Phenolic :
Antioxidants and Health Benefits

Dr. U. D. CHAVAN

Dr. U. D. Chavan obtained his B. Sc. and M. Sc. (Agri. in Biochemistry) degrees from Mahatma Phule Krishi Vidyapeeth Rahuri in 1985 and 1987 respectively. He received his Ph.D. degree in Food Science from Memorial University of Newfoundland St. John's Canada in 1999. He has done International training on "Global Nutrition 2002" at Uppsala University Uppsala, Sweden in 2002. He also attended follow-up International workshop on "Global Nutrition 2002" at Hanoi, Vietnam in 2002. Dr. Chavan visited Denmark, Finland, Ireland, France, Switzerland, Poland, Spain, Vietnam, Thailand, England, and U.S.A. under "Global Nutrition 2002" Programme sponsored by Swedish International Development Agency (SIDA). During this programme he worked on human nutritional diet and disorders as well as on genetically modified organisms (GMO).

Dr. Chavan worked as Senior Research Assistant in the Department of Biochemistry and Food Science and Technology at Mahatma Phule Krishi Vidyapeeth Rahuri from 1988 to 2000. During his Ph. D., he worked as Technician/Research Associate at Atlantic Cool Climate Crop Research Center and Agriculture and Agri-Food Canada. He received D.Sc. degree in 2006 from USA. He has guided 25 students for M.Sc. (Agri.) in Biochemistry and Food Science and Technology. From 2000 to 2004 he worked as Assistant Professor of Biochemistry at Mahatma Phule Krishi Vidyapeeth Rahuri.

Dr. Chavan received Korgavkar Trust Fellowship, under-graduate and post-graduate merit scholarship as well as Senior Research Fellowship from ICAR, New Delhi and United States Department of Agriculture U. S. A. He received University Graduate Fellowship from Memorial University of Newfoundland St. John's Canada for his Ph. D. programme. Dr. Chavan also received international Scholar Award and Excellentiam Award for his Ph. D. research work from Memorial University of Newfoundland St. John's Canada. He has received a "Certificate of Appreciation" from the U.S. Department of Agriculture in 1992 for his work on processing of groundnut under the guidance of Dr. S. S. Kadam (Principle Investigator). He has written 135 research papers and 140 popular articles. He has authored 19 books in Marathi, 23 books in English and eight book chapters in English. Dr. Chavan has visited 23 countries in the world for research and teaching programmes. He has been awarded **"Literary Award"** for his best book on **"Growth Regular"** on agriculture during 1997 by Government of Maharashtra. Dr. Chavan was selected as a best group leader and best presentation for "Global Nutrition 2002" by SIDA. He has been awarded "Life Time Achievement Award 2004" for his outstanding contribution in post-harvest technology of fruits and vegetables and allied fields by United Writers' Association of India. He has been elected as a Fellow of United Writers' Association of India in 2004. Dr. Chavan received pride of nation and Maharashtra Gunigan Ratna Award for 2006. He also received Jewels of India Award for 2006 for his contribution in the field of Food Science and Technology. He has been awarded **"Literary Awards"** for his best book on **"Sorghum Grain Processing"** on agriculture during **2009-2010** and **"Pulses cultivation to processing and value addition"** for **2012** by Government of Maharashtra. He is recipient of "Krishi Goorav Award-2011" from Bhartiya Krishak Samaj, Maharashtra State, Nasik. He also contributed in the development of crop varieties in wheat NIAW-917 and in Sorghum Phule Anuradha (RSV-423), Phule Chitra (RSV-1546), Phule Suchitra (RSV-1098), Phule Vasudha (RSV 423), Phule Revati (RSV-1006), SPV-1626, Phule Panchami (RSPOV-3) for Sorghum pops, Phule Godhan (SPV-2057) for forage single cut, CSH-50 (Sweet Sorghum), Phule Madhur (RSSGV–46) for sorghum *hurda* and Phule Rohini (RPASV-3) for sorghum *papad*. Now he is working as a Professor in the Department of Food Science and Technology, as well as a Senior Cereal Food Technologist in Sorghum Improvement Project and Foreign Student Advisor at Mahatma Phule Krishi Vidyapeeth, Rahuri, Dist, Ahmednagar, Maharashtra, India.

Phenolic :
Antioxidants and Health Benefits

U. D. Chavan
Professor
Department of Food Science and Technology
Mahatma Phule Krishi Vidyapeeth, Rahuri

Published by
SCIENTIFIC PUBLISHERS (INDIA)

Jodhpur –

5 A, New Pali Road
P.O. Box 91
Jodhpur - 342 001 INDIA

Delhi –

- 4806/24, Ansari Road
Daryaganj
New Delhi - 110 002 INDIA

ISBN: 978-93-86652-13-3

Visit the Scientific Publishers (India) website at
http://www.scientificpub.com

Printed in India

FOREWORD

The relationship between diet, moderate physical activity and health is of much current interest. Although initial evidence for this relationship was mainly epidemiological in nature, recent advances and developments in the area of nutraceuticals and functional foods have led to substantiation of claims by preclinical and clinical studies. Furthermore, availability of sophisticated analytical tools has led to the identification and appreciation of non-nutrient bioactives in the diet. Another motivating factor has been mounting health care costs that have emphasized a need for a preventive rather than a treatment approach by consumers, health professionals, researchers and perhaps some government departments in certain countries. The emerging findings in nutrigenomics and proteomics and the advent of nanotechnology for better delivery of bioactives have opened the way for individualized dietary regimes in the hopefully not too distant future. Over the past decade or two there has been a surge in the publication of research findings and developments in the field, although often in a non-targeted and fragmentary manner. Therefore, to fill the existing gap, the launch of a dedicated book for **"Phenolic: Antioxidants and Health Benefits"** was deemed necessary. This idea was first come in the mind of author and Mr. Tanay Sharma, Scientific Publishers (India), Jodhpur committed to publish this type of book. In this book mechanism of action of phenolic antioxidants, measurement of activity, classifications of antioxidants, natural sources and their health benefits as well as natural health products information are given. I think author has chosen the wright subject of the book and I hope this book will serve the purpose of students, teachers, readers, consumers and professional industrial processors.

J. K. Chavan
Ex-Head FST
MPKV, Rahuri

PREFACE

Latest advances are fast transforming raw materials into the processed products from a means of subsistence to an organized industry. This has stimulated a great deal of interest among widely diversified group of scientists. In this era of high yielding crop varieties and multiple cropping techniques, global competition in the quality of food products has put every country on the alert to improve upon nutritional quality, postharvest techniques and value addition sector.

Cereals, legumes, fruits and vegetables, oilseeds and plant oils, beverages, tree nuts, herbs and spices are important constituents of our diet which provide dietary fibers, vitamins, phytochemicals, and minerals in addition to proteins, carbohydrates and a small amount of lipids. These contribute more than 50% requirement of vitamin A. The dietary fibers aid in digestion, bowel movement and utilization of more concentrated foods in human diet.

Processing of agricultural produce and their postharvest management with additional phytochemical nutritional values are an emerging area in the present scenario encompassing a widely diversified group of scientists. India has now attained the stage where apart from enhanced production of crops, emphasis is being laid on processing and nutritional value addition of cereals, legumes, tree nuts, oilseeds, fruits and vegetables and herbs and spice crops.

Phenolic and polyphenolics compounds, ubiquitous in plants, are an essential part of the human diet and are of considerable interest due to their antioxidant properties and potential beneficial health effects. These compounds range structurally from a simple phenolic molecule to complex high-molecular-weight polymers. There is increasing evidence that consumption of a variety of phenolic compounds present in foods may lower the risk of health disorders because of their antioxidant activity. When added to foods, antioxidants control rancidity development, retard the formation of toxic oxidation products, maintain nutritional quality and extend the shelf-life of products.

The ever-increasing scientific research on the subject (Antioxidants and health benefits) has resulted in a large body of knowledge which can be found scattered in various scientific journals and technical reports and is not easily accessible. The available material needs to be assembled and briefed in order that the students/technologists/agriculturists/horticulturists/

engineers and others actively engaged in quantitative studies can have it as a ready reference. Antioxidants and health benefits book provides a comprehensive but concise treatise on the subject matter information.

U. D. Chavan

ACKNOWLEDGEMENT

Author is thankful to the Hon. Dr. S. S. Kadam, Ex. Vice-Chancellor, Marathwada Agricultural University, Parbhani and Ex. Head Department of Food Science and Technology, Mahatma Phule Krishi Vidyapeeth, Rahuri for his critical and fruitful suggestions while the preparation of **"Phenolic Antioxidants and Health Benefits"** book. I am thankful to Dr. Fereidoon, Shahidi, Dr. J.K. Chavan, Dr. Amarowicz Ryszard, Dr. J. V. Patil, Dr. V. D. Pawar, Dr. U. Y. Bhoite, Dr. K. D. Chavan, Dr. C. A. Nimbalkar, Mr. Ajay Jadhav, Mr. Kashinath Bagale, Mr. Suresh Shedage for helping during this manuscript preparation, as well as those scientists who worked on phenolic antioxidants and health benefits identification, utilization and usefulness for human beings and their contribution is acknowledged in this book as well their sources are cited in the reference section. Author is also thankful to all scientists and friends those helped directly or indirectly while preparation of this manuscript. I would like to express over profound thanks to my family members for their constant support, patience and devotion, which inspired me to complete this manuscript.

U.D. Chavan

CONTENTS

1.	Introduction	1-9
2.	Phenolic antioxidants	10-53
3.	Mechanism of action of phenolic antioxidants	54-76
4.	Measurement of antioxidant activity	77-159
5.	Health effect of phenolic antioxidants	160-183
6.	Classification of phenolic antioxidants	184-207
	Synthetic food phenolic antioxidants	185
	Natural phenolic antioxidants	191
7.	Degradation of phenolic antioxidants	208-210
8.	Natural sources of phenolic antioxidants	211-317
	Cereals and legumes	211
	Fruits and vegetables	243
	Oilseeds and plant oils	293
	Beverages	299
	Tree nuts	304
	Herbs and spices	312
9.	Effect of processing on antioxidant activity	318-343
10.	Future considerations	344-361
	References	362-499
	Appendix	500-503
	Subject Index	504-510

1 Introduction

Lipid oxidation is a major cause for food quality deterioration and generation of off odours and off flavours, decreasing shelflife, altering texture and colour, and decreasing the nutritional value of food (Alamed, Chaiyasit, McClements, and Decker, 2009). Numerous methods have been developed to control the rate and extent of lipid oxidation in foods, but addition of antioxidants is most effective. Antioxidants have become an indispensible group of food additives mainly because of their unique properties of extending the shelf-life of food products without any adverse effect on their sensory or nutritional qualities. Historically, gum guaiac was the first antioxidant approved for stabilization of animal fats especially lard in the 1930s (Nanditha and Prabhasankar, 2009). Halliwell, Aeschbacht, Loligert, and Aruoma (1995) reported that antioxidants are also of interest to biologists and clinicians because they may help to protect the human body against damage by reactive oxygen species (ROS). According to the United States Department of Agriculture (USDA) Code of Federal Regulation, "antioxidants are substances used to preserve food by retarding deterioration, rancidity or discoloration due to oxidation" (Shahidi and Wanasundara, 1992). Antioxidants for use in food system must be inexpensive, nontoxic and effective at low concentrations; highly stable and capable of surviving processing; have no odour, taste or colour of their own; easy to incorporate and have a good solubility in the product (Kiokias, Varzakas, and Oreopoulou, 2008). One of the primary pathways of lipid degradation is that of autoxidation. The process of autoxidation of polyunsaturated lipids in foods involves a free radical chain reaction that is generally initiated by exposure of lipids to light, heat, ionizing radiation, metal ions or metalloprotein catalysts. The enzyme lipoxygenase can also initiate oxidation (Shahidi, 2015; Shahidi and Naczk, 2004). The classic route of autoxidation includes initiation (production of lipid free radicals), propagation and termination (production of non radical products) reactions (Shahidi and Wanasundara, 1992). A general schematic pathway for autoxidation of polyunsaturated lipids is shown in Fig. 1.1. Antioxidants act at different levels in the oxidative sequence involving lipid molecules. They may decrease oxygen concentration, intercept singlet oxygen ($^1O_2$), prevent first-chain initiation by scavenging initial radicals such as hydroxyl radicals, bind metal ion catalysts, decompose primary products of oxidation to nonradical species and break chain reaction in order to prevent continued hydrogen bstraction from substrates (Shahidi, 2000a, 2000b, 2015; Shahidi and Naczk, 2004).

Initiation	RH→R• + H•	(1)
Propagation	R• + O2 → ROO•	(2)
	ROO• + RH→ R• + ROOH	(3)
	R• + R•	
Termination	R• + ROO• Nonradical products	(4)
	ROO• + ROO•	

Hydroperoxides are the primary products of lipid oxidation, but hydroperoxides, despite their deleterious effects on health have no effect on flavour quality of foods (Shahidi, 1998). However, these unstable molecules decompose readily to form a myriad of products such as aldehydes, ketones, alcohols and hydrocarbons, amongst others (Shahidi, 1998); these impart unpleasant flavours and odours to fats, oils and lipid containing foods. These ldehydes interact with sulphhydryl and amine groups in proteins and this may alter the functionality of proteins (McClements and Decker, 2007). Faustman, Liebler, McClure and Sun (1999) reported the ability of unsaturated aldehydes to react with histidine in myoglobin and accelerate the oxidation of oxymyoglobin. Phenolic compounds in foods originate from one of the main classes of secondary metabolites in plants (Naczk and Shahidi, 2004). At a low concentration, phenolics act as an antioxidant and protect food from oxidative rancidity (Karakaya, 2004). Phenolic antioxidants interfere with the oxidation process as free radical terminators and sometimes also as metal chelators. Phenols have been widely studied and confirmed to possess diverse bioactivities which could be beneficial to human health. They are known to reduce the risk of cancer, heart disease, and diabetes; to inhibit plasma platelet aggregation, cyclooxygenase (COX) activity, and histamine release, as well as to exert antibacterial, antiviral, anti-inflammatory, and anti-allergenic activities (Oak, El Bedoui, and Schini-Kerth, 2005; Shetty, 2004; Yang, Landau, Huang, and Newmark, 2001; Yao *et al.*, 2004).The benefits towards many of these conditions arise in part through the antioxidant characteristic of phenolics; therefore, it is important to quantify, identify and evaluate their antioxidant activities (Cevallos-Casals and Cisneros-Zevallos, 2010). This review summarizes both the synthetic and natural phenolic antioxidants, emphasizing their mode of action, health effects, degradation products and toxicology. In addition, sources of phenolic antioxidants are discussed in detail.

The photosynthetic process of green plants primarily supports life on earth by converting water and carbon dioxide into a variety of organic compounds. Traditionally, processes generating plant compounds have been grouped as primary and secondary metabolism. The chemical substances that have specific indispensable roles and that are found in broadly similar patterns in most, if not all, living cells are considered primary metabolites. Compounds having no specific functions in the cell metabolism and whose distribution is sporadic are referred to as secondary metabolites. One theory propounds that such compounds are waste products or accidents of metabolism (Muller, 1969). However, advances in plant physiology during the last 2 decades have clearly shown that plants do not randomly produce chemical compounds. Each metabolite is synthesized for a definite purpose, and all products are interrelated according to a complex process which conserves energy and scarce inorganic nutrients. Most secondary metabolites are now known to be essential to plant life; many of them providing a defense mechanism against bacterial, viral, and fungal attack analogous to the immune system of plants.

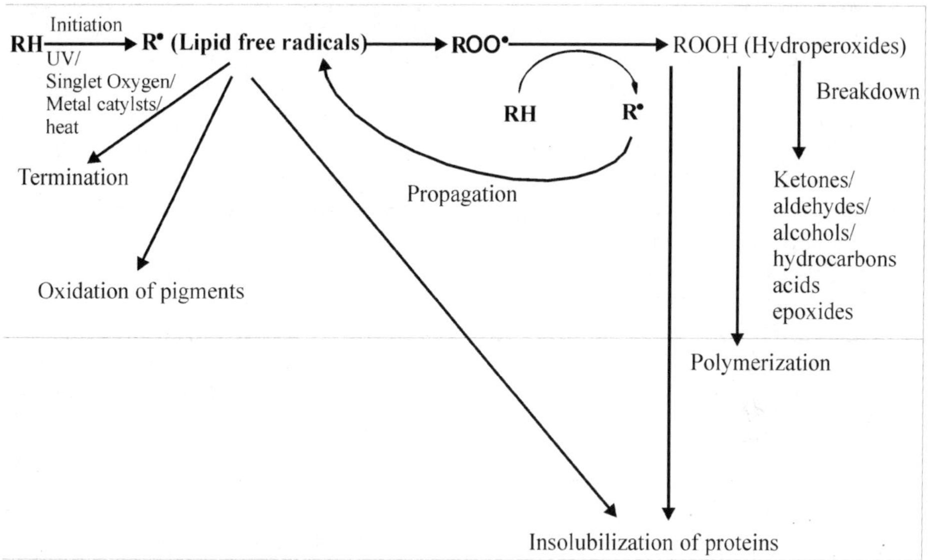

Fig. 1.1: General scheme for autoxidation of lipids containing polyunsaturated fatty acids (RH) and their consequences.

Phenolic compounds (phenolic acids, coumarins and flavonoids including tannins) are one such group of secondary metabolites. These are widely distributed in the plant kingdom, being nearly ubiquitous in nature. Although the pharmacological and therapeutic properties of many bioflavonoids are well documented in the literature, the antinutritional effects of certain flavonoids, like tannins, in the diet are only recently recognized. Plants provide nearly all the phenols to higher animals since the latter cannot synthesize these compounds. The occurrence of phenolic compounds is beneficial to plants, while it may or may not be beneficial to animals, including man. Anthocyanins are responsible for different colors of flowers and leave which aid in pollination by insects. Plant phenolics have been implicated in the resistance of plants against bird depredation (Hoshino *et al.*, 1979) insect attack (Woodhead *et al.*, 1980) preharvest seed germination (Harris and Burns, 1970) and diseases caused by fungi, bacteria, and viruses (Friend, 1981). The color of fruits, fruit juices, wines, beers, teas, or coffee extracts, and cooked beans or cooking liquors due to phenolic compounds gives immense aesthetic pleasure to man. Certain bioflavonoids have been implicated in stimulating antibacterial, antiathero-sclerolemic, antioxidant, and flavor-inducing effects (Singleton, 1981).

The phenolic compounds are also known to produce certain undesirable effects in foods. The browning reactions in foods are known to take place via enzymatic and/or nonenzymatic reactions with polyphenolic substances that affect consumer acceptability. Polyphenol oxidase is responsible for browning induced by mechanical or physiological injury and needs to be inactivated during the processing of fruits and vegetables for retention of color, flavor, and nutritive value (Mathew *et al.*, 1971). The o-quinones formed from phenolics further enhance the intensity of browning by oxidation of other substrates, complexing with amino acids and protein, and polymerization. The nonenzymatic discoloration was believed to involve metal-polyphenol complexes, as seen in the processed potato (Bate-Smith *et al.*, 1958), cauliflower (Donath, 1962) and asparagus (Deeds and Couch, 1948). The conversion of leucoanthocyanidins to pink anthocyanidins in the processed broad bean (Dikinson *et al.,* 1957), green-bean puree (Roseman *et al.*, 1957) and canned Bartlett pear (Luh *et al.*, 1960) has been documented. Another problem related to the phenolics is their considerable reactivity towards the animal body, as exerted by gossypol (isoprenoid) in cotton seed, tangeretin (methylated flavonoid) in citrus, rhein (anthraquinone) in rhubarb (Singleton and Kratzer, 1973), isoflavones in legumes (Harborne, 1967) and phlorizon (dihydrochalcone glycoside) in apples (Durkee and Poapst, 1965). Although many low molecular weight phenols are readily metabolized in the bodies of higher animals, especially in the kidney, heart and liver (Durkee and Poapst, 1965), the metabolic fate of phenolic compounds of flavonoid origin is not yet clearly known. The sheer number of naturally occurring flavonoid compounds in plant foods further complicates the problem in human nutrition. Common dietary plant phenols are not considered toxicants under normal amounts and conditions. The polymeric phenols, often called vegetable tannins, are one possible exception. By definition, vegetable tannins are polyphenols of 500 to 3000 mol wt and possess the ability to complex and precipitate proteins from aqueous solution. This property of polyphenols and their occurrence in certain foods has attracted the attention of nutritionists due to their possible antinutritional influence in animal and human diets.

On the basis of their structure, vegetable tannins are grouped into two classes, namely hydrolyzable and nonhydrolyzable or condensed tannins (Chapon *et al.*, 1961; Freudenberg, 1920). The hydrolyzable tannins consist of polyhydric alcohol (glucose) surrounded by phenolic acids, usually gallic or ellagic acid. They are readily hydrolyzed by acids, bases, and certain enzymes. The condensed tannins are oligomers of flavan-3-ols and flavan-3,4-diols (Haslam and Gupta, 1978). These are resistant to hydrolysis. The acid treatment promotes condensation of polyphenols as well as hydrolysis (Guesnel, 1968). Both types of vegetable tannins exhibit the ability to complex and precipitate proteins. Thus, both are nutritionally hazardous. Most of the reports related to the anti nutritional effects of tannins before 1970 are based on tannic acid and other hydrolyzable tannins. However, now most agree that hydrolyzable tannins, if present, are only in traces, and the condensed tannins are the major polyphenols of commonly consumed foods.

The condensed tannins are widespread in fruits, vegetables, forage plants, cocoa, red wine, and certain food grains such as sorghum, finger millet, and legumes (Guesnel, 1968). The dietary intake 'of tannin will vary among people depending on their dietary habits and sources of foods. Dietary tannins may not be a nutritional problem for those whose diets include animal foods and cereals, such as rice, wheat, corn, or barley. However, tannins may be of concern where diets are based on sorghum, millets, and pulses. Rao and Prabhavati (1982) reported a range of 1.5 to 2.5 g daily intake of dietary tannins in different regions of India. Tannin intake may be considerably higher for the people of certain parts of Africa where high tannin sorghum cultivars are grown and used for human consumption. The daily intake of tannin through tannin-containing foods like red wine, fruit juices, beetle quid or arecanut can be lowered to a minimum by consuming lesser quantities of such food items. However, tannins cannot be avoided when they are present in foods which are important components of a daily diet.

Sorghum, millets, and food legumes (pulses) are important food crops of several countries in semiarid regions of Asia and Africa (Table 1. 1). In certain countries of Asia, Africa, the Near East and the Middle East, almost the entire production of sorghum, millet, and pulses goes for human consumption and forms a major source of dietary energy and protein to people, particularly in rural areas and farming communities (Salunkhe *et al.*, 1985). Certain genotypes of sorghum, millet, and legumes are known to contain appreciable levels of condensed tannins which can be nutritionally harmful (Hulse, 1980; Salunkhe *et al.*, 1982). Tannins are reported to be responsible for decreases in feed intake, growth rate, feed efficiency, net and metabolizable energy, and protein digestibility (Hulse, 1980; Salunkhe *et al.*, 1982; Deshpande *et al.*, 1984). Such effects are exerted through the formation of complexes with proteins, starch, and digestive enzymes. Sorghum tannins have been reported to cause leg abnormalities in hens and delays in physical or sexual maturity or even death in hamsters (Butler *et al.*, 1986). Dietary tannins have been linked to the incidence of cancer of the esophagus (Morton, 1978; Kapadin *et a*l., 1983).

Table 1.1: World production of human food materials.

Region	Production (million t)					
	Cereals	Coarse grains	Pulses	Oilseeds	Fruits	Vegetables
World	2476.41	1121.93	68.82	170.27	608.92	1044.38
Africa	164.53	116.57	15.07	10.14	84.59	71.15
Americas	641.49	490.82	14.57	54.30	139.32	81.15
North America	447.32	353.06	7.94	24.73	26.85	37.92
Asia	1228.81	303.93	31.07	82.33	310.26	794.27
Europe	407.03	198.85	6.10	21.54	68.03	94.22
Oceania	34.53	11.74	2.01	1.95	6.70	3.56

From Food and Agricultural Organization, Production Yearbook.
Food and Agricultural Organization, Rome, Italy, 2013.

Attempts have also been made to remove or inactivate tannins in foods and feeds. The technologies developed for this purpose involve the identification of low tannin cultivars in cereals and legumes; pearling or dehusking the grains; soaking of grains in water and various chemical solutions; supplementation of high tannin diets with nutritional and/or nonnutritional tannin-binding proteins like gelatin and methyl group donors like methionine or its analogues; and cooking, sprouting and drying of grains (Deshpande *et al.*, 1984). However, every method has certain limitations in commercial applications.

A symposium on polyphenols in cereals and legumes was held during the 36th Annual Meeting of the Institute of Food Technologists at St. Louis, MO, in 1979 (Hulse, 1980). On the basis of the recommendations of this symposium, several investigators have attempted to study different aspects of polyphenols in sorghum, millets, legumes, and forages. These include the nature of phenolics in these grains; methods of analyses; changes during grain development, germination, cooking, drying, and storage; mechanism and specificities of tannin protein binding; identification of agronomically beneficial and nutritionally harmless forms of phenolics in these grains; and physiological effects of tannin consumption on the functioning of certain salivary glands. Attempts have also been made to develop simple and economical methods to remove or inactivate tannins in these grains before consumption. During the past decade, several reports and critical reviews have been published on these aspects of dietary tannins (Deshpande *et al.*, 1984; Butler *et al.*, 1986; Kapadin *et al.*, 1983; Reddy *et al.*, 1985; Butler *et al.*, 1984; Price and Butler, 1980). In this book, an attempt has been made to update and surnmerize the available literature on chemical nature, occurrence, composition of tannins in various foods and forages, methods of analyses, dietary consequences, and technologies developed to remove and/or inactivate them. The agronomical advantages of plant phenolics in crop growth are also reviewed, and future research needs are addressed.

Tree nuts are dry fruits with one seed in which the ovary wall becomes hard at maturity (Alasalvar and Shahidi, 2008). Common edible tree nuts include almond, Brazil nut, cashew, hazelnut, macadamia, pecan, pine nut, pistachio, and walnut, but the consumer definition also includes peanut, which is botanically a legume, but has a nutrient profile similar to those of tree nuts and is thus identified as part of the nut food group (Alasalvar and Shahidi, 2008; Ros, 2015).

Nuts are nutrient dense foods and have been a regular constituent of mankind's diet (Alasalvar and Shahidi, 2008). Nuts are convenient and tasty snacks that contribute to a healthful dietary pattern. They are usually consumed as whole nuts (either as raw or roasted or salted) (Alasalvar and Shahidi, 2008). Nuts contain high amounts of vegetable protein and fat-soluble bioactives (unsaturated fatty acids, phytosterols, phospholipids, phytostanols, essential oils, sphingolipids, tocopherols, tocotrienols, terpenoids, and squalene). Besides, they are also dense in a variety of other nutrients and provide dietary fibre, vitamins (e.g., folic acid, niacin, vitamin B6, and vitamin E), minerals (e.g., calcium, magnesium, and potassium), and many other phytochemicals such as phenolic acids, flavonoids, lignin, hydrolysable tannins, proanthocyanidins (condensed tannins), carotenoids, alkaloids, coumestan, and phytates, among others (Alasalvar and Bolling, 2015; Alasalvar and Shahidi, 2008; Bolling, Chen, McKay, and Blumberg, 2011; Bolling, McKay, and Blumberg, 2010b; Chen and Blumberg, 2008; Lainas, Alasalvar, and Bolling, 2016; Maguire, O'Sullivan, Galvin, O'Connor, and O'Brien, 2004; Ryan, Galvin, O'Connor, Maguire, and O'Brien, 2006).

A healthy diet supplemented with one daily serving of nuts prevents cardiovascular events (Ros, 2015) and possibly the development of other chronic diseases, including type II diabetes, cancer, high blood pressure, and neurodegenerative diseases (Afshin, Micha, Khatibzadeh, and Mozaffarian, 2014; Grosso et al., 2015; Kendall, Josse, Esfahani, and Jenkins, 2010; Kochar, Gaziano, and Djoussé, 2010; Luo et al., 2014; Ros, 2010; Zhou et al., 2014). The mechanism for these health effects probably lies in the synergistic interaction of the many bioactive constituents of nuts, which may all favourably influence human physiology (Liu, 2003, 2004). Despite the fact that tree nuts are high-fat and energy-dense foods, epidemiological studies and clinical trials suggest that regular nut consumption is unlikely to contribute to obesity and may even help in weight loss when consuming an energy-restricted diet (Ros, 2010; Vadivel, Kunyanga, and Biesalski, 2012). For this reason, the American Heart Association (Lloyd-Jones et al., 2010; Stone et al., 2014), the Canadian Cardiovascular Society (Anderson et al., 2013), the US Food and Drug Administration (US Food and Drug Administration, 2003), and the US Department of Agriculture (US Department of Health and Human Services and US Department of Agriculture, 2015) have recommend the regular consumption of nuts to the general population, in the context of a healthy diet.

Tree nuts rank third behind spices and fruits for containing the most phytochemicals (Pérez-Jiménez, Neveu, Vos, and Scalbert, 2010). Tree nuts and their co-products (skin or testa, hard shell, green leafy cover, hull, and leaf, among others) are rich sources of phytochemicals that possess multifunctional properties such as antioxidant activities, anti-carcinogenic, antimutagenic effects as well as anti-proliferative potential (Alasalvar, Hoffman, and Shahidi, 2008; Alasalvar and Shahidi, 2008; Oliveira et al., 2007; Shahidi, Zhong, Wijeratne, and Ho, 2008). These phytochemicals directly or indirectly provide protection against harmful free radicals and may reduce the risk of non-communicable diseases (NCDs) associated with oxidative stress. Therefore, inclusion of bioactive phytochemicals from tree nut co-products into the diet is of great interest as these may provide inexpensive sources of natural antioxidants for use as functional ingredients and nutraceuticals (Alasalvar et al., 2008; Alasalvar and Shahidi, 2008, 2009).

In the last decade, several different phenolic compounds were characterised and identified in nuts and their co-products (skin, hard shell, and hull/pellet) (Arcan and Yemeniciog˘lu, 2009; Chandrasekara and Shahidi, 2011; Lainas et al., 2016; Shahidi, Alasalvar, and Liyana-Pathirana, 2007; Slatnar, Mikulic-Petkovsek, Stampar, Veberic, and Solar, 2014, 2015; Wijeratne, Abou-Zaid, and Shahidi, 2006). The health effects of phenolic compounds depend on the amount consumed and their bioavailability (Landete, 2012; Manach, Scalbert, Morand, Rémésy, and Jiménez, 2004; Shahidi and Ambigaipalan, 2015). Comprehensive reviews on tree nut phytochemicals (phenolic compounds and fatsoluble bioactives), antioxidant efficacies, and health benefits have been published (Alasalvar and Bolling, 2015; Alasalvar and Shahidi, 2009; Bolling et al., 2011). Close scrutiny of the literature data, however, shows a lack of comprehensive review on the phenolic compounds and antioxidant capacity of natural (raw) and roasted nuts as well as their co-products (skin, hard shell, and hull/pellet). In this contribution, various phenolic compounds reported in popular tree nuts and their co-products are reviewed in terms of their processing (roasting) effect, antioxidant activity, bioavailability, and health benefits. This information is intended to stimulate large-scale commercial exploitation and generate interest among researchers in various scientific fields to study nuts and their co-products as functional food ingredients.

THE IMPORTANCE OF POLYPHENOLS AS ANTIOXIDANTS

From the above reports, it can be deduced that plants produce thousands of phenolic and polyphenolic compounds. Polyphenols constitute one of the most numerous and widely distributed groups of natural products in the plant kingdom. It includes an ample variety of molecules with phenol ring structure. At present, there is not any universal classification for polyphenols, but the most current classification divides this type of molecules in different groups according to their chemical structure (Motilva *et al.*, 2013). There are three main groups: non-flavonoids, flavonoids, and tannins. The non-flavonoids are molecules that have at least one phenolic ring with different reactive groups (hydroxyl, nitrosyl, SH, etc.); this group includes simple phenolic acids, phenyl alcohols, stilbenes, and chalcones (Fig. 1.2). The flavonoids are characterized by a phenyl chromane structure of C_{15} formed by two aromatic rings bound with a carbon chain (C_6–C_3–C_6) and include antocyanidins, flavonols, flavones, flavonons, isoflavons, flavan-n-ols, etc. The third group, called tannins, is subdivided into two groups: condensed tannins and hydrolysable tannins. Condensed tannins are flavonoid polymers type A (C_7–C_2 and an ether bond) and type B (C_4–C_8 or C_4–C_6) and hydrolysable tannins are phenolic acids polymers bound to a five or six carbon ring (Khanbabaee and Van Ree 2001). Polyphenols are essential for the physiology of plants. These compounds have been implicated in diverse functional roles, including structure, pigmentation, lignification, plant resistance against microbial pathogens, and animal herbivores such as insects (antibiotic and anti-feeding actions), protection against solar radiation (screen against DNA-damaging UV-B light), and they were probably important during early terrestrial plant evolution (Hatier and Gould 2008; Lattanzio *et al.*, 2008). The majority of polyphenols are synthesized by the highly branched phenylpropanoid pathway, which is responsible for the biosynthesis of a large number of chemical compounds with considerable structural diversity (Duthie *et al.*, 2003). Plant polyphenols exhibit a wide range of bio-physicochemical properties bundled within the phenol functional group, which makes them remarkably versatile metabolites and unique and intriguing natural products. The major characteristic of polyphenols is their capability to scavenge ROS, as well as oxidatively generate free radicals $RO^\bullet$ and $ROO^\bullet$ such as those derived from biomolecules like the low-density lipoproteins proteins (LDLs) Neudfforffer *et al.*, 2004) and oligonucleic acids (DNA and RNA) (Li *et al.*, 2000). In its most elementary structural form, namely a benzene ring bearing a hydroxyl group (PhOH), a phenol function constitutes an amphiphilic moiety that combines the hydrophobic character of its planar aromatic nucleus with the hydrophilic character of its polar hydroxy substituent, which can act either as a hydrogenbond donor or as an acceptor. Hydrophobic p-stacking (van der Waals) interactions and the formation of hydrogen bonds are seemingly dichotomic, yet are often complementary effects that plant phenolics can use to interact physically with other biomolecules, among which proteins are often first in line (Dangles and Dufour 2008).

Polyphenols are also able to chelate transition metals through their multiple hydroxy groups on a phenyl ring open (Andjelkovic *et al.* 2008). By chelating metal ions, such as iron(II)/copper(I) and iron(III)/copper(II) that are involved in the conversion of $O_2^{\bullet-}$ and H_2O_2 into highly reactive hydroxyl radicals (HOl), polyphenols can act as DNA protective agents from free radicals damages (Heim *et al.* 2002; Perron and Brumaghin 2009; Leopoldini *et al.* 2011). This capacity for metal chelating also contributes to plant pigmentation (Lattanzio *et al.* 2008) as well as cationic nutrient (e.g., Ca, Mg, Mn, Fe, Cu) cycling through plant–litter–soil interactions. Polyphenols act as scavengers of free radicals and ROS, when these are unable to be subdued by the regular action of other antioxidants such as glutathione (GSH), glutathione peroxidase, or superoxide dismutase or by vitamins (e.g., vitamins E and C, carotenoids). Two main intimate antioxidative mechanisms have been proposed for polyphenols (Wright *et al.* 2001). The first is based on the capacity of the phenol functional group to donate a hydrogen atom to a free radical $R^{\bullet}$ such as peroxy radicals $LOO^{\bullet}$ generated during lipid (LH) autoxidation (peroxidation; LH $\longrightarrow L^{\bullet}$, then $L^{\bullet} + 3O_2 \longrightarrow LOO^{\bullet}$) such as peroxy radicals $LOO^{\bullet}$ generated during lipid peroxidation, cting as chain-breaking antioxidants. The second mechanism is the single-electron transfer from the phenolic antioxidant (ArOH) to a free radical $R^{\bullet}$ with formation of a stable radical cation $ArOH^{\bullet+}$.

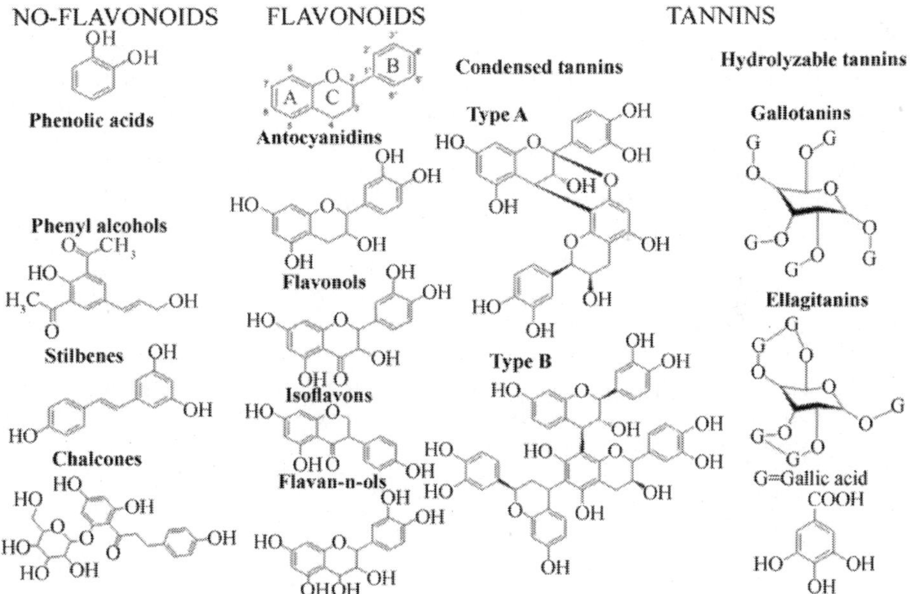

Fig. 1.2: Main groups of polyphenols according to their chemical structure. G gallic acid

2 Phenolic antioxidants

PHENOLIC ANTIOXIDANTS

Phenolic compounds are classified as primary antioxidants which are mainly free radical scavengers (FRS) that delay or inhibit the initiation step or interrupt the propagation step of lipid oxidation, thus decreasing the formation of volatile decomposition products (e.g., aldehydes and ketones) that cause rancidity (Alamed *et al*., 2009; Kiokias et al., 2008; Naczk and Shahidi, 2004; Nanditha and Prabhasankar, 2009; Shahidi and Naczk, 2004; Shahidi, Wanasundara, and Amarowicz, 1994).

Natural antioxidants from residual sources

The growing interest in the substitution of synthetic food antioxidants by natural ones has fostered research on vegetable sources and the screening of raw materials for identifying new antioxidants. Oxidation reactions are not an exclusive concern for the food industry, and antioxidants are widely needed to prevent deterioration of other oxidisable goods, such as cosmetics, pharmaceuticals and plastics. Polyphenols are the major plant compounds with antioxidant activity, although they are not the only ones. In addition, other biological properties such as anticarcinogenicity, antimutagenicity, antiallergenicity and antiaging activity have been reported for natural and synthetic antioxidants. Special attention is focused on their extraction from inexpensive or residual sources from agricultural industries. The aim of this review, after presenting general aspects about natural antioxidants, is to focus on the extraction of antioxidant compounds (mainly polyphenols) from agricultural and industrial wastes, as well as to summarize available data on the factors affecting their antioxidant activity and stability, and, in some cases, the reported major active compounds identified.

Natural sources of antioxidant compounds

Many of the antioxidants other than vitamin C, vita-min E and carotenoids occur as dietary constituents (Figure 2.1). Wang, Cao and Prior (1996) and Kalt, Forney, Martin and Prior (1999) published works about strong anti-oxidant compounds found in fruits. For example, anti-oxidants with important activity have been found in berries (Abuja, Murkovic and Pfannhauser, 1998; Heino-nen, Lehtonen and Hopia,

1998; Heinonen, Meyer and Frankel, 1998; Prior *et al*., 1998), cherries (Wang, Nair, Strasburg, Booren and Gray, 1999; Wang, Nair, Stras-burg, Chang, Booren and Gray, 1999; Wang, Nair, Strasburg, Chang, Booren, Gray and DeWitt, 1999), citrus (Saleh, Hashem and Glombitza, 1998) and in kiwi fruit (Dawes and Keene, 1999) prunes (Donovan, Meyer and Waterhouse, 1998) and olives (Romani, Mulinacci, Pinelli, Vincieri and Cimato, 1999). High activity anti-oxidants were found in olive oil (Blekas and Boskou, 1998; Papadopoulus and Boskou, 1991) and also in fruit juices (Chambers, Lambert, Plumb and Williamson, 1996; Spa-nos and Wrolstad, 1990, 1992; van Buren, de Vos and Pilnik, 1976; Wen, Wrolstad and Hsu, 1999). Recently a compre-hensive review summarised the role of phenolic compounds in the oxidative process of fruits (Robards, Prenzler, Tucker, Swaitsitang and Glover, 1999). The effects of processing and storage were evaluated on the changes and content of polyphenols in strawberry (Gil, Holcroft and Kader, 1997), plum (Raynal, Moutounet and Souquet, 1989), olive oil (Angerosa and di Giovacchino, 1996; Caponio, Allogio and Gomes, 1999), grape juice (Spanos and Wrostad, 1990, 1992), onions, beans and peas (Ewald, Fjelkner-Modig, Johansson, SjoÈ holm and A Ê Kesson, 1999).

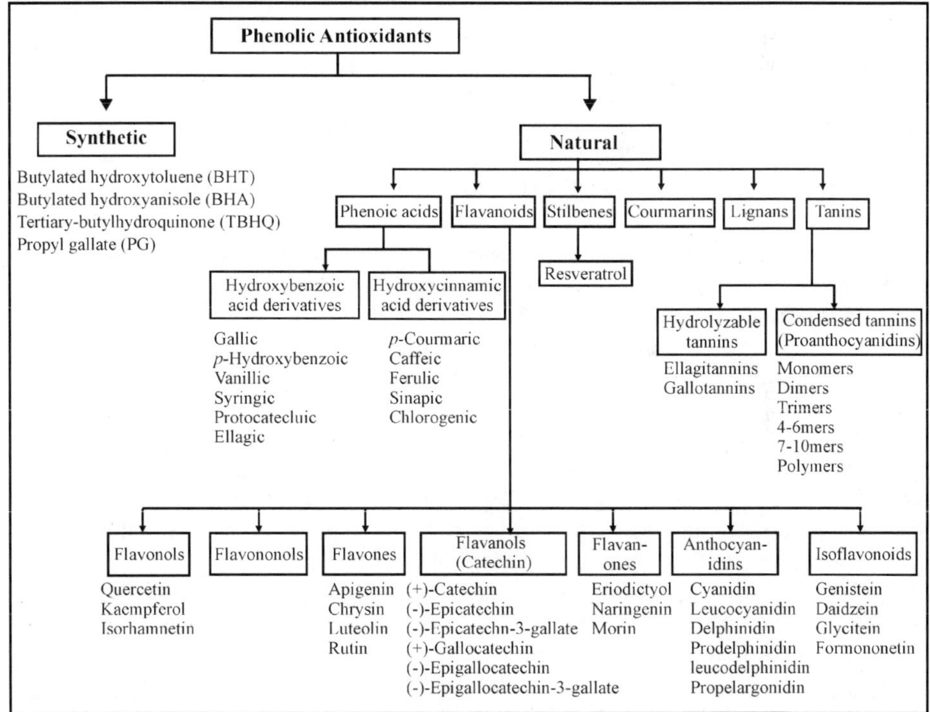

Figure 2.1: Classification of phenolic antioxidants

Several studies have analysed the antioxidant potential of a wide variety of vegetables (Furuta, Nishiba and Suda, 1997;Gazzani, Papetti, Massolini and Daglia, 1998; Hertog, Hollman and Katan, 1992; Vinson, Hao, Su and Zubik, 1998), and particularly, of cacao beans (Sanbongi, Osa-kabe, Natsume, Takizawa, Gomi and Osawa, 1998), potato (Al-Saikhan, Howard and Miller, 1995; Friedman, 1997; Ramamurthy, Maiti, Thomas and Nair, 1992), tomato (Abushita et al., 1997), spinach (Gil, Ferreres and TomaÂ s-BarberaÂ n, 1999), legumes such as Phaseolus vulgaris (Ganthavorn and Hughes, 1997; Tsuda, Ohshima, Kawakishi and Osawa, 1994) or vegetables such as paprika (Markus, Daood, KapitaÂny and Biacs, 1999; Matsufuji, Nakamura, Chino and Takeda, 1998). Both natural extracts and commercial products from garlic and ginger (Aruoma et al., 1997), from rosemary (Che Man and Tan, 1999; GuÈ ntensperger, HaÈmmerli-Meier and Escher, 1998; Jaswir and Che Man, 1999; OÈ zcan, 1999), from dietary supplements (Chambers et al., 1996), from smoke flavourings containing lignin dimers (GuilleÂn and Ibargoitia, 1998), or from drinks (Cano, Acosta and BanÄ oÂ n, 1998) were evaluated for antioxidant activity. Also antioxidants from seashore plants (Masuda et al., 1999) and seaweeds (Nakayama, Tamura, Kikuzaki and Nakatani, 1999) were studied.

Wines contain a variety of polyphenolic compounds, the most abundant being anthocyanins (Fogliano, Verde, Randazzo and Ritiene, 1999; Ghiselli, Nardini, Baldi and Scaccini, 1998; Heinonen, Lehtonen and Hopia, 1998; Hurtado, CalduÂ , Gonzalo, Ramon, MõÂnguez and Fiol, 1997; Lapidot, Harel, Akiri, Granit and Kranner, 1999; Larrauri, Sanchez-Moreno, RupeÂrez and Saura-Calixto, 1999; Saint-Cricq de Gaulejac et al., 1999; Sato, Ramarathnam, Suzuki, Ohkubo, Takeuchi and Ochi, 1996; Simonetti, Pietta and Testolin, 1997); antioxidant activity was also reported in whiskeys (McPhail, Gardner, Duthie, Steele and Reid, 1999), sake (Kitagaki and Tsugawa, 1999), Jerez-Sherries (Monedero, Olalla, MartõÂn-Lagos, LoÂ pez and LoÂ pez, 1999) and cavas (SatueÂ-Gracia, AndreÂs-Lacueva, Lamuela-RaventoÂ S and Frankel, 1999).

Green and black teas have been extensively studied for antioxidant properties since they can contain up to 30% of the dry weight as phenolic compounds (Lin, Lin, Liang, Lin-Shiau and Juan, 1998). Among studies of the antioxidant activity and identi®cation of polyphenols in green and in fermented teas are those of Cao, Sofic and Prior (1996), Chambers, Jimbin and McDonald (1988), Chen, Wang, Chan, Zhang, Chung and Liang (1998), Frankel, Huang and Aeschbach (1997), Lin, Juan, Liang and Lin (1996), Singh et al. (1999), Yen, Chen and Peng (1997), Wanasundara and Shahidi (1998), Yokozawa et al. (1998). Rooibos tea has also been investigated (von Gadow, Joubert and Hansmann, 1997a, b; Yen and Hsieh, 1998). Benzie and Szeto (1999) correlated the antioxidant activity with total phenolics content of the tea and found higher activity for green tea than for oolong or black tea. Among the major components (ÿ)epigallocatechin 3-gallate, which was thoroughly studied by Copeland, Clifford and Williams (1998), (ÿ)epigallocatechin, (-)epicatechin 3-gallate, (-)epicatechin, (+)gallocatechin and (+)catechin were also identified.

Some of the active principles of some medicinal pro-ducts are polyphenolic compounds. Thus, flavones that possess antimutagenic activity (Nakasugi and Komai, 1998), flavanones and xanthones, that exhibit antiviral, antimicrobial and antiin flamatory activities, and iso-flavones and coumestans that present important physiological effects in humans, have antioxidant action. A number of studies deal with the antioxidant activity of extracts from herbs, medicinal plants and spices (De la Torre Boronat and LoÂ pez Tamames, 1997; Duh and Yen, 1997b; Jung, Kim and

Kim, 1999; Kim, Kim, Kim, Oh and Jung, 1994; Madsen, Sùrensen, Skibsted and Bertelsen, 1998; Nieto et al., 1993; Pietta, Simonetti and Mauri, 1998). The antioxidant activity of sage components has been widely studied (Abdalla and Roozen, 1999; Aruoma, 1999; GuilleÂn and Manzanos, 1999; Marinova and Yanishlieva, 1997; Wang, Li, Rangarajan, Shao, LaVoie, Huang and Ho, 1998; Wang, Shao, Li, Zhu, Rangarajan, LaVoie and Ho, 1999; Wang, Yieh and Shih, 1999; Weinberg, Akiri, Potoyevski and Kanner, 1999; Yanishlieva, Marinova, Gordon and Raneva, 1999). Also ginger (Kikuzaki and Nakatani, 1993), Ganoderma species (Yen and Wu, 1999), dittany (Mùller, Madsen, Aaltonen and Skibsted, 1999), green pepper (Bandyopadhyay, Narayan and Variyar, 1990), Visnea mocanera L.F. (HernaÂ ndez-PeÂrez, HernaÂndez, GoÂ mez-CordoveÂ S, Estrella and Rabanal, 1996), Chrysanthemum (Chuda, Ono, Ohnishi-Kameyama, Nagata and Tsushida, 1996; Duh, 1999), Honeybush (Ferreira et al., 1998) or drugs (Ogata, Hoshi, Shimotohno, Urano and Endo, 1997) are anti-oxidants. Selection of clonal lines with high polyphenol content was studied for lavender (Al-Amier, Mansour, Toaima, Korus and Shetty, 1999). A number of studies focused on the composition of rosemary due to its potent antioxidant action applied either to the retarded oxidation in oil (Cuvelier, Richard and Berset, 1996; Houlihan, H. O. and Chang, 1984; Madsen et al.; Wu, Lee, Ho and Chang, 1982) or to the reduction of the loss of colour of carotenoids (Osuna-GarcõÂa, Wall and Waddell, 1997, 1998). The elucidation of the antioxidant mechanisms of its components has also been addressed (Hall, Cupett and Dussault, 1998).

Other potential vegetable sources, such as trees, have been evaluated for antioxidant compounds (Chung et al., 1999; HernaÂndez-PeÂrez et al., 1996; Venkatamuru, Patel and Rao, 1983; Yen and Hsieh, 1998). Among the different parts of the plants, leaves deserve especial attention, e.g. those from green barley (Osawa, Katsu-zaki, Hagiwara, Hagiwara and Shibamoto, 1992), *Pelargonium* sp., *Thalictrum flavum, Nerium oleander* L. (Mallet, Cerrati, Ucciani, Gamisans and Gruber, 1994), several willow species (Julkunen-Tiito, 1985), mulberry (Zhishen, Mengcheng and Jianming, 1999) or avocado (Torres, Mau-Lastovicka and Rezaaiyan, 1987). Roots (Yan, Suzuki, Ohnishi-Kameyama, Sada, Nakanishi and Nagata, 1999), buckwheat groats (Watanabe, 1998), cork from Quercus suber (CadahõÂa, Conde, FernaÂndez de SimoÂ n, GarcõÂa-Vallejo, 1998; Conde, CadahõÂa, Gar-cõÂa-Vallejo and FernaÂ ndez de SimoÂ n, 1998), bark from Fraxinus ornus (Marinova, Yanishlieva and Kostova, 1994), and sprouts from mung beans (Sawa et al., 1999) were also reported to contain antioxidants.

Seeds are another source of antioxidants as reported for tamarind (Tsuda, Watanabe, Ohshima, Yamamoto, Kawakishi and Osawa, 1994), canola (Krygier, Sosulski and Hogge, 1982; Naczk, Amarovicz, Sullivan and Shahidi, 1998; Wanasundara, Amarovicz and Shahidi, 1994), sesame (Shahidi, Amarovicz, Abou-Gharbia and Shehata, 1997), evening primrose (Balasinska and Troszynska, 1998; Wettasinghe and Shahidi, 1999), flaxseeds (Oomah, Kenaschuk and Mazza, 1995), lupinus seed (Tsaliki, Lagouri and Doxastakis, 1999), buckwheat (Pryzbylski et al., 1998), sunflower (Kubicka, JcÎedrychowski and Amarowicz, 1999) and *Rosa rubiginosa* and *Gevuina avellana* (Moure, Franco, Sineiro, DomõÂnguez, NuÂnÄ ez and Lema, 2000). Also in cereals (Baublis, Decker and Clydesdale, 2000; Lehtinen and Laakso, 1998), as in a recent study on corn kernel (Kurilich and Juvik, 1999), antioxidant activity was detected.

Hulls contain compounds with antioxidant activity (Shahidi and Naczk, 1995). Active compounds were detected in hulls from peanut (Yen, Duh and Tsai, 1993; Yen and Duh, 1994; Yen and Duh, 1995; Duh and Yen, 1995 and 1997a; Xing and White, 1997), mung bean (Duh, Yen, Du and Yen, 1997) and buckwheat (Watanabe, Ohshita and Tsushida, 1997). During the extraction of oil from oilseeds, the antioxidant compounds present in the hulls could be incorporated in the oil, as reported for peanut oil extracted from the coated seeds, which contained higher oxidative stability than the oil from dehulled seeds (Shahidi, Amarovicz, Abou-Gharbia and Shehata, 1997). The bran fraction has been reported to have more antioxidant activity than other fractions, as observed for durum wheat by Onyeneho and Hettiarachchy (1992), or in the coat of tamarind seeds, with strong oxidation-inhibiting activity, whereas no activity was detected in the germ (Tsuda, Ohshima, Kawakishi and Osawa, 1994). Also, in red and black bean seed coat of Phaseolus vulgaris, pro-oxidant species were found in the germ and not in the hulls (Muanza, Robert and Sparks, 1998). Baublis et al. (2000) found higher inhibition of iron-accelerated oxidation of phosphatidylcholine liposomes for the water-solubles from high-bran wheat than for re®ned wheat. The outer layers usually contain a greater amount of polyphenolic compounds, as expected from their protective function in the plants.

Agricultural and industrial residues are attractive sources of natural antioxidants. Potato peel waste (RodrõÂguez de Sotillo, Hadley and Holm, 1994a, b), rape of olive (Sheabar and Neeman, 1988), olive mill waste waters (Visioli et al., 1999), grape seeds (Gabrielska, Oszmianski and Lamer-Zarawska, 1997; Karakaya and Nehir, 1999; PekicÂ , KovacÏ , Alonso and Revilla, 1998; Pietta et al., 1998; Saint-Cricq de Gaulejac et al., 1999; Saura-Calixto, 1998; Wulff, 1997; Yamaguchi, Yoshimura, Nakazawa and Ariga, 1999) and grape pomace peels (Bonilla, Mayen, Merida and Medina, 1999; Larrauri, SaÂnchez-Moreno and Saura-Calixto, 1998; Lu and Foo, 1999; Meyer, Jepson and Sùrensen, 1998) have been studied as cheap sources of antioxidants and recently increased antioxidant activity in rat plasma after oral administration of grape seed extracts was reported (Koga et al., 1999). Identification of polyphenolic compounds from apple pomace (Lu and Foo, 1997), grape pomace (Lu and Foo, 1999, 2000), citrus seeds and peels (Bocco, Cuvelier, Richard and Berset, 1998), carrot pulp waste (Chen and Tang, 1998), old tea leaves (Zandi and Gordon, 1999), cocoa by-products (Azizah, Nik Ruslawati and Swee Tee, 1999), non-volatile residue from orange essential oil (Vargas-Arispuro, Sanz, MartõÂnez-TeÂ llez and Primo-YuÂ fera, 1998), and soybean molasses (Hosny and Rosazza, 1999) has also been reported. Spent ground coffee oil from the residue from the production of instant coffee was used to obtain an antioxidant product useful for food preservation and for aroma stabilisation, the antioxidant activity being due to the 5-hydroxytryptamide carboxylic acids (10-75% dry wt. of the product) (Bertholet, Kusy, Rivier and Colarow, 1998).

Scarce literature exists on studies with by-products other than those of plant origin, e.g. shrimp shell waste (Li, Seymour, King and Morrissey, 1998; Seymour, Li and Morrissey, 1996; Wang, Yieh and Shih, 1999). Other compounds such as the dipeptide carnosine (beta-alanyl l-histidine) (Kansci, Genot, Meynier and Gandemer, 1997; Lee and Hendricks, 1997; Lee, Hendricks and Cornforth, 1998) showed antioxidant potential. Protein (Roch, Dreyer, Lacan, Baccou and Ginoux, 1998), protein hydrolysates (Amarovicz and Shahidi, 1997), soluble elastin peptides (Hattori et al., 1998), water-soluble proteins (Okada and Okada, 1998) and pressure treated b-lactoglobulin (Mùller, Stapelfeldt and Skibsted, 1998) were also reported as antioxidant agents. Essential oils (Zygadlo, Lamarque, Maestri and Grosso, 1995), conjugated linoleic acids (Chen, Chan, Kwan and Zhang, 1997) and phospholipids (Bandarra, Campos, Batista, Nunes and Empis, 1999; Chu and Hsu, 1999; Saito and Ishihara, 1997) present antioxidant activity. Palm oil b-carotene (Farombi and Britton, 1999) and capsaicin, responsible for the pungent effect of hot chilli peppers (Henderson, Slickman and Henderson, 1999) are antioxidants, although this latter study evaluates the purified compound. Maillard reaction products were also reported as antioxidant agents (Alfawaz, Smith and Jeon, 1994; Bersuder, Hole and Smith, 1998; Lingnert and Waller, 1983; Nakamura, Ogawa, Nakai, Kato and Kitts, 1998; Pischetsrieder, Rinaldi, Gross and Severin, 1998; Tubaro, Micossi and Ursini, 1996; Wijewickreme and Kitts, 1998; Wijewickreme, Krejpcio and Kitts, 1999). The derivation of natural products with antioxidant activity from brewing seeds, grains and/or germs has been claimed by Niwa and Motoyama (1991) and Watanabe (1999). Also some microorganisms can produce antioxidants (Lin and Yen, 1999; Shimoni, Ampel, ZaÈhner and Neeman, 1998).

Few studies deal with the antioxidant activity of the bound phenolic compounds, linked to lignin or arabinoxylans (Cruz, DomõÂnguez, DomõÂnguez and ParajoÂ , 1999; Lehtinen and Laakso, 1998; Ohta, Yamasaki, Egashira and Sanada, 1994), even though their antioxidant activity in barley and malt is reported to be two-fold higher than that of free phenolic compounds (Maillard and Berset, 1995; Maillard, Soum, Boivin and Berset, 1996). However, other authors have found that the antioxidant activity of citrus peels and seed extracts is not directly related to the free or bound phenolic compounds (Bocco *et al.*, 1998).

POTENTIAL ANTIOXIDANT ACTIVITY FROM RESIDUAL WASTES

Several reports summarize on (Table 2.1) the antioxidant activity of extracts from residual sources and, when reported, the antioxidant activity of natural or synthetic antioxidants is given for comparative purposes. The concentrations tested, for both the natural extracts and for the standard compounds used for comparative purposes, are also indicated. As a general rule the extracts from vegetable materials of residual origin showed antioxidant activity, in some cases comparable to that of synthetic antioxidants, and their extraction and use could be an alternative for obtaining natural antioxidants. Even when the natural extracts are less efficient, the use of some of them as food antioxidants can be advantageous. Maximum levels established for synthetic food additives need not be applicable to naturally occurring compounds, e.g. those from grape marc (Bonilla et al., 1999).

Table 2.1: Antioxidant activities of extracts from residual materials from agro-industrial origin[a]

Residue (solvent)	Antioxidant activity assay	Activity (conc. antioxidant)	Reference
Rape of olives (A+ E)	Refined olive oil oxidation	PV (Control), 52 meq/kg PV (1000 ppm), 18 meq/kg AV (Control), 175	Sheabar and Neeman (1988)
	Refined olive oil oxidation	AV (1000 ppm), 60	
Durum wheat bran (E)	Soy oil oxidation	PV (0.05%), 37.6-42.0 meq/kg PV (0.05% BHA-BHT), 22.0 meq/kg PV (control), 129.0 meq/kg	Onyeneho and Hettiarachchy (1992)
Fraxinus oxinus bark (E)	Triacylglycerols of lard and sunflower oil oxidation	F(0.05%), 3.6-4.8 F(0.1%), 4-6.1 ORR (0.05%), 0.6-0.28 ORR (0.1%), 0.5-0.28	Marinova *et al.* (1994)
Corn bran hemicellulose fragments (FA sugar esters)	Lipid peroxidation of rat liver microsomes	ABS510nm (control), 0.23 ABS510nm (Toc 0.5 mM), 0.03 ABS510nm (0.5 mM), 0.065	Ohta *et al.* (1994)
Potato peel waste (W)	Sunflower crude oil oxidation	PV (200 ppm) ,37.47 PV (200 ppm BHA), 37.47 PV (Control), 49.15	RodrõÂguez de Sotillo *et al.* (1994b)
Tamarind seeds coats (E) (EA)	Linoleic acid oxidation	ILP (a-tocopherol 0.2 mg), 90% ILP (0.2 mg), 90% ILP (0.2 mg), 98%	Tsuda, Watanabe *et al.* (1994)
Wild rice hulls (M)	-	TBARS (0.1%), 2.4 TBARS (0.2%), 0.9 TBARS (Control), 3.8 TBARS (0.02%), 0.3	Wu *et al.* (1994)
Peanut hulls (M)	DPPH radical scavenging	IP (extract 1.5 mg/mL), 89.3% IP (BHA 240μM), 92.6% IP (catechin 8mM), 89.3%	Yen and Duh (1994)
Peanut hulls (M)	Linoleic acid oxidation	AA (9.6 mg), 96.1-96.8%	Yen and Duh (1995)
Mung bean hulls (M)	Linoleic acid peroxidation	ABS500 nm (100 ppm), 0.20 ABS500 nm (BHA, 100ppm), 0.23 ABS500 nm (control), 0.95	Duh *et al.* (1997)

Residue (solvent)	Antioxidant activity assay	Activity (conc. antioxidant)	Reference
	Soybean oil oxidation	PV (100 ppm),70 PV (BHA, 100 ppm),55 PV (control),105	-
	Soybean oil oxidation	ABS535nm (100 ppm), 0.7 ABS535nm (BHA, 100ppm), 0.75 ABS535nm (control), 1.4	-
Peanut hulls (M)	Soybean and peanut oil oxidation	O.S. (0.48%), 194 O.S. (1.20%), 292 O.S. (0.01% BHA), 143 O.S. (Control), 107	Duh and Yen (1997b)
Grape seeds (EA - C pptn)	Lecithin liposome oxidation	ILLO (0.1 mM),86% ILLO (0.1 mM BHT), 88.5% ILLO (+)catechin,40%	Gabrielska *et al.* (1997)
Rose hips (EA-C pptn)	Lecithin liposome oxidation	ILLO (0.1 mM),68% ILLO (0.1 mM BHT), 88.5% ILLO (+)catechin,40%	Gabrielska *et al.* (1997)
Buckwheat hulls/ (E-separation by cromatography)	Methyl linoleate oxidation	PV (Control),3.5 mM PV (0.042 g/l),1.75-2.5 mM PV (BHA), 0.75 mM	Watanabe *et al.* (1997)
Oat groats (M)	Soybean oil oxidation	IO (0.3%), 98.4%	Xing and White (1997)
Oat hulls (M)	Soybean oil oxidation	IO (0.3%), 96.4%	Xing and White (1997)
Lemon peel (M)	Citronellal oxidation	AOP, 0.16 L/g	Bocco *et al.* (1998)
Mandarin seeds (M)	-	AOP, 0.5 L/g	-
Sour orange peel (M)	-	AOP, 0.27 L/g	-
Sweet orange (M)	-	AOP, 0.20 L/g	-
Eucalyptus wood acid hydrolysates (EA)	Linoleic acid and b-carotene oxidation	AAC (400 mg),588 AAC (BHT 400 mg),916	Cruz *et al.* (1999)
Grape pomace (enzyme treatment, A)	Cu-induced human LDL oxidation	NPIT (catechin 3μM GAeq), 110.4 min NPIT (catechin 3μM GAeq), 62.2 min	Meyer, Jepsen *et al.* (1998b)
Grape seed extract (E)	Antiulcer activity (200 mg/kg)	Lesion length (control), 111mm Lesion length (catechin), 88mm Lesion length (extract), 4-20 mm	Saito *et al.* (1998)

Residue (solvent)	Antioxidant activity assay	Activity (conc. antioxidant)	Reference
Grape pomace (M,W-A,W)	DPPH radical scavenging	IC50,0.2 g (4 mg extractable polyph) IC50 (D,L, a-tocopherol), 0.02 g	Saura-Calixto (1998)
	Linoleic acid oxidation	IC50 ,0.7 g (14 mg extractable polyph) IC50 (D,L, a-tocopherol), 0.3 g	-
Olive mill waste water (W-E)	Cu-induced human LDL oxidation	LOOH (control) ,0.35 nm	Visioli *et al* (1999)
Olive mill waste water (EA)	-	LOOH (20 ppm), 0.1-0.2 nm	-
Shrimp shell waste (95% E)	b-carotene bleaching	34.09% BHA/BHT/citric acid, 11.36%	Li *et al*. (1998)
Lentil seed coat (MW)	50% inhibiton in the reduction of the nitro blue tetrazolium (NBT)	IC50,4.52 μg/ml IC50 (ascorbic acid), 5.5 mg/ml IC50 (catechin), 1.9 mg/ml	Muanza *et al.* (1998)
Brown lentil husk (M-W) (EA) (B)	DPPH radical scavenging	IC50,12.58-14.83 μg/ml IC50,6.62 μg/ml IC50,4.04 μg/ml IC50 (catechin), 5.46 μg/ml IC50 (quercetin), 1.73 μg/ml IC50 (gallic acid), 0.63 μg/ml IC50 (ascorbic acid), 2.58 μg/ml	Muanza *et al.* (1998)
Non-volatile residue from orange essential oil	Olive oil 70ºC, 48 h	Oxidation (%)(3000 ppm), 52-64% Oxidation (%)(BHA, 200 ppm),72%	Vargas-Arispuro *et al*. (1998)
Red grape marc (EA)	Refined olive oil oxidation	IP (100 mg/kg),22 h IP (100 mg/kg BHA),26 h IP (100 mg/kg BHT),24 h	Bonilla *et al.* (1999)
Grape seed (W)	Cu-mediated oxidation of rat plasma	Lag phase, 95 min (control) Lag phase 175 min (proanthocyanidin, incubation 30 min)	Koga *et al.* (1999)
Lemon peel (M)	Hemoglobin catalysed	IC50,122.0 ppm	Kuo *et al*. (1999)
Orange peel (M)	peroxidation of linoleic acid	IC50,68.8 ppm	

Residue (solvent)	Antioxidant activity assay	Activity (conc. antioxidant)	Reference
Peanut hull	-	IC50,111 ppm IC50 (BHA),0.65 ppm	-
Gevuina avellana hulls (E)	β-carotene bleaching	AAC (4000 mg/l),767 AAC (230mg BHA/l),901 IO (1 g/l),73.1%	Moure *et al.* (1999)
G. avellana hulls (M)	crude soybean oil oxidation	IO (0.01 g BHA/l),12.8% IO (0.01 g BHT/l),34.8%	-
Old tea leaves/(M)	Rapeseed oil 60°C, 20 days	(control); AV,80.7; PV, 290 meq/kg (OTL 0.25%); AV ,9.5; PV, 30 meq/kg (TBHQ 0.02%); AV,2.7; PV, 5 meq/kg (Rosemary 0.1%); AV,4.8; PV, 25 meq/kg	Zandi and Gordon (1999)

[a]A, Acetone / C, Chloroform / B, Butanol / E, Ethanol / EA, Ethylacetate / M, methanol / W, Water; IO, Inhibition of soybean oil oxidation, after 20 days at 60 ^{0}C in the dark, (PVcontrolÿPVtreatment)/ PVcontrol) *100; AA, Antioxidant activity (thiocyanate method), calculated as percentage of inhibition of peroxidation of linoleic acid; AAC, Antioxidant Activity Coeffcient (Miller, 1971), (Absorbance of extract120h-Absorbance of control120h)/ (Absorbance of control0h-Absorbance of control120 h); A.O.P., Reciprocal of the concentration required to double the half-life time of citronellal (l/g dry matter peel or seed); ABS500nm, Measure of linoleic acid oxidation by the thiocyanate method after 12 days; ABS510nm, Measure of lipid peroxidation in rat liver microsome; ABS535nm, Formation of TBARS (Thiobarbituric acid reactive sustances) on soybean oil after accelerated oxidation; BHA, butylated hydroxyanisole; BHT, butylated hydroxytoluene; DPPH, a a-diphenyl-b-picrylhydrazil; F (protection factor),IPinh/IP0, IPinh is the induction period in the presence of an inhibitor and IP0 is the induction period of the non- inhibited system; ILLO, Inhibition of lecithin liposome oxidation as increase in ABS535nm; ILP, Inhibition of linoleic acid peroxidation respect to a control (thiocyanate method); I.P., Induction Period, the time required for the fuorescence spots on a silica gel TLC plate sprayed with a 3% solution of linoleic acid in hexane, which was considered the induction period for lipid oxidation; IC50, Inhibitory concentration for 50% inhibiton in the reduction of oxidation; Lag phase of CE-OOH accumulation in $CuSO_4$-induced oxidation of rat plasma (min); LOOH, lipid peroxide production; NPIT, Net Prolongation of Induction Time (min) for conjugated diene hydroperoxide formation; O.R.R. (Oxidative Rate Ratio), Winh/Wo, Winh is the initial oxidation rate in the presence of an inhibitor, Who is the initial oxidation rate of the non-inhibited system; OS, Oxidative stability (h); AV, Anisidine value; PV, Peroxide value; TBHQ, tertiary butyl hydroquinone.

PRACTICAL APPLICATIONS

The natural antioxidants from residual sources may be used for increasing the shelf life of food by preventing lipid peroxidation and protecting from oxidative damage. Increasing the oxidation stability of vegetable oils is important for industrial practice, and many antioxidant tests are based on this ability to retard or inhibit the oil rancidity. Many tests of antioxidant activity use pure triacylglycerols or fatty acid methyl esters and many others use either crude, refined or commercial vegetable oils. These assays are carried out under extreme conditions; however, more practical assays have been used for food during storage or processing (Table 2.2). The stability of processed foods has also been assessed after addition of antioxidants. Both synthetic and natural phenols enhanced the oxidative stability of freeze-dried, ground extruded corn starch - soybean oil mixtures (Camire and Dougherty, 1998). Zandi and Gordon (1999) evaluated the methanolic extracts of old tea leaves (OTL) during deep-fat frying of potato crisps at 180°. The authors found that OTL extracts are slightly more active antioxidants than rosemary extract (both at 0.1%) during the frying stages. In this kind of assay, rosemary was more active than BHA or BHT (Che Man and Tan, 1999). The antioxidant compounds from residual sources could be used for increasing the stability of foods by preventing lipid peroxidation and also for protecting oxidative damage in living systems by scavenging oxygen radicals. Studies to incorporate the crude extracts or the whole vegetable material in foods (meats) as `antioxidant ingredients' are scarce. Wu, Zhang, Addis, Epley, Salih and Lehrfeld (1994) reported the beneficial effects of using wild rice as an antioxidant, particularly after particle size reduction and cooking. Water extracts are advantageous in relation to certification for food (Mùller *et al.*, 1999). Maillard reaction products were evaluated as antioxidants in cooked ground beef during storage at 4°C (Alfawaz *et al.*, 1994) or natural antioxidants in cooked, minced turkey (Bruun-Jensen, Skovgaard, Madsen and Skibsted, 1996; Bruun-Jensen, Skovgaard, Madsen, Skibsted and Bertelsen, 1996). Improvement in colour stability for different species of rock fish was observed in the presence of antioxidant extracts from shrimp shell waste. Carotenoid degradation was decreased by inhibiting autooxidation and/or lipoxygenase activity, this activity being increased with increasing concentrations of shrimp shell waste (Li *et al.*, 1998). Antioxidants have also been proposed for preventing loss or improving the stability of pigments from red beet juice in the food industry (Han, Kim, Kim and Kim, 1998), as well as for aroma protection and stabilization (Bertholet *et al.*, 1998) and for use in oral and topical pharmaceutical and cosmetic compositions (Wulf, 1997). Chambers et al. (1988) reported the use of antioxidant extracts for inhibiting the warmed-over flavour commonly associated with cooked roast beef which has been reheated. Probably the dietary supplementation of antioxidants in feed improves the antioxidative stability of processed meat, as recently reported for supplementation of vitamin E and β-carotene (Ruiz, PeÂrez-Vendrell and Esteve-GarcõÂa, 1999). The organoleptic characteristics of the extracts must be suitable for incorporation into food products without conferring the intense herb flavour that may limit some applications, as occurs for natural rosemary extract which has excellent antioxidant properties. In addition to colour and flavour, aspects such as production cost, antioxidant activity and toxic or pathogenic activity should be considered. The search for cheap, renewable and abundant sources of antioxidant compounds is attracting worldwide interest. Much research is needed in order to select raw materials; those of residual origin are especially promising due to their lower costs. However, extensive research on potential sources, optimisation of extraction processes, knowledge of the mechanisms of the in vivo action and assimilation, are still required.

Table 2.2: Assays for determining antioxidant activity[a]

Assay	References
Oxidation in hydrophobic, hydrophilic and emulsions of	
Vegetable and marine oils	Frankel *et al.* (1996); Han et al. (1990); Nieto *et al.* (1993).
Fatty acids, fatty acid methyl esthers, triacylglycerols	Foti *et al.* (1996); Heinonen *et al.* (1997); Ponginebbi, (1999)
Phospholipids	Farombi and Britton, (1999); Kuo *et al.* (1999); Marinova and Yanishlieva (1996)
Citronellal	Bocco et al. (1998),
β-Carotene oxidation in linoleic acid emulsion	Al-Saikhan *et al.* (1995); Mallet *et al.* (1994); Marco (1968); von Gadow *et al.* (1997a)
Lard	Kim *et al.* (1994); von Gadow *et al.* (1997c)
Scavenging of radicals	
$ABTS^+$ (metmyoglobin assay)	Hagerman *et al.* (1998); Rapisarda, Tomaino, Lo Cascio, Bonina, De Pasquale and Saija (1999)
Alkyl peroxyl radical	Milicǐ *et al.* (1998); Sawa *et al.* (1999)
a,a-diphenyl-b-picrylhydrazil radical scavenging activity	Brand-Williams *et al.* (1995); Duh (1998); Larrauri *et al.* (1998); Okada and Okada (1998); von Gadow *et al.* (1997 b,c),
Hydroxyl radical scavenging activity	Duh (1998); Watanabe *et al.* (1997)
Oxygen radical absorbance capacity	Cao *et al.* (1996); Kalt *et al.* (1999); Lin *et al.* (1996); Pior *et al.* (1998)
Superoxide scavenging activity	Aruoma *et al.* (1993); Nakamura *et al.* (1998), Yen and Duh (1994); Okada and Okada; Yamaguchi *et al.* (1999)
Others	
Redox potential	Hagerman *et al.* (1998)
Reducing power (reducing potassium ferricyanide)	Duh *et al.* (1997); Duh (1998)
Degradation rate of phenolic compounds as a consequence of their antioxidant activity	Chimi *et al.* (1991)
Chelating activity of Fe^{+2} or Cu^{+2}	Arora and Strasburg (1997); Chen and Ahn (1998); Hudson and Lewis (1983); Lin and Yen (1999); Okada and Okada (1998); Yen and Duh (1994); Yen and Wu (1999)
Oxidation in biological membrane models, cell assays and in vivo assays	

Assay	References
Liposome membranes (UV induced lipid peroxidation)	Gabrielska *et al.* (1997)
Oxidation of rat liver mitochondria	Yen and Hsieh (1998)
Lipid oxidation in egg lecithin microsomes	Yen and Hsieh (1998)
NADPH/iron-induced peroxidation in liver microsomes	Chambers *et al.* (1996); Yen and Hsieh, (1998)
Phosphatidylcholine liposome oxidation	Aruoma, (1997); Baublis *et al.* (2000)
Human plasma radical scavenging capacity	Carbonneau *et al.* (1998)
Cu-induced or AAPH-induced human LDL oxidation	Abuja *et al.* (1998); Carbonneau *et al.* (1998); Meyer, Heinonen *et al.* (1998); Meyer, Jepsen *et al.* (1998); Visioli *et al.* (1999); Zhu *et al.* (1999)
DNA oxidation and fragmentation assays, bleomycin-dependent DNA damage	Aruoma (1999); Aruoma *et al.* (1997); Aruoma *et al.* (1993); Hagerman *et al.* (1998); Saint-Criq de Gaulejac *et al.* (1999); Yen *et al.* (1997); Yokozawa *et al.* (1998)
Rat plasma antioxidant capacity, with Cu-induced or AAPH-induced formation of cholesteryl ester hydroperoxides	Koga *et al.* (1999)
Oxidation of blood plasma	Miyake and Shibamoto (1998); Nakagawa *et al.* (1999)
Oxidative susceptibility of ILDL in animals	Wakabayashi (1999)
Animal experiments, reduction of uremic toxins in the blood of rats	Yokozawa *et al.* (1998)
Inhibition of oxidation induced apoptosis in healthy human cells	Muanza *et al.* (1998)
Antiulcer activity	Saito *et al.* (1998)
Guinea pig complement serum hemolysis	Nakagami *et al.* (1995)
Leukotriene production by human neutrophiles	Visioli *et al.* (1999)
Food stability	
Storage of meat	Kanatt, Paul, D' Souza and Thomas, (1998)
Lipid stability in extruded corn	Camire and Dougherty (1998)
Lipid stability in precooked meat	Alfawaz *et al.* (1994); GuÈ ntersperger *et al.* (1998); Ramezanzadeh *et al.* (1999)
Lipid oxidation in our-lipid mixtures	Lehtinen and Laakso (1997); Wijewickreme and Kitts (1998)
Stability of oils during frying	Che-Man and Tan, (1999); Jaswir and Che-Man (1994); Zandi and Gordon (1999)

Assay	References
Colour stability and lipid oxidation of Rock®sh	Li *et al.* (1998); Seymour *et al.* (1998)
Storage of ground beef	Lee and Hendricks (1997), Lee *et al.* (1998)
Colour stability of carotenoids containing materials (paprika, carrot, etc.)	Han *et al.* (1998); Osuna-GarcõÂa *et al.* (1997); Talcott and Howard (1999)
Oxidative stability of turkey thigh meat homogenate	Mùller *et al.* (1999)

[a]Abbreviations: AAPH, 2, 2'-azobis (2-aminopropane) dihydrochloride; ILDL, intermediate and low density lipoproteins; NADPH, nicotinamide adenine dinucleotide phosphate in hydrogenated form.

In this book it has been emphasized that many residues are antioxidant sources; perhaps by-products of grape processing, such as seeds and peels are the most promising, together with vegetable residues such as tea leaves. Extract production is a key step for obtaining antioxidants with an acceptable yield. To select a solvent, comparative studies are required for each substrate. Besides conventional extraction with solvents as ethanol, methanol, ethylacetate, other methods such as supercritical extraction must be assayed, because they offer a good yield and preserve the properties of the antioxidants: Natural antioxidants often shown antioxidant powers lower than those of synthetic ones, but they are not law-limited in quantity. In any case, a detailed economic study, with reliable consideration of their potential toxicity must be done before any possible application on a practical scale.

Plant Phenolics: Structure, Classification and Biosynthesis

Phenolic compounds include a large group of chemierr1 substances containing at least one aromatic ring with one or more hydroxyl groups, in addition to a number of other substituent's. Numerous reports are available on the structure, classification, and possible biosynthetic pathways of phenolic compounds (White, 1957; Gupta and Haslam, 1980). However, the precise structure of polymeric phenols such as condensed tannins are not yet completely elucidated due to problems associated with their isolation and purification. The biosynthetic pathways reported in the literature are often speculative. The use of word "tannin" for polyphenols in food grains has been recently questioned (Hulse, 1984). The precise nomenclature for the phenolic substances in seeds of important food grains, such as sorghum, millets, and grain legumes, is yet to be developed. This section briefly discusses the structure, classification and biosynthesis of plant phenolics in foods and feeds.

STRUCTURES AND CLASSIFICATION

Harborne and Simmonds (Harborne and Simmonds, 1964) have classified plant phenolics into 15 major groups. However, the major or most common phenolic constituents of plants can be divided broadly into two main groups: phenolic acids and coumarins (C_6-C_1 and C_6-C_3 structures), and flavonoids, including anthocyanidins (C_6-C_3-C_6 types).

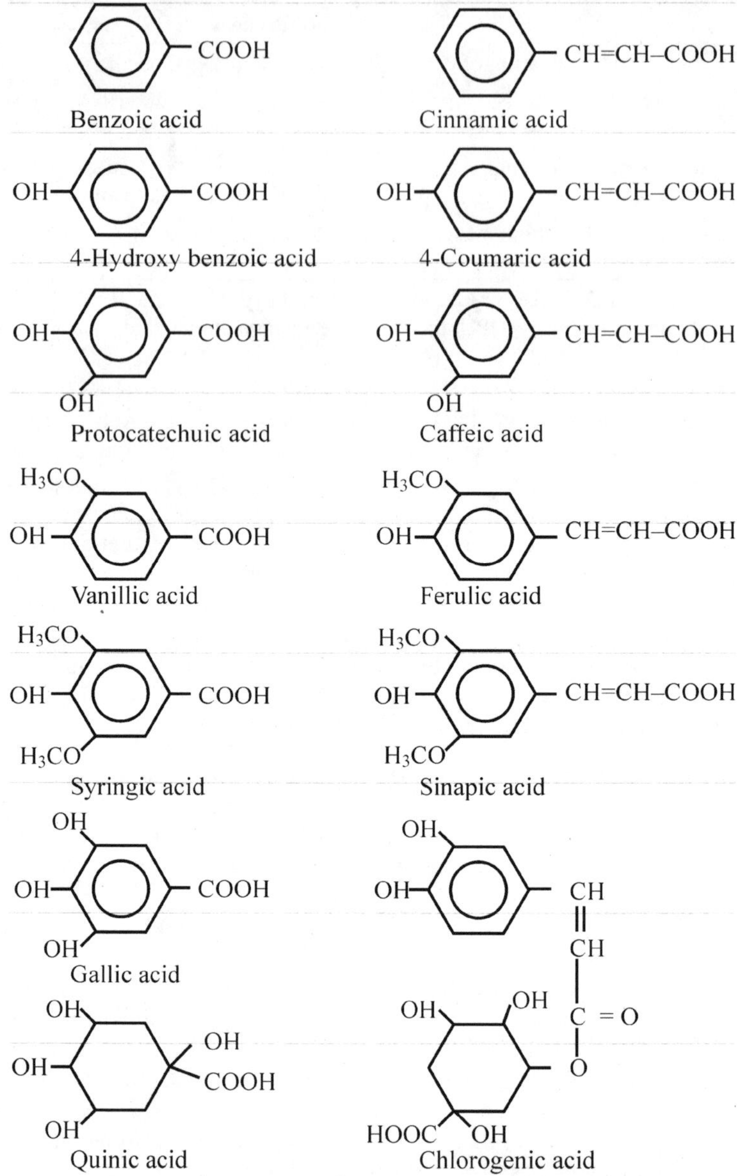

Figure 2.2: Structures of common phenolic acids found in plants.

A. Phenolic acids

Two families of phenolic acids are widely distributed in plants - a range of substituted benzoic acid (C_6-C_1) derivatives and those derived from cinnamic acid (C_6-C_3). Both types of phenolic acids usually occur in conjugated or esterified form. The simpler type of benzoic acid derivatives includes 4-hydrobenzoic, protocatechuic, vanillic, gallic and syringic acids and the o-hydroxy salicylic and gentisic acids (Figure 2.2). The cinnamic acids - 4-coumaric, caffeic, ferulic and sinapic are found in most oilseeds and occur frequently in the form of esters with quinic acid or sugars. The chlorogenic acid, which commonly occurs in sunflowers, coffee beans and cooked potatoes, is an ester of caffeic and quinic acid, and is found in several isomeric and derivatized forms. The phenolic acids such as caffeic and protocatechuic, commonly present in mushrooms, potatoes, sweet potatoes, tobacco leaves, apples, tomatoes, bananas, and tea leaves, play an important role in the development of enzymatic browning reactions. These acids have appropriate configurations for the phenolase activity. The quinones formed polymerize into red-brown polymers often called melanins, which appear at a point of mechanical injury in a potato tuber or at the surface of an apple. Chlorogenic acid is often responsible for the postcooking blackening sometimes observed in potatoes.

Aromatic amino acids like tyrosine and phenylalanine arc the most commonly observed C_6-C_3-type phenolic acids in all living cells and arc vital components of cell proteins. They are also involved in the biosynthesis of a number of other phenolic compounds. Glycosides of dihydoxyphenylalanine (DOPA) occur in bean plants. DOPA is also a precursor of the melanin pigments of animals. The cinnamic acid derivatives (R-CH=CH-COOH) which exist not as free acids but in bound form, such as esters, are of widespread occurrence. Ferulic acid and 4-coumaric acid esters are widely distributed in vascular plants. Common esters are chlorogenic acid p-coumaryl-D-glucose ester and the corresponding ferulic acid glucose ester. Cinnamic acids also occur as amides such as subaphylline or egelin. The ferulic acid has also been reported to form esters in wheat flour with xylan of glycoprotein (Fausch *et al.*, 1963).

B. Coumarins

Coumarins are lactones of *cis-o-hydroxy* cinnamic acid derivatives (Figure 2.3). They are of widespread occurrence in vascular plants. Neish (Quesnel, 1968) has listed over 30 naturally occurring coumarins. In addition to the usual hydroxyl and methoxyl substitutes, furocoumarins such as psoralen and angelicin are also found. The suberosin and braylin appear to have been alkylated by an isoprenoid group. Dicoumarol is a hermorraegic factor found in spoiled sweet clover hay. Coumarins are less widely distributed in seeds and occur primarily as glycosides. They exhibit limited chemical reactivity and have little effect on the organoleptic or nutritive value of food.

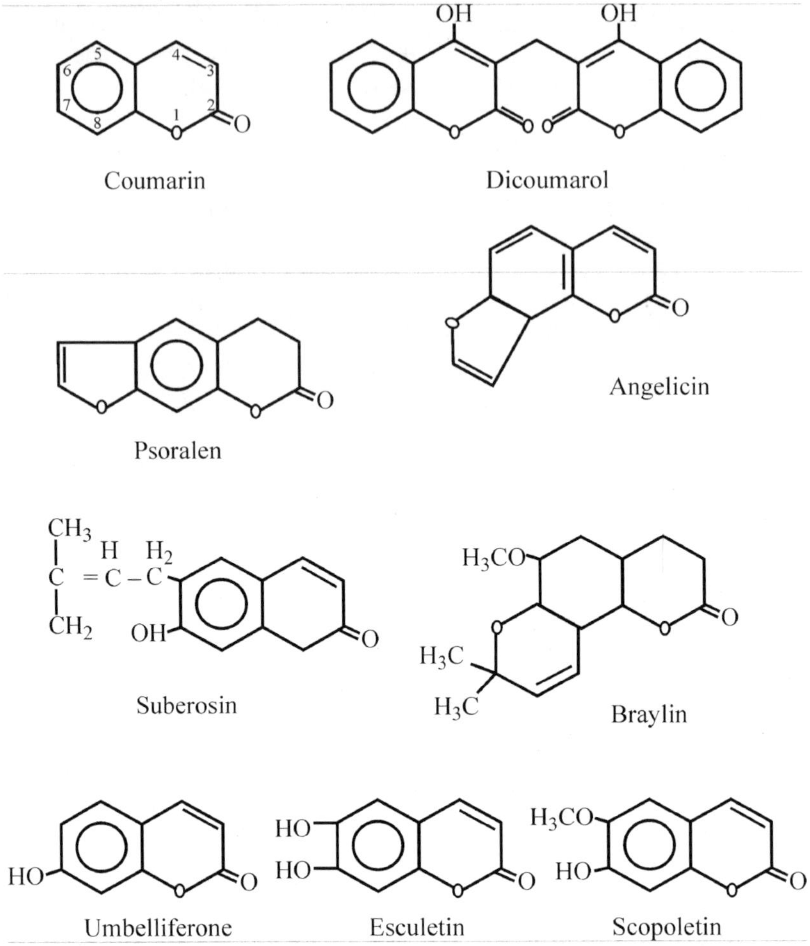

Coumarin Dicoumarol

Psoralen Angelicin

Suberosin Braylin

Umbelliferone Esculetin Scopoletin

Figure 2.3: Structures of some natural coummarins.

C. Flavonoid compounds

Flavonoids include the largest and most diverse range of plant phenolics (Harborne, 1965; Walker, 1975; Harhorne, *et al.*, 1975). They consist of scarlet, crimson, and purple anthocyanin pigments of flowers; ivory or pale yellow flavones, flavonols, flavonols less commonly found yellow chalcones and aurones and colorless flavonones and isoflavones. All flavonoids share a basic C_6-C_3-C_6 or the CIS skeleton of flavones (Figure 2.4A*)*, the ring A being derived from three acetate units condensed head-to-tail, while -ring B and the three carbon atoms of the central ring are derived from cinnamiç acid. Although more than one hundred different flavonoid aglycones have been isolated from plants, only II of these are most common (Figure 2.4B). The anthocyanidins are very widely present in colored flowers and fruits. Flavonols usually accompany anthocyanidins in flowers and also occur with great frequency in leaves.

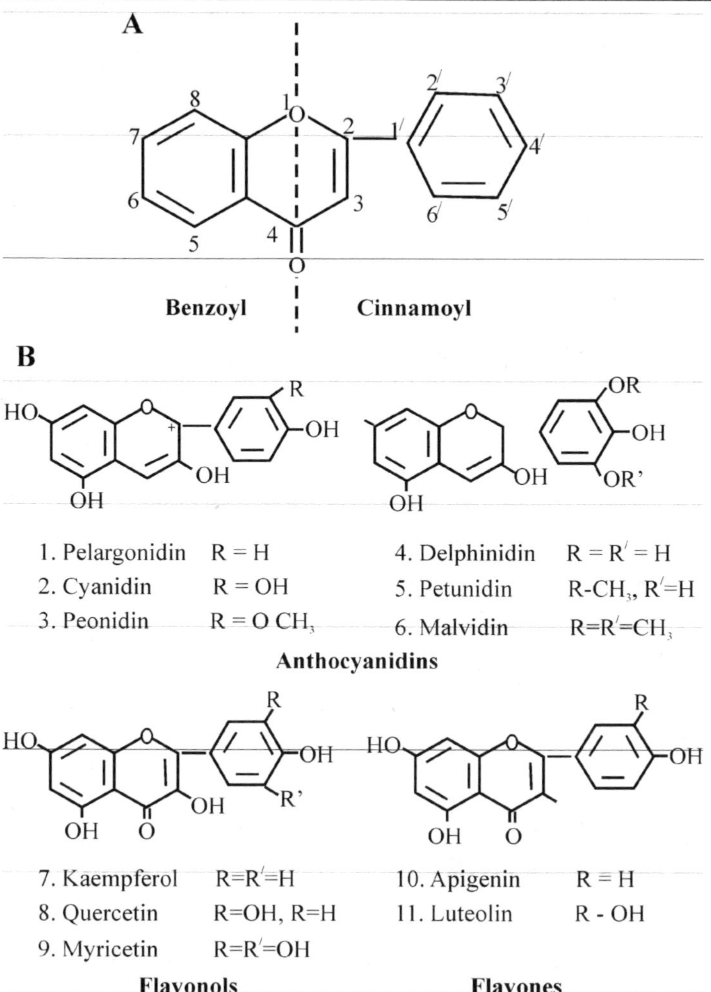

Figure 2.4: (A) Basic flavones, (B). Aglycones of common flavonoids: anthocyanidins (1 to 6); flavonols (7 to 9) and flavones (10 and 11).

Depending upon the presence or absence of an OH group in the side chain of the C. unit of the CIS skeleton, two branches of flavonoid families are generated. If an OH group is absent (as in the case of cinnamic, p-coumaric, or caffeic acid), the condensation of C. and C_6 units results in the formation of 3-desoxyflavonoids (Figure 2.5), whereas the presence of OH in the side chain of the C. unit (as in the case of the enol form of phenyl pyruvate or after the reaction of water with the quinoid radical of cinnamic acid) results in the formation of 3-hydroxyflavonoids (Figure 2.6). In both flavonoid families, accessory hydroxyl and methyl groups, as well as different sugar residues in O- and C-glycosidic linkages, can be introduced at different positions of the molecule. The positions, number, and combinations of these substituents contribute to the enormous complexity of the flavonoid spectrum. Both families are independent from each other in their origin as well as in occurrence, while flavanones and flavones are often found together. They are connected with each other and chalcones by specific enzymes, yet there is a certain mutual exclusion between flavones and flavonols in many plant families. Anthocyanins are almost absent in flavanone-rich plants (Harborne, 1967). On the other hand, the production of flavonols and anthocyanins is under a common and specific gene control. The interactions of the individual flavonoid subgroups are shown in Figure 2.7.

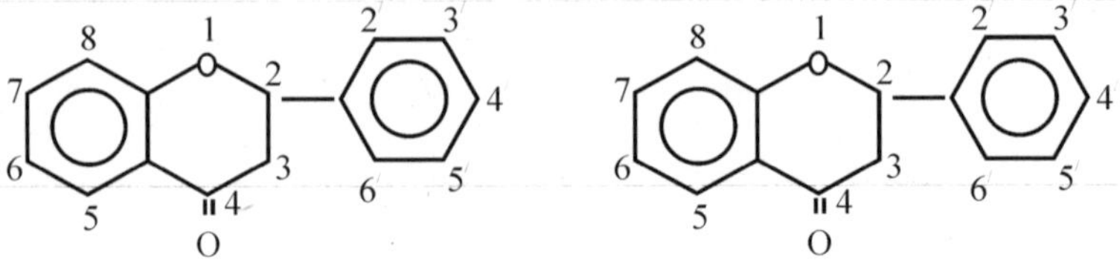

Figure 5. 3-Desoxyflavonoids

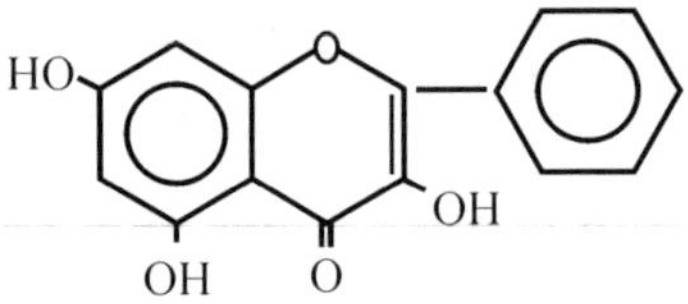

Flavonol

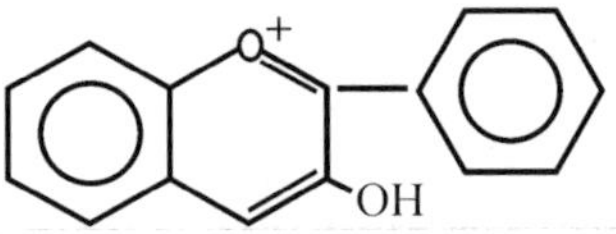

3-Hydroxyflavylium (Anthocyanidin)

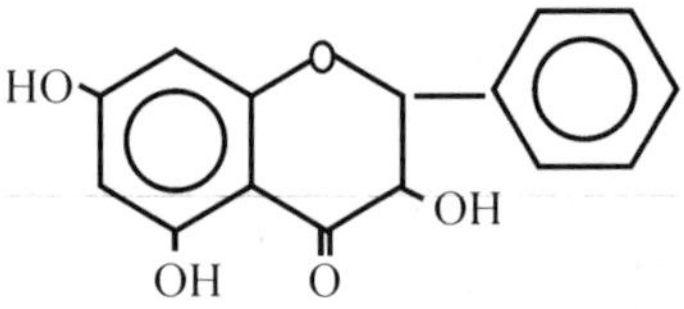

Flavan - 3,4-DIOL (Leucoanthocyanidin)

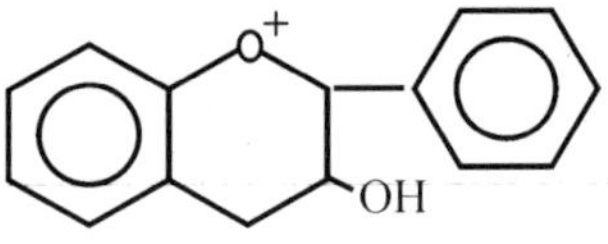

Flavan - 3-OL (Catechin)

Figure 2.5, 6: 3-Hydroxylflavonoids.

Most flavonoids occur as glycosides in which the C_6-C_3-C_6 aglycone part of the molecule is linked with a number of different sugars. Depending on whether the linkage of sugar to flavonoid aglycone is through an OH group or through a carbon-carbon bond, they are called O-glycosylflavonoids or C-glycosylflavonoids, respectively. The binding of sugars imparts sap solubility, stability to light, and resistance to enzymic degradation of flavonoids. The types of sugars and position of attachment vary greatly. The glycosidic linkage is usually (Harborne, 1965) except in rhamnosides and arabinosides. The linkages are usually 1 -> 2 or 1 -> 6 types, and the sugar at the reducing end is always glucose. Anthocyanins always have a sugar at the 3-hydroxyl group, although 3, 5-diglycosides are also common. Flavonols usually have a sugar at the 3-position, but also carry sugars at 7-, 3'-, and 4' as monoglucosides, and 3, 7-, 3, 4', and 7-4' –diglucosides.

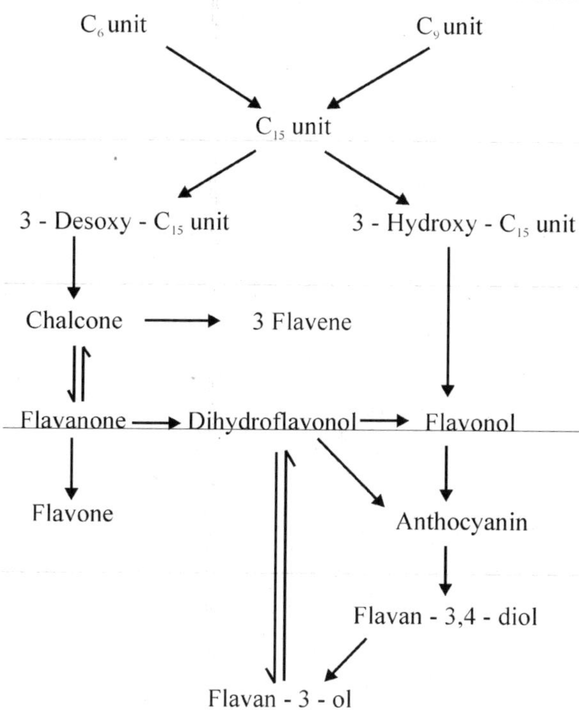

Figure 2.7: Interactions of flavonoids subgroups.

Some flavonoids among the flavones, isoflavones, and flavonones are C-C-linked glycosides. The number of sugars may be one (mono) or two (di-C-glycosylflavonoids). Among the mono-C-glycosylflavonoids, C-glycosylflavones, C-glycosylflavonols, C-glycosylflavonones, C-glycosylisoflavones, and C-glycosyldi-hydro-chalcones are well defined, and only C-glycosylflavones are the most important group.'·The well-characterized C-glycosylflavones arc vitexin, isovitexin, glucosylvitexin, orientin, isoorientin, and glucosylorientin (Figure 2.8). The vitexin, glucosylvitexin, and glucosylorientin are principal flavonoids of pearl millet grain.

Flavanones and dihydroflavonols are the simple reduction products of flavones and flavonols and serve as precursors of the more highly oxidized flavonoids in plants. Isoflavones arc isomenc with flavones and are produced from the same C_{15} skeleton by aryl migration. Flavanols are unique in that they do not occur as glycosides, but show reactivity through polymerization into condensed tannins.

D. Tannins

It is difficult to give a concise and complete definition of the word "tannin." The term "tannin," commonly used for the phenolics of cereals and legumes, has been questioned (Hulse, 1984). It has been recommended that the use of the word "tannin" be avoided as far as possible. Originally, this word was coined by Seguin (1980) to describe the substances present in vegetable extracts which are responsible for converting animal skins into stable product leather. The phenomenon is called the tanning process and the substances responsible to convert skin into leather are called tannins. In plant extracts, these substances are polyphenols of varying molecular size and complexity. In fact, many substances which exist in plants, other than polyphenols, have all the properties of tannins; however, they have not been tested for their ability to leather hides (Harborne, 1967). Furthermore, substances like systems, which can leather hides, are not of plant origin. Thus, it is clear that there can be a large group of substances of diverse origin and chemical makeup which can exhibit a tanning property.

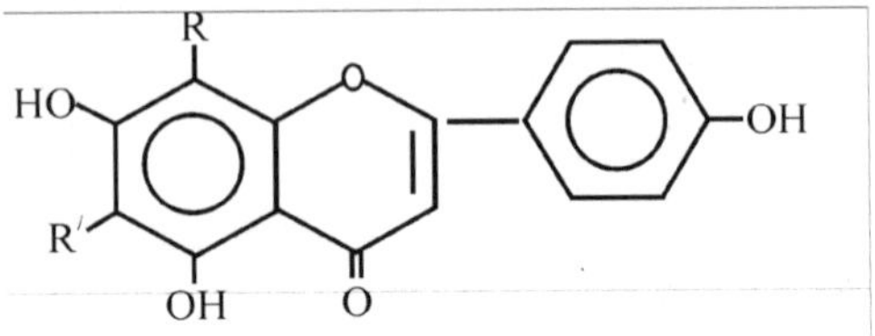

R = C - Glucosyl, R′=H,   Vitexin
R = H, R′ = C - Glucosyl   Isovitexin
R = C - Glucoglucosyl, R′=H,   Glucosylvitexin

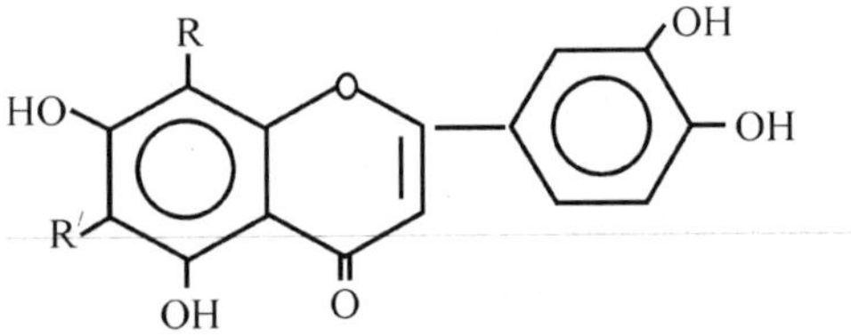

R = C - Glucosyl, R′=H,   Orientin
R = H, R′ = C - Glucosyl   Isoorientin
R = C - Glucoglucosyl, R′=H,   Glucosylorientin

Figure 2.8: Structures of important C-gycosylflavones.

All the phenolics (phenolic acids, coumarins, and flavonoids) of plants do not possess a tanning property. Among the polymeric phenols, only the polyphenols of certain molecular weight exhibit such phenomenon. The polyphenols with a molecular weight between 500 and 3000, containing a sufficiently large number of phenolic hydroxyl or other suitable groups (1 to 2 per 100 mol wt) can only form effective cross-links between protein and other macromolecules. These are often referred to as vegetable tannins. Phenolic compounds of a lower molecular weight do not form stable cross-links with proteins and compounds of high molecular weight are ineffective as tanning agents because they are too large to penetrate collagen fibrils. Therefore, polyphenols having a tanning property constitute only a proportion of the total phenols In a tissue of plants. White therefore suggests that much of the botanical data on the occurrence of tannins in plants is of doubtful validity since it is based on tests which arc sufficiently nonspecific. Several reports on the tannins in grains or food products are based on the general phenolic reaction with certain reagents which often estimate total phenolics in tissue rather than true vegetable tannins.

Bate-Smith and Swain''' defined vegetable tannins as water-soluble phenolic compounds having molecular weights between 500 and 3000. Besides giving usual phenolic reactions they have special properties such as the ability to precipitate. Alkaloids, gelatin and other proteins. Gupta and Haslam (1980) have further added to this definition that vegetable tannins are normal metabolic products and are not the products formed by in vitro transformation by chemical or other means. The precise nomenclature is, therefore, essential to describe the protein-precipitating polyphenols of plant tissues

The ability of vegetable tannins to complex with proteins through hydrogen bonds, covalent linkages, and hydrophobic interactions makes these tannins unique metabolic products. The formation of substantial quantities of free tannin molecules within the cell would cause the precipitation of metabolic proteins. It is similar to several antibiotics that are more toxic to the producing organism than are their precursors. The presence of tannins in plant tissue has several important consequences. The interaction of tannins with salivary proteins and glycoproteins in the mouth renders tissue astringent to the taste such as fruits like blackberry; strawberry, cranberry, and apple psychologically improve the acceptability of certain fruits. Because of their protein-binding property, tannins are of considerable importance in food processing, fruit ripening, and wine manufacturing. It is believed that changes in the palatability of many fruits that occur upon ripening are associated with concomitant changes in the chemical forms of polyphenols present in fruit.

Tannins have been classified by Freudenberg (1920) into two groups based on structural types: hydrolyzable tannins and non hydrolyzable or condensed 'tannins. This is the most widely accepted classification so far proposed (White, 1957; Swain, 1965; Haslam, 1966; Walker, 1975). Of the two groups, the condensed tannins are more widely distributed in higher plants (Schanderl, 1970).

1. Hydrolyzable tannins

These tannins contain a central core of polyhydric alcohol, such as glucose, and hydroxyl groups which are esterified either partially or wholly by gallic acid (gallotannins) or hexahydroxydiphenic acid (ellagitannins). The hexahydroxydiphenic acid is isolated as its stable dilactone ellagic acid. These types of tannins are readily hydrolyzed by acids, bases, or certain enzymes. Upon hydrolysis, they yield glucose or some other polyhydric alcohol and gallic acid or phenolic acids related to it.

a. Gallotannins

The present knowledge of the structure of gallotannins is based on the original work of Emil Fischer and Karl Freudenburg on Chinese and Turkish tannin. Chinese gallotannin (tannic acid) contains a glucose and gallic acid, in which each glucose molecule is linked to an average of nine to ten gallic acid residues. A possible structure of typical gallotannin (Chinese tannin) is shown in Figure 2.9. Some of the other known natural gallotannins include Turkish tannin (Figure 2.10), obtained from *Quercus infectora; Tara* tannin (Figure 2.11), and acidic tannin from the seed pods of *Caesalpinia spinosa; Acer* tannin (Figure 2.12), obtained from the leaves of Korean maple *(Acer ginnala);* and *Hamamelis* tannin (Figure 2.13), isolated from hazel *(Hamamelis virginiana)* (Swain, 1965).

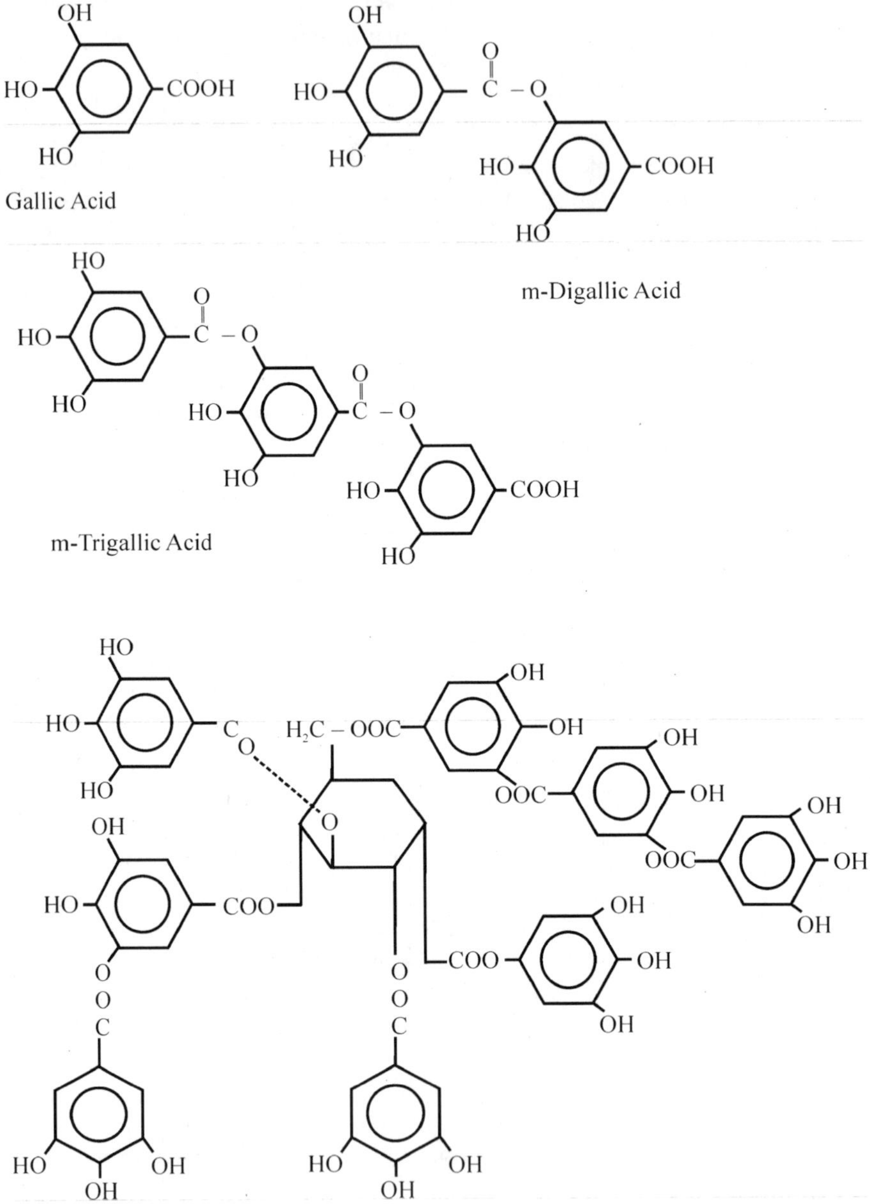

Gallic Acid

m-Digallic Acid

m-Trigallic Acid

Figure 2.9: Structures of gallotannins and its constituents

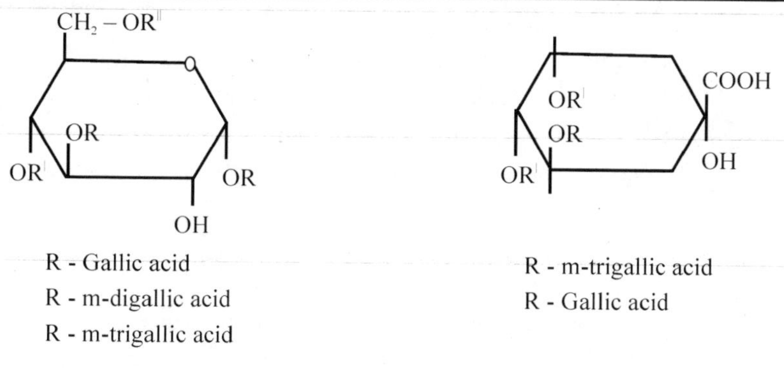

R - Gallic acid
R - m-digallic acid
R - m-trigallic acid

Figure 2.10. Turkish tannin

R - m-trigallic acid
R - Gallic acid

Figure 2.11. *Tara* tannin

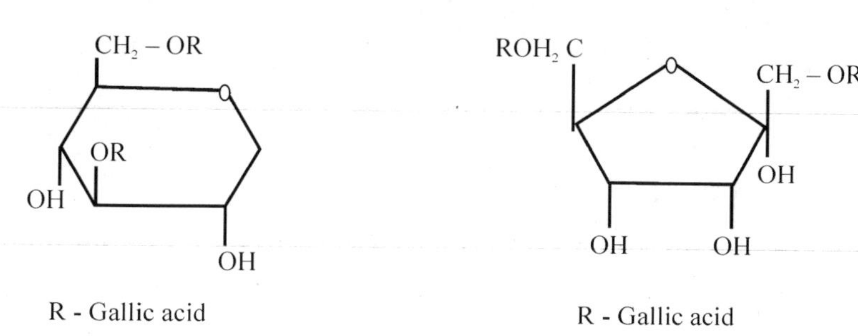

R - Gallic acid

Figure 2.12. Acer tannin

R - Gallic acid

Figure 2.13. Hamumekis tannin

Figure 2.10, 11, 12 and 13: Turkish tannin, Tara tannin, Acer tannin, and Hamumekis tannin.

b. Ellagitannins

In these types of tannins, a molecule of hexahydroxydiphenic acid is linked with glucose as a diester in addition to gallic acid. Upon hydrolysis, the hexahydroxy-diphenic acid undergoes lactonization to produce ellagic acid (Figure 2.14). The simplest ellagitannin, corilagin, occurs in *Caesalpinia. Coria ria*, *Terminalia chebula, Schinopsis* species, and *Eucalyptus sieberiana* (Swain, 1965). The levorotatory form of corilagin is known to occur in *Eucalyptus gigalllia.* The other important ellagitannins include chebulinic acid (Figure 2.15) and chebulagic acid (Figure 2.16), obtained in murobalan extract *(Terminalia chebula).* The brevilagins obtained from the extracts of *Caesalpinia brevifolia* are reported to be structurally related to these ellagitannins. The hydrolyzable tannins of sweet chestnuts contain dehydrodigallic acid, and that from valonia *Quercus aegilops* has the hexahydroxydiphenic congener (valoneaic acid) isolated as dilactone. Algarobilla contains, in addition to brevilagins, tannins which yield brevifolin carboxylic acid upon hydrolysis. The hydrolyzable tannins occur in seed pods; bark and wood; fruits; and leaves or galls of plants belonging to the family Leguminosae, Fabaceae, Combretaceae, Anacardiaceae; their concentrations range from 15 to 64%.

Hexahydroxy Diphenic Acid

Ellagic Acid

Ellagitannin (Corilagin)

Figure 2.14: Structure of typical ellagitannin.

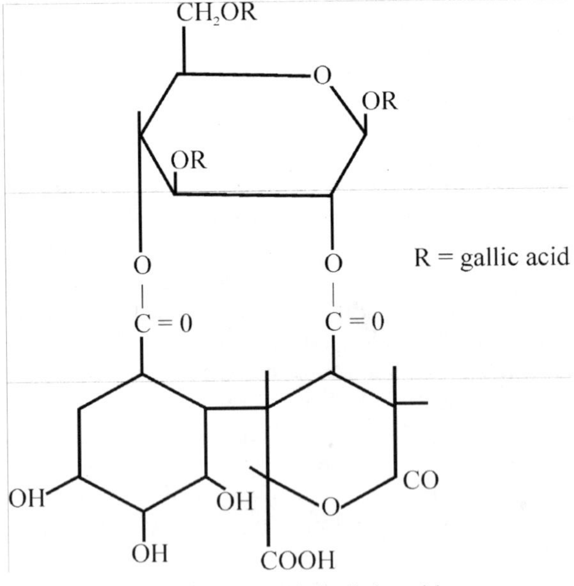

Figure 2.15: Chebulinic acid

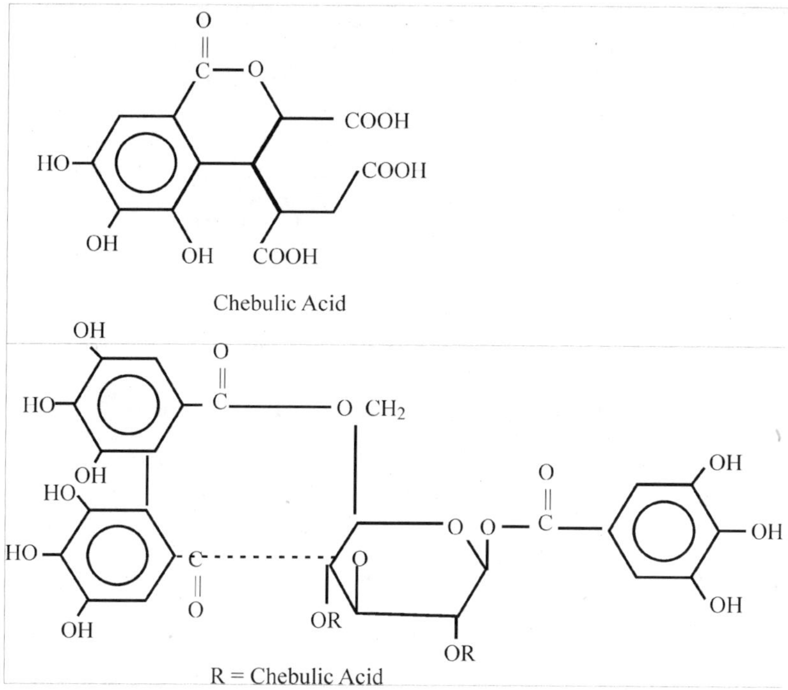

Chebulic Acid

R = Chebulic Acid

Figure 2.16: Chebulagic acid.

2. Condensed Tannins

These types of polymeric phenolics are present, often in substantial quantities, in many dicotyledons and in the most primitive of vascular plants, such as ferns and gymnosperms. The principal tannins present in sorghum, grain legumes, wine, and several fruits and forage plants are of this type. Hence, their structure, properties, biosynthesis, possible harmful effects to man and animals, physiological functions and postharvest processing technology to remove these compounds from food/feed before consumption or inactivate them has received the considerable attention of biochemists, nutritionists, food technologists, and food processors throughout the world. They are structurally more complex and their complete structures from various sources are yet to be determined. The extraction, isolation, purification, and characterization of monomers of these polymeric phenolics have shown that they are mainly the polymerized products of flavan-3-ols and flavan-3, 4-diols, or a mixture of the two. The polymer, referred to as flavolans, analogus to that of polysaccharide, araban, or dextran (Swain, 1965) is popularly called "condensed tannins".

a. Flavan-3-ols

Flavan-3-ols are often referred to as catechins. The molecular structure of different catechins is given in Figure 2.17. Since catechin molecules possess two asymmetric carbon atoms at C-2 and C-3, four isomers exist. They are (+) and (-) catechin in which 2-phenyl and 3-hydroxyl groups are *trans* and (+) and (-) epicatechin, in which these groups are *cis*. Only (+) catechin and (-) epicatechin are common in nature. The (-) epicatechin was first isolated from cacao beans. The (-) epigallocatechin and its 3-gallate are major phenolic constituents of green tea leaves. Besides (+) gallocatechin, (+) catechin, (-) epicatechin and its 3-gallate have been also shown to be present in tea

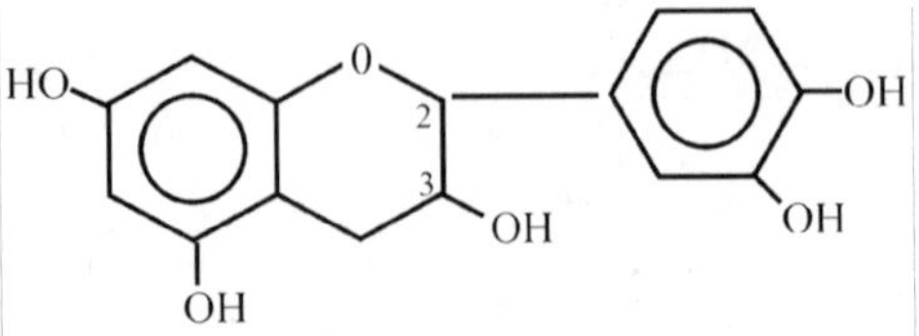

(+) - Catechin (2,3 - trans)
(-) - Catechin (2,3 - cis)

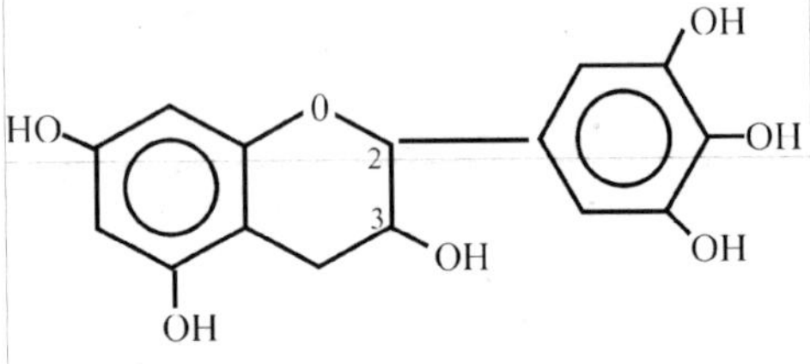

(+) - Gallocatechin (2,3 - trans)
(-) - Gallocatechin (2,3 - cis)

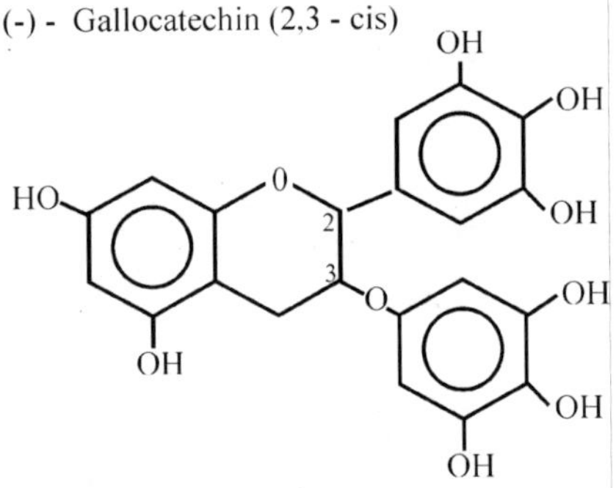

(-) Epigallo catechin-3-Gallate

Figure 2.17: Some natural catechins.

b. Flavan-3, 4-diols

These phenolics belong to the class of compounds called leucoanthocyanins because upon heating in acid solution. They not only polymerize to phlobaphene-like products, but also produce anthocyanidin. For example, leucocyamdin yields cyanidin (Figure 2.18). A flavan-3, 4-diol molecule possesses three asymmetric carbon atoms. C-2, C-3 and C-4, hence eight isomers arc possible. Flavan-3, 4-diols having hydroxylation patterns corresponding to flavan-3-ol are also known to occur. The structures of some known tlavan-3, 4diols obtained from various Indian plants are shown in Figure 2.18. The gallates and glycosides of tlavan-3.-diols are not found in nature (Swain, 1965).

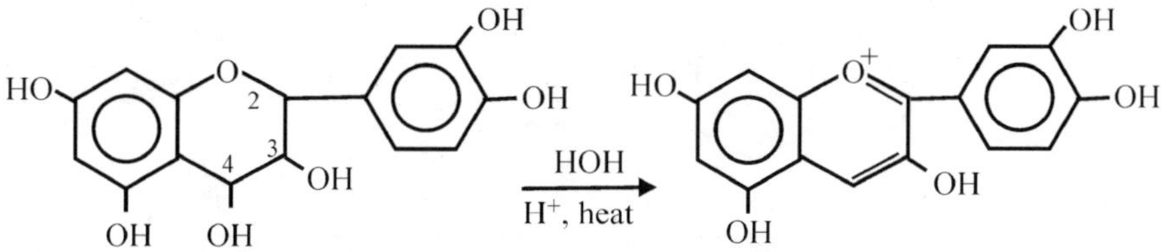

HOH

H^+, heat

Lecocyanidin
(Flavan 3,4-Diol)

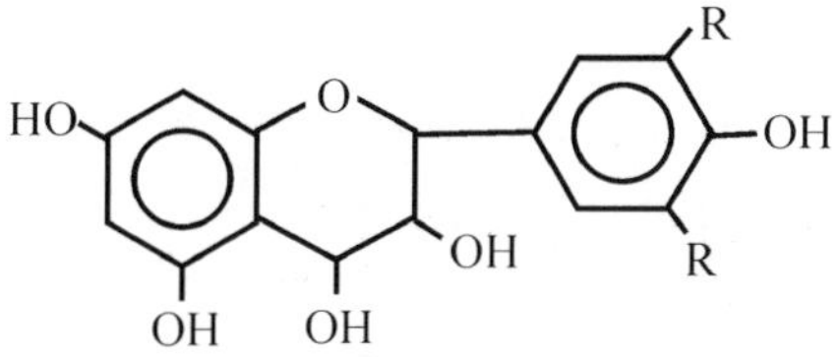

R = H, Leucopelargonidin
R = OH, Leucodelphinidin

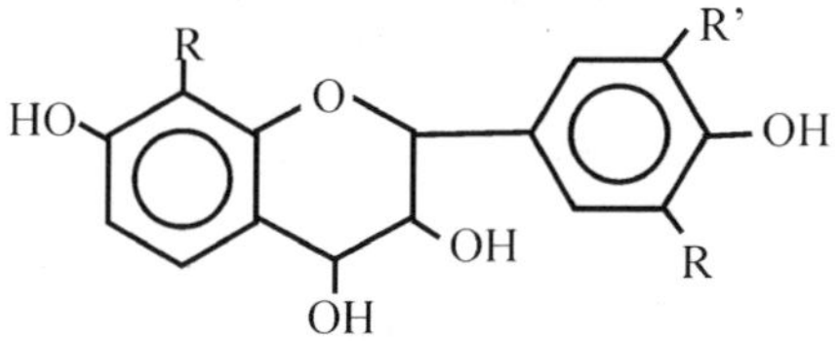

R = R' = R" = H; (+) Guibourtacacidin
R = H, R" = R = OH; (+) Leucorobinetinidin
R = R' = OH; R" = H; (-) Melacacidin
R = OH, R'=R" = H; (-) Teracacidin

Figure 2.18: Some naturally occurring flavan-3, 4-diols.

The chemical term "proanthocyanidins" has been used for those natural products which are convened to anthocyanidins upon heating with acid (Freudenberg and Weinges, 1962). Flavan-3-ols and flavan-3, 4-diols are among those proanthocyanidins which are widely distributed in plants. Weinges *et al*, (1969) implicated the term "leucoanthocyanidin" to represent monomeric flvan-3, 4-diols and condensed proanthocyanidin or procyanidin for flavan-3-ols dimers and oligomers. It is now clear that the condensed tannins, also referred to as procyanidins are mostly flavolans. The flavolans are polymers of flavan-3-0Is alone or a mixture of flavan-3-ols and flavan-3, 4-diols. Both flavan-3-ols and flavan-3, 4-diols are readily converted by dehydrogenase enzymes or even by a very dilute mineral acid at room temperature into flavonoid tannins (Weinges, 1968). Heating in an acid solution converts flavan-3, 4-diols to the corresponding anthocyanidins and brown phlobaphene-like polymers (Swain and Hillis, 1959). The condensed tannins are not readily degraded by acid treatment, but polymerize to foml amorphous phlobaphenes or "tannin-reds (Haslam, 1966). A typical structure of major condensed tannin, isolated from sorghum is shown in Figure 2.19 (Harborne, 1965).

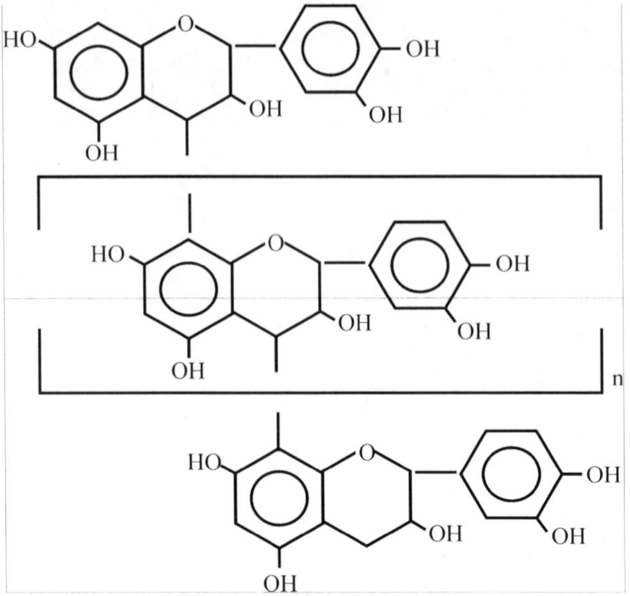

Figure 2.19: Possible structure of grain-sorghum condensed tannins.

A. Biosynthesis of phenolic acids

Phenolic acids in plants are synthesized through the shikimic acid pathway (Deshpande, *et al.*, 1984; Neish, 1965). Quinic acid, gallic acid and protocatechuic acid are formed from intermediates of the shikimic acid pathway. Tryptophan, tyrosine, and phenylalanine are end products of the shikimic aid pathway. Tyrosine and phenylalanine serve as precursors for the formation of cinnamic acid, which is further converted to various phenolic acids (Figure 2.20).

Basically, the shikimic acid pathway involves the initial condensation of erythrose-4 phosphate derived from the oxidative pentose phosphate cycle with phosphoenolpyruvate (PEP) from the glycolysis process. A series of enzymatic reactions lead to shikimic acid, which is then phosphorylated. The phosphorylated shikimic acid combines with a second molecule of PEP to yield prephenic acid via a chorismic acid intermediate. Prephenic acid is then decarboxylated to form phenyl pyruvate or p-hydroxyphenylpyruvate. On transamination these two compounds yield phenylalanine and tyrosine, respectively. The pathway contains certain branch points. The dehydroquinic acid can be converted either to 3-dehydroshikimate or quinic acid. Protocatechuic acid and gallic acid are synthesi7.ed by the direct aromatization of 3-dehydroshikimate, although they can also be formed by the hydroxylation of cinnamic acid and benzoic acid derivatives.

Common phenylpropane derivatives cinnamic acid, p-coumaric acid caffeic acid. Ferulic acid, 5-hydroxyfaulic acid, and sinapic acid are synthesized from phenylalanine. Benzoic acid derivatives, such as 4-hydroxy benzoic acid, vanillic acid, and syringic acid, are generally synthesized from corresponding cinamic acid derivatives by the β-oxidation of the propenyl side chain after ester formation with coenzyme A. Widespread chlorogenic acid is an ester of quinic acid. a product of the shikimic acid pathway and caffeic acid. Phenylpropanes are also important intermediates m the biosynthesis of lignin and flavonoids.

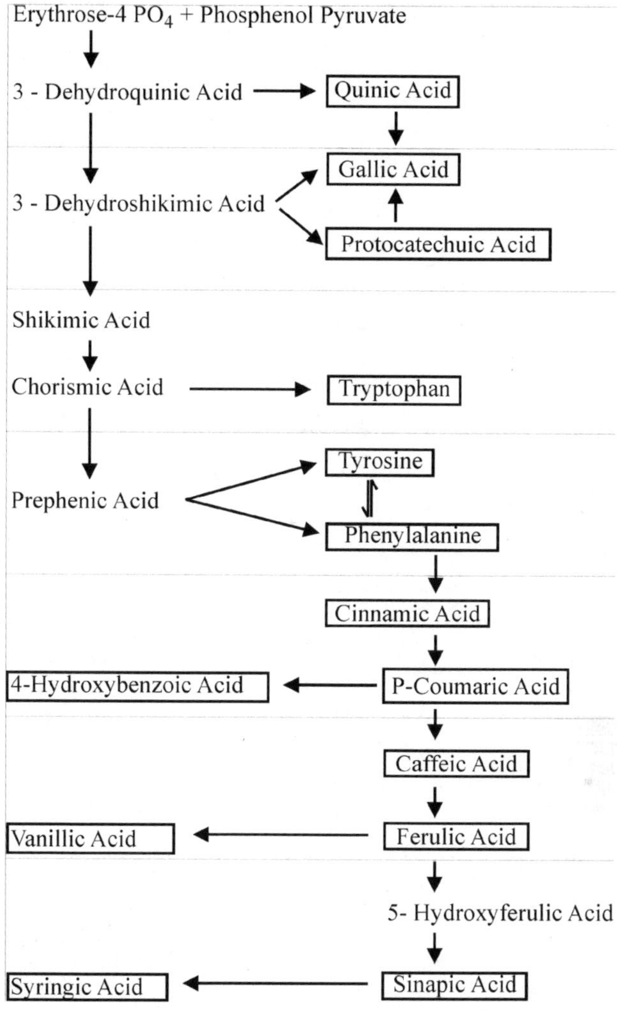

Figure 2.20: Outline for biosynthesis of some phenolic acids through the shikimic pathway.

B. Biosynthesis of coumarins

Coumarin is produced from cinnamic acid, which is synthesized from tyrosine and phenylalanine. A scheme for the biosynthesis of coumarin is shown in Figure 2.21 (Deshpande, *et al.*, 1984). Cinnamic acid is first hydroxylated at the ortho-position. The o-hydroxyl group subsequently undergoes glucosylation, which facilitates the isomerization of *trans-cinnamic* acid derivatives to the *cis* form. The hydrolysis of sugar then leads to spontaneous cyclization. Coumarins such as umbelliferone, esculetin and scopoletin, produced from p-coumaric acid, caffeic acid, and ferulic acid, respectively, through similar steps described for coumarin formation, are widely distributed in higher plants.

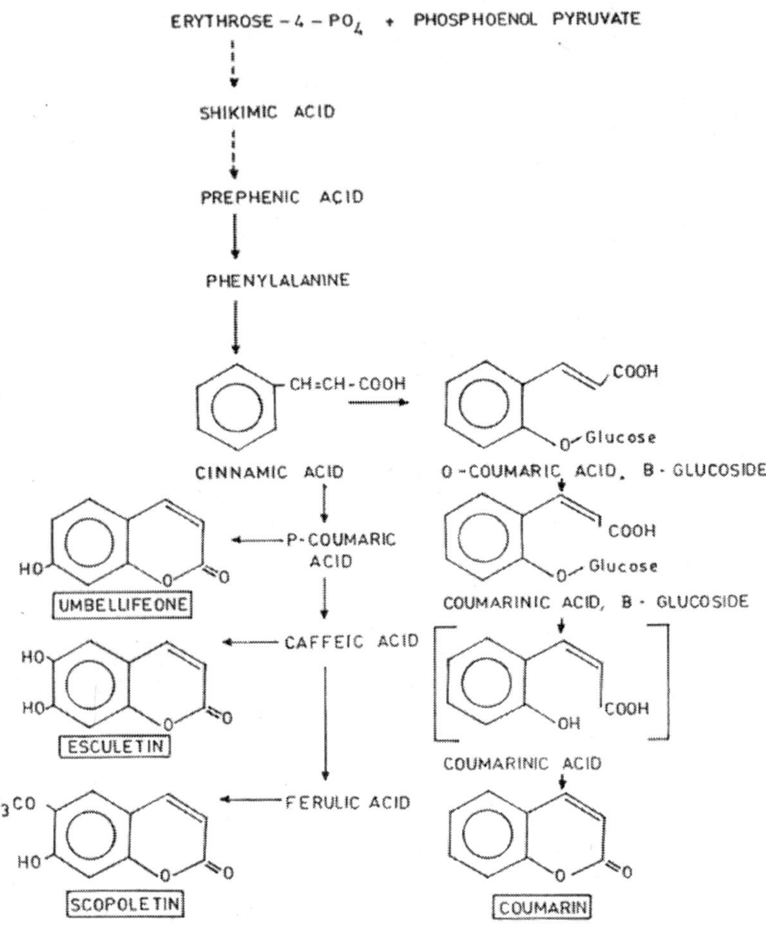

Figure 2.21: Biosynthesis of some natural coumarins from phenylalanine through cinnamic acid.

C. Biosynthesis of flavonoids

The biosynthetic pathways of different flavonoids have been described (Harhorne, *et al.*, 1975; Hahlbrock and Grisebach, 1975). The flavonoid and isoflavonoid ring structures are of mixed biosynthetic origin; ring A being derived from three acetate units condensed head-to-tail, while ring B and three carbon atoms of the central ring are derived from cinnamic acid. As the acetate units are first converted to malonyl CoA, both the acetate-malonate and the shikimic acid pathways contribute to flavonoid biosynthesis. Several flavones compounds occurring in the plant kingdom differ slightly from one another by the number and orientation of hydroxyl, methoxyl, glycosyl, and other substituent's in the aromatic rings A, C, and B (Figure 2.22).

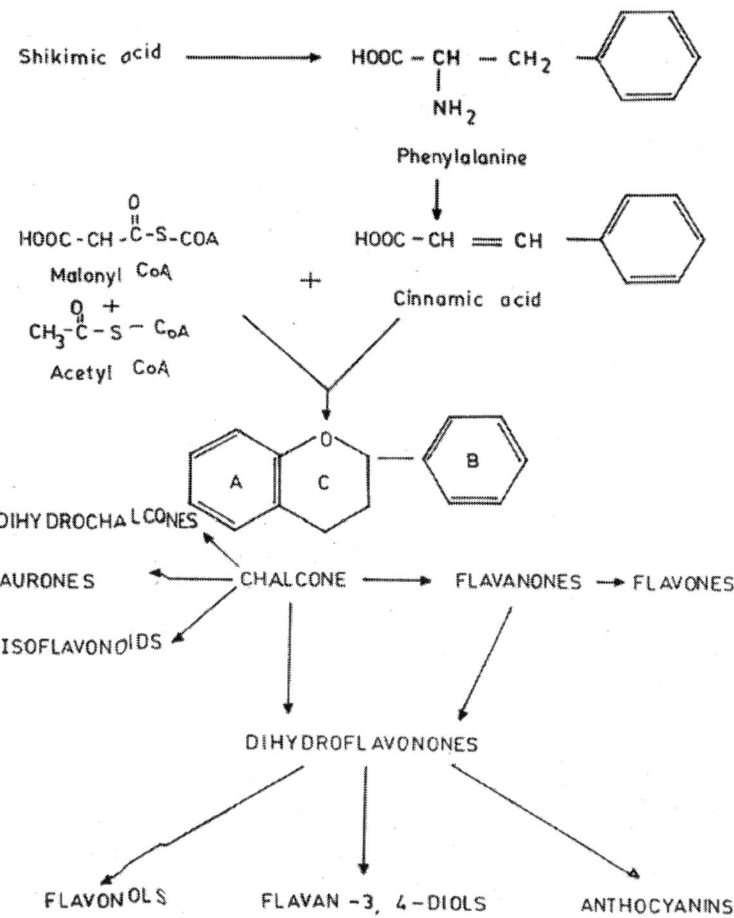

Figure 2.22: Biogenetic relationship of flavonoids.

The first product of the addition reaction between malonyl CoA and p-coumarin is a chalcone derivative. Chalcones can be considered as the precursor of all other classes of flavonoids. In general, all flavonoid compounds originate from a C_6-C_3-C_6 chalcone unit of which the central C_2 moiety shows structural variations leading to the formation of various types of flavonoids. The number of hydroxyl groups introduced in ring A at the stage of cyclization depends upon the degree of reduction of the triacetate ions. The first OH (Roux and Paulus, 1961) in ring B, introduced even before the formation of the C_6-C_3-C_6 unit, comes from the shikimic acid pathway. However, the second hydroxylation occurs at the chalcone level (Grisebach, 1967). Similarly, C-glycosylation also occurs at this level. Based on these steps, the biosynthetic pathway for C-glycosylflavones is outlined in Figure 2.23. More detailed reviews covering tracer work and enzymatic studies on the formation of the different classes of flavonoids have appeared in the literature (Hahlbrock and Grisebach, 1975; Ribereau, Gayon, 1972; Wong, 1976).

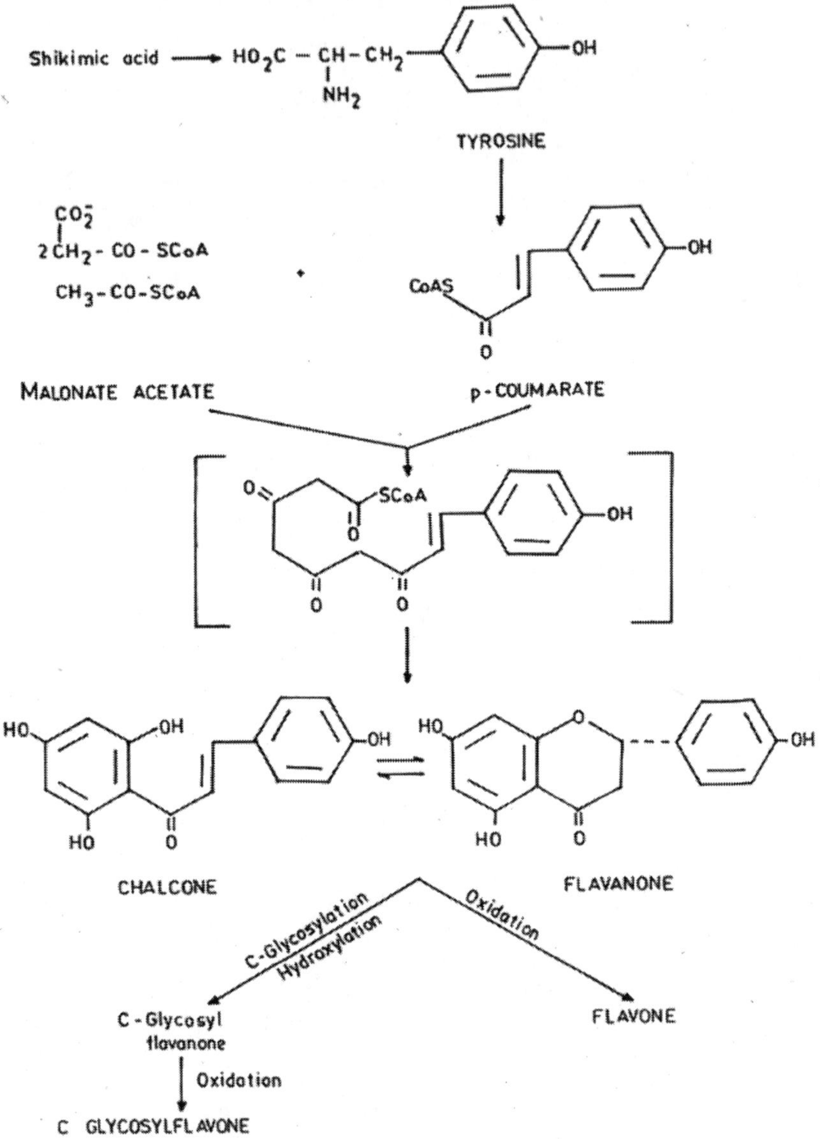

Figure 2.23: Biogenesis of C-glycosylflavones.

D. Biosynthesis of hydrolyzable tannins

The possible route for the biosynthesis of gallic acid and other phenolic acids through the shikimic acid pathway was shown in Figure 2.20; 3-Dehydroshikimic acid can be converted to both gallic acid and protocatechuic acid. The latter is known to be converted to gallic acid. Cinnamic acid produced from phenylalanine can also act as a precursor for gallic acid (Swain, 1965). Hexahydroxydiphcnic acid (a precursor of ellagic acid) is more likely formed by the oxidative coupling of gallic acid followed by the further oxidative cleavage of one of the aromatic rings to give chebulic acid. The coupling reactions between phenols are well documented. It has also been suggested that the hexahydroxydiphenic acid residue, in tannins such as corilagin are formed by the enzyme-catalyzed coupling of gallic acid residues already attached to the sugar. The linkage of the various phenolic acids to the sugars possibly goes via the corresponding coenzyme A derivative or acyl adenylate. The mechanism of esterification depends upon the specificity of the enzyme system involved, which needs to be elucidated.

E. Biosynthesis of condensed tannins

Condensed tannins are synthesized by the condensation of two or more molecules of flavonoids such as flavan-3-ols or a mixture of flavan-3-ols and flavan-3.4-diols. The mechanism of flavan polymerization is not clearly known, although several theories have been put forward. According to Freudenberg and Weinges (Freudenberg and Weinges, 1962) the flavan-3.4-diols can produce a carbonium ion at the reactive benzylic C_4-OH. The ion then reacts with the nucleophilic centers of another flavan to give a C_4-C_6 or C_4-C_8 linkage. Hathway *et al*, (1957) suggested that condensed tannins are either formed by autoxidation or formed enzymatically by polyphenol oxidase. The products formed in each case seem to be very similar and indistinguishable from each other. Polyphenol oxidase brings about an oxidation of a flavan to an o-quinone that can react with the A or B ring of another oxidized molecule. As a result, catechin condenses to give a "head-to-tail" polymer, while gallocatechin polymerizes in a "tail-to-tail" manner (Hathway and Seakins, 1957; Hathway, 1958). Malan and Roux (Malan and Roux, 1975) considered that the condensation patterns of catechin and flavan-3, 4-diols found in several plant sources follow most likely, electrophilic substitution pathways. Roux and Paulus (Roux and Paulus, 1961) suggested that the reactive C_4-hydroxyl group may be involved in the biochemical condensation offlavan-3, 4-diols. The knowledge of structure, stereochemistry, distribution, and metabolism of plant procyanidins or flavan-3-ol dimers or oligomers (Thompson *et a*l., 1972; Jacques and Haslam, 1974; Jacques *et al.*, 1977; Fletcher and Porter, 1977; Haslam *et al.*, 1977) led Haslam (1977) to propose the biosynthetic pathway of plant procyanidins and associated flavan-3-ols (Figure 2.24).

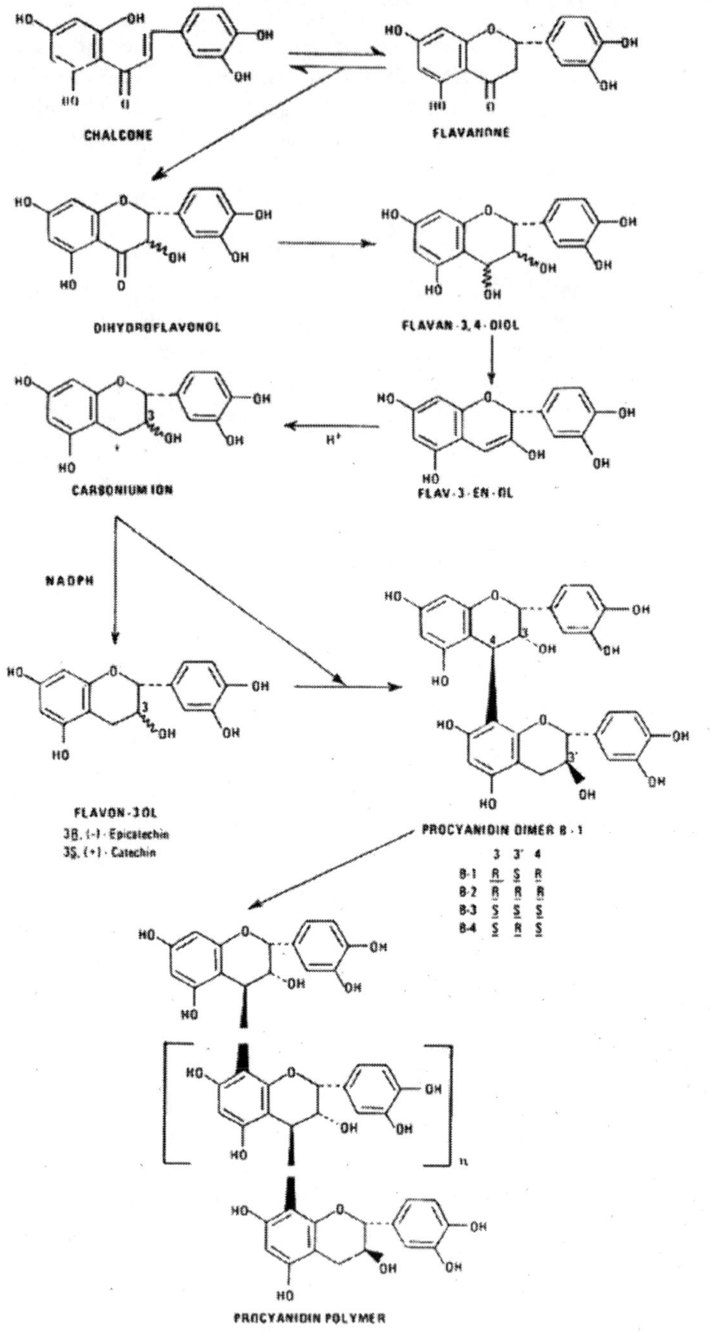

Figure 2.24: Biosynthesis of flavan-3-ol and procyanidin polymer in sorghum.

The chalcone-flavanone pair is the first intermediate, but the sequence of chemical changes in the C_3 unit that results in the formation of individual flavonoids themselves remains poorly defined. A scheme from the chalcone ← → flavanone pair to the flavan-3-ols [(+) catechin and (-)-epicatechin] is presented in Figure 2.24. A key intermediate is believed to be flavan-3-en-3-ol and a two-step sequence (stereospecific protonation followed by reduction) is postulated to yield flavan-3-ols. The procyanidins are thought to arise as byproducts in this two-stage process when the supply of biological reductant (NADPH) is rate limiting. The various distinctive patterns of procyanidins found in plants are then thought to arise by the reaction of one or both of the intermediate carbocations that escape from the active site of the enzyme with one or both of the flavan-3-ols utilizing their nucleophilic reactivity at carbon atom C_5 or C_8. The dimeric forms are first produced, but these react similarly with further carbocations to produce trimers, tetramers, and higher oligomers (Gupta and Haslam, 1980). The degree of polymerization of the procyanidins appears to be a characteristic of a specific plant. Any plant produces a range of procyanidins of differing molecular weight. Only those flavan-3-ols with molecular weights up to about 3000 are soluble. Flavan-3-ols with higher molecular weights are either insoluble or attached in some way to skeletal tissues of plants. The latter types of procyanidins represent a major fraction in plant tissue. Gupta and Haslam (Gupta and Haslam, 1980) reported that the polyphenol metabolism of this type does not continue over a long time in plants but there is a relatively short "burst" of procyanidin synthesis after which the level of product remains approximately the same in plant tissue until senescence occurs.

Major Procyanidins in sorghum grain are oligomers of flavan-3-ols. However, the precise structures of procyanidins from other sources such as forage plants, fruits, and wines need to be elucidated with respect to the nature of monomers and the size and shape of molecules. The tannin-protein interactions are reported to be both tannin-specific as well as protein specific (Asquith and Butler, 1986). Thus, it is clear that condensed tannins from different sources are structurally and biologically different. It is important to understand the structure of major tannins from various sources of foods and feeds in order to elucidate their nutritional and biological significance in diets.

PHENOLICS IN NATURAL AND ROASTED NUTS

The polyphenol content of many foods has been documented in nutrient databanks, such as the Phenol-Explorer and United States Department of Agriculture (USDA) flavonoid and proanthocyanidin databases (Pérez-Jiménez et al., 2010; USDA, 2015a, 2015b). Different classes and levels of phenolic compounds have been reported for different nuts and have also been included in Phenol-Explorer and USDA database. However, information concerning the individual phenolic compounds of both non-flavonoid and flavonoid compounds of nuts is still scarce. Using the Spanish cohort of the PREDIMED (PREvención con DIeta MEDiterránea) trial, Tresserra-Rimbau et al. (2013) determined the total dietary polyphenol intake of 7200 men and women aged 55–80 years and found total polyphenol intakes at 820 ± 323 mg/d per person with 10 mg/d intake from nuts, 66 and 34% of which is from walnut and other nuts, respectively. Furthermore, 6.4 and 3.6 mg/d per person of flavonoids and phenolic acids intake were reported (Tresserra-Rimbau et al., 2013). Of the 19.1 mg/d per person of hydroxycinnamic acids intake, 10% was derived from walnut.

Proanthocyanidins (condensed tannins) and hydrolysable tannins are generally the most abundantly found polyphenols in a majority of nuts. Besides, nuts also have a significant amount of phytates together with flavonoids, phenolic acids, stilbenes, and tyrosols (Alasalvar and Bolling, 2015; Bolling et al., 2011). Phenolic acids in foods occur in the free, esterified, glycosidic, and insoluble-bound forms. Free phenolic acids are known to contribute to the taste of foods. Flavonoids are another group of phenolic compounds that can be classified into seven groups: flavanones, flavones, isoflavones, anthocyanidins, flavonols, flavononols, and flavanols (flavan-3-ols) (Shahidi and Ho, 2005). Phenolic acids and flavonoids reported in nuts are predominantly conjugated with sugars or other polyols *via O*-glycosidic bonds or ester bonds. Flavonoids, which are the most common and widely distributed group of plant phenolics, are increasingly appreciated as being an important component of the human diet. Depending on their structures, tannins are defined as hydrolysable (gallotannins and ellagitannins) or condensed (monomers, dimers, oligomers, and polymers of flavan-3-ols). More importantly, the contents of phenolic compounds of nuts are affected not only by the type of nut (genotype) and its cultivar but also the harvest year, orchard location, processing steps (pasteurisation, bleaching, irradiation, and roasting), and storage conditions (temperature and duration). The detailed information of these factors has been reviewed by Bolling et al. (2011).

Phenolic compounds and antioxidant activities in processed (roasted) nuts

Nuts may be consumed as natural (raw) or preferably roasted. The main purpose of roasting is to improve the desirable flavour, colour, crispiness, and the crunchy texture of nuts (Hojjati, Calín-Sánchez, Razavi, and Carbonell-Barrachina, 2013; Pelvan, Alasalvar, and Uzman, 2012; Schmitzer, Slatnar, Veberic, Stampar, and Solar, 2011). The roasting process involves microstructural and chemical changes, such as the decrease in moisture content, lipid modifications, and changes in colour as well as the formation of compounds responsible for the typical roasted nut flavour, mainly due to Maillard reaction products (Alamprese, Ratti and Rossi, 2009; Amaral, Casal, Seabra and Oliveira, 2006; Saklar, Katnas and Ungan, 2001). However, microstructural and lipid modifications might lead to an enhanced susceptibility of roasted nuts to lipid oxidation (Alamprese et al., 2009), compared to raw nuts. Besides, changes in the content of metabolites (e.g., antioxidants) may also happen due to roasting (Amaral et al., 2006). Therefore, the roasting process may influence both the formation of health promoting components and those with potentially adverse health effects. To the best of our knowledge, limited studies are available on the effect of roasting on phenolic compounds of nuts. Hence, more research should be carried out to fill this knowledge gap. The impact of processing (roasting and stir-frying) on phenolic compounds of nuts demonstrates a rising trend at lower temperatures or over shorter heating durations, but a declining trend at higher temperatures or over longer heating periods (Bolling et al., 2011; Chandrasekara and Shahidi, 2011; Lin et al., 2016; Pelvan et al., 2012; Schlörmann et al., 2015;

Schmitzer et al., 2011; Vinson and Cai, 2012; Yang et al., 2015). However, the impact of roasting on antioxidant activities of nuts and their co-products is inconsistent mainly due to the use of different processing (roasting and stir-frying) duration and temperature, as well as different types of nuts.

Yang et al. (2015) have demonstrated that stir-frying pecan at a low temperature over a short time leads to an apparent increase of its extractable chlorogenic acid, gallic acid, and antioxidant activity. However, these values decreased with increases in heating time and temperature. Hence, it is only beneficial to stir-fry pecans at certain conditions during the processing. Pecan phenolics, with the exception of gallic acid, were always the highest in whole unroasted kernel and were significantly reduced after skin removal. Similarly, pecan TPC and antioxidant activity decreased when its skin was removed. Roasting had a significant negative effect on individual phenolics but not on the TPC and antioxidant activity of kernels (Yang et al., 2015).

In hazelnut, phloretin-2′-*O*-glucoside and flavan-3-ols (catechin and epicatechin) content decreased significantly when roasted, where significant differences were observed between the raw nut with skin and roasted nut without skin. Thermal treatment had a negative effect on the content of procyanidin dimers and trimers in hazelnut (Schmitzer et al., 2011). Again, significant differences existed between natural and roasted hazelnuts, suggesting that the stability of these compounds is indeed affected by temperature. Moreover, no polyphenols have been detected in roasted hazelnut from which the skin was removed. This shows that polyphenols in hazelnut are mostly present in the skin (Schmitzer et al., 2011). Another study on pistachio nut also reported a significant decrease in several phenolic compounds after raw nuts were roasted. Chlorogenic acid and daidzein were detected only in roasted pistachio, not in natural pistachio (Seeram et al., 2006). Roasting had an ambiguous effect on the content of protocatechuic acid in roasted hazelnut (Monagas et al., 2009). Similarly, roasting had a significant positive effect on the content of gallic acid in roasted hazelnut compared to natural hazelnut. This could be due to degradation of polymerised polyphenols, specifically hydrolysable tannins, and the hydrolysis of other glycosylated flavonoids (Monagas et al., 2009). Other factors may also affect the phenolic acid contents upon roasting, such as the content of phenolic acids in the skin, their thermal decomposition, and their liberation from esters, glucosides, and bound forms (Pelvan et al., 2012). Catechin and ellagic acid are also susceptible to heat, possibly by their highly hydroxylated structure that is vulnerable to redox reactions, causing losses when subjected to heat treatment. Therefore, it seems that roasting reduces the levels of phenolic compounds in most nuts, not only because of the removal of the skin but also due to chemical degradation of many phenolic compounds. Most of them are highly unstable and may be lost during processing, particularly when heat treatment is involved.The thermolability of phenolic compounds depends not only on their structure but also on the food matrix (Jinap, Jamilah, and Nazamid, 2004).

Enhanced roasting conditions led to a significant decrease of the antioxidant activity in hazelnut, macadamia, and walnut, whereas in almond and pistachio the antioxidant activity remained stable or was slightly enhanced. Furthermore, antioxidant activity was restored in hazelnut, almond, and walnut at the most intense roasting conditions (Schlörmann et al., 2015). Similar results were found by Açar,

Gökmen, Pellegrini, and Fogliano (2009), who observed a decline of antioxidant activity in several types of nuts by roasting at 150°C for 10–30 min and enhanced activities after 60 min. This effect can be explained by the loss of polyphenols due to thermal treatment which is counteracted by the formation of antioxidant-active compounds due to Maillard reactions as also discussed by Açar et al. (2009). Most of the antioxidative compounds of nuts, such as polyphenols, are located in the skin and its removal is accompanied by a loss of antioxidant activity (Arcan and Yemeniciog˘ lu, 2009; Schmitzer et al., 2011).

Vinson and Cai (2012) determined the antioxidant activities of nut and peanut butter polyphenols as measured by the concentration to reduce the oxidation of LDL + VLDL by 50%. The order of decreasing efficacy (increasing IC50) for natural (raw) nuts is walnut > cashew > hazelnut ~ pecan ~ almond ~ macademia > pistachio > Brazil nut > peanut. For roasted nuts and peanut butters, the order is almond ~ macademia ~ pecan ~ peanut butter crunchy > peanut butter creamy > peanut > pistachio > hazelnut ~ walnut > Brazil nut > cashew. However, roasting caused a statistically insignificant decline in antioxidant efficacy. The average IC50 for raw nuts was 4.3 ± 2.1 mM and for roasted nuts (not including peanut butters) was 6.9 ± 3.9 mM. Raw walnut had the highest antioxidant activity among all nuts. Besides, the efficacy of nut polyphenols to bind to LDL, thus protecting the LDL from oxidation, was also determined [almond (2.1 mM), peanut (5.6 mM), and walnut (1.0 mM)]. That means walnut-bound polyphenol antioxidants are twice as effective as almond and 5-fold effective compared to peanut in lipoprotein binding.

Schlörmann et al. (2015) have evaluated the effect of roasting (125°C/25 min to 185.5°C/25 min) on antioxidant capacities of almond, hazelnut, macadamia, pistachio, and walnut using TBARS, hydrophilic and lipophilic antioxidant assays. The TBARS levels significantly increased with roasting conditions (both temperature and time). The increase in TBARS levels was relatively low in hazelnut (from 0.94 ± 0.02 in raw up to 1.73 ± 0.1 μg malondialdehyde equivalents (MDAE)/g) and macadamia (from 0.40 ± 0.04 in raw up to 0.69 ± 0.05 μg MDAE/g) compared to pistachio (from 1.41 ± 0.15 in raw up to 3.51 ± 0.36 μg MDAE/g). In contrast, the highest increase of TBARS levels was observed in walnut (from 0.95 ± 0.05 in raw up to 16.3 ± 3.16 μg MDAE/g) and in walnut roasted under extreme conditions (185.5°C/25 min). These results are in agreement with those of Yaacoub, Saliba, Nsouli, Khalaf, and Birlouez-Aragon (2008) in roasted almond and pistachio. For hydrophilic antioxidant activity, no effect on the antioxidant activity was found in almond and pistachio, whereas antioxidant activity significantly decreased in hazelnut and macadamia after roasting. In contrast, the antioxidant activity significantly increased after roasting (171°C) in macadamia. In walnut, the hydrophilic antioxidant activity significantly decreased after roasting. In addition, the lipophilic antioxidant activity decreased in pistachio and walnut due to roasting (Schlörmann et al., 2015). According to Xu and Chang (2008), the thermal treatment applied to foods of plant origin by heating or roasting cause's evaporation of intracellular water, triggering chemical reactions that can change the lignocellulosic structure, and promotes protein denaturation, which might result in a greater availability of plant phenolic compounds in the matrix. Enhancement of antioxidant activity might be attributed to the release of some bound antioxidant phenolic compounds acting as free radical scavengers from the cell matrix, and increase phenolic extractability after roasting (Lin et al., 2016; Schmitzer et al., 2011).

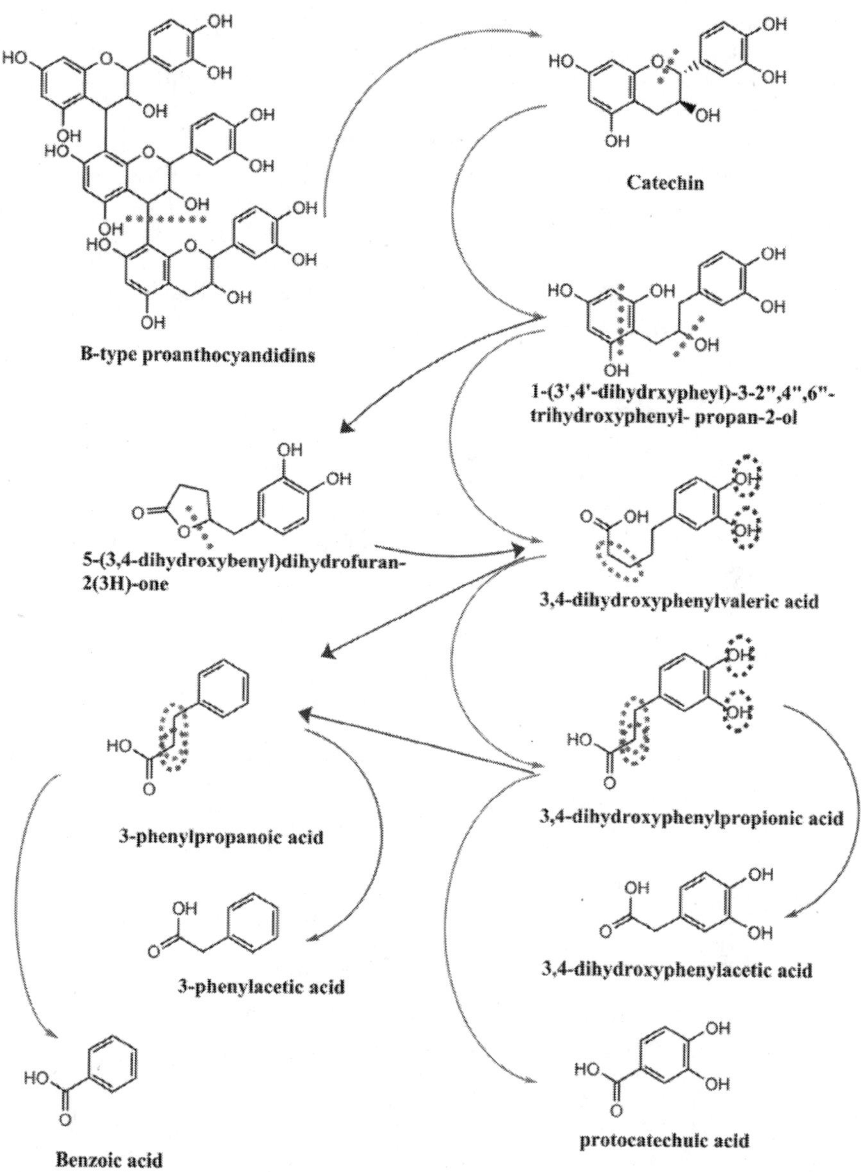

Figure 2.25: Metabolism of B-type nut proanthocyanidins. B-type proanthocyanidins, which are catabolised by gut microbes and small phenolic compounds (e.g., valerolactones and benzoic acids), are subsequently absorbed and subjected to glucuronidation, sulphation, and/or methylation.

Bioavailability of nut and co-product polyphenols

Many nut co-products are rich sources of polyphenols. Polyphenols are extensively metabolised after consumption. Thus, the bioavailability of nut polyphenols should be considered when evaluating the potential health benefits of nut co-products. Nut tannins are extensively metabolised by gut microbiota. Proanthocyanidins are metabolised to valerolactone intermediates and hydroxybenzoic acids (Figure 2.25). Nut ellagitannins are hydrolysed to ellagic acid and converted to urolithins by microbial metabolism (Figure 2.26; Cerda, Tomás-Barberán, and Espin, 2005). The extent of urolithin produced after consumption of nut ellagitannins varies between individuals, and circulating urolithins exist as primarily glucuronides and sulphates (González-Sarrías et al., 2010). Urolithin metabolites enter tissues, and urolithin A and B glucuronides were identified in human prostate tissue after walnut consumption (González-Sarrías et al., 2010).

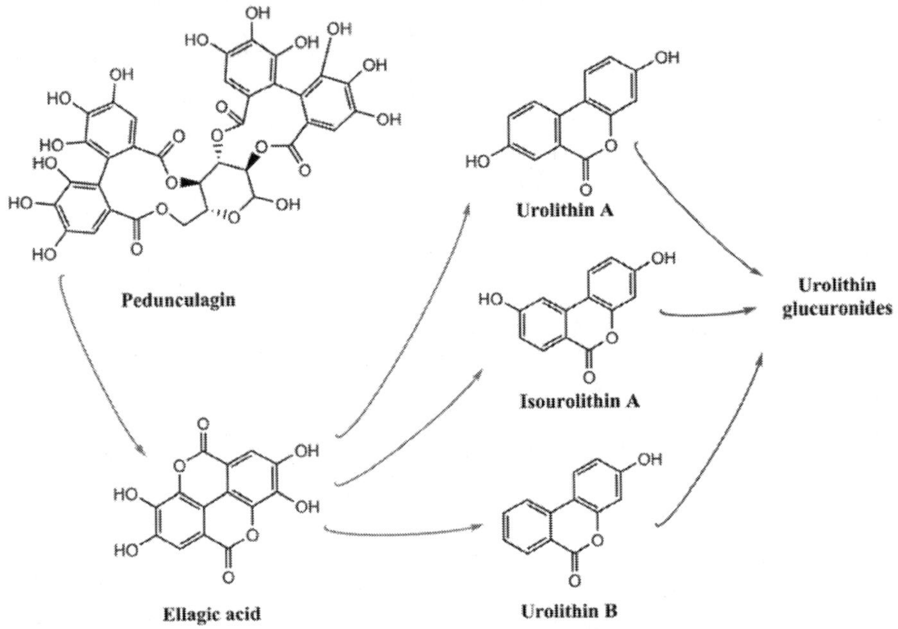

Figure 2.26: Metabolism of nut ellagitannins. Gut microbes catalyse the release of ellagic acid from nut ellagitannins, which are subsequently metabolised to urolithins in the colon. Urolithins are absorbed and further subjected to phase II metabolism.

The bioavailability of almond, pecan, and walnut polyphenols has been determined in humans after consumption of whole nuts or crude extracts. Almond skin polyphenols and metabolites were characterised in plasma and urine in two individuals that consumed 884 mg of almond skin polyphenols (Urpi-Sarda et al., 2009).Valerolactone, hydroxycinnamic acid, and hydroxybenzoic acid metabolites were significantly increased in urine after consumption of almond skin extract, but plasma polyphenol concentrations were modestly increased, in the nanomolar range. Low-micromolar concentrations of epicatechin, naringenin, and isorhamnetin phase 2 metabolites (e.g., glucuronidated, sulphated, or methylated metabolites) were recovered in urine (Urpi-Sarda et al., 2009). A subsequent placebo-controlled study further characterised almond skin polyphenol bioavailability in 12 adults (Garrido et al., 2010). Urinary epicatechin sulphate and naringenin *O*-glucuronide reached maximum concentrations 2–6 h after almond skin extract intake. Urinary excretion of microbial polyphenol catabolites increased progressively from 10 to 24 h after intake. Among other nuts, consumption of 90 g walnut with a control meal increased plasma catechin 1 h after consumption (Haddad, Gaban-Chong, Oda, and Sabate, 2014). Urinary excretion of urolithin A and 4-methoxyphenylacetic acid also increased at 12–24 h after walnut consumption (Haddad et al., 2014). Similarly, a 90 g dose of pecan increased plasma epicatechin, epigallocatechin gallate, and epigallocatechin (Hudthagosol et al., 2011). Epigallocatechin gallate reached a maximum concentration of 192 nmol/l in plasma at 3.3 h after pecan consumption (Hudthagosol et al., 2011). Urinary 3-methoxy- 4-hydroxyphenylacetic acid, a microbial phenolic metabolite, also increased 13–24 h after pecan consumption (Hudthagosol et al., 2011).

Polyphenol metabolites may also be useful biomarkers of intake of some nuts. For example, 5-hydroxyindole-3-acetic acid, a metabolite of serotonin, has been identified as a marker for walnut intake (Andersen et al., 2014; Tulipani et al., 2011). Tulipani et al. (2012) also discovered that urinary urolithins A and B were increased in individuals with metabolic syndrome that consumed 30 g/day of mixed walnut, almond, and hazelnut for 12 weeks. Urinary walnut polyphenolic metabolites, particularly urolithins A and B and pyrogallol phase 2 conjugates could help differentiate walnut consumers from non-consumers (Garcia-Aloy et al., 2014). Experiments in rodents indicate that nut polyphenols are distributed to various tissues. Almond skin polyphenol consumption increased liver epicatechin and kaempferol in hamsters (Chen, Milbury, Lapsley, and Blumberg, 2005). In rats fed hazelnut skin polyphenols, polyphenol metabolites were detected in multiple tissues. Protocatechuic acid in liver, heart, lung, intestine, and kidney were increased after chronic hazelnut polyphenol consumption (Serra et al., 2011). *p*-Hydroxyphenylpropionic acid also increased by 15 nmol/g in rat brain (Serra et al., 2011). In a subsequent study that considered the kinetics of rat tissue accumulation after hazelnut extract consumption, the microbial polyphenol metabolites *p*-hydroxybenzoic acid and *o*-hydroxyphenylacetic acid increased in liver 1.5 h after consumption. Ferulic acid sulphate and catechin glucuronide peaked 2 h in rat kidney after hazelnut extract consumption (Serra et al., 2013). Based on these studies, the tissue concentrations of nut polyphenol catabolites appear to be transient and require chronic consumption to maintain tissue levels (González-Sarrías et al., 2010; Serra et al., 2013).

The extent of polyphenol bioaccessibility from the nut or nut co-product matrix also impacts polyphenol bioavailability. While nearly 70–95% of almond skin polyphenols were bioaccessible in simulated human digestion, flavan-3-ols and hydroxycinnamic acids were more accessible than flavonols and flavanones (Mandalari et al., 2010d). In addition, cashew fibre polyphenols were not as bioaccessible as cashew apple juice polyphenols during *in vitro* simulated digestion (de Lima et al., 2014).Therefore, nut co-product polyphenol bioavailability may depend on the matrix in which it is consumed.

Mandalari et al. (2013) evaluated the release of polyphenols from pistachios (raw pistachio, roasted salted pistachio, andmuffins made with raw pistachio) during simulated human digestion using an *in vitro* model. More than 90% of the pistachio polyphenols were released during simulated digestion of more than 3 h. No significant differences existed between the natural and roasted pistachios, with an initial slower release for the hydroxybenzoic acids compared with isoquercetin, eriodictyol, and luteolin. Similar trends were observed with pistachio muffins for most of the identified polyphenols, with the exception of protocatechuic acid and luteolin, whereas a slower release was observed in the gastric and small intestinal environments, as a result of the food matrix (muffins). Releases of 64.7 and 78.3% were detected for protocatechuic acid in pistachio muffins after gastric (G) and gastric + duodenal (G + D) digestion, respectively, whereas lower releases were found for luteolin, 27.0 and 35.8%, after G and G + D digestion, respectively. Phenolic acids, such as gallic and chlorogenic acids are generally absorbed in the upper gastrointestinal (GI) tract within 1 to 2 h after their intake.

It has been shown that absorption of flavonols is affected by the attached sugars and the presence of fat, whereas epimerisation reactions occurring during processing could affect the absorption of flavanols such as catechins (Scholz and Williamson, 2007). More than 90% of the pistachio polyphenols were released in the gastric compartment, with little or no increase in the duodenal phase as demonstrated by Mandalari et al. (2013). A similar trend was also observed with polyphenols from almond. The absorption of phenolic acids seems to be affected by the attached sugar, which can covalently link these compounds to the food matrix (Scholz and Williamson, 2007). Hence, the rate of absorption of polyphenols in the small intestine may be significantly affected by the food matrix. Relatedly, the release of protocatechuic acid and luteolin from pistachio was affected by the presence of the muffin, and this may be due to the matrix composition and processing conditions. However, roasting had no effect on polyphenol bio-accessibility (Scholz and Williamson, 2007). It has been reported that roasting significantly increases the antioxidant capacity of nut extracts, such as almond skin, compared with blanching and oven drying because of the degradation of polymerised polyphenols and the hydrolysis of glycosides (Garrido et al., 2008). In this study, a slight decrease of total polyphenols in roasted pistachio compared with its natural counterpart, although some compounds were found only after roasting.

Relatedly, most phenolics (catechin, epicatechin, rutin, eriodictyol, cyanidin-3-*O*-glucoside, and cyanidin-3-*O*galactoside) found in pistachio were not readily accessible during the simulated *in vitro* digestion with only 5% phenolics being released except protocatechuic and gallic acids, as determined by Liu, Blumberg, and Chen (2014a). This result is in contrast to the results obtained by Mandalari et al. (2013) in which gallic and procatechuic acids, luteolin, and eriodyctiol were >90% bioaccessible. Besides, it is also interesting to note that quercetin glycosides dominantly present in the skin were not bioaccessible during the model digestion. Nieman et al. (2014) have evaluated the effect of pistachio nut consumption (3 oz or 85 g/d) on exercise performance and 21-h post exercise recovery from inflammation among cycling athletes for two weeks. Surprisingly, they demonstrated elevations in several metabolites related to amino acid metabolism (leucine, indoleacetate, 2-aminooctanoate, γ-glutamylleucine, γ-glutamyltyrosine, phenylacetate, and isobutyrylcarnitine) and fatty acid metabolism (2-hydroxydecanoate), and may have reflected the ingestion of pistachio before and during exercise. Besides, gut-derived raffinose, sucrose, and *myo*-inositol metabolites that are related to leukotoxic effects and oxidative stress were also present in the circulation of endurance athletes after the combination of 2-weeks pistachio ingestion and intensive exertion. Pistachio ingestion impairs a 75-km cycling performance in overnight fasted cyclists (Nieman et al., 2014).

A major challenge in evaluating the role of individual health promoting components in natural foods is the lack of information on their behaviour in the GI tract and especially the factors that influence their bioavailability. In fact, bioavailability is influenced by bioaccessibility, which is defined as the relative amounts of nutrients or phytochemicals "released" from a complex food matrix in the lumen of the GI tract, and therefore, available for absorption into the body (Kamiloglu, Pasli, Ozcelik, and Capanoglu, 2014; Mandalari et al., 2010d). This process is highly influenced by the physical and chemical properties of the food matrix and the way these change during digestion (Kamiloglu et al., 2014). The choice of food matrix is therefore crucial for testing the bioaccessibility of several compounds under simulated GI digestion. Cell walls play a crucial role in regulating nutrient bioaccessibility from almond and polyphenols from almond skin are potentially available for absorption during digestion (Kamiloglu et al., 2014; Mandalari et al., 2010d).

Even though a simulated *in vitro* GI tract digestion model cannot directly mimic the human *in vivo* conditions, this model is helpful for investigating the effect of food matrix and enzymes on polyphenol bioaccessibility. Furthermore, studies on bioaccessibility using *in vitro* models were well-correlated with the results from human studies and animal models (Kamiloglu et al., 2014). More research should be carried out to determine the metabolites of bioactive compounds from nuts (e.g., Brazil nut, cashew, hazelnut, macadamia, pecan, peanut, and pine nut) especially their co-products after urinary excretion coupled together with cellular models in order to obtain more qualitatively well-correlated results with human/clinical studies. These data are essential for demonstrating the true biological relevance of these compounds in the context of nutrition and human health (Chang, Alasalvar, and Shahidi, 2016).

■■■■

3 Mechanism of Action of Phenolic Antioxidants

MECHANISM OF ACTION OF PHENOLIC ANTIOXIDANTS

The antioxidant potential of phenolic compounds depends on the number and arrangement of the hydroxyl groups in the molecules of interest (Cao, Sofic, & Prior, 1997; Sang, Lapsley, Jeong, *et al.*, 2002). Phenolic antioxidants (AH) can donate hydrogen atoms to lipid radicals and produce lipid derivatives and antioxidant radicals (Reaction 5), which are more stable and less readily available to promote autoxidation (Kiokias *et al.*, 2008). The antioxidant free radical may further interfere with the chain-propagation reactions (Reactions 6 and 7).

$$R\bullet\ RO\bullet\ ROO\bullet + AH \rightarrow A\bullet + RH\ ROH\ ROOH \quad (5)$$

$$RO\bullet\ ROO\bullet + A\bullet \rightarrow ROA\ ROOA \quad (6)$$

$$ROO\bullet + RH \rightarrow ROOH + R\bullet \quad (7)$$

As bond energy of hydrogen in a free radical scavenger decreases, the transfer of hydrogen to the free radical is more energetically favourable and thus more rapid (McClements & Decker, 2007). Any compound that has a reduction potential lower than the reduction potential of a free radical (or oxidized species) is capable of donating its hydrogen atom to that of the free radical unless the reaction is kinetically unfeasible. For example, FRS including α-tocopherol ($E^{\circ\prime}$ = 500 mV) which have reduction potential below that of peroxyl radicals ($E^{\circ\prime}$ = 1000 mV) are capable of donating their hydrogen to the peroxyl radical to form a hydroperoxide (McClements & Decker, 2007). The phenoxyl radical is stabilized by delocalization of its unpaired electron around the aromatic ring (Fig. 3.1), which participates in the termination reaction. Gorden (1990) reported that substitution at the *para* position with an ethyl or n-butyl group rather than a methyl group improves the activity of the antioxidant; however, the presence of chain or branched alkyl groups in this position decreases the antioxidant activity. The stability of the phenoxyl radical is further increased by bulky groups in the 2 and 6 positions as in 2,6-di-*t*-butyl-4-methylphenol (BHT) (Miller & Quakenbush, 1957), since these substituents increase the steric hindrance in the region of the radical and thereby further reduce the rate of propagation reactions involving the antioxidant radical (Reactions 8, 9, 10).

$$A\bullet + O2 \rightarrow AOO\bullet \quad (8)$$

$$AOO\bullet + RH \rightarrow R\bullet + AOOH \quad (9)$$

$$A\bullet + RH \rightarrow AH + R\bullet \quad (10)$$

The effect of antioxidant concentration on autoxidation rates depends on many factors, including the structure of the antioxidant, oxidation conditions, and the nature of the sample being oxidized (Shahidi & Naczk, 2004). Often phenolic antioxidants lose their activity at high concentrations and behave as prooxidants (Gorden, 1990) by involvement in initiation reactions (Reactions 11, 12). Phenolic antioxidants are more effective in extending the induction period when added to any oil that has not deteriorated to any great extent. However, they are ineffective in retarding decomposition of already deteriorated lipids (Mabarouk & Dugan, 1961). Thus, antioxidants should be added to foodstuffs as early as possible during processing and storage in order to achieve maximum protection against oxidation (Shahidi & Wanasundara, 1992).

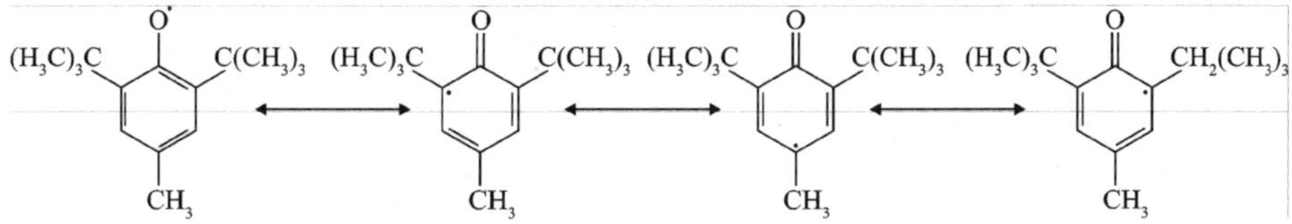

Fig. 3.1: Resonance stabilization of phenoxyl radical.

Polyphenols and antioxidant activity of barley

The beneficial effects of beverages containing polyphenols such as teas, coffees, fruit juices, wines and beers are today well recognized (Gorinstein *et al*., 2000; Kaur and Kapoor, 2002). As far as beer is concerned, a number of studies have outlined its phenolic content (Fantozzi et al., 1998), and reported correlations between the level of beer polyphenols and the antioxidant activity (Gorinstein *et al*., 2007; Gorjanovic *et al*., 2010), between its alcohol content and the absorption of polyphenols (Bourne *et al*., 2000), and between beer consumption and human health (Preedy, 2009). In addition to their physiological properties, beer polyphenols are technologically crucial: they are involved in foam maintenance, physico-chemical stability and shelf-life of beer (Mikyska *et al*., 2002).

Beer polyphenols are derived from cereal grains (e.g. barley, rice, wheat, sorghum, oat, etc.) and hop; and their levels depend on malt and hop varieties (Agu, 2002; Derdelinckx, 2008; Liu and Yao, 2007). Barley and its derived malt are the focus of attention due to their high phenolic content (benzoic and cinnamic acids, proanthocyanidins, tannins, flavonols, chalcones, flavones, flavanones, and amino phenolic compounds) and the associated antioxidant activity (Goupy *et al*., 1999; Hernanz *et al*., 2001; Liu and Yao, 2007; Qingming *et al*., 2010).

Yet, the levels of polyphenols in beer also depend on the malting and brewing processes. Data showed that the brewing process had a considerable impact on the phenolic content and the antioxidant activity of beer, mainly due to reactions undergone by polyphenols (Derdelinckx, 2008; Fantozzi *et al.*, 1998; Gorjanovic *et al.*, 2010; Pascoe *et al.*, 2003). In most cases however, the antioxidant activity is studied without separation of beer's individual compounds, which takes into account the matrix complexity. Yet, from an analytical point of view, this hampers the identification of antioxidant compounds and the changes they individually undergo during the processing steps. In a recent study, we used a Chromatographic Online Antioxidant Detection System (COADS) to assess the effects of brewing, boiling and fermentation on beer polyphenols (Leitao *et al.*, 2011). It has been shown that with global determinations, the interactions occurring, while more or less representative of the food system, affect the assessment of the actual phenolic activities.

As far as malting is concerned, several studies investigated its effect on total polyphenols and antioxidant activity (Friedrich and Galensa, 2002; Lu *et al.*, 2007; Qingming *et al.*, 2010; Samaras *et al.*, 2005). It has been found that malt had a higher antioxidant activity and phenolic content than the corresponding unmalted barley, which suggested that the malting process was of significance for this increase, especially the later stages of germination and subsequent kilning. Yet, the effect of malting on the fate of individual polyphenols and their corresponding antioxidant activities remains to be determined. The novelty of the present study is that it set out to determine the fate of individual phenolic compounds during the malting process. A further monitoring during the subsequent brewing is also presented.

Experimental

Extraction of phenolic compounds

The phenolic compounds extraction procedure used was adapted from Samaras *et al.* (2005) with slight modifications. Barley/malt samples were finely ground in liquid nitrogen (Air Liquide, Paris, France) during 9 min with a cryogenic ground (6870 Freezer/Mill, Spex CertiPrep, Stanmore, U.K.) at a rate of 20 impacts/sec. Powdered samples (1.63 g) were homogenized in 10.72 mL of acetate buffer (50 mM ammonium acetate, pH 5.4) for 2 min using a pestle and mortar (kept on ice). This corresponded to 10 mL of wort at 13 ^{0}Plato (g of sugar/100 g of wort). After centrifugation of the extracts (3000 rpm, 10 min), supernatants were filtered through a Whatman No.1 filter paper, to remove particulate material. Samples were adjusted to pH 2.0 by the addition of HCl (37%), and 0.5 g of sodium chloride was added. As previously described (Leitao et al., 2011), extraction was carried out in 50 mL Corning centrifuge tubes with 10 mL of ethyl acetate (three times, for periods of 15 min) on an orbital shaker at 200 rpm. The extracts were centrifuged (5000 rpm, 10 min), and the supernatants were collected and evaporated to dryness under vacuum (30°C, 80 mbar). Samples were then redissolved in 1 mL methanol/water (50/50, v/v), membrane-filtered (0.45 μm, MachereyeNagel, Hoerdt, France) and injected (20 μL) in the chromatographic system.

HPLC analysis

The conditions and equipment used for COADS method were as previously described (Leitao *et al.*, 2011). Briefly, separation was carried out using a Waters HPLC system (Waters, Saint- Quentin-Fallavier, France), a hypersil BDS C18 HPLC column (5 μm, 250 x 4.6 mm i.d., ThermoScientific, Gometz-le-Châtel, France), and a mobile phase consisting of a water (0.1% formic acid, v/v)/methanol gradient delivered at 1 mL/min. Detection was done at 254 nm for phenolic compounds and at 412 nm for their respective antioxidant activities after post-column reaction with $ABTS^{\bullet +}$ prepared as previously described (Leitao *et al.*, 2011). Identification of phenolic compounds with LC-ESI-MS Identification of phenolic compounds was performed with electrospray ionization (ESI) mass spectrometer after online HPLC separation. High-resolution mass spectra were obtained with a 6520 Accurate Mass Q-TOF spectrometer (Agilent, Massy, France), with separation on a C18 HPLC column (1.9 μm, 100 x 1 mm) and an isocratic water/acetonitrile mobile phase containing 0.01% (v/v) formic acid and delivered with a 1200 solvent delivery module (Agilent). The sampling, extraction chromatographic determination, phenol content and their antioxidant activity are shown in Fig. 3.2; 3.3; and 3.4.

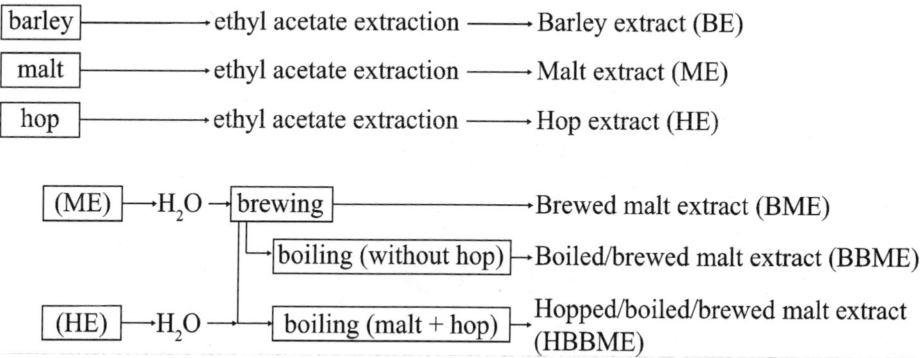

Figure 3.2: Sampling and extraction at different stages of malting and brewing.

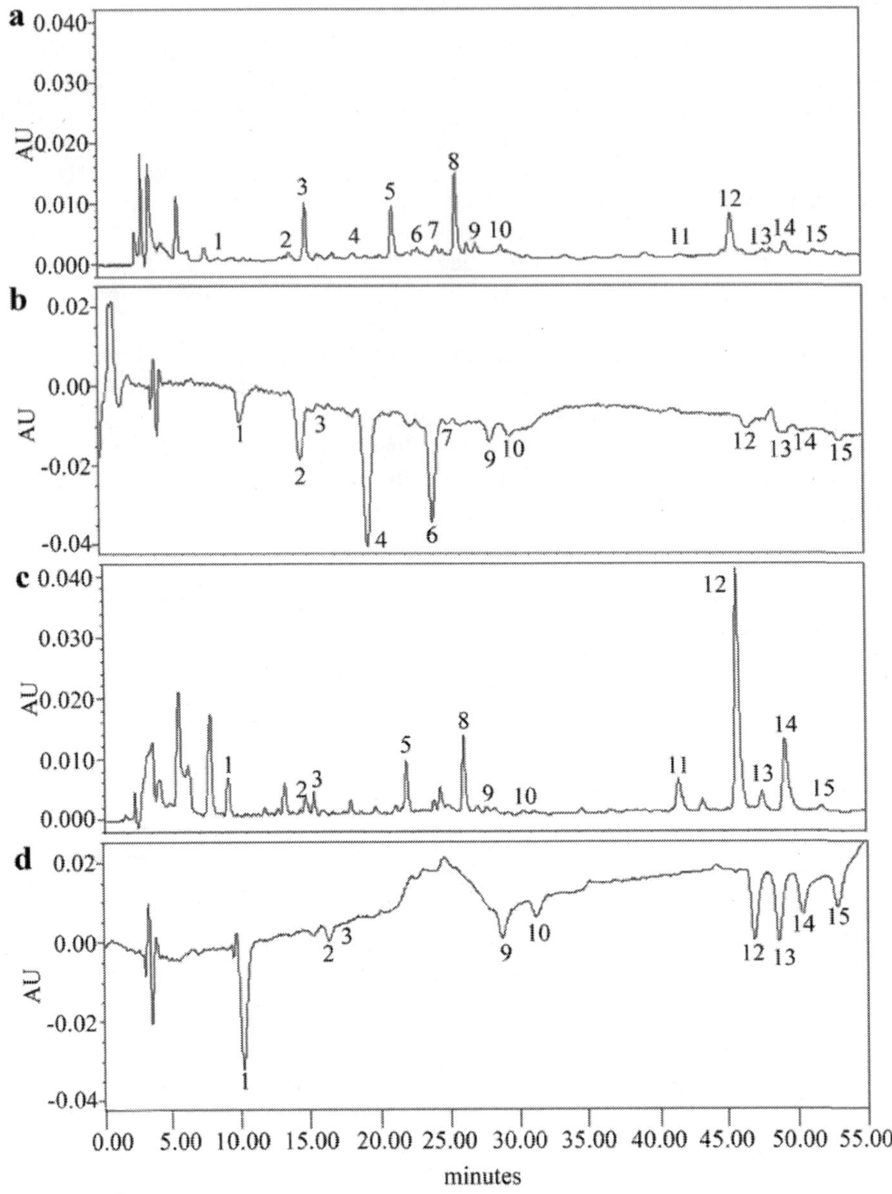

Figure 3.3: Chromatographic determination of phenolic compounds (a, c) and their corresponding antioxidant activities (b, d) in extracts of barley (a, b) and malt (c, d). Peaks are as follows: 1, unidentified compound; 2, prodelphinidin B3; 3, protocatechuic acid; 4, procyanidin B3; 5, p-hydroxybenzoic acid; 6, catechin; 7, chlorogenic acid; 8, vanillic acid; 9, caffeic acid; 10, epicatechin; 11, p-coumaric acid; 12, ferulic acid; 13, sinapic acid; 14, unidentified compound; 15, unidentified compound.

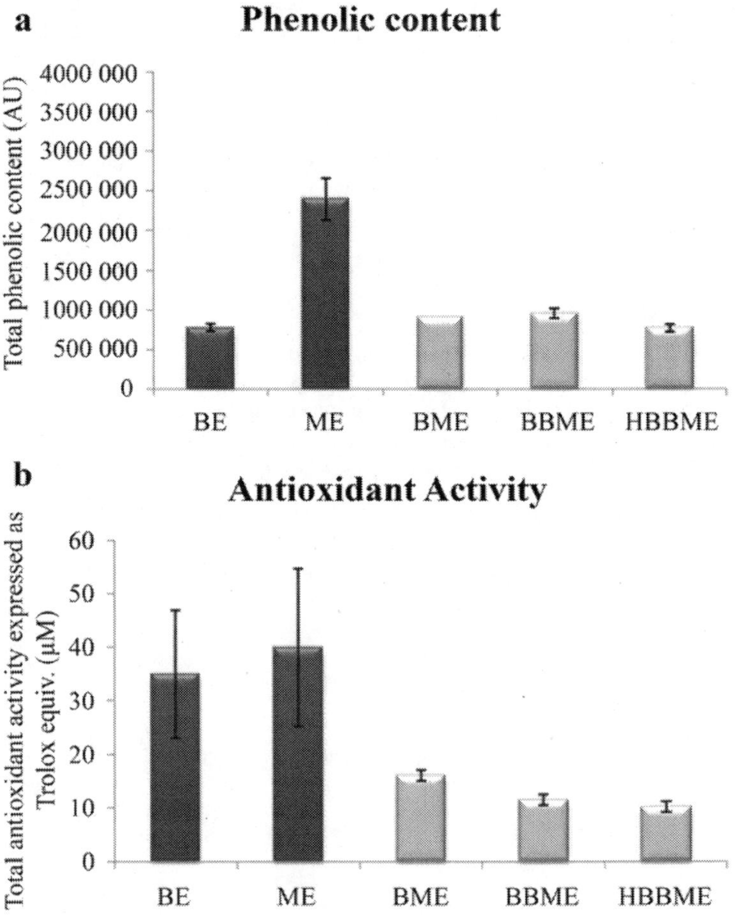

Figure 3.4: Total phenolic content (a) and antioxidant activity (b) of barley extract (BE) and malt extract (ME) at different steps of brewing: BME, brewed malt extract; BBME, boiled and brewed malt extract; and HBBME, hopped, boiled and brewed malt extract.

Legume Tannins and Phenolic Acids

Tannins are complex polyphenolic compounds present in a wide variety of plant materials. Polyphenolic compounds are classified as phenolic acids and derivatives, tannins, and flavonoids. The flavonoids are subclassified into anthocyanins, flavones, flavonols, and related substances. Anthocyanin pigments are known to be responsible for nearly all the pink, scarlet, red, mauve, violet, and blue colours of flowers, leaves, fruits, fruit juices, and wines, but nonetheless flavones and flavonols do make a significant contribution, either as yellow pigments or as copigments of anthocyanins (Harborne, 1965). According to Horowitz (1964) the taste of flavanones has a definite relation to their chemical structures. Astringency is closely connected to the tanning reactions of condensed tannins (Rossi and Singleton, 1966a, b) and their ability to inhibit enzymes. Based on their structural features, tannins are classified either as hydrolyzable tannins, which generally have gallic acid as their monomers, or condensed tannins such as catechin, gallocatechin, leucocyanidin, leucodelphinidin, and dimeric procyanidins (Naczk *et al.*, 1992b) (Figure 3.5). The condensed tannins are dimers, oligomers and polymers of polyhydroxyflavan-3-ol monomer units linked by $4 \rightarrow 6$ or $4 \rightarrow 8$ bonds. During the maturation of seeds/ fruits, increased condensation of phenolics (condensed tannins) (Goldstein and Swain, 1963) is accompanied by a decrease in the total phenolics and astringencey (Craft, 1961). This decrease in the content of phenolics during maturation may be due to polymerization of existing polyphenolic compounds to high molecular weight insoluble polymers such as lignins (Kadam *et al.*, 1982).

Sosulski and Dabrowski (1984) fractionated the phenolic constituents of defatted flours and hulls of ten legume species and classified them into free acids, soluble esters, and insoluble-bound residues. They reported that the total content of free phenolic acids (trans-ferulic, trans-*p*-coumaric, and syringic acids) in legume flours varied from 1.8 to 16.3 mg/100g of sample (Figure 3.6).

Polyphenolic compounds in legumes have received considerable attention largely as a result of their possible influence on the nutritional and aesthetic qualities of foods, biochemical and physiological functions, and their pharmacological and toxicological implications (Jadhav *et al.*, 1989).

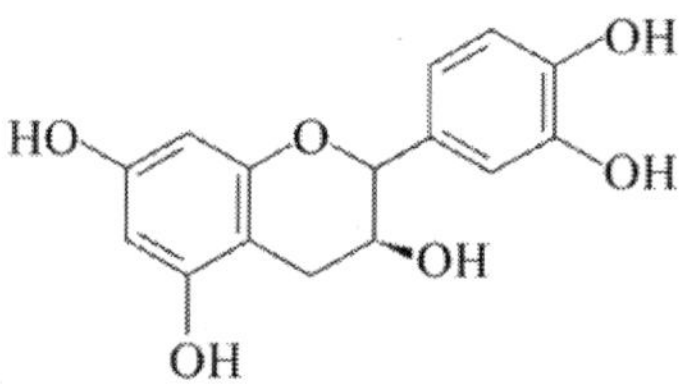

Catechin

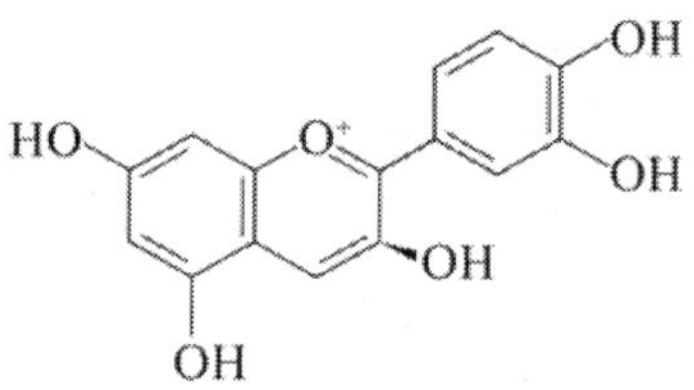

Leucocyanidin

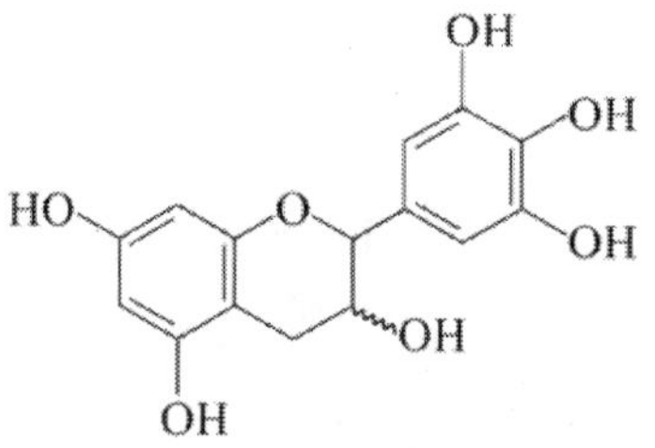

Gallocatechin

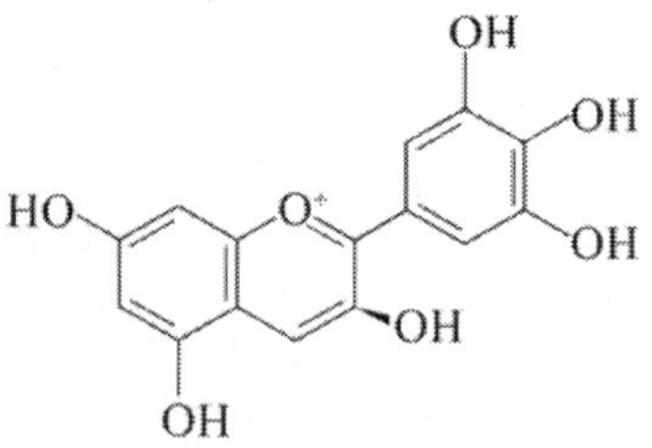

Leucodelphinidin

Figure 3.5: Structures of polyphenols of pea seed meals.

Benzoic acid derivatives

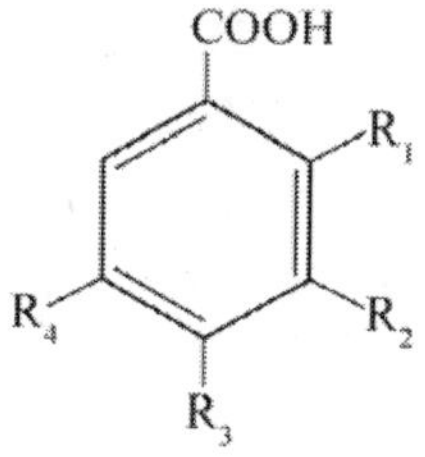

R_1	R_2	R_3	R_4	
H	H	OH	H	*p*-Hydroxybenzoic acid
H	OH	OH	OH	Gallic acid
OH	H	H	H	Salicylic acid
H	H	OH	OH	Protocatechuic acid
H	OCH_3	OH	OCH_3	Syringic acid

Cinnamic acid derivatives

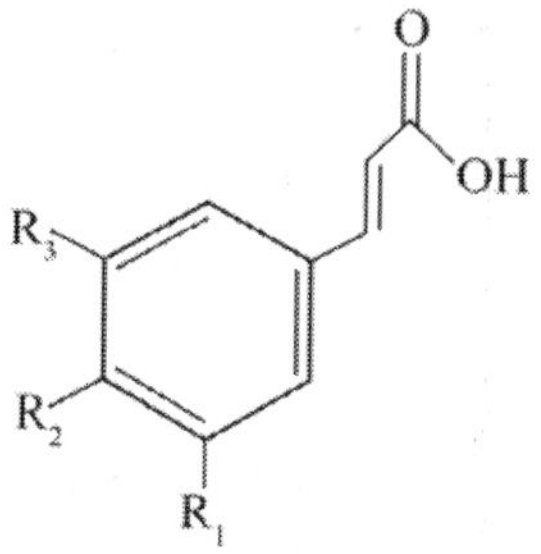

R_1	R_2	R_3	
H	OH	H	*p*-Coumaric acid
H	OH	OCH_3	Ferulic acid
OCH_3	OH	OCH_3	Sinapic acid

Figure 3.6: Phenolic acids of pea meals.

Chemistry and biogenesis

The term "tannin" was historically used to describe a chemically heterogeneous group of compounds which precipitated proteins. Molecular weights of tannins ranged between 500 and 3000 Da and, besides giving the usual phenolic reactions, had special properties such as the ability to precipitate alkaloids, gelatin and other proteins (Swain and Bate-Smith, 1962). The condensed tannins, or proanthocyanidins (Hagerman and Butler, 1994), are flavanol-based compounds. Condensed tannins react in alcoholic solutions of strong mineral acids to release the corresponding anthocyanidin, with a characteristic colour. Structural diversity of the proantho-cyanidins is a consequence of the substitution patterns and stereochemistry of the flavanol subunits. Structural complexity also results from the diversity of positions for interflavan bond formation, and from the stereochemical variation in the interflavan bond.

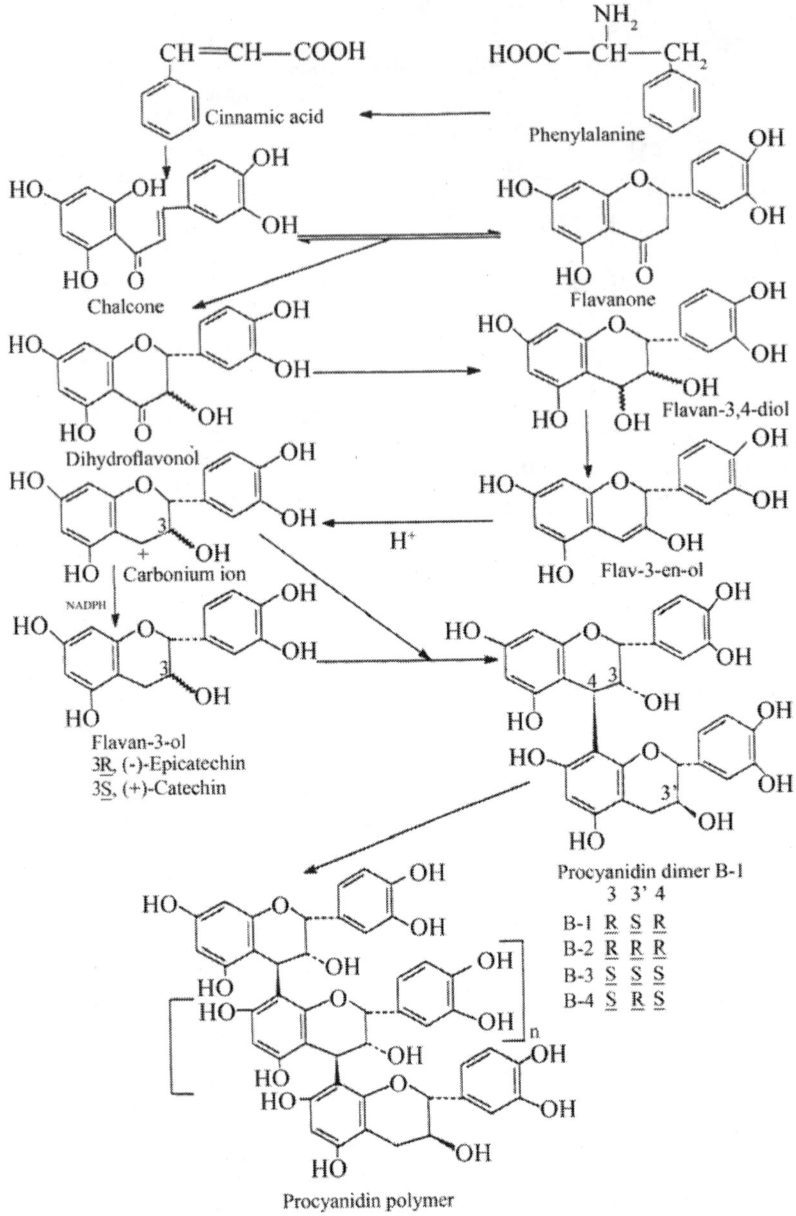

Figure 3.7: Biosynthesis of polyphenols [Tannins] (adopted from Haslam, 1977).

The hydrolyzable tannins may be hydrolyzed to a carbohydrate, usually glucose, and a phenolic, either gallic acid or its dimer ellagic acid. On the other hand, condensed tannins mainly ellagitannins, yields several derivatives of gallic acid besides gallic acid itself. The resistance of the condensed tannins towards hydrolysis arises from the condensation of two or more molecules of flavonoids such as flavan-3-ols (catechins) or flavan-3, 4-diols or a mixture of the two. The chemical term "proanthocyanidins" has been used for those colourless natural products which are converted to anthocyanidins on heating with acids (Freudenberg and Weinges, 1962). Weinges *et al.* (1969) implicated the term "lecuoanthocyanidin" to represent monomeric proanthocyanidins such as flavan-3,4-diols and condensed proanthocyanidin or procyanidin for flavan-3-ol dimers and oligomers.

Freudenberg and Weinges (1962) proposed that flavan-3, 4-diol is capable of producing a carbonium ion at the reactive benzylic C_4 -OH which could react with the nucleophilic centres of another flavan to give a $C_4 \rightarrow C_6$ or $C_4 \rightarrow C_8$ linkage. According to Hathway (1958), the condensed tannins are formed either by autoxidation or enzymatic reaction involving polyphenoloxidase. Polyphenoloxidase combines an oxidation product of a flavan, namely an *O*-quinone, with the A or B ring of another oxidized flavan. As a result, catechin condenses to give a head to tail polymer, while gallocatechin polymerizes in a tail to tail manner (Hathway and Seakins, 1957). The degree of polymerization of the procyanidins appears to be characteristic of specific plants controlled by metabolic activities during maturity (Haslam *et al.*, 1977). The structure, stereochemistry, distribution, metabolism, and biosynthetic pathways of plant procyanidins (condensed tannins) and associated flavan-3-ols are well explained by Haslam *et al.* (1977) (Figure 3.7).

In addition to tannins, plants contain a wide variety of nontannin phenolics (Hagerman *et al.*, 1997). These phenolics, which are usually relatively low molecular weight (<500Da), are distinguished from tannins by their inability to precipitate proteins. Dietary nontannin phenolics have different metabolic fates from tannins because of their different reactivity.

Chemical analyses

Chemical methods for assessing tannins require their extraction from tissues, separation, and isolation from other compounds and detection or quantification by spectrophotometric methods. The solvents widely used for extraction of tannins are methanol (Naczk *et al.*, 1992b), ethanol (Shahidi *et al.*, 1994), and acetone (Amarowicz *et al.*, 1996a). These solvent systems destroy the cell membrane and simultaneously dissolve the phenolic compounds. Sephadex LH-20 or G-50 chromatography has proven to be a more successful and widely used technique for fractionation of tannins (Davis and Hoseney, 1979). Compounds that are more sensitive but do not allow recovery of the intact tannins, involve the formation of coloured products with vanillin in concentrated HCl or *p*-toluenesulphonic acid and precipitation with gelatin, neutral lead acetate, or ferric chloride. Thin-layer and paper chromatographic techniques have also been employed to check the purity of the isolated condensed tannins (Jones *et al.*, 1976; Leung *et al.*, 1979). In general, the methods used for tannin analysis are based on general phenolic reactions, protein precipitation, functional group reactions, or HPLC. A large proportion of the current assays involve spectrophotometry of tannins or their chromogen. UV spectrophotometry appears to be a simple, fast, and convenient procedure (Sharp *et al.*, 1978) as it excludes partial purification of tannins encountered by Haslam and Gupta (1978).

The Prussian blue method (Price and Butler, 1977) or the Folin-Denis method (Ribereau-Gayon, 1972) may be used to measure both tannin and nontannin phenolics in plant extracts. Methods based on the formation of coloured phenolic-metal ion complexes are also useful for measuring total phenolics (Hagerman and Butler, 1978). The property of tannins to form insoluble complexes with proteins has been adopted for quantification of tannins from plant materials (Hagerman and Butler, 1978). The insoluble complexes can be dissolved in SDS-triethanolamine solution and treated with ferric chloride to form a violet-coloured complex with condensed tannins. The biological activity of condensed tannins could be reported as percentage inhibition of porcine pancreatic α-amylase (Davis and Hoseney, 1979).

Mechanisms of antinutritive action of tannins

Tannins in peas have been determined by various methods and expressed as catechin- or tannic acid-equivalents. The tannin contents of several pea varieties have been reported to range from 0 to 4% (Reddy *et al.*, 1985). The antinutritional effects of certain types of tannins include high toxicity to some animal and avian species (Singleton, 1981). However, varying effects have been observed in different biological systems. Since polyphenols (tannins) form insoluble complexes with proteins, carbohydrates and other food components, their presence in foods interferes with the utilization of dietary proteins in nonruminants (Martin-Tanguy *et al.*, 1977) and ruminants (Kumar and Singh, 1984).

The tannin-protein interaction may depend on the size, conformation, and charge of the protein molecule. Tannins may reversibly complex with proteins via hydrogen bonding of their hydroxyl groups with carbonyl functionality of peptide bonds in proteins, or irreversibly by oxidation to quinones which may in turn combine with reactive groups (e.g. $-NH_2$) of protein molecules. The phenol-protein complex may also be stabilized by ionic bonds between the phenolate anion and the cationic site of the protein molecules and/or hydrophobic interactions between aromatic ring structure of tannins and hydrophobic regions of proteins (Kumar and Singh, 1984). Tannins are also considered potent inhibitors of digestive enzymes due to their capacity to bind with enzyme proteins as well as other substrates. The phenolic compounds can affect the enzymes by either reducing their solubility due to the formation of an insoluble protein-phenolic complex or by inhibiting the enzyme-inhibitor complex. For example, tainting of eggs laid by hens fed on tannin-containing seed meals (tainting effect) may be a manifestation of the formation of a tannin-trimethylamine oxidase complex. The formation of the enzyme-inhibitor complex prevents the conversion of trimethylamine to the odourless and water-soluble trimethylamine oxide (Fenwick *et al.*, 1984).

The other important property of condensed tannins is related to their metal-precipitating ability due to the presence of *ortho*-dihydroxyl groups in the flavonoid ring. The precipitation of metal ions inhibits the absorption of dietary minerals. The inhibition may be due to the formation of insoluble metal-tannin complexes in the gastrointestinal tract, thus making the metal ions unavailable for absorption.

The affinity of tannins for polysaccharides is strongly dependent on the molecular size, conformational mobility and shape, as well as water solubility of polyphenols. Thus, an increase in molecular size and conformational flexibility of tannins enhances their affinity for carbohydrates. Davis and Hoseney (1979) reported that 40-60% tannins may bind starch depending on the source of tannins as well as the type of starch.

Biological significance

Bate-Smith (1958) apparently made the first connection between phenolic substances and textural quality of plant products. It was shown, microscopically, that lignin is not the only material that takes part in the encrustation and toughening of plant cell walls. Leucoanthocyanins, polymerized to form condensed tannins, are apparent in the toughness. The biochemical function of tannins has long been debated. The importance of tannins to the plant is believed to be due to their effectiveness as repellents to predators, whether animal or microbial. This function depends on the combined properties of astringency which renders the plant tissue unpalatable, and protein precipitation, which denatures salivary proteins of predators and inactivates microbial extracellular enzymes (Gupta and Haslam, 1980; Rhodes, 1985). Zucker (1983) explained that, due to their large molecular size and structure, condensed tannins are tightly complexed to proteins, pectins and cellulose of the cell wall in such a way to deny microbial extracellular enzymes binding sites to these substrates, thus providing protection against microbial attack. The function of tannins may be related to pea seed hardness since this defect is thought to be a physiological response to environmental stress (Hincks and Stanley, 1986). Lignin-like materials are deposited around pea/bean cotyledon cells and promote hardening, both as a result of their own mechanical strength as well as their action in preventing water imbibition and swelling (Hincks and Stanley, 1986, 1987). Ma and Bliss (1978) and Deshpande *et al.* (1982) examined whole and dehulled seeds from nearly 39 cultivars of common legumes for tannins and reported that the ratio of tannin content of testa to cotyledon was about 4:1. High concentrations of tannins in grains during the milk stage of development resist bird attack, while tannins in mature grains are responsible for decreased preharvest seed sprouting (Salunkhe *et al.*, 1982b).

ANTIOXIDANT ACTIVITY OF PHENOLIC FRACTIONS OF BEACH PEA HULL EXTRACT

Separation of phenolic fractions from beach pea hull extract

Three phenolic fractions (Fraction I-III) were separated from acetone extract (70%, v/v, containing 1% concentrated HCl) of beach pea hulls via Sephadex LH-20 column chromatography and each fraction was tested for the presence of condensed tannins (Figure 3.8). One major peak (III) and two minor peaks (I and II) were observed when absorbances were measured at 280 nm (Figure 3.8). For beach pea hulls, the highest content of phenolic compounds and condensed tannins was observed in fraction III. Furthermore, the crude extract possessed considerably more phenolic compounds than fractions I and II, while condensed tannins were not detected in fractions I and II (Table 3.1). Phenolic compounds from Polish white bean, pea, everlasting pea, lentil, broad bean and faba bean seed hulls showed higher amount of crude extract compared with the whole seeds but less than beach pea hulls (Amarowicz *et al.*, 1996b). They also had high antioxidative properties, compared with those of seeds from mustard, canola, rape, and flax (Amarowicz *et al.*, 1996a). These authors have also reported that the extract obtained from seed coats contained 2.5 - 13 times higher amounts of total phenolic than those from whole seeds.

Table 3.1: Percentage recovery of beach pea hulls extract and their total phenolic and condensed tannin contents[1]

Fraction	Relative content (% of total)	Total phenolic (% of extract)[2]	Condensed tannins (% of extract)[3]
Crude extract	21.53 ± 1.23[c]	54.15 ± 1.26[b]	156.73 ± 1.53[b]
Fraction-I	23.87 ± 0.87[b]	3.11 ± 0.41[d]	ND
Fraction-II	19.22 ± 0.42[d]	19.21 ± 0.23[c]	ND
Fraction-III	56.89 ± 1.03[a]	79.06 ± 1.06[a]	220.19 ± 3.83[a]

[1]All fractions separated on Sephadex LH-20 column. Results are means of three determinations, ± standard deviation. Means followed by different superscripts in each column are significantly ($p<0.05$) different from one another. ND, Not detected.

[2]As sinapic acid equivalents.

[3]As catechin equivalents.

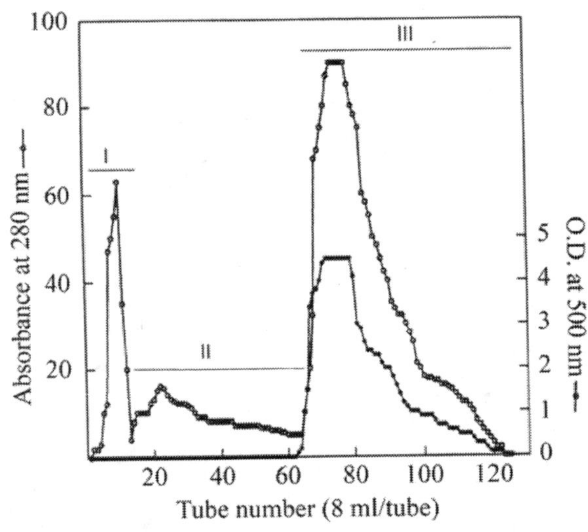

Figure 3.8: Separation of phenolic fractions of beach pea hulls extracts by Sephadex LH-20 column chromatography: UV absorbance of phenolics (280 nm) and condensed tannins (500 nm) following colour development (Fraction I and II ethanol mobile phase and III fraction with acetone-water, 50:50 (v/v) mobile phase).

The absorption maxima of separated fractions (Figure 3.9) occurred mainly in the range of 280 to 290 nm. Fractions I, II and III had only one maximum at 284, 284 and 282 nm, respectively. This suggests that flavonoids are potentially the main phenolics present in beach pea hulls extracts. Mabry *et al.* (1970) reported that flavones and flavonols produced two major absorption peaks in the 240 to 400 nm range in methanol. Amarowicz and Raab (1997) separated five phenolic fractions from the acetone extract of everlasting pea, faba bean and broad bean using Sephadex LH-20 column chromatography. They also reported that UV spectra of separated fractions from these legumes had most of their absorption bands in the 270 to 280 nm range.

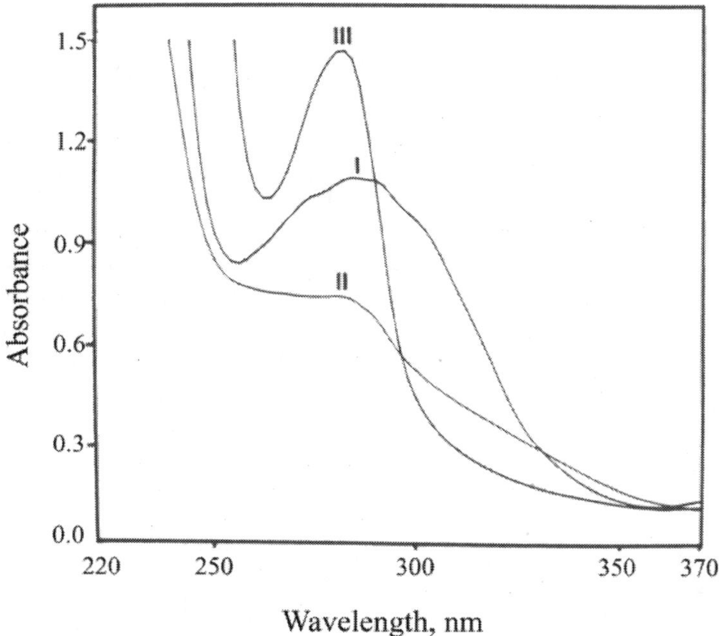

Figure 3.9: UV spectra of individual fractions of beach pea hull extracts separated on a Sephadex LH-20 column (I and II fractions from ethanol as mobile phase and fraction III acetone-water (v/v) as mobile phase).

TLC separation of phenolic fractions of beach pea hull extract

The TLC plates after development of the three isolated fractions and crude extract from beach pea hulls following spraying with different reagents are shown in Figure 3.10. Thin layer chromatography indicated that the separated fractions contained several phenolic compounds. Compounds from fraction III of beach pea hulls extract, close to the solvent front in both developing systems, possessed maximum antioxidant activity. Several phenolic compounds were visualized on the TLC plate. Therefore, fraction III with the highest antioxidant activity may contain several antioxidative phenolics. The total number of hydroxyl groups present in compounds of fraction III may be higher than those of compounds in other fractions or different additional compound(s) may be present; this may partially be responsible for better antioxidative properties of this fraction as shown in the inhibition of the bleaching of β-carotene.

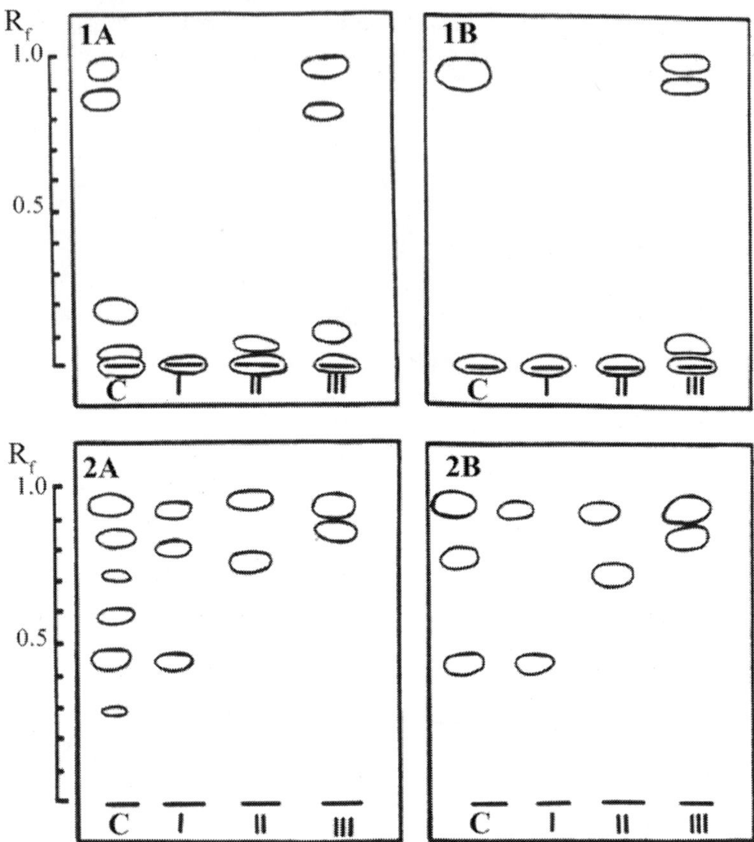

Figure 3.10: TLC of phenolic fractions separated from beach pea hull extract; chromatograms were developed using (1) acetic acid-petroleum ether-diethyl ether (1:20:80, v/v/v) and (2) acetic acid-water-n-butanol (10:10:30, v/v/v); plates were sprayed with a solution of ferric chloride (A) to give spots of phenolic compounds and (B) β–Carotene-linoleate in order to evaluate antioxidant activity of crude extract (C) and fractions (I-III).

Antioxidant activity of phenolic fractions of beach pea hull extract

The antioxidant activity of each fraction and crude extract, as compared with BHA is presented in Figure 3.11. Fraction III exhibited the highest antioxidative activity against bleaching of β-carotene. The crude extract had a better antioxidative effect than fractions I and II. The activity of fraction III was higher than that of fractions I, II and the crude extract. Fraction III which exhibited the best antioxidative activity, contained 79.1% phenolics, as sinapic acid equivalents, and 220.2 % condensed tannins, as catechin equivalents (Table 3.1). This indicates that the amount of phenolic compounds and their molecular structures play an important role in their antioxidative activity (Wanasundara *et al.*, 1996).

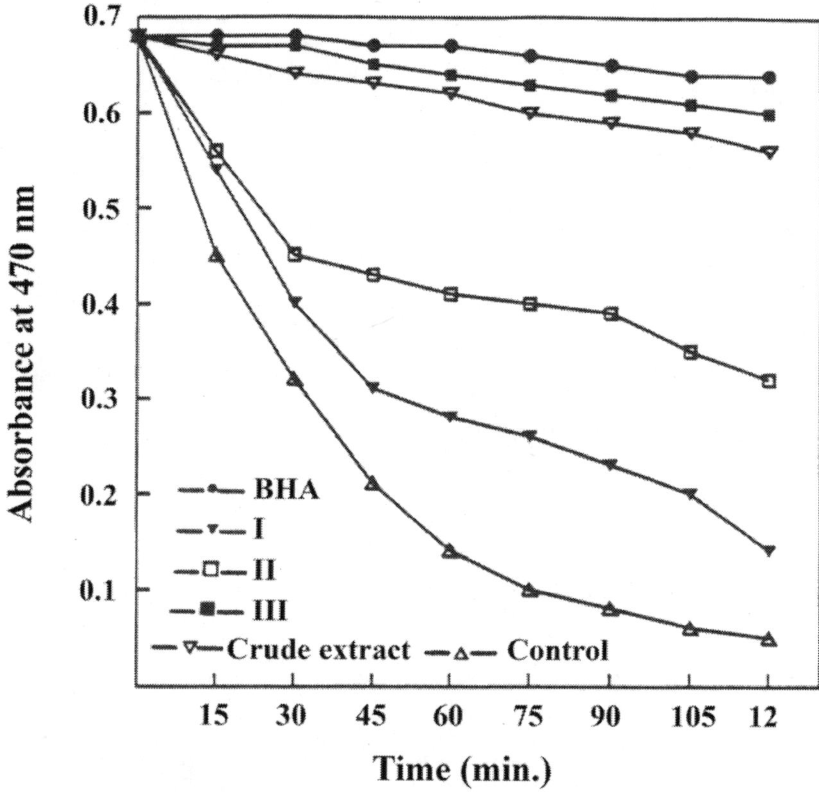

Figure 3.11: Inhibition of bleaching of β–Carotene in a model β–Carotene-linoleate system containing the crude extract or individual fractions of beach pea hull extracts separated on a Sephadex LH-20 column.

Detection of phenolic/tannin compound present in fractions of beach pea hull extract

Fraction III of the beach pea hull extract which contained the highest amount of phenolics, condensed tannins and possessed strong antioxidant activity, was further separated on a semi-preparative HPLC. Presence of (+) catechin and (-) epicatechin as main phenolic compounds in this fraction was detected (Figure 3.12).

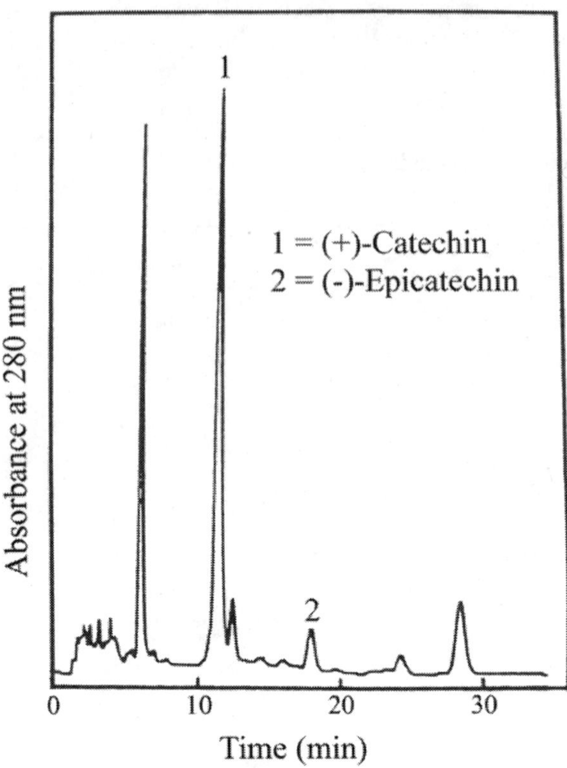

Figure 3.12: Chromatogram of HPLC separated (+) catechin and (-) epicatechin from Sephadex LH-20 fractions.

Effect of different solvent extractions on phenolic compounds, tannins and sugars of beach pea seeds

The extraction capability of different solvents of phenolic compounds, tannins and sugars with time of extraction is presented in Figure 3.13 and Table 3.2, respectively. UV data after consecutive stages of extraction at 280 nm show that acetone-water (80:20, v/v) is most effective in extracting a maximum amount of phenolic compounds and tannins from beach pea. Acetone-water mixture extracted almost 1.5 times more phenolics and condensed tannins from beach pea seeds than methanol-water or ethanol-water mixtures. However, methanol-water was most effective in extracting a maximum amount of sugars from beach pea seeds only at 15 min extraction. This might be due to simple sugars and oligosaccharides which dissolve more easily in methanol-water than in ethanol-water or acetone-water. Price and Spiro (1985) and Price and Spitzer (1993) reported that the highest extraction of phenolic compounds from plant material with methanol-water (80:20, v/v) was achieved during the first stage of extraction. Acetone-water (80:20, v/v) was most effective in extracting the phenolic compounds from lentil seeds (Amarowicz *et al.*, 1995), but this solvent system was less effective in the removal of sugars.

Table 3.2: Comparison of extraction efficiency of methanol, ethanol and acetone on beach pea as percentage of total amounts of sugars, phenolics and tannins[1].

Extraction, solvent and time (min)	Soluble sugars	Phenolic compounds[2]	Condensed tannins[3]
Methanol			
15	57.65	32.00	31.98
30	23.86	26.90	27.08
60	10.54	23.32	24.43
90	7.95	17.78	16.51
Ethanol			
15	42.03	32.96	30.15
30	30.63	27.35	27.85
60	19.24	22.78	23.55
90	8.10	16.91	18.45
Acetone			
15	38.19	49.34	52.51
30	31.21	30.32	21.35
60	19.92	14.39	17.55
90	10.68	5.95	8.59

[1]Results are means of two determinations. [2]As sinapic acid equivalents. [3]As catechin equivalents.

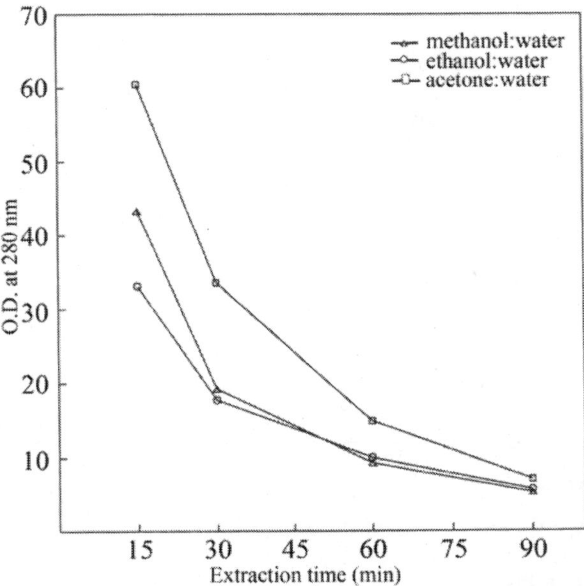

Figure 3.13: Comparison of extracting capability of the solvents used: extraction of compounds with UV absorbance at 280 nm; -Δ- methanol-water; -O- ethanol-water; -□- acetone-water extracts; all solvents at 80:20, (v/v) ratio.

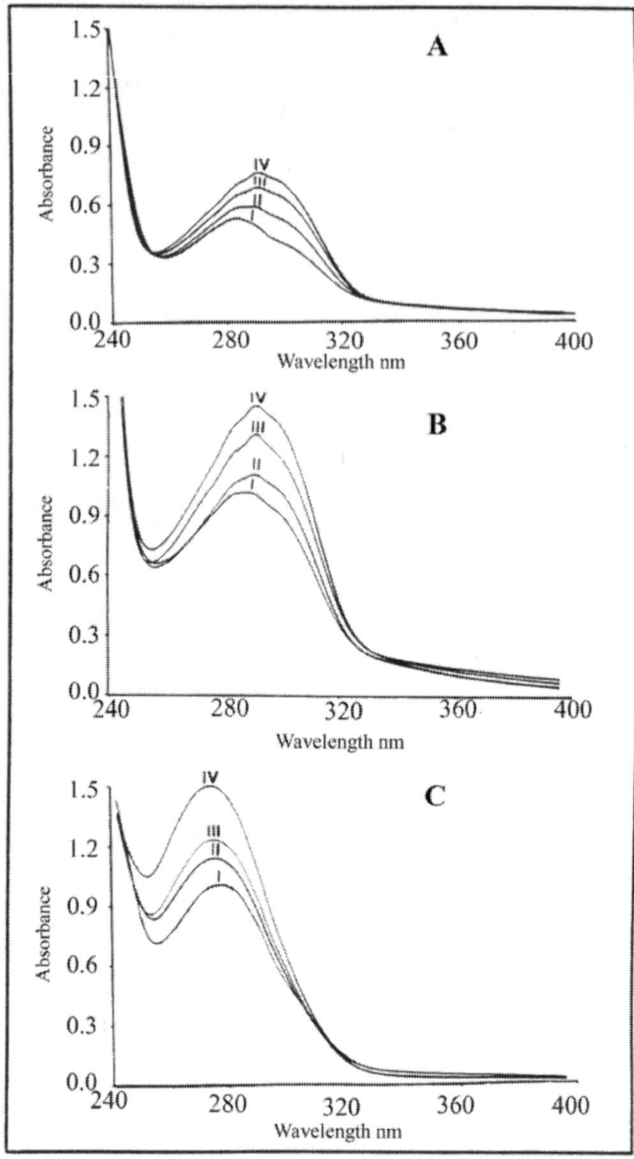

Figure 3.14: UV spectra of each extraction (I-IV); A, methanol-water (80:20, v/v); B, ethanol-water (80:20, v/v), C, acetone-water (80:20, v/v) from beach pea seed extracts.

UV spectra of beach pea seed extracts

The beach pea seeds were extracted 4 times (I-IV) with methanol:water (A) ethanol:water (B) and acetone:water (C); their UV spectra are shown in Figure 3.14. Four extractions (I-IV) of each solvent combined with their UV spectra recorded between 240 and 400 nm are shown in Figure 3.15 (ethanol : water, 1; methanol : water, 2; acetone : water, 3). The UV spectra of ethanol-water extract showed a maximum at 292 nm. The first methanol-water extraction showed UV absorption maximum at 284 while the next three extractions and combined extracts (I-IV) exhibited a maximum at 292 nm. When acetone-water was used for the extraction, a maximum was observed at 284 nm. These UV spectral data indicate that methanol-water and ethanol-water solvent systems extract the same types of compound, but in case of acetone-water different compounds may be extracted from the seeds. Frejnagelt *et al.* (1994) reported that the absorption maxima of phenolic compounds obtained from the faba bean seed coats ranged from 264 to 280 nm. Amarowicz *et al.* (1995) used similar conditions for extraction of phenolic compounds and sugars from lentil and reported that UV spectrum of the acetone extract had a maximum at 274 nm and compounds in methanol and ethanol extracts had a maximum at 272 nm. Acetone-water (8:2, v/v), used for the extraction of phenolic compounds from everlasting pea, faba bean, and broad bean afforded absorption maximum in the range of 260 to 282 nm (Amarowicz *et al.*, 1996b).

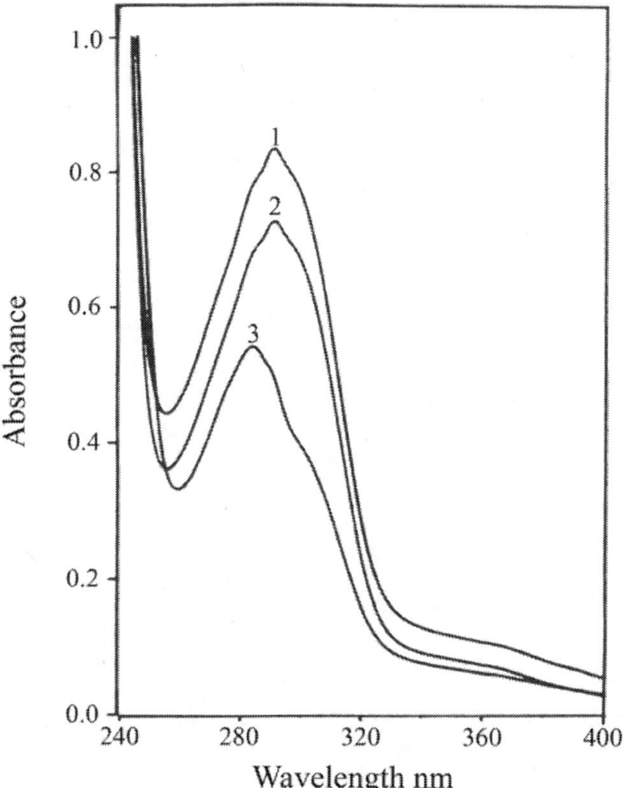

Figure 3.15: UV spectra of combined extracts (Extractions I-IV) of beach pea seed obtained using (1) ethanol-water (80:20, v/v); (2) methanol-water (80:20, v/v) and (3) acetone-water (80:20, v/v) solvents.

Total extracts and the content of phenolics, condensed tannins and sugars extracted by different solvent systems

The amounts of extract recovered with methanol (80:20, v/v), ethanol (80:20, v/v) and acetone (80:20, v/v) are shown in Table 3.3. The amount of extract recovered by acetone as a solvent was significantly ($p<0.05$) higher than when methanol and ethanol solvent systems were used. Extraction of sugars was highest when methanol was employed and the content of phenolic compounds extracted from beach pea seeds by acetone were twice that recovered with methanol or ethanol. Extraction of condensed tannins with acetone-water was also ten time more effective than methanol or ethanol. However, the sugar content in methanol extract was higher than that in ethanol and acetone extract. These results indicate the importance of choice of solvent in quantification of different components of the extracts.

Table 3.3: The content of extract (%), sugars (%), phenolic compounds (mg/100g) and condensed tannins (mg/100g) in beach pea seeds using different solvent systems[1]

Extraction solvent	Extract	Sugars	Phenolic compounds[2]	Condensed tannins[3]
Methanol (80:20, v/v)	16.62±0.60[ab]	5.03±0.35[a]	619.64±0.31[b]	748.93±2.29[b]
Ethanol (80:20, v/v)	15.01±0.65[b]	4.87±0.29[a]	421.06±0.54[c]	381.06±1.70[c]
Acetone (80:20, v/v)	19.32±0.90[a]	2.95±0.15[b]	1283.87±3.90[a]	7485.74±3.01[a]
Difference significance	A=M, A>E, M=E	M=E>A	A>M>E	A>M>E

[1]Results are means of triplicate determinations, ± standard deviation. Means followed by different supercripts in each column are significantly ($p<0.05$) different from one another. M, Methanol; E, Ethanol; and A, Acetone.
[2]As sinapic acid equivalents. [3]As catechin equivalents.

TLC separation of beach pea seed extracts

TLC plates, with beach pea seed extracts developed using different solvent systems and sprayed with different reagents are presented in Figure 3.16. Chromatograms of extracts developed with acetic acid-water-n-butanol (10:10:30, v/v/v), a high polar solvent system, showed three intense spots from ethanol extract close to the solvent front. Methanol extract also showed three intense spots, one very close to the origin and the other two at R_f values of 0.3 and 0.9. Acetone extract also showed two intense spots at R_f values of 0.3 and 0.9 (Figure 3.16A). Chromatograms developed with acetic acid-petroleum ether-diethyl ether (1:20:80, v/v/v), a non-polar solvent (Figure 3.16B), and sprayed with a ferric chloride solution showed only two intense spots for ethanol extract at R_f values of 0.8 and 0.9. The same extracts used for chromatogram development in water-methanol-chloroform (10:35:65, v/v/v), a polar solvent (Figure 3.16C), and sprayed with 10% sulphuric acid and heated at 120°C showed four very intense spots close to the solvent front for ethanol extract, two spots for the methanol extract at R_f values of 0.1 and 0.8, respectively, for the presence of sugars and no intense spots were observed for acetone extract. Ferric chloride reacts with phenolic compounds and gives blue or greenish coloured spots while sulphuric acid char the sugars at higher temperatures to yield black spots.

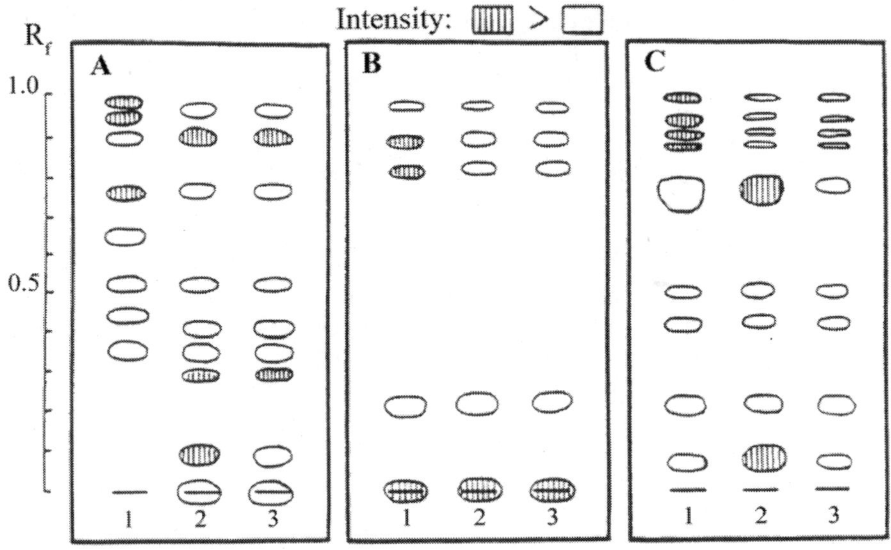

Figure 3.16: TLC beach pea seed extracts; developed using A: acetic acid-water-n-butanol (10:10:30, v/v/v); B: acetic acid-petroleum ether-diethyl ether (1:20:80, v/v/v); C: water-methanol-chloroform (10:35:5, v/v/v, Lower phase). Plates were spray with, A and B-ferric chloride solution and C-10% sulphuric acid and heated at 120°C.

Separation of phenolic/tannin fraction of beach pea extract

The phenolics of beach pea extracted by acetone-water were further separated using Sephadex LH-20 column chromatography. UV absorbances of extracted phenolics at 280 nm and condensed tannins at 500 nm, following colour development, are presented in Figure 3.17. Major phenolics were found in tube numbers 10-23 and condensed tannins in tube numbers 23-30. The condensed tannins fraction, separated on Sephadex LH-20, was further fractionated by semi-preparative HPLC. The major fraction (tube numbers 23-30) of condensed tannins of beach pea seed extract contained (+) catechin and (-) epicatechin as its major compounds (Figure 3.18). Similar separation of (+) catechin and (-) epicatechin were reported for pea and bean extract (Tsuda *et al.*, 1993).

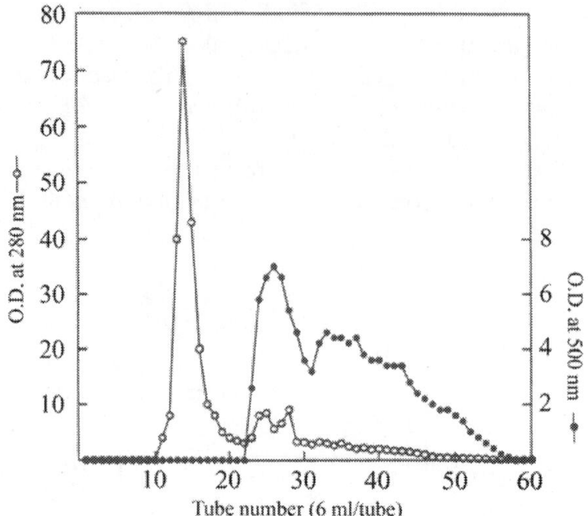

Figure 3.17: Separation of phenolic fractions of beach pea seed extracts by Sephadex LH-20 column chromatography: UV absorbance of phenolics (280 nm) and condensed tannins (500 nm) following colour development.

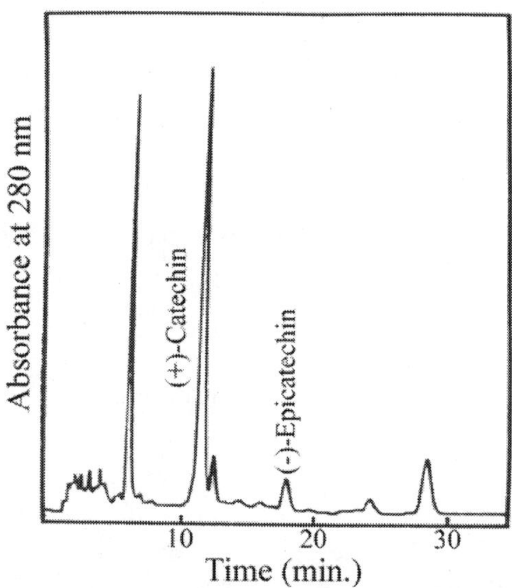

Figure 3.18: Chromatogram of the analytically separated pure (+) catechin and (-) epicatechin obtained from Sephadex LH-20 fractions (tube numbers 23-30, Figure 4.15) followed by semi-preparative HPLC separation.

■■■■■

4 Measurement of Antioxidant Activity of Phenols

Measurement of antioxidant activity

The need to measure antioxidant activity is well documented; these are carried out for meaningful comparison of foods or commercial products and for provision of quality standards for regulatory issues and health claim (Shahidi and Ho, 2007). Lipid oxidation is conventionally studied by determination of peroxide value (PV), thiobarbituric acid reactive substances (TBARS), conjugated dienes (CD), or anisidine value (AV) or by assessing volatile compounds (Kristinova, Mozuraityte, Storrø, and Rustad, 2009). There are numerous methods for measuring antioxidant activity; these may be classified into two categories. The first category measures the ability of antioxidants in inhibiting oxidation in a model system by monitoring the associated changes using physical, chemical or instrumental means. Radical scavenging assays include methods based on hydrogen atom transfer (HAT) or single electron transfer (SET) mechanisms. Oxygen radical absorbance capacity (ORAC), total radical trapping antioxidant parameter (TRAP) and crocin bleaching assays are the major methods that measure HAT whilst trolox equivalent antioxidant capacity (TEAC), ferric reducing antioxidant power (FRAP) and 2, 2-diphenyl-1-picrylhydrazyl (DPPH) assays represent SET-based methods (Shahidi and Ho, 2007). HAT-based methods measure the classical ability of an antioxidant to quench free radicals by hydrogen donation whilst SET-based methods detect the ability of a potential antioxidant to transfer one electron to reduce any compound, including metals, carbonyls and radicals (Shahidi and Zhong, 2005, 2007).

Table 4.1 summarizes the methods commonly used to measure antioxidant activity and the units they carry. However, a detailed discussion of these methods is beyond the goal of this review and has been communicated elsewhere (Shahidi and Zhong, 2015).

Table 4.1: Antioxidant activity measurement methods and units.

Methods	Measurement unit
Peroxide value (PV)	Milliequivalents of oxygen per kilogram of sample (meq/kg)
Conjugated dienes and trienes	Conjugable oxidation products (COPs)
Thiobarbituric acid reactive substances (TBARS) assay	Milligrams of malondialdehyde (MDA) equivalents per kilogram sample or micromoles of MDA equivalents per gram of sample (meq/g)
p-Anisidine value (*p*-AnV)	Absorbance of a solution resulting from the reaction of 1 g of fat in isooctane solution (100 ml) with *p*-anisidine
Electrical conductivity method	Oil stability index (OSI) value, which is defined as the point of maximal change of the rate of oxidation
Oxygen radical absorbance capacity (ORAC) assay	µmol of trolox equivalents
Total radical-trapping antioxidant parameter (TRAP) assay	µmol per litre peroxyl radical deactivated
Trolox equivalent antioxidant capacity (TEAC) assay	mM trolox equivalent to 1 mM test substance
2,2-Diphenyl-1-picrylhydrazyl (DPPH) assay	EC50 (Concentration to decrease concentration of test free radical by 50%); TEC50 (Time to decrease concentration of the test free radical by 50%); AE (Antiradical efficiency (1/EC50) TEC50)
Ferric reducing antioxidant power (FRAP) assay	Absorbance of Fe^{2+} complex at 593 nm produced by antioxidant reduction of corresponding tripyridyltriazine Fe^{2+} complex

PROTOCOLS

1. Determination of antioxidant activity by the DPPH$^{\bullet}$ test

The DPPH$^{\bullet}$ test is based on the ability of the stable 2, 2-diphenyl-1-picrylhydrazyl free radical to react with hydrogen donors. The DPPH$^{\bullet}$ radical displays an intense UV-VIS absorption spectrum. In this test, a solution of radical is decolourized after reduction with an antioxidant (AH) or a radical (R$^{\bullet}$) in accordance with the following scheme: DPPH$^{\bullet}$ + AH → DPPH$^{\bullet}$-H + A$^{\bullet}$, DPPH$^{\bullet}$ + R$^{\bullet}$ → DPPH$^{\bullet}$-R. This method is very simple and also quick for manual analysis.

Reagent preparation: 0.95 mmol·L-1 solution of radical DPPH$^{\bullet}$ (m = 0.00374 g/100 mL). First, this amount is dissolved in 50 mL of DMSO and after dissolution made up to a volume of 100 mL with ACS water. The solution can be used for 7 days when stored at 4 °C and in the dark.

Measurement procedure for an automated analyzer: A 200 μL volume of reagent is incubated with 20 μL of sample (gallic acid, Trolox®). Absorbance is measured after 1,520 seconds at λ = 510 nm. To calculate the antioxidant activity, the values determined before decrease of the absorbance (224th second of measurement – A224) and the last measurement value (1520th second of measurement – A_{1520}) are used. Resulting value is calculated in accordance with the following formula: $A = A_{1520}-A_{224}$.

2. Determination of antioxidant activity by the ABTS test

The ABTS radical method is one of the most used assays for the determination of the concentration of free radicals. It is based on the neutralization of a radical-cation arising from the one-electron oxidation of the synthetic chromophore 2, 2'-azino-bis (3-ethylbenzothiazoline-6-sulfonic acid) ($ABTS^{\bullet}$): $ABTS^{\bullet}$ – e- $ABTS^{\bullet+}$. This reaction is monitored spectrophotometrically by the change of the absorption spectrum. Results obtained using this method are usually recalculated to Trolox® concentration and are described as "Trolox® Equivalent Antioxidant Capacity" (TEAC). For chemically pure compounds, TEAC is defined as the micromolar concentration of Trolox® equivalents demonstrating the same antioxidant activity as a tested compound (at 1 mmol·L-1concentration).

Reagent preparation: Seven mmol·L-1 $ABTS^{\bullet}$ (m = 0.03841 g/10 mL) and 4.95 mmol·L-1 potassium peroxodisulphate (m = 0.01338 g/10 mL) are mixed and dissolved in ACS water. The solution is then diluted with ACS water in a 1:9 v/v ratio (10 mL is quantitatively transferred into 100 mL calibrated flask and diluted). The solution is incubated for 12 hours in the dark and the reagent is usable for seven days if stored in the dark at 4 °C.

Measurement procedure for an automated analyzer: A 245 μL volume of reagent is pipetted into a plastic cuvette with subsequent addition of 5 μL of sample (gallic acid, Trolox®). Absorbance is measured at λ = 670 nm after 1,520 seconds. For calculating the antioxidant activity, values before decrease of the absorbance (224th second of measurement – A_{224}) and the last measurement value (1,520th second of measurement – A_{1520}) are used. Resulting value is calculated in accordance with following formula: $A = A_{1520}-A_{224}$.

3. Determination of antioxidant activity by the FRAP method

The FRAP method (Ferric Reducing Antioxidant Power) is based on the reduction of complexes of 2,4,6-tripyridyl-*s*-triazine (TPTZ) with ferric chloride hexahydrate ($FeCl_3 \cdot 6H_2O$), which are almost colourless, and eventually slightly brownish. This chemical forms blue ferrous complexes after its reduction. The method has its limitations, especially in measurements under non-physiological pH values (3.6). In addition, this method is not able to detect slowly reactive polyphenolic compounds and thiols.

Reagent preparation: Solution 1: 10 mmol·L-1 solution of TPTZ (m = 0.07802 g/25 mL), in 40 mM of hydrochloric acid. Solution 2: 20 mM solution of ferric chloride hexahydrate (m = 0.13513 g/25 mL) in ACS water. Solution 3: 20 mM acetate buffer, pH 3.6 (weight of sodium acetate trihydrate is 0.27216 g in 100 mL ACS water, adjusted to the desired pH using HCl). These three solutions (TPTZ, $FeCl_3$, acetate buffer) are mixed in a 1:1:10 ratios. Reagent can be used for seven days if stored at 4°C in the dark.

Measurement procedure for an automated analyzer: A 245 μL volume of reagent is pipetted into a plastic cuvette with subsequent addition of a 5 μL sample (gallic acid, Trolox®). Absorbance is measured for 1,520 seconds at primary $\lambda = 578$ nm and secondary at $\lambda = 630$ nm wavelengths. For calculating the antioxidant activity, the differential absorbance (primary absorbance – secondary absorbance) is used.

4. Determination of antioxidant activity by the DMPD method

The compound *N,N*-dimethyl-1,4-diaminobenzene (DMPD) is converted in solution to a relatively stable and coloured radical form by the action of ferric salt. After addition of a sample containing free radicals, these are scavenged and as a result of this scavenging, the coloured solution is decolourized.

***Reagent preparation*:** Solution 1: acetate buffer (0.2 mol·L-1, pH 5.25); 1a) 2.17 g of sodium acetate trihydrate is dissolved in 80 mL of ACS water; 1b) 300 μL of concentrated acetic acid (>99.5 %, v/v) is diluted to a volume of 20 mL with ACS water. These two solutions are mixed to reach pH = 5.5. Solution 2: 0.74 mmol·L-1 ferric chloride: 1 mg of $FeCl_3 \cdot 6H_2O$ is dissolved with ACS water to a volume of 5 mL. Solution 3: 36.7 mmol·L-1 DMPD: 25 mg of DMPD is dissolved in 5 mL of ACS water. This solution must be prepared at the time of use due to its low stability. These three solutions (solutions No. 1, 2 and 3) are mixed in a 20:1:1 (*v/v/v*) ratio.

***Measurement procedure for an automated analyzer*:** A 200 μL volume of reagent is pipetted into a plastic cuvette. Then, 5 μL of sample (gallic acid, Trolox®) is added. Absorbance is measured at $\lambda = 510$ nm for 1,520 seconds. For calculating of the antioxidant activity, values before decrease of absorbance (224th second of measurement – A_{224}) and last (1520th second of measurement – A_{1520}) are used. Resulting value is calculated in accordance with following formula: Differential absorbance $A = A_{1520} - A_{224}$.

5. Determination of antioxidant activity by the Free Radicals method

This method is based on ability of chlorophyllin (the sodium-copper salt of chlorophyll) to accept and donate electrons with a stable change of maximum absorption. This effect is conditioned by an alkaline environment and the addition of catalyst.

***Reagent preparation*:** 5 mL of reaction buffer (100 mmol·L-1 HCl) is diluted with 45 mL of ACS water. To this solution, 100 μL of chlorophyllin is added. After its complete dissolution, 0.25 mL of catalyst is added. Reaction buffer is stable for one month when stored at 2 - 8 °C in the dark.

Measurement procedure for an automated analyzer: Into a plastic cuvette, a 200 μL volume of prepared reagent is pipetted. Next, 8 μL of sample (gallic acid, Trolox®) is added. Absorbance is measured at $\lambda = 450$ nm for 560 seconds. For calculating of the antioxidant activity, values before decrease of the absorbance (224th second of measurement – A_{224}) and the last value of measurement (560th second of measurement – A_{560}) are used. The resulting value is calculated in accordance with the following formula: $A = A_{560}-A_{224}$.

6. Determination of antioxidant activity by the Blue CrO_5method

Chromium peroxide (CrO_5) is very strong pro-oxidant produced in an acidic environment by ammonium dichromate in the presence of H_2O_2. It is a deep blue potent oxidant compound, miscible and relatively stable in polar organic solvents that can be easily measured by spectrometry.

Reagent preparation: Solution 1: 1a) 10 mL of solution of sulphuric acid (25 mmol·L-1; 13.4 µL of 98.8 % sulphuric acid diluted with ACS water to a volume of 10 mL); 1b) 10 mL of 20 mmol·L-1 ammonium dichromate solution (50.4 mg of ammonium dichromate dissolved in ACS water); 1c) 30 mL of 99.5 % DMSO (*v/v*). These three solutions are mixed in a 1:1:3 ratios (*v/v/v*). Solution 2: 1.6 mol·L-1 solution of hydrogen peroxide.

Measurement procedure for an automated analyzer: 400 µL of solution 1 is mixed with 4 µL of sample (gallic acid, Trolox®, α-tocopherol). This solution is incubated for 192 seconds. The first absorbance is measured at the 416th second (A_{416}). Subsequently, 40 µL of solution 2 is added and the solution is incubated for 192 seconds with measurement of the second absorbance (A_{608}). Measurements are carried out at $\lambda = 546$ nm. Resulting absorbance was calculated in accordance with the formula $A = A_{608} - A_{416}$.

UV-VIS SPECTROMETRIC PROTOCOLS

1. DPPH

A volume of 1,000 µL of reagent was pipetted into plastic cuvettes. Subsequently, a volume of 100 µL of the standard (gallic acid, Trolox®) was added. Absorbance was measured at $\lambda = 510$ nm. For calculation value of absorbance of reagent itself (AR) and value determined after 25 minutes of incubation A_{25} was used. Resulting value was calculated in accordance with following formula: $A = A_{25}-A_R$.

2. ABTS

A volume of 980 µL of reagent and subsequently a volume of 20 µL of measured sample (gallic acid, Trolox®) was added into plastic cuvettes. Absorbance was measured at $\lambda = 670$ nm. For calculation of the absorbance of the reagent itself (AR) and value determined after 25 minutes of incubation A_{25} was used. Resulting value was calculated in accordance with following formula: $A = A_{25}-A_R$.

3. FRAP

A volume of 980 µL of reagent and subsequently a volume of 20 µL of measured sample (gallic acid, Trolox®) was pipetted into plastic cuvettes. Absorbance was measured after 25 minutes of incubation at primary $\lambda = 578$ nm and secondary $\lambda =$

630 nm wave lengths. The obtained values of absorbance were subtracted (primary absorbance - secondary absorbance = differential absorbance). For calculating of antioxidant activity, differential absorbance was used.

4. DMPB

A volume of 1,000 μL of reagent was added into plastic cuvettes with subsequent addition of 25 μL of sample (gallic acid, Trolox®). Absorbance was measured at λ = 510 nm. For calculation of the absorbance value of the reagent itself (AR) and value of absorbance determined after 25 minutes of incubation (A_{25}) was used. Resulting value was calculated in accordance with following formula: differential absorbance = $A_{25} - A_R$.

5. Free radicals

A volume of 1,000 μL of prepared reagent was pipetted into plastic cuvettes with immediate addition of 40 μL of sample (gallic acid, Trolox®). Absorbance was measured after 9 minutes at λ = 450 nm. For calculation of the absorbance value of the reagent itself (AR) the absorbance determined after 25 minutes of incubation (A_{25}) was used. Resulting value was calculated in accordance with following formula: $A = A_{25}-A_R$.

6. Blue CrO_5

Solution No. 1 (1,200 μL) was pipetted into plastic cuvettes and mixed with sample solution (12 μL, gallic acid, Trolox®, δ-tocopherol). Incubation took 3 minutes. After that, the first value of absorbance (A1) was determined. Subsequently, a volume of 120 μL of reagent No. 2 was added and incubated for 3 minutes. After this incubation, a second absorbance value was determined at λ = 546 nm (A_2). Resulting absorbance was calculated according to the formula $A = A_2 - A_1$.

Cereal grains and products have gained popularity in contributing to healthy eating behaviour because of their antioxidant properties associated with protection against chronic diseases. In this book, notable studies on the *in vitro* and *in vivo* antioxidant activity of commonly consumed cereal grains are summarized. Cereals contain phytochemicals or certain minor components with antioxidant properties. The antioxidant potential of cereals depends on their bioaccessibility, absorption in the gastrointestinal and their bioavailability utilization *in vivo*. The *in vitro* gastrointestinal digestion and fermentation of cereals increased their antioxidant potentials which are significantly correlated with their total phenolic contents. Most studies performed *in vivo* have been concerned with the antioxidant properties of coloured rice, wheat bran and rye products. There are inadequate *in vitro* and *in vivo* studies on antioxidative potentials of fermented versus unfermented cereals. Therefore, further studies are necessary to maximize possible health benefits of cereal antioxidative phytochemicals.

Epidemiological and experimental studies have suggested that diet plays a crucial role in the prevention of chronic diseases such as heart disease, cancer, diabetes, and Alzheimer's disease (Temple, 2000; Willet, 1994). Consumption of cereal grains has gained popularity with whole grain products being regarded as "healthy foods" because of their potential protection against life-style and diet related disorders such as obesity, diabetes (Fardet et al., 2008; Lui *et al.*, 2000; Meyer et al., 2000), cardiovascular diseases (Fardet *et al.*, 2008; Kelly *et al.*, 2007) and cancers (Fardet *et al.*, 2008; Schatzkin *et al.*, 2007). This effect is thought to be partly due to their phytochemicals that combat oxidative stress. Overproduction of oxidants can cause oxidative damage to large biomolecules such as lipids, DNA and proteins resulting in increased risk for cancer, cardiovascular and other diseases (Temple, 2000; Ames *et al.*, 1993; Wagner *et al.*, 1992). Dietary antioxidants may reduce such oxidative damages to biomolecules through several mechanisms (Duthie *et al.*, 1996; Frag *et al.*, 1991).

The importance and health benefits of regular cereal grain consumption in the prevention of chronic diseases are a focus for many research laboratories. Current research data suggest that whole-grains contain more antioxidant phytochemicals than was previously reported (Liu, 2007). The aim of this review is to critically discuss research data pertaining to the antioxidant activities of cereal grains produced by *in vitro* and *in vivo* studies. Such information has been summarized here for a better understanding of antioxidant properties of commonly consumed cereal grains (wheat, corn, rice, barley, sorghum, rye, oats and millet).

ANTIOXIDANTS IN CEREAL GRAINS

Cereal grains have a long history as major sources of staple foods worldwide. Whole grain cereals are good sources of phenolic compounds which include derivatives of benzoic and cinnamic acids, anthocyanidins, quinines, flavonols, chalcones, flavones, flavanones, and amino phenolic compounds (Bellido and Beta, 2009; Jang and Xu, 2009; Hosseinian *et al.*, 2008; Lloyd *et al.*, 2000; Shahidi and Naczk, 1995; Thompson, 1994). Grains contain tocotrienols and tocopherols (Thompson, 1994), and oryzanols (Lloyd *et al.*, 2000). This wide range of phytochemicals has been recognized to support overall health through their antioxidant potential. The protective function of antioxidants in the body is through balanced oxidative stress free radicals. Cereal grain antioxidants are thought to act as direct free radical scavengers, cofactors of antioxidant enzymes or as indirect antioxidants (Fardet *et al.*, 2008). Free radical compounds result from normal metabolic activities as well as from the diet and environment. Excess amounts of these reactive substances can cause oxidative damage to biomolecules such as lipids, protein and DNA (Temple, 2000) resulting in an increased risk of chronic disease and contribution to general inflammatory response and tissue damage (Slavin, 2003; Young and Woodside, 2001). Reactive oxygen species (ROS) including superoxide anion, hydrogen peroxide, peroxyl radical, hydroxyl radical, singlet oxygen and peroxynitrite are produced during cellular metabolic processes (Sánchez-Moreno, 2002). Several studies suggested a strong association between elevated levels of these molecules and the pathogenesis of several chronic diseases through the process of oxidative stress (Leavy, 2014; Nogueira and Hay, 2013; Sugamura and Keaney Jr, 2011). ROS are involved in a variety of physiological and pathological processes, including cell signaling transduction, cell proliferation, differentiation, apoptosis, as well as ischemia-reperfusion, inflammation and many neurodegenerative disorders (Ames, Shigenaga and Hagen, 1993; Bland, 1995). The process of ROS generation/ production occurs when there is an imbalance between pro-oxidants and antioxidants leading to oxidative stress (Qingming *et al.*, 2010). The process of oxidation in cells is regulated by antioxidants, which delay or prevent cellular damage. Antioxidant protection is normally achieved through a balance between prooxidants and endogenous and/or dietary antioxidants (Sobha and Andallu, 2013; Hancock *et al.*, 2001).

The dietary antioxidants (minerals, trace elements, vitamins, carotenoids, polyphenols, alkylresorcinols, betaine, choline, sulphur amino acids, phytic acids, lignans and avenanthramides) of whole grain cereals are primarily located in the bran and germ fractions (Fardet, 2010). Different antioxidant mechanisms have been attributed to cereal grain phytochemicals. These include prevention of oxidation of polyunsaturated lipids (vitamin E), reduction of the concentration of plasma homocysteine (vitamin B9, betaine, choline), acting as cofactor of antioxidant enzymes such as superoxide dismutase, glutathione peroxidase and thioredoxine (Zn, Fe, Se, Cu, Mn), or stabilization and delocalization of unpaired electrons (vitamin E, polyphenols, alkylresorcinols). Phenolic acids (γ-coumaric acid, ferulic, syringic, sinapic and vanillic) are also believed to chelate transition metals as well as activate or repress particular genes, while sulphur-containing amino acids (cysteine and methionine) contribute to the synthesis of a major endogenous antioxidant glutathione (Cukelj *et al.*, 2010; Fardet et al., 2008). Whole cereal grain antioxidants are water and fat soluble, providing protection through the entire digestive tract (Slavin, 2003).

In vitro cereal grain antioxidant activity

The antioxidant capacity of different cereals is evaluated based on two chemical antioxidant assays, namely the electron transfer (ET) and hydrogen atom transfer (HAT) methods. The ET method is characterized by change in colour during the reduction of oxidants. HAT involves competition between the antioxidant and substrate (probe) for free radicals (Huang *et al.*, 2005). Examples of ET assays include trolox equivalent antioxidant capacity (TEAC), the ferric reducing ability of plasma (FRAP) assay, copper reduction assay (CUPRAC), and 2,2,-diphenyl-1-picrylhydrazyl (DPPH) assay. The HAT assay includes the crocin bleaching assay, oxygen radical absorbance capacity (ORAC) and total peroxyl radical trapping antioxidant parameter (TRAP) (Huang et al., 2005). Most methods are based mainly on solvent extraction and quencher procedures which may largely underestimate total antioxidant capacity of cereals. However, both solvent extraction and quencher procedures are performed prior to digestion. In this section, antioxidants released by solvent extraction methods, quencher procedure, enzymatic digestive extraction and solid-state fermentation extraction will be described.

Wheat

Yu *et al* (2013) investigated the antioxidant properties of refined and whole wheat flour and their resultant bread. Their data showed that whole wheat flour and bread were superior to refined flour and bread in regard to their *in vitro* antioxidant properties. Žilic *et al* (2012) studied the distribution of phenolic compounds and yellow pigments in wheat grains and their relation to the total antioxidant capacity of bran and debranned flour. The colour intensity of yellow pigments, the activity of lipoxygenase (LOX) and peroxidase (POX) enzymes were also measured. Their results showed that bran fraction contains significantly higher concentrations of phenolic acids, flavonoids and yellow pigments. The LOX activity was concentrated in endosperm and embryo, while the POX activity was mostly concentrated in the bran fraction. Therefore, their results suggest that the bran fraction of wheat would potentially provide natural antioxidants. Liu et al (2010) used the DPPH assay to investigate antioxidant activities of various coloured wheat grains and their phenolic compounds. The results showed that Charcoal purple wheat had the highest antioxidant activity (up to 6899 μmol/100 g) followed by red Fife wheat and yellow Luteus wheat; while white AC vista wheat had the lowest antioxidant activity. The antioxidant activity was positively correlated to phenolic contents in these grains. The major phenolic composition identified in wheat grains consisted of phenolic acids, flavonoids, and anthocyanins. Ferulic acid was reported as the most dominant phenolic acid (74-87 mg/100g). Vanillic, ρ-coumaric, and sinapic acids were found in moderate levels (about 2 mg/100g), whereas caffeic acid was present in the least amount (< 1 mg/100g). Purple wheat, compared with coloured wheat grains, was distinguished by higher contents of vanillic acid (> 2.58 mg/100g) and ferulic acid (> 81.38 mg/100g). The flavonoids ranged from 21.59-102.95 mg/100g in purple wheat, 13.44 mg/100g and 10.72 mg/100g for yellow and red wheats respectively, and 9.6 mg/100g for white wheat. Anthocyanin contents were only present in purple wheats ranging from 2.5 to 23.5 mg/100g. Verma et al (2008) measured the free, bound, and total phenolic contents and antioxidant activities in the bran of 51 wheat cultivars belonging to eight Western Canadian spring wheat market classes grown in a replicated trials at Saskatoon, Saskatchewan, Canada. The free phenolic content ranged from 854.1±265.1 to 1754.9±240.3μg/g of bran gallic acid equivalent (GAE). Saponification followed by a liquidliquid solvent extraction released bound phenols ranged from 2304.9±438.0 to 5386.1±927.5 μg/g of bran GAE, contributing 66-82% of the total wheat bran phenolic content. Total phenolic acids ranged from 3406.4±32.3 to 6702.7±19.6 μg/g of bran GAE, with an average of 5197.2±804.9 μg/g of bran GAE. Antioxidant activity was determined as % discolouration = (1-[(absorbance of sample at 30 min)/(absorbance of control at time 0)]) and ranged from 11.86±2.59 to 20.12±0.51%, while the overall average was 15.6±2.2%. Based on varietal means, antioxidant activity correlated with free, bound and total phenolic contents (TPC; $r = 0.8$, $P < 0.05$).

Anson *et al* (2010) investigated the antioxidant and anti-inflammatory capacity of bioaccessible compounds from wheat fractions after gastrointestinal digestion. In their study, the bioaccessible compounds from aleurone, bran and flour were obtained from a dynamic *in vitro* model of the upper gastrointestinal tract. They found that bioaccessible compounds from aleurone had the highest antioxidant capacity and provided a prolonged anti-inflammatory effect as compared to bran and flour. Bhanja *et al* (2009) investigated the TPC and antioxidant potential of ethanolic extract of non-fermented and fermented wheat grains with two filamentous fungi

(*Aspergillus oryzae* and *Aspergillus awamori* nakazawa). They found that the total phenolic content and antioxidant property of wheat (54% ethanol extraction) was drastically increased when fermented with *Aspergillus oryzae* and *Aspergillus awamori* nakazawa. Moore *et al* (2007) evaluated the potential of solid-state yeast fermentation for their ability to improve the extractable antioxidant properties of wheat. The results demonstrated that solid state yeast treatment significantly increased releasable antioxidant properties ranging from 28 to 65, 0 to 20, 23 to 19, 0 to 25, 50 to 100 and 3 to 333% for scavenging capacity against ORAC, ABTS, DPPH TPC, and phenolic acids, respectively.

Rice

Shao *et al* (2014) investigated total phenolic contents, antioxidant capacity by using DPPH and ORAC assays, and phenolic acids in fractions of white (unpolished), red, and black rice after flowering and maturity, as well as the distribution of phenolic acids and anthocyanins in endosperm, embryo, and bran of white, red, and black rice. Their study showed that the total phenolic contents (TPC) of white rice (14.6-33.4 mg/100 g) and red rice (66.8-422.2 mg/100 g) were higher at 1-week development than at late stage, and black rice (56.5-82.0 mg/100 g) had a higher antioxidant activity at maturity (Shao et al., 2014). Shao et al (2014) also demonstrated that rice bran had a higher TPC (7.35 mg GAE/g average) and contributed to 60%, 86% and 84% of phenolics in white, red and black rice whole grain, respectively. The average TPC of the embryo and endosperm were 2.79 and 0.11 mg GAE/g accounting for 17% and 23%, 4% and 10% and 7% and 9% in white, red and black rice, respectively. The antioxidant capacity followed a similar trend to TPC. Deng *et al* (2012) used spectrophotometric methods to investigate the antioxidant properties and lipophilic and hydrophilic phenolic contents in 24 cereal grains from China. Their results showed that cereals have diverse antioxidant capacities. The phenolic compounds gallic acid, kaempferol, quercetin, galangin and cyanidin 3-glucoside were widely found in the cereals tested. The pigmented cereals, such as black rice, red rice, and purple rice had the highest antioxidant capacities and total phenolic contents among tested cereals.

Qiu *et al* (2009 and 2010) used DPPH and ORAC assays to evaluate antioxidant properties of wild rice. In both studies, their results showed that wild rice had a higher antioxidant activity than white rice (control), most likely due to their high phenolic acid contents. Zhang *et al* (2010) determined the phytochemical profiles and antioxidant activity of rice bran samples from 12 diverse varieties of black rice. Their data showed that there are significant differences in phytochemical content and antioxidant activity among the different black rice varieties. Black rice bran has a higher content of phenolics, flavonoids, and anthocyanins as well as higher antioxidant activity compared to white rice. The phenolic, flavonoids and anthocyanins of black rice bran are mainly present in the free form.

Bhanja *et al* (2008) investigated a self-designed new bioreactor (NB) for the improvement of phenolics and antioxidant activity in rice koji and compared the results with solid-state fermentation (SSF). Rice fermented in the NB resulted in a higher yield of phenolics and DPPH compared to the SSF and control. This might be due to higher titre values of β-glucosidase (62.7%) and α-amylase (40.7%) in the extraction media of NB compared to SSF.

Other cereal grains:

Kljak and Grbeša (2015) evaluated the relationship between carotenoid content and antioxidant activity in six Croatian commercial high-yield corn hybrids using TEAC, ABTS and thiobarbituric acid reactive substances (TBARS) systems. Their data showed that the corn hybrids varied in carotenoid content and antioxidant activity. Antioxidant activity in both assays increased linearly with total carotenoid content. Ndolo and Beta (2013) investigated carotenoid antioxidant activity using DPPH scavenging activity on aleurone, endosperm and germ fractions of barley (purple and regular), yellow corn, oats and wheat (purple and regular). The data suggested that antioxidant activity of carotenoids from the germ fraction showed higher scavenging activity compared to aleurone and endosperm. Malunga and Beta (2015) analysed the antioxidant capacity of water-extractable nonstarch polysaccharides (NSP) and feruloylated arabinoxylans (WEAX) from commercial barley, wheat and wheat fractions (germ, bran, and aleurone) using DPPH, ABTS and ORAC assays. Their data showed that NSP and WEAX exhibited antioxidant activity. The results demonstrated that arabinoxylan content, TPC, xylose and degree of substitution influenced the NSP and WEAX antioxidant capacity. Gamel and Abdel-Aal (2012) investigated phenolic acid composition and antioxidant capacity against DPPH and ABTS radicals, and inhibition of oxidation of human low density lipoprotein (LDL) cholesterol of selected Canadian and Egyptian barley whole grain flours and four pearling fractions. The data showed significant variations among barley wholegrain flour and pearling/milling fractions in terms of phenolic acid composition and antioxidant capacity.

Bhanja Dey and Kuhad (2014) evaluated the antioxidant potential of various filamentous fungi-fermented products derived from different whole grain cereals (wheat, brown rice, oats and maize). The fermented products showed a high efficiency for the improvement of water soluble TPC and antioxidant properties including ABTS and DPPH. Fermented wheat showed a 14-fold improvement in TPC, and 6.6 and 5-fold enhancement of DPPH and ABTS respectively. This study demonstrates that fermented wheat can be a powerful source of natural antioxidants. Gong *et al* (2013) compared the *in vitro* procedure of antioxidant extraction with the Quencher method and water as an extracting solvent. Total antioxidant capacity values of cereal grains obtained using an *in vitro* procedure were 1.8-10.3 times higher than the Quencher procedure and 3.5-10.5 times higher than the water extract. This indicates that the *in vitro* gastrointestinal digestion procedure is more useful in the screening of grains, and assessing for their health benefits compared to the Quencher procedure and water extraction. Prajapati *et al* (2013) evaluated the total phenol content, flavonoids, flavonol and antioxidant activity of five cereals (wheat, pearl millet, rice, maize and sorghum) by *in vitro* digestion and chemical extraction along with the effect of cooking. Their results showed higher values of total phenolics, flavonoids, flavonols as well as increase in antioxidant activity measured by the FRAP method in *in vitro* digesta (enzymatic extract) of cereals compared to their chemical extracts. Cooking of cereal also resulted in increased total antioxidant capacity and enhanced phenolics, flavonoid and flavonol content. Based on this study, antioxidant components like phenolic compounds, flavonoids and flavonols of selected cereals were affected by *in vitro* digestion. Pazinatto *et al* (2013) evaluated the effect of different process-defatting, protein concentration, thermal treatment, hydrolysis with Alcalase and *in vitro* digestion on the antioxidant capacity of amaranth seeds. The combination of protein concentration and hydrolysis with Alcalase led to a product with high antioxidant activity as evaluated by DPPH and ORAC. After *in vitro* digestion, protein concentrate and its hydrolysate showed similar antioxidant capacity. Cai *et al* (2014 and 2012) evaluated the potential of solid-state fermentation for the improvement of TPC and antioxidant potential of oat. They found that fermentation increased TPC and the antioxidant activity. Đor_evic *et al* (2010) evaluated the influence of fermentation by *Lactobacillus rhamnosus* and *Saccharomyces cerevisiae* on antioxidant activities and total phenolic content of 4 cereals (buckwheat, wheat germ, barley and rye) and compared them to their unfermented counterparts. Total phenolic content increased significantly and antioxidant activities were enhanced through fermentation by lactic acid bacteria and yeast.

Intracellular antioxidant cereal grains

Masisi *et al* (2015) investigated antioxidant potential of carotenoids from aleurone, germ, and endosperm of barley, corn, and wheat. The antioxidant properties using DPPH, ABTS and ORAC assays revealed significantly higher antioxidant activity in the germ than in the aleurone and endosperm fractions. 3-(4,5-dimethylthiazolyl-2)-2,5-diphenyltetrazolium bromide (MTT) assay, and 2,2-azobis (2-amidinopropane) dihydrochloride (AAPH)-induced cell loss was effectively reduced by pre-incubating Caco-2, HT- 29, and FHs 74 Int cells with carotenoid extracts. The data showed that carotenoids of germ, aleurone, and endosperm fractions improve biochemical and intracellular antioxidant activity. Hirawan et al (2011) investigated the properties of whole purple wheat, unpolished red rice and partially polished red rice infant cereals. ORAC, TPC, total anthocyanin contents (TAC) and cellular antioxidant activity (CAA) were measured. Home- and laboratory-made unpolished red rice had higher TPC and peroxyl radical scavenging activities than purple wheat infant cereals; the latter had higher TAC. Pigmented infant cereals were found to have higher TPC, TAC and ORAC than commercial ones. Anthocyanins were identified in whole purple wheat but not in unpolished red rice. Purple wheat infant cereal had higher CAA than unpolished rice. Whole purple wheat infant cereals showed higher antioxidant activities than the commercial infant cereals. Stein et al (2010) investigated the effects of fermented wheat aleurone on the expression of genes involved in stress response, toxicity, activity of drug-metabolising enzymes and anti-genotoxic potential. They also investigated the protection against H_2O_2-induced DNA damage in HT29 cells. The data showed that fermented aleurone significantly induced mRNA expression in *CAT*, *GSTP1* and *SLUT2B1* (HT29) and *GSTP1* (epithelial stripe). The enzyme activity of GST (HT29) and CAT (HT29, epithelial stripes) were also increased. DNA damage induced by H_2O_2 was significantly reduced by the fermented aleurone.

In vivo cereal grain antioxidant activity

Although *in vitro* experiments indicate great antioxidant abilities of the whole grain cereals, it is questionable if the methods underestimate physiological antioxidant capacity. Several studies have been done in order to evaluate the bioactivity of cereal grain phytochemicals, mostly on animals and human.

Animal studies

Qingming *et al* (2010) evaluated the antioxidant activity of malt extract from barley *in vivo* (male Kunming mice). Scavenging effects on the hydroxyl and superoxide radicals, and protection against reactive oxygen species induced lipid, protein and DNA damage were evaluated. The Kunming mice were induced with D-galactose to evaluate the ability of malt extract to behave as an antioxidant. The extract exhibited high antioxidant activities in both *in vitro* and *in vivo*, evidenced by its ability to scavenge hydroxyl-and superoxide-radicals, high reducing power, and protection against macromolecular (lipid, protein and DNA) oxidation damage. Malt extract prevented the decrease of antioxidant enzyme activities, decreased liver and brain malondialdehyde levels and carbonyl contents and improved total antioxidant capacity in D-galactose-treated mice. Wang *et al* (2010) investigated the protective effect of wheat feruloyl oligosaccharides (FOs) against oxidative stress in rat plasma. The oxidative markers (oxidized glutathione and malondialdehyde) and the activity of antioxidant enzymes (superoxide dismutase, catalase, and glutathione peroxidase) in the plasma of rats fed a standard diet supplemented with 1% FOs were evaluated. The anti-radical capacity of rat plasma after ingestion of 0.5 mg FOs was measured using AAPH as the free radical inducer. The results suggested that FOs enhanced plasma antioxidant enzyme activity and lowered oxidized glutathione and malondialdehyde levels for the group supplemented with FOs compared to the control. Moreover, the plasma of rats after ingestion of FOs were resistant to AAPH-induced hemolysis than was the control group.

Zhang *et al* (2009) studied the effect of replacing white rice and processed wheat starch with wild rice as the chief source of dietary carbohydrates in rats fed a high fat diet/cholesterol diet. The rats fed wild rice were found to suppress the build-up of oxidative stress by improving antioxidant capacity, increasing superoxide dismutase (SOD) activities and reducing malondialdehyde concentrations, both in serum and liver tissue of the rats. Another study on the potential cardiovascular benefits of wild rice in male and female LDL-receptor-deficient (LDL-r-KO) mice showed that a wild rice diet has protective roles in improving the plasma lipids profile in LDL-r-KO mice (Surendrian *et al.*, 2013). Guo *et al* (2007) evaluated the effect of an anthocyanin-rich extract from black rice on hyperlipidemia and insulin resistance in fructose-fed rats. Dietary supplementation with the anthocyanin-rich extract (5 g/kg of high-fructose) prevented the development of fructose-induced insulin resistance. In addition, rats supplemented by the extract exhibited lower oxidative stress than the fructose-fed control animals, as indicated by the lower concentrations of plasma TBARS and blood oxidized GSH. Fardet *et al* (2007) investigated the metabolic responses of 2 groups of rats fed 2 diets (whole-grain wheat flour, WGF and refined wheat flour, RF) for 2 weeks each in a crossover design. Their metabolic approach showed that consumption of WGF produced metabolic changes, some of which may protect the organism against oxidative stress. Álvarez *et al* (2006) fed healthy mice for 5 weeks with a cereal fraction (wheat germ, buckwheat flour, and rice bran and wheat middlings) to investigate supplementation effect. The results showed that all cereal fractions caused improvement of the leukocyte parameters including chemotaxis capacity, microbicidal activity, lymphoproliferative response to mitogens, interleukin-2 (IL-2) and tumor necrosis factor (TNF) release, as well as oxidized glutathione (GSSG), GSSG/reduced glutathione (GSH) ratio, CAT activity and lipid oxidative damage. The observation was similar among the cereal fractions. These effects may be due to the antioxidant activity of polyphenols naturally present in cereals.

A study on black rice by Chiang *et al* (2006) showed that black rice extract significantly increased superoxide dismutase and catalase activities by 161.6% and 73.4%, respectively. Hsieh *et al.* (2005) evaluated the effect of oxidative damage to DNA by feeding rice bran oil to streptozotocin-induced diabetes rats. They measured the levels of 8-hydroxy-2'-deoxyguanosine (8-OhdG) oxidative DNA damage in various tissues of rats with streptozotocin-induced diabetes. There were significant reductions in mtDNA 8-OhdG levels in the liver, kidney, and pancreas of rats treated with rice bran oil as compared to those in diabetic rats without intervention. Another study using rabbits showed that a black rice diet significantly decreased aortic 8-OhdG and serum and aortic malondialdehyde, hence indicating decreased oxidative stress (Ling et al., 2002). In a study conducted by Hegde et al (2005) alloxan-induced diabetic rats were fed a diet supplemented with 55% (w/w) finger millet and kodo millet. Blood glucose, cholesterol, enzymatic and nonenzymatic antioxidants, lipid peroxides in plasma, and glycation of tail tendon collagen were measured. Feeding diabetic rats diet supplemented with millet restored the levels of enzymatic (GSH, vitamin E and C) and nonenzymatic antioxidants (SOD, CAT, GPx and glutathione reductase) and lipid peroxide which were reduced by their diabetic condition. Blood sugar and total cholesterol were lowered in rats fed diets formulated with millet. Another confirmation for postulating that the beneficial effects could involve mechanisms related to freeradical scavenging activity is the inhibition of glycation of rat-tail tendon collagen in the milletfed diabetic rats. Chen *et al* (2004) examined the bioavailability of avenanthramides and phenolics from oats using BioF1B hamsters. Absorbed oats phenolics did not change the resistance of hamster LDL particles against Cu2+-induced oxidation until ascorbic acid was introduced to the assay mixture.

Human studies

The effects of a diet high in wheat aleurone on plasma antioxidants status, markers of inflammation and endothelial function, were studied by Price *et al* (2012). Seventy-nine healthy, older, overweight participants incorporated either aleurone-rich cereal products (27 g aleurone/d), or control products balanced for fibre and macronutrients, into their habitual diets for 4 weeks. Fasting blood were collected at baseline and on day 29. The results showed that, compared to the control, consumption of aleurone-rich products provided substantial amounts of micronutrients and phytochemicals which may function as antioxidant. However, no changes were found in other markers of inflammation, antioxidant status or endothelial function. Söderholm *et al* (2012) evaluated the effect of none versus high intake of rye bread on the oxidation resistance of LDL in healthy humans while otherwise on habitual diet. Their data showed that intake of rye in four-weeks improved significantly resistance of LDL. However, they were unable to identify rye-originating substances in LDL responsible for the enhancement. The antioxidant ability of capsules containing oats avenanthramides was evaluated in 120 healthy individuals (60 women and 60 men). The materials were randomly assigned to 4 groups, which consumed 4 basal corn oil capsules (placebo), 4 capsules containing 1.56 mg of oat avenanthramides-enriched extract (OAEs) and 8 capsules containing 3.12 mg of OAEs or without treatment (control) for 1 month. Plasma lipid peroxides and antioxidant status were measured. For the 8 capsules group, the level of serum SOD and reduced glutathione hormones were significantly increased by 8.4 and 17.9%, respectively (P<0.05) and malondialdehyde level significantly decreased by 28.1%. The total cholesterol, triglyceride, and low density lipoprotein cholesterol levels were lowered by 11.1, 28.1 and 15.1%, respectively (P<0.05), while the high density lipoprotein cholesterol level was increased by 13.2%. Oat extracts containing avenanthramides possess a high antioxidative activity in humans (Liu *et al.*, 2011). A study on the bioavailability and bioactivity of avenanthramides in six healthy older adults showed an increased antioxidant capacity after consuming skimmed milk containing avenanthramides (Chen *et al.*, 2007). A study by Wang *et al* (2007) investigated the influence of black rice pigment fraction (BRF) supplementation in sixty patients with coronary heart diseases (CHD) for six months. After six months of intervention, compared to white rice pigment fraction supplementation, BRF supplementation greatly enhanced plasma total antioxidant capacity. The physiological consequences of using white rice or mixed rice were evaluated in overweight Korean women. The data showed that plasma GPx activity increased by 15% when subjects consumed brown and / or black rice for 6 wk (Kim *et al.*, 2008).

Price *et al* (2008) evaluated total phenolics (TP) and antioxidant potential (AOP) in plasma and urine of humans following consumption of a single meal of unprocessed wheat bran or a refined cereal (ground white rice). Their data showed that wheat bran phenolics are relatively well absorbed and may enhance antioxidant status. The potential antioxidative effect of rye bran intervention was investigated by measuring LDL susceptibility to copper oxidation *ex vivo*. Rye bran intervention had no influence on lag time or propagation rate of the LDL oxidation *ex vivo* (Harder *et al.*, 2004). Table 4.2 summarizes the findings on i*n vivo* antioxidant activity of cereal grains.

Cereal grains remain a staple component of diets worldwide. They provide significant levels of bioactive phytochemicals including folates, phenolic acids, carotenoids, betaine, choline, sulphur amino acids, phytic acid, lignins, avenanthramides and alkylresorcinols. These phytochemicals are unevenly distributed between the various tissues and cell types of the grains. *In vitro* and *in vivo* studies have shown that cereal grain phytochemicals may improve antioxidant capacity, and thus have the potential to mitigate oxidative stress hence delaying the onset of some chronic diseases. However, there are inadequate studies in humans in which real, complex cereal foods are used.

Table 4.2: *In vivo* antioxidant activity of cereal grains.

Cereal grains	Potential bioactive compounds	Observed biological activity	Animal model	References
Millet	Phenolics	Scavenged hydroxyl-and super-oxide-radicals, high reducing power, and protect against macromolecular (lipid, protein and DNA oxidation damage.	D-galactose-induced mouse aging model	Qingming *et al* 2010
	Phenolics, tannins and phytates	Improvement in the enzymatic (glutathione, Vitamin E and C) and non-enzymatic antioxidants (SOD, CAT, GPx, glutathione reductase)	Alloxan-induced diabetic rats	Hedge *et al* 2005
Oats	Avenanthramides and Phenolics	Absorbed oats phenolic did not change the resistance of hamster LDL particles against CA^{2+-} induced oxidation until ascorbic acid was introduced.	BioF1B hamster	Chen *et al* 2004
		Serum SOD and reduced glutathione hormones increased by 8.4 and 17.9% respectively; malondialdehyde decreased by 28%; TC, TAG, and LDL cholesterol were lowered by 11.1, 28.1, 15.1% respectively	Healthy human	Liu *et al* 2011
		Increased antioxidant capacity.	Healthy old adults	Chen *et al* 2007
Black rice	Anthocyanin	Prevented development of fructose-induced insulin resistant; lowered plasma TBARS concentration and blood oxidized glutathione	Fructose-fed rats	Guo *et al* 2007
		Increased SOD and CAT activities by 161.1% and 73.4% respectively	C57BL/6 mice	Chiang *et al* 2006
		Decreased aortic 8-OhdG and serum and aortic malondialdehyde	Rabbits	Ling *et al* 2002
		Enhanced plasma TAC, reduced plasma levels of vascular cell adhesion molecule-1, soluble CD40 ligand and high sensitive C-reactive protein.	CHD human	Wang *et al* 2007

Cereal grains	**Potential bioactive compounds**	**Observed biological activity**	**Animal model**	**References**
Brown rice		Plasma GPx activity increased by 15%	Over weight Korean women	Kim *et al* 2008
Wild rice		Increased SOD and reduced malondialdehyde concentration, in serum and liver	Male sprague–dawley rats	Zhang *et al* 2009
Rice bran	Polyphenol	Increased SOD and CAT activities	LDL-receptor-deficient mice	Surendrian *et al* 2013
		Improved GSSG, GSSG/ GSH ratio, catalase activity and lipid oxidative damage	Healthy mice	Álvarez *et al* 2006
Rye	Ferulic acid	Improved oxidation resistance of LDL	Healthy human	Söderholm et al 2012
		Significant reduction in mtDNA 8-OhdG levels in liver, kidney and pancreas	Streptozotocin -induced diabetes rats	Hseih *et al* 2005
		Increased urine ferulic acid but did not change measurable antioxidative effect on the subject's LDL	Healthy human	Harder *et al* 2004
Wheat (Bran germ and flour)	Phenolics	Decreased transit time and Increased TBARS	Obese rat	Kim *et al* 2012
		Consumption of aleurone-rich products provided substantial amounts of micronutrients and phytochemicals which may function as antioxidant	Overweight human	Prince *et al* 2012
		Improved GSSG, GSSG/ GSH ratio, catalase activity and lipid oxidative damage	Healthy mice	Álvarez et al 2006
		Increased SOD and reduced malondialdehyde concentr-ation, in serum and liver	Male sprague–dawley rats	Zhang *et al* 2009
		WGF produced metabolic change	Rats	Fardet *et al* 2007
		Enhanced antioxidant status	Healthy human	Price *et al* 2008

(Adopted from Masisi, K. *et al* 2015).

The beneficial health effects are associated with the additive and/or synergistic action of these phytochemicals as have been supported by results from studies on the activity of single or combination of a number of antioxidants. However, the antioxidant capacity of cereal grains have probably been underestimated, since most methods are based on the solvent extraction and quencher procedures which are both performed before digestion. The antioxidant potential of cereals depends on their bioaccessibility, absorption in the gastrointestinal tract and their bioavailability *in vivo*. The antioxidant potential of cereals may be improved *in vitro* by the action of fermentation and *in vitro* gastrointestinal digestion of cereals which significantly correlated with their total phenolic contents. Thus fermentation and *in vitro* gastrointestinal digestion offers a tool to further increase the bioactive potential of cereals. Further studies are necessary on fermented and *in vitro* gastrointestinal digesta of other cereal antioxidative phytochemicals. Most studies performed *in vivo* have been concerned with the antioxidant properties of coloured rice, wheat bran and rye products. However, there are inadequate *in vivo* studies on other cereal grains and antioxidant potential of fermented versus unfermented cereals. *In vivo* studies are therefore needed to further explore the antioxidant potential of other cereal grains and fermented versus unfermented cereals so as to maximize possible beneficial health effects of cereal phytochemicals. Polyphenols (mainly ferulic and phenolic acid) are the main contributors in these antioxidant properties of cereal grains in both *in vitro* and *in vivo* studies. Correlations between antioxidant activities and other types of components remain to be investigated. Few clinical studies on the antioxidant potential of cereals in animals and humans have been carried out. The mechanisms, however, are not fully established. Therefore, more information on the mechanisms involved is necessary in order to provide strong, convincing arguments about their health benefits and new health claims in the future. Few studies deal with cereal grain consumption on the induction/repression of genes coding for antioxidants compounds. Therefore, further studies are necessary to explore this new area of research using the most recent comprehensive analysis of genomic and transcriptomic techniques. Moreover, cereal grain consumption can be investigated through metabolomics which will allow further investigation on how complex antioxidant-rich cereal grains can modify metabolism and elucidate metabolic pathways affected by cereal grains antioxidants. Such findings will provide novel information on the health benefits of cereal grains and the development of healthy functional cereal grains products.

METHODS OF TANNIN ANALYSIS

The methods used for the determination of phenolic compounds in plant products are mostly based on the polarity of phenolic hydroxyl groups. Some methods estimate all phenols as a group, while few methods are used to determine specific phenolic substance or class of phenolics. The acidity of hydroxyl groups is often used as a means of separating phenolic constituents. The activated benzene ring reacts with a variety of reagents to yield specific colors. Phenols also absorb light in the UV region. Excellent reviews on the analysis of various phenolic compounds have been published. Markham has described isolation techniques for flavonoids using various chromatographic methods. Unless completely esterified, etherified, or glycosylaled, phenolic compounds are usually soluble in polar solvents. These compounds can be dissolved in solutions of sodium hydroxide and sodium carbonate, but only a few are soluble in bicarbonate. Most phenolic glycosides are soluble in water, but the corresponding a glycones are usually insoluble. Butanol and methanol are offen used as solvents whereas phenols with only a few hydroxyl groups are soluble in nonpolar solvents, i.e., ether, benzene, chloroform, and ethyl acetate. In this section, only those methods specific in the dietary tannins of plant products are summarized.

General considerations

Vegetable tannins form colored solutions and precipitate with iron and other metals. These compounds possess the ability to form insoluble, nonputrescible precipitates with albumin, gelatin, collagen, and other skin proteins. These can be precipitated by alkaloids, pyridines, and quinones. These properties of vegetable tannins are often used for their analysis. Tannins are insoluble in nonpolar solvents, i.e., ether, chloroform and benzene, and are sparingly soluble in ethyl acetate. Because of the presence of a large number of polar groups, these tannins can be readily dissolved in water and alcohol to form colloidal sols. The extraction of tannins by aqueous mixtures of polar solvents depends upon the random transfer of individual bonds from the substrate to competing sites on the solvent molecule (McLeod, 1974). The release of bound tannins depends upon a number of bonds involved and usually proceeds in the order of increasing molecular weight. The actual ratio of solvent soluble to solvent-insoluble tannin is determined by the nature and concentration of the organic solvent (Htock, *et al.*, 1958). Cold acetone (40%) in aqueous solution extracts almost the same quantity of tannins as doe's hot water. It also avoids the oxidation of labile phenolic compounds.

Tannins are easily oxidizable by phenolases present in plant tissues into highly colored compounds. This tendency also causes practical problems during sample preparation and extraction. The fission products of tannins and accompanying flavonols are, in many cases, similar. In fact, based on the nature of the fission products obtained by acid hydrolysis. Russell (Russell, 1935), classified natural tannins into three groups:

1. Tannins related to depsides
2. Tannins related to diphenyldimethyloids
3. Phlobaphene-produeing tannins or phloba tannins

The first two classes yield crystalline fission products gallic acid and glucose and ellegic acid and glucose respectively. They are called hydrolyzable tannins," while those of group three are called condensed tannins.

The quantitative extraction of condensed tannins from plant tissue is always difficult as it may remain behind attached to a carbohydrate or other matrix and may be quite insoluble due to a high degree of polymerization. As a general rule, plant material is extracted first with a nonpolar solvent to remove fats, waxes, chlorophylls, and carotenoids more or less completely. Subsequently, extraction is carried out with a suitable solvent, usually methanol (Burns, 1971; Haslam and Gupta, 1978), 1% HCl in methanol (Maxson and Rooney, 1972) or 70% acetone (Jones, *et al.*, 1976; Leung, *et al.*, 1979). The precipitation of condensed tannins by either is useful in the fractionation procedure (Martin-Tanguy *et al.*, 1976) but, more recently, Sephadex LH-20 and/or G-50 chromatography has proved to be the more successful and hence, more widely used technique. A number of workers (Haslam and Gupta, 1978; Leung, *et al.*, 1979; Strumeyer and Malin 1975; Marquardt *et al.*, 1975; Davis and Hoseney, 1979) have used this method for the separation of condensed tannins. It involves the adsorption of tannins onto the column, followed by exhaustive washing to remove low molecular weight polyphenols, and then elution with methanol or 70% acetone. Of the tests used for locating tannin fractions, absorption spectrophotometry is one of the more useful since it does not destroy tannins. Compounds that are more sensitive, but do not allow recovery of intact tannins, are involved in the formation of colored products with vanillin HCl or toluene sulphonic acid, and precipitate with gelatin, neutral lead acetate, or ferric chloride. Thin-layer and paper chromatography have also been employed to check the purity of isolated condensed tannins (Jones, *et al.*, 1976; Leung, *et al.*, 1979; Strumeyer and Malin 1975). Several procedures have been known to determine the molecular weight of condensed tannins, including the elevation of boiling points using methylated or acetylated derivatives, ultracentrifugation, colorimetry (Rollman *et al.*, 966) and filtration on molecular sieves (Diemair and Polaster, 1967; Lewak, 1968).

Schanderl (1970) classified the methods of tannin analysis based on the physical or chemical properties as follows:

1. Precipitation by inorganic reagents. e.g., copper acetate, salts of aluminum, zinc, lead, and copper
2. Precipitation by organic reagents, e.g., hides powder, cinchonine, gelatin, and formaldehyde
3. Formation of colored products, e.g. alkaline-phosphotungstate phosphomolybdate (Folin-Denis), ferric chloride, acid-vanillin, diazotized p-nitroaniline, and nitrous acid
4. Oxidation with acid permanganate or ferricyanide on a volumetric basis

The most common methods for analyzing tannins of legumes and cereals include the vanillin test (Bate-Smith and Lerner, 1954; Burns, 1963) the Prussian blue test, (Price and Butler, 1972) and the Folin-Denis test (Burns, 1963). The Prussian blue and Folin-Denis methods are based on the reducing power of phenolic hydroxyl groups. These methods are not very specific and detect all phenols (phenolic acids, fl avonoids, and tannins) with varying sensitivity. However, it is possible to quickly distinguish between zero-, low-, intermediate-, and high-tannin varieties of sorghum by the development of shades of yellow, green, and blue colors with the Prussian blue reagent (Price and Butler, 1972). Researchers (Price and Butler, 1972) justified the use of this redox method for Spectrophotometric tannin assay as the sorghum cultivars they studied seemed to contain negligible amounts of nontannin phenolics. The vanillin assay is widely used for the determination of condensed tannins or their monomeric components in plants. The advantage of this procedure lies in the fact that it is specific for a narrow range of fl avanols and dihydrochalcones which have a single bond at the 2.3 position or free moriented hydroxyl groups on the Bring (Sarkar and Haworth, 1976). Leucoanthocyanidins and proanthocyanidins react with vanillin in the presence of HCl to give a light red color. Although vanillin reacts primarily with fl avanols (leucoanthocyanidins), other tl al'onoid compounds give significant color development. Anthocyanidins are red in mineral acids and, therefore, a blank need to be used if only tannins are to be determined.

Vanillin Assay

The vanillin assay is widely employed as a method for the quantitative determination of condensed tannins (proanthocyanidins) in plant materials such as fruits (Goldstein and Swain, 1963); sorghum grain (Broadhurst and Jones, 1978) and forage legumes (Broadhurst and Jones, 1978). It is a sensitive, relatively simple assay specific for tlavan-3-ols, dihydrochalcones and proanthocyanidins (Gupta and Haslam, 1980). For convenience, catechin, a monomeric flavan-3-ol unit of condensed tannin is often used to standardize the assay, although this leads to a considerable overestimation of tannin content (Price *et al.*, 1978).

A. Principle

The basic reaction involved is the substitution (condensation) of aldehyde (vanillin) with a phenolic hydroxyl group of phenols (tannins) to yield a red-colored condensation product. Phenols have three limiting structures (**Figure 4.1**) with the electron density particularly high at *ortho* and *para* positions of phenolic hydroxyl. The substitution reactions, requiring a strong electron density are favored in these positions. The hydroxyl groups on phenols act as donors of electrons; they increase electron density on the benzene ring and enhance its ability to undergo substitution reactions. Polyphenols with several OH groups in altemate positions undergo substitution reactions very easily, especially so in the case of resorcinol and phor-oglucinol. However, it is not easy in the case of catechol and pyragallol (Figure 4.2), which the OH groups are adjacent to each other. Substitution reactions occur more easily in the A ring of natural flavans (e.g. catechins) than that of flavones and flavonols (e.g. quercetins). Although in each case, the A ring has a phloroglucinol pattern of substitution (with one OH group engaged in the heterocyclic bond); the A ring of catechin is particularly easily substituted in position 6 and 8, whereas in quercetin, the A ring is deactivated by the CO group.

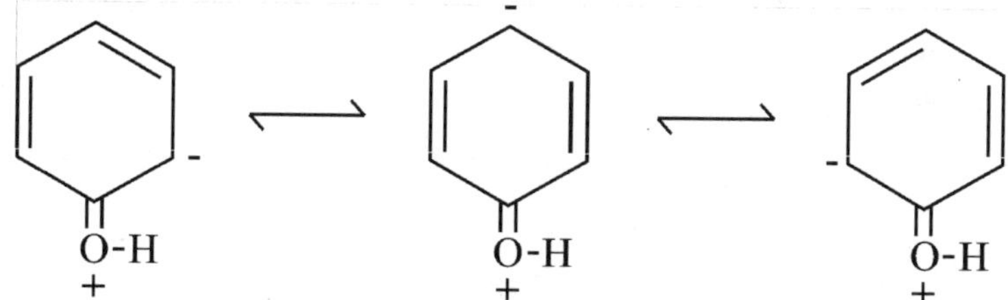

Figure 4.1: Mechanism of phenolic ring showing the limiting structures.

The substitution of the OH group on the A ring with aldehyde (vanillin) ha, been widely used in both qualitative and quantitative analyses of polyphenols (Figure 4.3). In the presence of strong acids, carbonyl compounds, especially aldehydes, acquire a proton to form an electrophilic radical which is able to attack the aromatic nucleus of a phenol to yield a condensation product. Vanillin (Vanillaldehyde) reactions with leucocyanidin are summarized in Figure 4.4. The electrophilic radical of vanillin in acid solutions reacts only on those benzene rings which are activated, i.e., resorcinol or phloroglucinol nuclei. Thus, vanillin reacts with the A ring of leucocyanidin (5, 6, 3', 4' -tetrahydroxyflavan-3, 4-dlol) at one of the positions, 6 or 8, activated by the presence of several OH groups. It does not react with the B ring. In the case of flavones and flavonols, which have a CO group in position 4, the A ring deactivated and does not neither react, nor does the B ring. The radical resulting from the protonation of aldehyde is substituted exclusively in positions 6 and 8 to give an intermediate which is readily dehydrated to the end product of the reaction. This condensation product is red in color with an absorbance maxima at 480 to 550 nm, depending on the maximum absorbance of the carbonium ion formed.

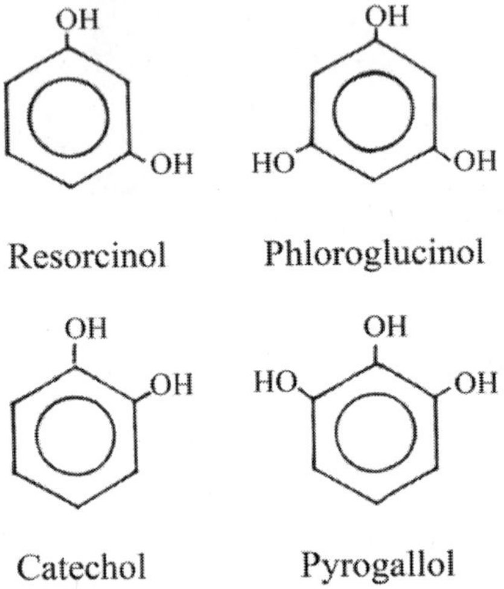

Resorcinol

Phloroglucinol

Catechol

Pyrogallol

Figure 4.2: Hydroxilation pattern of phenolic compounds that influence substitution reactions.

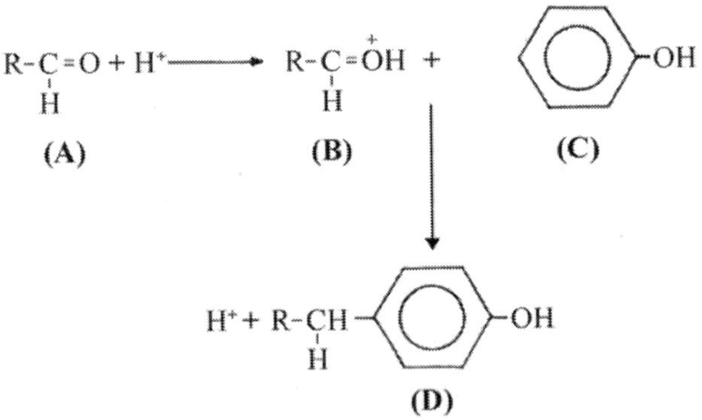

(A) (B) (C) (D)

Figure 4.3: Condensation of an aldehyde with aromatic nucleus of phenol ring (A) aldehyde, (B) electrophilic radical of aldehyde, (C) phenol and condensation product.

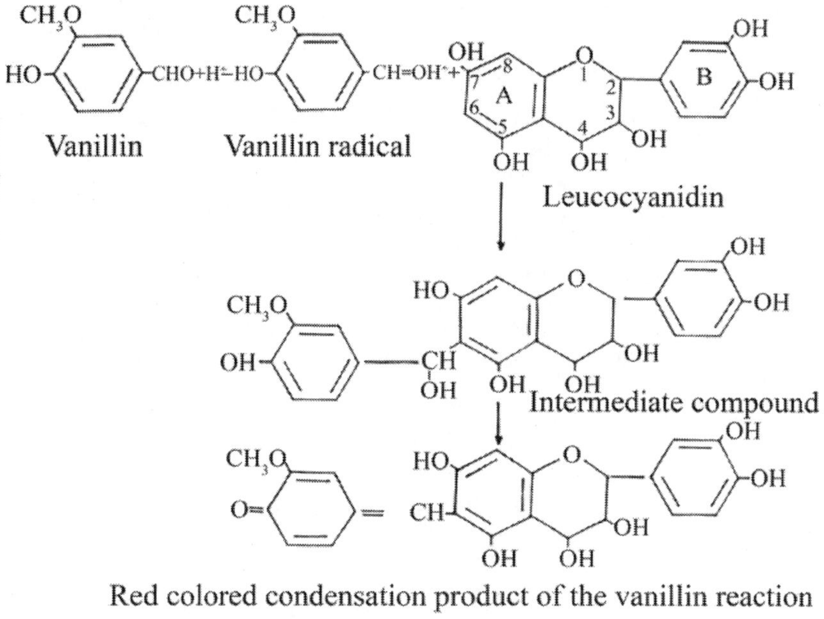

Vanillin Vanillin radical Leucocyanidin

Intermediate compound

Red colored condensation product of the vanillin reaction

Figure 4.4: Condensation reactions of vanilladehyde with flavons.

On the basis of a vanillin reaction, Goldstein and Swain classified plant phenolics into three groups:

1. Those which give high absorbance values with absorption maximas at 500 nm
2. Those which give lower absorbance values with absorption maximas at 520 nm, e.g., resorcinol, pyragallol and brazilin
3. Those which give no pigment, e.g., gallic acid and quercetin

The reaction was found to be approximately stoichiometric with compounds containing the phloroglucinol group. Both phloroglucinol and theaflavin gallate, which differ widely in molecular weight, showed nearly the same molecular absorptivity (3.5×10^4 and 3.0×10^4, respectively). On the other hand, the linkages in condensed tannins involve these reactive positions and, hence, any monomeric flavan which is part of such polymers becomes un-reactive to vanillin. Thus, on a weight basis, vanillin gives a stronger reaction with monomeric flavans than with condensed tannins. Also, catechins consistently give higher absorptivity values than their epimers. Goldstein and Swain did not observe any reactions with numerous phenolic compounds not containing meta-oriented hydroxyl groups, except for homocatechol. The latter might react through the methyl group. In both the phloroglucinol and resorcinol series of compounds, C-substitution reduced their absorptivity values. They have suggested that in the case of flavolans, a similar, if not greater, reduction might be expected, due to the formation of new C-C bonds, and that the reactivity of vanillin decreases as the molecular weight of the tannin increases.

B. Procedure

Bate-Smith and Lemer (1954) first used the vanillin reagent for the detection of leucoanthocyanidins and catechins. The first standard version of this assay for tannin determination was described by Swain and Hillis (1959). A freshly prepared solution of 1% vanillin in 70% concentrated sulfuric acid (w/v) was used for the determination of flavanols containing phloroglucinol or resorcinol A rings. It was reported that the vanillin-sulfuric acid reagent gave a more stable-colored carbonium compound than the vanillin-HCI reagent. Numerous modifications have been reported in the whole procedure of the vanillin assay with respect to vanillin concentration, extractant, time of reaction, standard, and extraction time to obtain reproducibility (Burns, 1971; Sarkar and Haworth, 1976; Goldstein and Swain, 1963; Gupta and Haslam, 1980; Price, et al., 1978; Swain and Hillis, 1959).

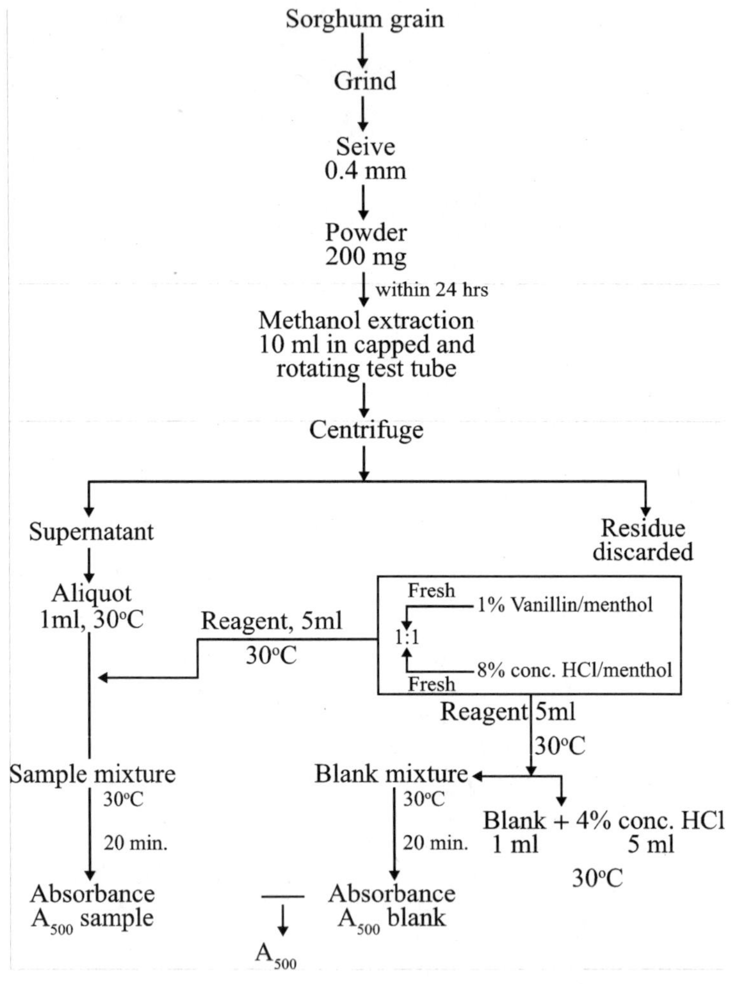

Figure 4.5: Revised procedure for the vanillin catechol.

Price *et al.* (1954) critically evaluated the vanillin reaction by examining several factors that influence the assay. Some of these include the extraction time and medium, background sample color, adequacy of catechin as a reference standard, vanillin and HCl concentration, temperature and reproducibility of the assay and the sensitivity of the results to time elapsed between grinding and analysis. Based on their results, a revised procedure for the vanillin assay for sorghum tannins, that offered a desirable reproducibility, has been suggested (Figure 4.5). This procedure is now most routinely followed for tannins of cereals and legumes. The vanillin assay is still not specific for polyphenols that are tannins. It does not discriminate the narrow range of reactive compounds on the basis of their molecular weight. However, within the narrow specificity of the reagent and with certain restrictions in regard to the choice of reaction conditions, the vanillin assay appears to be a workable and reproducible procedure.

The vanillin assay may be more useful if used in conjunction with some index which gives a value in direct relation to the degree of condensation. Goldstein and Swain (1963) developed a method for the determination of the degree of polymerization of tannins by using a vanillin/leucoanthocyanidin ratio (V/LA ratio or index). This determination only applies to tannins formed from flavan-3, 4-diols. It consists of applying two different analytical chemical reactions to the same sample, affected differently by the degree of polymerization. These reactions involve the transformation of the flavan-3,4-diols to an leucoanthocyanidin (LA) and their combination with vanillin (V) to give a colored product. There are two possible ways of using the leucoanthocyanin reaction for the determination of tannins, using either aqueous (Masquelier *et al.*, 1959) or alcoholic (Swain and Hillis, 1959; Pigman *et al.*, 1953) solutions. Although the latter gives a better yield of anthocyanidin, aqueous acids are generally preferred for more easily reproducible results (Ribereau-Gayon and Stonestret, 1966). The findings of Goldstein and Swain (1963) on the use of the V/LA index are summarized as follows.

1. In the less polymerized fractions extracted by absolute methanol, any increase in polymerization results in a more rapid reduction in the value of LA than V; the ratio of V/LA is thus increased.
2. In the more condensed fractions soluble in 50% methanol, any increase in polymerization has little effect on LA, but reduces the value of V; the ratio V/LA is thus lowered.

The variations in the values of LA and V as a function of polymerization are still inconclusive. The method also fails to account for catechins (flavan-3-ols). The latter react with vanillin and are thus included in any measure of V. However, they do not contribute to the LA determination. Some investigators have also suggested the use of other ratios, such as vanillin Folin-Denis *(V/FD)* or vanillin Folin-Ciocalteu (V/FC) (Rossi and Singleton (1966). Goldstein and Swain[34] pointed out that the vanillin reaction is suppressed when flavonols polymerize. Relative to the Folin-Ciocalteu values, this suppression is an indicator of the molecular size within a group of condensed tannin polymers. Rossi and Singleton (1966) reported that a relatively high ratio of vanillin to Folin-Ciocalteu value is indicative of a wine with high catechin content and a high browning tendency.

Recently, Butler *et a*l. (1982) reported a modification of vanillin assay for proanthocyanidins (condensed tannins) to estimate the relative, degree of polymerization. They used glacial acetic acid as an extractant in place of methanol. When the' reaction of flavanols with vanillin was carried out in glacial acetic acid, the absorbance produced was much greater than in methanol. In glacial acetic acid, the time course of the vanillin reaction with catechin was similar to that of tannin. The absorbance produced in glacial acetic acid was approximately proportional to the concentration of flavan-3-ol end groups and thus measured the concentration of oligomeric molecules rather than the total concentration of flavan-3-0Is, as is the case with methanol. It was found that, by use of this reaction in glacial acetic acid, the degree of polymerization of purified tannin samples can be determined. It is, however, indicated that methanol is the solvent of choice for the determination of tannin content because in methanol the reaction is much less sensitive to monomer units such as catechin than it is to the polymeric tannins.

The vanillin assay in glacial acetic acid, although useful in measuring the degree of polymerization, apparently measures only a single terminal flavan-3-ol unit in oligomers containing 1 to 6 or more of these units. Butler' later reported a procedure to determine the degree of polymerization of condensed tannins in extracts of developing sorghum seed. It was based on the ratio of total flavan-3-ol residues (estimated by the formation of a thocyanidins in HCl/butanol) to terminal flavan-3-ol residues (estimated by color production in the vanillin reaction carried out in glacial acetic acid). The degree of polymerization of the tannin extractable from several high-tannin, bird-resistant sorghum cultivars was determined. Unlike the vanillin reaction, this assay does not detect flavan-3-ol monomers such as catechin. Hence, the degree of polymerization, as determined by the anthocyanidin formation/vanillin product ratio, is under estimated, especially in extracts rich in catechin and/or epicatechin. It was, therefore, suggested to estimate the relative degree of polymerization, instead of absolute degree of polymerization of flavan-3-ols, in seed extracts to compare their tannin contents.

Among other aldehydes, formaldehyde has also been used as a reagent for tannin determinations. Formaldehyde attacks the benzene ring in several positions with the removal of water and results in the formation of polymers of high molecular weight phenol formaldehyde resins. The phenolics present in plant extracts can similarly be quantitatively removed by precipitating them with formaldehyde in acid solution. The formaldehyde-HCl method is based on Stiasney's reagent and was first suggested by Clarke *et al.* (1939). The sample extract is reacted with Stiasney's reagent. The mixture is then boiled under an air condenser for 0.5 to 3 h. The red precipitate obtained is filtered off on a weighed crucible, dried, and weighed. Because of its complexity, the equipment needed, nonspecificity, and the development of newer and easier methods, Stiasney's reagent is now seldom used for the tannin analysis of plant products.

Oxidation-Reduction Methods

The oxidation of phenols involves chemical reactions of extraordinary complexity. The first step in the oxidation process involves the removal of a hydrogen atom from phenolic hydroxyl. The free radical with a single bound oxygen atom carries an unpaired electron. Being highly unstable, it rapidly changes with one which has in the *ortho* or *para* position, a trivalent carbon atom carrying an unpaired electron (Fieser and Fieser, 1956). The radical thus formed can either dimerize or react with another radical, forming, in order of importance, C-C, C-O, or O-O bonds (Thomson, 1964). In all probability, the chemical and enzymatic oxidation of phenols operates in the same way. Forsyth and Quesnel (Forsyth and Quesnel, 1957a; Forsyth and Quesnel, 1957b) and Forsyth *et al.* (1960) showed that in the case of catechol, these two types of oxidations lead through intermediate radicals to a mixture containing several tetrahydroxybiphenyls and a quinone (Figure 4. 6). Oxidizing agents are employed for the determination of phenols and for their detection in chromatography. Commonly used oxidizing agents include ammoniacal silver nitrate, a mixture of phosphotungstic and phosphomolybdic acids (Folin-Denis reagent), acid or alkaline permanganate, ferric sulfate and ferricyanide. Phenols also reduce Fehling's solution, but the latter is seldom used in analytical methods for tannins.

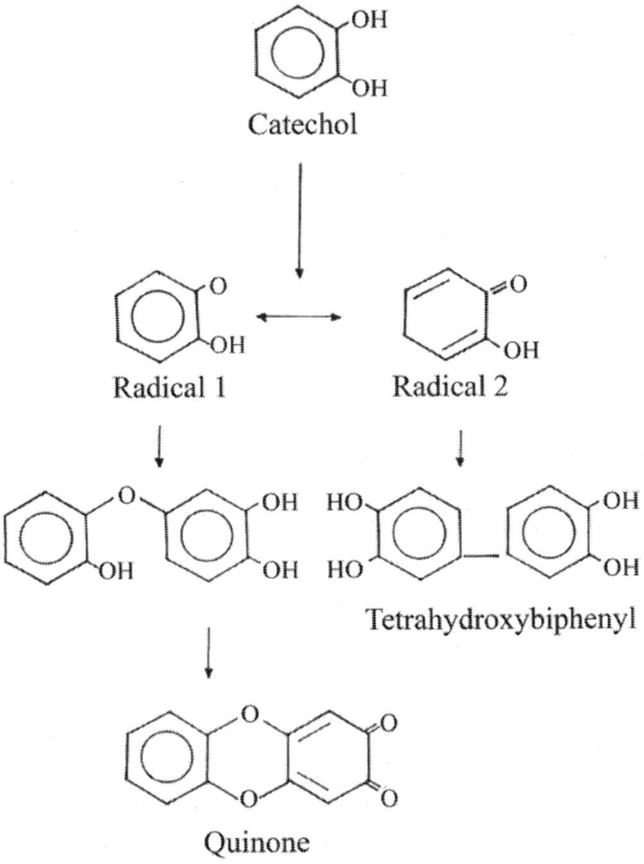

Figure 4.6: Oxidation of typical phenol-catechol.

Ammoniacal silver nitrate is routinely used in paper chromatography for the detection of phenolic compounds. It gives brown to black spots with compounds containing either pyrogallol or catechol nuclei (Roux and Maihs, 1960). In the cold, the reagent is instantly reduced by the vicinal phenolic hydroxyl groups of ortho-dihydroxy and trihydroxy groups, while mono- and metahydroxyphenols reduce the reagent very slowly. The metallic gray to black to brown colorations produced depends on the nature of the phenolic compounds present in the sample. The use of other commonly used oxidizing agents in quantitative tannin analysis is described below.

A. Folin-Denis Method

The Folin-Denis method is the most widely used method for the determination of total phenols in a wide range of plant products and beverages. In the U.S., it is the official method for wine analysis (AOAC, 1965). The original colorimetric method (Folin and Denis, 1912) was later applied to the determination of tyrosine in proteins" and of vanillin in vanilla extract (Folin and Denis, 1912).

1. Principle

Phosphomolybdic-phosphotungstic acid (Folin-Denis reagent) is reduced to a blue-color complex in alkaline solution by phenols. The intensity of the blue color produced is measured with a spectrophotometer at 725 to 760 nm. Tannic acid is often used as standard to calculate the concentration of phenolics in the sample and is expressed as a tannic acid equivalent. The method estimates total phenols and does not distinguish between the type of phenolics.

2. Procedure

The method of the Association of Official Agricultural Chemists (AOAC, 1980) for tannins, using the Folin-Denis reagent is outlined in Figure 4.7. However, for the routine analysis of a large number of samples, a modification, suggested by Swain and Hillis (38) is often used. In this modification, a suitable aliquot of the solution under test (usually 0.5 to 1.0 ml, containing about 5 ml of methanol or ethanol) is diluted with distilled water to about 7 ml, followed by the addition of 0.5 ml of Folin-Denis reagent. Exactly 3 min later, 1 ml of standard sodium bicarbonate solution is added, and the mixture is made up to 10 ml with water. The optical density is recorded after 1 hr at 725 nm using water and reagents only as the instrument blank. The scaled-down volumes in this procedure, therefore, make it easier to handle a large number of samples on a daily basis.

Christcnsen (1974) suggested a modification to the AOAC method. Tannic acid was added to each sample to make an internal standard curve. This proved to be necessary as the internal standard was different and depended upon sample material, even with material of the same origin. In this method, a 1 g finely ground sample is weighed in four flasks and mixed with 150 ml of water. An internal standard curve is prepared by adding 0, 10, 15 and 20 ml of tannic acid (100 mg tannic acid/100 ml water) to the flasks respectively. The contents are heated for 30 min while maintaining the original volumes. Later, clear supernatants are collected after centrifuging the contents at 2000 rpm and filtering them through cotton. The filtrates containing tannins are diluted to 250 ml. The reaction mixture for color development contained 10 ml of filtrate, 75 ml of water, 2.5 ml of Folin-Denis reagent, and 5 ml of saturated sodium carbonate. The color intensity is read at 740 nm after 30 min. A standard curve is prepared by using tannic acid as a standard. A line is drawn through four points, and where this line cuts the Y-axis, is taken as concentration of tannin in sample. This value (E 740) is doubled and read on the curve to calculate the concentration of tannin in a sample.

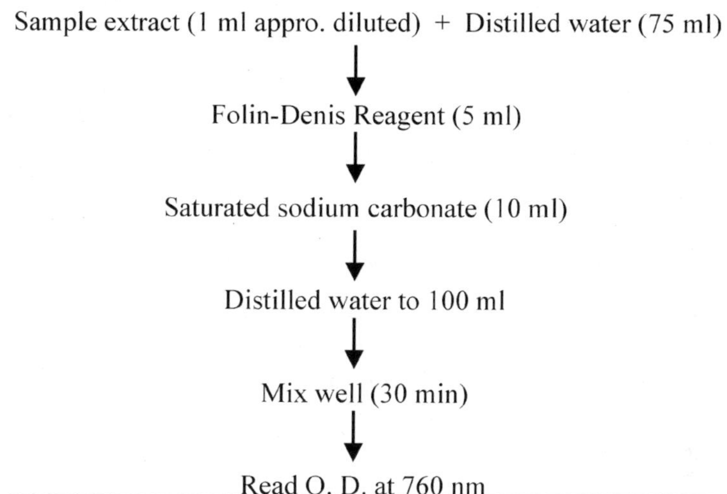

Figure 4.7: Official Association of Official Agricultural Chemists procedure for tannin analysis by the Folin-Denis method.

The Folin-Denis reagent has been used for the determination of tannin in a number of food products, including wine, whiskeys, and a variety of fruit products (Craft, 1961; Guadagni *et al.*, 1949; Pro, 1955; Rosenblatt and Peluso, 1941; Scott, 1922; Sherman *et al.*, 1953). Swain and Hillis (1959) considered the Folin-Denis reagent as the most convenient for the analysis of total phenols in fruits and their method, as described earlier, has been widely adopted for this purpose. However, they observed that most of the methods for the estimation of total phenols are rather empirical. Goldstein and Swain'· later reported that unlike the vanillin reagent, the Folin-Denis reagent does not react in a stoichiometric manner with phenolic compounds, even when the number of hydroxyl groups are taken into account. It is, thus similar to other oxidizing agents; the differences presumably arising from the relative oxidation-reduction potentials of different phenoic groups. Goldstem and Swam (1963) also observed that the molecular absorptivities of some complex molecules are approximately the same as obtained by the addition of coefficients of separate phenolic nuclei. Polymeric tannins will therefore give approximately the same amount of color with this reagent as the monomers unless the hydroxyl groups are involved in linkage or are transferred into stabilized quinoid groups during polymerization.

Folin-Denis reagent is often substituted with the Folin-Ciocalteu reagent (Folin and Ciocalteau, 1927). The Sensitivity of the reagent to reduction is improved by increasing the molybdate content and by the addition of lithium sulfate to prevent precipitation of sodium-complex salts. Even then, with absolute methanolic extracts of several legumes, a fine, cloudy precipitate is often formed up on the addition of a sodium carbonate solution, necessitating filtration or centrifugation prior to the reading of the optical density of the test samples (Deshpande, 1985). The use of the Folin-Cioealteu reagent was investigated and modified further for wines by Singleton and Rossi (Singleton and Rossi, 1965). They suggested the use of selected alkali-time-temperature conditions -and gallic acid as a reference standard. Singleton and Rossi (Singleton and Rossi, 1965) found that the molar color yield under improved conditions was reasonably predictable for the series of phenols they investigated. Monomeric phenolic compounds gave similar molar color yields, except that phenols with suppressed ionization such as salicylic acid, gave somewhat lower color yields than normal. The color yield of polyphenols tested was equal to that predicted from the number of phenolic groups present which may be affected by quinoidal reaction possibilities.

One of the major drawbacks of the Folin-Denis reagent is that it also reacts with other plant constituents, i.e. xanthine, amino acids, and proteins (Lowry, *et al*., 1951). Therefore, with certain cereals and legumes containing alcohol-soluble proteins, this method yields higher estimations of tannins. The method is also subjected to interference by several reducing agents occurring naturally m plant products, i.e., ascorbic acid, hydrogen sulfide, sulfurous acid hydrogen peroxide and ferrous ions (Snell and Snell, 1953; Snell *et al*., 1961).

B. PRUSSIAN BLUE ASSAY

This method has been suggested by Price and Butler (1972) for the qualitative and quantitative detem1matIon of condensed tannins in grain sorghum.

1. Principle

Ferric iron is reduced to ferrous iron by tannins and other phenolics to form a ferricyanide ferrous color complex commonly known as Prussian blue. The color intensity is measured at 720 nm.

2. Procedure

A suitable aliquot (usually 6 to 10 ml) of extract is mixed with 50 ml of distilled water. To this, 3 ml of 0.1 M ferric chloride in 0.1 N HCl is added, followed by the timed addition of 3 ml of 0.08 M potassium ferricyanide reagent. The optical density is read at 720 nm after 10 mm. Increasing amounts of tannins result in the production of an increasing amount of blue color, which absorbs at the red end of the spectrum. The solution however appears green because the blue end of the spectrum is still masked by unreacted ferricyanide. A similar approach has also been used for quick, visual estimation of sorghum tannins. Accordingly to Price and Butler (1972), the series of colors produced by these reagents can be more easily distinguished visually than when the vanillin-HCl reagent is used. The latter produces only one color of varying intensities.

Budini *et al.* (1980) modified the procedure of Price and Butler to improve its sensitivity for determining total phenols in strawberries. In the modification, 3 ml of an appropriately diluted sample extract is mixed with 200 ml each of the ferricyanide and ferric chloride reagents. The color intensity is read at 700 nm after 5 min. Ascorbic acid is the main interfering compound. Hence, the use of an ascorbic acid blank is suggested (Budini *et al.* 1980). The color development in this modification was found to be independent of other naturally occurring substances. The method was found to be 20 times more sensitive than the acid-vanillin assay and three times more sensitive than the titanium method" when used as a ripening index of strawberries.

The Prussian blue assay, however, makes no distinction between condensed tannins and other phenolic compounds. Also, polyphenolics with varying hydroxyl at ion patterns and degree of polymerization react differently to the reagent. For example, quercetin yields 84% greater molecular absorptivity than catechin. Price and Butler (1972) also suggested that the actual amount of tannins present in sorghum may be more accurately measured by the aqueous Prussian blue test when corrected by the salt-extraction method or by the modified vanillin assay (Price *et al.*, 1972). Salt solutions precipitate condensed tannins, whereas anthocyanidins, which also respond to the Prussian blue test can be' quantitatively extracted with aqueous salt solutions (Goto *et al.*, 1976). The difference in solubility in water and salt solutions, therefore, could be used for a Prussian blue test which is specific for condensed tannins.

A simple and rapid quantitative procedure for determining total phenols in agricultural products, based on their color complex formation with titanium, has been reported (Eskin *et al.*, 1978). The ethanol-extracted solids are mixed with 5 ml of acetone containing 0.25 ml of a titanium reagent (20% $TiCl_2$ in concentrated HCl). The color is read at 405 to 450 nm, depending upon the absorption maxima. It is claimed that the color complex is stable for a longer period of time. However, different molar absorptivities are obtained when phenolics of varying hydroxylation patterns.

C. POTASSIUM PERMANGANATE TITRATION METHOD (LOWENTHAL METHOD)

The potassium permanganate titration method is one of the old methods still used for the determination of tannins in beverages. The standard official permanganate-reducing method of AOAC (1980) is used for tea, cloves, and spices. The determination is based on the reduction of potassium permanganate by phenols in the cold. The aqueous extract of samples is titrated with a potassium permanganate solution.

The oxidation of tannins in an acid solution by permanganate is slow. A large excess of indigocarmine indicator is added to achieve an accurate end point and to avoid the oxidation of substances other than phenols in wine (Ribereau-Gayon and Peyand, 1958). Since the permanganate oxidizes many compounds other than polyphenols, the Lowenthal method is not very specific. To improve the sensitivity of this method, Pro suggested that after the sample titration by permanganate, another separate titration should be performed in which tannins are removed using precipitating agents like gelatin, hide powder, or kaoline. This accounts for compounds other than phenolics, such as ascorbic acid, sugars, alcohol, lactic acid, and oxalic acid to be often present in wine and also contribute to the permanganate-titer value (Singleton, 1964; Singleton and Draper, 1962).

At equivalent concentrations, different phenolic compounds do not reduce the same amount of permanganate. This is a major drawback of this method. It has been reported that only orthodiphenols and paradiphenols react, while monophenols and metadiphenols do not react with permanganate (Bate-Smith, 1954). Hence, the method yields results for any given tannin that are comparable among them, but greatly vary with tannins from different sources. The accuracy of the permanganate method was found to be only ± 10%. The detailed procedure of this method is summarized in several reports (Joslyn, 1950). The method has limitations for tannin assay in cereals and legumes.

D. FERRIC CHLORIDE AND FERRIC AMMONIUM CITRATE METHOD

Balavoigne (1934) first used ferric chloride in an acid solution for the determination of tannins in wine. Ferric chloride was later replaced with ferric ammonium citrate (Czerniwski, 1958). Phenols are oxidized by the iron reagent to produce a violet color. The procedure involves the mixing of the iron reagent with 5 ml of phosphate buffer (pH 8) and 50 ml of aliquot of a sample extract. The colour intensity is read at 725 nm after 30 min. Pyrogallol and rutin are used as tannin standards.

This method is more specific than either the Folin-Denis or fomaldehyde-HCl method and gives coloration characteristic of phenols and hydroxamic acid. However, the method is not very sensitive to low tannin concentrations. The results obtained by this method and that of the Folin-Denis method on tannins in forage crops of *Sericea lespedeza* and certain tick clover species were found to be similar (Burns, 1963).

E. PROTEIN PRECIPITATION METHODS

Tannins are known to form insoluble precipitates with proteins and protein-like precipitants. Based on this property of tannins, several methods for their quantitation have been developed, depending on the type of protein or precipitant used (Bate-Smith, 1973; Swain, 1965). These methods measure the biological activity of tannins that cannot be obtained with chemical assays. Similarly, only protein-precipitating tannins and not total phenols are measured. Gelatin, casein, bovine serum albumin, hemoglobin and different enzymes have been used in the assay by various investigators. The protein-like precipitants used include polyamide, polyvinylpyrrolidone (PVP), formaldehyde-HCl and various inorganic and organic reagents. The formation of a tannin-precipitant complex is primarily by hydrogen bonding (Singleton, 1967a; Singleton, 1967b).

Gelatin and casein precipitation has long been used to separate total phenols into true protein-precipitable tannins and non-precipitable polyphenols (Hathaway and Seakins, 1958; Zitko and Rosik, 1962). The official AOAC method for tea tannins is based on the precipitation of gelatin. Bate-Smith (1983) used hemoglobin as tannin precipitant. The method involved the colorimetric estimation of hemoglobin remaining after the reaction between tannins and proteins of the hemolyzed blood. The results are expressed in terms of relative astringency, the ratio of the concentration of tannic acid standards to that of tannin, which affects the same degree of precipitation. Saponins and other plant metabolites are known to interfere with the assay (Bate-Smith 1977). The method is inconvenient for routine use because it requires large amounts of freshly drawn animal blood. Schultz *et al.* (1981) suggested that this reaction may be used for the quantitative analysis of both hydrolyzable and condensed tannins.

The tannins and polyphenolic nontannins in crude extracts of biological origin were separated by passing the extract through a small column of immobilized bovine serum albumin on Sepharose (Hoff and Singldon, 1977). Only protein-bound tannins are held on the column, and other phenolics are allowed to pass through the column. Later, the tannins are eluted with dimethylformamirle while monitored under UV absorbance or determined by the Folin-Ciocalteu method, the results obtained by this method were more closely related to the physiological effects observed mammal feeding studies than those obtained by the vanillin-HCl method (Hoff and Singldon, 1977). The major disadvantage of this method is those tannins which occur naturally in association with proteins or those which bind to proteins during the course of extraction with an acidic buffer are probably not detected. Hoff and Singleton (1977) suggested that the extraction of residue from acidic extracts with an alkaline buffer, and then the determination of tannins in the clear extracts, would eliminate this problem.

Hagerman and Butler (1978) developed a method for tannin determination based on the coloration of tannins precipitated by bovine serum albumin with ferric chloride. The aqueous or metabolic extracts (up to 1 ml of 1 mg/ml tannins) are added to 2 ml of a standard protein solution (1 mg BSA/ml in 0.2 *M* acetate buffers, pH 5 containing 0.17 *M* NaCl). The combined solutions are mixed and allowed to stand at room temperature for 15 min and then centrifuged. The washed precipitate is dissolved in 4 ml of sodium dodecylsulfate (1%) triethanolamine (5%, *v/v)* solution. After the addition of 1 ml of 0.01 *M* ferric chloride solution in 0.01 N HCl, the absorbance of the violet-colored complex is read at 510 nm after 15 to 30 min using appropriate reagent blanks. Catechin, which gives a positive test with acid-vanillin, does not precipitate proteins and, hence, is not precipitated by this method. However, at higher concentrations, Hagerman and Butler (1978) observed an apparent increase of 10% in the amount of tannin detected in the precipitate. This was attributed to the nonspecific binding of catechin to the tannin-protein complex. A radio-labeled bovine serum albumin was reacted with tannins, and the activity was measured in the resulting precipitate (Hagerman and Butler, 1980). However, this technique requires the use of expensive and hazardous radioactive reagents. Several attempts have been made to estimate tannin concentration by its inhibition of various enzymes (Hagerman and Butler, 1980; Griffiths and Jones, 1977; Daiber, 1975; Davis and

Hoseney, 1979; Deshpande *et al.*, 1982; Becker and Martin, 1982). Enzymes that were used among others include β-glucosidase (Goldstein and Swain, 1965), cellulases (Griffiths and Jones, 1977) and amylases (Deshpande *et al.*, 1982). These assays measure either substrate loss (Maxson *et al.*, 1983) or product formation as a function of time. However, the correlation between tannin concentration and degree of inhibition has been reported to be unsatisfactory (Earp *et al.*, 1981; Bullard *et al.*, 1981). This may be due to the retention of variable activity in enzyme-tannin complexes (Armstrong, 1983; Buller *et al.*, 1984). Besides, these methods are time consuming and require considerable attention. Results are often difficult to interpret, since the relationship between enzyme activity and the formation of insoluble complexes is not fully understood·

Precipitants of tannins other than proteins have also been used for tannin determination. Inorganic or organic reagents, i.e., copper acetate, salts of aluminum, zinc, lead and copper, cinchonine, and trivalent utterbium are employed as tannin precipitants (Reed *et al.*, 1985). Cinchoninetannate procedures have long been used for cacao products (Chapman, 1908; Duthie, 1928; Jensen, 1985). Tannins in cacao products are precipitated as cinchonine-tannate. The precipitate is collected, dried, and weighed by using a specially derived factor; the weight of the precipitate is converted into the corresponding amount of tannins. A gravimetric method based on the precipitation of tannins with trivalent utterbium for forage crops has been reported (Maxson *et al.*, 1973).The solutions of utterbium precipitated phenolics are then used for further studies, i.e. enzyme inhibition, protein precipitation, UV absorbance, and colorimetric assays.

Asquith and Butler (1985) used a dye-labeled protein as the Spectrophotometric assay for tannins. Bovine serum albumin was covalently labeled with Remazol brilliant blue R to provide a substrate for a Spectrophotometric assay of tannins. The blue protein was very useful for measuring protein precipitation by vegetable tannins because its absorption maxima are at a wavelength where plant pigments exhibit minimum absorption. Blue BSA was used to determine, by competition assays, the relative affinity of various proteins for tannins. The sorghum tannins with higher degrees of polymerization precipitate about twice as much blue BSA as an equal weight of quebracho tannin. Porter and Woodruff (1984) reported that the procyanidin chain length is important in tannin astringency. Tannins with different protein precipitating capacities cannot be directly compared in competition assays at the same tannin concentration.

UV SPECTROPHOTOMETRY

Simple phenols have absorption maximas in the UV region of around 220 and 280 nm (Owades *et al.*, 1958a, b; Ribereau-Gayon, 1972). This property of phenolics is very useful in their identification. In natural complexes, however, these peaks may be displayed and more peaks may appear due to the presence of a chain of carbon atoms linked to one or more benzene rings. The wide variations in molecular absorptivity of even very c1oscly related phenolics makes results difficult to compare. Vuataz *et al.* ((1959) reported that such values vary between 1285 and 3120 for different catechins, and between 11,500 and 13,500 for their esters with gallic acid. The absorption of natural phenols is also affected by many factors, including the nature of the solvent and pH. However, such change in absorption with a change in the nature of a solvent can be used in the determination of a single component or a class of components in a complex mixture (Bate-Smith and Rasper, 1969; Maranville and Goldschmidt, 1954). The pH, therefore, may be changed or metal complexes may be formed. Such methods measure the absorption before and after the addition of a specific reagent for the particular substance or class (Ribereau-Gayon, 1972).

Phenolics in tea and beer have been estimated by UV absorption technique (Owades *et al.*, 1958a, b). Phenols are extracted from acidified sample extracts with ethyl acetate. The solvent is evaporated, and the residue is heated to 80°C to remove traces of the solvent. The residue is redissolved in acidified methanol, and the UV absorbance is read at 270 nm for beer and 275 nm for tea. In the case of beer, interfering substances, such as isohumulone, are removed by preliminary extraction with isooctane. Absorption peaks, however, are subjected to interference by many constituents of wines and other carbonyl compounds that also absorb in the same region. Hence, the method is not suitable for wine analysis. Lipis *et al.* (1964), however, reported that the 280 nm absorption of diluted wines, compared with calibration data with tannic acid, gave reproducible results for total wine phenols.

Tannins of cereals and legumes have also been measured by UV spectrophotometry (Davis, 1982). Polyphenols of sorghum could be selectively adsorbed on a Sephadex LH-20 column in ethanol and then removed by elution with methanol. The UV absorption of separated polyphenols is measured at 280 nm. Although the method is claimed to be relatively specific for tannins in sorghum, it is time consuming and tedious. When applied to legume tannins, it was observed that a number of coloring compounds interfered with the assay of tannins by this technique.

A large number of different approaches have been attempted to develop a simple and satisfactory method for the determination of vegetable tannins. However, none has been found completely satisfactory. This can be attributed to the complex nature of phenolics within one sample and from samples of different origin, extraction methods, and sample particle size, location of tannins in the plant tissue, age of sample, and difficulty in one specific standard for the assay. The influence of these variables is briefly described below.

FACTORS INFLUENCING TANNIN ANALYSIS

A. NATURE OF PLANT PHENOLICS

Plant phenolics, with varying degrees of hydroxylation and polymerization, pose a unique problem for their determination. In any plant tissue, a wide range of phenolics are present together in varying quantities. This makes the estimation of one class of phenolics by a single method very difficult. Phenolic compounds are often associated with a large number of potentially interfering compounds with the exception of flavans (progenitors of condensed tannins) and substances that are included with the flavonoids, such as fully alkylated hydroxyflavones (Ribereau-Gayon, 1972; Geissman, 1955). Seeds often contain free phenolics. Except in isolated cases, aglycones in plant extracts are present only as a result of partial or complete hydrolysis during extraction.

The most common linkages found in phenolic compounds are ether (acetal) type (-C-O-C- as in the case of O-glycosides), ester linkages (-CO-O-C-) of phenolic acids and aglycone sugar linkage (-C-C-) of C-glycosylflavones. C-glycosides are more resistant to hydrolysis, causing additional problems in their analysis. In anthocyanins, both glycosidic and ester linkages may be present; the latter in the case of acylated anthocyanins in which a cinnamic acid residue is esterified with a glycosidic sugar. A great number of closely related phenolic compounds in plant extract also present practical problems in colorimetric estimation because most of these compounds absorb at closely similar wave lengths. Thus, it becomes almost impossible to determine the concentration of anyone compound by the measurement of absorbance. Often in such cases, absorbance has to be determined at any given wave length before and after treatment with a reagent specific for that compound or for the group of substances to which it belongs.

Condensed tannins of plants are a mixture of several polymers which are condensed products of flavans. Every sample of natural tannin, therefore, contains several different flavolans of varying molecular weights. It is the relative proportion of each of these which determines the overall properties of a tannin mixture. Attempts to determine the molecular weight of tannins include the elevation of the boiling point using methylated or acetylated derivatives, ultracentrifugation, and colorimetry and gel filtration. Food grains, such as sorghum, and legumes contain mainly condensed tannins. Nutritionists are often interested only in these tannins, but not in total phenols. Hence, the separation of other phenols becomes important.

Non-phenolic and non-polar substances like waxes, fats, terpenes and chlorophylls are potential interfering compounds in tannin analysis. They need to be removed first with nonpolar solvents. The desired phenolic compounds then could be extracted with acetone, ethanol, methanol, or water. The choice of solvent usually depends upon the number of hydroxyl groups present on the molecule. Some of the other factors that influence the estimation of condensed tannins are described below.

B. EXTRACTION METHODS

There is no single, completely satisfactory method for the quantitative extraction of condensed tannins from plant tissue. Some portion of such tannins may remain behind, attached to a carbohydrate or protein matrix, while some fraction may be quite insoluble due to a high degree of polymerization. It has been reported that the processing of high tannin grains decreases their nutritional value even though the tannin level has been reduced significantly (Price and Butler, 1980). This has been attributed to the binding of tannins to proteins during processing. Such bound tannins become insoluble, cannot be extracted and, therefore, escape measurement.

The sample is extracted with a nonpolar solvent to remove the interfering substances before tannin assay. Geissman (1955) however, observed that methylated flavones in leaves and fruits could be easily extracted by light petroleum ether. Absolute methanol, acidified methanol, acetone and ethanol and their aqueous solution, and water have been used to extract tannins. Methanol and ethanol can be easily removed by vacuum evaporation, particularly if further purification and characterization of tannins are intended. The time of extraction is also important. In literature, extraction times of 20 min to 24 h are used. Many phenols tend to oxidize during sample preparation and extraction. Hence, a longer extraction time may result in a lower estimation of tannins. The addition of suitable reducing agents to the extraction medium has been found to increase the yield of phenolic compounds (Khanna *et al.,* 1968). Price et al. (1978) observed almost a complete extraction of tannins from grain sorghum using absolute methanol within 20 min. It was further reported that when the extraction time was extended to 24 h, there was a decrease of 10 to 15% in assayable tannins.

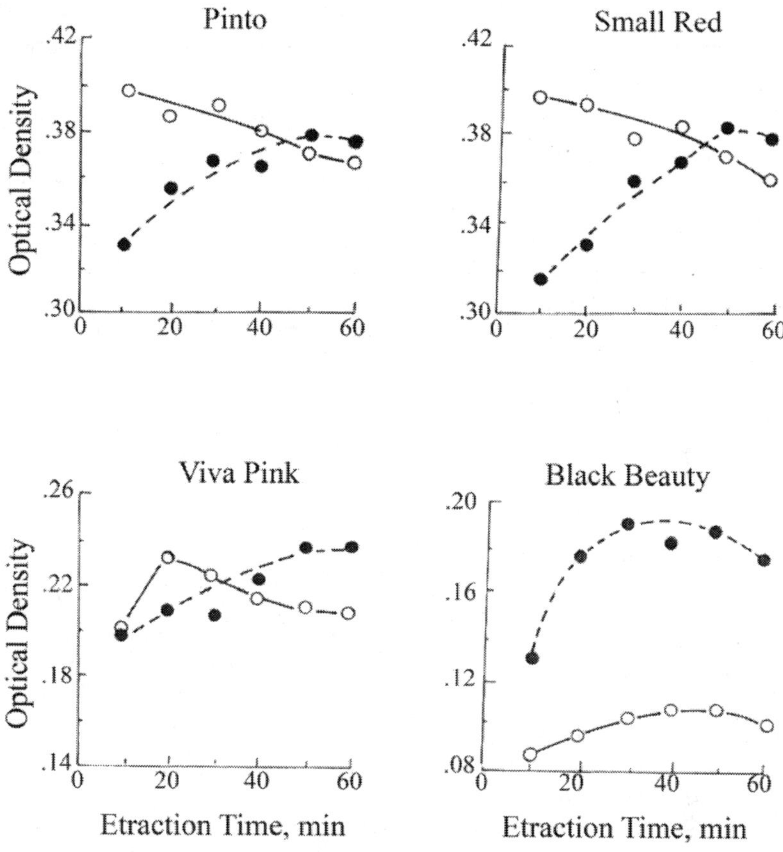

Figure 4.8: Effects of extraction time on the optical density of tannin extracts: Curve A= MeOH extract and Curve B= HCl-MeOH extract.

Maxson and Rooney (1872) extracted sorghum tannins in acidified methanol (1% HCl in methanol) for 24 h. However, Price *et al.* (1978) observed a gradual decrease in tannins with extraction times of up to 2 h, using the same solvent. Extraction for 24 h gave overall 40 to 70% lower estimates of tannins for all the sorghum varieties they studied as compared to a I h extraction. Such a decrease in extraction times has been attributed to the destruction of tannins by HCI since, even after the extracts were separated from the grain, the absorbance continued to decrease with time. Deshpande (1985) recently observed that the tannin extraction pattern of dry beans *(Phaseolus vulgaris* L.) is markedly affected by the solvent used. With absolute methanol, maximum absorbance at 500 nm, using a 0.5% vanillin assay of Price *et al.* (1978) was obtained within 10 to 20 min of extraction (Figure 4.8) and the values slowly decreased with longer extraction times, presumably due to tannin oxidation. When acidified methanol was used as an extractant, the time required to extract maximum tannin was at least 50 to 60 min. Among the varieties studied, the extraction pattern was quite different for Black Beauty beans. The absorbance was almost twice as high in acidified methanol as in methanolic extracts. Such differences were attributed to the nature of phenolic compounds present in this variety.

Maxson and Rooney (1972) used acidified methanol for the extraction of phenolic compounds in sorghum varieties that were thought to have no detectable tannins when absolute methanol was employed for extraction. This MV-HCl method, however, consistently estimated more sorghum tannins than the vanillin method of Bums (Table 4.3) (1971). Later, Cummings and Axtell (1968) classified sorghum into three groups based on the differences between the results of these two assays. Grains having catechin equivalent (CE) values of less than one by both methods were classified as Group I, those with CE less than one by the vanillin and greater than two by the MV-HCl method as Group II and the remaining varieties were classified as Group III. In the MV-HCl assay, an implicit assumption is made in that the higher CE values indicate that more tannin is extracted in acidified methanol than in methanol itself (Price and Butler, 1972). When corrections were applied for the large amount of background color present in the extract by using appropriate sample blanks, both the vanillin (Burns, 1971) and MV-HCl assays gave comparable results. Only the Group II sorghum varieties showed a distinctly different behavior. These varieties, in spite of a colored testa and tannins, were found to be nutritionally equivalent to low tannin varieties.[124] Based on their observations, Price *et al.* (1972) suggested that Group II tannins may be nutritionally inert because they are covalently bound to some grain component and, unless released, do not dissolve and, therefore, do not precipitate proteins. It was further suggested that Group II tannins may be monomeric flavonoids which are attached to some grain component. The extraction of these compounds seems to be dependent on an acid-catalyzed reaction to release these tannins, which then give a vanillin positive test.

Ethanol, propanol, water, acetone, and dimethylfonnamide are some of the other solvents tried for the extraction of sorghum tannins. Gupta and Haslam'6 found absolute methanol to be the most suitable solvent with regards to the subsequent purification and analyses of tannins. However, methanol may not completely extract the tannins. Such data on legume tannins is lacking in the literature.

Price and Butler (1972) using the differential solubility of condensed tannins' in aqueous and salt solutions, studied the specificity of the Prussian blue assay for tannins alone. Along with tannins, the assay also measures both anthocyanidins and other low molecular weight phenols. Salts are known to precipitate condensed tannins from solutions, whereas anthocyanidins can be quantitatively extracted (Feeny and Bostock, 1968; Goto, *et al*., 1978). Using the Prussian blue assay, they observed lower CE values for high tannin sorghum when 1 M NaCl was used for extraction as compared to those of water extraction. Salt-extracted polyphenolics apparently varied only within the narrow range, and their concentrations never approached those in true high-tannin Group III sorghums. Some sorghum varieties appeared to be quite sensitive to low salt concentrations. However, all varieties contained phenolics, other than tannins, whose extractability was unaffected by up to 1M NaCl concentrations. A good correlation (r = 0.957) between the CE values of sorghum tannins from water, minus salt extractions and those from the corrected vanillin-HCl assay, has been observed (Price and Butler (1972). However, CE values for the latter were almost three to four times higher than those measured by the former approach. Based on their studies, Price *et al.* (1972) further suggested a need to determine appropriate extraction times for the grinding and extraction techniques employed within each laboratory.

Table 4.3: Comparison of CE values for sorghum tannins using the vanillin-HCl and MV-HCl methods.

Variety colour	Mg CE/100 mg	
	Modified	Unmodified
Brown	4.32	4.54
Brown	4.20	4.34
Brown	3.25	2.97
Brown	3.23	1.64
Brown	2.83	1.66
Red	0.67	0.04
White	0.31	0.03
White	0.23	0.01
White	0.14	0.00
White	0.11	0.00
White	0.08	0.00

CE=Catechin equivalent.

C. PARTICLE SIZE OF SAMPLE

The degree of grinding or milling of a sample significantly affects the tannin assay. However, the effects seem to vary with the type of sample under consideration. Chopping or macerating fresh plant material or wilting in the field to reduce the tannin content is commonly followed in the case of *Lespedeza cueala* (1954). Bate-Smith (1954) found it difficult to extract tannins from the leaves of some species of forage legumes. A fine grinding of grains, passed through a 0.4 mm sieve, has been found to yield consistent and reproducible results for sorghum tannin (Price *et al.*, 1972; Hagerman and Butler, 1978). On the contrary, Deshpande and Cheryan (1985) observed a consistent reduction in assay able tannins of legumes when the particle size of the sample flour was reduced from 820 to 250 μm (Table 4.4). Assayable tannins of four dry bean varieties studied decreased by about 25 to 49% with a 2%

vanillin assay and by 36 to 50% with a 0.5% vanillin assay. In most cases, the differences between particle sizes were statistically significant. The 2% vanillin assay of Bums, 15 which does not include a correction for the background color of a sample, consistently gave higher values than the 0.5% vanillin assay, as suggested by Price *et al.* (1972). In fact, with the former method, Black Beauty beans yielded higher values for tannins with decreasing particle size. However, when corrections were applied for the background color, the same relationship as described here was observed between assayable tannins and particle size.

Table 4.4: Effects of particle size on tannin analysis of dry beans.

Minimum particle size (μm)	2% Vanillin assay				5% Vanillin assay			
	Pinto	Small Red	Viva Pink	Black Beauty	Pinto	Small Red	Viva Pink	Black Beauty
820 (20)b	337	304	169	154	328	248	162	60
420 (40)	269	262	111	172	251	210	118	45
350 (45)	245	250	105	213	226	191	99	36
250 (60)	223	227	87	258	186	159	95	30
LSD (p = 0.05)			9.8			5.6		

The differences in seed coat color also affect the tannin assay. The varieties with light colored seed coats (Pinto and Yiva Pink beans) did not exhibit significant differences in tannin content by either of the two assays with 820 μm particle size. However, the tannin content sharply differed in the Black Beauty variety (dark-colored seed coat) when assayed b\ both methods. This may presumably be due to the presence of monomeric anthocyanin like compounds which turn red under the acidic conditions employed for the assay. Unless corrected for by a sample blank, this leads to erroneous estimations of tannin in these two types of beans. The differences between the old and corrected vanillin assays were also more pronounced with decreasing particle size. Apparently, such compounds are attached to some seed components and yield the characteristic red coloration of anthocyanins under acidic conditions only when separated from them. However, these appeared to be quick different than the monomeric flavans of Group II sorghum tannins (Price *et al.*, 1972).

The tendency of many phenols to get oxidized during sample preparation and extraction is well-known. The loss of assayable condensed tannins appears to be directly related to the time and extent of milling of the seeds. The heat generated during grinding for a prolonged period of time to achieve fineness may create a microenvironment in which tannins are either easily oxidized or can complex with other seed proteins. The significant reduction in assayable tannins with the fineness of bean flours may also be related to the high protein content of legumes as compared to that of sorghum. The effect of particle size on assayable tannins of beans, therefore, may be a function of the exposed surface area of flour particles which contain protein that can bind with tannins (Table 4.5). The surface area and volume ratios for bean flours were calculated using the average percent particle-size distribution within each mesh size. Assayable tannins were directly related to the volume to surface area ratios of varying particle sizes of flour (r = 0.96 to 0.99). Davis and Hoseneyn.98 observed that the biological activity values calculated as percent a-amylase activity for sorghum tannins were much higher when expressed on a per kernel basis as compared to those expressed on a flour weight basis. They also found that the amount of condensed tannins per kernel of grain sorghum decreased with maturity. Since beans vary in size, both within and between varieties, this may also influence assayable tannins.

Table 4.5: Relationship between surface area and volume of bean flours and assayable tannins (0.5% vanillin assay).

Bean	Flour particle size (μm)	Surface area (μ^2 x 10^6)	Volume (μ^3 x 10^6)	Volume/ surface area (μ)	Assayable tannin
Pinto	820	1.69	2.23	t31.8	328
	420	0.36	0.22	60.5	251
	350	0.30	0.17	54.8	226
	250	0.14	0.05	36.7	186
			r = 0.97		
Small Red	820	1.22	1.50	122.7	248
	420	0.38	0.24	62.7	210
	350	0.25	0.13	51.0	191
	250	0.t3	0.04	34.7	159
			r = 0.96		
Viva Pink	820	1.42	1.81	127.2	162
	420	0.42	0.27	64.3	118
	350	0.23	0.11	49.2	99
	250	0.15	0.05	37.5	95
			r = 0.99		
Black Beauty	820	1.69	2.22	131.3	60
	420	0.42	0.27	63.8	45
	350	0.27	0.14	52.8	36
	250	0.16	0.06	38.3	30
			r = 0.97		

D. LOCATION OF TANNINS IN SEEDS

Tannins are located primarily in the seed coats of cereals and legumes (Hahn *et al.*, 1984; Ma Yu and Bliss, 1978). However, tannins are usually extracted from whole seed flours, particularly of cereals, since it is often tedious to scrap seed coats. Seed coats of legumes are, however, relatively easy to remove. Laurena *et al.* (1984) analyzed the tannin contents in the seed coat, whole seeds, and cooking broths of several varieties of cowpeas. The seed coats and cooking broths contained seven to ten times' greater values for condensed tannins as compared to those for the whole seeds. The large difference in tannin content was attributed to the interaction of tannins with seed proteins during the analysis of tannins in the whole seeds, thereby decreasing their activity. The seed coat proteins of cowpeas (9 to 16%) may also interfere with the analysis of condensed tannins.

In order to avoid tannin-protein complex formation during extraction, tannins were extracted from intact beans (Deshpande and Cheryan, 1985). However, only 10 to 20% of the total tannins were detected in absolute methanol with a 48 h extraction period. It was realized that some grinding is essential to release tannins from intact seed coat cells. The seed coats were removed manually and analyzed for assayable tannins (Table 4.6). The seed coats contained approximately 0.5 to 7% tannins on a weight basis. When considered on a total seed basis, these values were 1.1 to 2.5 times higher as compared to when extractions were made on the bean flours. This further confirmed that seed proteins are the major interfering components with the tannin assay.

Table 4.6: Dry bean seed coat tannin extracted with absolute methanol.

Variety	Assayable tannin		
	Seed coat (%)	**G CE/100g of seed coat**	**Mg CE/100g of beans**
Pinto	8.2	6.85	562
Small Red	8.1	5.95	482
Viva Pink	8.0	6.48	518
Black Beauty	8.1	0.51	44

Separately ground batches of seeds consistently cause a greater variation in tannin content than do the replicate assays within a batch of ground grain (Price *et al.*, 1972), indicating a strong possibility of the inherent color variation causing such discrepancies. Deshpande and Cheryan (1985) investigated the effects of such color differences within a given sample of bean seeds. The light- and dark-colored small red beans (approximately 20:80 distributions) were subjectively separated and ground in different batches prior to extraction. Tannin contents, as determined by the 0.5% vanillin assay, were 228 ± 9 to 254 ± 14 mg CE per 100 g of beans in the light - and dark-colored seeds, respectively. The inherent color variation in a sample of seeds, therefore, may also influence tannin determination. Deshpande and Cheryan (1985) conducted analyses using all the optimum conditions of extraction and assay. Although significant differences in tannin content were observed as determined by the optimum and 20-min assay, no definite pattern with regard to solvent used could be established. The distinct varietal differences in the properties of condensed tannins were, however, observed. This emphasizes a need for the standardization of assay conditions within each laboratory and for different grain samples.

E. EFFECTS OF STORAGE

Tannin content was found to be steadily decreased in sorghum when ground samples were stored for 1 to 18 days (Table 4.7). After 18-days storage, assayable tannins were decreased by about 15%. Similar observations for bean tannins have also been reported (Cummings and Axtell, 1973). Such decreases in assayable tannins were found to be characteristic of a particular variety.

Table 4.7: Reproducibility of vanillin assay[a] and Effects of Time Elapsed between Grinding and Analysis of BR-54 Sorghum

Sample	Day 1	Day 4	Day 8	Day 18
	0.676	0.643	0.642	0.574
Sample I[b]	0.676	0.646	0.636	0.574
	-	0.640	0.639	0.567
Mean	0.676	0.643	0.639	0.572
	0.730	0.692	0.687	0.623
Sample II[b]	0.710	0.694	0.684	0.625
	-	0.697	0.679	0.614
Mean	0.720	0.694	0.683	0.621

[a]Results reported as $A_{500}/_{200}$ mg of grain after subtracting the blank. Each result is for a separate extraction with 1 ml of methanol extract and 5 ml of 2% vanillin in 4% concentrated HCl in methanol incubated in a 30°C water bath 20 min before reading.

[b]12 g of cleaned BR-54 was divided into 4-g (l) and 8-g (I) sample and ground separately.

The anthocyanin-like pigments responsible for the background color in the vanillin assay apparently seem to be more prone to oxidation than the polymeric-condensed tannins. It is, therefore, recommended that samples should be preferably analyzed for tannins within 6 to 8 h after milling,

The relative humidity of storage also influences assay able tannins, depending on the time before analysis (Deshpande *et al.*, 1984; Mitaru *et al.*, 1983). High-moisture storage decreases the extractable tannins of high tannin sorghum to a level similar to that of a low-tannin variety (Figure 4.9). The greatest reduction in assayable tannins in each of the four sorghum varieties studied occurred in the first 10 days of storage and then gradually fell to lower values at 20 days. The moisture content of the grains in this study was raised by adding 25% (w/w) distilled water prior to their storage at 25°C. It was suggested (Mitaru *et al.*, 1983) that the process of a decrease of tannins under high-humidity storage conditions probably is similar to that of the polymerization of tannins during the ripening of a seed. An increase in moisture may provide a medium in which the natural process of polymerization occurs. The polymerized tannins are highly insoluble and are too large to form complexes with proteins. However, under normal conditions, the relative humidity of the environment should not be a major influencing factor in tannin analysis.

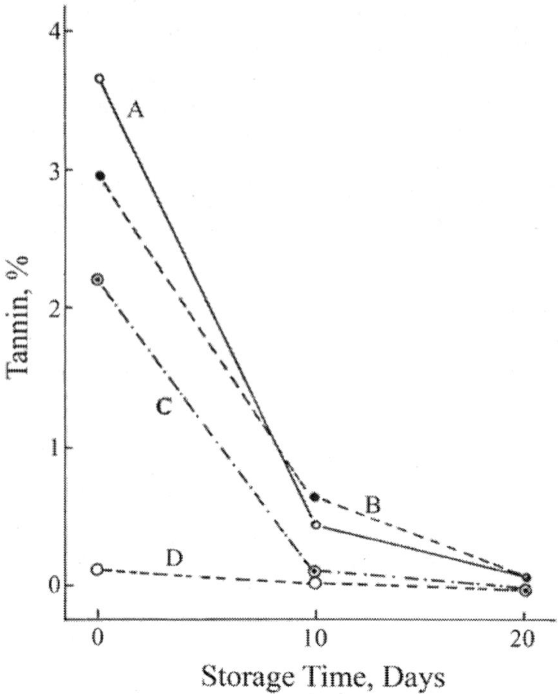

Figure 4.9: Effects of reconstitution on tannin deactivation in these high and one low tannin sorghum: (A) P570; (B) AKS 614; (C) AR 3003 x TX 430 and (D) Yellow No. 2.

F. SELECTION OF A STANDARD

Phenolics are an extremely complex class of naturally occurring substances, and the use of a specific standard for their determination is not possible. Variations in the molecular reactivity of phenolic compounds with different reagents cause additional problems in the selection of a suitable standard for tannin analysis. Tannic acid is often used as an arbitrary standard in total phenol analysis, due to its structural relationship to natural tannins (Rosenblatt and Peluso, 1941). However, tannic acid is a mixture of four phenolic compounds, with varying proportions between samples (King and Pruden, 1970). Chlorogenic acid has also been used as a standard for total phenol analysis. However, none has been found suitable, since most methods of total phenol determination are based on redox principles. Several studies have shown differential oxidation-reduction behavior of phenolic classes. The number and position of hydroxyls on the phenolic ring and the degree of polymerization also influences the molar absorptions of phenols.

A number of model substances for the analysis of phenolic compounds using several different methods have been studied. Data on the molecular absorptivities of the colored products formed during reactions with various reagents arc summarized in Table 4.8. Vanillin reacts in an approximately stoichiometric manner with compounds containing a phloroglucinol group with an absorption maximum at 500 nm. The resorcinol type of compounds, on the other hand, although they react in a similar fashion, give higher molecular absorptivities at a slightly different wavelength (520 μm). Unlike vanillin, the Folin-Denis reagent does not react in a stoichiometric manner with phenolic compounds, even when the number of hydroxyl groups arc taken into account. Other nonselective substitution-type reagents. such as 4-aminophenazone, showed a marked decrease in reactivity with substituted phenols and give variable results with polymerized catechins and tannins, presumably due to the fact that they require an alkaline medium for the reaction.

Table 4.8: Molecular Absorptivities of Colored Products Formed from Various Model Compounds by the Folin-Denis Reaction with Vanillin, 4-Amino Phenazone, and Molybdate Reagents

Compound	$E \times 10^{-3}$			
	Folin-Denis	Vanillin	4-Amino phenazone	Molybdate
Catechol	20.2	0.04	6.1	1.8
Homocatechol	17.6	3.6	4.0	1.7
Tertiary butyl alcohol	9.7	0.05	2.0	1.3
Chlorogenic acid	21.8	0.0	-	-
Resorcinol	15.8	45.7	6.8	0.0
4-Elhylresorcinol	10.4	38.4	2.5	0.0
2.4-Dimethylresorcinol	9.3	38.4	1.6	0.0
Pyrogallol	10.9	44.6	0.50	5.3
2-Methylpyrogallol	13.2	21.6	0.20	-
Gallic acid	18.5	0.0	-	-
Phloroglucinol	7.7	35.0	2.6	0.0
2-Methylphloroglucinol	5.2	28.0	0.18	0.0
7-Hydroxyflavan	13.3	30.5	-	0.0
(+)-Catechin	30.7	32.9	3.3	1.8
(+)-Epicatechin	26.7	24.7	3.2	1.8
(+)-Gallocalechin	30.1	28.0	-	-
(-)-Epigallocatechin	16.7	24.0	2.5	2.4
(-)-Epigallocatechin gallate	43.5	32.1	4.0	10.7
Theaflavin gallate	49.8	30.0	-	-
Brazilin	36.8	39.1	-	-

Goldstein and Swain (1963) also observed the molecular absorptivities for the more complex molecules with the Folin-Denis reagent to be approximately the same as that obtained by the addition of coefficients of separate phenolic nuclei. It was presumed that polymeric tannins will approximately give the same color with the

Folin-Denis reagent as their monomers, unless hydroxyl groups are involved in linkage or are transformed into stabilized quinonoid groups during polymerization. Several polymers were then prepared by heating catechin or gallo-catechin under acidic conditions, and their behavior to the Folin-Denis and vanillin reagents was investigated (Deshpande and Cheryan, 1987). In each case, the color formed with the former reagent was less than that of catechin, although the reduction in extinction coefficient was not as great as that obtained with the vanillin reagent (Table 4.9). The ratios of these two values (V/FD) were, therefore, lower for polymers and could be taken as a measure of polymerization.

Table 4.9: $E^{1\%}_{1cm}$ values for various tannins and catechin polymers.

Tannins	$E^{1\%}_{1cm}$ X 10^{-2}		
	Folin-Denis	Vanillin	V/FD %
Catechin	10.6	11.4	108
Acid-catalyzed polymer	11.3	10.3	91
Autoxidation polymer	4.9	1.6	33
Gallocatechin	9.9	9.1	92
Acid-catalyzed polymer	9.9	8.2	83
Autoxidation polymer	2.2	0.9	42
Grape fruit	4.1	1.2	29
Grape leaf	4.6	1.8	39
Persimmon	5.6	3.4	61
Cacao	4.6	2.2	48
Quebracho bark	6.5	2.5	38
Wattle bark	6.1	3.1	51

Table 4.10: Molar absorptivities for the Prussian Blue Reaction.

Phenolic compound	Molar absorptivity x 10^{-4}	Beer's plot (r^2)
L-Epicatechin	1.71	1.00
Orcinol	0.0	-
Hydroquinone	1.80	1.00
Catechol	1.82	0.99
Gallic acid	3.62	1.00
Resorcinol	0.0	-
3,4-Dihydroxycinnamic acid	1.66	0.99
Pyrogallol	2.82	1.00
Quercetin	8.00	1.00
Cyanidin chloride	2.64	0.99

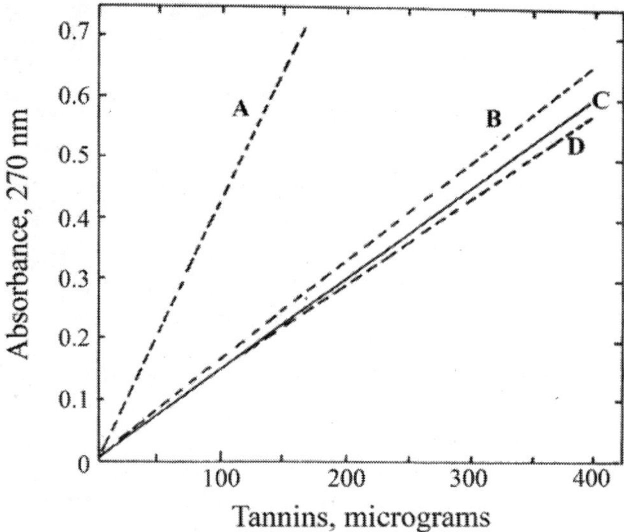

Figure 4.10: Standard curve using hop and malt tannins for the UV sepectrophotometric determination of beer tannins (A) digallic acid; (B) average of three hops, (C) standard curve and (D) average of four malts.

The varying response of even closely related phenolics to the same reagent has, therefore, been the major problem in the selection of an appropriate standard for the estimation of total phenols using the redox approaches. Similar no specificity of other redox methods of tannin analysis including the Prussian blue assay (Table 4.10) and phenol-titanium redox system (Table 4.11) as well as tannin analysis using UV spectrometry (Figure 4.10), have been reported. The wide variation in the relative oxidation-reduction potential of the reference standard to those mixtures of phenolics present in extracts often makes the data on total tannin or phenol content difficult to interpret. It may be more appropriate, therefore, to use a partially purified phenol extract from the source itself as a reference standard

Table 4.11: Spectrophotometric data for phenolic-titanium systems.

Phenolic compound	Max (nm)	Molar absorptivity	Measured at nm	Beer's law (r')
Phenol	410	2.4 X 10^3	410	0.97
Catechol	430	1.9 x 10^3	430	1.00
Resorcinol	410	1.6 x 10^3	410	1.00
Quinol	440	4.1 x 10^3	440	1.00
Phloroglucinol	410	1.1 x 10^3	410	0.96
P-Hydroxybenzoic acid	410	1.0 x 10^3	410	1.00
4-Hydro.y-3-methoxybenzoic acid	405	1.2 x 10^3	405	0.99
P-hydroxycinnamic acid	410	4.6 x 10^3	410	0.99

Phenolic compound	Max (nm)	Molar absorptivity	Measured at nm	Beer's law (r')
3,4-Dihydroxycinnamic acid	450	2.0×10^3	450	1.00
3,5-Dimethoxy-.4-hydroxycinnamic acid	405	9.0×10^3	405	0.94
Chlorogenic acid	450	2.1×10^3	450	1.00
Catechin	430	3.5×10^3	430	1.00
Apigenin	410	9.6×10^3	410	0.99
Naringenin	410	8.8×10^3	410	0.99

The choice of standard for the determination of condensed tannins in cereals or legumes using the vanillin assay is relatively less problematic. Catechin (flavan-3-ol), which is one of the two basic flavan constituents of tannins, has been a standard of choice for the vanillin assay. However, the selectivity of the vanillin assay is known to be affected by the degree of polymerization of condensed tannins. Hence, the adequacy of catechin as a standard for such analysis has been questioned. The use of catechin has been reported to over estimate tannins in sorghum (Price *et al.*, 1972). The standard curve for catechin deviated considerably from linearity and the overestimation varied from about twofold when A was less than 0.2 to threefold at A_{500} = 1.4 (Figure 4.11). Gupta and Haslam (1980) also observed different reaction kinetics for the color development of various substrates in the vanillin assay (Table 4.12).

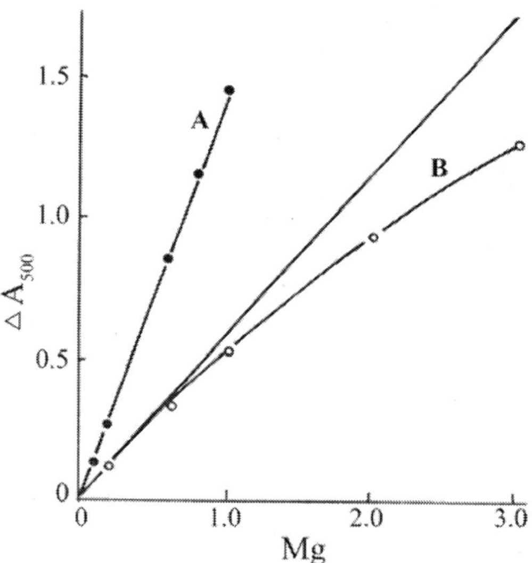

Figure 4.11: Standard curve for the 0.5% vanillin assay, A_{500} after a 20 min incubation at 30°C vs. a milligram sample using partially purified BR-54 tannin (A) and catechin (B) in methanol.

Table 4.12: Tannin content of sorghum seeds using catechin and native polymer as reference standards.

	Vanillin assay		UV method		Cyanidin method	
Variety	Catechin	Polymer	Catechin	Polymer	B-2	Polymer
IS 3648	1100	2.50	2.97	238	5.80	2.80
IS 4225	2.50	0.66	0.70	0.55	1.60	085
IS 9115	1.64	0.53	0.68	0.54	1.33	071
BR 44	2.87	0.76	1.20	0.85	1.50	0.80
BR 54	6.00	1.61	1.42	1.14	3.14	1.68

Depending on the method used the overestimations varied from about 1.2 to 4 times when catechin was used as a standard. Reaction kinetics in the vanillin assay is quite different for the monomeric and polymeric flavans and a true quantitative estimation of tannin may be obtained only when the native polymer is used as a standard.

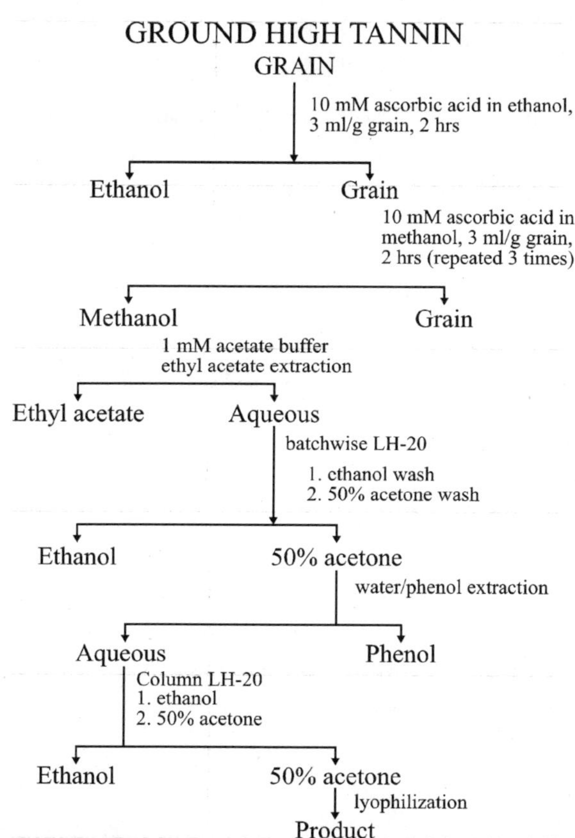

Figure 4.12: Scheme for the purification of tannin from sorghum grain.

It is, therefore essential to develop a method to purify sorghum procyanidin polymer practically free from bound proteins and other organic materials. The sorghum tannins prepared and used as standards earlier (Price *et al.*, 1972) contained up to 20% protein. Recently, Hagerman and Butler (1980) have described a procedure to purify sorghum tannins (Figure 4.12). In this procedure, the co-purification of tannins and proteins is minimized. The first ethanol extraction removes most of the nontannin activity and ethanol-soluble proteins of grain, preventing the simultaneous extraction and subsequent co-purification of the proteins with the tannin. This step can be omitted for beans as they contain practically no alcohol-soluble proteins. Another important step is phenol extraction of proteins. Seed proteins, including the alcohol-soluble prolamins, are completely extracted from water by phenols, but flavonols are not. The nontannin contaminants in tannin-containing extract, such as low molecular weight phenolics, tannin precursors, and added ascorbic acid to prevent oxidation during extraction, are removed by extracting in ethyl acetate and later by adsorption chromatography with Sephadex LH-20. The non-tannins are eluted in ethanol washing, and the bound tannins are recovered later by elution with aqueous acetone. The sorghum tannin obtained by this procedure contained no more than 2 to 3% residual proteins. Such material can serve as a more appropriate standard for the tannin assay.

Table 4.13: Tannin analysis of Phaseolus beans by four different methods.

Sample	Protein precipitable phenols (%)	Modified vanillin assay	Folin-Denis assay (tannic acid equ. g/100g)	Uncorrected vanillin assay (catechin equ. g/100g)
1.	0.44	o.68	0.91	1.59
2.	0.25	0.57	0.77	0.66
3.	0.26	0.60	0.86	0.98
4.	0.35	0.84	0.77	1.30
5.	0.52	1.12	0.91	2.54
6.	0.16	0.32	0.70	0.40
7.	0.20	0.37	0.87	0.56
8.	0.38	1.04	0.95	2.10
9.	0.37	0.73	0.92	1.48
10.	0.28	0.59	0.89	1.10
11.	0.30	0.60	0.81	1.05
12.	0.30	0.79	0.93	1.62
13.	0.38	0.56	0.95	1.78

A measure of the actual tannins present in legumes is probably more accurately provided by the vanillin assay when corrected for the background color, although the monomeric proanthocyanidins, which are nutritionally inert, are also detected. The UV-Spectrophotometric determination of tannins of yellow peas, lentils, and white beans was reported by Davis (1982). However, such an approach has limitations due to the presence of several coloring compounds in dry beans. The protein-precipitation and enzyme-inhibition assay correlate most highly with the nutritional quality of foods. The results, however, are often variable and not quite quantitative. In spite of the availability of such a large number of different approaches for tannin analyses, none has been found completely satisfactory.

Bressani *et al.* (1983) studied the condensed tannins and related phenols in 13 samples of red Guatemalan common beans by four chemical assays (Table 4.13) and calculated their correlations. The four assays differed in their capacity to detect various types of polyphenols, although the values were highly correlated. Deshpande and Cheryan (1963) investigated the phenolic compounds of ten genotypes of dry beans using four different methods. Significant differences were observed for the tannins measured by the vanillin and the protein-precipitable phenol tests (Table 4.14). Of the two redox assays, the Prussian blue assay gave significantly higher total phenols than the Folin-Ciocalteu assay. Both reagents were found to be more sensitive to tannic acid than to catechin. Although the tannin values varied two- to threefold for a given variety, an excellent correlation was observed between the different assays employed for bean-tannin analysis (Table 4.15).

Table 4.14: Tannin determination by vanillin assay and protein-precipitation method in dry beans.

Variety	0.5% Vanillin assay (mg catechin equ./100g)	Protein precipitable phenols (mg tannic acid equ./100g)
Sanilac	ND	ND
Small white	ND	ND
Great Northern	ND	ND
Cranberry	82	53
Viva Pink	160	92
Pinto	306	218
Light Red Kidney	172	155
Dark Red Kidney	116	124
Small Red	257	187
Black Beauty	49	8
LSD	7.14	10.9

Two major factors preventing meaningful comparisons between laboratories using similar methods appear to be the solvent used for tannin extraction and the choice of the reference standard. Extraction in acidified methanol consistently yielded higher values, and unless corrected for the sample blanks as suggested by Price *et al.* (1978) may result in erroneous estimations of tannins in dry grains. This may also be a major contributing factor to the controversy over the possible relationship between seed color and tannin content assayed by the uncorrected vanillin assay first described by Bums. Since the redox assays do not only detect all phenols, but are also influenced by several interfering compounds, the choice of a standard solvent. such as absolute methanol, and the use of a corrected vanillin-HCl assay, in conjunction with a protein-precipitation assay using the purified natural tannin of respective source as a standard, may yield more meaningful data on tannins in food grains, fruits, wine, and forages.

Table 4.15: Simple correlation coefficients between different assays used for dry bean tannins

Assay	Folin-Ciocalteu	Prussian Blue	Protein precipitable phenols
Vanillin	0.92	0.87	0.94
Folin-Ciocalteu	-	0.95	0.79
Prussian Blue	-	-	0.67

CHEMICAL ASSAYS FOR ANTIOXIDANT ACTIVITY

Radical/ROS scavenging methods

A large group of assays are available for direct measurement of hydrogen atom or electron transfer from potential antioxidants to free radicals in simple "lipid free" systems. The antioxidant activities reported by this group of methods are generally associated with their scavenging capacity against specific types of radical species, some of which may be artificial and biologically irrelevant. Therefore, they have been criticized for not reflecting the situation in an oxidizing food or an *in vivo* situation. However, the hydrogen atom or electron donating capacity data obtained by these methods provide important information on their intrinsic antioxidant potential with minimum environmental interference. These assays require no lipid substrate and usually employ a chemical system containing an oxidant (free radicals or other ROS), an oxidizable probe (not necessary for some assays), and antioxidants under investigation. Depending on the chemical reactions involved, these assays fall into two categories of hydrogen atom transfer (HAT) reaction based assays and single electron transfer (ET) reaction-based assays. Antioxidants can scavenge radicals or other ROS (e.g. hydrogen peroxide and lipid peroxides) by HAT and ET leading to the same end results, regardless of the mechanism involved, although kinetics and potential for side reactions vary (Prior, Wu and Schaich, 2005). Proton-coupled ET and HAT reactions may also occur in parallel, and in this case the dominating mechanism in a given system is dependent on antioxidant structure and properties, solubility and partition coefficient, as well as the solvent system (Prior et al., 2005). Antioxidant activities are either expressed as inhibition against ROS-mediated oxidation of the probe, or equivalents of a selected reference antioxidant such as trolox, ascorbic acid or another compound. Oxidation of the probe is measurable by various detection technologies, which are advancing with availability of up-to-date instrumentation.

These include spectrophotometric, fluorometric, chemiluminescent, EPR (electron paramagnetic resonance), FT-IR (Fourier transform-infrared), NMR (nuclear magnetic resonance), and amperometric methods, among others. Chemical assays (summarized in Table 4.16) for antioxidant activity measurement are simple, quick, and usually automated, and are widely used in initial screening and evaluation of novel antioxidant compounds or extracts of natural products/by-products.

Table 4.16: Major reactive oxygen species (ROS) scavenging- and redox potential-based assays.

Name of assay	Dominating mechanism	Oxidant	Probe	Detection
ROS scavenging assays				
ORAC	HAT	Peroxyl radical generated by AAPH	Fluorescein	Fluorometry
Chemiluminescence	HAT	Hydrogen peroxide	Luminol	Fluorometry
DPPH scavenging	ET	DPPH radical	DPPH radical	Spectrophotometry or EPR
TEAC		ABTS radical cation	ABTS radical cation	Spectrophotometry
Redox potential assays				
FRAP	ET	Fe^{3+}	Ferricyanide	Spectrophotometry
CUPRAC	ET	Cu^{2+}	Neocuproine	Spectrophotometry
Ag^+ reducing	ET	Ag^+	Ag nanoparticle	Surface plasmon resonance
Au^{3+} reducing	ET	Au^{3+}	Au nanoparticle	Cyclic voltammetry
CERAC	ET	Ce^{4+}	Indigo carmine dye	Spectrophotometry
CHROMAC	ET	Cr^{6+}	Cr3+ complex	Spectrophotometry

AAPH, 2, 2′-azobis (2-amidinopropane) dihydrochloride; DPPH, 2,2-Diphenyl-1-picrylhydrazyl; ABTS, 2,2′-azinobis(3-ethylbenzothiazoline-6-sulphonic acid.

OXYGEN RADICAL ABSORBANCE CAPACITY (ORAC) ASSAY

The ORAC assay measures the radical chain breaking ability of antioxidants by monitoring the inhibition of peroxyl radical induced oxidation. Peroxyl radicals are the predominant free radicals found in lipid oxidation in foods and biological systems under physiological conditions. Hence, ORAC values are considered by some to be of biological relevance as a reference for antioxidant effectiveness. In this assay, the peroxyl radical produced by a generator reacts with a fluorescent probe resulting in the loss of fluorescence, which is recorded with a fluorometer. A set of fluorescence decay curves can be constructed in the absence or presence of antioxidants and the net integrated area under the decay curves (area gain in the presence of antioxidants compared to that of a blank run without antioxidants) can be calculated as an indicator of the peroxyl radical scavenging capacity of the antioxidants. A standard antioxidant, usually trolox, is used as reference and ORAC values of the tested antioxidants are reported as trolox equivalents. The ORAC method measures hydrogen atom donating ability of antioxidants, and is therefore an HAT-based method. The assay is readily automated using a multichannel liquid handling system coupled with a fluorescence microplate reader. Commonly used peroxyl radical generators in ORAC assays include a group of azo compounds such as the lipophilic AIBN (α,α-azobisisobutyronitrile), ABAP (2,2-azobis(2-amidinopropane) hydrochloride) and AMVN (2,2′-azobis(2,4- imethylnaleronitrile)) and the hydrophilic AAPH (2,2′-azobis(2-amidinopropane) dihydrochloride) (Becker, Nissen, and Skibsted, 2004). Reaction of peroxyl radical generation is simplified as given below:

$$R{-}N{=}N{-}R \xrightarrow{O2} N_2 + 2ROO^{\bullet}$$

The azo compound AAPH, for example, undergoes thermal decomposition in the presence of oxygen to yield peroxyl radicals as oxidants that further react with the probe and antioxidant. Generation of peroxyl radical is sensitive to temperature, and has been reported to be one of the major contributing factors that interferes with the results and inconsistency of the assay (Prior et al., 2005). It is therefore very important to control and monitor the temperature during the assay. However, this issue has been well mitigated by development of highly automated fluorescence plate readers equipped with an incubator. Incubation of the reaction buffer at 37 °C prior to the dissolution of AAPH is recommended to reduce the intra-assay variability (Prior et al., 2005). A number of fluorescent materials has been identified and used as probes for ORAC assay. B-phycoerythrin (B-PE), a protein isolated from *Porphyridium cruentum*, was the fluorescent probe initially employed in the early ORAC studies until it was found to suffer from several disadvantages. Being a natural component derived from red algae of various species, it shows large lot-to-lot variability in reactivity with peroxyl radicals. In addition, its nonspecific binding to polyphenol antioxidants and photobleaching by excitation light can all introduce interference to the reaction and hence false interpretation of antioxidant activity (Huang, Ou, and Prior, 2005; Prior et al., 2005). Alternative synthetic non-protein probes have been identified as replacement of B-PE, among which fluorescein is the most frequently selected fluorescent probe for ORAC assay in decades. However, fluorescein has been demonstrated to undergo undesired fluorescence quenching and side reactions (Apak et al., 2013), and novel fluorescent probes have been proposed. A long wavelength fluorophore Nile blue was used as alternative probe to determine ORAC values of fruit juices and wines, which were in good agreement with results obtained by fluorescein method (Godoy-Navajas, Aguilar-Caballos, and Gomez-Hens, 2011). Barahona, Chandia, Encinas, Matsuhiro, and Zuniga (2011) compared ORAC values measured by using c- phycocyanin and fluorescein probes for soluble polysaccharides from seaweeds. Even though use of c-phycocyanin probe resulted in higher ORAC values than fluorescein in general, the relative ORAC values and trend among different samples were similar for both probes. Ortiz et al. (2012) investigated the use of pyrogallol red probe for ORAC assay, which showed good linearity, precision and accuracy. Guclu, Kibrislioglu, Ozyurek, and Apak (2014) reported ORAC values of liver and kidney samples measured using *p*-aminobenzoic acid (PABA) as the fluorescent probe. The authors suggested replacement of fluorescein with PABA in ORAC assay for amino acids, plasma and thiol-type antioxidants as well as albumin, as inconsistent results have been reported for thiols such as glutathione and cysteine.

Although itwas originally developed for measurement of hydrophilic antioxidants, ORAC assay has also been adapted to detect both lipophilic and hydrophilic antioxidants by altering the radical source and solvent. Huang, Ou, Hampsch-Woodill, Flanagan, and Deemer (2002) developed and validated a modified ORAC assay for lipophilic antioxidants using randomly methylated β-cyclodextrin (RMCD) as solubility enhancer. RMCD at 7% in an acetone/water (50:50, v/v) mixture was suggested to efficiently solubilize tocopherols and other lipophilic phenolic antioxidants in phosphate buffer. This modified ORAC assay, using cyclodextrin as solubilizer, has been successfully employed for measuring ORAC values of tocopherols, tocotrienols, γ-oryzanol, resveratrol, lipophilized tea catechins, and various food and biological samples (Huang et al., 2002; Lucas-Abellan, Mercader-Ros, Zafrilla, Gabaldon, and Nunez-Delicado, 2011; Prior et al., 2003; Zhong, Ma, and Shahidi, 2012). It has been argued by Nkhili and Brat (2011) that further modification of ORAC method was needed to eliminate the effect of metal ions in the test systems on measured ORAC values of antioxidants. Antioxidants can form complexes with metal ions, leading to underestimated peroxyl radical scavenging capacity.The metal chelator, ethylenediaminetetraacetic acid (EDTA) can be helpful to mitigate the interference, and significantly increased measured ORAC values were reported (Nkhili and Brat, 2011). Other similar HAT-based methods sharing the same principle and common features of ORAC assay include total radicaltrapping antioxidant parameter (TRAP) and crocin bleaching assays (Shahidi and Zhong, 2007). Similar to ORAC, TRAP and crocin bleaching consist of an azo compound as peroxyl radical generator (e.g. AAPH or ABAP), a molecular fluorescence (e.g.

R-phycoerythrin or dichlorofluorescin diacetate) or UV–vis probe (crocin), and an antioxidant. They both collect kinetic curves of fluorescence or absorbance decay for quantification of antioxidant activity, and express results as trolox equivalents. Like the ORAC assay, they are all readily adapted to high-throughput automated methodologies using microplate reader. However, all these three assays suffer from one limitation, i.e. naturally existing pigments and fluorophores in antioxidant extracts that may interfere with absorbance and fluorescence readings. Despite its demonstrated usefulness in the evaluation of antioxidant activity, USDA website removed information relevant to the listing of foods based on their ORAC values. The reason given was that they did not realistically reflect the *in vivo* situation (USDA, 2010).

Chemiluminescence assay

In addition to peroxyl radicals, several other ROS contribute to oxidative stress of food and biological systems and are targets of antioxidant action. Singlet oxygen ($^1O_2$), superoxide radical anion ($O_2^{\bullet -}$), hydroxyl radical ($^{\bullet}OH$) and hydrogen peroxide are highly reactive oxygen-derived species that are short-lived and difficult to evaluate. Highly sensitive and reproducible methods have been developed for measurement of antioxidant activity in quenching these ROS, such as the chemiluminescence assay for hydrogen peroxide quenching capacity. Chemiluminescence is a phenomenon observed when a chemical reaction produces an electronically excited species that either luminesces (direct chemiluminescence) or transfers the energy to a fluorophore that generates emission (indirect chemiluminescence). A majority of chemiluminescence assays for antioxidant activity is direct chemiluminescence-based and generally consists of a chemiluminescent species, an oxidant (hydrogen peroxide) in the presence or absence of a metal or enzymatic catalyst, and an antioxidant or extract.A variety of chemiluminescent reagents such as luminol, lucigenin, oxalates, pyrogallol or Cypridina luciferin, among others have been proposed, with luminol being the most commonly used aqueous chemiluminescent reagent (Bezzi, Loupassaki, Petrakis, Kefalas, and Calokerinos, 2008; Papadopoulos, Triantis, Yannakopoulou, Nikokavoura, and Dimotikali, 2003; Stepanyan, Arnous, Petrakis, Kefalas, and Calokerinos, 2005). Luminol reacts with the potent oxidizing agent hydrogen peroxide to yield 3-aminophthalate in an excited electronic state, which is a light-producing emitter. Antioxidants can quench hydrogen peroxide by hydrogen atom donation and inhibit hydrogen peroxide-induced chemiluminescence. Therefore, this assay is based on the competition between antioxidants and luminol for hydrogen peroxide and presence of antioxidants results in a decrease in light emission intensity.The antioxidant of interest is tested over a range of concentrations, and antioxidant activity is usually expressed as EC50, i.e. the effective concentration that reduces chemiluminescence by 50%. Although hydrogen peroxide is the oxidizing agent used in chemiluminescence assays, it is not the only form of ROS reacting with luminol and being quenched by antioxidants. In many cases, a catalyst is added to magnify the chemiluminescence signal, by converting the relatively weak oxidant hydrogen peroxide into stronger ROS superoxide radical anion and hydroxyl radical. A metal (e.g. iron, copper, manganese, and cobalt) containing compound or enzyme can catalyse production of hydroxyl radical and superoxide radical anion from hydrogen peroxide through Fenton reaction. Even though adding catalyst can significantly increase chemiluminescence signal of luminol and linearity of the assay, it may complicate the interpretation of the results by introducing multiple ROS due to preference of antioxidants to scavenge particular ROS (Giokas, Vlessidis, and Evmiridis, 2007; Sariahmetoglu, Wheatley, Cakici, Kanzik, and Townshend, 2003). Given this concern, the hydrogen peroxide–metal catalyst–chemiluminescent reagent is still the most frequently used system for this assay. Oxidation and chemiluminescence of luminol may also be enhanced by using enzymes such as horseradish peroxidase and sensitizers such as *p*-iodophenol (Minioti and Georgiou, 2008) as in enhanced chemiluminescence. However, limitations with luminol-based antioxidant assays have also been reported. Luminol molecule is negatively charged and its signal may be quenched by non-antioxidant quenchers in the sample, leading to overestimation of antioxidant activity (Kruglov et al., 2010). The excited luminol molecule may also be quenched by antioxidants by a non-antioxidant mechanism, which interferes with the interpretation of their ROS-scavenging capacity (Zargoosh, Ghayeb, Aeineh, and Qandalee, 2014). Different sources of chemiluminescent species have been explored for their potential use in the evaluation of antioxidants of varying hydrophilicity (or lipophilicity). For example, lucigenin–hydrogen peroxide chemiluminescence in 2-propanol has been developed to evaluate the activity of both hydrophilic and lipophilic antioxidants (Christodouleas, Fotakis, Papadopoulos, Yannakopoulou, and Calokerinos, 2009). While alternative chemiluminescent species have been explored, indirect chemiluminescence methods have also become an option for antioxidant assessment. Zargoosh et al. (2014) reported peroxyoxalate–hydrogen peroxide–imidazol–fluorophore system for evaluation of antioxidants in olive oils and honey samples. A furan dicarboxylate derivative was used as the fluorophore.The mechanism is believed to involve highly energetic intermediates from peroxylate–hydrogen peroxide reaction that can transfer the energy and excite the fluorophore. A similar method was proposed by Nikokavoura, Christodouleas, Yannakopoulou, Papadopoulos, and Calokerinos (2011) who developed a lucigenin–hydrogen peroxide–fluorophore system for assessing various hydrophilic and lipophilic antioxidants and extracts.The fluorophore used in this method is *N*-methylacridone, which is emitted as a result of the lucigenin–hydrogen peroxide reaction. Chemiluminescence method has been used as a sensitive, low-cost and robust tool for assessing antioxidant activity, particularly quenching capacity of hydrogen peroxide-derived ROS. More recently, it has been automated in the forms of flow based assays, including flow-injection analysis (FIA), sequentialinjection analysis (SIA), multisyringe FIA (MS-FIA) and multicommutation (Fassoula, Economou, and Calokerinos, 2011). These automated chemiluminescence assays facilitate fast screening of antioxidant activity in food and natural health products.

2, 2-Diphenyl-1-picrylhydrazyl (DPPH) radical scavenging assay

DPPH radical scavenging assay is among the most frequently used methods and offers the first approach for evaluating antioxidant activity. It is an ET-based method with HAT mechanism being only a marginal reaction pathway in the assay (Prior et al., 2005). DPPH is a stable chromogen radical with a deep purple colour. It is commercially available and does not need to be generated prior to the assay. The DPPH scavenging assay is based on electron donation of antioxidants to neutralize DPPH radical. The reaction is accompanied with colour change of the DPPH measured at 517 nm, and the discolouration acts as an indicator of the antioxidant efficacy. The antioxidant activity by DPPH scavenging method is often reported as EC50 which is defined as the effective concentration of the antioxidant necessary to decrease the initial DPPH concentration by 50%. In addition, TEC50 which is the time needed to reach the steady state with EC50 may be used. Occasionally, antiradical efficiency (AE) is reported which combines EC50 and TEC50 into one parameter according to the equation AE = (1/EC50) TEC50. The DPPH assay is a simple technique and requires only a UV spectrophotometer or an EPR (electron paramagnetic resonance) spectrometer. However, it has been argued that the scavenging of DPPH does not mimic the radical scavenging mechanism of antioxidants in real food or biological systems due to lack of oxygen radical in the assay (Benzie and Strain, 1999). Hence, this method is largely based on the assumption that antioxidant activity is equal to its electron donating capacity or so-called reducing power. Plank et al. (2012) conducted a collaborative study to determine antioxidant activity of a variety of foods and beverages using DPPH assay and validated the method for certain types of foods but challenged its suitability for others. Even though the DPPH assay is simple and does not require special sample treatment, its sensitivity may be affected by a number of factors, such as the type and amount of solvent used, presence and concentration of hydrogen and metal ion and freshness of DPPH reagent (Dawidowicz,Wianowska, and Olszowy, 2012; Kedare and Singh, 2011; Musa, Abdullah, Kuswandi, and Hidayat, 2013). Modifications to the assay have been proposed by different authors in an attempt to minimize the aforementioned problems and further simplify and automate the method (Kedare and Singh, 2011). Musa et al. (2013) investigated the use of DPPH dry agent as an alternative to freshly prepare DPPH in an automated microplate-based assay. The modified method was validated with the classic DPPH assay, and according to the authors, it showed excellent sensitivity while being more convenient and simple to operate with minimal solvent required.

One major limitation of the spectrophotometric DPPH assay is the overlapped spectra of compounds that absorb in the same wavelength range as DPPH.For example,anthocyanins have strong absorption over thesamewavelength range (500–550 nm)as DPPH, and thus may introduce interference with the results and their interpretation. Thus, EPR spectroscopy which directly measures the concentration of DPPH radical at submicromolecular levels may be employed.The EPR detection has advantages over the classic spectrophotometric detection, not only for highly coloured samples, but also for samples appearing turbid in the chosen solvent (Gardner, McPhail, and Duthie, 1998). The DPPH radical, having an unpaired electron, generates a characteristic and distinguishable EPR spectrum in amagnetic field under microwave radiation.The peak intensity in the EPR spectrum,which is proportional to the concentration of DPPH, decays with time in the presence of antioxidants as a result of DPPH radical scavenging by the antioxidants.The rate of DPPH signal decay or signal intensity drop at an end pointmay serve as an indicator of scavenging capacity of the antioxidant against DPPH radical (Zhong and Shahidi, 2011). In addition to EPR spectroscopy, scavenging of DPPH radical can be monitored by amperometric detection. At an applied voltage, DPPH radical generates a constant electric current, and the decreased amount of radicals as a result of scavenging by antioxidants leads to a drop of the amperometric signal.Thus, quantification of antioxidants is accomplished via amperometric detection of the residual concentration of unreacted DPPH radical. Amatatongchai, Laosing, Chailapakul, and Nacapricha (2012) developed an amperometric flow injection (FI) method using a carbon nanotube modified-glassy carbon electrode as the working electrode to detect the DPPH radical continuously flowing through, while antioxidant samples are injected into the carrier stream. The authors demonstrated good linearity and precision of this method for evaluation of DPPH scavenging capacity of numerous antioxidants including gallic acid, catechin, quercetin, caffeic acid and trolox. Sireerat and Schulte (2012) determined the antioxidant activity of tea infusions, fruit juices and vegetable extracts using an automated amperometric DPPH assay, in which antioxidant capacity measurementswere achieved using DPPH radical as an amperometric redox indicator, with a pencil Pb working electrode, in conjunction with a Pt counter-electrode and Ag/AgCl reference electrode. This method was validated by synthetic antioxidants of known concentrations.

DPPH radical scavenging assay has also been used in combination with other techniques for particular purposes. DPPH assay can be coupled with online HPLC for rapid screening of a large number of antioxidant samples which reduces the loss of antioxidants during purification processes. Online HPLCDPPH methods are based on post-column reaction of antioxidant eluates with DPPH radicals. Crude antioxidant samples are first subjected to HPLC (normally reversed-phase) separation, and the eluate corresponding to each peak is reacted with DPPH radical at a certain flow rate for a constant time period (e.g. 60 seconds), resulting in loss of DPPH radical and the antioxidant itself, which can be recorded by UV detector (Kedare and Singh, 2011). Some methods monitor decrease of DPPH absorbance at 517 nm (Zhang et al., 2013), while others monitor decrease of antioxidant peak area (Qiu et al., 2012) after the postcolumn reaction. HPLC-DPPH methods can be further coupled with identification of the components showing scavenging capacity towards DPPH radical. Nuengchamnong et al. (2005) combined separation, antioxidant activity determination and structural identification of complex antioxidant mixtures in one assay, in which the eluates after column separation are split into two flows with the major portion going through the ESI/ MS/MS analysis and the minor portion reacting with DPPH radical for antioxidant activity evaluation. Qiu et al. (2012) screened and identified antioxidant components in peanut shell using DPPH-HPLC-DAD-TOF/MS method.These methods are rapid and efficient as a means of antioxidant screening; however, they require sophisticated instruments which may not be readily available in every laboratory. More recently, the DPPH assay has been re-evaluated by Xie and Schaich (2014) for its suitability for antioxidant activity assessment through kinetic and stoichiometric approaches. The authors questioned the application of DPPH assay for ranking antioxidants, and suggested its use to distinguish the active reaction mechanism of an antioxidant, namely ET and HAT, which exhibit different reaction rates and patterns in different solvents.

Trolox equivalent antioxidant capacity (TEAC) assay

TEAC assay was first developed by Miller et al. (1993) as a simple and convenient method for total antioxidant capacity (TAC) determination. The assay measures the ability of antioxidants to scavenge the stable radical cation $ABTS^{\bullet+}$ (2,2′-azinobis(3-ethylbenzothiazoline-6-sulphonic acid)), a blue-green chromophore with maximum absorption at 734 nm which decreases in its intensity in the presence of antioxidants. $ABTS^{\bullet+}$ can be generated from ABTS in the presence of strong oxidizing agents.Antioxidants can neutralize the radical cation$ABTS^{\bullet+}$ by either direct reduction via electron donation or by radical quenching via hydrogen atom donation,and the balance of these twomechanisms is generally determined by antioxidant structure and pH of the medium (Prior et al., 2005). Hence, although TEAC assay is usually classified as an ET-based method, the HAT mechanism also applies. The extent of discolouration of the blue-green colour, quantified as a decrease in absorbance at 734 nm, depends on the duration of reaction, the intrinsic antioxidant activity and concentration in the sample. Similar to other radical scavenging assays discussed above,TEAC assay can either monitor the absorbance decay with time, or the drop at a certain end point. When operated as an end-point assay, TEAC method requires reliable determination of end point of the reaction for quantitative evaluation of antioxidant activity (Huang et al., 2005). Results obtained by TEAC assay are reported as trolox equivalents. In the original TEAC assay, metmyoglobin and hydrogen peroxide are used to generate an intermediate ferrylmyoglobin radical, which then reacts with ABTS to produce $ABTS^{\bullet+}$. The oxidizing agent was later replaced by peroxide or persulphate. Potassium persulphate is the most commonly used oxidant for $ABTS^{\bullet+}$ generation. Modifications of the original assay have been proposed in many aspects, including the methods for generation of $ABTS^{\bullet+}$, wavelengths for monitoring the reaction, and the means of quantification, among others (Prior et al., 2005). Durmaz (2012) prepared a ready-to-use radical powder of $ABTS^{\bullet+}$ by incubating ABTS with potassium peroxodisulphate followed by freeze-drying of the resultant radical solution. This freeze-dried radical powder was then used as a replacement for the freshly prepared radical solution and its linearity and accuracy of the modified assay was demonstrated by the authors. In another version of the assay, the radical $ABTS^{\bullet-}$ anion was used as the target radical instead of the cation. $ABTS^{\bullet-}$ was generated via oxidation of $ABTS2^{-}$ by potassium persulphate, and the radical $ABTS^{\bullet-}$ anion was reduced back by antioxidants to the colourless $ABTS2^{-}$ product (Huang et al., 2005). The TEAC assay has been used to measure the total antioxidant activity of pure substances, body fluids and plant materials. The TEAC assay, similar to other radical scavenging methods, can be automated and adapted to microplates and flow injection techniques (Milardovic, Kerekovic, and Rumenjak, 2007). It may also be coupled with HPLC by including a post column reaction with ABTS radical cation to facilitate the search of individual antioxidants from a complex mixture. HPLC-TEAC provides an efficient and rapid method for separation and identification of bioactive compounds fromthe source material (Gong, Plander, Xu, Simandi, and Gao, 2011; Raudonis, Bumblauskiene, Jakstas, Pukalskas, and Janulis, 2010). Better sensitivity and precision of TEAC assay can be achieved when coupled with other detection techniques, such as amperometric (Chan-Eam, Teerasong, Damwan, Nacapricha and Chaisuksant, 2011; Milardovic et al., 2007) and FTIR (Cerretani et al., 2010), among others. Many of these modified TEAC assays employ online enzymatic generation of$ABTS^{\bullet+}$, especially those using continuous flowsystems. For example, Milardovic et al. (2007) reported the generation of $ABTS^{\bullet+}$ by glucose oxidase and peroxidase separately immobilized in tubular flow-through reactors for flowinjection analysis of alcoholic beverages for their TEAC values. The TEAC assay is simple and can be performed as a routine analysis in almost any laboratory without the need for sophisticated equipments. $ABTS^{\bullet+}$ can be solubilized in both hydrophilic and lipophilic media and is not affected by ionic strength of the medium (Arnao, 2000). However, limitations of the TEAC assay have been noted, including the effect of incubation time, antioxidant/$ABTS^{\bullet+}$ ratio and end-point determination on TEAC values (Roginsky and Liss, 2005). The limitation of the applicability of TEAC assay to rank antioxidant activity was recently reported due to its failure to evaluate radical scavenging rate (Tian and Schaich, 2013). Moreover, like the DPPH scavenging assay, the TEAC assay has also been challenged for its lack of biological relevance due to use of the artificial ABTS radical cation that is not found in food or biological systems. Nevertheless, the usefulness of these assays for ranking of antioxidant activity of similar substrates under comparable test conditions cannot be ignored.

Non-radical redox potential-based methods

Antioxidants are also known as reductants. Their electron donating capability can not only allow scavenging of free radicals and other oxygen derived ROS, but also reduce higher valent elements to their lower valence state. The redox potential or reducing power of antioxidants is an important indicator of their antioxidant efficacy, and is measured through redox reaction with various metal ions, such as iron, copper, chromium, and cerium, among others. This group of methods can be classified as ET-based methods and are conducted in simple chemical assays of less biological relevance, compared to ROS scavenging measurement.

Ferric reducing antioxidant power (FRAP) assay

The FRAP assay is a typical ET-based method that measures the reduction of ferric ion (Fe^{3+})–ligand complex to the intensely blue coloured ferrous (Fe^{2+}) complex by antioxidants in acidic media. Antioxidant activity is determined as increase of absorbance at 593 nm, and results are expressed as micromolar Fe^{2+} equivalents or relative to an antioxidant standard (Antolovich et al., 2002). Unlike other ET-based methods, FRAP assay is carried out under acidic pH conditions (pH 3.6) in order to maintain iron solubility and more importantly drive electron transfer. This will increase the redox potential, causing a shift in .dominant reactionmechanism (Hegerman et al., 1998). The original FRAP assay uses tripyridyltriazine (TPTZ) as the iron-binding ligand, while alternative ligands have also been employed for ferric binding, such as ferrozine for ascorbic acid reducing power evaluation (Molina-Diaz, Ortega-Carmona, and Pascual-Reguera, 1998). More recently, potassium ferricyanide has been the most popular ferric reagent used in FRAP assays. In the latter case, Prussian blue is produced as the end product which is quantified spectrophotometrically and indicates the reducing power of the antioxidants tested. Production of Prussian blue may be through two different routes with the same outcome. Antioxidants can either reduce the Fe^{3+} in the solution to Fe^{2+}, which binds the ferricyanide to yield Prussian blue, or reduce the ferricyanide to ferrocyanide, which binds the free Fe^{3+} in the solution and forms Prussian blue. The simplified scheme for these two reactions is given blow (Berker, Gueclue, Tor, Demirata, and Apak, 2010).

$$\text{Antioxidant} + Fe^{3+} \rightleftarrows Fe^{2+} + \text{oxidized antioxidant}$$

$$Fe^{2} + Fe\,(CN)_6^{3-} \rightleftarrows Fe\,[Fe\,(CN)_6]^{-} \text{ or}$$

$$\text{Antioxidants} + Fe\,(CN)6^{3-} \rightleftarrows Fe(CN)_4^{4-} + \text{oxidized antioxidant}$$

$$Fe\,(CN)_6^{4-} + Fe^{3+}\, Fe[Fe(CN)_6]^{-}$$

One drawback of this ferricyanide FRAP assay is the tendency of the Prussian blue to precipitate, forming a suspension and staining the measuring cuvette. Therefore, time to add Fe^{3+} ($FeCl_3$) is essential and may introduce error to result interpretation. In order to stabilize the Prussian blue against precipitation, Berker et al. (2010) proposed addition of the surfactant sodium dodecyl sulphate, while adjusting the optimal pH to 1.7 to maintain the redox activity of ferric ion without hydrolysis. The authors suggested that this modification also allows evaluation of antioxidants whose redox potential do not exceed that of the Fe^{3+}/Fe^{2+} in conventional FRAP assay (i.e. ferric reagent two weak to oxidize the antioxidants), such as thiol type physiological nonenzymatic antioxidants.The FRAP assay was further modified by choosing water/acetone as a solvent in the absence of solubility enhancer RMCD to enable simultaneous measurement of both hydrophilic and lipophilic antioxidants, the latter of which was restricted by conventional FRAP assay (Berker, Demirata, and Apak, 2012). The FRAP assay is simple, fast and cost-effective and does not require specialized equipment. It was originally employed to measure reducing power in plasma, but its use has been extended for assessing antioxidant activity in other biological fluids, foods, and plant extracts. However, Pulido, Bravo, and Saura-Calixto (2000) reported that FRAP results might vary depending on the analysis time as observed for the reaction between antioxidants and Fe^{3+}, which ranged from several minutes to several hours.Therefore, a single-point absorption endpointmay not represent a complete reaction, since different antioxidants require different reaction times for detection (Prior et al., 2005). More recently, the FRAP assay has adopted electrochemical detection techniques for better sensitivity, accuracy and reproducibility. A coulometric titration method has been developed to determine the FRAP value of different antioxidant materials (Abdullin, Turova, Budnikov, Ziyatdinova, and Gajsina, 2002; Ziyatdinova, Nizamova, and Budnikov, 2011; Ziyatdinova Salikhova, and Budnikov, 2014). In this modified FRAP method, antioxidant reacts with coulometric titrants (electrogenerated ferricyanide ions), and the quantity of electricity consumed for the titration till the end point (when initial current is resumed) serves as indicator for reducing power of the antioxidant (Ziyatdinova et al., 2014). The FRAP assay by coulometric titration has been attractive as a highly sensitive, reliable and simple test for reducing power evaluation. Another electrochemical version of the FRAP assay is the chronoamperometric measurement of antioxidant reducing power. As demonstrated by Brainina, Varzakova, and Gerasimova (2012), a shift of potential in the K_3 [Fe $(CN)_6]$ $/K_4[Fe(CN)_6]$ mediator system is indirectly correlated with the reducing power of the antioxidant present. More specifically, at certain potentials (higher than 0.35 V), oxidation process takes place only forming K_3 $[Fe(CN)_6]$, and concentration of $K_4[Fe(CN)_6]$ is entirely from the reaction of $K_3[Fe(CN)_6]$ with antioxidant and is proportional to the reducing power of the antioxidant (Brainina et al., 2012). This chronoamperometric method has been proposed for assessing antioxidant activity of biological samples owing to its very low detection limit (e.g. $2 \times 10-6$ M for ascorbic acid). The FRAP assay, in general, as an ET-based non-radical method, has been argued to have little relationship with radical quenching process (HAT mechanism) occurring in lipid systems, and has poor correlation with other antioxidant activity measurements. It is therefore suggested that this assay could be used in combination with other methods in distinguishing dominant mechanisms for different antioxidants (Prior et al., 2005).

Cupric reducing antioxidant capacity (CUPRAC) assay

The CUPRAC assay is a copper reduction assay developed as a variant of the FRAP assay. It uses copper as the oxidant instead of iron in the FRAP assay. The method measures the reducing power of antioxidants to convert cupric (Cu_2^+) to cuprous (Cu^+) ion. Similar to the FRAP assay, a ligand is employed to form a copper–ligand complex to facilitate absorbance measurement. Neocuproine (Nc; 2,9-dimethyl-1,10-phenanthroline) is the ligand commonly employed in CUPRAC assay. The Cu_2^{+-} neocuproine complex can be reduced by antioxidants to Cu^{+-} neocuproine, which is a chromophore with maximum absorption at 450 nm (Apak, Güçlü, Özyürek, and Karademir, 2004). The original CUPRAC assay has been modified in order to accommodate different samples in diverse applications. For example, acetone/water medium with the aid of methyl-β-cyclodextrin has been employed for simultaneous determination of hydrophilic and lipophilic antioxidants (O″ zyürek, Bektas¸ogˇgˇlu, Güçlü, Güngör, and Apak, 2008). In order to evaluate antioxidant phenolics in both free and bound forms in the food matrix without preliminary extraction and the time and labour intensive hydrolysis process, a forced solubilization of bound antioxidants by the CUPRAC reagent was proposed, which takes advantage of the surface reaction phenomenon between solid (bound antioxidants) and liquid (soluble CUPRAC reagent) material (Tufan, Celik,O″ zyürek, Güçlü, and Apak, 2013).This modified method was suggested to be applicable to relatively insoluble food matrices as well as to insoluble cosmetic products, such as creams, balms, and powders. In another modified CUPRAC assay, an optical sensor containing immobilized chromogenic redox reagent was adopted for measuring the reducing power of liquid samples without any pretreatment (Bener, O″ zyürek, Güçlü, and Apak, 2010). The CUPRAC reagent Cu^{2+}–neocuproine complex was immobilized onto a cation-exchanger film of Nafion, and the complex was reduced by antioxidants to Cu^{+-} neocuproine, giving rise to changes of absorbance at 450 nm, which was monitored spectrophotometrically. Another version of optical sensor aided CUPRAC assay used a miniature reflectance spectrometer to measure reflectance changes at 530 nm instead of absorbance (Bener, O″ zyürek, Güçlü, and Apak, 2013). Use of optical sensor was suggested to significantly simplify the operation, exhibit reliability and robustness, and enable the *in situ* estimation of antioxidant capacity of various food extracts and biological samples.

OTHER REDUCING POWER ASSAYS

Metal ions other than those derived from iron and copper have been used as target of redox reaction of antioxidants. Such sources of metal ions include silver, gold, cerium, and chromium, among others. Reduction of Ag^+ to spherical Ag nanoparticles by antioxidants in the presence of trisodium citrate and Ag seeds produces a very intense surface plasmon resonance (SPR) absorption band of Ag nanoparticles at 423 nm, which allows quantitative detection of antioxidants (O″ zyürek, Güngör, Baki, Güçlü, and Apak, 2012). A similar method was developed utilizing resonance light scattering of gold nanoparticles, a reduction product from the reaction of Au^{3+} with antioxidants in the presence of cetyltrimethylammonium bromide (Andreu-Navarro, Fernandez-Romero, and Gomez-Hens, 2011). Detection of gold nanoparticles can also be achieved by an electrochemical method, such as cyclic voltammetric detection of enlarged gold nanoparticle on electrode surface, as a result of tetrachloroauric acid reduction by antioxidants (Wang, Zhou, Zhu, Huang, and Li, 2007). Reduction of Ce4+ to Ce3+ by antioxidants can be employed for reducing power evaluation, as in the CERAC (cerium reducing antioxidant capacity) assay. In this assay, the remaining unreacted Ce^{4+} (sulphate salt) is measured spectrophotometrically at 320 nm, which is inversely related to the reducing power of the antioxidant under investigation (Ozyurt, Demirata, and Apak, 2010). In another cerium-based assay, the unreacted Ce^{4+} reacts with a known amount of chromogenic probe indigo carmine dye followed by measuring the unreacted indigo carmine dye solution at 610 nm (Padmarajaiah et al., 2012). In the recently developed CHROMAC (chromium reducing antioxidant capacity) assay, formation of Cr^{3+} complex from reduction of Cr^{6+} complex by antioxidants gave rise to absorbance increase at 540 nm (Isik, Sahin, and Demir, 2013). Quaranta et al. (2013) proposed an enzyme-based method for evaluating reducing power of antioxidants. This method does not involve metal ions, but rather indirectly measures the reduction of azobismethylene quinone to syringaldazine by antioxidants.The reduction product syringaldazine is reoxidized by laccase accompanied by oxygen consumption, which is measured by an optical oxygen sensor. This assay was coupled with HPLC for online monitoring and identification of antioxidant components with distinguished reducing power. The variety of reducing power assays allows selective evaluation of a broad range of antioxidants with varying redox potential while minimizing interference from reductants out of the redox potential range. In other words, it is important to select the proper metal or non-metal oxidizing agent that is strong enough to oxidize the antioxidants of interest but not too strong to oxidize other reductants such as citric acid and reducing sugars in the food matrix.

CYCLIC VOLTAMMETRY

Electrochemical methods for evaluating antioxidant activity have emerged in the past decade, among which cyclic voltammetry has attracted much attention as an alternative method to conventional chemical assays. Cyclic voltammetry measures electron donation capability (redox potential) of antioxidants, which respond to a voltametric scan according to their redox potential.A cyclic voltammogram is obtained which indicates the oxidation potential of a specific antioxidant compound or functional group (usually phenyl group) and its concentration. The antioxidant activity is determined by three major parameters:

(1) the anodic peak current (Ia), usually the y-axis on a cyclic voltammogram; (2) the peak oxidation potential (Epa), usually the x-axis on a cyclic voltammogram; and (3) the area under anodic current wave (Q) (Zhang et al., 2011). The Epa value indicates redox potential or electron donating capability of antioxidants. The closer the Epa is to zero, the higher capability of the compound to donate an electron (Ragubeer, Beukes, and Limson, 2010). The Ia as the anodic current at maximum current response reflects the concentration of the antioxidant compound in the mixture. The Q value is related to the amount of charge transferred and has been proposed as indicator of overall antioxidant activity.

When more than one antioxidant component is present in a mixture, the Q value provides the total antioxidant activity of the mixture. Polyphenols with multiple phenol rings may show multiple peaks in the cyclic voltammogram. Zhang et al. (2011) analysed antioxidant activity of 14 flavonoid compounds by cyclic voltammetry and differentiated the peaks corresponding to the sequential oxidation of the hydroxyl groups at different positions, e.g. hydroxyl groups in different aromatic rings of quercetin, isorhamnetin, catechin, epicatechin, genistein and daidzein, among others. Cyclic voltametric measurement of antioxidant activity as a redox-based method has many advantages over conventional chemical assays, because it is rapid and simple and can be used for initial screening of antioxidants based on reducing power. Cyclic voltammetry is also clean since it does not require sophisticated chemical reagents or solvents. No special sample preparation is required for the cyclic voltametric method, and it can be used to assess samples of any hydrophilicity or lipophilicity, as well as turbidity, which are restricted in many spectrophotometric methods (Ragubeer et al., 2010). The cyclic voltametric method has been used to determine the activity of antioxidants from various origins, including edible oils (Gulaboski, Mirceski, and Mitrev, 2013), chocolate (Brcanovic et al., 2013), dry fruits (Zielinski, Zielinska, and Kostyra, 2012), and marine algae (Keyrouz et al., 2011; Ragubeer et al., 2010), among others. However, similar to some other reducing power assays, cyclic voltametric determination of antioxidant activity may experience interference from other non-antioxidant reducing substances, such as reducing sugars.

METAL CHELATION CAPACITY

Transition metal ions are known to stimulate lipid oxidation via Fenton reaction and also by decomposing lipid hydroperoxides into more reactive peroxyl and alkoxyl radicals. Some antioxidants such as flavonoids are powerful metal chelators. These antioxidants can easily deactivate prooxidant metal ions and thus prevent or retard metal ion-induced lipid oxidation. The antioxidant property of metal chelators is assessed when a complex is formed between the antioxidant and the metal, in such a way that metal ions can no longer act as an initiator of lipid oxidation. Therefore, metal chelation capacity is also used as an indicator of antioxidant activity, usually in combination with other antioxidant assays. In general, metal chelation capacity is determined by measuring the chelating effect of antioxidants for ferrous ion. Ferrous sulphate and ferrozine are the most commonly used sources of ferrous ion. A loss of absorbance at 485 nm (for ferrous sulphate) or 562 nm (for ferrozine) after the addition of antioxidants represents the formation of metal–antioxidant complex, and metal chelation capacity of the added antioxidant can be quantified spectrophotometrically. In most cases, metal chelation capacity of antioxidants is expressed as EDTA (ethylenediaminetetraacetic acid) equivalents. EDTA is a standard metal chelator used in many food and pharmaceutical applications. Its regulatory status as a synthetic chemical prompts its replacement with other natural metal chelating antioxidants. Metal chelation capacity has been investigated for a number of antioxidants and extracts (Chandrasekara and Shahidi, 2010; Karawita et al., 2005; Wettasinghe and Shahidi, 2002; Wijeratne, Abou-Zaid, and Shahidi, 2006; Zhong et al., 2012).

TOTAL PHENOLIC CONTENT (TPC)

Total phenolic content (TPC) is another important parameter of total antioxidant capacity (TAC), and widely used for evaluation of antioxidant extracts, including extracts from herbs, spices and fruits, cereals and legumes, among others.The Folin–Ciocalteu assay is the well-known method for determination of TPC.The Folin–Ciocalteu assay was initially designed for the analysis of proteins, which takes advantage of phenolic amino acid tyrosine in proteins (Folin and Ciocalteu, 1927). It was later adopted by Singleton, Orthofer, and Lamuela-Raventos (1999) to analyse phenolic components in wine, after which it became a routine analysis for antioxidant assessment of food and plant extracts. The Folin–Ciocalteu assay is based on the reduction of the Folin–Ciocalteu reagent by phenolic compounds under alkaline condition.The exact chemical nature of the Folin–Ciocalteu reagent is not clearly defined, but it is believed to contain phosphomolybdic/phosphotungstic acid complexes that are reduced to yield a blue-coloured chromophore with maximum absorption at 765 nm (Magalhaes, Segundo, Reis, and Lima, 2008). The molybdenum centre in the complexes is generally accepted as the reduction site, where the Mo_6^+ ion is reduced to Mo_5^+ by accepting an electron donated by the phenolic antioxidant. Thus, the Folin–Ciocalteu assay is as an ET-based assay and associated with the reducing power of phenolic antioxidants. Gallic acid is the commonly used reference standard and results of TPC is usually expressed as gallic acid equivalents. However, TPC results are also occasionally expressed as catechin, caffeic acid, chlorogenic acid, or ferulic acid equivalents, necessitating standardization of the reported results (Karadag, Ozcelik, and Saner, 2009). The Folin–Ciocalteu assay for TPC determination has many advantages, including its simplicity, reproducibility and robustness. It does, however, suffer from several drawbacks. First, the test is sensitive to pH, temperature and reaction time (Karadag et al., 2009), hence proper selection of reaction condition is necessary for consistent and reliable results. Secondly, overestimation of TPC is a major concern for Folin–Ciocalteu assay, due to the contribution of non-phenolic reducing agents present in the system to the reduction of the Folin–Ciocalteu reagent (Blasco, Rogerio, Gonzalez, and Escarpa, 2005; Escarpa and González, 2001). Examples of such contaminants include reducing sugars and certain amino acids. Thus, results for TPC determination may be over-estimated by an order of magnitude compared to those obtained by HPLC methods. The assay is also carried out in aqueous systems and its application for lipophilic phenolics is limited unless modifications on solvent system are applied. Berker, Olgun, Ozyurt, Demirata, and Apak (2013) proposed NaOH-containing isobutanol/water medium for simultaneous measurement of hydrophilic and lipophilic phenolics. TPC measured by alternative methods other than traditional Folin–Ciocalteu assay have been reported. Bonanni, Campanella, Gatta, Gregori, and Tomassetti (2007) demonstrated the use of a biosensor to measure the content of phenolic components, which were oxidized to quinones by tyrosinase while consuming oxygen. Oxygen consumption was recorded amperometrically and correlated with the content of phenolic compounds.The same authors also proposed an OXY-adsorbent method that utilizes reaction of phenolic antioxidants with hyperchlorous acid (HClO) while quantifying theTPC by measuring the unreacted excess hyperchlorous acid spectrophotometrically through a chromogenous agent (Bonanni et al., 2007).

ANTIOXIDANT EVALUATION IN FOOD MODEL SYSTEMS

Since food is a major application area of antioxidants, their performance in the food matrix is important for selection and optimized usage level of antioxidants in food. For this purpose, model food systems are used for antioxidant activity evaluation. Normally, most assessments of antioxidant activity are performed in oil, giving sensible prediction for the activity in oil or water-in-oil emulsions, whereas the results may be misleading for oil-in-water emulsions (Gordon, 2001). Similarly, antioxidant effectiveness in meat (fresh or cooked) differs from that in oil. Localization of antioxidants in the food matrix is believed to play an important role in their functional behaviour and hence efficiency. Antioxidants of different lipophilicity have different distribution pattern in the lipid-containing food system; some are preferentially distributed in either the water or lipid compartment while others tend to be localized at the interface (Porter, 1993).The interfacial phenomenon was widely accepted to provide a strong support for the so-called "polar paradox"theory.However, this theory was recently re-visited and its limitations noted as more factors were found to contribute to the complicated antioxidant behaviour patterns (Shahidi and Zhong, 2011). Thus, effectiveness and efficacy of antioxidants are only representative when a proper model system is selected. It should also be ensured that the starting lipid does not contain high levels of oxidation products (Decker, Warner, Richards, and Shahidi, 2005). Several oxidation products of lipids such as hexanal and 2,4-decadienal have been reported to exhibit pro-oxidative effects (El-Magoli, Karel, and Yong, 1980). Furthermore, stripping of oils may be necessary in such evaluations because the endogenous antioxidants in non-stripped oils are found to enhance the oxidative stability of test samples, thus giving rise to erroneous results about the efficacy of antioxidants under investigation (Khan and Shahidi, 2000; Povilaityte, Cuvelier, and Berset, 2001). In addition to oils and fats, lipid substrates used for testing antioxidant activity may include fatty acids, fatty acid ethyl esters or triacylglycerols, and β-carotene (Amarowicz, Naczk, and Shahidi, 2000; Shahidi, Desilva, and Amarowicz, 2003). Determination of antioxidants activity in food model systems is usually carried out under accelerated oxidation conditions. The combination of increased temperature and oxygen supply, addition of metal catalysts, and exposure of the reactants to light can reduce the oxidative stability of the food model system and shorten the time frame of the test to a large extent. Nevertheless, the elevated temperature may bring about changes in the oxidation mechanism thus causing difficulties in the prediction of antioxidant activity at low temperatures when compared with those at high temperatures (Antolovich et al., 2002; Gordon, 2001). Therefore, high oxidation temperatures (>60 °C) are not recommended for testing antioxidant activity (Decker et al., 2005), unless similar thermal treatment is involved in the processing of the food system under investigation. After the substrate is oxidized under standard conditions, the oxidation is monitored by chemical, instrumental or sensory methods. An appropriate measure of end point is essential for assessing antioxidant activity. Analytical strategies for end point determination include measurement at a fixed time point, measurement of reaction rate, lag phase measurement, and integrated rate measurement (Antolovich et al., 2002).

MEASUREMENT OF ANTIOXIDANT ACTIVITY IN OIL SYSTEMS

Bulk oil, triacylglycerols and free fatty acids or their alkyl ester are typical food model systems used for antioxidant evaluation. Methods vary depending on the particular scenario, but they all share the common principle, i.e. assessing the current oxidative state of the model system in the presence and absence of an antioxidant. The extent that the oxidation process is inhibited by added antioxidant indicates its antioxidant activity. The process of lipid oxidation can be monitored through a number of chemical and physico-chemical procedures, including measuring the substrate, oxidant/initiator, intermediates or final products. Quantification of the loss of reactants, depletion of oxygen (oxygen consumption) and formation of primary or secondary oxidation products may provide the most appropriate markers depending on the stage of oxidation. Methods for measuring lipid oxidation have been reviewed elsewhere (Shahidi and Zhong, 2005).

MEASUREMENT OF OXYGEN CONSUMPTION

Consumption of oxygen during the initial stages of lipid oxidation results in an increase in oil weight or a drop of headspace oxygen pressure (or concentration). Therefore, changes in oxygen consumption caused by antioxidants serve as an indicator of antioxidant activity. The oxygen consumption is usually determined by means of monitoring weight gain or headspace oxygen uptake. Heating an oil and periodically testing for weight gain is one of the oldest methods for evaluating its oxidative state (Antolovich et al., 2002). To avoid the influence of mass change by volatiles, samples can be preheated in an inert atmosphere. The induction period can be obtained by plotting weight gain against storage time. While this method offers advantages such as simple operation and low instrumentation cost, it suffers from drawbacks of compromised reproducibility associated with discontinuous heating and weighing. It was suggested that this technique may be extended to more sophisticated continuous monitoring of mass and energy changes as in thermogravimetry (TG)/differential scanning calorimetry (DSC) (Antolovich et al., 2002). This method is most suitable for evaluation of highly unsaturated oils, such as marine oils and vegetable oils containing a high content of polyunsaturated fatty acids as the oxidizable substrates. In addition to the weight gain method, oxygen consumption can be measured directly by monitoring the drop in oxygen pressure. Using headspace oxygen method, oil samples with and without antioxidants are placed in a closed vessel containing a fixed amount of headspace oxygen at elevated temperatures, commonly around 100°C.The pressure reduction in the vessel which is due to the oxygen consumption is monitored continuously and recorded automatically. The induction period as the point of maximum change in rate of oxygen uptake can be calculated (Velasco and Dobarganes, 2002). Pressure change caused by oxygen consumption can be measured electronically by a commercial instrument Oxidograph equipped with pressure transducers (Velasco and Dobarganes, 2002). Oxygen consumption may also be measured electrochemically by detecting changes in oxygen concentration. The use of a semi-automatic polarographic method has been proposed as an improvement for evaluation of lipid oxidation by determination of oxygen consumption (Genot, Kansci, and Laroche, 1994). This method is based on the use of two oxygen meters with microcathode oxygen electrodes, coupled to a computerized data collection and a processing unit.

MEASUREMENT OF CHANGE IN OXIDATION SUBSTRATE

Oxidative state of an oil model system may also be assessed quantitatively by measuring the loss of initial substrates. Unsaturated fatty acids are the main reactants whose composition changes significantly during oxidation. Changes in fatty acid composition provide a direct measure of the extent of lipid oxidation and in turn antioxidant efficacies, as antioxidants exert a significant inhibitory effect on oxidation of unsaturated fatty acids. Fatty acid methyl esters (FAME) are commonly used for determination of fatty acid composition by gas chromatography (GC). Similar to the oxygen consumption method, it is recommended that oils rich in highly unsaturated oils be used as the model system. Selection of markers for fatty acid profile change is essential for determination of the degree of oxidation and antioxidant activity.While some methods use the ratio of different fatty acid groups; others monitor the change of a particular fatty acid. Kim, Yeo, Kim, Kim, and Lee (2013) presented antioxidant activity of sesamol in thermally treated lard and soybean oil by measuring changes in the ratio of total saturated fatty acids over total unsaturated fatty acids as well as the concentration of a specific fatty acid, caprylic acid in particular. Chung, Lee, and Choe (2006) used linolenic acid level changes in combination with conjugated dienes and *p*-anisidine value of fried products as indicators of lipid oxidation during frying in soybean oil and reported improved oxidative stability by addition of sesame oil in the frying oil. Decrease in linoleic acid and increase in oleic acid contents was correlated with oxidation of pine nut oil and antioxidant activity of tocopherols and carotene; the content of saturated fatty acids was not altered significantly during oxidation (Kim, Rhee, and Cheigh, 1995).

MEASUREMENT OF OXIDATION PRODUCTS

Oxidation of fats/oils, triacylglycerols and free fatty acids or their alkyl esters leads to continuous formation of primary and secondary oxidation products. Antioxidants suppress oxidation reaction and inhibit the formation of oxidation products, which can be measured by different methods using a variety of oxidation markers, which are summarized in Table 4.17. The peroxide value (PV) is an indicator of the initial stages of oxidative change, and inhibition of formation and/or action of these unstable species by antioxidants can be used for assessing antioxidant activity. A number of methods have been developed for determination of PV,among which the iodometric titration, ferric ion complex measurement spectrophotometry, and infrared spectroscopy are most frequently used (Yildiz, Wehling, and Cuppett, 2003). Iodometric titration assay, which is based on the oxidation of the iodide ion (I^-) to iodine by hydroperoxides, is the basis of current standard methods for PV determination.The liberated iodine (I_2) is titrated against a standardized solution of sodium thiosulphate and starch as an end point indicator. The PV is calculated and reported as milliequivalents of oxygen per kilogram of sample (meq/kg). Although iodometric titration is the most commonly practised method for PV measurement, it is time consuming and labour intensive and introduces difficulties in the titration end point determination. Furthermore, result interpretation may be confounded by possible absorption of iodine across unsaturated bonds and oxidation of iodide by dissolved oxygen (Shahidi and Wanasundara, 2002). Modifications such as colourimetric determination at 560 nm (Asakawa and Matsushita, 1978), potentiometric end point determination (Hara and Totani, 1988), and spectrophotometric determination of the I_3^- chromophore at 290 or 360 nm (Gebicki and Guille, 1989; Hicks and Gebicki, 1979) have been reported.An electrochemical technique has been proposed as an alternative to the titration step in order to increase the sensitivity for determination of low PV by reduction of released iodine at a platinum electrode maintained at a constant potential (Shahidi and Wanasundara, 2002).

Table 4.17: Antioxidant activity assays measuring oxidation products.

Name of assay	Oxidation marker	Reagent	Detection	Antioxidant activity expression
Peroxide value	Hydroperoxide	Potassium iodide	Titration	% inhibition
		Thiocyanate or xylenol orange	Spectrophotometry	% inhibition
		Triphenylphosphine	FTIR	% inhibition
Conjugated dienes	Conjugated dienes		Spectrophotometry	% inhibition
TBARS	MDA equivalent	Thiobarbituric acid	Spectrophotometry	% inhibition
p-anisidine value	2-alkenals and 2,4-alkadienals	Anisidine	Spectrophotometry	% inhibition
Total carbonyls	Carbonyl compounds	2,4-Dinitrophenyl-hydrazine	Spectrophotometry	% inhibition
Headspace volatile	Volatile compounds, e.g. hexanal, propanal		Headspace GC	% inhibition
Oil stability index (OSI)	Volatile organic acids		Conductivity measurement	Induction period, protection factor
Free radical measurement	Free radicals	α-Phenyl-*tert*-butylnitrone (PBN) spin trap	EPR	% inhibition

Alternatively, PV can be determined spectrophotometrically by measuring the ability of lipid hydroperoxides to oxidize ferrous to ferric ions, which are complexed by either thiocyanate or xylenol orange to form red-violet or blue-purple chromophores, respectively (Eymard and Genot, 2003; Jiang, Woollard, andWolff, 1991). The spectrophotometric methods are rapid, simple and more sensitive than iodometric titration method. However, because many factors such as the amount of oil sample, solvent used and source of reagents such as xylenol orange may affect the absorption coefficient, knowledge of the nature of hydroperoxides present in the oil sample and careful control of the conditions used are required for accurate measurements (Dobarganes and Velasco, 2002). Lipid hydroperoxides can also be quantitatively determined by IR spectroscopy via measurement of their characteristic O—H stretching absorption band (Sedman, van de Voort, and Ismail, 1997). An absorption band at about 2.93 μm indicates the generation of hydroperoxides, whereas the replacement of a hydrogen atom on a double bond or polymerization accounts for the disappearance of a band at 3.20 μm. Arapid Fourier transform infrared (FTIR) spectroscopy method based on the stoichiometric reaction of triphenylphosphine (TPP) with hydroperoxides has been developed and successfully applied for PV determination (Dong, Ma, van de Voort, and Ismail, 1997). The hydroperoxides present in oil samples react stoichiometrically with TPP to produce triphenylphosphine oxide (TPPO), which has an intense absorption band at 542 cm^{-1} in the mid- IR spectrum (Riuz, Ayora-Canada, and Lendl, 2001). The band intensity is measured and converted to peroxide value. By using tert-butylhydroperoxide spiked oil standards and evaluation of the band formed at 542 cm^{-1} a linear calibration graph covering the range of 1–100 PVwas obtained (Riuz et al., 2001). The FTIR method can also be coupled with flow injection analysis, which offers exact and highly reproducible timing of sample manipulation and reaction aswell as a closed environment with oxygen and light being easily excluded (Riuz et al., 2001). The FTIR spectroscopy is a simple, rapid and highly precise method. It shows excellent correlation with the iodometric titration method and avoids the solvent and reagent disposal problems associated with the standardwet chemical method (Dong et al., 1997; Riuz et al., 2001). It provides an automated, efficient and low-cost means of evaluating oxidative state of lipid and efficiency of antioxidants in the model oil system undergoing accelerated oxidation induced by thermal stress (Innawong, Mallikarjunan, Irudayaraj, and Marcy, 2004; Kiritsakis, Kanavouras, and Kiritsakis, 2002). However, there is a need to characterize the spectral changes, assign wavelengths to more common molecular species produced and access potential spectral cross interferences (Kiritsakis et al., 2002).An improved FTIR attenuated total reflectance (FTR-ATR) method using the whole FTIR spectral data instead of particular wave numbers has been proposed (Innawong et al., 2004). In addition to the hydroperoxide marker, conjugated dienes and trienes can be used as indicators of the degree of oxidation and inhibition by antioxidant. Due to the fact that polyunsaturated fats and oils are generally methylene interrupted in nature, conjugated dienes and trienes are formed upon oxidation, which can be measured spectrophotometrically at 234 and 268 nm in the ultraviolet (UV) region, respectively (Shahidi andWanasundara, 2002). An increase in UV absorption theoretically reflects the formation of primary oxidation products in fats and oils. Therefore, antioxidant activity can be assessed by lag phase measurements and percentage inhibition in the presence of a certain antioxidant (Antolovich et al., 2002). Good correlations between conjugated dienes and peroxide value have been reported (Wanasundara,

Shahidi, and Jablonski, 1995). Ultraviolet detection of conjugated dienes is simple, fast, and requires no chemical reagents and only small amounts of samples are needed. However, this method has less specificity and sensitivity than PV measurement. Furthermore, the result may be affected by the presence of compounds absorbing in the same region, such as carotenoids (Shahidi and Wanasundara, 2002). To avoid these interferences, an alternate spectroscopic method measuring conjugable oxidation products (COPs) has been proposed. In this method, hydroperoxides and some decomposition products are converted to more conjugated chromophores by reduction and dehydration. The concentrations of the resultant conjugated trienes and tetraenes are determined from their respective absorption at 268 and 301 nm and expressed as COP values (Gordon, 2001; Shahidi and Wanasundara, 2002).

The PV and conjugated dienes are primary oxidation products. They are unstable and decompose to more stable secondary oxidation products, such as a variety of carbonyl compounds, including aldehydes and ketones. Hence, primary oxidation products alone are less evidential and are usually reported in combination with secondary oxidation products measurement for a more comprehensive picture of the oxidation process. Among the secondary oxidation products produced frombreakdown of hydroperoxides, TBARS (thiobarbituric acid reactive substances) are awell-known marker of oxidation process both for the early appearance as oxidation occurs and for the sensitivity of the analytical method (Cesa, 2004).TBARS are reported as malonaldehyde or malondialdehyde (MDA) equivalents.MDA, a degradation product of polyunsaturated fatty acids, reacts with TBA (thiobarbituric acid) and forms a pink MDA–TBA complex that has maximum absorption at 530–535 nm. Since certain oxidation products other than MDA also react with TBA and contribute to the colour intensity, the term TBARS is used to include an array of substances mainly carbonyl compounds that react with TBA in similar mechanism as that of MDA. These substances include alkanals, 2-alkenals, 2,4-alkdienals, ketones, ketosteroids, acids and esters,among others. Despite the lack of specificity and sensitivity (de las Heras, Schoch, Gibis, and Fischer, 2003), TBARS is among the most frequently used tests for assessing oxidation in fats and oils. In more complicated food matrix system, interference may arise from other non-lipid components such as proteins, sucrose, urea, pyridines, and pyrimidines, among others (Guillen-Sans andGuzman-Chozas, 1998; Jardine, Antolovich, Prenzler, and Robards, 2002). For improved specificity and sensitivity, TBARS test has been modified by using FTIR technology for direct detection of MDA and HPLC for separation and characterization of individual species of TBARS (Jardine et al., 2002). For oil systems containing antioxidants, a decrease in the amount of total TBARS or a specific species can be quantified for oxidation inhibition and in other words efficiency of the antioxidants.

The *p*-anisidine value (*p*-AnV) is another measure of the content of aldehydic oxidation products (principally 2-alkenals and 2,4-alkadienals) generated during decomposition of hydroperoxides. It is based upon the colour reaction of *p*-methoxyaniline (anisidine) and the aldehydic compounds under acidic conditions which yield yellowish products that absorb at 350 nm (Doleschall, Kemeny, Recseg, and Kovari, 2002). The *p*-AnV is a reliable indicator of oxidative rancidity in fats and oils and thus a good approach to measure inhibition of oxidation by antioxidants. A highly significant correlation between *p*-AnV and PV has been found, although its significance may depend on a number of factors such as storage time and temperature, among others (List, Evans, Kwolek, Warner, and Boundy, 1974). Nevertheless, the same oil model system should be used when evaluating antioxidant activity by *p*-AnV method, since the initial AnV varies among oil sources, being higher for unsaturated oils than for saturated oils, due to stronger absorption of coloured products from unsaturated aldehydes than that from saturated aldehydes (White, 1995) at this wavelength.

The analysis of total carbonyl compounds provides an alternative approach to measure the extent of lipid oxidation in fats and oils and hence antioxidant activity in fat/oil model systems.This method is based on the absorbance of the carbonyl derivatives through a colourimetric 2,4-dinitrophenylhydrazone formation reaction. The carbonyl compounds formed during lipid oxidation are reacted with 2,4-dinitrophenylhydrazine (DNPH) followed by reaction of the resulting hydrazones with alkali. The final coloured products are then analysed spectrophotometrically (Melton, 1983). Many variations of this method using an alternate solvent, reagent, wavelength, orworkup have been reported. The determination of total content of carbonyls has been used in different oxidative stability studies.However, it isworth noting that in food model systems other than fats and oils, carbonyls produced from oxidation of other components such as proteinsmay also contribute to the colourimetric reaction giving rise to higher total carbonyl values than expected. Meanwhile, high temperature must be avoided because some short-chain carbonylsmay be removed and hence affecting the total carbonyls.

Some of the secondary oxidation products are volatile with varying threshold values, which contribute to the off-flavour or aroma of oxidized oils through different routes and mechanisms depending on nature of the oil and oxidation conditions. Measurement of headspace volatile compounds as total volatiles or specific compounds, usually volatile aldehydes, has been an effective tool for assessment of lipid oxidation and antioxidants in lipid-containing model systems. For oils rich in omega-6 fatty acids such as majority of vegetable oils, hexanal is the major volatile compound formed upon oxidation, while marine oils rich in omega-3 fatty acids produce mainly propanal (Zhong, Lall, and Shahidi, 2007). Gas chromatography (GC) equipped with a headspace sampler is used in determining secondary volatile oxidation products. More recently, headspace solid-phase microextraction (HS-SPME) technique has been used for analysis of volatile oxidation products. In this technique, the analytes are directly adsorbed from an aqueous or gaseous phase onto the fibre coating and later thermally desorbed onto the GC injector and transferred to the capillary column (Gioti, Fiamegos, Skalkos, and Stalikas, 2007). HS-SPME has been used for measurement of oxidation and antioxidant performance in model oil systems. Gioti et al. (2007) evaluated antioxidant activity of St. John's wort extract in a sunflower oil model system by simultaneous detection of MDA, pentanal and hexanal using HS-SPME technique. Lee, Kim, Chang, and Lee (2007) suggested the use of HS-SPME to determine antioxidant activity in a mixed free fatty acid model system using total volatiles or 2-heptenal as oxidation markers.

In addition to volatile aldehydes, volatile organic acids, mainly formic and acetic acids are produced as secondary volatile oxidation products at high temperatures, simultaneously with hydroperoxides (Kiritsakis et al., 2002; Schwarz et al., 2001). Furthermore, other secondary products including alcohols and carbonyl compounds can be further oxidized to carboxylic acids (Kiritsakis et al., 2002). The formation of volatile acids can be measured by monitoring the change in electrical conductivity when effluent from oxidizing oils is passed through water (Gordon, 2001).This method provides the oil stability index (OSI) value, which is defined as the point of maximal change of oxidation rate, attributed to conductivity increase by the formation of volatile organic acids during lipid oxidation (Miura, Kikuzaki, and Nakatani, 2002).Two commercially available equipment, the Rancimat (Metrohm Ltd.) and the Oxidative Stability Instrument (Omnion Inc.) are employed for determining the OSI value. This method requires a somewhat higher level of oxidation (PV > 100) to obtain measurable results than other methods in which hydroperoxides are the most important products formed and detected (De la Presa-Owens, Lopez-Sabater, and Rivero-Urgell, 1995).Therefore, for some oils that are stable under normal conditions, the oxidation process is accelerated by exposing them to elevated temperatures in the presence of excess air or oxygen (Canizares-Macias, Garcia-Mesa, and Luque de Castro, 2004; Tan, Che-Man, Selamat, and Yusoff, 2002). In the OSI assay, a flow of air is bubbled through heated oil, usually at 100 °C or above. For marine oils temperatures as low as 80 °C are often used. The OSI assay enables continuous monitoring of the oxidation process and excellent correlation between Rancimat and conjugated dienes has been reported (Canizares-Macias et al., 2004). The OSI assay has been practised as industrial standard for oxidative stability of oils and effectiveness of antioxidant products. However, its application for evaluation of thermosensitive antioxidant compounds is questionable. Furthermore, it is not recommended for assessment of volatile antioxidants which may be swept out of the oil by the air flow under test conditions (Gordon, 2001).

Oxidation level of fats and oils in the absence and presence of antioxidants can also be measured directly by detecting the formation of radicals by using EPR spectroscopy. However, the short life times and low steady state concentration of the highly reactive lipid-derived radicals make it difficult to detect these radicals at concentrations lower than the 10−9M minimum detectable concentration (Andersen and Skibsted, 2002). To overcome this problem, various approaches have been used, including pulse radiolysis and UV photolysis, continuous flow systems and spin trapping, among which spin trapping has been the most widely employed procedure (Antolovich et al., 2002). Spin trapping technique allows the accumulation of detectable concentrations of longer-lived radicals by addition to samples of a spin trapping agent, which reacts with free radicals to form more stable spin-adducts (Andersen and Skibsted, 2002; Antolovich et al., 2002; Velasco, Anderson, and Skibsted, 2004). Nitroso compounds and nitrones are the most common spin traps, both leading to nitroxyl type spin-adducts such as α-phenyl-*tert*-butylnitrone (PBN)-adducts (Andersen and Skibsted, 2002). EPR spectroscopy is of great value for the study of the early stages of lipid oxidation and prediction of oxidative stability of fats and oils as well as evaluation of antioxidant activity. It has high sensitivity, and allows mild conditions by applying low temperatures and requiring little sample preparation. Nevertheless, spin traps used in the EPR assay have been reported to exhibit widely differing trapping efficiencies for different radicals and show both prooxidant and antioxidant effects (Halliwell and Gutteridge, 1999; Lee, 1994), thus may give erroneous results of antioxidant activity.

Measurement of system change during oxidation process

While a majority of methods measure the formation of oxidation products or consumption of reactants (substrates and oxygen), a few methods monitor the change of the whole system during the oxidation process without targeting particular species. These include NMR spectroscopy and DSC. High-resolution ^{1}H NMR spectroscopy has been used to evaluate oxidative deterioration of fats and oils (Shahidi and Wanasundara, 2002). The principle of NMR is that hydrogen atoms in a strong magnetic field absorb radiation energy in the radiofrequency range, depending on their molecular environment, which changes during the oxidation process (Shahidi and Wanasundara, 1999). These changes may be monitored by NMR spectroscopy as a reflection of oxidation level of the oil or fatty acids. More specifically, the hydrogen atoms (proton, ^{1}H) located at different positions in the triacylglycerol or free fatty acid molecules can be calculated from the NMR spectra and ratios of aliphatic to olefinic protons (Rao) and aliphatic to diallylmethylene protons (Rad) can be obtained (Shahidi and Wanasundara, 1999). Both ratios increase steadily during lipid oxidation, as the allylic and bisallylic positions are the reactive sites for oxidation. Attenuation of this ratio increment suggests antioxidant effectiveness in the oil or fatty acid model systems. NMR spectroscopy has been used by many researchers, and the changes in Rao and Rad measured by NMR correlated well with both conjugated dienes and TBA values (Senanayake and Shahidi, 1999).

In addition to ^{1}H NMR, ^{13}C NMR and ^{31}P NMR are also powerful tools to predict oxidative stability of oils. ^{13}C NMR enables direct observation of carbon atoms. The selectivity and dispersion of ^{13}C NMR spectra are very high (Diehl, 1998). ^{13}C NMR assesses lipid oxidation by monitoring the changes of carbon chains in triacylglycerol or fatty acid molecules, revealing the specific sites where oxidative degradation occurs (Medina, Sacchi, Giudicianni, and Aubourg, 1998). However, because the abundance of ^{13}C isotope is only 1.12% of that of ^{12}C, the sensitivity of ^{13}C NMR is usually much lower than that of ^{1}H NMR (Diehl, 1998).

NMR spectroscopy is a rapid, non-destructive and reliable technique for assessing lipid oxidation. It simultaneously measures both the primary and the secondary oxidative changes in the oil or fatty acid model systems, and provides specific information on oxidative regions in the substrate molecules. Using NMR for antioxidant test, a reduction in oxidation of the model system relative to a control sample is used as an index of antioxidant activity.

DSC is another tool to monitor the oxidation process in terms of change in energy. During lipid oxidation, fats and oils reveal a number of thermally-induced transitions such as the transfer of oxygen molecules to unsaturated fatty acids (exothermic process) (Tan and Che Man, 2002). Therefore, thermal analysis can be used in accelerated oxidation tests. The DSC technique, which is based on thermal release of oxidation reactions, has the potential as a non-chemical method for assessing oxidative stability of fats and oil model systems as affected by antioxidants (Velasco et al., 2004). It provides unique energy profile information, which specifically measures the temperature and heat flows associated with lipid oxidation as a function of time and temperature (Tan and Che Man, 1999). The method uses isothermal or non-isothermal conditions and a flow of oxygen as purge gas, with a calorimeter measuring the heat flow into (endothermic) or out of (exothermic) an oil model system undergoing oxidation changes (Tan and Che Man, 2002; Velasco et al., 2004). The oxidation curves of the sample are obtained with different heating time, and a dramatic increase for the evolved heat can be observed with the appearance of a sharp exothermic curve during initiation of oxidation. The end point is taken at the time where a rapid exothermic reaction between oil and oxygen occurs, and induction period (IP) determined automatically by intersection of extrapolated baseline and tangent line (leading edge) of the exotherm. The DSC also measures oxidation onset temperature, the temperature at maximum reaction, and the ending temperature (Tan and Che Man, 2002). The isothermal and non-isothermal DSC show good agreement, suggesting that both isothermal and nonisothermal DSC are suitable for oxidative stability studies of oils (Litwinienko, Daniluk, and Kasprzycka-Guttman, 1999). Addition of antioxidants can improve the oxidative stability of the model system. This improvement can be determined by DSC method and used to indicate the antioxidant activity. The DSC technique has been reviewed by Tan and Che Man (2002).

Measurement of antioxidant activity in model emulsion systems

Antioxidants are found to behave differently when used in different media. Hence, their efficacy in bulk oil may not necessarily reflect that in oil/water emulsions. Moreover, it has been recognized that high surface-to-volume ratio of emulsions are the natural conditions, while low surface-to-volume bulk lipid is more like an artefact that is less common in foods and biological systems. Many food lipids exist as oil-in-water emulsions (e.g. milk, mayonnaise, dressings, dips, sauces, beverages, ice cream, etc.), which are often more susceptible to oxidation than bulk oils due to their higher surface areas that promote interaction of the oil with prooxidants in the aqueous phase. Therefore, it is important to also include information on effectiveness of an antioxidant in oil/water emulsion systems for a more comprehensive assessment of antioxidant activity.

The β-carotene bleaching assay is one of the oldest and commonly practiced methods of estimating antioxidant activity in oil-in-water emulsion. In this method, the oxidative state of an emulsion composed of β-carotene, linoleic acid and water is measured spectrophotometrically in the absence and presence of the antioxidant. Free radicals generated from oxidation of linoleic acid in the emulsion lead to discolouration (bleaching) of β-carotene, which can be monitored with time at 470 nm. The presence of antioxidants can reduce the loss and hence discolouration of β-carotene during the coupled oxidation of linoleic acid and β-carotene in the emulsified aqueous system. Usually, a kinetic curve of absorbance is obtained, which suggests how a certain antioxidant can attenuate the decrease of absorbance. The β-carotene bleaching assay in different versions has been applied in the assessment of various synthetic and natural antioxidant extracts. It was recently suggested that standardization of this method is necessary as the results can be significantly influenced by a number of factors, such as type and amount of solvent in which the antioxidants are dissolved, and presence of hydrogen and metal ions, among others (Dawidowicz and Olszowy, 2010).

In addition to the well-established β-carotene bleaching assay, there are many other emulsion model systems for antioxidant evaluation tailored for specific need. Such model systems include vegetable oil-in-water emulsion (e.g. sunflower oil, or soybean oil-in-water emulsions) (Abdalla and Roozen, 1999), fish oil-in-water emulsion (Farvin et al., 2014), or fatty acids/water emulsions (Wardhani, Fucinos, Vazquez and Pandiella, 2013). Most oxidation markers discussed above for bulk oil model systems are also generally suitable for oil/water emulsions. Typical tests include measurement of PV, TBARS, and volatile hexanal or propanal, among others. Someemulsion model systems involve use of emulsifiers to stabilize the emulsion against lipid/water phase separation. However,emulsifier introduced in the system may affect antioxidant effectiveness and cause false interpretation of results.The quantity and type (cationic, anionic and non-ionic) of emulsifier in a heterogeneous system can drastically change the antioxidant distribution in the emulsified medium and in turn the antioxidant activity. Moreover, emulsifiers in high quantities (above critical micelle concentration) form micelles, which may trap antioxidants in these self-assembled structures and carry them to the water phase. Therefore, mass transfer of antioxidants in emulsions between micelles and the oil–water interface must be taken into consideration (Shahidi and Zhong, 2011).

Measurement of antioxidant activity in muscle food model system

Lipids are not only important components of bulk oils such as cooking oil and emulsions such as salad dressings and margarines, they are also incorporated into processed whole tissue foods such as in many fast foods and are responsible, to a large extent, for their quality deterioration. The lipids in processed whole tissue foods are characterized by a high surface-tovolume ratio and are usually considered a very polar medium (Porter, 1993). These foods generally require a larger amount and different types of antioxidants due to their predisposition to oxidation during precooking and long storage time under high surface-to-volume conditions (Porter, 1993). Cooked ground muscle foods provide an excellent model for assessing effectiveness of antioxidants in thermally processed whole or modified tissue foods, where heating causes rapid oxidation of lipid and meat flavour deterioration as well as development of "warmed-over flavour" (Rubin and Shahidi, 1988).

Lipid oxidation in muscle foods proceeds by different mechanisms and interaction with other components such as protein, haem, metal and salt as well as processing and storage conditions play important roles in their oxidative stability (Ladikos and Lougovois, 1990). For example, disruption of the muscle membrane by homogenization allows oxygen to penetrate inside the tissue and, in turn, accelerate lipid peroxidation (Pikul, 1992). Similarly, cooking causes breakdown of haem constituents and the release of iron. Non-haem iron is an active catalyst responsible for rapid oxidation of lipids in heated meat (Stodolak, Starzynska, Czyszczon, and Zyla, 2007). Proteins in meat also undergo oxidation during thermal treatment, and produce carbonyl compounds leading to quality deterioration. Addition of antioxidants can inhibit oxidation of both lipid and proteins, and the overall effect is considered as their efficiency in quality preservation.

Muscle food model systems used for antioxidant assessment vary depending on the ultimate application of the antioxidants. More general antioxidant screening typically utilizes ground pork, beef, poultry meat or fish, ground or fillet, in fresh, refrigerated, or cooked forms. Antioxidants may be added in different processing steps and their efficiency is evaluated by measuring the difference in levels of oxidative markers with and without antioxidants.The per cent inhibition in oxidation by antioxidants compared to a control without antioxidants can be calculated and reported as antioxidant efficiency. As already noted for other system,TBARS and headspace volatiles (hexanal, propanal and others) are common oxidative markers used formuscle foods. Measurement of TBARS in muscle foods differs slightly fromthat of oil, owing to the matrix structure that entraps the carbonyl lipid and protein oxidation products. The TBA reagent reacts directly with the meat sample, followed by extraction of the TBA-MDA and related adducts prior to the spectrophotometric analysis. Alternatively, the MDA equivalent TBARS are extracted either by steam distillation or solvent extraction and the extract is then reacted with the TBA reagent for subsequent absorbance measurement of the coloured product.Tarladgis,Watts, Younathan, and Dugan (1960) reported that the distillation of an acidified sample was essential to liberate MDA fromprecursor or bound forms, to produce maximal colour development and, especially, to separate TBARS from the food matrix (de las Heras et al., 2003). However, distillation method is generally considered to lack accuracy and reproducibility, and hence methods involving direct reaction ofTBA reagent with food samples are becoming more popular for determining TBARS in meat. Like in oil model systems, TBARS in meat can also be detected by techniques other than spectrophotometry. Papastergiadis, Mubiru, van Langenhove, and de Meulenaer (2012) detected TBARS in pork sausage and cooked fish by spectrophotometric and HPLC methods and concluded that spectrophotometric method tended to overestimate the content of MDA and related products.

Headspace hexanal for meat and poultry and propanal for fish is frequently used oxidation marker and indicator for antioxidant activity (Hammond and Skonberg, 2012; Vuorela et al., 2005). Headspace volatile measurement is simple, rapid, and accurate with minimum sample treatment requirement;muscle tissue can be directly measured for the headspace volatile without extraction or colourimetric reaction. However, special attention needs to be paid to possible loss of volatiles during meat processing and storage or conversion into tertiary oxidation products. It is recommended that the headspace hexanal/ propanal method be used in combination with other oxidation tests for a full picture of antioxidant effectiveness.

Measurement of antioxidant activity in biological model systems

In addition to the use in food products as additives for quality preservation, antioxidants are of great interest to health professionals for their protective roles in the body against oxidative stress and the associated diseases and health disorders. *In vitro*, *ex vivo* and *in vivo* biological model systems are employed for assessment of antioxidants in fighting oxidative stress. Similar to the food model systems, biological systems used differ in the type of substrate, oxidation initiators, reaction conditions and oxidation markers. Typical biological systems for antioxidant potential evaluation are discussed in this section and summarized in Table 4.18.

Table 4.18: Evaluation of antioxidant activity in biological model systems.

Model system (substrate)	Oxidant	Oxidation marker	Detection
LDL-cholesterol	Cu2+ or AAPH-generated peroxyl radical	Conjugated dienes	Spectrophotometry
Supercoiled DNA	Fe2+/hydrogen peroxide-generated hydroxyl radical, or AAPH-generated peroxyl radical	Open circular form of DNA, or linear form DNA	Gel electrophoresis
Red blood cells	Hydrogen peroxide, or AAPH-generated peroxyl radical	Haemolysis	Spectrophotometry, or turbidity measurement
HepG2 cells	ABAP-generated peroxyl radical	2′,7′-dichlorofluorescin	Fluorometry

AAPH, 2,2′-azobis(2-amidinopropane) dihydrochloride; ABAP, 2,2′-azobis (2-amidinopropane) hydrochloride.

LDL-cholesterol oxidation inhibition assay

Oxidative modification of LDL-cholesterol plays a key role in the pathogenesis of atherosclerosis and heart disease (Sajilata, Bajaj, and Singhai, 2008). Oxidized LDL containing both lipid oxidation products and oxidized apoprotein particles contribute to all stages of the atherosclerosis process, thus promoting the atherogenic effects (Shahidi, Zhong, and Chandrasekara, 2012). Antioxidants such as many polyphenols can inhibit LDL oxidation by scavenging free radicals and other ROS, chelating prooxidant metals, and binding with the apo-lipoprotein B, which promotes the access of antioxidant to the lipids and prevents interaction with prooxidants (Satue-Gracia, Heinonen, and Frankel, 1997). Antioxidants are potential anti-atherogenic agents in cardiovascular diseases risk reduction, and their potencies can be assessed by the LDL-cholesterol oxidation inhibition assay.

In the LDL-cholesterol assay, oxidation of LDL-cholesterol can be induced by transition metals such as copper ion (Cu^{2+}) or peroxyl radical generated from thermal decomposition of AAPH. AAPH-generated peroxyl radical is considered a preferred oxidation initiator due to its biological relevance over the copper ion (Magalhaes et al., 2008). The oxidation substrate (LDL-cholesterol), initiator metal ion or peroxyl radical and antioxidants are incubated at physiological temperature of 37 °C, and formation of conjugated dienes is monitored spectrophotometrically at 234 nm periodically or at a certain end point. Antioxidant activity is reported as per cent inhibition of conjugated dienes formation as compared to a control without antioxidants. Alternatively, oxidative state of LDLcholesterol can be measured by the formation of TBARS as MDA equivalents, which is more time-consuming and labour intensive. This assay is generally carried out in aqueous buffer but also occasionally in organic solvents and micelles. Reaction in micelles requires special sample treatment since direct spectrophotometry is not suitable (Karadag et al., 2009). The LDL-cholesterol oxidation inhibition assay has been used by many researchers for antioxidant activity and potential antiatherogenic effect of various antioxidant compounds and extracts (Chandrasekara and Shahidi, 2011; Madhujith and Shahidi, 2007; Zhong and Shahidi, 2012b).

Similar to conjugated dienes method for antioxidants in oil model systems, food components that also absorb at 234 nm will interfere with the antioxidant activity assessment in the LDL-cholesterol model system. This interference may be minimized or eliminated by including a blank without the LDL substrate. Another drawback of this assay is that isolation and purification of LDL-cholesterol is a lengthy procedure and may introduce inconsistency which leads to compromised reproducibility of the assay.

Inhibition of DNA oxidation and nicking

Oxidative stress in cells also leads to DNA damage, which is implicated inmutagenesis and carcinogenesis, and other pathological processes as well as ageing. Oxidative damage to the DNA may occur at both the phosphate backbone and the nucleotide bases, resulting in a wide variety of modifications, including strand scission, sister chromatid exchange, DNA–DNA and DNA–protein cross-links as well as base modification (Davies, 1995). Antioxidant activity of many polyphenol compounds and extracts has been evaluated in DNA model systems for their potential as antimutagenic agents in cancer prevention and/or treatment (Chandrasekara and Shahidi, 2011; Zhong and Shahidi, 2012b).

Effect of antioxidants on free radical induced DNA strand scission has been determined in a supercoiled plasmid pBR322 DNA model system, where DNA strand breaking is induced by hydroxyl or peroxyl radicals (Chandrasekara and Shahidi, 2011; Zhong and Shahidi, 2012b). Hydroxyl and peroxyl radicals are selected because they are a major source of biologically relevant ROS responsible for oxidative damage of DNA, especially the mitochondrial DNA (Perron, Hodges, Jenkins, and Brumaghim, 2008). Hydroxyl radical can be generated by the Fenton reaction between ferrous ion and hydrogen peroxide, while peroxyl radical is usually generated by AAPH. After incubation with radicals and antioxidants at 37 °C, the DNA fractions are separated by gel electrophoresis and bands identified after visualization. Both open circular form of the DNA from single strand cleavage and linear form from double strand cleavage may be observed as results of DNA oxidation, with the open circular form being dominant. The concentration of the native (supercoiled) and nicked DNA fractions is obtained from densitometer as indicated by the intensity or density of the corresponding bands. Antioxidants inhibit DNA scission possibly through a combination of radical scavenging and ferrous ion chelation mechanisms, and their inhibition efficiency can be calculated as DNA retention (per cent of DNA retained un-oxidized and supercoiled). The DNA oxidation inhibition assay provides valuable information on the efficacy of antioxidants in protecting DNA from oxidative stress, which may suggest their anti-mutagenic and anti-carcinogenic potentials.

Cellular assays for evaluation of antioxidant activity

Development of the *in vitro* chemical assays and use of biologically relevant macromolecule model systems have enabled evaluation of intrinsic antioxidant activity of a great number of synthetic bioactive compounds and natural extracts. However, these tests fail to predict the antioxidant activity *in vivo* in many circumstances, due to the cellular absorption, metabolism and antioxidant behaviour in the more complicated cellular environment that varies from one substance to another. Since it is not always an easy approach to measure antioxidant effectiveness directly in animal or human subjects for initial antioxidant screening, it is suggested that cellular models be used for a more comprehensive understanding of antioxidant activity at cellular level and better correlation with their performance *in vivo*. Cell culture models provide cost-effective and relatively fast approach that addresses the issues of uptake, distribution, and metabolism of antioxidant compounds (Wolfe and Liu, 2007).

HAEMOLYSIS INHIBITION ASSAY

Plasma serves as an excellent substrate of oxidation and model system for antioxidant evaluation. It is biologically relevant and suitable for both hydrophilic and lipophilic antioxidants (Omata, Yoshida, and Niki, 2010). The erythrocytes (red blood cells, RBC) are constantly exposed to oxidative stress owing to their specific role as oxygen carrier in the blood. The RBC, lacking nucleus, mitochondrial and other typical cellular organelles as well as sufficient self-repair mechanisms, provide a simplified, yet sensitive, model for antioxidant assessment (Giao et al., 2010). The ROS, especially hydrogen peroxide as a normal cell metabolite released from enzymatic and nonenzymatic reactions, can initiate oxidation of protein (haemoglobin) and lipid (mainly cholesteryl ester), leading to destruction of the cell shape and membrane structure and ultimately haemolysis (Giao et al., 2010).

Capability of antioxidants to inhibit haemolysis can be measured for evaluation of antioxidant activity. Hydrogen peroxide and AAPH generated peroxyl radical have been used as common oxidation initiators, among others. After incubation with the oxidation initiator and antioxidant at 37°C, the degree of haemolysis can be determined periodically by centrifugation and spectrophotometrically measuring the concentration of released haemoglobin in the supernatant solution at 524 nm (Giao et al., 2010; Tai, Sawano, Yazama, and Ito, 2011). Inhibition of haemolysis by antioxidants can then be calculated by comparing with a control containing no antioxidant.The haemolysis inhibition assay has been used to evaluate efficacy of a number of antioxidant compounds and extracts. The original spectrophotometric determination of the haemolysis is timeconsuming and labour intensive, and has recently been modified using the automated microplate reader technology which allows simultaneous measurement of a large number of antioxidant samples. In this modified method, the degree of haemolysis is determined by turbidity (through light scattering detection) of the erythrocyte suspension without centrifugation (Takebayashi, Iwahashi, Ishimi, and Tai, 2012).

CELLULAR ANTIOXIDANT ASSAY (CAA)

Cellular antioxidant assay (CAA) was recently developed to quantitatively measure capability of antioxidants in inhibiting oxidation in cells (Song et al., 2010; Wolfe et al., 2008; Wolfe and Liu, 2007). Different from haemolysis inhibition assay, CAA takes advantage of both cell culture models and chemical assays by introducing a special fluorescent probe into the cells. In this method, the 2′,7′-dichlorofluorescin diacetate (DCFH-DA) probe is taken up by the cultured HepG2 human hepatocarcinoma cells, where the absorbed DCFH-DA is activated to DCFH through cellular esterases-catalysed deacetylation. DCFH is sensitive to ROS such as hydrogen peroxide and peroxyl radical and RNS (reactive nitrogen species) such as nitric oxide and peroxynitrite, and is converted to its oxidation product DCF which is fluorescent upon excitation. The CAA specifically uses ABAP as peroxyl radical generator to initiate oxidation of DCFH to DCF. The fluorescence intensity measured is proportional to the level of oxidation. Antioxidants absorbed into the cells can scavenge the peroxyl radical resulting in lower degree of oxidation observed as attenuated fluorescence increase. Per cent inhibition of oxidation by antioxidants can be calculated as indicator of antioxidant activity. In-depth description of this method and its use to assess different types of antioxidants is available (Wolfe and Liu, 2007).

The CAA assay is believed to offer many advantages for antioxidant assessment in biological model systems. These include use of biologically relevant substrate under physiological conditions, low level of oxidants (ABAP concentration), cellular uptake of antioxidants being considered that correlates with bioavailability *in vivo*, among others (Wolfe and Liu, 2007). Like most chemical assays utilizing fluorescence probes, CAA is automated with microplate reader technology and net area under the curve (AUC) integration calculation, allowing kinetic study of all oxidation reaction stages including lag time.

The CAA has been used to evaluate efficacy of a variety of antioxidant compounds and extracts, including extracts of common fruits, vegetables, and dietary supplements (Song et al., 2010; Wolfe et al., 2008; Wolfe and Liu, 2007). It was proposed as a standardized method for measuring cellular antioxidant activity in order to minimize cross laboratory variations caused by difference in types of cell lines and oxidants, cell treatment procedure and quantification method for oxidation and antioxidant activity, etc. The CAA is attracting more attention from researchers as a result of growing bioavailability issue with many bioactive phytochemicals. However, stability of the cell line and other issues has raised concern about reproducibility of results and opportunity to compare results from different laboratories with one another, among other concerns. It should also be noted that proper choice of cell lines may well address this shortcoming, e.g. use of Caco2 or other cells instead of HepG2 cells.

CONCLUSION

Antioxidants are of growing interest to food scientists, health professionals and the general public owing to their protective roles in food products against oxidative deterioration and in the body against oxidative stress-mediated pathological processes. Natural and potent antioxidants are in demand for food preservatives and nutraceuticals / pharmaceuticals. Effective search for sources of naturally occurring antioxidants and design of novel antioxidant compounds require reliable methods of antioxidant activity evaluation. A number of chemical assays, food and biological model systems have been developed to measure at molecular or cellular level the radical scavenging capacity, reducing power, and other specific attributes of antioxidants as well as overall oxidation inhibition in more complex food and biological systems. These methods vary in terms of antioxidant mechanism, substrate type, oxidation initiator, result expression and ease of operation. Proper method selection or combination of methods is important for valid assessment of antioxidant activity and ultimately potential of the antioxidant as food preservative or health promoting agents.

■■■■

5 Health Effect of Phenolic Antioxidants

A. INTRODUCTION TO PHENOLIC ANTIOXIDANTS

Dietary intake of phenolics is greatly affected by the eating habits and preferences of individuals (Shahidi and Naczk, 2004). The average daily intake of dietary polyphenols is approximately 1 g per person; the main sources are beverages, fruits and, to a lesser extent, vegetables and legumes (Scalbert andWilliamson, 2000). Simple phenolics such as hydroxycinnamic acid conjugates and flavonoids are important constituents of fruits, vegetables and beverages.These compounds showa wide range of antioxidant activities *in vitro* (Rice-Evans, Miller, Bolwell, Bramley, and Pridham, 1995) and are thought to exert protective effects against major diseases such as cancer and cardiovascular diseases (Boudet, 2007). Oxidative stress imposed by reactive oxygen species (ROS) indeed plays a crucial role in the pathophysiology associated with neoplasia, atherosclerosis and neurodegenerative diseases. The potential mechanism of the protective effects of phenolic compounds, including flavonoids, is thought to be due to their direct scavenging of free radicals (Heim, Tagliaferro, and Bobilya, 2002). Lee and Lee (2006) showed the efficacies of antioxidant therapies, based on antioxidative phenolics, which decreased ROS levelswere equivocal at best. Some antioxidants exhibit prooxidant activity under certain conditions and potential carcinogenicity under others, thus dietary supplementation with large amounts of a single antioxidant may be deleterious to the health (Boudet, 2007). One of the most crucial conclusions at the present time is that low risk of cancer is more closely related to a diet rich in multiple antioxidants than one supplemented with an individual antioxidant.Acombination of antioxidants with different modes of action is believed to increase efficacy and minimize toxicity (Lee and Lee, 2006).The additive and synergistic effects of dietary phytochemicals obtained from fruits and vegetables are likely responsible for their anticancer activities in a more efficient way than dietary supplements (Liu, 2004). Liu *et al.* (2000) found that a combination of fruits such as orange, apple, grape and blueberry, renders a synergistic effect on antioxidant activity *in vitro*; the half maximal effective concentration (EC50) of these combined fruitswas five times lower than their individual EC50, suggesting a synergistic effect due to the combination of the four fruits (Jaganath and Crozier, 2010). Huang, Liu, Dushenkov, Ho, and Huang (2009) found that oral feeding of orange peel extract, black tea extract and caffeine had anti-obesity effects by suppressing bodyweight gain and adipose tissue formation, suggesting that orange peel extract acted synergistically with black tea extract and caffeine to render the most effective antiobesity action. The health implications of dietary phenolic compounds, including flavonoids, are also dependent on the composition of the components of the diet and the bioavailability of the individual compounds under investigation. Increasing evidence shows that hydroxycinnamic acid derivatives and flavonoids can be absorbed into the human body in amounts that are, in principle, sufficient to exert antioxidant or other biological activities *in vivo* (Olthof, Hollman, and Katan, 2001; Scalbert and Williamson, 2000). Dietary polyphenols are substrates for β-glucosidases, UDP-glucuronosyltransferase, or catechol-*O*methyltransferase in the small intestine aswell as for a number of phase I and II enzymes in the liver (Rechner *et al.*, 2002; Scalbert and Williamson, 2000). In addition, ingested polyphenols are subjected to hydrolysis and degradation in the colon due to the activity of enzymes of the colonic microflora (Booth, Emerson, Jones, and Deeds, 1957). Rechner *et al.* (2002) demonstrated that intact conjugated polyphenolswere present atmuch lower levels than their degradation products resulting from the action of colonic bacterial enzymes and subsequent metabolism in the liver. Consideration should also be given to these degradation products (Boudet, 2007). Phenolic compounds exhibit a wide range of physiological properties, such as anti-allergenic, anti-artherogenic, anti-inflammatory, antimicrobial, antioxidant, anti-thrombotic, cardioprotective and vasodilatory effects (Benavente-Garcia, Castillo, Marin, Ortuño, and Del Río, 1997; Manach, Mazur, and Scalbert, 2005; Middleton, Kandaswami, and Theoharides, 2000; Puupponen-Pimiä *et al.*, 2001). Recent research findings indicate that tea polyphenols can protect against different stages of carcinogenesis (Khan and Mukhtar, 2010).EGCG (epigallocatechin-3-gallate), the main catechin in green tea, serves as a cancer chemopreventive agent (lungs, liver, gastrointestinal tract, skin and prostate cancer), as well as antiobesity and cardiovascular protective compound (Khan and Mukhtar, 2010; Klaus, Pultz, Thone-Reineke, and Wolfram, 2005; Yang and Wang, 1993). The antioxidant activity and beneficial health effects of the main polyphenol of green tea, epigallocatechin gallate (EGCG),was enhanced upon conjugation with docosahexaenoic acid (DHA) and the tetraester so formed was able to arrest colon cancer effectively (Zhong, Chiou, Pan, Ho, and Shahidi, 2012). Polymethoxyflavones, the major components of orange peel, have been found to render health benefits, including anti-inflammatory, anti-carcinogenic, anti-viral, antioxidant, anti-thrombogenic and anti-atherogenic properties (Huang et al., 2009; Lai et al., 2007; Li, Wang, Li, Li, and Wang, 2009; Middleton *et al.*, 2000). Caffeinewas also effective in suppressing body weight gain by stimulating thermogenesis, extending sympathetic stimulation, suppressing food intake and reducing adipose tissue mass (Kobayashi-Hattori, Mogi, Matsumoto, and Takita, 2005; Tremblay, Masson, Leduc, Houde, and Despres, 1988). In addition, hydroxytyrosol, one of the major phenolic constituents in olive oil, was reported to reduce the risk of coronary heart disease and atherosclerosis by itself (Salami, Galli, De Angelis, and Visioli, 1995; Tuck and Hayball, 2002). It has also been postulated that hydroxytyrosol inhibits arachidonic acid metabolism (Petroni et al., 1997) and platelet aggregation (Petroni *et al.*, 1994).

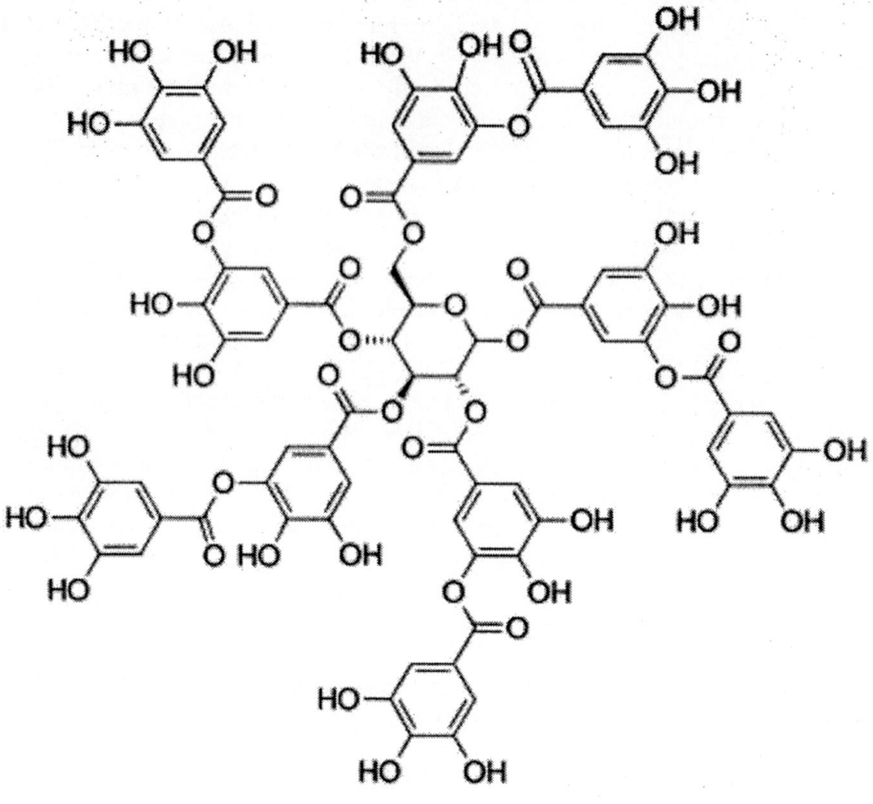

Polyphenols

It is presumed that hydroxytyrosol penetrates into cell membranes and consequently inhibits the production of leukotriene B4 (LTB4) effectively from endogenous arachidonic acid (Kohyama, Nagata, Fujimoto, and Sekiya, 1997; Tuck and Hayball, 2002). Considerable evidence exists from epidemiological and experimental studies for preventive effects of soy or its isoflavones against chronic diseases including cancer (breast, prostate, colorectal, lung), osteoporosis, cardiovascular disorders and menopausal symptoms, but this has not always been consistent (Franke, Halm, Kakazu, and Li, 2010; Levis, Strickman-Stein, Doerge, and Krischer, 2010; Rimbach *et al.*, 2008). Miyake *et al.* (2005) reported that high intake of soy and isoflavones may be associated with reducing the prevalence of allergic rhinitis.The structural similarity of isoflavones to steroidal oestrogens and the potent binding of genistein, one of the most predominant soy isoflavones, to the oestrogen receptor-beta, including its transactivation, are believed to be the basis for many beneficial effects of soy intake (Cotroneo, Wang, Fritz, Eltoum, and Lamartiniere, 2002; Kuiper *et al.*, 1997). Resveratrol is a polyphenolic compound with potent antioxidant activity. It is found in a number of plants, notably grapes, pistachio, peanuts and berry fruit, and is attracting increased attention due to its health benefits, especially incommon age-related diseases such as cancer, type 2 diabetes, cardiovascular disease and neurological conditions (Marques, Markus, and Morris, 2009). Resveratrol first received the attention of the scientific community in 1990s, when it was shown to be responsible for the cardioprotective effects of red wine (Wu *et al.*, 2001). Rayalam, Yang, Ambati, Della-Fera, and Baile (2008) reported that resveratrol may be more effective in lowering body weight at higher doses, and could be a potential treatment for obesity. It can decrease fat mass by inhibiting adipogenesis and can induce apoptosis in adipocytes by affecting expression of genes that modulate mitochondrial function.

Phenolic compounds

The carcinogenesis process involves two major stages: initiation and promotion. The initiation stage involves the intracellular formation of reactive electrophiles that can attack chemical nucleophiles in the cell, including DNA, thus causing its chemical or structural alteration. The promotion stage is the malignant transformation of the damaged cells leading to mutagenesis or carcinogenesis (Newmark, 1984; Namiki, 1990). The initiator may be the agent itself or a product of its metabolic activation by mixed function oxidases. Thus, to block carcinogenesis, a substance may act as follows: (a) prevent the formation of carcinogen (e.g. N-nitroso compounds, active oxygen species) in foods or endogenously, (b) alter the enzymic activation of the potential initiator to ultimate carcinogen, (c) stimulate carcinogen detoxifying systems, or (d) block or trap the ultimate carcinogen electrophiles in a nucleophilic chemical reaction to form innocuous products, or inhibit by arachidonic acid metabolites via the Iipoxygenase or cyclooxygenase pathways (Wattenberg, 1983; Newmark, 1984).

Some recent reviews have identified phenolic compounds in foods as antimutagens and anticarcinogens, perhaps functioning in one or more of the above ways (Wattenberg, 1983; Newmark, 1984; Stich and Rosin, 1984; Cheng and Ho, 1988; Hayatsu *et al.,* 1988; Williams *et al.,* 1989; Namiki, 1990; Caragay, 1992). Several phenolics (e.g. chlorogenic acid, gallic acid, caffeic acid, tannic acid, catechin) have been shown to inhibit the mutagenic effects of both direct-acting carcinogens (e.g. benzo(a)pyrene diol epoxide) and carcinogens that require metabolic activation (e.g. aflatoxin B_1) and to trap nitrite, thus reducing the nitrosat-ing species and preventing the endogenous formation of carcinogenic nitroseamines (Stich and Rosin, 1984). In mice fed diets containing 1·1% caffeic or ferulic acid, about 40% reduction in benzo(a)pyrene induced stomach cancer was observed (Wattenberg *et al.,* 1980). Ellagic acid produced significant reduction in mice skin tumour risk promoted by 12-0-tetradecanoyl-phorbol-13acetate (Chang *et al.,* 1985) or induced by 3-methylcholanthrene (M ukhtar *et al.,* 1986). Ellagic acid, quercetin or robinetin injection also inhibited the incidence of lung tumours in mice (Lesca, 1983; Chang *et al.,* 1985) although other studies showed no significant effect of ellagic acid on lung or skin tumours (Smart *et al.,* 1986). Quercetin and rutin caused reduction in azoxymethanol-induced colon cancer (Deschner *et al.,* 1991). Green tea polyphenols, particularly epigallocatechin gallate, have been shown to reduce the mutagenicity of AFB1, 4-NQO, BaP, MelQ and cigarette smoke condensate, to induce the activity of glutathione-S-transferase and superoxide dismutase, and to reduce the development of skin, liver and lung cancer (Cheng and Ho, 1988).

B. ANTHOCYANIN ACT AS ANTIOXIDANTS

Edible berries, a potential source of natural anthocyanin antioxidants, have demonstrated a broad spectrum of biomedical functions. These include cardiovascular disorders, advancing age-induced oxidative stress, inflammatory responses, and diverse degenerative diseases. Berry anthocyanins also improve neuronal and cognitive brain functions, ocular health as well as protect genomic DNA integrity. This chapter demonstrates the beneficial effects of wild blueberry, bilberry, cranberry, elderberry, raspberry seeds, and strawberry in human health and disease prevention. Furthermore, this chapter will discuss the pharmacological benefits of a novel combination of selected berry extracts known as OptiBerry, a combination of wild blueberry, wild bilberry, cranberry, elderberry, raspberry seeds, and strawberry, and its potential benefit over individual berries. Recent studies in our laboratories have demonstrated that OptiBerry exhibits high antioxidant efficacy as shown by its high oxygen radical absorbance capacity (ORAC) values, novel antiangiogenic and antiatherosclerotic activities, and potential cytotoxicity towards Helicobacter pylori, a noxious pathogen responsible for various gastrointestinal disorders including duodenal ulcer and gastric cancer, as compared to individual berry extracts. OptiBerry also significantly inhibited basal MCP-1 and inducible NF-jb transcriptions as well as the inflammatory biomarker IL-8, and significantly reduced the ability to form hemangioma and markedly decreased EOMA cell-induced tumor growth in an in vivo model. Overall, berry anthocyanins trigger genetic signaling in promoting human health and disease prevention.

Selected anthocyanidins and their substitutions

Basic structure	Anthocyanidin	$R_{3'}$	$R_{4'}$	$R_{5'}$	R_3	R_5	R_6	R_7
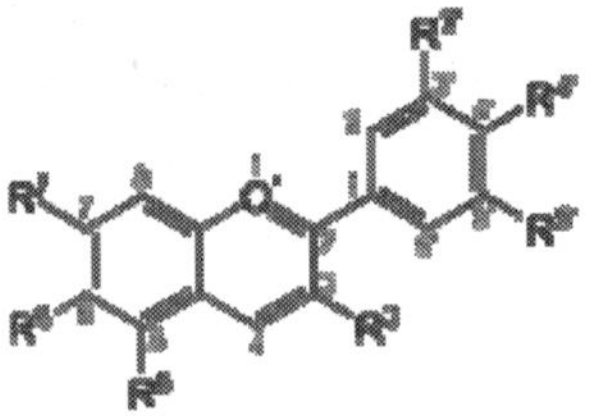	Aurantinidin	-H	-OH	-H	-OH	-OH	-OH	-OH
	Cyanidin	-OH	-OH	-H	-OH	-OH	-H	-OH
	Delphinidin	-OH	-OH	-OH	-OH	-OH	-H	-OH
	Europinidin	$-OCH_3$	-OH	-OH	-OH	$-OCH_3$	-H	-OH
	Pelargonidin	-H	-OH	-H	-OH	-OH	-H	-OH
	Malvidin	$-OCH_3$	-OH	$-OCH_3$	-OH	-OH	-H	-OH
	Peonidin	$-OCH_3$	-OH	-H	-OH	-OH	-H	-OH
	Petunidin	-OH	-OH	$-OCH_3$	-OH	-OH	-H	-OH
	Rosinidin	$-OCH_3$	-OH	-H	-OH	-OH	-H	$-OCH_3$

Role of anthocyanins in human health

In addition to coloring properties, anthocyanins exhibit strong anti-oxidant activity which helps to prevent neuronal diseases, cardiovascular illnesses, cancer, diabetes, inflammation and many such others diseases. Anthocyanins are reported to have effect on treatment of cancers (Nichenametla *et al.,* 2006) and human nutrition

(Stintzing and Carle, 2004). They are reported to be effective in suppressing tumor growth by arresting cell growth between S phase and G2 phase of the cell cycle (Koide *et al.,* 1996). Ability of anthocyanins to counter oxidants makes them atherosclerosis fighters. Anthocyanins are found to relax blood vessels and protect the integrity of the endothelial cells that line the blood vessel walls. Based on animal experiments, strawberry has been shown to have inhibitory effect against esophageal cancer and reverses the neuronal and behavioral aging in these experimental animals (Torronen and Maatta, 2002). This therapeutic activity of strawberries has been correlated with anthocyanin content in these fruits. Other health benefits of anthocyanins include allergy relief, healthy heart (Basu *et al.,* 2010; Wallace, 2011), better eyesight (Ghosh and Konishi, 2007), ulcer treatment and cognitive function (Moskovitz *et al.,* 2002). They have been found to have positive effects in the treatment of various diseases resulting from capillary fragility (Tamura, *et al.,* 1994). For instance preventing cholesterolinduced atherosclerosis and inhibiting platelet aggregation. Therefore, due to the interesting coloring and heath properties, researchers are involved in exploring the natural potential of anthocyanins. Large number of reports are found in the literature regarding the techniques for purification and separation of anthocyanins application of anthocyanins in food (Giusti and Wrolstad, 2003), identification and distribution in plants (Matera *et al.,* 2012), stability (Cavalcanti *et al.,* 2011; Durge *et al.,* 2013), quantitative analysis using chromatographic and electrophoretic techniques (Huang *et al.,* 2009) and their degradation kinetics (Reyes and Zevallos, 2007).

Bioavailability of anthocyanins

The bioavailability is the proportion of a particular nutrient that is digested, absorbed, and metabolized through normal pathways. Bioavailability of anthocyanins is a major issue regarding their biological effects. To perform their multiple effects the bioavailability of anthocyanins present in different fruits and vegetables is important, but it still remains not so well understood issue. Anthocyanins need to be ingested and distributed within the body successfully. However, as anthocyanins are usually ingested in combination with different food sources, the effect of food matrixes on their absorption makes bioavailability studies more complex. Various studies have been carried out to investigate the bioavailability of anthocyanins (Clifford, 2000; McGhie and Walten, 2007). The bioavailability of anthocyanins is less than 1% (Bub *et al.,* 2001; Matsumoto, *et al*., 2001; Manach *et al.*, 2005) Anthocyanins containing different glycosides have different bioavaliabilities. In general, non-acylated anthocyanins are better absorbed than acylated ones (Tsuda, *et al.,* 1996; Zhang, *et al.,* 2005). Bub *et al.* (2001) while studying malvidin-3-glucoside (M3G), an anthocyanin, occurring in red wine and red grape juice reported that M3G is poorly absorbed and not anthocyanins themselves rather yet not defined anthocyanin metabolites or other polyphenols might be responsible for the observed antioxidant and health effects. Anthocyanins are rapidly absorbed and eliminated and that they are absorbed with poor efficiency (Manach *et al.,* 2005). As such, continuous intake of anthocyanins may be needed for their systemic benefits.

Anthocyanins are important components present naturally in most of the fruits, vegetables and in some cereals. In addition to the coloring properties they provide a number of health benefits but are very sensitive to environmental conditions during processing and storage. Encapsulation can be used to improve stability of anthocyanins. Till now no substantial work has been done on microencapsulation of anthocyanins. Only a few researchers have worked on this area and use of spray drying method for encapsulation has been reported. However, use of other techniques for encapsulation of anthocyanins is still an unexplored field of research. Researchers have also used maltodextrin as the coating material for microencapsulation of anthocyanins. Different coating materials can be exploited to with stand different conditions of processing and for target delivery of anthocyanins. Therefore, ample opportunities exist to explore this field of research and to take it from its infancy to a well examined stage.

C. CITRIC ACID ACT AS AN ANTIOXIDANTS

Health Benefits of Citric Acid

Juicy grapefruits, lemons, and oranges are delicious to eat and can also help you out health wise. There are many health benefits of citric acid that will help you clear free radicals out of your system, get rid of painful kidney stones, and keep the acid levels in your body at a normal level. Unfortunately, the delectable fruits that contain citric acid aren't always in season but you can work around this by taking citric acid supplements and getting the same benefits all year round. Consult your physician before doing this and find out how citric acid can improve your overall health.

It's an anti-oxidant

Because citric acid is an anti-oxidant, it can reduce the harmful impacts free radicals have on your body. What are free radicals you ask? Free radicals are compounds that build up inside you after you have been exposed to harmful chemicals. These can accumulate after you have been around cigarette smoke, cleaning with chemicals, or around a factory that works with noxious substances. The health benefits of citric acid are many, and its' power as an anti-oxidant will keep harmful materials from getting in the way of your body's regular routines.

Keeps kidney stones at bay

Kidney stones are one of the most uncomfortable and painful medical situations known to man, rivalled only by giving birth. The acid in citrus fruits and citric acid supplements can help prevent you from landing up in the hospital with these painful little stones. Citric acid has alkalizing properties that effectively break up the mineral compounds that build up and cause kidney stones. When it comes to the health benefits of citric acid, this may be the most valuable one for many individuals.

Controls acid balance

You might think it's ironic to fight acid with acid, but the acid levels in your body can be regulated with a little help from citric acid. There is a condition known as metabolic acidosis that happens when your blood has high levels of acid in it. The causes of this disease are numerous, and it can make a person overly tired, mentally unaware, and eventually lead to an unfortunate death. This can be prevented by a little more citric acid in your diet through more citric fruits or citric acid supplements.

The many health benefits of citric acid might make eating your morning grapefruit a little more tolerable. If you were looking for a way to help out your kidneys, break up free radicals in your body, and control the acid levels inside yourself, then citric acid is the answer.

Used for centuries to preserve meat and fish, citric acid has innumerable health benefits. It increases the acidity of the bacterial environment, making it difficult for the microbes to survive and, more importantly, to reproduce. Being an acid, citric acid is an alkalizing agent that helps decrease acidity levels in body fluids. It also binds the excess calcium and allows it to leave the body.

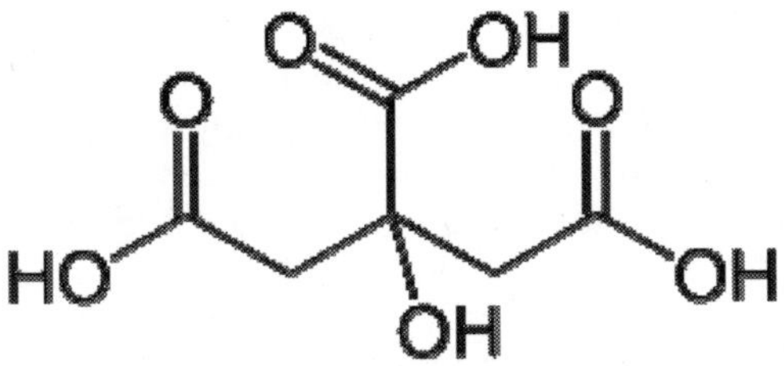

Citric acid

Citric Acid: Naturally Healthy

Citric acid can be obtained both naturally (in plants) and artificially (in human tissues and body fluids). It's present in large quantities in juicy fruits such as limes, oranges, grapefruits, kiwis, strawberries, apples, pears, cherries and raspberries and vegetables such as mushrooms, potatoes, tomatoes, peas and asparagus. Citric acid is often used for proper mineral supplementation of food, as an acidity regulator, and as a flavouring ingredient. Though not an antioxidant itself, citric acid enhances the activity of many beneficial antioxidants. All these properties offer a wide range of health benefits.

- **Boosting Renal Health:** Citric acid ensures proper functioning of the kidneys and prevents the formation of kidney stones. Its alkalizing properties and the ability to bind to calcium are useful in treating other kidney ailments, too.
- **Fighting Free Radicals:** As an antioxidant, citric acid can neutralize the damaging effects of free radicals, which are unstable compounds that trigger the growth of cancerous tumours.
- **Improving Mineral Absorption:** Experts say that citric acid mixes with minerals and metals easily, facilitating their absorption in the body.
- **Revitalizing The Skin:** Citric acid is a common ingredient in face packs and skin products. It helps in the regeneration of skin tissue, slowing down the

aging process. Creams rich in citric acid can have a depigmenting effect and can lighten freckles.

- **Soothing a Sore throat:** Citric acid can reduce the inflammation caused by tonsillitis. Gargling with a combination of citric acid and water clears throat infection by killing the infection-causing germs. (Related Article: 6 Natural Sore Throat Remedies)
- **Curbing Nausea:** Citric acid can alleviate nausea by reducing gastric acidity.

D. VITAMIN E ACTS AS ANTIOXIDANTS

Vitamin E refers to a group of compounds that include both tocopherols and tocotrienols. Of the many different forms of vitamin E, γ-tocopherol is the most common form found in the North American diet. The γ-Tocopherol can be found in corn oil, soybean oil, margarine, and dressings. The α-tocopherol, the most biologically active form of vitamin E, is the second-most common form of vitamin E in the diet. This variant can be found most abundantly in wheat germ oil, sunflower, and safflower oils. As a fat-soluble antioxidant, it stops the production of reactive oxygen species formed when fat undergoes oxidation. Regular consumption of more than 1,000 mg (1,500 IU) of tocopherols per day may be expected to cause hypervitaminosis E, with an associated risk of vitamin K deficiency and consequently of bleeding problems.

Vitamin E has many biological functions, the antioxidant function being the and best known. Other functions include enzymatic activities, gene expression, and neurological function(s).

- As an antioxidant, vitamin E acts as a peroxyl radical scavenger, disabling the production of damaging free radicals in tissues, by reacting with them to form a tocopheryl radical, which will then be reduced by a hydrogen donor (such as vitamin C) and thus return to its reduced state. As it is fat-soluble, it is incorporated into cell membranes, which protects them from oxidative damage. Vitamin E has also found use as a commercial antioxidant in ultra-high molecular weight polyethylene (UHMWPE) used in hip and knee replacements, to help resist oxidation.
- As an enzymatic activity regulator, for instance, protein kinase C (PKC), which plays a role in smooth muscle growth, can be inhibited by α-tocopherol. The α-Tocopherol has a stimulatory effect on the dephosphorylation enzyme, protein phosphatase 2A, which in turn, cleaves phosphate groups from PKC, leading to its deactivation, bringing the smooth muscle growth to a halt.
- Vitamin E also has an effect on gene expression. Macrophages rich in cholesterol are found in the atherogenetic tissue. Scavenger receptor CD36 is a class B scavenger receptor found to be up-regulated by oxidized low density lipoprotein (LDL) and binds it. Treatment with α-tocopherol was found to downregulate the expression of the CD36 scavenger receptor gene and the scavenger receptor class A (SR-A) and modulates expression of the

connective tissue growth factor (*CTGF*). The *CTGF* gene, when expressed, is responsible for the repair of wounds and regeneration of the extracellular tissue lost or damaged during atherosclerosis.

- Vitamin E also plays a role in neurological functions and inhibition of platelet aggregation.
- Vitamin E also protects lipids and prevents the oxidation of polyunsaturated fatty acids.

So far, most human supplementation studies about vitamin E have used only α-tocopherol. This can affect levels of other forms of vitamin E, e.g. reducing serum γ- and δ-tocopherol concentrations. Moreover, a 2007 clinical study involving α-tocopherol concluded supplementation did not reduce the risk of major cardiovascular events in middle-aged and older men.

Form	Structure
alpha-Tocopherol	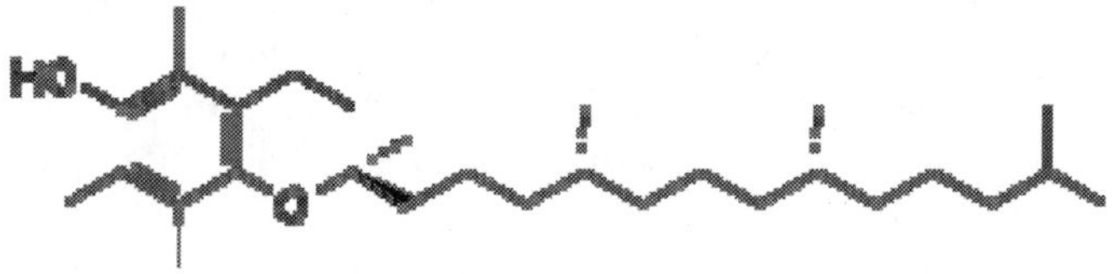
beta-Tocopherol	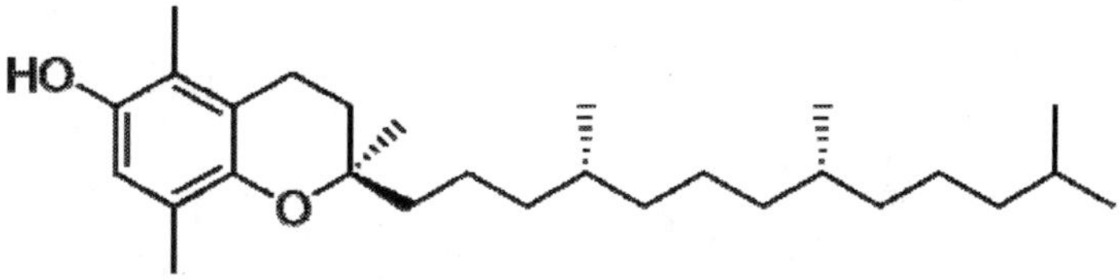
gamma-Tocopherol	
delta-Tocopherol	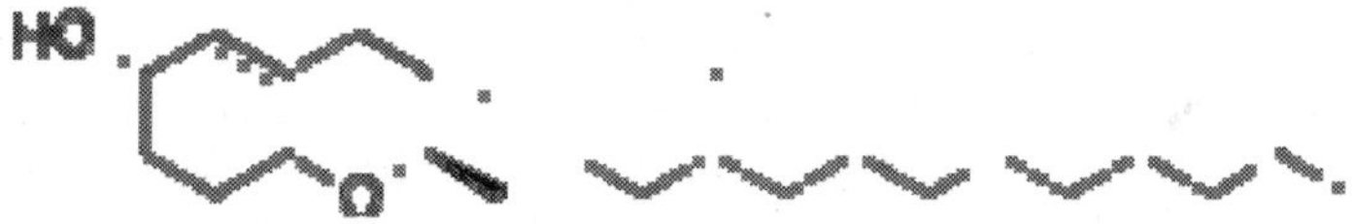

E. LYCOPENE AT AS ANTIOXIDANTS

Lycopene is a naturally present carotenoid in tomatoes. Among the carotenoids, lycopene is a major component found in the serum. High levels of lycopene have also been found in the testes, adrenal glands, and prostate. Several recent studies including cell culture, animal and epidemiological investigations have indicated the effect of dietary lycopene in reducing the risk of chronic diseases such as cancer and coronary heart disease. Although, the antioxidant properties of lycopene are thought to be primarily responsible for its beneficial properties, evidence is accumulating to suggest other mechanisms such as intercellular gap junction communication, hormonal and immune system modulation and metabolic pathways may also be involved.

Lycopene, a predominant carotenoid in tomatoes, exhibits highest antioxidant activity among all dietary carotenoids. Its potential role in human health is beginning to be recognized. The main supporting evidences come from three different research directions: (a) tissue culture studies using different cell lines to investigate anti cell proliferative activities, (b) use of animal models to study anti tumorogenic properties of lycopene and (c) epidemiological studies done with control and at risk population.

Dietary guidelines for the prevention of cancer and coronary heart disease recommend increased consumption of plant foods that include fruits, vegetables, cereals and legumes. Presence of micronutrient phytochemical compounds are now being recognized as playing an important role in the disease prevention properties of the plant foods, possibly through their effect on hormonal metabolism and oxidative damage. Lycopene, a carotenoid antioxidant, has been shown in recent epidemiological and experimental studies to protect against oxidative damage and contribute towards reducing the risk of cancer and coronary heart disease. In addition to its antioxidant mediated activity, other metabolic effects of lycopene have also been demonstrated.

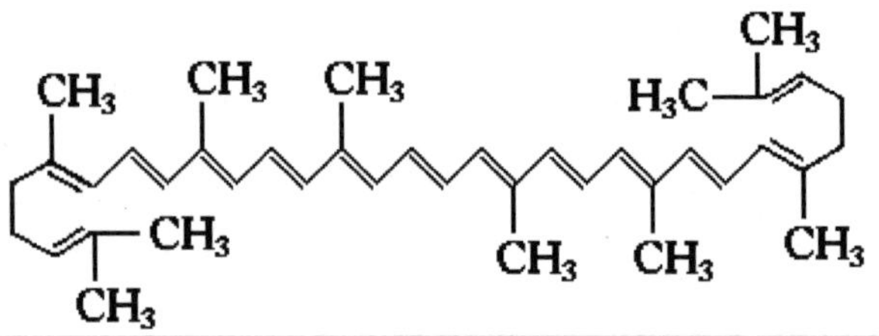

Lycopene

Lycopene's function in the human body

Although not considered an essential nutrient, research has shown that lycopene may have various benefits for human health. As a major carotenoid in human blood, lycopene protects against oxidative damage to lipids, proteins, and DNA. Lycopene is a potent quencher of singlet oxygen (a reactive form of oxygen), which suggests that it may have comparatively stronger antioxidant properties than other major plasma carotenoids. Lycopene has been found to be a potent and specific inhibitor of cancer cell proliferation, which is regulated by an elaborated cellular process called "cell cycle." Rapid and uncontrolled cell division is a hallmark of cancer cell metabolism; lycopene's activity in retarding cell cycle progression may explain its demonstrated activity in retarding the spread of certain types of cancer. Lycopene may prevent malignant transformation (the cellular process which describes the transformation of a normal cell into a cancer cell). Contact inhibition is one of the mechanisms that control excessive cell division. By this mechanism, a cell, in crowded surroundings, will stop multiplying. Special structures in the cell membrane, termed a "gap-junction," function as communication channels between cells. Normal cells are both contact-inhibited and have a functional gap-junction whereas most tumour cells exhibit fewer of these structures. Lycopene was found to induce the formation of the protein connexion 43, one of the major building blocks of these channels, and thereby to restore gap junctions. Lycopene induces Phase II enzymes which help to eliminate carcinogens and toxins from the body. The change of the levels of so many regulatory proteins is related to lycopene's ability to modulate various transcription factors which are key players in the process of new cellular protein synthesis.

F. BETA CAROTENE ACT AS ANTIOXIDANTS

Beta-carotene, the most common form of carotene, is an important organic compound derived in abundance from fruits and vegetables. There are ten types of carotenes that are synthesized by plants, but alpha and beta carotene are the only types that the human body can use as a precursor to Vitamin A. And within the hundreds of types of carotenoids created by plants, only 10% give us the important provitamin form of vitamin A carotenoid.

Beta carotene is further typified as a terpenoid, a kind of red to yellowish-orange pigment required for the production of vitamin A (retinol) in the body. In order for us to have healthy eyes, skin, and an optimally functioning heart, the body needs Vitamin A, and the beta-carotene that supports its productions. What is more, beta-carotene is another form of anti-inflammatory and free-radical-fighting antioxidant. It is from these antioxidant capacities that beta carotene aids in stopping harmful cancerous tumor growth.

Mechanisms of Action in Beta Carotene

Beta carotene is a fat-soluble (lipophilic) hydrocarbon that is naturally synthesized by plants. It is converted into retinol in the human body. Beta carotene may act to inhibit changes cellular modulation and cell cycle regulatory proteins, alterations in insulin-like growth factor-1. These processes stop a cell from changing into a cancerous growth by preventing the process of oxidative DNA damage. What is more, this hydrocarbon antioxidant can induce glutathione production in our cells, a process which also increases antioxidant activity. Beta carotene may also increase the amount and functioning of cancer-fighting enzymes in the cells. Beta carotene also stimulates increased levels of T-helper lymphocytes, (NK) cell cytotoxicity (a natural form of cancer-killing cells), as well as macrophage action.

Form	Structure
Beta carotene	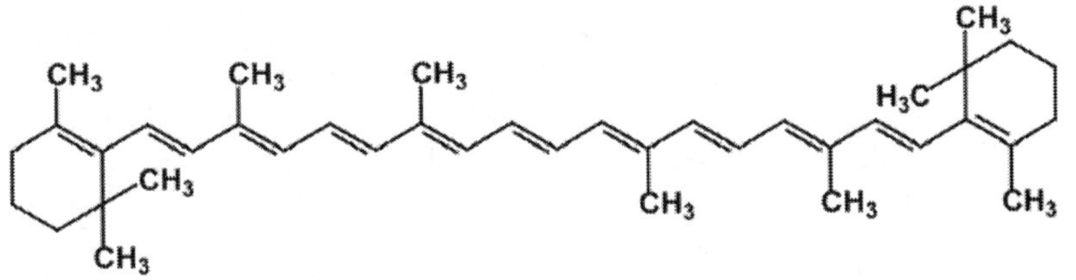 CH_3 CH_3 CH_3 CH_3 H_3C CH_3 CH_3 CH_3 CH_3 CH_3

Alpha carotene

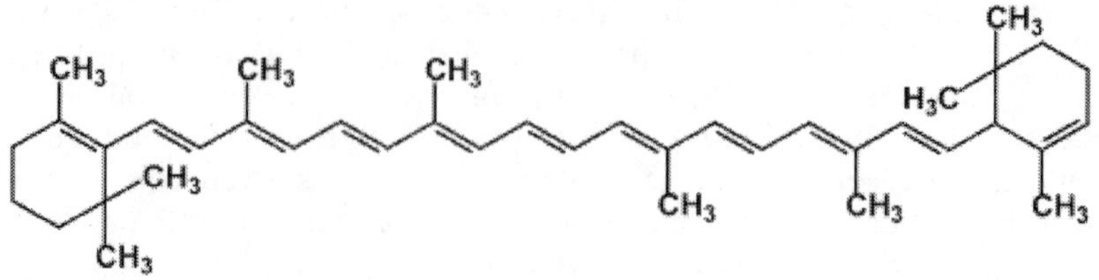

Health Benefits of Beta Carotene

The health benefits of this vitamin A precursor are numerous. The following highlights some of the most important and effective benefits that the antioxidant may provide:

- May prevent night blindness
- May prevent skin disorders
- Acts as a natural immune system booster
- Protects the body against cancer
- Reduces chances of developing colds, flu, and infections
- Has important protective mechanisms for the formation of bones and teeth
- May reduce the chances of developing lupus
- May ease the symptoms of arthritis
- May aid to decrease risks for certain types of cancer when taken in the form of consistent fruits and vegetables
- Acts as a powerful antioxidant
- May prevent or treat certain types of cancer such as cervical cancer, but may actually increase lung and prostate cancers
- May treat HIV
- May treat or prevent heart disease
- May in treatment and prevention of cataracts, as well as macular degeneration

G. HEALTH BENEFITS OF NUT CO-PRODUCTS

While many human intervention studies have evaluated the potential health benefits of nuts, few studies have included nut co-products. A larger evidence base for the function of nut polyphenols or nut co-products in animal models of human disease is available. Given the extensive metabolism of nut polyphenols discussed in details and in vitro experiments of crude extracts are not included in this review. To the best of our knowledge, reports on the function of Brazil nut, macadamia, and pine nut co-products for human or animal health are not available. Health benefits of almond, cashew, hazelnut, pistachio, pecan, and walnut co-products are reviewed in detail.

Almond

An almond blanch-water extract was incorporated into a skin treatment to prevent ultra-violet (UV)-B-induced erythema in humans (Mandalari et al., 2013). Application of a 2% blanches water extract in an oil-in-water emulsion reduced UV-Binduced erythema by 50%, whereas the tocopheryl acetate positive control reduced erythema by 22% (Mandalari et al., 2013). Almond skin or its polyphenol-rich extract have been applied to animal models of disease. Consumption of 50 mg/kg body weight (bw) almond skin polyphenol extract increased hamster ex vivo LDL oxidation resistance (Chen et al., 2005). Consumption of 30 mg/kg bw almond skin reduced inflammation after spinal cord injury in CD1 mice (Mandalari et al., 2011a, 2011b). However, polyphenol-depleted almond skin obtained after blanching whole almond did not confer protection in the same model (Mandalari et al., 2011b). The same 30 mg/ kg bw dose of almond skin reduced colitis symptoms and colonic cytokines after rectal injection of dinitrobenzene sulphonic acid in CD1 mice (Mandalari et al., 2011a). An oral gavage of 10 mg/kg of isolated almond skin procyanidins to mice reduced acetaminophen-induced hepatotoxicity (Truong et al., 2014).

Mandalari et al. (2010b) evaluated the prebiotic potential of almond skin using an in vitro gastric and duodenal digestion, followed by colonic fermentation using mixed faecal bacterial cultures. Their results showed that almond skin exhibited the potential to be used as a novel source of prebiotics, increasing the populations of bifidobacteria and Clostridium coccoides/Eubacterium rectale group and decreasing the number of the Clostridium hystolyticum group. In this connection, Mandalari et al. (2010a) demonstrated that almond skin fractions inhibited gram-positive bacteria (Staphylococcus aureus and Listeria monocytogenes) with the raw skin being more active than blanched skin. Liu et al. (2014b) evaluated the prebiotic effects of almond and almond skin intake in healthy humans. Almond and almond skin ingestion may lead to an improvement in the intestinal microbiota profile and a modification of the intestinal bacterial activities, by inducing changes in bacterial enzyme activities, specifically a significant increase in faecal β-galactosidase activity and decreases in faecal β-glucuronidase, nitroreductase, and azoreductase activities. These would induce the promotion of health beneficial factors and the inhibition of harmful factors. In fact, the presence of various flavonoids and phenolic acids can contribute to the prebiotic properties of almond skin (Mandalari, 2012; Mandalari et al., 2010a, 2010b, 2010d).

Cashew

Consumption of cashew-derived anacardic acids prevented ethanol-induced gastric damage, pH-induced erosion of tooth enamel, and improved cognitive performance in rodents. Acute oral consumption of anacardic acids from cashew liquid, 10, 30, or 100 mg/kg, prevented subsequent gastric damage from ethanol administration to mice (Morais et al., 2010). Daily application of anacardic acids isolated from cashew liquid to bovine teeth prevented pH-induced erosion ex vivo (Silveira et al., 2014). Supplementation of 100 mg/kg/d cardanol acetate for 2 months to young adult Wistar rats reduced their errors on a radial maze (Tobinaga et al., 2012)

Hazelnut

Consumption of a polyphenol-rich hazelnut skin extract prevented increases in LDL and total plasma cholesterol after the consumption of high-fat diets by Golden Syrian hamsters (Caimari et al., 2015). However, liver lipids, cholesterol, and triacylglycerols (TAG) were not different among high-fat treatments (Caimari et al., 2015). The hazelnut skin extractenriched diet also reduced plasma TBARS represented as MDAE as a marker of oxidative stress (Caimari et al., 2015).

Pecan

A polyphenol-rich aqueous acetone pecan extract was fed to Wistar rats that consumed a high-fat diet for 9 weeks (Dominguez-Avila et al., 2015). Pecan polyphenol-supplemented rats had lower plasma lipids than the high fat control group (Dominguez-Avila et al., 2015). An aqueous pecan shell extract containing gallic acid, rutin, and other polyphenols was also applied to several rodent models of chemical-induced toxicity. Consumption of the pecan shell extract for 30 days, as 5% ad libitum, protected against cyclophosphamide-induced toxicity and reduced oxidative stress and lipid peroxidation (Benvegnu et al., 2010; 2013). Pecan shell extract consumption also prevented ethanol-induced hepatic damage to rats and cigarette smoke-induced oxidative stress in mice (Muller et al., 2013; Reckziegel et al., 2011). Acute consumption of 300 and 1000 mg pecan shell extract/kg reduced acetic acidinduced writhing in Swiss mice, as well as hydrogen peroxide and cinnamaldehyde-induced paw oedema (Trevisan et al., 2014).

Pistachio

A polyphenol-rich pistachio skin extract was incorporated into a topical formulation for preventing UV-B skin damage in adults (Martorana et al., 2013). Application of the pistachio skin extractfortified product inhibited skin erythema by 65%, compared to 23% inhibition by the tocopheryl acetate control (Martorana et al., 2013). In rats, topical application of a pistachio-hull extract accelerated wound healing (Farahpour, Mirzakhani, Doostmohammadi, and Ebrahimzadeh, 2015). In contrast, acute oral consumption of aqueous or ethanolic extracts of pistachio skin did not inhibit carrageenan-induced hind paw oedema in mice (Orhan, Kupeli, Aslan, Kartal, and Yesilada, 2006). Pistachio polyphenol-supplemented atherogenic diets also reduced the development of aortic lesions and lowered plasma TBARS represented as MDAE in rabbits (Marinou et al., 2010).

Walnut

Consumption of 5.6 g of walnut skin with a gelatine matrix lowered reactive hyperemia index from baseline, but did not affect plasma total thiols, FRAP or TBARS in 15 adult participants (Berryman et al., 2013). Mice that consumed high-fat diets fortified with 100 or 200 mg/kg bw walnut skin polyphenols for 13 days had less serum and liver TAG than mice consuming a control diet (Shimoda et al., 2009). A walnut polyphenol fraction was orally administered to C57BL/KsJ-db/db diabetic mice at 200 mg/kg for 4 weeks and found to reduce urinary 8-hydroxydeoxyguanosine and serum TAG, but not glucose (Fukuda, Ito, and Yoshida, 2004). An acute oral dose of walnut skin polyphenols at 200 mg/kg bw inhibited carbon tetrachloride liver injury in mice (Shimoda et al., 2008).A chronic 100 mg/kg bw dose of walnut extract reduced pulmonary oedema formation from cigarette smoke extract inWistar rats (Qamar and Sultana, 2011). Walnut husk is rich in juglone, which prevented high-fat diet induced weight gain in rats and also prevented diet-induced increases in serum endotoxin and TBARS contents expressed as MDAE (Peng, Nie, Wu, Huang, and Cheng, 2015).

H. NUTRITIONAL IMPORTANCE OF MILLETS IN HUMAN HEALTH

The key element in maintaining overall physical well-being is directly proportional to nutritional wellbeing showing as sustainable force for health, development and maximization of human genetic potential. In the food industry, cereal grains and plant nutrients are largely used as a major source of dietary nutrients worldwide. The richness in calcium, dietary fiber, polyphenol and protein contents in millets make them unique among the cereals. Generally, millets show significant amounts of amino acids (methionine and cysteine) and also have high fat content than maize, rice and sorghum.

Foxtail millet which is rich in essential amino acid lysine can be used as a supplementary protein source to most cereals and the millet protein characterization showed that its protein concentrate is a potential functional food ingredient. When compared to the protein of wheat and grain, proso millet showed significantly higher content 11.6% (dry matter) and was also rich in essential amino acids (leucine, isoleucine and methionine) than wheat protein. The average of nutrient composition of some millet grains and other grains is summarized in Table 1. The amino acid contents of various millets are given in Table 2.

About 5-8% of protein is present in finger millet, 65-75% carbohydrates, 15-20% dietary fiber and 2.5-3.5% minerals. The naked caryopsis of finger millet with brick red coloured seed coat is generally used in the form of whole meal in the traditional food preparations such as *roti,* muddle and *ambali* (thin porridge). Regular consumption of whole grain cereals and their products have shown in epidemiological studies that they can protect against risk of *diabetes mellitus,* gastro-intestinal diseases and cardiovascular risks. The use of millets as whole grain makes the essential nutrients such as dietary fiber, minerals, phenolics and vitamins concentrated in the outer layer of the grain or the seed coat form the part of the food and offer their nutritional and health benefits.

The edible component of millet kernel is the rich source of phytochemicals, such as dietary fiber and polyphenols (0.2-0.3%). Also millets contribute to antioxidant activity with phytates, polyphenols and tannins present in it having important role in aging and metabolic diseases. The highest calcium content is present is finger millet with 344 mg/100g among the cereals; Also rich in phytates 0.48g/ 100g, polyphenols, tannins 0.61%.

In addition, black finger millet contains 8.71 mg/g dry weight fatty acid and 8.47g/g dry weight protein. Kodo millet and little millet were also reported to have 37% to 38% of dietary fiber, which was once considered as 'anti-nutrient' and is now termed as a nutraceutical and highest among cereals. Thus, it makes millets a complete food ingredient suitable for large scale utilization as processed products, snacks, baby foods etc. and also plays a major role in propagating food security among under developed and developing countries.

Table 5.1: Nutritive value of cereals/millets

Cereals/ Millets (%)	Protein %	Fat (%)	Crude Fiber (%)	Ash (%)	Carbohydrate (%)	Total dietary fiber (%)	Total phenol (mg/100 g)
Wheat	14.4	2.3	2.9	1.9	64	12.1	20.5
Rice	7.5	2.4	10.2	4.7	77.2	3.7	2.51
Maize	12.1	4.6	2.3	1.8	62.3	12.8	2.91
Sorghum	11	3.2	2.7	1.8	73.8	11.8	43.1
Barley	11.5	2.2	5.6	2.9	58.5	15.4	16.4
Oats	17.1	6.4	11.3	3.2	52.8	12.5	1.2
Rye	13.4	1.8	2.1	2	68.3	16.1	13.2
Finger millet	7.3	1.3	3.6	3	59	19.1	102
Pearl millet	14.5	5.1	2	2	60.5	7	51.4
Proso millet	11	3.5	9	3.6	56.1	8.5	NA
Foxtail millet	11.7	3.9	7	3	59.1	19.11	106
Kodo millet	8.3	1.4	9	3.6	72	37.8	368

Source: (a) Kamara *et al.* (2009); (b) Bagdi *e. al.*, 2011; (c) Saldivar (2003); (d) Devi *et al.* (2011).

Table 5.2: Amino acid profiles of different millets

Amino acids (g/100g)	Foxtail millet (defatted flour) (a)	Proso millet (dehulled grain) (b,c)	Pearl millet (true prolamine) (c)	Finger millet (native grain) (d)
Essential amino acid				
Iso leucine	4.59	4.1	5.1	4.3
Leucine	13.6	12.2	14.1	10.8
Lysine	1.59	1.5	0.5	2.2
Methionine	3.06	2.2	1	2.9
Phenylalanaine	6.27	5.5	7.6	6
Threonine	3.68	3	3.3	4.3
Valine	5.81	5.4	4.2	6.3
Histidine	2.11	2.1	1.7	2.3
Tryptophan	NA	0.8	1.2	NA

Amino acids (g/100g)	Foxtail millet (defatted flour) (a)	Proso millet (dehulled grain) (b,c)	Pearl millet (true prolamine) (c)	Finger millet (native grain) (d)
Non essential amino acid				
Alanine	9.3	10.9	8.1	6.1
Arginine	3	3.2	0.9	3.4
Aspartic acid	7.71	6.2	6.2	5.7
Cysteine	0.45	NA	0.8	NA
Glutamic acid	22	21.3	22.8	23.2
Glycine	2.91	2.1	0.7	3.3
Serine	4.56	6.3	5.4	5.3
Tyrosine	2.44	4	2.7	8.6
Proline	5.54	7.3	8.2	9.9
*PER(b)	0.8	1.1	1.6	2

*PER-Protein Efficiency Ratio; NA-Not available.

Health benefits of millets

Millets have potential health benefits and epidemiological studies have shown that consumption of millets reduces risk of heart disease, protects from diabetes, improves digestive system, lowers the risk of cancer, detoxifies the body, increases immunity in respiratory health, increases energy levels and improves muscular and neural systems and are protective against several degenerative diseases such as metabolic syndrome and parkinson's disease. The important nutrients present in millets include resistant starch, oligosaccharides, lipids, antioxidants such as phenolic acids, avenanthramides, flavonoids, lignans and phytosterols which are believed to be responsible for many health benefits.

Cardiovascular diseases

Being rich sources of magnesium, millets help in reducing blood pressure and risk of heart strokes especially in atherosclerosis. Also, the potassium present in millets helps in keeping blood pressure low by acting as a vasodilator and help to reduce cardiovascular risk. Also, the plant lignans present in millets have the ability to convert into animal lignans in presence of mircroflora in digestive system and protect against certain cancers and heart disease. The high fiber presents in millets play a major role in cholesterol lowering eliminating LDL from the system and increasing the effects of HDL. Improved plasma levels of adiponectin, high density lipoprotein (HDL) cholesterol in genetically obese type-2 diabetic mice under high fat conditions were observed on feeding proso millet. The finger millets and proso millets have also shown to lower significantly the concentrations of serum triglycerides than white rice and sorghum fed rats. Finger millet and proso millet may prevent cardiovascular disease by reducing plasma triglycerides in hyperlipidemic rats.

Diabetes mellitus

Diabetes mellitus is a chronic metabolic disorder characterized by hyperglycemia with alterations in carbohydrate, protein, and lipid metabolism. It is considered as the most common endocrine disorder and results in deficient insulin production (type 1) or combined resistance to insulin action and the insulin-secretory response (type 2). The efficiency of insulin and glucose receptors in the body is increased by the significant levels of magnesium content present in millets and help in preventing diabetes. Finger millet based diets have shown lower glycemic response due to high fiber content and also alpha amylase inhibition properties which are known to reduce starch digestibility and absorption. Finger millet also had shown significant results in dermal wound healing process. Few studies done on rats also showed that it improves antioxidant status and controlled blood sugar levels. Barnyard millet has been reported to be beneficial for type 2 diabetics especially the de hulled varieties, as the glycemic index for dehulled millet (50.0) and heat treated is millet 41. 7.

The rapidly digestible carbohydrate and slowly digestible carbohydrates were reduced significantly in the *in-vitro* studies performed in the extruded products made from amaranth, buck wheat and millet combination food products. The polyphenols of finger millets in a study were found as major antidiabetic and antioxidant components, when evaluated for aldose reductase (AR) - inhibiting activity the reason being the phenolic with an OH group present at the 4th position which are responsible for this inhibitory activity. The gallic, protocatechic, p-hydroxy benzoic, p-coumaric, vanillic, syringic, ferulic, trans-cinnamic acids, quercetin inhibited cataract eye lens effectively. Therefore, these studies show strong evidence for finger millets protein in inhibiting the cataractogeneisis in humans. The strong inhibition on α-glucosidase and pancreatic amylase were found by few other phenolic compounds from the millet seed coat. The aqueous extracts of foxtail millets have excellent anti hyperglycemic activity activity. Proso millet had shown to improve the glycemic. Therefore, millet grains have the potentials to be useful in preventing diabetes and for treatment of diabetics.

Gastrointestinal disorders

Regulating digestive process can increase nutrient retention and reduce chances of more serious gastrointestinal conditions like gastric ulcers or colon cancer. Fiber content in millets helps in eliminating disorders like constipation, excess gas, bloating and cramping. An immune mediated enteropathic disease called celiac disease which is usually triggered by the ingestion of gluten in susceptile individuals. A gluten free diet primarily plays a major role in affecting food consumption in the grain food group. Replacing cereals like wheat, barley, rye-based foods with foods made from gluten free grains, including rice, corn, sorghum, millet, amaranth, buck wheat, quinoa, wild rice may help people adhering to gluten free diet. As millets are gluten free, they have considerable potential in foods and beverages and can meet the growing demand for gluten free foods and will be suitable for individuals suffering from celiac disease.

Cancer

Millet grains based on literature values are known to be rich in phenolic acids, tannins, and phytate. These nutrients reduce the risk for colon and breast cancer in animals. The fiber present in sorghum and millet and also the phenolics have been attributed for lower incidences of esophageal cancer than those consuming wheat or maize. Recent research has revealed that fiber as one of the best and easiest ways to prevent the onset of breast cancer in women. The chances of breast cancer can be reduced by more than 50% by consumption of more than 30 gm. of fiber every day.

Detoxification (Anti-oxidant properties)

Many of the antioxidants found in millet have beneficial impact on neutralizing the free radicals, which can cause cancer and clean up other toxins from body such as those in kidney and liver. Quercetin, cucurmin, ellagic acid and various other beneficial catechins can help to clear the system on any foreign agents and toxins by promoting proper excretion and neutralizing enzymatic activity in those organs. Therefore, tremendous attention has been given to polyphenol due to their roles in human health. The antioxidant, metal chelating and reducing powers are shown by the soluble and insoluble bound phenolic extracts of several varieties of millet (kodo, finger, foxtail, proso, pearl and little millets). Foxtail millet contains 47mg polyphenolics/100 g and 3.34 mg tocopherol/100 g (wet basis); however, proso millet contains 29 mg polyphenolics/100 g and 2.22 mg tocopherol/100 g (wet basis). In addition, a positive and significant correlation (R^2 =0.9973, $P<0.01$) between polyphenolic content and radical cation scavenging activity was observed. Over 50 phenolic compounds belonging to several classes, namely, phenolic acids and their derivatives, dehydrodiferulates and dehydrotriferulates, flavan-3-ol monomers and dimers, flavonols, flavones, and flavanonols in 4 phenolics fractions of several whole millet grains (kodo, finger, foxtail, proso, little, and pearl millets) were positively or tentatively identified using HPLC and HPLC-tandem mass spectrometry (MS). Therefore, millet grains can be used as functional food ingredients and as sources of natural antioxidants. Kodo millet, finger millet, little millet, foxtail millet, barnyard millet, and sorghum bicolor grown in India and their white varieties were screened for free radical quenching of I-ldiphenyl-2-picrylhydrazyl (DPPH) by electron spin resonance. Furthermore, finger millet extracts were found to have a potent radical-scavenging activity that is higher than those of wheat, rice and other species of millet. In addition, defatted foxtail millet protein hydrolysates also exhibited antioxidant potency. Thus, millets may serve as a natural source of antioxidants in food applications and as a nutraceuticals and functional food ingredient in health promotion and disease risk reduction.

Millets contain many nutritional and health benefits. Nutritionally important components in millets such as dietary fiber, minerals, vitamins, and phytochemicals that include phenolic compounds are important contributors for positive health benefits in humans. The antioxidant and scavenging active compounds in the millets play a major role in preventing cancer and protecting the human system from cancer. Thus, further studies on processing and changes in nutrients should focus on nutritional and health benefits of millets. Also novel approaches and models should be proposed to improve the effectiveness of the millet consumption. Preparation and distribution of millets and millet products should be propagated more to promote food security which includes nutritional security.

Bioactivities of millet phenolics

Foods rich in phenolic compounds have been shown to possess antioxidative properties (Benavente-Garc□a, Castillo, Lorente, Ortun□o, and Del Rio, 2000; Visioli and Galli, 1998). Natural antioxidative substances occurring in foods have gained more interest over the synthetic counter parts due to the safety concerns for the latter. A number of studies have reported the antioxidant activity of the extracts of several millet grains (Chandrasekhar a and Shahidi , 2011b;Pra deep andGuha,2011; Pushparaj and Urooj, 2011; Kim, Hyun,andKim,2010;Ra gaee and Abdel-Aal, 2006).In these studies antioxidant activity of grain extracts displayed through their free radical scavenging, reducing power , ferrous chela ting properties and inhibition of reactive oxygen species was demonstrated, all of which play a major role in the inhibition of oxidation of lipids. Chandrasekara and Shahidi (2011c) showed that millet grain phenolics were effective in inhibiting the formation of conjugated dienes and hydroperoxides in phosphatidyl choline (PC) liposome membranes which were oxidized under UV A light. Ultraviolet radiation is known to damage lipid membranes directly or via a process mediated by RO S (Bose, Agarwal, and Chatter gee, 1989; Einspahr, Stratton, Bowden, and Alberts, 2002; Krutmann, 2000). A relationship existed between exposure to sunlight and incidence of skin cancers suggesting the involvement of free radicals in photocarcinogenesis (Black, 1987; Freeman and Crapo, 1982). Oxidative damage could modify DN A permanently, leading to several pathological conditions, including cancer and aging (Valko, Izakovic, Mazur, Rhodes, and Telser, 2004). Millet grain extracts showed effective, concentration dependent inhibition ability against hydroxyl and peroxyl radical mediated DN A scission (Chandrasekara and Shahidi, 2011c). Furthermore, in the same stud y millet phenolic extracts showed suppressing action against proliferation of HT -29 human colon adeno - carcinoma cells. The degree of antiprolifiration was time- and dose-de pendent and ranged from 30%to 100% at a concentration of 0.5 mg/mL at the end of a 96 h study. Flavones such as kaempherol, apigenin, luteolin, and quercetin in millets extracts may contribute to the observed antiprolifer active activities in vitro (Manthey and Guthrie, 2002; Wang et al., 2000). In other studies Chandrasekara and Shahidi, 2011d, 2012a showed that millet grain phenolics were effective in the inhibition of copper-catalyzed human low density lipoprotein (LDL) cholesterol oxidation in vitro. It has been shown that free radical-induced or metal ion-de pendent oxidation of LDL cholesterol is an important step in deve loping atherosclerotic lesions and leads to cardiovascular disease s (CVDs) (Halliwell, 2007). Copper catalyzes the oxidation of unsaturated fatty acid moieties in the cholesterol molecule. The mechanisms which inhibit the LDL cholesterol oxidation by phenolics could account for chelation of cupric ions as well as scavenging of peroxyl radicals formed (Decker, Ivanov, Zhu, and Frei, 2001). Dietary antioxidants such as millet phenolics that inhibit LDL oxidation may help reducing the occurrence of CVDs. Furthermore, Chethan, Dharmesh, and Malleshi (2008) showed that phenolics of finger millet seed coat inhibit reversibly aldose reductase by non-competitive inhibition. The constituents of seed coat phenolics such as gallic, pro to-catechuic, p-hydroxybenzoic, p-coumaric, vanillic, syringic, ferulic, and trans-cinnamic acids and quercetin were identified as major compounds responsible for inhibiting cataract of the eye lens. Cataract genesis is one of the major causes of blindness in the world and is a common complication of diabetes mellitus. Shobana et al. (2009) showed that finger millet seed coat phenolics were effective non-competitive inhibitor s of pancreatic amylase and intestinal a-glucosidase enzyme s in a dose- de pendent manner. Amylase and a-glucosidase are important enzymes which convert complex carbohydrates into simple glucose for the absorption in the human gut. In the treatment of diabetes mellitus, one therapeutic approach is decreasing of postprandial hyper glycaemia by blocking the action of enzymes needed for hydrolysis of complex carbohydrates. Therefore, amylase and a-glucosidase inhibitors may help to reduce postprandial hyper glycaemia which is the ear - lays metabolic abnormality occurring in type 2 diabetes. Phenolic compound s in the outer layers of plant organs act as barriers for their physical invasion by microorganisms. Polyphenols of millets showed antimicrobial activity (Viswanath, Urooj, and Malleshi, 2009). Finger millet seed coat phenolic extract had higher antibacterial and antifungal activities compared to that of finger millet flour (Viswanath et al., 2009). Phenolic acids such as caffeic, p-coumaric, ferulic and proto catechuic acids have been reported to process antifungal effects (Dragl and, Senoo, Wake, Holte, and Blomhoff, 2003; Moure et al., 2001). Several studies have demonstrated anti-microbial activity of finger millet phenolics against Escherichia coli, Bacillus cereus, and Aspergillusniger, among others (Antony, Moses, and Chandra, 1998; Chethan and Malleshi, 2007b; Viswanath et al., 2009). Recently Banerjee, Sanjay, Chethan, and Malleshi (2012) reported antibacterial activity of finger millet crude phenolic extracts as well as isolated individual phenolic compounds against several pathogenic bacteria as deter - mined by agar diffusion bioassay. Finger millet phenolics exhibited effective proliferation inhibitor y activities on *E. coli, B. cereus, Staphylococcusaureus, Listeria monocytogenes, Streptococcus pyogenes, Proteus mirabilis, Pseudomonas aeruginosa, Serratiamar cescens, Klebsiella pneumonia*, and *Yersinia enterocolitica*. High phenolic content of finger millet that comprised of both phenolic acids and their derivative s as well as flavonoids may inhibit microbial enzymes and oxidation of microbial membranes leading to inhibitory activities of proliferation of bacterial cells. 7. Health benefits of millets A number of in vivo studies have demonstrated the beneficial role of millet grain consumption in the reduction of diseases associated with oxidative stress. Lakshmi Kumari and Sumathi (2002) have shown that consumption of finger millet-based diets results in significantly lower plasma glucose levels in individuals with non-insulin dependent diabetes mellitus. A protective effect of feeding finger millet and kodo millet whole grains meal against hyper glycemic status and alloxan induced oxidative stress in Wistar rats was reported (Hegde, Anitha, and Chandra, 2005). The effect of finger millet feeding was also examined on the dermal wound healing process of alloxan induced diabetic rats (Rajasekaran, Nithya, Rose, and Chandra, 2004). The results showed that finger millet feeding for 4 weeks to the diabetic animals controlled the glucose levels and improve d the antioxidant status as well as the wound healing process. In another study, Hegde, et al. (2005) demonstrated that topical application of an aqueous paste of kodo millet and finger millet flour was effective in rat der - mal wound healing. Recently, Lee, Chung, Cha, and Park (2010) reported that foxtail and pro so millet grain feeding decreased the triacylglycerollevel and foxtail millet reduced C-reactive proteins which is an inflammation related indicator in hyperlipidemia rats suggesting their potential application in lowering the risk for CVD. Furthermore, Shobana et al. (2009) showed that feeding of 20% finger millet seed coat had hypoglycaemic, hypocholester olaemic, nephroprotective and anti-cata- ractogenic properties in streptozotocin-induced diabetic rats.

Bioavailability of millet phenolics

Bioavailability of a compound can be defined as the quantity that passes through the cell membranes in the intestine and is available for action within the cells. Several studies have demonstrated the bioavailability and bio efficacy of phenolics using both animal models and human volunteers (Adam et al., 2002; Bourne and Riv e-Evans, 1998; Chen et al., 2004; Lafay et al., 2006; Manach, Williamson, Morand, Scal-bert, and Remesy, 2005; Olthof, Hollman, Buijsmanvan Amelsvoort, and Katan, 2003; Rondini et al., 2004; Williamson and Manach, 2005; Zhao et al., 2004). Very few studies have been carried out on the bioavailability of hydroxycinnamic acids which are abundant in cereals. Esterification to arabinoxylans of the grain cell walls hampered the absorption of ferulic acids in rats (Adam et al., 2002; Zhao et al., 2004). However, Rondini et al. (2004) showed that bound ferulic acid from bran was more bioavailable than pure ferulic acid fed to rats. It has been shown that quercetin is more available when ingested from onions than as a pure compound in healthy ileostomic subjects (Hollman, de Vries, van Leeuween, Mengelers, and Katan, 1995). Furthermore, presence of glucose moiety may enhance the absorption of quercetin in onions in humans (Hollman et al., 1995). Bourne and Rive-Evans (1998) reported that urinary excretion of total free ferulic acid and feruloyl glucuronide was 11–25%of that ingested from tomatoes, pro viding approximately 21 - 44mg of ferulic acid. Enzymatic digestion and colonic fermentation in the digestive tract release phenolic compounds from food matrices and are available in the gut and may potentially be bioavailable. Recently, Chandrasekara and Shahidi (2012c) re-ported that simulated gastrointestinal digestion and colonic fermentation in-vitro released millet grain phenolics which showed antioxidant activity thro ugh different mechanisms such as free radical scavenging, ferrous ion chela ting, and reducing activity. Furthermore, antioxidant activity de pended on the type of millet considered. Accor ding to this stud y the total phenolic content (TPC) and total flavonoid content (TFC) of millet extracts obtained following enzymatic digestion in the gastric phase ranged from 10.2 to 26.9 μmol ferulic acid equivalents (FAE)/g cooked grain (dw) and from 0.64 to 2.74 μmol catechin equivalents (CE)/g cooked grain (dw), respectively. This was 2–5times more than the total amount of released phenolics than those released from the corresponding aqueous extracts of samples. In the process of digestion, both TPC and TFC increased for all millet grains employed (Chandrasekara and Shahidi, 2012c).

Millets are underutilized cereals in the developed countries while they serve as staple foods for millions of people in parts of Africa and Asia. In addition to serving as a source of macro and micronutrients, similar to other principal cereals in the world, millets are rich in bioactive phytochemicals, particularly phenolics. Moreover, millets are staple food substitutes for celiac patients as they are gluten-free. Millets are climatic change compliant crops that can be cultivated round the year. It is well known that pre vent able diet- and lifestyle-r elated chronic diseases such as hyper tension, cardiovascular disease, and type II diabetes impose an enormous social and economic bur den in a country. Phenolic content as well as composition of different millet grains and available evidences for beneficial health effects emphasize their significance as a functional food ingredient in non-communicable disease (NCD) reduction and improving health.

The research findings clearly showed that millet grain phenolics are released during gastrointestinal digestion and colonic fermentation. The compounds so produced were soluble in the medium, thus they are considered accessible for absorption into the body and may exhibit bioactivity. Several antioxidant activity assays employed in the present work demonstrated that bio-accessible phenolic compounds had a considerable antioxidant effect and exerted their activities through different mechanisms such as free radical scavenging, ferrous ion chelating, and reducing activity. The findings of this study further demonstrated that different millet types may possess varying antioxidant activities under physiological conditions. Furthermore, it was shown that phenolic compounds bound to the insoluble fibre released during colonic fermentation may exert their health benefits locally as well as systemically upon absorption. Nevertheless, identification of phenolics in the digest as well as fermented end products is warranted in order to determine their fate in millet grains in vivo.

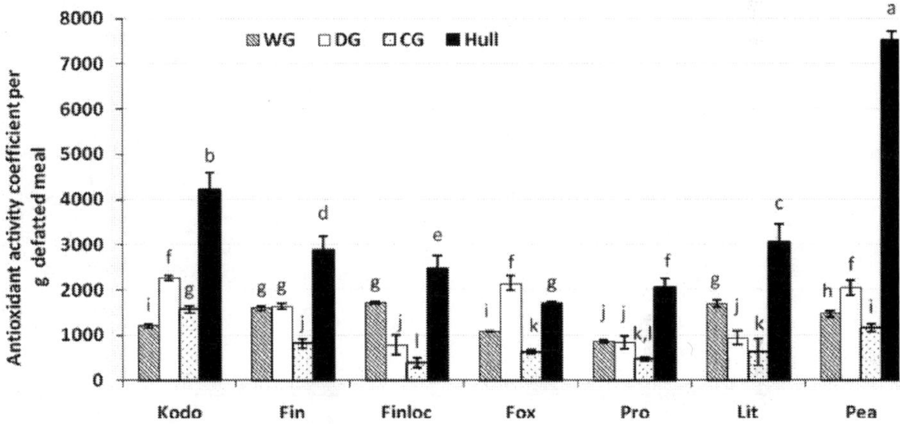

Figure 5.1: Antioxidant activity of whole (WG), dehulled (DG), cooked (CG) grains and hulls of millets in beta carotene/linoleate emulsion system. Bars with different letters have significantly different ($p < 0.05$) mean values. The letter 'a' represents the highest value. Kodo, kodo; Fin, finger (Ravi); Finloc, finger (local); Fox, foxtail; Pro, proso; Lit, little; Pea, pearl millets.

■■■■

6 Classification of Phenolic Antioxidants

CLASSIFICATION OF PHENOLIC ANTIOXIDANTS

Phenolic antioxidants can be classified as synthetic and natural (Fig. 6.1), based on their origin.

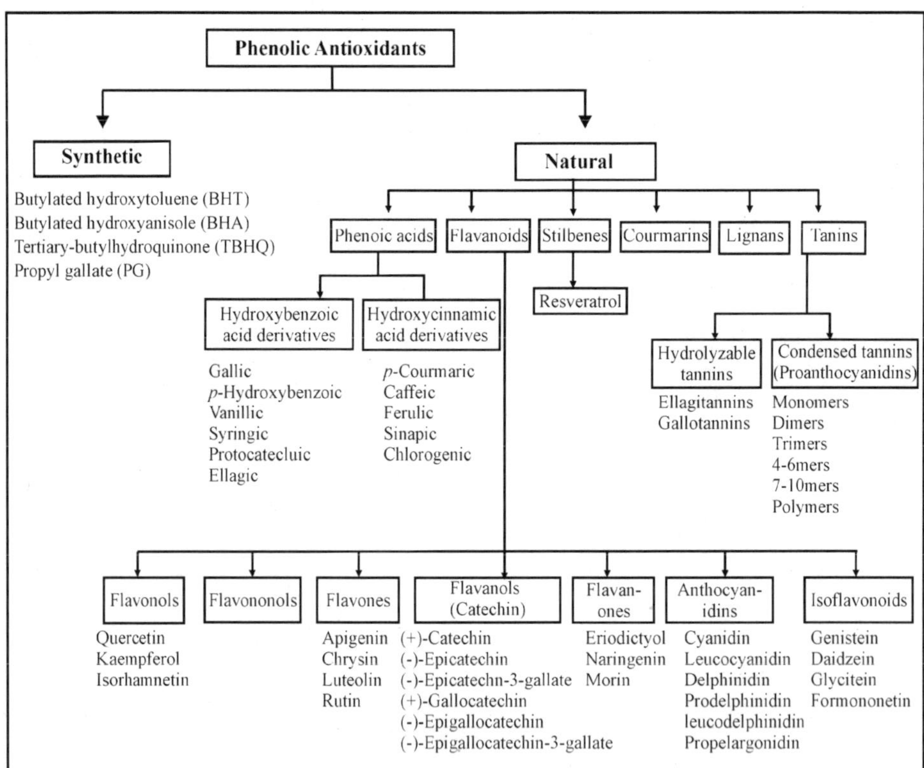

Fig. 6.1: Classification of phenolic antioxidants.

A. SYNTHETIC FOOD PHENOLIC ANTIOXIDANTS

Synthetic phenolic antioxidants currently permitted for use in foods are butylated hydroxyanisole (BHA), butylated hydroxytoluene (BHT), propyl gallate (PG) and tertiary-butylhydroquinone (TBHQ). In addition octyl gallate (OG) and dodecyl gallate (DG) are also used as a synthetic antioxidant (Makahleh, Saad, and Bari, 2015). These synthetic phenolic antioxidants are deliberately added to products in order to prevent or delay the onset of lipid oxidation during processing and storage of fats, oils and lipid-containing foods and have been used by the food industry for some 60 years or so (Saad *et al.*, 2007). Berdahl, Nahas, and Barren (2010) reported that the food and beverage market for antioxidants is a $500 billion industry and is growing at 5–7% per year, and is expected to grow at that rate through the year 2017. However, Verhagen, Schilderman, and Kleinjans (1991) reported that synthetic antioxidants may not be very effective in preventing oxidative deterioration of fats and oils of vegetable origin.

Table 6.1: Chemical and physical properties of synthetic food antioxidants.

Characteristic	BHA	BHT	Gallate		TBHQ
			Propyl	Dodecyl	
Melting point (°C)	50–52	69–70	146–148	95–98	126–128
Boiling point (°C)	268[a]	265[a]	181 ± 20.8[a]	-	291 ± 20.0[a]
Density (g/cm3) at 20 °C	-	0.89[a]	1.36 ± 0.06[a]	-	1.09 ± 0.06[a]
Carry-through properties	Very good	Fair-good	POOR	Fair-good	Good
Synergism	BHT and gallates	BHA	BHA	BHA	-
Solubility (w/w%) in					
Water	0	0	<1	<1	<1
Propylene glycol	50	0	6.1	4	30
Lard	30-40	50	1	-	5-10
Corn oil	30	40	0	0	10
Glycerol	1	0	25	-	<1
Methyl linoleate	Very soluble	Very soluble	1	1	>10

Source: Adapted from Shahidi and Wanasundara (1992); [a]Andre *et al.* (2010).

Table 6.1 summarizes the physical and chemical properties of synthetic antioxidants.The use of synthetic phenolic antioxidants in foodstuffs are strictly regulated by governments due to their potential toxicity effects. BHA and BHT have been suspected of being responsible for liver damage and carcinogenesis when used at high levels in laboratory animals (Biparva, Ehsani, and Hadjmohammadi, 2012; Grice, 1986; Rodil, Quintana, Basaglia, Pietrogrande, and Cela, 2010; Wichi, 1988). It has been reported that amongst the various food additives BHA, TBHQ, 2-tert-butyl-4-methylphenol (TBMP) and PG have potential to form molecular complexes with nucleic acid structure and cause damage to double helical structure of DNA (Dolatabadi and Kashanian, 2010).The maximumpermitted levels of synthetic antioxidants are given in Table 6.2. Synthetic phenolics are effective in numerous food systems; however, their use in the food industry has recently declined owing to safety concerns and consumer demand for all natural products (McClements and Decker, 2007). Berdahl *et al.* (2010) reported that the market is 50% larger for naturals than it is for synthetics, where the market for ascorbates and erythorbates is as large as the entire synthetic market.

Table 6.2: Maximum usage levels (Codex General Standards) permitted by Codex Alimentarius Commission for synthetic antioxidants

Food category	**Maximum usage level (mg/kg)**			
	BHA	**BHT**	**PG**	**TBHQ**
Bakery wares	200	200	200	-
Beverage whiteners	100	100	–	100
Breakfast cereals, including rolled oats	200	100	200	-
Butter oil, anhydrous milk fat, ghee	175	75	100	-
Chewing gum	400	400	1000	400
Cocoa and chocolate products	200	200	200	200
Confectionery including hard & soft candy, nougats etc.	200	200	200	200
Decorations (e.g., for fine bakery wares), toppings (nonfruit) and sweet sauces	200	200	200	200
Desserts – dairy based/fruit based/cereal and starch based/egg bas	-	-	90	-
Dried vegetables (including mushrooms and fungi, roots and tubers, pulses and legumes, and aloe vera), seaweeds, and nuts and seeds	200	200	50	-
Edible ices, including sherbet and sorbet	200	100	-	200
Fat emulsions mainly of type oil-in-water, including mixed and/or flavoured products based on fat emulsions	200	200	200	200
Fat spreads, dairy fat spreads and blended spreads	200	200	200	200
Fat-based desserts excluding dairy-based dessert products	200	200	200	200
Food supplements	400	400	400	-
Frozen battered fish, fish fillets, and fish products, including molluscs, crustaceans, and echinoderms	200	200	-	-
Frozen fish, fish fillets, and fish products, including molluscs, crustaceans, and echinoderms	200	200	-	-
Fully preserved, including canned or fermented fish and fish products, including molluscs, crustaceans, and echinoderms	200	200	-	-

Herbs, spices, seasonings and condiments (e.g., seasoning for instant noodles)	200	200	200	200
Imitation chocolate, chocolate substitute products	-	200	200	-
Lard, tallow, fish oil, and other animal fats	200	200	200	200
Milk powder and cream powder (plain)	100	200	200	-
Mustards	-	-	-	200
Pre-cooked pastas and noodles and like products	200	200	100	200
Processed comminuted meat, poultry, and game products	200	100	200	100
Processed meat, poultry, and game products in whole pieces or cuts	200	100	200	100
Processed nuts, including coated nuts and nut mixtures (with e.g., dried fruit)	200	-	200	-
Ready-to-eat savouries	-	200	-	200
Sauces and like products	200	100	200	200
Semi-preserved fish and fish products, including molluscs, crustaceans, and echinoderms	200	200	-	-
Smoked, dried, fermented, and/or salted fish and fish products, including molluscs, crustaceans, and echinoderms	200	200	100	-
Snacks – potato, cereal, flour or starch based (from roots and tubers, pulses and legumes)	200	-	200	-
Soups and broths	200	100	200	200
Vegetable oils and fats	200	200	200	200
Water-based flavoured drinks, including "sport", "energy", or "electrolyte" drinks and particular drinks	-	-	1000	-
Whole, broken, or flaked grain, including rice	-	-	100	-
Yeast and like products	200	-	-	-

Source: Adapted from Codex General Standard for Food Additives (GSFA) online database. Joint FAO/WHO Codex Alimentarius Commission (Anonymous, 2009).

1. Butylated hydroxyanisole (BHA)

Butylated hydroxyanisole is a monophenolic antioxidant. Commercial BHA is available as white waxy flakes, which is a mixture of two isomers, 3-tertiary-butyl-4-hydroxyanisole (90%) and 2-tetiary-butyl-4-hydroxyanisole (10%) (Fig. 6.2). BHA is very soluble in fats and oils and insoluble in water and a more effective antioxidant in preventing deterioration of flavour and colour of essential oils (Hettiarachchy and Kalapathy, 2000). BHA is slightly better than BHT in its carry-through properties (Table 6.2). BHA is particularly effective in controlling the oxidation of short-chain fatty acids such as those found in coconut and palm kernel oils that are used typically in cereal and confectionery products (Shahidi andWanasundara, 1992).A synergistic effect of BHA with other antioxidants such as BHT, TBHQ or PG provides greater antioxidant potency than what might be expected from the contribution of each individual antioxidant. BHA (0.01%) mixed with dodecyl gallate (0.005%) is more effective than BHA alone in stabilizing margarine (Nanditha and Prabhasankar, 2009). An examination of the scientific literature reveals some paradoxical properties that BHA shares with other antioxidants. Thus, BHA may be referred to as an antioxidant, a prooxidant, an anticarcinogen, a carcinogen, and a tumour promoter (Iverson, 1999). It is hypothesized that 1.5–2% BHA gives rise to tumour formation in rodent forestomach by inducing heritable changes in DNA. Evidence indicates that reactive oxygen species (ROS), in particular hydroxyl radicals, may play a crucial role (Verhagen *et al.*, 1991). However, dietary BHA at levels of 0.25 and 0.75% was found to protect young female mice against the acute toxicity of monocrotaline (Miranda, Reed, Cheeke, and Buhler, 1981).Williams, Iatropoulos, and Whysner (1999) concluded that BHA is a rodent carcinogen, which is species-specific for all practical purposes, and not relevant to humans.

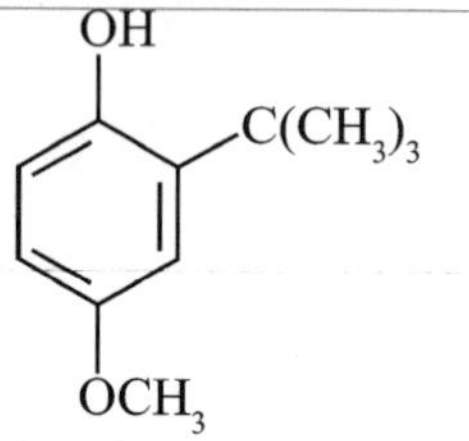

OH, $C(CH_3)_3$, OCH_3

3-tertiary-butyl-4-hydroxyanisole
2-tertiary-butyl-4-methoxyphenol

OH, $CH(CH_3)_3$, OCH_3

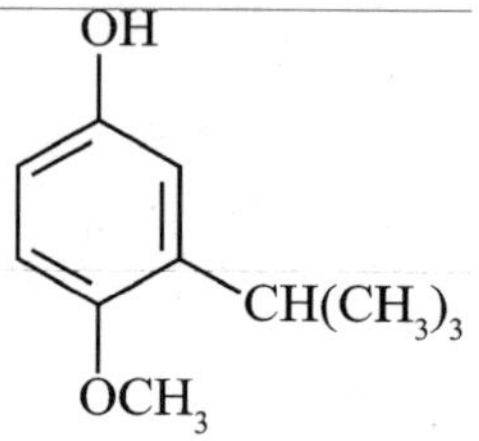

2-tertiary-butyl-4-hydroxyanisole
3-tertiary-butyl-4-methoxyphenol

OH, $(H_3C)_3C$, $C(CH_3)_3$, CH_3

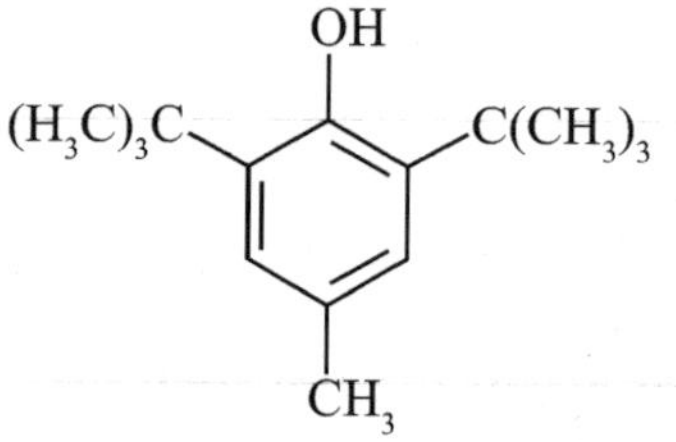

3,5-di-tertiary-butyl-4-hydroxytoulene
2,6-di-tertiary-butyl-4-methylphenol

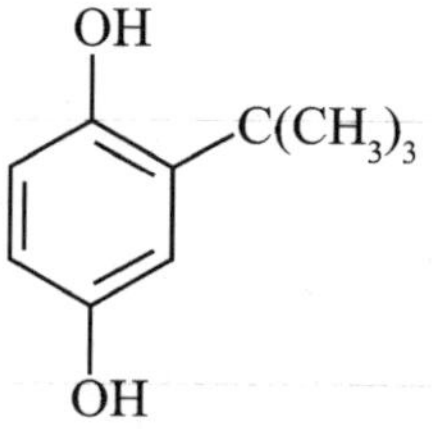

OH, $C(CH_3)_3$, OH

tertiary-butylhydroquinone

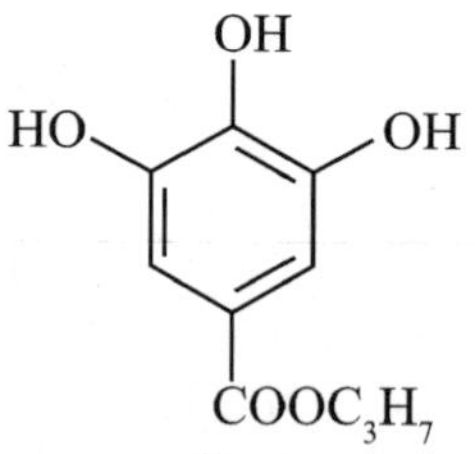

OH, HO, OH, $COOC_3H_7$

3,4,5-trihydroxypropylbenzoate (propyl gallate)

Fig. 6.2: Chemical structures of synthetic antioxidants.

This supports the decisions of authorities that have reviewed these data to recommend continued use of BHA (Iverson, 1995; JECFA, 1996) despite the international agency for research on cancer evaluation that BHA is possibly carcinogenic to humans (IARC, 1987). The potential dietary intake of BHA in Canada was estimated using dietary recall data on food consumption and maximum permitted use levels for this antioxidant. Based on these estimates, it was concluded that the dietary intake of BHA and other permitted phenolic antioxidants (BHT and propyl gallate) is unlikely to exceed 1 mg/kg body weight/day and on average is less than 0.4 mg/kg body weight/day (Kirkpatrick and Lauer, 1986). Joint FAO/WHO Expert Committee on Food Additives (JECFA) reported that the acceptable daily intake (ADI) of BHA is 0 to 0.5 mg/kg (Leclercq, Arcella, and Turrini, 2000; Shahidi and Wanasundara, 1992). Furthermore, at concentrations as low as 125 ppm, which is closer to food additive levels, BHA exhibits anticarcinogenic properties (Williams *et al.*, 1999). BHA is used in baked and fried foods as well as in dry cereals, potato products, dessert mixes and beverages (Dolatabadi and Kashanian, 2010; Makahleh *et al.*, 2015).

2. Butylated hydroxytoluene (BHT)

Butylated hydroxytoluene is also a monophenol and commercially available as a white crystalline compound. However, BHT is not as effective as BHA mainly because of the presence of two tert-butyl groups, which offer greater steric hindrance than BHA to the molecule (Nanditha and Prabhasankar, 2009). BHT is soluble in fats and oils and insoluble in water (Table 6.2) and more effective in suppressing oxidation of animal fats than vegetable oils (Dziezak, 1986). Shahidi and Naczk (2004) reported that BHT can produce radical intermediates with moderate resonance delocalization.The tertiary butyl groups of BHT do not generally allow involvement of the radical formed from it after hydrogen abstraction in other reactions. Thus, a lipid peroxyl radical may join the molecule of BHT in the *para* position to the phenoxy group. The volatile nature of BHA and BHT makes them important additives in packaging materials because they can migrate into foods. For this purpose, these antioxidants are added directly to the wax used in making inner liners or applied to the packaging board as an emulsion (Dziezak, 1986; Shahidi and Wanasundara, 1992). BHT functions synergistically with BHA. A combination of 3-BHA and BHT showed a higher antioxidant activity than either of them used singly in soybean oil, lard, and methyl oleate (Nanditha and Prabhasankar, 2009). Chevillard, Nouhi, Anna, Paquet, and Blank (2010) found the first evidence that BHT exposure not only affects lung function but also leads to impaired adipogenesis in adipocytes. A number of studies have shown that BHT may cause internal and external haemorrhaging at high doses that is severe enough to cause death in some strains of mice and guinea pigs. This effect is due to the ability of BHT to reduce vitamin K dependent blood clotting factors (Ito *et al.*, 1986). Babich (1982) suggested that BHT may exert a positive effect on life span; it exerts a negative effect on the lungs, kidneys, myocardial cells, liver metabolism of lipids, and clotting factors, and is a potential behavioural and developmental teratogen. Williams et al. (1999) reported that BHT is not genotoxic or reproducibly carcinogenic, although at high doses of 250 mg/kg/day or greater, it was associated with some unconfirmed increases in spontaneous neoplasms and, like BHA, has some tumour-promoting activity. The overall evaluation of IARC (International Agency for Research on Cancer) was that BHT is not carcinogenic to humans (IARC, 1987). According to the Joint FAO/WHO Expert Committee on Food Additives (JECFA), the acceptable daily intake (ADI) for BHT is 0 – 0.3 mg/kg body weight (Leclercq *et al.*, 2000).

3. Tertiary-butylhydroquinone (TBHQ)

Tertiary-butylhydroquinone (TBHQ, Fig. 6.2) is a highly effective preservative for unsaturated vegetable oils, many edible animal fats and meat products. It does not cause discolouration even in the presence of iron, and does not change the flavour or odour of the material to which it is added (Kashanian and Dolatabadi, 2009). Metabolically, TBHQ is formed from 3-*tert*-butyl-4-hydroxyanisole (BHA), another widely used food additive, by *O*-demethylation (Okubo, Yokoyama, Kano, and Kano, 2003). TBHQ may be considered as an alternative to hydrogenation for increasing oxidative stability (Dziezak, 1986; Shahidi and Wanasundara, 1992). TBHQ is commercially available as a beige coloured powder and may be used alone or in combination with BHA or BHT at a maximum concentration of 0.02% or 200 ppm, based on the fat content of foods, including essential oils (Shahidi and Naczk, 2004). Chelating agents such as citric acid and monoacylglycerol citrate can further enhance lipid stabilizing properties of TBHQ, which in combination is primarily used in vegetable oils and shortenings but not extensively for animal fats (Shahidi and Wanasundara, 1992). Khan and Shahidi (2001) reported that amongst synthetic antioxidants, TBHQ was more effective than BHA and BHT and served as the strongest antioxidant in borage and evening primrose oil triacylglycerols (TAG). The two *para*hydroxyl groups in TBHQ are responsible for its antioxidant activity (Nanditha and Prabhasankar, 2009). TBHQ reacts with peroxyl radicals to form a semiquinone resonance hybrid.The semiquinone radical intermediate may undergo different reactions to form more stable products.They can also react with one another to produce a dimer or react with one another to produce a quinone and a hydroquinone molecule or add to a lipid peroxyl radical to produce a semiquinone (Shahidi and Naczk, 2004). In high doses, TBHQ has some negative health effects in laboratory animals, such as precursor to stomach tumours and damage to DNA.A number of studies have shown that TBHQ causes DNA cleavage *in vitro* and the formation of 8-hydroxydeoxyguanosine in calf thymus DNA due to the generation of ROS such as superoxide radical anion and hydrogen peroxide (Kashanian and Dolatabadi, 2009). TBHQ and its metabolite 2-*tert*-butyl-1, 4-benzoquinone (TBQ) were both cytotoxic in human monocytic leukaemia U937 cells, TBQ being more strongly cytotoxic (Okubo *et al.*, 2003). Increase of antioxidative potential by TBHQ protects against cell death associated with 6-hydroxydopamine-induced oxidative stress in neuroblastoma SH-SY5Y cells (Hara, Ohta, Ohta, Kuno, and Adachi, 2003). The acceptable daily intake (ADI) of TBHQ is 0 – 0.2 mg/kg body weight (Nanditha and Prabhasankar, 2009; Shahidi and Wanasundara, 1992). Recent studies suggest that addition of TBHQ could prevent biodiesel oxidation and increase the storage time (De Araujo, Barbosa, Viana, and Ferreira, 2011; Goulart, Teixeira, Ramalho, Terezo, and Castilho, 2014).

4. Propyl gallate (PG)

Propyl gallate (PG, Fig. 6.2) has been used since 1948 to stabilize cosmetic, and food-packaging materials, and foods containing fats, and as an additive in edible fats, oils, mayonnaise, shortening, baked products, pressure-sensitive adhesives, lubricating oil additives and transforming oils (Zurita et al., 2007). Propyl gallate is commercially prepared by esterification of gallic acid with propyl alcohol followed by distillation to remove the excess alcohol and is available as a white crystalline powder. It is soluble in ethanol but practically insoluble in water (Shahidi and Naczk, 2004). With a melting point of 148°C, PG loses its effectiveness during heat processing and is therefore not suitable in frying applications that involve temperatures exceeding 190°C. PG chelates iron ions and forms an unappealing blue-black coloured complex. Hence, PG is always used with chelators such as citric acid to eliminate the pro-oxidative iron and copper catalysts (Shahidi and Naczk, 2004). Good synergism between PG and BHA and BHT is obtained; however, their coapplication with TBHQ is not permitted (Buck, 1984; Shahidi and Wanasundara, 1992). Propyl gallate and its metabolite, gallic acid, have been shown to exhibit liver toxicity and enhance carcinogenesis (Eler, Peralta, and Bracht, 2009; Kim, Kang, Lee, Lee, and Lee, 2008). PG is cytotoxic to isolated rat hepatocytes because it impairs mitochondria, leading to ATP depletion (Nakagawa, Nakajima, Tayama, and Moldeus, 1995). PG inhibits the growth of microorganisms by inhibiting respiration and nucleic acid synthesis; it also decreases hepatic microsomal hydrolase and demethylase activities and inhibits the activity of some redox enzymes (Han and Park, 2009). A number of studies have demonstrated that the antioxidative and cytoprotective properties of propyl gallate may change to prooxidative, cytotoxic and genotoxic in the presence of Cu (II) (Jacobi, Eicke, and Witte, 1998; Jacobi, Hinrichsen, Weβ, and Witte, 1999). The acceptable daily intake (ADI) of PG is 0 – 0.25 mg/kg body weight (Nanditha and Prabhasankar, 2009).

B. NATURAL PHENOLIC ANTIOXIDANTS

In recent years, the possible toxicity of synthetic antioxidants has been considered. Thus, the potential of plant products to serve as antioxidants to protect against various diseases induced by free radicals has been explored (Hou et al., 2003). In addition to scavenging free radicals, the multiple activities of antioxidants include inactivating metal catalysts by chelation, reducing hydroperoxides into stable hydroxyl derivatives, and interacting synergistically with other reducing compounds (Frankel and Finley, 2008). Increased popularity of natural food additivesmay prompts more food manufacturers to replace synthetic antioxidants with ingredients containing natural antioxidative compounds. Therefore, research on natural additives has gained momentum as they are perceived as posing no health risk to the consumers (Shahidi andWanasundara, 1995). Naturally-occurring antioxidant compounds are flavonoids, phenolic acids, lignans, terpenes, tocopherols, phospholipids and polyfunctional organic acids, amongst others. Sources of natural antioxidants are primarily plant phenolics that may occur in all parts of plants.They can be found in fruits, vegetables, nuts, seeds, leaves, flours, roots and barks (Wanasundara, Amarowicz, and Shahidi, 1996). There have been numerous studies on the biological activities of phenolics, which are potent antioxidants and free radical scavengers (Naczk and Shahidi, 2004, 2006; Tung, Wu, Kuo, and Chang, 2007).

1. Tocopherols and tocotrienols

Tocopherols and tocotrienols (collectively known as tocols) are monophenolic compounds, and comprise a group of eight chromanol homologues that possess vitamin E activity in the diet (Blekas, Tsimidu, and Bosku, 1995). Tocopherols consist of a 6-chromanol group and an apolar phytal chain, having the prefix α-, β-, γ-, or δ-, depending on the number and position of methyl groups attached to the chromane rings (Dziezak, 1986; Kiokias et al., 2008; Shahidi and Wanasundara, 1992). The tocotrienols differ from tocopherols in having unsaturated side chains at positions 3′, 7′, and 11′ (Seppanen, Song, and Csallany, 2010). The chemical structures and vitamin E activity related compounds are given in Fig. 6.3.Theymay retard the development of precancer lesions and tumours (Lee *et al.*, 2009) and combat free radical reactions that can cause DNA mutations (Gester, 1993). Tocopherols are present in oilseeds, leaves, and other green parts of higher plants. α-Tocopherol is present mainly in the chloroplasts of plant cells, whilst β-, γ- and δ-homologues are usually found outside these organelles. In contrast, tocotrienols are not found in the green parts of the plants but, rather, in the bran and germ fractions of certain seeds and cereals (Kamal-Eldin and Appelqvist, 1996). Tocopherols and tocotrienols play two main roles as antioxidants, firstly as scavengers of lipid peroxyl radicals, and secondly as singlet oxygen quenchers and chemical scavengers (Munne-Bosch and Alegre, 2002). The antioxidant activity of tocols is mainly due to their ability to donate their phenolic hydrogen to lipid free-radicals (Kamal-Eldin and Budilarto, 2015).With regard to vitamin E activity, α-tocopherol is the most potent member of this family (Shahidi and Naczk, 2004) whilst the other tocols are less effective, being 15–40% for β-tocopherol, 1–20% for γ-tocopherol and 15–30% for δ-tocopherol. Meanwhile, the corresponding values for tocotrienols are 1% for α-tocotrienol, 15–30% for β-tocotrienol, 1–5% for γ-tocotrienol and 1% for δ-tocotrienol (Shahidi, 2000a). Although it is generally agreed that the relative antioxidant activity of tocopherols *in vivo* is in the order of $\alpha > \beta > \gamma > \delta$, there is a widespread confusion concerning their relative potency *in vitro* (Burton and Ingold, 1981). Hamam and Shahidi (2006) reported that γ- and δ-tocopherols were stronger antioxidants *in vitro*. The chemical structures of the tocopherols and tocotrienols support a hydrogen-donating power in the order $\alpha > \beta > \gamma > \delta$. This order was also obtained when the activity of the four tocopherols was compared in a homogeneous solution of dichlorobenzene (Burton and Ingold, 1981), but a reverse order ($\alpha < \beta \approx \gamma < \delta$) was obtained when relative antioxidant potencies were compared in fats, oils, and lipoproteins *in vitro* (Gottstein and Grosch, 1990; Hamam and Shahidi, 2006; Kamal-Eldin and Appelqvist, 1996; Kamal-Eldin and Budilarto, 2015; Lea andWard, 1959; Olcott and Van Der Ven, 1968). The reason behind this reverse order is not yet clearly understood. However, it is now known that the "absolute" and "relative" *in vitro* activities of the tocopherols are not only dependent on their absolute chemical reactivities towards hydroperoxyl and other free radicals, but also on many other possible factors such as temperature and light, type of substrate and solvent, and by other chemical species acting as prooxidants and synergists in the system (Kamal-Eldin and Budilarto, 2015).

(a)

Tocopherol

Tocotrienol

Compound	R_1	R_2	R_3
5,7,8-Trimethyl tocopherol (α-tocopherol) 5,7,8-Trimethyl tocotrienol (α-tocotrienol)	CH_3	CH_3	CH_3
7,8-Dimethyl tocopherol (β-tocopherol) 7,8-Dimethyl tocotrienol (β-tocotrienol)	H	CH_3	CH_3
5,8-Dimethyl tocopherol (γ-tocopherol) 5,8-Dimethyl tocotrienol (γ-tocopherol)	CH_3	H	CH_3
8-Methyl tocopherol (δ-tocopherol) 8-Methyl tocotrienol (δ-tocotrienol)	H	H	CH_3

(b)

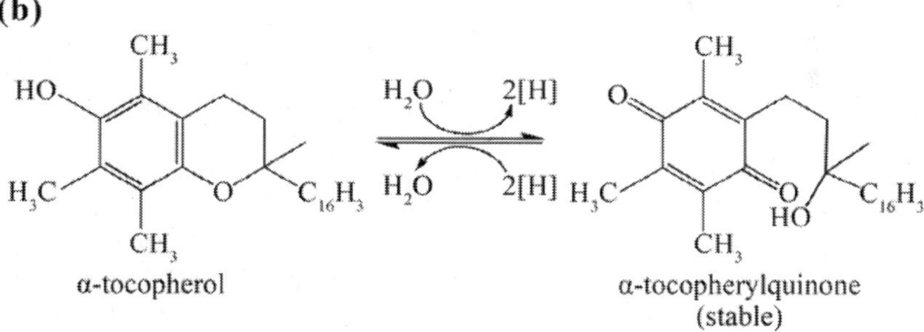

Fig. 6.3: (a) Chemical structures of tocopherols and tocotrienols, (b) α-Tocopherol/α-tocopherylquinone redox system (adapted from Shahidi and Naczk, 2004).

The antioxidant activity of α-tocopherol is based mainly on the tocopherol–tocopherylquinone system (Fig. 6.3) (Shahidi and Wanasundara, 1992).Tocopherols (AH2) are radical scavengers and quench lipid radicals (R•), thus regenerating RH molecules as well as producing a tocopheryl semiquinone radical. Two tocopheryl semiquinone radicals (AH) may form a molecule of tocopheryl quinone (A) and a regenerated tocopherol molecule (Reactions 11 and 12) (Schuler, 1990).

$$R\bullet + AH_2 \rightarrow RH + \bullet AH \quad (11)$$

$$2\,\bullet AH \rightarrow A\bullet + AH_2 \quad (12)$$

However, in the absence of antioxidants such as ascorbic acid and glutathione, which recycle tocopheroxyl and tocotrienoxyl radicals (Munne-Bosch and Alegre, 2002), the radicals may undergo radical–radical coupling with other lipid peroxyl radicals to form adducts, may disproportionate to form quinones or may undergo self-coupling with other tocopheroxyl and tocotrienoxyl radicals to form dimers and/or trimers (Kamal-Eldin and Appelqvist, 1996; Kamal-Eldin and Budilarto, 2015; Munne-Bosch and Alegre, 2002). Several authors have claimed that adding tocopherols to vegetable oils (even to the refined oils, where some tocopherols are removed during deodourization) hardly shows any improvement in oxidative stability due to the fact that an optimum concentration of these compounds is still present (Abuzaytoun and Shahidi, 2006; Hamam and Shahidi, 2006; Kiokias et al., 2008). Tocotrienols are present especially in palm oil and in cereal grain oils, especially rice bran oil (Cerretani, Lerma-Garcia, Herrero-Martinez, Gallina-Toschi, and Simo-Alfonso, 2010). The relative concentrations of tocopherols and tocotrienols vary widely from one oil to another, which can be used to distinguish them according to their botanical origin. Thus, α-tocopherol is the most representative antioxidant in olive oil (Tasioula-Margari and Okogeri, 2001) whilst γ- and δ- tocopherol contents are high in soybean and sunflower oils, with soybean oil particularly rich in γ-tocopherol (Cerretani et al., 2010). Tocopherol and/or tocotrienol concentrations have been used to detect the adulteration of olive oil with red palm and hazelnut oils and soybean oil with linseed oil (Dionisi, Prodolliet, and Tagliaferri, 1995; Manandhar, Nagao, and Yamazaki, 1986). Approximate content of tocopherols and tocotrienols found in vegetable oil are given in Table 6.3. In food manufacturing practice, it is recommended to keep the amount of total tocopherol (natural or added) at levels between 50 and 500 ppm (Kiokias et al., 2008). Kamal-Eldin and Appelqvist (1996) reported the synergistic effect of tocols with other antioxidants. Kancheva et al. (2014) observed synergism for the binary mixtures of α-tocopherol with ascorbyl palmitate and dehydrozingerone, which is one half of the molecule of curcumin. α-Tocopherol and β-carotene may play complementary roles or even show synergism in systems of low oxygen pressure. Parkhurst, Skinner, and Strum (1968) found that mixed tocopherols (α-, γ-, and δ-) gave better protection to lard than individual tocopherols and suggested some synergistic interaction. Synergism between tocopherols and phospholipids has also been reported in the literature (Hudson and Mahgoub, 1981; Kamal-Eldin and Appelqvist, 1996; Shahidi, 2000a), which are mainly due to the metal-chelating properties of the latter (Linow and Mieth, 1976).

Table 6.3: Tocopherol and tocotrienol contents (mg/kg) of vegetable oils.

Oil	Tocopherol				Tocotrienol			
	α	β	γ	δ	α	β	γ	δ
Borage[c]	–	–	150	1350	–	–	–	–
Camelina[a]	38	0.9	720	15	–	–	–	–
Coconut[b]	5–10	–	5	5	5	Trace	1–20	–
Corn[a]	180	11	440	22	9.4	–	13	2.6
Cottonseed[b]	40–560	–	270–410	0	–	–	–	–
Evening primrose[c]	160	–	420	65	–	–	–	–
Linseed[a]	12	Trace	520	9.5	–	–	–	–
Maize, grain[b]	60–260	0	400–900	1–50	–	0	0–240	0
Maize, germ[b]	300–430	1–20	450–790	5–60	–	–	–	–
Olive[b]	1–240	0	0	0	–	–	–	–
Palm[b]	180–260	Trace	320	70	120–150	20–40	260–300	70
Peanut[b]	80–330	-	130–590	10–20	–	–	–	–
Rapeseed/canola[b]	180–280	-	380	10–20	–	–	–	–
Safflower[b]	340–450	–	70–190	230–240	–	–	–	–
Sesame[a]	79	4.1	360	12	Trace	–	3.4	–
Soybean[b]	30–120	0–20	250–930	50–450	0	0	0	–
Sunflower[b]	350–700	–	10–50	1–10	–	–	–	–
Walnut[a]	560	20–40	590	450	–	–	–	–
Wheat grainb	560–1200	660–810	260	270	20–90	80–190	–	–
Wheat germ[c]	1100	440	330	110	–	110	–	–

Sources: Adapted from [a]Schwartz, Ollilainen, Piironen and Lampi (2008); [b]Shahidi and Naczk (2004); [c]Shahidi (2004).

Kago and Terao (1995) suggested that phospholipids enhance the antioxidant potency of tocopherols through a physical rather than a chemical action. It was postulated that the phospholipids form reverse micelles (or microemulsions) when dissolved in organic solvents or bulk oils and that tocopherols are solubilized and positioned in these microemulsions with their active phenolic group near the polar region where peroxyl radicals are concentrated (Kamal-Eldin and Appelqvist, 1996). Maillard reaction products (called melanoidins) produced by reactions of reducing sugars with amino acids are good inhibitors of oxidation, and act as synergists with tocopherols (Kamal-Eldin and Appelqvist, 1996; Yamagushi and Fujimaki, 1974). Ascorbic acid was mentioned to regenerate α-tocopherol from its tocopheroxyl radical *in vivo* and *in vitro* and thereby restoring its antioxidant activity (Han, Yi, and Shin, 1991; Iglesias, Pazos, Torres, and Medina, 2012; Kamal-Eldin and Appelqvist, 1996; Kamal-Eldin and Budilarto, 2015; Shahidi, 2000a; Sharma and Buettner,

1993). Glutathione is also capable of regenerating α-tocopherol from its radical (Niki, Tsuchiya, Tanimura, and Kamiya, 1982). The commercial synthesis of α-tocopherol involves the condensation of 2, 3, 5-trimethylhydroquinone with phytol, isophytol, or phytal halogenides (Shahidi and Wanasundara, 1992). Extraction of vitamin E from natural sources has received increasing interest due to the high antioxidant activity associated with this family of compounds. Besides its well-known antioxidant activity, studies have demonstrated that synthetic vitamin E is less effective than natural vitamin E (Hadolin, Skerget, Knez, and Bauman, 2001). Tocopherols are now obtained mainly by vacuum distillation of deodourizing-step residues generated in the refining of vegetable oils. Throughout this process, that includes several steps such as solvent recovery and purification, copious amounts of organic solvents and energy are required, and thermal degradation of tocopherols is commonly encountered (de Lucas, Martinez de la Ossa, Rincón, Blanco, and Gracia, 2002). Interest in the application of supercritical fluid extraction using carbon dioxide (SFE-CO_2) has grown continuously because SFE-CO_2 shows several advantages over classical extraction processes with organic solvents. Under the supercritical fluid conditions, supercritical fluid extraction (SFE) technology is suitable to decrease volatility and thermal degradation during compounds extraction (de Lucas *et al.*, 2002; Gelmez, K□ncal, and Yener, 2009; Nyam *et al.*, 2010). The total tocopherol content of the extracts is usually between 30 and 80%, but higher in γ- and δ-tocopherols.To obtain stable α-tocopheryl acetate; the mixture is methylated and subsequently acetylated (Shahidi and Naczk, 2004). The α-tocopheryl acetate is the commercially available form of vitamin E, which is not an antioxidant because it's active -OH group is esterified. However, under acidic aqueous conditions, tocopherol is released by hydrolysis and the released tocopherol may then act as an antioxidant (Schuler, 1990).

2. Phenolic acids (hydroxybenzoic and hydroxycinnamic acids)

Phenolic acids, known to serve as multipurpose bioactive compounds, are widely spread throughout the plant kingdom. Most of them are an integral part of the human diet and are also consumed as medicinal preparations. Many of the health protective effects of phenolic compounds have been ascribed to their antioxidant, antimutagenic, anticarcinogenic, antiinflammatory, antimicrobial, and other biological properties (Xu, Ye, Liu, Ma, and Chen, 2008). Substituted derivatives of hydroxybenzoic and hydroxycinnamic acids are the predominant phenolic acids in plants, with hydroxycinnamic acids being the more common (Fig. 6.4). These derivatives differ in the pattern of the hydroxylation and methoxylation in their aromatic rings (Mattila and Hellström, 2007; Shahidi and Naczk, 2004). Technically speaking, only benzoic acid derivatives are phenolic acids and cinnamic acid derivatives are phenylpropanoids.The basic pathway for synthesis of phenolic acids in plants begins from sugars through to aromatic amino acids – phenylalanine, and, in some rare cases, tyrosine. The formation of *trans*-cinnamic acid from phenylalanine and *p*-hydroxycinnamic acid from tyrosine is catalysed by phenylalanine ammonia lyase (PAL) and tyrosine ammonia lyase (TAL), respectively (Amarowicz, Carle, *et al.*, 2009; Amarowicz, Zegarska, *et al.*, 2009; Shahidi, 2015). Phenolic acids are present in some plant foods mostly in the bound form. The most common hydroxycinnamic acids are caffeic, *p*-coumaric and ferulic acids, which frequently occur in foods as simple esters with quinic acid or glucose. Probably the most well-known bound hydroxycinnamic acid is chlorogenic acid, which is combined caffeic and quinic acids. Unlike hydroxycinnamates, hydroxy-benzoic acid derivatives are mainly present in foods in the form of glucosides; *p*-hydroxybenzoic, vanillic and protocatechuic acids are the most common forms (Herrmann, 1989; Mattila and Hellström, 2007; Shahidi and Chandrasekara, 2010; Shahidi, McDonald, Chandrasekara, and Zhong, 2008; Shahidi and Naczk, 2004; Yeo and Shahidi, 2015). Phenolic acids behave as antioxidants, due to the reactivity of their phenol moiety (hydroxyl substituent on the aromatic ring). Although there are several mechanisms, the predominant mode of antioxidant activity is believed to be radical scavenging via hydrogen atom donation.

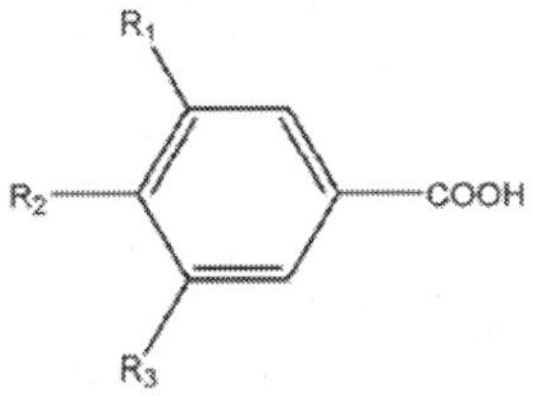

Acid	R_1	R_2	R_3
p-Hydroxybenzoic	H	OH	H
Protocatechuic	OH	OH	H
Vanillic	OCH_3	OH	H
Syringic	OCH_3	OH	OCH_3
Gallic	OH	OH	H

Hydroxybenzoic acid

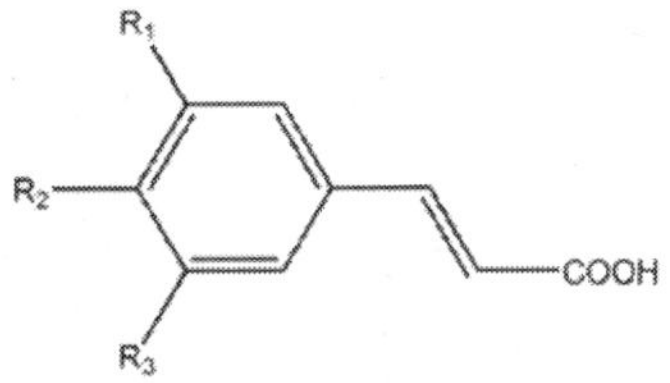

Acid	R_1	R_2	R_3
p-Courmaric	H	OH	H
Caffeic	OH	OH	H
Ferulic	OCH_3	OH	H
Sinapic	OCH_3	OH	OCH_3

Hydroxycinnamic acid

Fig. 6.4: Chemical structures of naturally occurring phenolic acids and related compounds.

Other established antioxidant, radical quenching mechanisms are through electron donation and singlet oxygen quenching (Shahidi and Wanasundara, 1992). Substituents on the aromatic ring affect the stabilization and therefore affect the radical-quenching ability of these phenolic acids. Different acids therefore have different antioxidant activities (Rice-Evans, Miller, and Paganga, 1996). The antioxidant behaviour of the free, esterified and glycosylated phenolics has been reported (Robbins, 2003). There is an awareness and interest in the antioxidant behaviour and potential health benefits associated with these simple phenolic acids. It is their role as dietary antioxidants that have received the most attention in recent literature (Rice-Evans *et al.*, 1996; Robbins, 2003). Because of their ubiquitous presence in plant-based foods, humans consume phenolic acids on a daily basis. The estimated range of consumption is 25 mg^{-1} g per day depending on the diet consumed (fruit, vegetables, grains, teas, coffees, spices) (Clifford, 1999). Caffeic acid, one of the most prominent naturally occurring cinnamic acids, is known to selectively block the biosynthesis of leukotrienes, components involved in immunoregulation diseases, asthma and allergic reactions (Koshihara et al., 1984). Other studies have reported that caffeic acid and some of its esters might possess antitumour activity against colon carcinogenesis (Olthof *et al.*, 2001; Robbins, 2003).

3. Flavonoids

Typically, when discussing phenolics in plant foods, flavonoids are the predominant class described, because they account for approximately two-third of the dietary phenols (Robbins, 2003). Flavonoids are cyclized diphenylpropanes that commonly occur in plants and particularly plant foods (Cao *et al.*, 1997). More than 6000 flavonoids have been identified (Harborne and Williams, 2000). The immediate family members of flavonoids include flavones, flavonol, isoflavones, flavanones, flavanonol, flavanol and anthocyanidin. Flavanones undergo a series of transformations affecting the heterocyclic carbon ring to give rise to anthocyanins and catechins (Das, 1994). Flavonoid derivatives vary in their structure around the heterocyclic oxygen ring, but all have the characteristic C_6—C_3—C_6 carbon skeleton (Fig. 6.5) (Shahidi and Naczk, 2004; Yao *et al.*, 2004). In general, all flavonoids are derivatives of the 2-phenylchromone parent compound composed of three phenolic rings (A, B and C in Fig. 6.5), all of which exhibit various levels of hydroxylation and methoxylation (Yao et al., 2004). The biochemical activities of flavonoids and their metabolites depend on their chemical structures and the relative orientation of various moieties in the molecules. Isoflavonoids are found mainly in the Leguminosae family, consisting of a phenyl ring (the A-ring, Fig. 6.5) fused with the six-membered heterocyclic C-ring and another phenyl ring (the B-ring) at the C-3 position, whereas for flavonoids, the B-ring is substituted to the C-2 position. Despite the subtle structural differences, some isoflavonoids are more active as antioxidants than their corresponding flavonoids (Han et al., 2009). Flavones and flavonols occur as aglycones in foods; approximately 200 flavonols and some 100 flavones have been identified in plants. These compounds possess a double bond between C-2 and C-3. Flavonols are different from flavones in that they possess a hydroxyl group in the 3-position and can be regarded as 3-deoxyflavonols (Fig. 6.5) (Shahidi and Naczk, 2004).

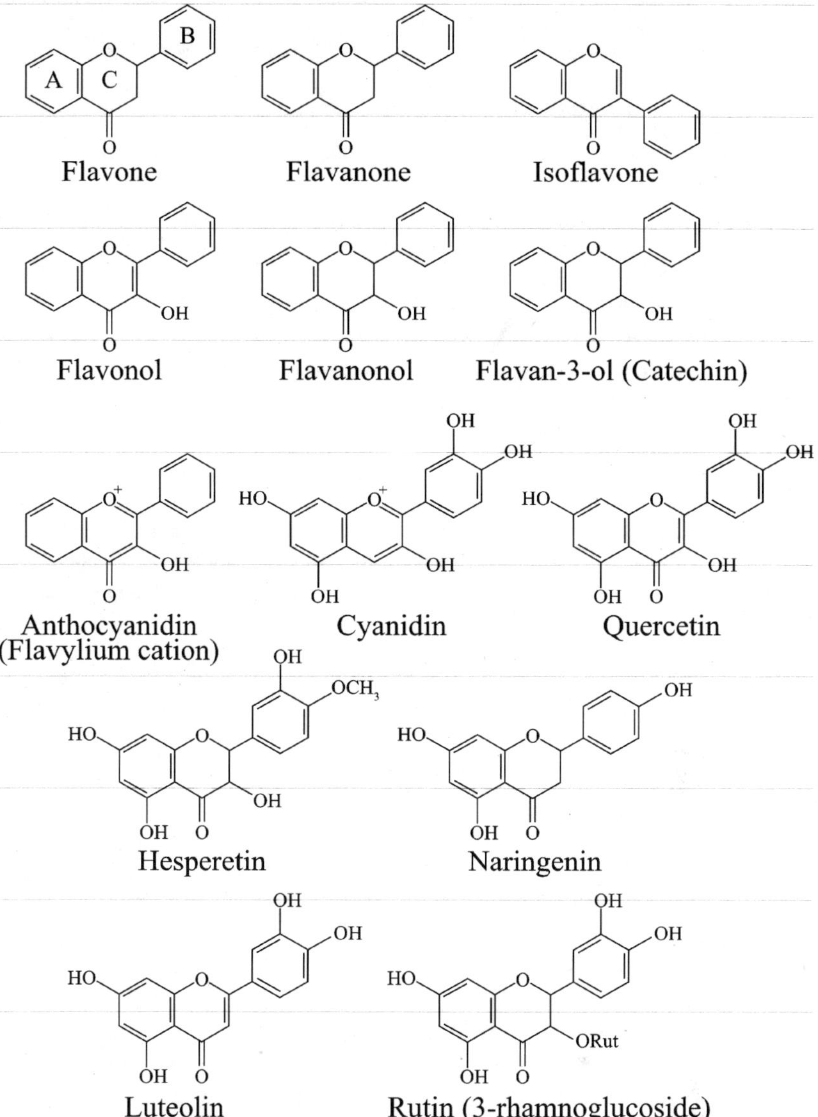

Fig. 6.5: Chemical structures of major classes of flavonoids and some selected flavonoids and flavonoid related compounds.

Meanwhile, flavonones and flavononols are characterized by the presence of a saturated C_2—C_3 bond and an oxygen atom (carbonyl group) in the 4-position. Thus, flavonones may be referred to as dihydroflavones. Flavononols differ from flavonones by having a hydroxyl group in the 3-position and are often referred to as 3-hydroxyflavonones or dihydroflavonols (Shahidi and Naczk, 2004). Amongst flavonoids, anthocyanins and catechins, knowncollectively as flavans because of lack of the carbonyl group in the 3-position, are important; flavan-3-ols and flavan-3, 4-

diols belong to this category (Shahidi and Naczk, 2004). Examples of some naturally occurring flavonoids in foods and food components are provided in Table 6.4.

Table 6.4: Different classes of flavonoids and dietary sources.

Class	Name	Dietary source
Flavone	Chrysin	Fruit skins
	Apigenin	Parsley, Celery
Flavonone	Naringin	Citrus, grapefruit
	Naringenin	Citrus
	Taxifolin	Citrus
	Eriodictyol	Lemons
	Hesperidin	Oranges
	Isosakuranetin	Citrus
Flavonol	Kaempferol	Leek, broccoli, endives, grapefruit, black tea.
	Quercetin	Onion, lettuce, broccoli, tomato, tea, berries, apples, olive oil, cranberry.
	Rutin	Buckwheat, citrus, red, pepper, red wine, tomato skin.
Flavononol	Engeletin	White grape skin
	Astilbin	White grape skin
	Genistin	Soybean
	Taxifolin	Fruits
Isoflavone	Genistein	Soybean
	Daidzin	Soybean
	Daidzein	Soybean
Flavanol	(+)-Catechin	Tea
	(+)-Gallocatechin	Tea
	(−)-Epicatechin	Tea
	(−)-Epigallocatechin	Tea
	(−)-Epicatechin gallate	Tea
	(−)-Epigallocatechin gallate	Tea
Anthocyanidin	Epigenidin	Stored fruits
	Cyanidin	Cherry, raspberry, strawberry, grapes
	Delphinium	Dark fruits
	Pelargonidin	Dark fruits

Source: Adapted from Shahidi and Naczk (2004).

Flavonols are represented mainly by quercetin, kaempferol and myricetin, of these, quercetin is most ubiquitous. In a recent study by Sultana and Anwar (2008), the highest concentrations of flavonolwere detected in the medicinal plant, moringa (*Moringa oleifera*), followed by strawberry, peepal, spinach and cauliflower amongst 22 plant materials (Jaganath and Crozier, 2010). Unlike flavonols, flavones are not widely distributed with significant concentrations being reported in only celery, parsley and artichoke (Jaganath and Crozier, 2010). Flavan-3-ols are found abundantly in fruits such as apricots, sour cherries, grapes and blackberries whereas flavanones are exclusively found in citrus fruits in their glycosidic forms (Jaganath and Crozier, 2010). The most widespread anthocyanin in fruits is cyanidin-3-glucoside (Kong, Chia, Goh, Chia, and Brouillard, 2003).Anthocyanins occur in abundance in berries where they provide the fruits with their distinctive and vibrant palate of colours. Cranberry, blackberry and elderberry contain derivatives of only one type of anthocyanin (i.e., cyanidin), whilst a wide array of anthocyanins occur in blueberry and blackcurrant (Jaganath and Crozier, 2010). Meanwhile, soybeans are almost the sole dietary source of isoflavones.Common isoflavones such as genistein, daidzein and glycitein, also occur, albeit in lowlevels, in black beans and green peas (Jaganath and Crozier, 2010). Delmonte, Perry, and Rader (2006) reported the presence of isoflavones in red clover and a perennial vine kudzu (*Pueraria lobata*).

In general, the ability of flavonoids to be effective antioxidants depends on three factors: (i) the metal-chelating potential that is strongly dependent on the arrangement of hydroxyls and carbonyl group around the molecule, (ii) the presence of hydrogen-/electron-donating substituents able to reduce free radicals, and (iii) the ability of the flavonoid to delocalize the unpaired electron leading to the formation of a stable phenoxyl radical. Both known modes of the antioxidant action, i.e., preventive mechanism and chain-breaking mechanism, are postulated to be responsible for the high activity of flavonoids (Musialik, Kuzmicz, Pawłowski, and Litwinienko, 2009).There are numerous studies devoted to the importance of flavonoid structure for their antiradical activity as chain-breaking antioxidants (Cao et al., 1997; Rice-Evans *et al.*, 1996; Zhou, Miao, Yang, and Liu, 2005). It is generally accepted that the excellent antioxidant properties of flavonoids are due to the presence of catechol hydroxyl groups in the B ring (Musialik *et al.*, 2009; Rice-Evans *et al.*, 1996; Zhou *et al.*, 2005). Shahidi and Wanasundara (1992) reported that the *o*-dihydroxylation of the B ring contributes to the antioxidant activity.The *p*-quinol structure of the B ring has been shown to impart an even greater activity than *o*-quinol; however, *p*- and *m*-hydroxylation of the B ring does not occur naturally. All flavonoids with 3′, 4′-dihydroxy configuration possess antioxidant activity.The TEAC (trolox equivalent antioxidant activity) assay, originally described by Miller, Rice-Evans, Davies, Gopinathan, and Milner (1993), is based on scavenging of long-lived radical anions (Scott, Chen, Bakac, and Espenson, 1993). In this assay radicals are generated through the peroxidase activity of metmyoglobin in the presence of hydrogen peroxide and can easily be detected spectrophotometrically at 734 nm or by electron paramagnetic resonance (Shahidi and Zhong, 2007; van den Berg, Haenen, van den Berg, and Bas, 1999). Trolox equivalent antioxidant activity of flavonoids is given in Fig. 6.6. Quercetin has an identical number of hydroxyl groups in the same positions as catechin, but also contains the 2,3-double bond in the C ring and the 4-oxo group which allows delocalization between the A and B rings stabilizing the aryloxyl radical after hydrogen donation (Fig. 6.5). This structural advantage confers an enhancement of the TEAC value to 4.7 ± 0.10 mM (Rice-Evans *et al.*, 1996). Cyanidin has *o*-dihydroxy structure in the B ring as quercetin, and the same number and arrangement of the five hydroxyl groups. Thus, the same TEAC value ($4.44 \pm$

0.01 mM) as quercetin. Dehydroxylation to a monophenol in the B ring as in pelargonidin (1.3 ± 0.1 mM) gives much the same value as the equivalent flavan-3-ol, kaempferol.Two flavanones, naringenin and hesperitin (Fig. 6.5), have only one hydroxyl group on the B ring and possess little antioxidant activity.The glycosylation of flavonoids reduces their activity when compared to the corresponding aglycones (Shahidi and Wanasundara, 1992). Blocking the 3-hydroxyl group in the C ring of quercetin as a glycoside (whilst retaining the 3′,4′-dihydroxy structure in the B ring) as in rutin or removing the 3 - OH group in the C ring as in luteolin decreases the antioxidant activity to a value of 2.40 ± 0.12 and 2.10 ± 0.05 mM, respectively (Fig. 6.5).

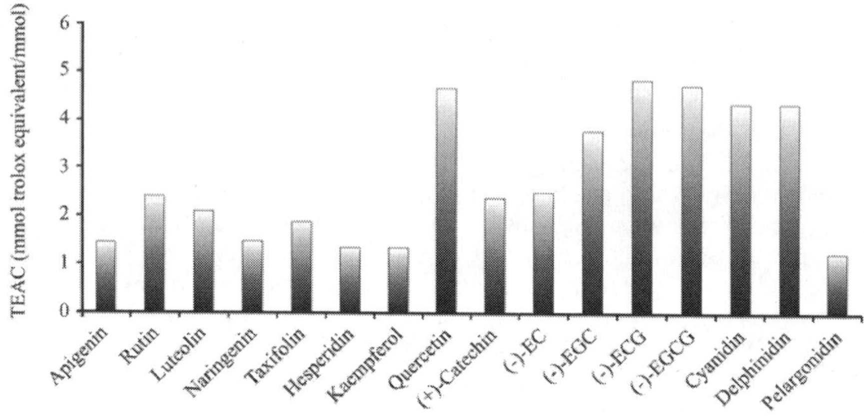

Fig. 6.6: Trolox equivalent antioxidant activity of flavonoids (adapted from Rice-Evans et al., 1996).

Thus, the maximum effectiveness for radical scavenging apparently requires the 3 –OH group attached to the 2, 3-double bond and adjacent to the 4-carbonyl in the C ring (Rice-Evans et al., 1996). Han et al. (2009) reported that the isoflavonoid genistein was a better antioxidant than the isomeric flavonoid apigenin in phosphatidyl liposomes at pH 7.4. In the isoflavone, it is clear that both hydroxyl groups in 4′ and 5′ positions are needed for significant antioxidant activity as in genistein. Even 6, 7, 4′-trihydroxyisoflavone is marginally active, when compared to analogous flavone apigenin, which is inactive as an antioxidant.

Genistein is particularly active.The resonance-stabilized quinoid structures show that for isoflavone the carbonyl group at position four remains intact and can interact with the 5-hydroxy group, if present; however, in flavone, the carbonyl group at position four loses its functionality. This may explain the superior antioxidant activity of genistein compared with that of apigenin (Shahidi and Naczk, 2004). The current data on the antioxidant properties of naturally occurring isoflavones and their corresponding metabolites indicate that the metabolism of genistein and daidzein to the oxidative metabolites 3′-hydroxygenistein and 3′- 6-, and 8-hydroxyldaidzein as well as to the bacterial metabolite equol enhances their antioxidant properties. These compounds are more effective antioxidants than the known antioxidants, quercetin and ascorbic acid (Rufer and Kulling, 2006).

Flavonoids with high antioxidant activity have the following characteristic structures: (1) a 3′,4′-dihydroxyl group in the B ring, (2) the 3 -OH moiety in the C ring, (3) the C_2—C_3 double bond in the C ring conjugated with a 4-keto group causing electron delocalization from the B ring, and (4) both 3 -OH group in C ring and 5 -OH group in A ring combined with a 4-carbonyl group and C_2—C_3 double bond (Sun and Powers, 2007). In addition to antiradical efficiency, flavonoids act as metal chelators. The chelating capacity of flavonoids for Cu and Fe ions has been investigated (Brown, Khodr, Hider, and Rice-Evans, 1998; Engelmann, Hutcheson, and Cheng, 2005; Mira et al., 2002). The ortho-hydroxyl groups on the B ring, the 3-hydroxyl combined with the 4-carbonyl group or the 5-hydroxyl combined with the 4-carbonyl group are the metal-complexing sites. However, Cao et al. (1997) observed that flavonoids efficiently scavenged peroxyl or hydroxyl radicals but were prooxidant with Cu2+. Shahidi and Naczk (2004) reported that such complexations with Cu may contribute to the antioxidative action of flavonoids. Chelation of metal ions renders them catalytically inactive.The antioxidant mechanisms of flavonoids might include synergistic effects. The antioxidant activity of flavonoids usually increases with an increase in the number of hydroxyl groups and a decrease in glycosylation. EC and ECG (epicatechin and epicatechin gallate), respectively with a vicinal diphenol structure in the B ring and a saturated C ring exhibit the strongest effects (Yao et al., 2004). Salah et al. (1995) showed that the total antioxidative activity and the order of effectiveness of green tea polyphenols as radical scavengers is ECG > EGCG (epigallocatechin gallate) > EGC > gallic acid > EC = catechin. In addition, lyophilized derivatives of green tea catechins exhibited strong antioxidant acitivities (Zhong *et al.*, 2012; Zhong and Shahidi, 2011). However, the oxidation of low density lipoproteins is inhibited by catechin, EC, ECG, and EGCG to a similar degree, not as much as in the presence of EGC or gallic acid. Pekkarinen, Heinonen, and Hopia (1999) reported thatmyricetin and rutin exhibit a synergistic effect with α-tocopherol. Myricetin, quercetin and rutin protect α-tocopherol from decomposition, where myricetin is more protective than quercetin and rutin (Pekkarinen *et al.*, 1999). The interaction between flavonoids and α-tocopherol might be resulting from their different localization in methyl linoleate. The synergistic effect of (−)-epicatechin and quercetin with α-tocopherol in egg yolk phosphatidylcholine was partly explained by the different localization of antioxidants in the substrate (Terao, Piskula, and Yao, 1994).The remarkable synergistic effect of quercetin with α-tocopherol in a fish oil-bile salt emulsion, has been shown to be due to different localization of these antioxidants (Pekkarinen *et al.*, 1999). The daily intake of flavonoids (reported at above 100 mg) is almost at the same level as the sum of the daily doses of other antioxidants including β-carotene (2–3 mg), vitamin C (70–100 mg), and vitamin E (7–10 mg) (Musialik *et al.*, 2009). These data correspond to the reports of many health benefits originating from dietary intake of flavonoids. Some flavonoids have been found to possess antilipoperoxidant (Terao *et al.*, 1994), antitumoural (Deschner, Ruperto,Wong, and Newmark, 1991), antiplatelet (Tzeng, Ko, Ko, and Teng, 1991), anti-ischaemic, anti-allergic, and anti-inflammatory (Middleton and Kandaswami, 1992) activities (Cao *et al.*, 1997). One interesting example of flavonoid activity is the so-called "French paradox", that is, despite high fat intake, mortality from coronary heart disease is lower in some regions of France, a fact attributed to the regular drinking of red wine which contains high levels of flavonoids (approximately 200 mg per glass) and resveratrol (0.1–15 mg/l) (Zhou et al., 2005). In contrast to their beneficial effects, some flavonoids have also been found *in vitro* to be mutagenic (Ahmad, Fazal, Rahman, Hadi, and Parish, 1992; Cao *et al.*, 1997; Popp and Schimmer, 1991; Rahman et al., 1992; Sahu and Gray, 1993, 1994). Flavonoids, such as quercetin and kaempferol, have been shown to induce nuclear DNA damage and lipid peroxidation in the presence of transition metals (Ahmad et al., 1992; Rahman *et al.*, 1992;

Sahu and Gray, 1993, 1994). These harmful effects were suspected to result from the prooxidant rather than antioxidant action of the related flavonoids (Hanasaki, Ogawa, and Fukui, 1994; Sahu and Gray, 1994). The rapid metabolic inactivation of mutagenic flavonoids catalysed by catechol-*O*-methyltransferase has been demonstrated by Zhu, Ezell, and Liehr (1994) *in vivo.* Cao *et al.* (1997) suggested that the action of catechol-*O*methyltransferase could be a major reason for the lack of carcinogenic activities of some flavonoids *in vivo.* Transition metal-initiated prooxidant action of a flavonoid may also be responsible for changes in regulation of enzyme activities by the flavonoid (Cao *et al.*, 1997).

4. Stilbenes

Stilbenes, in particular *trans*-resveratrol and its glucoside, are beneficial to health, having antioxidative, anticarcinogenic, and antitumour properties (Burns, Yokota, Ashihara, Lean, and Crozier, 2002; Jung *et al.*, 2009; Torres, Poveda, Jimenez-Barbero, Ballesteros, and Plou, 2010). In plants a major form of resveratrol is *trans*-resveratrol-3-*O*-β-D-glucoside, often referred to as piceid or polydatin (Fig. 6.7) (Jensen, Wertz, and O'Neill, 2010). Resveratrol (3,5,4′-trihydroxystilbene) is produced by plants in response to damage, particularly in grapevines (Langcake and Pryce, 1976), pines, and legumes (Soleas, Diamandis, and Goldberg, 1997).The main dietary stilbene is resveratrol from red wine and peanuts (Burns *et al.*, 2002) with lesser amounts found in pistachio, berries, red cabbage, spinach and certain herbs (Jaganath and Crozier, 2010). *Trans-Resveratrol* has gained significant worldwide attention because of its ability to inhibit or retard a wide variety of animal diseases, including cardiovascular disease and cancer (Jaganath and Crozier, 2010). Gülçin (2010) reported that resveratrol is an effective antioxidant in different *in vitro* assays including total antioxidant activity, reducing power, $DPPH^{\bullet}$, $ABTS^{\bullet+}$, $DMPD^{\bullet+}$ and $O_2^{\bullet-}$ scavenging, hydrogen peroxide scavenging, and metal chelating activities, when compared to standard antioxidant compounds such as BHA, BHT and α-tocopherol (Gülçin, 2010). Resveratrol may also be used in minimizing or preventing lipid oxidation in pharmaceutical products, retarding the formation of toxic oxidation products, maintaining nutritional quality, and prolonging the shelf life of pharmaceuticals. Hubbard *et al.* (2013) confirmed that resveratrol is an effective allosteric activator of SIRT1, an enzyme that belongs to the sirtuin family of nicotinamide adenine dinucleotide (NAD+)-dependent deacetylases as well as responsible for gene silencing, DNA repair, and metabolic regulation, which are all related to ageing. Pterostilbene, a derivative of resveratrol with only one hydroxyl group, is found to be more effective in rendering health benefits than resveratrol itself (Chang et al., 2012; Larrosa, Tomás-Barberán, and Espín, 2003; Stivala *et al.*,

2001). McCormack and McFadden (2012) reported that pterostilbene inhibits cancer growth through alteration of the cell cycle, induction of apoptosis, and inhibition of metastasis *invitro*. Furthermore, Pterostilbene has been shown to inhibit vascular endothelial growth factor (VEGF) production in human breast cancer [MDA-MB-231] cells (Sanghani, Kandra, Bavadekar, and Vansal, 2013). More recently, Siddalingappa, Benson, Brown, Batty, and Chen (2015) synthesized and evaluated the ester and ether-based poly-(ethylene glycol) succinylamide resveratrol and polylactide succinyl ester resveratrol polymer conjugates (soluble and polymeric micelles) and demonstrated that polymeric conjugates could be an effective approach to improve the pharmacokinetic profile of resveratrol as well as stability *in vitro* and *in vivo*.

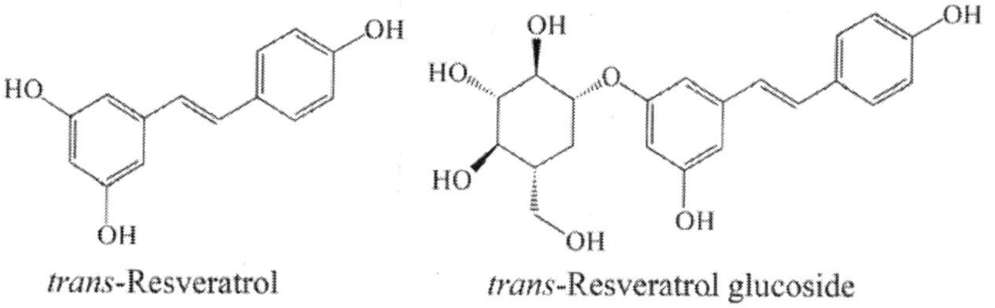

trans-Resveratrol   *trans*-Resveratrol glucoside

Fig. 6.7: Chemical structures of resveratrol and its glucoside (adapted from Jensen *et al.*, 2010).

5. Tannins

Tannins are defined as hydrolysable or condensed (proanthocyanidins), depending on their chemical structures (Fig. 6.8) (Shahidi and Naczk, 2004). The condensed tannins are oligomers and polymers of flavonoids, specifically flavan-3-ols, whilst hydrolysable tannins are glycosylated gallic acid (Ferreira and Li, 2000; Khanbabaee and van Ree, 2001). The phenolic groups of tannins bind very tightly with the —NH groups of peptides and proteins; they prevent their hydrolysis and digestion in the stomach and therefore are known to be anti-nutritional in nature (Shahidi and Naczk, 2004). In general terms, proanthocyanidins are principally found in fruits, especially berries, cocoa and some beverages like wine, beer and tea. Berries, legumes and leafy vegetables are the major sources of hydrolysable tannins (Serrano, Puupponen-Pimi, Dauer, Aura, and Saura-Calixto, 2009). Tannins are known to inhibit lipid peroxidation and lipoxygenases *in vitro*, and are able to scavenge radicals such as hydroxyl, superoxide, and peroxyl, which are known to be important in cellular prooxidant state (Gyamfi and Aniya, 2002). Systemic effects related to mechanism of antioxidation have been debated since verification *in vivo* has not been successful due to inadequate biomarkers of effects (Halliwell, Rafter, and Jenner, 2005; Huang, Boxin, and Prior, 2005; Prior,Wu, and Schaich, 2005). Riedl, Carando, Alessio, McCarthy, and Hagerman (2002) have demonstrated that both hydrolysable and condensed tannins scavenge free radicals in a kinetically complex fashion involving both a fast and a slow scavenging step. Using the structure–activity studies for monomeric and polymeric phenolic compounds shows that 4 moles of radical are scavenged per *ortho*-substituted diphenol group (Riedl *et al.*, 2002).

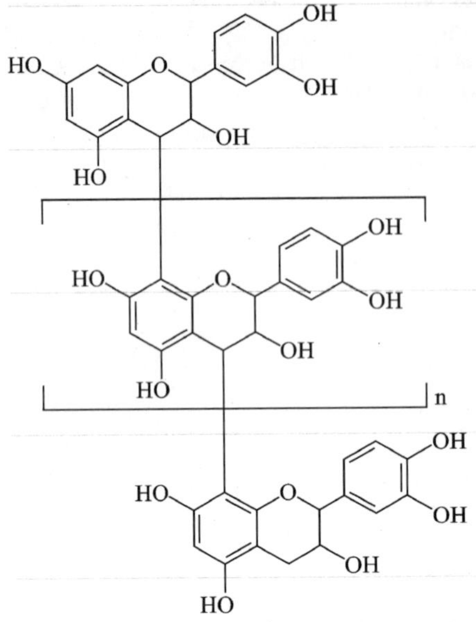

Condensed Tannin

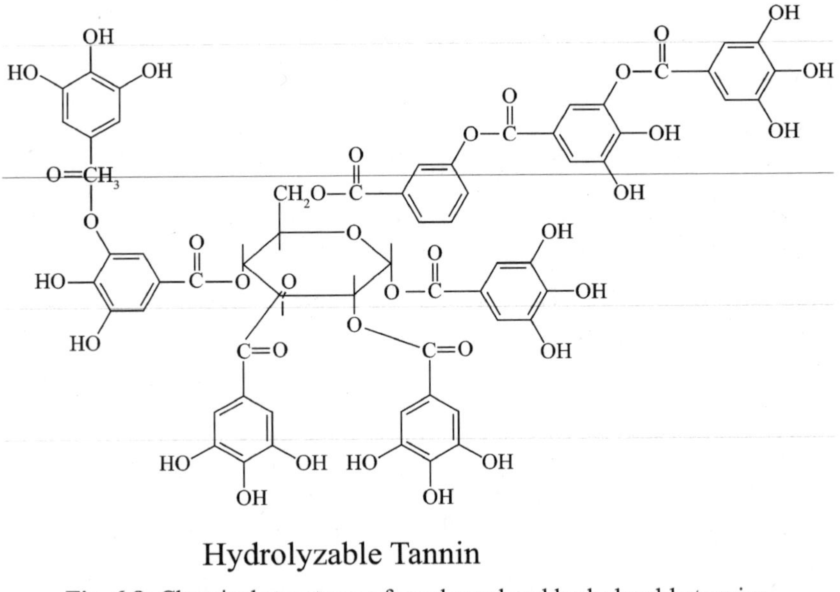

Hydrolyzable Tannin

Fig. 6.8: Chemical structures of condensed and hydrolysable tannins.

The fast scavenging reaction is inhibited by complexation of the tannin with protein, but the overall capacity of the tannin–protein complex for scavenging is similar to that of the free tannin (Riedl *et al.*, 2002). Scalbert (1991) reported that tannins seem to affect bacterial growth in several mechanisms, such as inhibition of extracellular microbial enzymes, deprivation of the substrates required for microbial growth or direct action on microbial metabolism through inhibition of oxidative phosphorylation. In addition, several hydrolysable tannins have been shown to significantly inhibit the cytopathic effects of human immunodeficiency virus, HIV, and the expression of HIV antigen in human lymphotropic virus type I-positive MT-4 cells (Nakashima *et al.*, 1992). Amarowicz and Pegg (2013) reported antiproliferative activities in a concentration-dependent manner of hydrolysable tannins against five carcinoma cell lines, namely MCF-7 (oestrogen receptor-positive breast carcinoma), DU-145 (androgen receptor-negative prostate carcinoma), HT-29 (colon carcinoma), SK-MEL-5 and MDA-MB-435 (melanoma; skin carcinoma). Yoshida, Nakata, and Okuda (1999) reported that raspberry extracts containing ellagitannins sanguiin H6 and lambertianin exhibits antitumour activities.

■■■■

7 Degradation of Phenolic Antioxidants

DEGRADATION OF PHENOLIC ANTIOXIDANTS

Phenolic antioxidants generally undergo degradation during oxidative processes such as those of fats and oils and produce a range of products, especially dimers of the antioxidants (Shahidi and Naczk, 2004). These dimers may be produced by the formation of phenoxyl radicals followed by radical rearrangement and a coupling reaction with another radical (Kikugawa, Kunugi, and Kurechi, 1990; Shahidi andWanasundara, 1992). Moreover, most oxidation products of antioxidants retain some antioxidant activity, which may influence the potential of the parent antioxidant during the course of its degradation (Kikugawa *et al.*, 1990). Amongst the breakdown products of BHT shown in Fig. 7.1a, products through possess antioxidant properties whilst products and are not effective as antioxidants (Kikugawa *et al.*, 1990; Shahidi and Naczk, 2004). The antioxidant activities of these breakdown products were determined as the ratio of the induction period of the oil containing these compounds to that of the oil containing the parent antioxidant (Kikugawa *et al.*, 1990). The degradation products of BHA shown in Fig. 7.1b are less effective than BHA itself, in the order of BHA > (7) > (6) (Shahidi and Naczk, 2004). All oxidation products of TBHQ retained some antioxidant effect, but compounds and exhibited a greater antioxidant activity than that ofTBHQ (Fig. 7.1c). As shown in Fig. 7.2a, irradiation of propyl gallate in ethanol produces ellagic acid with excellent antioxidant activity (Shahidi and Naczk, 2004). Degradation of mixtures of BHA, BHT and PG produced heterodimers between different antioxidant components (Fig. 7.2b). Although product was new, the heterodimer (15) and heterodimer (16) exhibited activities comparable to that of PG (Shahidi and Naczk, 2004). Studies have shown that extended exposure to certain temperatures, several other factors such as oxygen, alkali, light, minerals, and hydroperoxides degrade α-tocopherol (Sabliov *et al.*, 2009). It has been shown that UV light initiates formation of α-tocopherol oxy radical (breaking O - H bond) and can be dissociated to make a semiquinone anion-radical. The new semiquinone anion-radical could interact with oxygen and convert into a di-radical (Sabliov *et al.*, 2009).

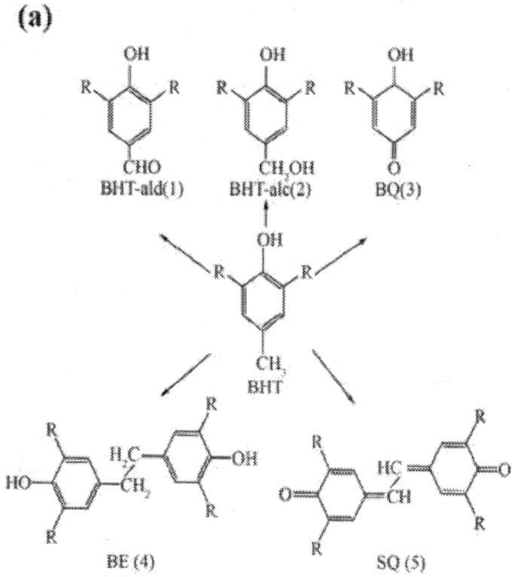

(a)

R = tertiary-butyl
(1) 3,5-ditertiary-butyl-4-hydroxybenzaldehyde (BHT-ald)
(2) 3,5-ditertiary-butyl-4-hydroxybenzyl alcohol (BHT-alc)
(3) 2,6-ditertiary-butylbenzoquinone (BQ)
(4) 3,5,3',5'-tetra-tertiary-butyl-4,4'-dihydroxy-1,2-diphenylethane (BE)
(5) 3,5,3',5'-tetra-tertiary-butylstilbenequinone (SQ)

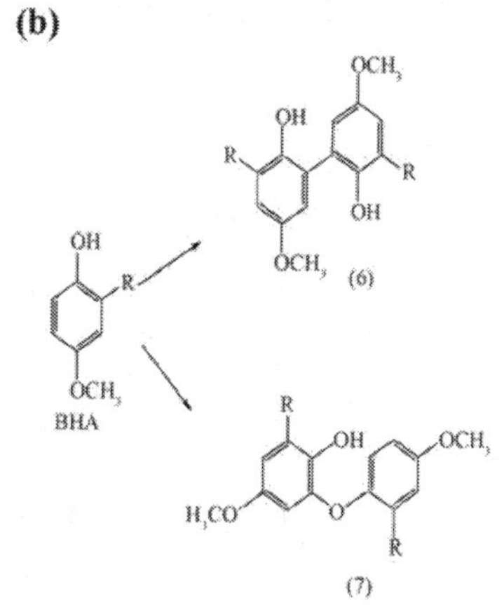

(b)

R = tertiary-butyl
(6) 2,2'-dihydroxy-5,5'dimethoxy-3,3'-di-tertiary-butylbiphenyl
(7) 2',3-ditertiary-butyl-2-hydroxy-4',5-dimethoxybiphenyl ether

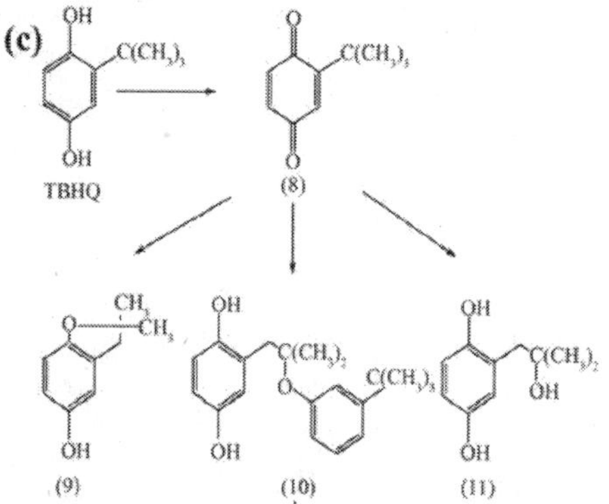

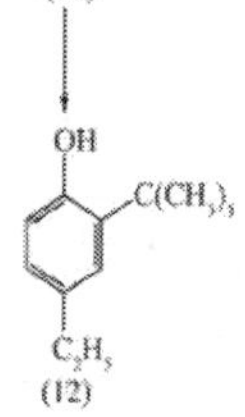

(c)

(8) 2-tertiary-butyl-*p*-benzoquinone
(9) 2,2'-dimethyl-5-hydroxy-2,3-dihydrobenzo[6]furan
(10) 2-[2-(3'-tertiary-butyl-4'-hydroxy-phennoxy-2-methyl-1-propyl]hydroquinone
(11) 2-(2-hydroxy-2-methyl-1-propyl)hydroquinone
(12) 2-tertiary-butyl-4-ethoxyphenol

Fig. 7.1: (a) Degradation products of BHT, (b) BHA, and (c) TBHQ.

(a)

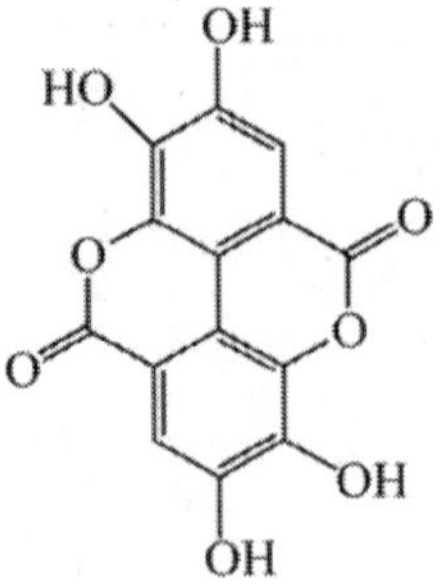

OH HO O O O O OH OH

(13) Ellagic acid

(b)

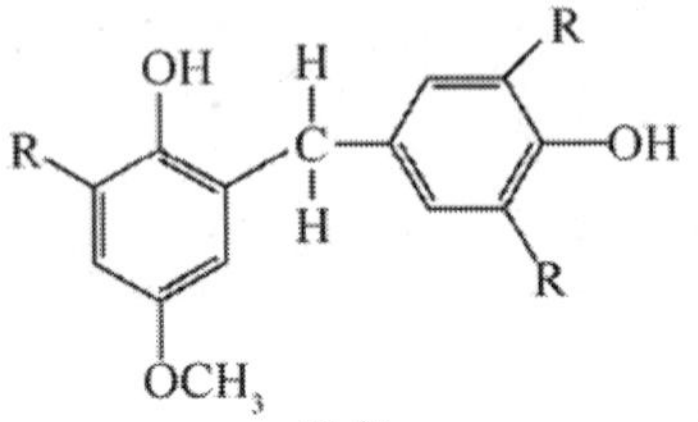

(14) 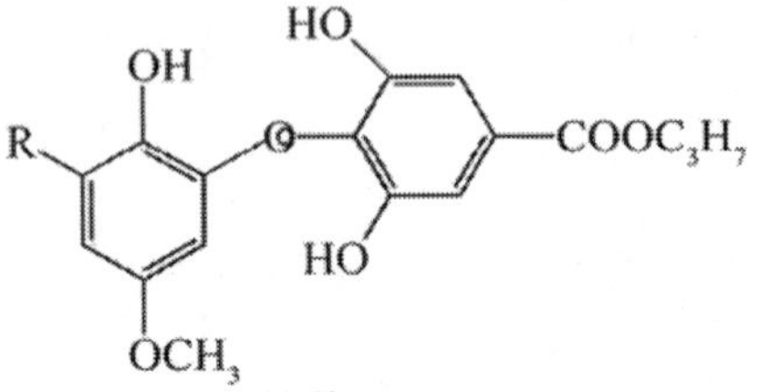 (15) 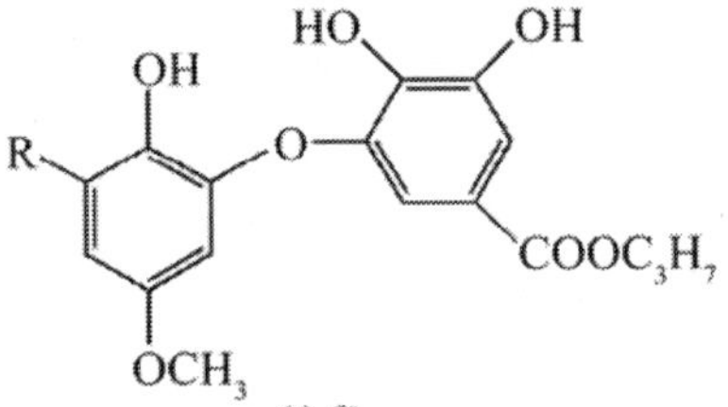 (16)

(14) 3,3',5'-tri-tertiary-butyl-5-methoxy-2,4'-dihydroxydiphenylmethane
(15) propyl-3,5-dihydroxy-4-(2'-hydroxy-5-methoxy-3'-tertiary-butyl phenoxy) benzoate
(16) propyl-3,4-dihydroxy-5-(2'-hydroxy-5'-methoxy-3'-tertiary-butyl phenoxy) benzoate

Fig. 7.2: (a) Degradation product of PG and (b) mixture of BHA and PG.

■■■■

8 Natural sources of Phenolic Antioxidants

A. CEREALS AND LEGUMES

1. CEREALS

Whole grains are recommended for healthy diets and are recognized sources of dietary fiber and antioxidant compounds (Ragaee, Abdel-Aal, and Noaman, 2006). Phenolic acids and flavonoids are present in cereals in the free and conjugated forms. The outer layers of cereal grains (husk, pericarp, testa and aleurone cells) contain the greatest concentrations of total phenolics, whereas their concentration is considerably lower in the endosperm layers (Kähkönen et al., 1999; Liyana-Pathirana and Shahidi, 2005, 2006; Madhujith and Shahidi, 2006, 2007a, 2007b). For instance, free and esterified phenolic acids were found in wheat, corn, rice, and oat (Sosulski, Krygier, and Hogge, 1982). Some researchers have reported that phenolic acids are concentrated in the cell walls of the outer layers, where they are mainly esterified to the arabinose side groups of arabinoxylans. However, others have indicated that phenolic acids are present mainly in the aleurone layer and endosperm (Yu, Vasanthan, and Temelli, 2001). Naczk and Shahidi (2006) reported that the highest concentration of phenolic acids and flavonoids is in the aleurone layer of cereal grains, but these compounds are also found in embryos and seed coat of grains. Total antioxidant capacities of cereals are shown in Figure 8.1. Water and ethanol extract of buckwheat showed the highest TEAC value (93.6 ± 1.6) and millet and corn had the lowest TEAC values (9.1 ± 3.0) (Serpen, Gökmen, Pellegrini, and Fogliano, 2008). Wheat extracts have shown potential antioxidant properties as wheat phenolics appear to serve as powerful antioxidants through radical scavenging and/or metal chelation (Liyana-Pathirana, Dexter, and Shahidi, 2006; Liyana-Pathirana and Shahidi, 2006), whilst barley contains substantial amounts of phenolic antioxidants that effectively scavenge peroxyl, DPPH, and hydroxyl radicals, and control oxidation of LDL cholesterol, thereby having a great potential in the development of nutraceuticals rich in antioxidants (Madhujith and Shahidi, 2006, 2007a). The antioxidant activities and total phenolics of some cereals are shown in Table 8.1. Adom and Liu (2002) reported that corn had the highest total phenolic content (15.55 ± 0.60 μmol of gallic acid eq/g of grain) of the grains tested, followed by wheat (7.99

± 0.39 μmol of gallic acid eq/g of grain), oats (6.53 ± 0.19 μmol of gallic acid eq/g of grain), and rice (5.56 ± 0.17 μmol of gallic acid eq/g of grain). Recently, Ndolo and Beta (2014) showed that sinapic acid is more concentrated in the pericarp and germ of wheat, whereas isoferulic acid is concentrated in the germ of purple barley. Moreover, syringic and vanillic acids have been shown to be concentrated in the pericarp and sinapic acid in the aleurone layer of yellow corn (Ndolo and Beta, 2014).

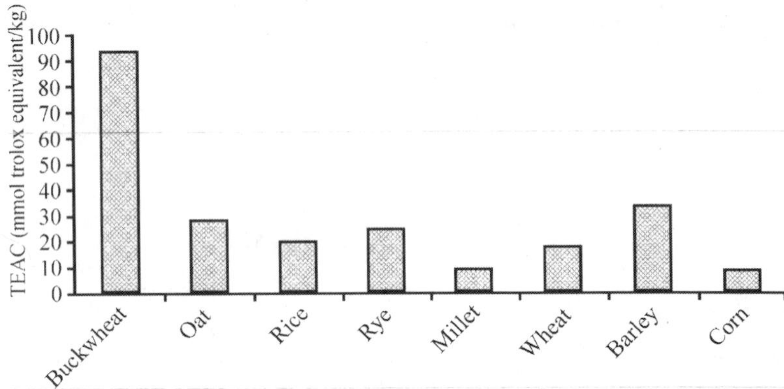

Figure 8.1: Total antioxidant capacities, expressed as mmol of Trolox equivalent antioxidant capacity (TEAC) per kg of cereals (adapted from Serpen *et al.*, 2008).

Table 8.1: Phenolic content and antioxidant activity of cereals.

Cereal	Total phenolic content[a]	DPPH scavenging (IC50) capacity[b]	FRAP[c]
Barley	16.4	>200	15.56
Buckwheat	50.7	76.7	49.43
Wheat	16.2	>200	12.15
Rye	13.2	>200	8.94

Source: Adapted from Đordevic, Šiler-Marinkovic and Dimitrijevic-Brankovic (2010).
[a]mg gallic acid equivalents/g dried extract; [b]μg/ml. [c] nmol Fe2+/mg dried extract.

a. Wheat (Wheat antioxidants)

Wheat kernels contain a number of phenolic compounds, namely ferulic, vanillic, gentisic, caffeic, salicylic, syringic, *p*-coumaric and sinapic acids as well as vanillin and syringaldehyde, none of which contribute to the antioxidant activity of the products (Liyana-athirana and Shahidi, 2005, 2006). Of these, ferulic acid was the primary phenolic acid in the grain, accounting for up to 90% of total phenolic acids (Liyana-Pathirana *et al.,* 2006; Liyana-Pathirana and Shahidi, 2005, 2006; Okarter, Liu, Sorrells, and Liu, 2010). A study on six varieties of whole wheat grain reported no vanillic acid in the free or bound fractions. Further, no syringic acid or *p*-hydroxybenzoic acid were detected in the free fraction, which is consistent with the results reported by Moore *et al.* (2005), but not consistent with that reported by

Mattila, Pihlava, and Hellstrom (2005) and Okarter *et al.* (2010). Whole wheat varieties tested contained mainly α- and β-tocopherols and α- and β-tocotrienols. β-Tocotrienol was the predominant form of vitamin E found in all varieties of whole wheat (Okarter *et al.,* 2010).Tricin (5, 7, 4-trihydroxy 3, 5-dimethoxyflavone) was found to be the dominant flavone pigment in wheat. In addition, two *C*-glycosyl-flavones, namely 6-*C*-pentosyl-8-*C*-hexosylapigenin and 6-*C*-hexosyl-8-*C*-pentosylapigenin were isolated from wheat bran (Naczk and Shahidi, 2006). Liyana-Pathirana and Shahidi (2007a) reported that phenolic compounds were concentrated in the outermost layers, namely the bran. Thus, consumption of wheat with bran in the form of whole grain may provide beneficial health effects. Asymmetrical distribution of antioxidative components in the wheat grain was prominent. The concentration of bioactive constituents was greater in the external layers; thus the bran fraction alone demonstrated a higher antioxidant activity than that of other milling fractions (Liyana-Pathirana and Shahidi, 2007b). Campestanyl ferulate and sitostanyl ferulate were the main components of steryl ferulates present in wheat grain and that rye and wheat and especially their bran fractions are comparable to corn as sources of steryl ferulate (Hakala et al., 2002). Wheat bran also contained ferulic acid dehydrodimers (DiFA), which strengthen the aleurone walls during the maturation of wheat grain by the formation of bridges between two arabinoxylan chains. These dehydrodimers are products of oxidative coupling of ferulic acid catalysed by peroxidase (Naczk and Shahidi, 2006).

Wheat (*Triticum aestivum* L.) is one of the most important cereal grains in the world because of the universal use of wheat for a wide range of products such as bread, noodles, cakes, biscuits, cookies, etc. Wheat kernel is composed of endosperm (81–84%), bran (14–16%), and germ (2–3%) (Pomeranz, 1988). Endosperm is the inner part playing a role as storage of energy (starch) and functioning protein (gluten components). Bran is the outer layer protecting the grain and germ is the kernel's reproduction system. Wheat endosperm contains mostly starch and protein, whereas bran and germ are rich in dietary fiber, minerals and phytochemicals which play important roles in nutrition and health benefits for humans (Pomeranz, 1988). Therefore, the customers are strongly recommended to consume whole wheat flour and its related products. The significant health benefits are derived from the important diets in bran and germ such as dietary fiber or phenolic acids. The results of TPCs in different milling fractions of wheat grains are shown in Figure 8.2. The TPC of bran/germ fractions (2867-3120 μmol of gallic acid equiv/100 g) was 15-18-fold higher ($p < 0.01$) than that of respective endosperm fractions (176-195 μmol of gallic acid equiv/100 g of flour) in the five wheat samples tested (Adom *et al.*, 2005). Recently, the gradual milling method has been developed to grade whole grains from outer to inner parts without removal of germ and bran using a modified Japanese rice-polisher (Hayashi *et al.*, 1991). The graded flour fractions containing large amounts of nutrients such as dietary fibers and minerals were considered as the sources of nutrition for human beings (Maeda, 2001). Hung *et al.* (2009) reported that the graded flour fractions milled from the outer parts of grain contained significantly higher amount of phenolics and exhibited significantly higher antioxidant capacity than did the whole grain. Likewise, the inner flour fractions milled from mostly endosperm part had significantly higher amount of phenolics and exhibited significantly higher antioxidant capacity than did the white flour, which was milled by a conventional milling method. In addition, the roller milling method with or without debranning have been reported to increase the amount of phenolic compounds and improve antioxidant activity of the pearled wheat fractions (Beta *et al.*, 2005). Phenolic acids existed in cereal grains in free, soluble conjugate and insoluble bound forms. Organic solvents such as ethanol, methanol and acetone may be employed to extract the free and soluble conjugate phenolic acids, whereas bound phenolics may be released by alkali, acid, or enzymatic treatment of samples prior to extraction (Sosulski *et al.*, 1982; Krygier *et al.*, 1982a; Krygier *et al.*, 1982b; Andreasen et al., 2001; Zupfer *et al.*, 1998; Bartolome, and Gomez-Cordoves, 1999). The previous studies reported that phenolic acids in wheat grains are mostly in the bound form and exist in bran associated with cell wall materials (Adom and Liu, 2002; Liyana-Pathirana *et al.*, 2006). The bound phenolics were considered to have more health benefits because they may escape from upper gastrointestinal digestion conditions along with cell wall materials and are absorbed into blood plasma during digestion of intestinal microflora (Andreasen *et al.*, 2001). Therefore, the increased consumption of the outer layer fractions of wheat grains might increase in an amount of the bound phenolics absorbed. During germination of wheat, free phenolic acids in wheat increased, whereas bound phenolic acids significantly decreased. However, the steeped, then incubated sub-samples, in two Canadian wheat classes, Canadian Western Red Spring (CWRS) and Canadian Western Amber Durum (CWAD), exhibited significantly greater TPCs than their corresponding non-germinated sub-samples (Hung *et al.*, 2011). Thus, the germinated wheats exhibits better nutritional properties than un-germinated wheat and could be used to improve the nutrition value in food products.

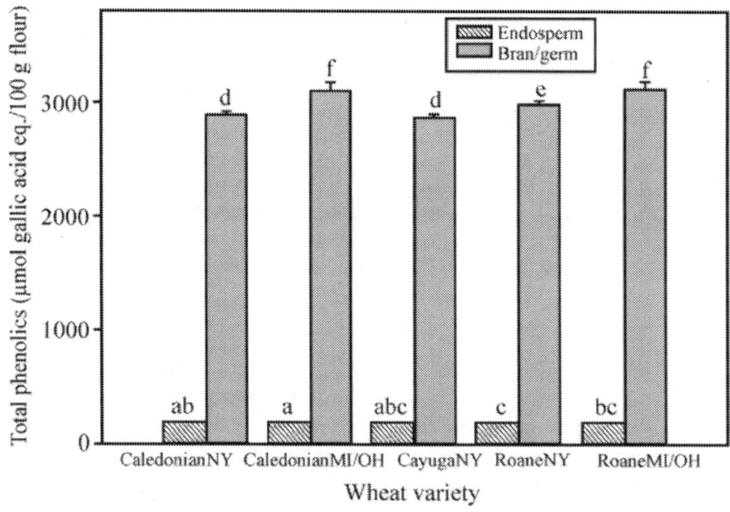

Figure 8.2: Total phenolic content of milled fractions of wheat varieties (Adapted with permission from Adom *et al.* (2005). Copyright ©2005 American Chemical Society)

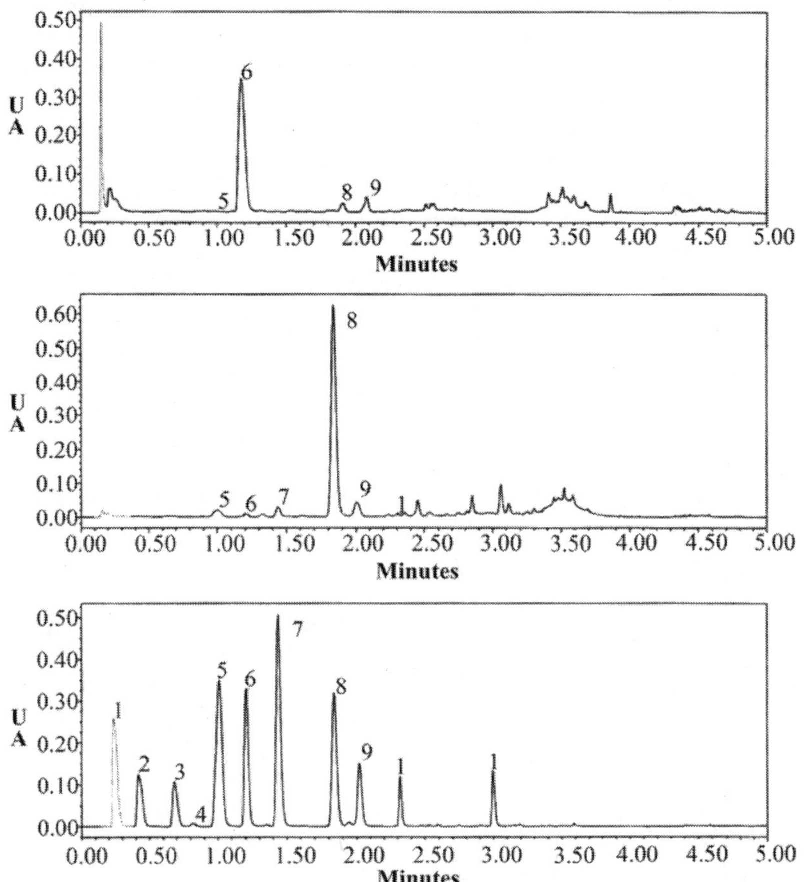

Figure 8.3: Typical chromatography for phenolic compounds from alcohol extract (A), alkaline extract (B) and a mixed standard (C) detected at 280 nm. 1, Gallic acid; 2, Protocatechuic acid; 3, 4-Hydroxybezoic acid; 4, Gentisic acid; 5, Vanillic acid; 6, Syringic acid; 7, p-Coumaric acid; 8, Ferulic acid; 9, Sinapic acid; 10, Rutin; 11, Quercetin.

Phenolic acids in cereal grain extracts can be detected by liquid chromatography system. A high pressure chromatography system (HPLC) has been commonly employed to detect the phenolic acids in wheat extracts (Kim *et al.*, 2006; Hatcher and Kruger, 1997). Recently, the first report on the application of an ultra-performance liquid chromatography (UPLC) for the rapid separation and quantification of simple phenolic acids in wheat extract replacing the traditional high pressure liquid chromatography system (HPLC) has been published (Hung *et al.*, 2011). The UPLC system was developed by Waters Corp. to use sub 3μm-particle chromatography columns resulting in analyses with short running time, high sensitivity and resolution. The UPLC system used includes a binary solvent manager which delivers up to 15,000 psi pressure, a photodiode array (PDA) detector with a spectra in a range of between 190 and 800 nm, a 1.8 lm particle column and a sample manager with small injection volume used (0.5–5 µl). Figure 8.3 shows the typical chromatography of phenolic compounds extracted from the wheat flours and detected by the UPLC system. Twelve standards were separately detected in the running program for only 5 min (Figure 8.3C) by using the UPLC system as compared to a 60-min running by a HPLC system reported by Kim *et al.* (2006). The main phenolic compound detected in wheats extracted by an aqueous of 80% ethanol was syringic acid, whereas the main phenolic compound detected in the extracts by alkaline hydrolysis was ferulic acid. These results confirm that the hydroxybenzoic acid derivatives can be easily extracted using alcohol, whereas the hydroxycinnamic acid derivatives are released from the bound form by alkaline hydrolysis. Using 80% aqueous ethanolic extraction, five phenolic compounds were found in wheat varieties with a decreasing order of syringic>sinapic>ferulic>coumaric>caffeic, whereas The phenolic profile of CWRS and CWAD flours after alkaline hydrolysis includes seven compounds with a decreasing order of ferulic>sinapic>vanillic>4-hydroxybenzoic> p-coumaric>syringic>ceffeic (Hung *et al.*, 2011). However, the phenolic acids in wheats were found to vary among the published reports which might be due to the different condition of extraction and chromatography system (Hung *et al.*, 2011; Liyana-Pathirana *et al.*, 2006; Hatcher *et al.*, 1997; Mattila *et al.*, 2005; Zhou *et al.*, 2004; Solsulski *et al.*, 1982).

The antioxidant activity of phenolic fractions was evaluated using Trolox equivalent antioxidant capacity (TEAC), 2,2-diphenyl-1-picrylhydrazyl radical scavenging (DPPH), reducing power, oxygen radical absorbance capacity (ORAC), inhibition of oxidation of human low-density lipoprotein cholesterol (LDL) and DNA, Rancimat, inhibition of photochemilumenescence (PCL), and iron(II) chelation activity (Liyana-Pathirana *et al.*, 2006). In the methods of ORAC, PCL, and Rancimat, wheat phenolics may act as free radical scavengers by donating a hydrogen atom. In assays that determine the ability to inhibit LDL and DNA oxidation by wheat phenolics, the antioxidative properties may be rendered by both hydrogen donation and/or metal chelation. Wheat phenolics appeared to serve as powerful antioxidants by radical scavenging and/or metal chelation in a consistent manner. The antioxidant capacity is positively correlated to the TPCs of wheat extracts. The extracts with higher phenolic contents exhibited the stronger antioxidant capacity than the lower one. Liyana-Pathirana *et al.* (2006) reported that the 10% byproduct from the pearling of wheat demonstrated the highest antioxidant activity. Subsequent removal of external layers resulted in a decrease in phenolic content with concurrent lower antioxidant activity values for higher degrees of pearling. Likewise, Hung *et al.* (2009) reported that all fractions contained a part of bran by gradual milling resulting in improvement of antioxidant capacity of these flours as compared to the white flour. In addition, the bound phenolic extracts of wheats exhibited higher antioxidant capacity than did the free phenolic extracts. The total antioxidant capacity of all graded milling fractions were mainly contributed by the bound phenolic extracts (72.5 - 83.2%), which mostly existed in the outer fractions rather than the inner fractions of wheat grain. As phenolic compounds are concentrated in the outermost layers, the bran fractions resulting from pearling may be used as a natural source of antioxidants and as value-added products in the preparation of functional food ingredients or for enrichment of certain products (Liyana-Pathirana *et al.*, 2006).

b. Barley (Barley antioxidants)

Barley phenolics include tyrosine, tyramine and its derivatives, phenolic acids, their esters and glycosides, anthocyanins, proanthocyanidins, lignans and substances related to lignin (Naczk and Shahidi, 2006). The free forms of salicylic, *p*-hydroxybenzoic, vanillic, protocatechuic, *o*-, *m*- and *p*-coumaric, syringic, ferulic and sinapic acids have been identified in barley grains (Madhujith, Izydorczyk, and Shahidi, 2006; Shahidi, 2009; Shahidi and Naczk, 2004). Yu *et al.* (2001) reported the presence of chlorogenic and protocatechuic acids in barley. Ferulic acid was the predominant free phenolic acid in barley seeds (Nordkvist, Salomonsson, and Aman, 1984) and barley bran (Renger and Steinhart, 2000). On the other hand, *p*-hydroxybenzoic acid was the major bound phenolic acid detected in barley extracts obtained by sequential treatment of grain with acid, α-amylase and cellulase (Yu *et al.,* 2001).

Madhujith and Shahidi (2009) found that the antioxidant and antiradical activities of insoluble-bound phenolic fraction, in general, were higher than those of soluble conjugates and free phenolic fractions. Madhujith and Shahidi (2006) reported total phenolic content ranging from 0.81 to 1.38 mg of ferulic acid eq/g in defatted barley flour. The coloured barley grains contained at least eight pigmented compounds that were identified as derivatives of cyanidin, delphinidin and pelargonidin, as well as cyanidin-3-arabinoside and, possibly, cyanidin-3-glucoside (Madhujith and Shahidi, 2009; Shahidi and Naczk, 2004). Barley grains also contain a range of flavanols such as (+)-catechin, (−)-epicatechin, dimeric prodelphinidin B3 and procyanidin B3, as well as trimeric procyanidin C_2 and three trimeric prodelphinidins (Goupy, Hugues, Boivin, and Amiot, 1999). Osawa, Katsuzaki, Hagiwara, Hagiwara, and Shibamoto (1992) and Kitta, Hagiwara, and Shibamoto (1992) isolated a novel flavone C-glycoside, 2″-*O*-glycosylisovitexin, from green barley leaves and reported that its antioxidative activity was similar to that exhibited by α-tocopherol. Sharma and Gujral (2010) reported that the antioxidant activity of barley significantly increased with duration of germination up to 24 hours and also a strong positive correlation existed between total phenolic content and antioxidant activity of barley. Phenolic extracts from whole barley kernel possessed high antioxidant, antiradical, and antiproliferative potentials.Therefore, inclusion of whole barley into the daily diet may render beneficial health benefits (Madhujith and Shahidi, 2007a, 2007b). Another study showed that breads made by replacing 40% of wheat flour with barley flour increased the antioxidant properties of the breads (Holtekjølen, Bævre, Rødbotten, Berg, and Knutsen, 2008).

Barley (*Hordeum Vulgare* L.) is one of the first agricultural domesticates together with wheat, pea, lentils dating from about 10,000 years ago (Smith, 1998). It is the fourth most important cereal in the world in terms of world production after wheat, rice and corn. Barley has been known to be used in the beer industry, malting and animal feed with 80-90% of barley production is used for malting and animal feedstocks (Baik, and Ullrich, 2008). Recently, barley is considered to be used as an ingredient for production of functional foods (Holtekjolen *et al.*, 2008) due to its high contents of bioactive compounds such as β-glucans, tocopherols, tocotrienols (Jadhav *et al.*, 1998) and many classes of phenolic compounds such as benzoic and cinnamic acid derivatives, proanthocyanidins, quinones, flavonols, chalcones, flavones, flavanones and amino phenolic compounds (Goupy *et al.*, 1999; Hernanz *et al.*, 2001). The abundant content of phenolic compounds in barley reveals that it may serve as an excellent dietary source of natural antioxidants with antiradical and antiproliferative potentials for disease prevention and health promotion (Madhujith and Shahidi, 2007; Zhao *et al.*, 2008).

Like other cereals, the content of insoluble-bound phenolics was significantly higher than those of soluble conjugate and free phenolic fractions among all barley extracts tested as evaluated by Madhujith and Shahidi (2009). Free phenolic content ranged from 0.18 to 0.42 mg ferulic acid equivalents per gram defatted material while soluble conjugates ranged from 0.42 to 0.81 mg ferulic acid equivalents per gram of defatted material and insoluble-bound fraction ranged from 2.03 to 3.36 mg ferulic acid equivalents per gram of defatted material. The TPC of barley grains is also concentrated in bran other than the endosperm part. The TPCs of the pearling byproducts decreased when moving from the outermost layer to the center of the barley grain. The outermost layer (F1), which is basically the bran, contained the highest TPC while F2 also showed a significantly higher TPC compared to the rest of the inner fractions (Madhujith *et al.*, 2006). Sharma and Gujral (2010) also reported that the TPC in barley bran was highest followed by whole flour and refined flour in all the germinated and ungerminated samples. In whole flour the TPC was less compared to the bran as the endosperm diluted the concentration of phenolic compounds from the outer layers. A similar effect was also observed for the refined flour that showed the lowest TPC as compared to the bran and the whole flour.

Phenolic acids from 30 barley varieties (combination of hulled/hulless/two-row/sixrow/regular/waxy) were investigated by HPLC following four different sample treatments have been reported by Yu *et al.* (2001). Seven major phenolic acids including the benzoic acid (*p*hydroxybenzoic, vanillic, and protocatechuic acids) and cinnamic acid derivatives (coumaric, caffeic, ferulic, and chlorogenic acids) were found in the barley varieties. *p*-Hydroxybenzoic acid was the major phenolic acid in the 30 barley varieties studied by using acid, α-amylase, and cellulase hydrolysis treatments, followed by ferulic acid. By hot water extraction, the protocatechuic acid and chlorogenic acid were the two major compounds that existed in free form in barley grain even though their concentrations were very low (<2.9 μg/g for protocatechuic acid and <16.3 μg/g for chlorogenic acid). After acid hydrolysis, the concentrations of protocatechuic, vanillic, chlorogenic, and ferulic acids could be quantified. On the other hand, *p*-hydroxybenzoic, caffeic, and *p*-coumaric acids could be detected but not quantified because of poor resolution or low levels of those compounds. After acid hydrolysis with α-amylase hydrolysis, a better separation for all phenolic acids in the complex medium was obtained. All six phenolic acids could be separated except for protocatechuic acid. The treatment combining acid, α-amylase, and cellulase yielded the highest concentration for most phenolic acids, indicating that most phenolic acids in barley are primarily bound to other grain components (for example, starch, cellulose, β-glucan, pentosans, and others) (Yu *et al.*, 2001). In another study, Kim *et al.* (2007) has found 17 phenolic acids and 7 flavonoids present in 127 lines of coloured (black, blue, and purple) barley (Table 8.2). In the tested barley, chlorogenic acid and phloroglucinol were present in hulled and unhulled groups as the major compounds, respectively. All barley groups had very low concentrations of syringic acid, *o*hydroxyphenylacetic acid, ferulic acid, and 3,4-dimethoxybenzoic acid. The concentrations of phloroglucinol acid, vanillic acid, syringic acid, *o*-hydroxyphenylacetic acid, 3, 4-dimethoxybenzoic acid, salicylic acid, and *o*-coumaric acid in the blue and purple groups were significantly higher than those in the black groups.

Table 8.2: Total concentrations (μg/g) of phenolic compounds in coloured barley groups.

Phenolic acids	**Hulled**				**Unhulled**			**LSD (0.05)**
	Black 1	**Black 2**	**Black 3**	**Purple**	**Black**	**Blue**	**Purple**	
A. Phenolic acids								
Phloroglucinol	2.4±7.0	0.8±2.8	3.6±10.4	19.3±27.3	46.3±21.4	52.8±59.6	34.2±16.5	15.41
Gallic acid;	14.6±0.6	15.2±0.9	15.0±1.0	14.6±0.5	15.6±0.9	15.7±13.3	15.7±1.6	2.42
Pyrogallol	15.7±8.0	19.7±4.4	23.3±11.0	15.2±0.4	21.8±11.0	22.2±3.1	22.5±9.0	8.96
Homogentisic acid	5.5±3.3	7.8±4.3	10.3±4.5	7.2±1.4	11.3±5.1	11.9±0.4	9.8±2.7	4.24
Protocatechuic acid	8.1±6.6	7.7±5.4	13.2±5.7	8.6±1.8	9.31.7	12.2±7.4	10.7±7.0	6.28
Chlorogenic acid	30.0±7.1	33.3±9.4	38.8±12.7	25.0±2.6	29.2±6.2	35.6±6.1	30.6±6.1	10.16
β-resorcylic acid	4.5±1.8	4.0±1.0	5.1±4.3	6.3±2.0	4.0±1.3	19.5±18.7	4.1±2.0	4.18
Vanillic acid	0.5±0.8	0.4±0.5	0.3±0.8	0.4±0.6	0.2±0.3	9.3±9.0	1.1±1.4	1.71
Syringic acid	-	0.0±0.0	-	-	-	1.4±1.1	0.0±0.0	0.20
Caffeic acid	10.3±2.3	10.3±2.5	8.4±4.8	11.1±0.5	11.6±1.1	11.0±0.4	8.1±5.1	3.86
o-hydroxy phenylacetic acid	-	0.1±0.0	0.1±0.5	-	-	0.1±0.1	0.2±0.7	0.43
p-coumaric aicd	3.4±0.5	3.0±0.4	2.9±0.3	3.3±0.3	2.6±0.1	3.5±0.9	2.7±0.3	0.41
Ferulic acid	0.2±0.3	0.2±0.2	0.1±0.2	-	-	2.6±3.5	-	0.62
3,4-dimethoxy-bezoic acid	0.1±0.3	0.1±0.3	0.2±0.4	1.9±2.6	-	0.9±1.3	0.2±0.4	0.60
Salicylic acid	10.9±6.9	11.0±7.0	14.2±4.3	15.8±0.7	16.4±0.6	28.1±15.3	15.5±4.1	6.43
Benzoic acid	2.0±3.5	4.4±7.6	1.9±1.6	12.4±9.5	4.6±6.3	9.2±13.1	2.3±5.4	6.11
o-coumaric acid	4.1±1.0	3.81.3	4.51.6	10.4±5.7	4.3±0.2	5.2±1.2	7.5±4.7	2.46
Total	**112.3±23.6**	**121.8±17.3**	**142.0±30.6**	**151.5±20.0**	**177.2±24.4**	**241.3±62.2**	**165.2±30.6**	**28.74**
B. Flavonoids								
Catechin	17.7±6.0	17.6±6.7	16.7±6.9	13.1±3.3	16.8±6.1	23.8±13.3	20.0±10.3	7.90
Rutin	3.2±1.7	3.0±1.4	3.7±2.2	4.6±1.2	0.7±1.1	4.6±2.9	3.4±2.7	2.10
Naringin	0.0±0.0	0.1±0.2	0.1±0.2	-	-	-	0.1±0.2	0.17
Hesperidin	0.4±1.3	0.2±0.7	0.2±0.8	2.4±3.4	-	1.0±1.5	0.4±1.0	1.13
Myricetin	28.0±1.1	27.7±1.0	27.7±1.1	39.1±1.6	33.4±8.6	42.6±16.0	31.4±2.2	4.29
Quercetin	17.4±1.1	17.1±3.0	17.4±0.7	19.3±2.0	14.8±6.0	66.6±69.8	17.9±1.6	11.76
Kaempferol	12.7±2.0	12.3±0.6	12.2±0.9	13.4±2.1	11.9±0.2	23.9±16.9	13.1±2.3	3.14
Total	**79.3±8.4**	**78.1±7.9**	**77.9±7.9**	**91.9±19.9**	**77.6±10.2**	**162.5±98.3**	**86.3±12.8**	**18.99**
Grand Total	**191.6±26.2**	**199.9±20.2**	**220.0±34.7**	**243.4±15.2**	**254.7±26.1**	**403.8±144.9**	**251.4±38.0**	**39.02**

There are a number of publications reported on the antioxidant capacity of phenolic extracts from barley in the literature (Goupy *et al.*, 1999; Kim *et al.*, 2007; Madhujith *et al.*, 2006; Madhujith *et al.*, 2007; Madhujith *et al.*, 2009; Zhao *et al.*, 2008). The insoluble phenolic fraction exhibited significantly higher TEAC and ORAC values and DPPH radical scavenging capacity than the free, soluble conjugate phenolic fractions of barley extracts. A similar trend was observed against inhibition of LDL cholesterol oxidation and radical-induced DNA breakage (Madhujith *et al.*, 2009). The antioxidant and antiradical activities are mainly concentrated in the outer fractions (F1-F3 fractions), which are counted approximately up to 25% of the grain on weight basis (Madhujith *et al.*, 2006). In colored barley, the DPPH radical scavenging activities varied from 46.4 to 86.3%. Average radical scavenging activity in the unhulled barley groups (66.5%) was higher than that of hulled barley (63.5%). Although the total concentration of anthocyanins in the colored barley varied among the tested groups, the radical scavenging activity among colored barley groups was not significantly different. This results indicate that antioxidant capacity of the colored barley is related to the concentration of phenolic compounds such as chlorogenic acid, 3,4-dimethoxybenzoic acid, homogentisic acid, protocatechuic acid, and rutin (Kim *et al.*, 2007). Zhao *et al.* (2008) also reported that the TPC showed strong correlations with DPPH radical scavenging activity, ABTS radical cation scavenging activity, and reducing power ($P < 0.01$), whereas its correlations with superoxide anion radical scavenging activity and metal chelating activity were poor ($P > 0.05$) using the Pearson correlation analysis.

c. Buckwheat

Buckwheat is categorized as a pseudo cereal and its seed serves as a rich source of flavonoids. Flavonols such as rutin, hyperin, quercitrin and quercetin and flavones such as vitexin, isovitexin, orientin, and isoorientin have been identified in immature buckwheat seeds (Guo *et al.,* 2012; Luo, Peng, Fei, Yang, and Fan, 2014; Shahidi and Naczk, 2004). Only six flavonoids have been identified in mature seeds and their hulls. Rutin and isovitexin were the only flavonoids of seeds, although the hulls contained rutin, quercetin, orientin, vitexin, isovitexin and isoorientin (Dietrych-Szostak and Oleszek, 1999). Four catechins, namely (−)-epicatechin, (+)-catechin 7-*O*-β-Dglucopyranoside, (−)-epicatechin 3-*O*-*p*-hydroxybenzoate and (−)-epicatechin 3-*O*-(3, 4-di-*O*-methyl) gallate were identified in the ethanolic extracts of buckwheat groats (Dietrych-Szostak and Oleszek, 1999). Of the isolated flavonoids, vitexin and isovitexin revealed almost no peroxyl radical-scavenging activity (Watanabe, Ohshita, and Tsushida, 1997). However, the isolation of isovitexin from long-life rice seeds Ramarathnam, Osawa, Namiki, and Kawakishi, 1989) suggests that this compound probably acts as a protectant in the seeds against oxidative damage. In buckwheat the content of ferulic and hydroxycinnamic acids is low. Bran–aleurone fraction of buckwheat contained bound syringic, *p*-hydroxybenzoic, vanillic and *p*-coumaric acids (Shahidi and Naczk, 2004).Watanabe et al. (1997) identified protocatechuic acid and 3, 4-dihydroxybenzaldehyde from buckwheat hulls. Presence of soluble condensed tannins based on pelargonidin and cyanidin structures has been confirmed (Shahidi and Naczk, 2004).Watanabe *et al.* (1997) found that condensed tannins of buckwheat were composed of a mixture of proanthocyanidins with various degrees of polymerization. Antioxidant activity of buckwheat seeds and leaves proved to be higher when compared with those of oats, barley, buckwheat straws and hulls (Holasova *et al.,* 2002). The antioxidant effects of two types of buckwheat sprouts on human hepatoma HepG2 cells revealed that both tartary buckwheat (TBS) and common buckwheat (CBS) sprout could decrease the production of intracellular peroxide and remove the intracellular superoxide anions in HepG2 cells, but TBS reduced the cellular oxidative stress more effectively than CBS, possibly because of its higher rutin (and quercetin) content (Liu, Chen, Yang, and Chiang, 2008). Substituting 15% of wheat flour in the bread formula with common buckwheat flour enhanced functional components, namely rutin and quercetin and its antioxidant activity (Lin, Liu, Yu, Lin, and Mau, 2009). Recently, Chitarrini *et al.* (2014) showed that the healthy antioxidant compounds present in buckwheat possess antimicrobial activity. Thus, the authors suggested that buckwheat antioxidants could be considered as markers for tolerance against mycotoxigenic pathogens and used for improving food safety (Chitarrini *et al.,* 2014). Phenolic extracts from the sprouts of buckwheat also have been shown to exhibit strong antioxidant activity (Brajdes and Vizireanu, 2012; Ren and Sun, 2014). A study has shown that buckwheat honeys could protect non-site-specific hydroxyl radical-mediated DNA damage and site-specific hydroxyl radical-mediated DNA strand breaks under *in vitro* conditions (Zhou *et al.,* 2012).

d. Corn/Maize (Corn antioxidants)

In corn, insoluble bound phenolic acids constitute the predominant fraction of phenolic acids present (Naczk and Shahidi, 2006).The embryo and aleurone layer of corn contain phenolic acids linked covalently to amine functionalities, such as feruoylputrescine, *p*-coumarylputrescine, diferuloylputrescine, di-*p*-coumarylputrescine, *p*-coumarylspermidine, diferuloylspermidine and diferuloylspermine (Sen *et al.,* 1994). Kurilich and Juvik (1999) reported the presence of tocopherols in corn kernel tissue. Monomeric flavan-3, 4-diols, namely leucopelargonidin and lecocyanidin and two favonols, kaempferol and quercetin, have also been reported in hydrolysates of aleurone tissue of corn (Shahidi and Naczk, 2004).

Anthocyanins cyanidin-3-glucoside, pelargonidin-3-glucoside and peonidin-3-glucoside were found in Chinese purple corn (Yang and Zhai, 2010). Mohsen and Ammar (2009) reported that the ethanolic extract of corn tassels was successfully utilized to retard the oxidation of sunflower oil. The present investigation depicted that corncob is a good source of potent antioxidants that can be used in the nutraceutical and functional food applications, and to protect vegetable oil (Sultana, Anwar, and Przybylski, 2007). These findings suggest that *p*-coumaric acid derivatives, which might play a beneficial role against oxidative damage, exist in corn steep liquor, thus this byproduct might be a useful source of phenolic antioxidants (Niwa, Doi, Kato, and Osawa, 2001).

Corn (*Zea mays* L.) is a major staple food in the world after wheat and rice, providing nutrients for humans and animals and serving as a basic raw material for the production of starch, oil and protein, alcoholic beverages, food sweeteners and, more recently, fuel. The kernels are often white or yellow in colour, although black, red and a mixture of colours is also found. There are a number of grain types, distinguished by differences in the chemical compounds deposited or stored in the kernel. White maize is biologically and genetically very similar to yellow maize, although there is a difference in appearance due to the absence of carotin oil pigments in the kernel which otherwise cause the yellow colour of the grain (FAO, 1992). In addition, mutant corn genotypes with different amylose contents (typical, waxy and high-amylose maize) have been genetically developed.

Corn has a greater total phenolic content (TPC) and total antioxidant activity than do wheat, oats, and rice. TPC of corn extract was 15.55 ± 0.60 mmol of gallic acid equiv/g of grain, which was higher than that of wheat (7.99 ± 0.39 μmol of gallic acid equiv/g of grain), oats (6.53 ± 0.19 μmol of gallic acid equiv/g of grain), and rice (5.56 ± 0.17 μmol of gallic acid equiv/g of grain). The bound phenolics were also found to be significantly higher than free phenolics in corn extract. About 69% of the total phenolics present in corn are in the insoluble bound forms, with ferulic acid being the major phenolic compound present. The bound phenolic content was also highest for corn (13.43 ± 0.59 μmol/g of grain), followed by wheat (6.10 ± 0.39 μmol/g of grain) and then oats (4.76 ± 0.14 μmol/g of grain) (Adom et al., 2002). Li *et al.* (2007) studied on the phenolic contents and antioxidant capacity of their compounds extracted from the typical and mutant corn genotypes and reported that the TPCs in the mutant corns were higher than the amount found in the typical corn in all extraction (methanol, HCl/methanol extraction and alkaline hydrolysates). The TPC was found to be greater in alkaline hydrolysates than in methanol and HCl/methanol extracts and high-amylose corn (F4-HA) had the highest TPC among the genotypes. The phenolic acids found in corn genotypes included *p*-hydroxybenzoic, vanillic, cafefic, syringic, *p*-coumaic, *m*-coumaric, *o*-coumaric, and ferulic acids, in which ferulic acid was the predominant phenolic acid present in all four types of corn. Other major phenolic acids were *o*-coumaric acid (126.53-575.87 mg/kg) and *p*-coumaric (97.87-211.03 mg/kg) present in corn. Free and bound ferulic acids in corn were higher than those of wheat, rice and oat (Adom *et al.*, 2002), whereas mutant corns were found to have greater ferulic acid levels than their typical parents. The total antioxidant activity of corn was the highest ($p < 0.01$) of those of the grains tested including corn, wheat, oat and rice. Free phenolic compounds were the major contributors to the total antioxidant activity in methanol and HCl/methanol extracts, whereas bound phytochemicals were responsible for the free radical scavenging activity of alkaline hydrolysates (Li *et al.*, 2007). The total antioxidant activity of the bound extraction highly correlated with phenolic content ($R2 = 0.991$, $p < 0.01$) and ferulic acid content ($R2 = 0.999$, $p< 0.01$), while there was a low correlation between parameters measured for free extracts (Adom *et al.*, 2002). For the mutant corn genotypes, the high-amylose corn had a better antioxidant capacity than did typical (nonmutant) corn genotypes (Li *et al.*, 2007). In addition, purple pigmented corn kernels are very rich in anthocyanins with well-established antioxidant and bioactive properties (Tsuda *et al.*, 2003). Also, the total antioxidant activity of sweet corn becomes elevated by 44% after thermal processing (Dewanto *et al.*, 2002).

e. Millets

Other cereals such as sorghum, oats, rye, and millet have also contributed to supply the major caloric and protein source for humanity and are main materials for food and non-food processing due to their high production in the world. In addition, these cereals have been reported to have significantly antioxidant capacity associated with the health benefits. Therefore, it is very important to evaluate TPCs of these cereals and their antioxidant capacity.

Millets are rich sources of phenolics and tannins, which can act as antioxidants (Hegde, Rajasekaran, and Chandra, 2005). Protocatechuic, gallic, and caffeic acids were the predominant free phenolic acids, whilst ferulic, caffeic and coumaric acids were the major bound phenolic acids found in finger millets (Subba Rao and Muralikrishna, 2002). Chandrasekara and Shahidi (2010) have shown that ferulic acid and *p*-coumaric acids are the major bound phenolics found in several varieties of millets (kodo, finger, foxtail, proso, pearl, and little millets). Viswanath, Urooj, and Malleshi (2009) identified diadzine, gallic, coumaric, syringic and vanillic acids as major phenolic acids from the extracted phenolics of finger millets. The only millet flavonoids reported were flavones, eight of which were identified in the leaves of finger millet, namely orientin, isoorientin, vitexin, isovitexin, saponarin, violanthin, lucenin-1 and tricin (Dykes and Rooney, 2006). The tiny millet grain has different colours, usually a dark brown seed coat, richer in polyphenols compared to other continental cereals such as barley, rice, maize and wheat. Finger millet seed coat is a good source of polyphenols with significantly higher antioxidant activity compared to the whole flour (Viswanath et al., 2009). Watanabe (1999) identified three antioxidative phenolic compounds, one serotonin derivative and two flavonoids were isolated from an ethanolic extract of Japanese barnyard millet (cv. Kurohie) grain by Sephadex LH-20 chromatography and preparative highperformance liquid chromatography. Their structures were established to be *N*-(*p*-coumaroyl) serotonin, luteolin, and tricin. Inclusion of millet in multigrain breads is now practiced in North America. Shahidi and Chandrasekara (2013) suggested that millet grain phenolics are bioaccessible, possess bioactivities against several pathophysiological conditions, as well as could serve as potential natural source of antioxidant in food and biological systems.

HEALTH BENEFITS OF MILLETS

A number of in vivo studies have demonstrated the beneficial role of millet grain consumption in the reduction of diseases associated with oxidative stress. Lakshmi Kumari and Sumathi (2002) have shown that consumption of finger millet-based diets results in significantly lower plasma glucose levels in individuals with non-insulin dependent diabetes mellitus. A protective effect of feeding finger millet and kodo millet whole grains meal against hyper glycaemic status and alloxan-induced oxidative stress in Wistar rats was reported (Hegde, Anitha, & Chandra, 2005). The effect of finger millet feeding was also examined on the dermal wound healing process of alloxan-in duced diabetic rats (Rajasekaran, Nithya, Rose, & Chandr a, 2004). The results showed that finger millet feeding for 4 weeks to the diabetic animals controlled the glucose levels and improve d the antioxidant status as well as the wound healing process. In another study, Hegde et al. (2005) demonstrated that topical application of an aqueous paste of kodo millet and finger millet flour was effective in rat der - mal wound healing. Recently, Lee, Chung, Cha, and Park (2010) reported that foxtail and pro so millet grain feeding decreased the triacylglycerol level and foxtail millet reduced C-re active proteins which are an inflammation related indicator in hyperlipidaemic rats suggesting their potential application in lowering the risk for CVD. Furthermore, Shobana et al. (2009) showed that feeding of 20% finger millet seed coat had hypoglycaemic, hypocholester olaemic, nephroprotective and anti-cataractogenic properties in streptozotocin-induceddiabetic rats.

Effect of Processing on Phenolic Content of Millet Grains

Cereal grains are usually subjected to some type of processing such as soaking, decortication, flaking, grinding, malting, fermentation and heat treatment. Whole cereals grains undergo consider able changes in composition during threshing (separation of grains from stalks), milling, fermentation and heat treatment.

Upon malting the content of free and bound phenolic acids in finger millet changed and the content of bound phenolic acid continuously decreased whereas free phenolic acid content increased upon malting for 96 h. Ferulic, caffeic and coumaric acids were found as major bound phenolic acids in native finger millet and upon malting for 96 h 1- to 2-fold decrease in their content s was reported (Rao & Muralikrishna, 2002).

Malting of finger millets for 96 h decreased the proto catechuic acid content from 45 to 16 mg/100 g. However, the content of other free phenolic acids, namely p-coumaric, gallic and ferulic acids was increased upon 96 h of malting by 2-, 4- and 10-fold, respectively (Rao & Muralikrishna, 2002). Accor ding to Hag, Tinay, and Yousif (2002), fermentation and dehulling of pearl millet reduced the total phenolic content as determine d by Folin–Denis method. The total phenolic content of standard cultivar of pearl millet decreased from 304 to 122 mg/100 g upon fermentation for 14 h. Dehulling, a process for removing the outer layers of the grain, also reduced the total phenolic content by 22% from its original value (Ha g et al., 2002). Opoku, Ohenhen and Ejiofor (1981) showed that germination of millet (*Pennisetum typhoides*) decreased the levels of tannins from 1.6% to 0.83%. Progressive decortication from 0% to 50% and cooking reduced the concentration of C-glycosylflavones (vitexin and orientin) of pearl millets (Akingb et al, 1991). Matuschek, Towo, and Savanberg (2001) reported that cooking, soaking and germination reduced the total phenol content of sorghum udo and finger millet.

Sripriya, Chandrasekaran, Murty, and Chandra (1996) reported that fermentation and germination of finger millet decreased their DPPH radical scavenging ability compared to the raw finger millet. Accor ding to Shobana and Malleshi (2007) hydrothermal treatment and decortication of finger millets reduced polyphenolic content by 14% and 74%, respectively. In addition, Towo, Svanberg and Ndossi (2003) showed that hydrothermal treatment of finger millet reduced the total phenolic content (TPC) by 1.7 times compared to that of the raw grain. Processing may increase the bioavailability of bioactive compounds in grains; however, processing may decrease their levels (Slavin, Jacobs, Marquart, & Wiemer, 2001). Several studies have shown that the outermost layers of the grains possess a high phenolic content and antioxidant activity (Gar cia-Conesa et al., 1997; Liyana-Pathirana & Shahidi, 2007)

A recent study showed that dehulling decreased the TPC of whole grain millets and this change was due to the rem ova l of the outer layers of the grain (Chandrasekara, Naczk, & Shahidi, 2012). Hulls had higher TPC than those of dehulled grains in kodo, foxtail, proso, pearl, little and finger millets. The same stud y further showed that TPC of dehulled and cooked grains of millet varieties did not differ significantly except for finger millets. Phenolic acids were mainly concentrated in the hulls of millet grains (Chandrasekara & Shahidi, 2011a). Kodo and pearl millet hulls had 34- and 4-folds more ferulic acid, respectively, compared to their dehulled counterparts. In addition, p-coumaric acid content in hulls of kodo and pearl millet grains was 7 times higher than their corresponding dehulled grains (Chandrasekara & Shahidi, 2011b).

Bioavailability of millet phenolics

Bioavailability of a compound can be defined as the quantity that passes through the cell membranes in the intestine and is available for action within the cells. Several studies have demonstrated the bioavailability and bioefficacy of phenolics using both animal models and human volunteers.

Very few studies have been carried out on the bioavailability of hydroxycinnamic acids which are abundant in cereals. Esterification to arabinoxylans of the grain cell walls hampered the absorption of ferulic acids in rats. However, Rondini et al. (2004) showed that bound ferulic acid from bra n was more bioavailable than pure ferulic acid fed to rats. It has been shown that quercetin is more available when ingested from onions than as a pure compound in healthy ileostomic subjects. Furthermore, presence of glucose moiety may enhance the absorption of quercetin in onions in humans. Bourne and Rive-Evans (1998) reported that urinary excretion of total free ferulic acid and feruloyl glucuronide was 11–25% of that ingested from tomatoes, pro viding approximately 21- 44 mg of ferulic acid. Enzymatic digestion and colonic fermentation in the digestive tract release phenolic compounds from food matrices and are available in the gut and may potentially be bioavailable. Recently, Chandrasekara and Shahidi (2012) reported that simulated gastrointestinal digestion and colonic fermentation in vitro released millet grain phenolics which showed antioxidant activity thro ugh different mechanisms such as free radical scavenging, ferrous ion chelating, and reducing activity. Furthermore, antioxidant activity de pended on the type of millet considered. Accor ding to this stud y the total phenolic content (TPC) and total flavonoid content (TFC) of millet extracts obtained following enzymatic digestion in the gastric phase ranged from 10.2 to 26.9 μmol ferulic acid equivalents (FAE)/g cooked grain (dw) and from 0.64 to 2.74 μmol catechin equivalents (CE)/g cooked grain (dw), respectively. This was 2–5 times more than the total amount of released phenolics than those released from the corresponding aqueous extracts of samples. In the process of digestion, both TPC and TFC increased for all millet grains employed (Chandrasekara and Shahidi, 2012).

Outcome from Millets

Millets are underutilized cereals in the developed countries while they serve as staple foods for millions of people in parts of Africa and Asia. In addition to serving as a source of macro and micro-nutrients, similar to other principal cereals in the world, millets are rich in bioactive phytochemicals, particularly phenolics. Moreover, millets are staple food substitutes for celiac patients as they are gluten-free. Millets are climatic change compliant crops that can be cultivated round the year. It is well known that pre vent able diet- and lifestyle-r elated chronic diseases such as hypertension, cardiovascular disease, and type II diabetes impose an enormous social and economic burden in a country. Phenolic content as well as composition of different millet grains and available evidences for beneficial health effects emphasize their significance as a functional food ingredient in non-communicable disease (NCD) reduction and improving health.

f. Sorghum

Sorghum is a good source of phenolic compounds (Hahn, Faubion, and Rooney, 1983). Protocatechuic, *p*-hydroxybenzoic, caffeic, *p*-coumaric and ferulic acids are found in the free and bound forms. However, cinnamic and vanillic acids are found in the free and/or bound forms only in some sorghum varieties. Gallic acid is present only in the bound form (Hahn et al., 1983). Recently, four phenolic acids (caffeic, *p*-coumaric, ferulic and sinapic acids) were detected in all pearling fine and pearled kernel fractions of two sorghum genotypes namely white (PR6E6) and red (PR6E14) (Luthria and Liu, 2013). Sorghum tannins readily associate with sorghum proteins (Shahidi and Naczk, 2004). Only varieties with a pigmented testa have condensed tannins. Sorghum tannins are excellent antioxidants, which slow hydrolysis in foods, produce naturally dark-coloured products and increase the dietary fiber levels of food products (Dykes and Rooney, 2006). The proanthocyanidins in sorghums are the B-type with (−)-epicatechin as extension units and catechin as terminal units (Gu *et al.,* 2002, 2003; Gupta and Haslam, 1978). The anthocyanins are the major class of flavonoids studied in sorghum. Unlike common anthocyanins, sorghum anthocyanins are unique since they do not contain the hydroxyl group in the 3-position of the C-ring and thus are called 3-deoxyanthocyanins (Dykes and Rooney, 2006). The antioxidant properties of the 3-deoxyanthocyanidins are similar to those of the anthocyanins (Awika, Rooney, and Waniska, 2004a). The two common 3-deoxy-anthocyanidins in sorghum are the yellow, apigeninidin, and the orange, luteolinidin (Awika, Rooney, andWaniska, 2004b). Other 3-deoxyanthocyanins identified in sorghum grains include apigeninidin 5-glucoside, luteolinidin 5-glucoside, 5-methoxyluteolinidin, 5-methoxyluteolinidin 7-glucoside, 7-methoxyapigeninidin, 7-methoxyapigeninidin 5-glucoside, 5-methoxyapigeninidin and 7-methoxyluteolinidin (Dykes and Rooney, 2006). Red pericarp sorghums have flavan-4-ol compounds, such as luteoforol and apiforol, which are produced from flavanones (i.e., naringenin and eriodictyol) and may be precursors of sorghum anthocyanidins (Dykes and Rooney, 2006). Other flavonoids isolated and identified in sorghum grains include the flavones apigenin and luteolin, which are predominant in tan-pigmented sorghums. Flavanones, eriodictyol and eriodictyol 5-glucoside have been reported.

The flavonol, kaempferol 3-rutinoside-7-glucuronide and the dihydroflavonols taxifolin, and taxifolin 7-glucoside have also been isolated (Dykes and Rooney, 2006). Kil *et al.* (2009) suggested that sorghum could be used as a natural ingredient with biological function for its antioxidant and antimicrobial properties in the food industry. Fernandez *et al.* (2009) reported that some sorghum varieties indigenous to West Africa (Nigeria and Niger) have a yellow endosperm, which is pigmented by carotenoids. The most abundant carotenoids identified are lutein, zeaxanthin and β-carotene (Taylor, Belton, Beta, and Duodu, 2014).

Sorghum (*Sorghum bicolor* M.) is the fifth leading cereal food crop in the world and is particularly important as a human food resource in Asia and Africa (Rooney and Waniska, 2000). Sorghum is a rich source of various phytochemicals including tannins, phenolic acids, anthocyanins, phytosterols and policosanols (Awika and Rooney, 2004). Tannin content of sorghum grain ranged from 10.0 – 68.0 mg/g (Awika *et al.*, 2004), in which only sorghums containing a pigmented testa had significant amounts of condensed tannins (Dykes *et al.*, 2005). Like other cereals such as wheat, corn and barley, the sorghum phenolic compounds are mostly concentrated in the bran and in bound forms. Ferulic acid is the most abundant bound phenolic acid in sorghum (Hahn *et al.*, 1983), while other phenolic acids including syringic, protocatechuic, caffeic, p-coumaric, and sinapic are also found in sorghum as the more abundant (Waniska *et al.*, 1989). The most common anthocyanins in sorghums are the 3-deoxyanthocyanidins, which include orange luteolinidin and yellow apigeninidin. Both compounds have good potential for use as natural colorants due to their pH stability (Dykes *et al.*, 2005). Recently, a number of studies on health-related benefits of sorghum reported that sorghum has significantly antioxidant activity (Choi *et al.*, 2006), antimicrobial activities (Kil *et al.*, 2009), anticarcinogenic effects (Kwak *et al.*, 2004), and cholesterol-lowering effects (Ha *et al.*, 1998), and can reduce the risk of cardiovascular disease (Cho *et al.*, 2000).

Phenolic acids are derivatives of benzoic and cinnamic acids (Figure 8.4) and are present in all cereals (Table 8.3). Other flavonoids found in fruits vegetables are also reported in cereals. For example the flavone apigenin a compound found in parsley and celery, is also reported in millet, oat and sorghum (Table 8.4). Flavonoids are reported to have antioxidant, anticancer, anti-allergic, anti-inflammatory, anticarcinogenic and gastroprotective properties. There are two classes of phenolic acids; hydroxybenzoic acids and hydroxycinnamic acids. Hydroxybenzoic acids include gallic, ρ-hydroxybenzoic, vanillic, syringic and protocatechuic acids. The hydroxycinnamic acids have a C_6– C_3 structure and include coumaric, caffeic, ferulic and sinapic acids. The phenolic acids reported in cereals occur in both free and bound forms. Sorghum and millet have the widest variety of phenolic acids. Free phenolic acids are located in the outer layer of the pericarp and are extracted using organic solvents. Bound phenolic acids are esterified to cell walls; acid or base hydrolysis is required to release these bound compounds from the cell matrix. The major phenolic acids in cereals are ferulic and ρ-coumaric acids. Phenolic acid levels vary among cereals; their brans concwntrate these compounds threefold (Table 8.4).

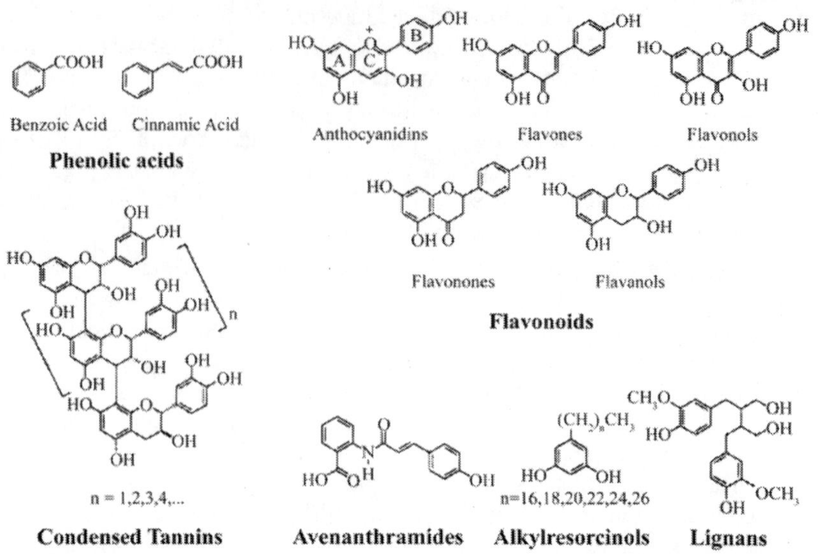

Figure 8.4: Chemical structure of classes of phenolic compounds in cereal grains.

Table 8.3. Phenolic acids and flavonoids reported in cereal grains.

Compound	**Grain**
A. Phenolic acid	
1. Hyderoxybenzoic acids	
Gallic	Millet, rice, sorghum
Protocatechuic	Barley, maize, millet, oat, rice, rye, sorghum, wheat
ρ-Hydroxybenzoic	Barley, maize, millet, oat, rice, rye, sorghum, wheat
Gentisic	Millet, sorghum
Salicylic	Barley, sorghum, wheat,
Vanillic	Barley, maize, millet, oat, rice, rye, sorghum, wheat
Syringic	Barley, maize, millet, oat, rice, rye, sorghum, wheat
2. Hydroxycinnamic acids	
Ferulic	Barley, maize, millet, oat, rice, rye, sorghum, wheat
Caffeic	Barley, maize, millet, oat, rice, rye, sorghum, wheat
α-Coumaric	Barley
m-Coumaric	Barley
ρ-Counaric	Barley, maize, millet, oat, rice, rye, sorghum, wheat

Compound	Grain
Cinnamic	Millet, sorghum, wheat
Sinapic	Barley, millet, oat, rice, rye, sorghum
B. Flavonoids	
1. Anthocyanins	
Apigeninidin	Sorghum
Apigeninidin-5-glucoside	Sorghum
Cyanidin	Barley
Cyanidin 3-galactoside	Maize, wheat
Cyanidin 3-glucoside	Barley, maize, rice, rye, wheat
Cyanidin 3-rutinoside	Maize, rice, wheat
Delphinidin	Barley
Delphinidin 3-glucoside	Wheat
Delphinidin 3-rytinoside	Rye, wheat
Luteolinidin	Sorghum
Luteolinidin 5-glucoside	Sorghum
5-Methoxyapigeninidin	Sorghum
7- Methoxyapigeninidin	Sorghum
7- Methoxyapigeninidin 5-glucoside	Sorghum
5-Methoxyluteolinidin	Sorghum
5-Methoxyluteolinidin 7-glucoside	Sorghum
7- Methoxyluteolinidin	Sorghum
Pelargonidin	Barley
Pelargonidin 3-glucoside	Maize
Pelargonidin glucosides	Barley, maize
Peonidin-3-glucoside	Maize, rice, rye, wheat
Petunidin 3-glucoside	Barley, wheat
Petunidin 3-rutinoside	Wheat
2. Flavones	
Apigenin	Millet, oat, sorghum
Apigenin glucosides	Wheat
Glucosylorientin	Millet
Glucosylvitexin	Millet
Luteolin	Millet, oat, sorghum
Isovitexin	Oat
Tricin	Millet, oat, wheat
Vitexin	Millet, oat
3. Flavanones	
Eriodictyol	Sorghum
Eriodictyol 5-glucoside	Sorghum
Homoeriodictyol	Oat
Naringenin	Sorghum

Compound	Grain
4. Flavonols	
Chrysoeriol	Barley
Kaempferol	Maize, oat
Kaempferol 3-rutinoside	Oat
Kaempferol 3-rutinoside-7- glucuronide	Sorghum
Quercetin	Maize, oat
Quercetin 3-rutinoside	Oat
5. Dihydroflavonols	
Taxifolin	Sorghum
Taxifolin 7-glucoside	Sorghum
6. Flavan-4-ols	
Apiforol	Sorghum
Luteoforol	Sorghum
7. Flavanols (monomers/dimers)	
Catechin	Barley, sorghum
Leucocyanidin	Barley, maize
Leucodelphinidin	Barley
Leucopelargonidin	Maize
Procyanidin B-1	Sorghum
Procyanidin B-3	Barley
Prodelphinidin B-3	Barley

Table 8.4: Phenolic acid content in cereal grains

Sample	Amount (µg/g)
A. Whole grains	
Barley	450-1346
Finger millet	612
Foxtail millet	3907
Maize	601
Oat	472
Pearl millet	1478
Rice	197-376
Rye	1362-1366
Sorghum	385-746
Wheat	1342
B. Brans	
Oat	651
Rye	4190
Wheat	4527

Figures 8.5 and 8.6 compare phenol and in vitro antioxidant activity levels of a wide array of cereal grains. The Folin-Ciocalteu and the 2, 2'-azinobis (3-ethyl-benzothiazoline-6-sulfonic acid (ABTS) assays were used. respectively using 1% HCI in methanol as the solvent. In general, tannin-containing grains (i.e., sorghum) and pigmented cereal grains had the highest levels of phenols and antioxidant activity in each grain category. Tannin sorghums and black rice had the highest levels of phenols and antioxidant activity whereas non-pigmented cereals (i.e., white rice, wheat, and waxy barley) had the lowest levels. These results suggest that condensed tannins and pigment-contributing compounds such as the anthocyanins increase phenols and antioxidant activity. There is a high correlation between total phenols and antioxidant activity (R^2 = 0.96), which suggests that the antioxidant activity is contributed by the phenolic compounds. In vitro methods used to measure antioxidant activity (i.e., ABTS, DPPH, and ORAC) do not give information about the bioavailability or metabolism of these compounds in biological systems. However, these methods are useful to screen and compare antioxidant activity levels among a wide variety of samples. To date, reports on human health benefits of cereal phenols are limited and, therefore, more research is needed in this area.

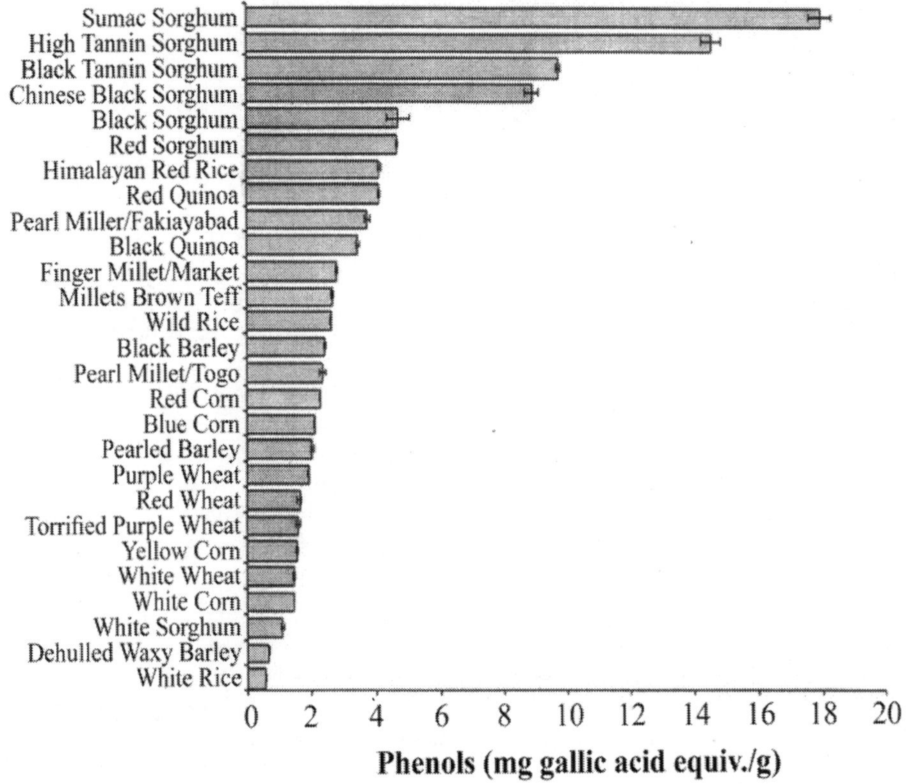

Figure 8.5: Total phenol levels of cereal grains (Adapted from Guajardo-Flores *et al.*, 2006).

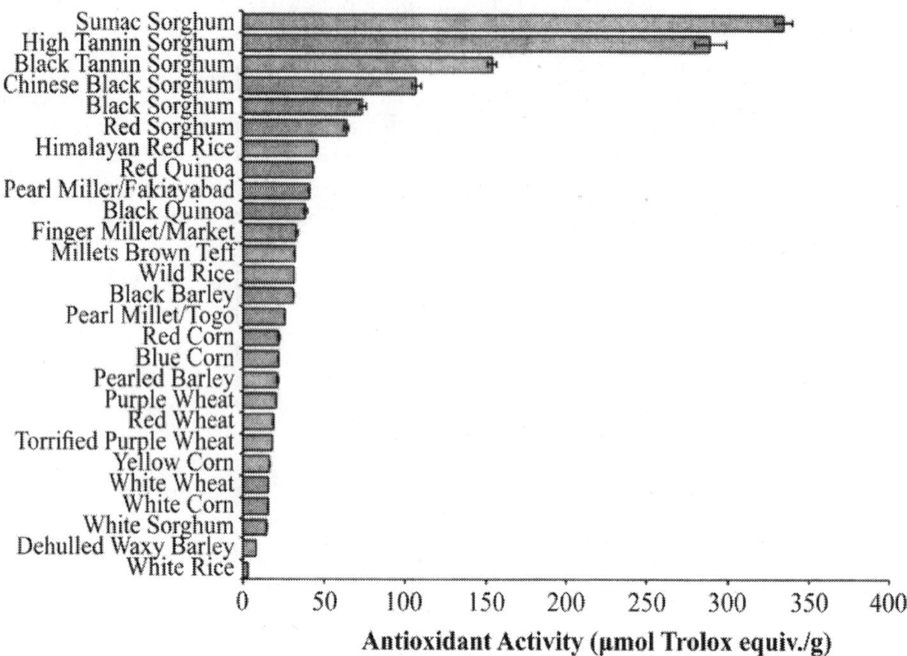

Figure 8.6: Antioxidant activity (ABTS) levels of cereal grains (Adapted from Guajardo-Flores *et al.*, 2006).

Sorghums containing condensed tannins have consistently shown the highest antioxidant activity in vitro, and they approach or exceed the antioxidant levels of fruits and vegetables (Table 8.5). Sorghums containing condensed tannins dominate production of grains in hot, humid regions of Africa because they have significantly improved resistance to grain molds and birds, which allow for their successful production. Tannin sorghums contain a pigmented testa, which contributes astringency during grain maturation and causes birds to utilize other food sources. When other grains are unavailable, birds consume the tannin sorghums. These tannin sorghums are grown extensively in east and southern Africa.

The condensed tannins cause depressed feed efficiency because they slow down and decrease digestion of the grain components. Thus, the tannin sorghums are discriminated against in sorghum markets. Very little tannin sorghums are grown in the United States since they are discounted in the grain markets. In Africa, the tannin sorghums are used in a wide variety of traditional foods including beer, porridges, unleavened breads, and other products. Special tannin sorghums are consumed when farmers are doing field work. The tannin sorghum porridges stay with the person longer and are said to be more satisfying, probably because of the reduced rate of digestion. The estimated glycemic index of ground whole sorghums with tannins is significantly lower than sorghums without tannins.

Table 8.5: Antioxidant activity (ORAC) of sorghum grains and brans compared to common fruits and vegetables.

Commodity	ORAC (µmol TE/g, dry wt.)
Tannin sorghum (grain)	868
Tannin sorghum (bran)	3124
Black sorghum (grain)	219
Black sorghum (bran)	1008
Red sorghum (grain)	140
Red sorghum (bran)	710
White sorghum (grain)	22
White sorghum (bran)	64
Blueberry, low bush	842
Strawberry	402
Plum	495
Watermelon	18
Apple red delicious	295
Orange navel	137
Broccoli	173
Carrot	108
Onion red	93
Sweet pepper green	105
Radishes	217
Potatoes russet	63

Adapted from Dykes and Rooney (2006).

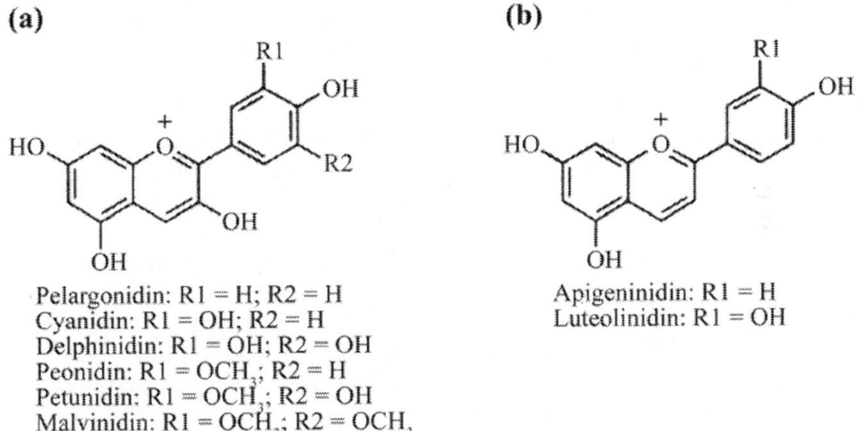

Figure 8.7: Chemical structure of A, the six common anthocyanidins and B, the 3-deoxyanthocyanidins.

Sorghum contains unique anthocyanins called 3-deoxyanthocyanins, which lack the hydroxyl group in the 3 position of the C-ring (Figure 8.7). This feature is believed to increase their stability at high pH compared to common anthocyanins, which could make them good natural food colorants. The two main sorghum 3-deoxyanthocyanidins are the yellow apigeninidin and the orange luteolinidin. Sorghum with a black pericarp has higher levels of 3-deoxyanthocyanins than red sorghums. Decortication of black sorghum grain to produce bran concentrates anthocyanin levels almost seven fold. For example the grain and bran of Tx430 black have anthocyanin levels of 944 (Table 8.6) and 6.695 g/g respectively (unpublished data).

Table 8.6. Anthocyanin content of pigmented cereal grains

Sample	Amount (μg/g)
Blue barley	4
Maize	
Pink	93
Red	558
Blue	225
Purple	965
Black rice	2283
Black sorghum	944
Wheat	
Blue	106-153
Purple	13-139

g. Oat

Oat is a source of a number of compounds that exhibit antioxidant activity. Vitamin E (tocols), phytic acid, phenolic compounds and avenanthramides are amongst the most abundant antioxidants in oat along with flavonoids and sterols. These antioxidants are concentrated in the outer layers of the kernel (Peterson, 2001). The distribution of antioxidants within oat kernels is important in implementing technologies to develop oat products with enhanced antioxidant capacity. Oat germ, the high-lipid compartment of kernels has concentrated levels of tocopherols (α- and γ-homologues), whereas tocotrienols are concentrated in endosperm and absent in the germ (Peterson, 1995). Oat bran is less effective in scavenging free-radicals compared with other cereal brans such as wheat, barley and rye (Martínez-Tomé et al., 2004). However, Lehtinen and Laakso (1997) reported that oat itself has greater antioxidant capacity than the same three cereals, with highest activity observed for its soluble fiber fraction. Avenanthramides (Figure 8.8) are a group of alkaloids in oat grain that consist of an anthranilic acid derivative linked to a hydroxycinnamic acid derivative by a pseudo peptide bond and exhibit antioxidant activity (Peterson, Hahn, and Emmons, 2002).The three most abundant avenanthramides are *N*-(4′-hydroxy-3′-methoxycinnamoyl)-5-hydroxyanthranilic acid, *N*-(4′-hydroxycinnamoyl)-5-hydroxyanthranilic acid, and *N*-(3′,4′-dihydroxycinnamoyl)-5-hydroxyanthranilic acid

(Peterson *et al.,* 2002). In general, avenanthramides show a higher antioxidant activity than ferulic acid, gentisic acid, *p*-hydroxybenzoic acid, protocatechuic acid, syringic acid, vanillic acid, vanillin, and phytic acid which are typical in cereals (Martínez-Tomé *et al.,* 2004). *p*-Hydroxybenzoic acid, vanillic acid, caffeic acid, vanillin, *p*-coumaric acid, and ferulic acid were identified in oat milling fractions (Emmons, Peterson, and Paul, 1999). Oat flour has been shown to consist of three flavones, apigenin, luteolin and tricin (Emmons et al., 1999; Shahidi and Naczk, 2004). Oat phenolics, including avenanthramides, have been shown to interact synergistically with vitamin C to protect LDL during oxidation (Chen *et al.,* 2004). In addition, oat phenolics have been shown to possess anti-inflammatory and anti-hypertensive activity (Chu *et al.,* 2013). Recently, Bindu and Krishnaveni (2013) showed that supplementation of oats porridge significantly decreased the low density lipoprotein (LDL), very low-density lipoprotein (VLDL) and blood pressure.

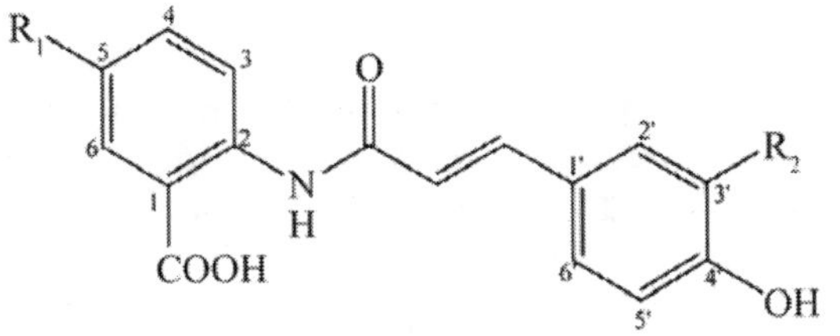

	R_1	R_2
A	OH	H
B	OH	HCH_3
C	OH	OH
D	H	H

Figure 8.8: Chemical structures of oat avenanthramide.

Oat (*Avena sativa* L.) and millet (*Eleusine coracana*) are also rich in phenolic compounds which possess antioxidant capacity. Phenolic acids of the oat are *p*-hydroxybenzoic acid, vanillic acid, caffeic acid, vanillin, *p*-coumaric acid, and ferulic acid (Emmons *et al.*, 1999), whereas diadzene, gallic, coumaric, syringic and vanillic acids were identified as major phenolic acids from the extracted phenolics of the millet (Viswanath *et al.*, 2009). In addition, three avenanthramides and an unidentified ferulate derivative were also detected in oat extracts (Emmons *et al.*, 1999). Phenolic compounds in oat and millet are also concentrated in outer layer (bran) of grains. Like other cereal grains, the extracts of oat and millet also exhibited significantly antioxidant capacity, which was correlated with measures of oxygen radical absorbance capacity and low-density lipoprotein oxidation. The millet extract also showed higher antimicrobial activity against *Bacillus cereus* and *Aspergillus flavus*. Total phenolic content of the phenolic extracts of both oat and millet was significantly positively correlated with antioxidant activity (Emmons *et al.*, 1999; Viswanath *et al.*, 2009). In addition, avenanthramides, substituted *N*-cinnamoyl-anthranilate alkaloids isolated only from oats, have also shown antioxidant activity in an *in vitro* linoleic acid oxidation system (Dimberg *et al.*, 1993).

h. Rice (Rice antioxidants)

Ferulic acid was found as the most abundant phenolic acid (up to 355 mg/kg) followed by sinapic acid in wild rice (Irakli, Samanidou, Biliaderis, and Papadoyannis, 2012; Qiu, Liu, and Beta, 2010).They both occur mainly in the insoluble form. Other monomeric phenolic acids present in wild rice consisted of *p*-coumaric, vanillic, syringic, and *p*-hydroxybenzoic acids, along with two phenolic aldehydes, *p*-hydroxy benzaldehyde and vanillin, present in both soluble and insoluble forms. Phenolic acid dehydrodimers are cell wall bound and only appear in the insoluble fractions featured by diferulic acids (DiFA) and disinapic acids (DiSA) (Qiu *et al.,* 2010). The antioxidant activity of wild rice was 30 times greater than that of the white rice (Qiu, Liu, and Beta, 2009). Meanwhile, rice bran contained a higher phenolic content (2.51–3.59 mg/g) than wheat bran (Iqbal, Bhanger, and Anwar, 2005; Lai, Li, Lu, and Chen, 2009). Furthermore, rice bran contained 100 different antioxidants, amongst the most powerful of which were oryzanols, tocopherols and tocotrienols (Iqbal *et al.,* 2005). Black rice is an especially economically important rice species and derives its name from its rich natural anthocyanin compounds, such as cyanidin 3-glucoside and peonidin 3-glucoside, which possess antioxidative and anti-inflammatory activities (Kong and Lee, 2010). Isovitexin, a flavonoid also known as *C*-glucosylflavone, with good antioxidant activity, has been reported in rice hulls (Osawa, Narasimhan, Kawakishi, Namdci, and Tashiro, 1985). Moreover, anisole, *m*-hydroxybenzaldehyde, vanillin (4-hydroxy-3-methoxybenzaldehyde), and syringaldehyde (4-hydroxy-3, 5-dimethoxybenzaldehyde) have been identified in methanolic extracts of wild rice hulls (Asamarai, Addis, Epley, and Krick, 1996). Cycloartenyl ferulate and 24-methylenecycloartanyl ferulate are major steryl ferulates in rice bran. Rice steryl ferulates (γ-oryzanol) have been shown to possess good antioxidant activity (Nyström, Mäkinen, Lampi, and Piironen, 2005). Butsat and Siriamornpun (2010) demonstrated that rice bran and husk can be considered as valuable sources of bioactive components with high antioxidant properties.

Rice is one of the oldest food crops in the world. Most of the rice varieties cultivated today belong to the species, *Oryza sativa* L. Rice harvested from the field is called paddy. The paddy consists of the hull and the rice caryopsis, also known as brown rice. The brown rice is the edible part of the rice grain and furthermore milled into white rice and bran. The white rice is starchy endosperm part of rice grain, which mostly contains carbohydrate and protein, whereas rice bran consists of germ, pericarp, aleurone cells and seed coat, which is rich in dietary fibers, minerals and B vitamins. Rice bran also contains high levels of several phytochemicals such as vitamin E and γ-oryzanol fractions that have antioxidant activities and health-related benefits. Figure 8.9 shows the structures of vitamin E and γ-oryzanol components in rice bran analyzed by liquid chromatography/mass spectrometry/mass spectrometry (Yu *et al.*, 2007). A natural form of vitamin E in rice bran consists of eight homologues, four of which are tocopherols and the other four comprising tocotrienols (Jariwalla, 2001). Gamma-oryzanol is a mixture of ferulic acid esters of triterpene alcohols and sterols (Evershed *et al.*, 1988; Diack, and Saska, 1994; Rogers *et al.*, 1993; Xu, and Godber, 1999). Both γ-oryzanol and vitamin E in rice bran had significant antioxidant activities, which protect cells from the oxidative damage of plasma very low-density lipoprotein, cellular proteins and DNA and from membrane degeneration (Akihisa *et al.*, 2000; Topinka *et al.*, 1989; Xu *et al.*, 2001). Among the vitamin-E homologues, α-tocotrienol has known to have more than three times the in vitro free radical scavenging activity of α-tocopherol (Packer, 1995). Tocotrienols have also reported to have significantly health benefits in inhibiting cholesterol synthesis, lowering serum-cholesterol levels in various animal models, and suppressing tumor-cell proliferation (Qureshi *et al.*, 2000). In addition, γ-oryzanol has been found to be 13 to 20 times (w/w) greater content in rice bran than total tocopherols and tocotrienols (Bergman and Xu, 2003) and has reported to decrease animal serum-cholesterol levels, have anti-inflammatory activity, and inhibit cholesterol oxidation in vitro (Rong *et al.*, 1997; Akihisa *et al.*, 2000; Xu *et al.*, 2001).

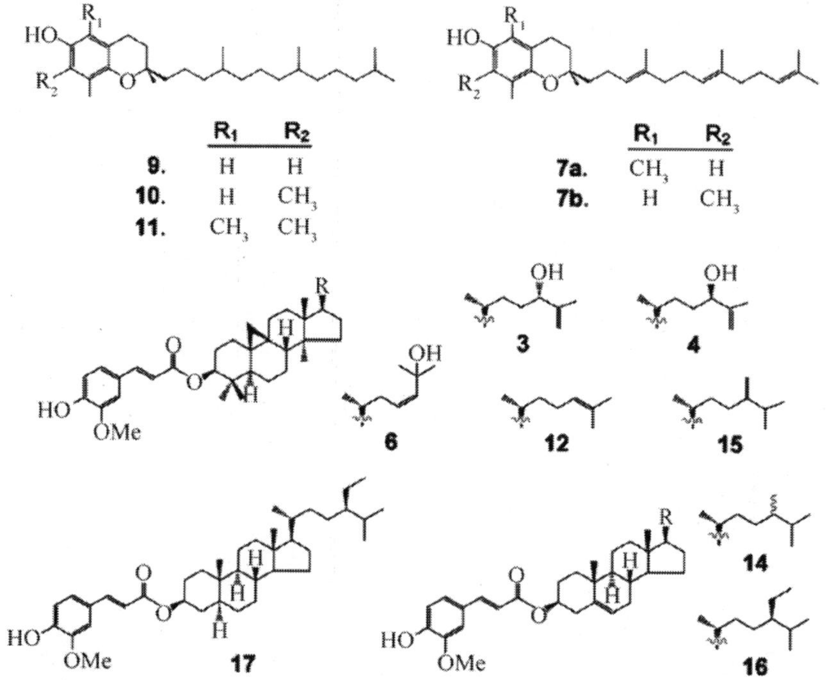

	R_1	R_2
9.	H	H
10.	H	CH_3
11.	CH_3	CH_3

	R_1	R_2
7a.	CH_3	H
7b.	H	CH_3

Figure 8.9: Structures of vitamin E and g-oryzanol components. 7a, β-tocotrienol; 7b, γ-tocotrienol; 9, δ-tocopherol; 10, γ-tocopherol; 11, α-tocopherol; 3, (24S)-cycloart-25-ene-3β, 24-diol-3β-trans-ferulate; 4, (24R)-cycloart-25-ene-3β,24-diol-3β-trans-ferulate; 6, cycloart-23Z-ene-3β,25-diol-3β-trans-ferulate; 12, cycloartenol trans-ferulate; 14, campesterol trans-ferulates; 15, 24-methylenecycloartanol trans-ferulate; 16, sitosterol trans-ferulate; 17, stigmastanol trans-ferulate (Adapted with permission from Yu et al. (2007). Copyright ©2007 American Chemical Society).

The phenolic profiles of both brown and milled rice were dominated by ferulic and pcoumaric acids with lesser amounts of gallic, vanillic, caffeic and syringic acids. The total phenolic acids in the ''bran'' (calculated as the difference in contents of brown and milled rice) typically contributed 70–90% of the total phenolic acids in the grain, depending on cultivar and the particular phenolic acid. These results suggest that the concentrations of phenolic acids increased from endosperm to the aleurone (Zhou *et al.*, 2004). The free phenolic content of rice was lower than that of corn but higher than those of wheat and oat. However, rice had the lowest bound phenolic content as compared to other grains (Adom *et al.*, 2002). Ferulic acid was the dominant phenolic acid, both in total and bound states, in brown and milled rice, followed by pcoumaric acid (Zhou *et al.*, 2004). The antioxidant capacity of the rice extracts, both in the bound and total phenolics, was the lowest compared to that of corn, wheat or oat. The lowest antioxidant activity of rice extracts compared to those of other cereals was due to the low amount of phenolic acids, ferulic and p-coumaric acids, in the bound phenolic extracts (Adom *et al.*, 2002). However, the total antioxidants in rice are significantly contributed by tocopherols, tocotrienol and γ-oryzanol, which has not been examined by Adom and Liu (2002). Thus, brown rice and rice bran containing large amount of phytochemicals derived from the outer layer of rice grain are considered to have significantly impact on human health. The brown rice should be consumed instead of the white rice and the extracts of rice bran and its oil could be used as a functional food or as additives to improve the storage stability of foods.

i. Rye

Ferulic acid is the most abundant hydroxycinnamic acid in rye, followed by sinapic acid and *para*coumaric acid. Amongst the benzoic acid derivatives only small amounts of vanillic acid, syringic acid and *para*-hydroxybenzoic acid have been reported in rye grain and products (Andreasen, Christensen, Meyer, and Hansen, 2000). The antioxidant activities of purified monomeric and dimeric hydroxycinnamates and of phenolic extracts from rye (whole grain, bran, and flour) were investigated using an *in vitro* copper-catalysed human LDL oxidation assay. The most abundant ferulic acid dehydrodimer (diFA) found in rye, 8-*O*-4-diFA, was a slightly better antioxidant than ferulic acid and *p*-coumaric acid. The antioxidant activity of the monomeric hydroxycinnamates decreased in the order of caffeic acid > sinapic acid > ferulic acid > *p*-coumaric acid (Andreasen, Landbo, Christensen, Hansen, and Meyer, 2001). Steryl ferulate extracts from wheat or rye bran were studied for their capability to inhibit hydroperoxide formation in bulk methyl linoleate and methyl linoleate emulsion and found to possess good antioxidant activity (Karamac, Amarowicz, Weinder, Abe, and Shahidi, 2002; Nyström *et al.*, 2005). The key difference in the steryl ferulates of rice from those of wheat and rye is the sterol structure. In rice, the sterols are 4, 4-dimethylsterols with two methyl groups in carbon C4; whereas the sterols in other cereals are principally desmethylsterols that have no methyl groups in their C4 position (Nyström *et al.*, 2005).

Rye (*Secale cereale* L.) is a grass grown extensively as a grain and as a forage crop with low requirements regarding soil and fertilization, as well as a relatively good overwinter ability. Therefore, it is grown primarily in Eastern, Central and Northern Europe (Salovaara and Autio, 2001). Rye grain, like other cereal grains, contributes significant quantities of energy, protein, selected micronutrients and non-nutrients to a human diet (Edge *et al.*, 2005). Therefore, it is widely used for flour, rye bread, rye beer, some whiskeys, some vodkas, and animal fodder. Phytochemicals of rye include tocopherols and tocotrienols (Zielinski *et al.*, 2007), phenolic acids and ferulic acid dehydrodimers (Andreasen *et al.*, 2000), and alkylresorcinols (Gliwa *et al.*, 2011). Tocopherol contents in rye ranged from 2.43 – 10.2 μg/g (db) with the main tocopherols were α-tocopherol and β-tocopherol. The whole grain contained higher tocopherol content than did endosperm with embryo or pericarb with testa. Tocotrienol contents with only α- tocotrienol and then β-tocotrienol ranged from 9.53 – 28.8 μg/g (db), in which tocotrienol contents of the whole grains and the fraction of endosperm with embryo of the investigated cultivars were twice as low as that of pericarb and testa fraction (Zielinski *et al.*, 2007). Total phenolic content of rye is in a range of 2.16 – 4.01 mg/g (db), in which the pericarb with testa fraction was about twice richer in TPC than was the endosperm with embryo fraction or the whole grain (Zielinski *et al.*, 2007). The phenolic acids existed in rye grain include ferulic, sinapic, *p*-coumaric, caffeic, *p*-hydroxybenzoic, protocatechuic, and vanillic acids. The concentration of ferulic acid, the most abundant phenolic acid, ranged from 900 to 1170 μg/g dry matter, followed by sinapic acid ranging from 70 to 140 μg/g dry matter, *p*-coumaric acid ranging from 40 to 70 μg/g dry matter, and caffeic, *p*-hydroxybenzoic, protocatechuic, and vanillic acids, which were all detected in concentrations less than 20 μg/g dry matter (Andreasen *et al.*, 2000). The highly positive correlation coefficient between the TPCs and antioxidant capacity of the whole grains, pericarb with testa fractions and endosperm with embryo fractions of rye were also observed. The results obtained indicate the possibility of using antioxidant contents and antioxidant properties of pericarb with testa fraction, rather than these of the whole grain, for the selection of rye variety of high technological quality (Zielinski *et al.*, 2007). Andreasen *et al.* (2001) reported that the most abundant phenolic acids in rye, ferulic acid, sinapic acid, and the dimer 8-*O*-4-diFA, exert antioxidant activity to inhibit low-density lipoprotein (LDL) oxidation in vitro. The rye bran extracts highly inhibited LDL lipid oxidation rather than the whole flour extract. Alkylresorcinols (ARs), compounds which were demonstrated to occur in cereal grains and related materials such as whole-grain cereal products, are similar to tocopherols except that they have a straight aliphatic hydrocarbon side chain and a single phenolic ring (Figure 8.10). ARs are well-known antioxidants, which also possess antimicrobial, antibacterial, antifungal, and antitumor activities (Kozubek and Tyman, 1999). In rye bran, and the presence of AR homologues C15:0, C17:0, C19:0, C21:0, C23:0, and C25:0 was confirmed, in which the AR homologue C19:0 was the most common and C15:0 was the least common (Gliwa *et al.*, 2011). Rye bran is known to have a high amount of ARs compared to the inner part of rye grain. ARs in rye bran is a range of 360 - 3200 μg/g, higher than those of triticale (580 - 1630 μg/g), wheat (317 – 1010 μg/g), and barley (44 - 500 μg/g) (Ross *et al.*, 2003). DPPH radical reduction varied from ~10% to ~60% for the rye bran alkylresorcinol homologues (C15:0–C25:0) at concentrations from 5 to 300 μM and was not dependent on the length of the alkyl side chain of the particular homologue. Values of EC_{50} for all the alkylresorcinol homologues were significantly higher than those for Trolox and α-, δ-, and γ-tocopherols, compounds with well-defined antioxidant activity and used as positive controls (Korycinska *et al.*, 2009). The antioxidant activity using oxygen radical absorbance capacity (ORAC) of rye extracts decreased from the outermost fraction to the innermost fraction. From a comparison of the results for antioxidant properties obtained using the ORAC assay and the total amount of ARs using GC-MS, it is confirmed that ARs are in fact the components that give rye its antioxidant activity (Gliwa *et al.*, 2011).

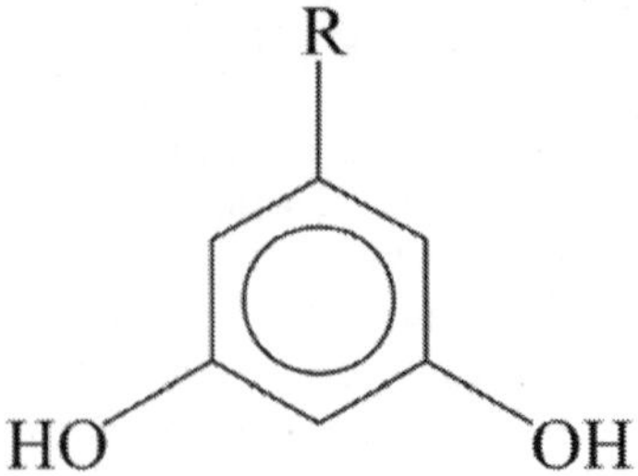

Figure 8.10: Chemical structure of the alkylresorcinols. R - the alkyl side chain. R = 15, 5-npentadecylresorcinol (C15:0); R = 17, 5-n-heptadecylresorcinol (C17:0); R = 19, 5-nnonadecylresorcinol (C19:0); R = 21, 5-n-heneicosylresorcinol (C21:0); R = 23, 5-ntricosylresorcinol (C23:0); R = 25, 5-npentacosylresorcinol (C25:0).

In all cereals, phenolic compounds are found mostly in the outer layer rather than in the endosperm. Therefore, the extracts from the bran always contained higher total phenolic contents as compared to the endosperm. The highly positive correlation between TPCs and antioxidant activity of the phenolic extracts were observed in literature. Ferulic acid was abundant acid in the bound form of the wheat, corn, rice, sorghum and rye, whereas *p*-hydroxybenzoic is the major acid in the barley and oat, and diadzene is dominant acid in the millet. The phenolic extracts from cereal grains exhibit significantly antioxidant capacity evaluated using trolox equivalent antioxidant capacity, 2,2-diphenyl-1-picrylhydrazyl radical scavenging, reducing power, oxygen radical absorbance capacity, inhibition of oxidation of human low-density lipoprotein cholesterol and DNA, Rancimat, inhibition of photochemilumenescence, and iron(II) chelation activity. Nowadays, the whole grain and whole-grain products are encouraged to consume instead of the white flour because they contain large amount of nutrients such as dietary fibers, vitamins, minerals and phytochemicals derived from the outer layer of grains. The consumption of whole grains is considered to have significantly health benefits in prevention from chronic diseases such as cardiovascular disease, diabetes, and cancer. In addition, the extracts from the bran is considered to be used as a source of natural antioxidants, which can be used to produce functional foods or used as additives in food storage.

2. LEGUMES

Amarowicz and Pegg (2008) reviewed legumes as a source of natural antioxidants. The major polyphenolic compounds of legumes consist mainly of tannins, phenolic acids and flavonoids (Campos-Vega, Loarca-Piña, and Oomah, 2010). Flavonol glycosides, tannins and anthocyanins are responsible for the colour of seed coat in dry beans (Madhujith, Naczk, and Shahidi, 2006; Madhujith and Shahidi, 2005; Shahidi and Naczk, 2004). Legumes with the highest polyphenolic content are the dark, highly pigmented varieties, such as red kidney beans (*Phaseolus vulgaris*) and black gram (*Vigna mungo*) (Campos-Vega et al., 2010). The distribution of phenolic compounds differs in the cotyledon and the seed coat, with non-flavonoid phenolic compounds, such as free and combined hydroxybenzoic and hydroxycinnamic acids, being located mainly in the cotyledon of lentils (Duenas, Hernandez, and Estrella,

2002). Flavonoids, such as glycosides of flavonols and flavones, were identified in the seed coat of lentils, together with *trans*-resveratrol-3-*O*-glucoside, and higher concentrations of proanthocyanidins (Duenas et al., 2002; Duenas, Sun, Hernandez, Estrella, and Spranger, 2003). According to the literature, total phenolic content is directly associated with antioxidant activity (Amarowicz, Troszynska, Barylko-Pikielna, and Shahidi, 2004; Awika, Rooney, Wu, Prior, and Cisneros-Zevallos, 2003; Shahidi and Naczk, 2004). Total phenolic contents (TPC) of the extracts (using acidic 70% acetone) from selected legumes are presented in Table 8.7. Lentils have the highest phenolic, flavonoid and condensed tannin contents (7.53 mggallic acid equivalents/g, 2.21 and 8.70 mgcatechin equivalents/g, respectively), followed by black beans and red kidney beans (Xu and Chang, 2007). Pulses with the highest total phenolic content (lentil, black, and red kidney beans) exert the highest antioxidant capacity as assessed by DPPH radical scavenging, FRAP and ORAC (Xu and Chang, 2007; Yeo and Shahidi, 2015). Condensed tannins (proanthocyanidins) have been quantified in hulls of several varieties of field beans (*Vicia faba*) (Amarowicz et al., 2004; Beninger and Hosfield, 2003; Campos-Vega *et al.,* 2010; Smulikowska *et al.,* 2001; Troszyn□ska, Estrella, López-Amóres, and Hernández, 2002). Tannins are located mainly in the testa and play an important role in the defence system of seeds that are exposed to oxidative damage bymany environmental factors (Troszyn□ska *et al.,* 2002). Beach pea hullswere found to contain very high amounts of condensed tannins compared to those reported for other legumes such as pigeon pea, chickpea, cowpea and green pea (Shahidi, Chavan, Naczk, and Amarowicz, 2001). Tannins in beans are linear polymers of flavan-3-ol (catechin and gallocatechin) and flavan-3, 4-diol (leucocyanidin and leucodelphinidin) units (Martin-Tanguy, Guillaume, and Kossa, 1977).Oomah, Luc, et al. (2011) have shownthat phenolic content linearly related to tannin content is the best predictor of antioxidant activity in low-tannin faba bean genotypes. More recently, Yeo and Shahidi (2015) introduced a new indicator, the ratio of insoluble bound phenolics to soluble phenolics, and suggested it as an effective means to monitor changes in the antioxidant activity of lentils during processing as exemplified for the germination process. There are, however, only a few reports on identification and quantification of flavonoids in food legumes (Xu and Chang, 2007). For example, only several reports on common beans (Beninger, Hosfield and Bassett, 1999; Beninger, Hosfield, andNair, 1998; Hempel and Bohm, 1996; Madhujith, Naczk, et al., 2006; Romani et al., 2004) and peas (Troszyn□ska et al., 2002) are available. Hempel and Bohm (1996) reported that 3-*O*glucuronides and 3-*O*-rutinosides of quercetin and kaempferol were the main flavonoid glycosides in six varieties of yellow and green French beans. Ferulic acid was the most abundant phenolic acid in common beans and *p*-coumaric and sinapic acids were also present (Luthria and Pastor-Corrales, 2006; Oomah, Cardador-Martinez, and Loarca-Pina, 2005; Shahidi and Naczk, 2004). Legumes, namely mung bean, field pea, faba bean, lentil and pigeon pea contained 18 to 31 mg of total phenolic acids (*trans*ferulic, *trans*-p-coumaric and syringic acids) per kilogram, whilst in navy bean, lupin, lima bean, chickpea and cowpea, the total phenolics content ranged from 55 to 163 mg/kg (Sosulski and Dabrowski, 1984).

Table 8.7: Phenolic content and antioxidant activity of legumes.

Legumes	Total phenolic[a]	Total flavonoid[b]	Condensed tannin[c]	DPPH scavenging capacity[d]	FRAP value[e]	ORAC value[f]
Green pea	1.13	0.26	0.91	0.59	1.03	6.17
Yellow pea	1.28	0.29	1.02	1.36	1.34	13.30
Chickpea	1.57	2.87	1.21	0.94	0.81	7.41
Lentil	7.53	2.21	8.70	19.09	10.65	38.60
Yellow soybean	2.23	0.41	0.85	1.40	0.34	44.08
Black soybean	6.18	2.57	4.09	17.58	9.93	122.75
Red kidney	5.90	2.93	5.37	18.94	9.22	23.2
Black bean	6.89	3.21	6.74	18.33	11.03	51.54

Source: Adapted from Xu and Chang (2007).

[a] mg gallic acid eq/g; [b] mg catechin eq/g; [c] mg catechin eq/g; [d] µmol Trolox eq/g; [e] mmol Fe2+ eq/100 g; [f] µmol Trolox eq/g.

Oomah, Caspar, Malcolmson, and Bellido (2011) reported that green and red lentil hulls are excellent sources of potent phenolic antioxidants.Long-Ze, Harnly, Pastor-Corrales, and Luthria (2008) analysed twenty four common bean samples and reported that common beans tested contained the same hydroxycinnamic acids, whereas they had different flavonoid components. Black beans contained primarily the 3-*O*-glucosides of delphinidin, petunidin, and malvidin, whilst kaempferol and its 3-*O*-glycosideswere present in pinto beans. Light red kidney beans had traces of quercetin 3-*O*-glucoside and its malonates, but pink and dark red kidney beans contained the diglycosides of quercetin and kaempferol. Small red beans contained kaempferol 3-*O*-glucoside and pelargonidin 3-*O*-glucoside, whilst flavonoids were undetected in great northern and navy beans. According to the USDA (2002) survey on isoflavone content, lentils do not contain any significant amount of isoflavones. Chickpeas contained daidzein, genistein and formononetin (0.04, 0.06, and 0.14 mg/100 g, respectively), and approximately 1.7 mg/100 g biochanin (Campos-Vega *et al.,* 2010). Dini Schettino and Dini (1998) identified two newisoflavonoid derivatives (Figure 8.11), mutabilin (3′-methoxy-5-hydroxy-7-*O*-β-D-glucosyliso-flavone) andmutabilein (3′-methoxy-5, 7-dihydroxyisoflavone) in seeds of *Lupinus mutabilis* (Fabaceae) (Shahidi and Naczk, 2004). Furthermore, formononetin, genistein and the phytoestrogen secoisolariciresinolwere found in seeds of *L. mutabilis* (23, 2420, and 3.1 µg/100 g, respectively) (Mazur, Duke, Wahala, Rasku, and Adlercreutz, 1998). Delphinidin 3-glucoside is the major anthocyanin in beans of CanadianWonder cultivar, whilst cyaniding 3-diglucoside, 3, 5-diglucoside, pelargonidin 3-glucoside and 3, 5-diglucoside are the minor anthocyanins (StantonandFrancis, 1966). The anthocyanins in seed coats of beans were identified as delphinidin 3-glucoside 65.7%, petunidin 3-glucoside 24.3%, and maldivin 3-glucoside 8.7% (Salinas-Moreno, Rojas-Herrera, Sosa-Montes, and Pérez-Herrera,

2005) which are mainly responsible for the colour of black and purple seed coat beans (Shahidi and Naczk, 2004). Chickpeas contain a wide range of polyphenolic compounds, including flavonols, flavone glycosides, flavonols, and oligomeric and polymeric proanthocyanidins (Sarma, Singh, Mehta, Singh, and Singh, 2002; Zia-Ul-Haq et al., 2008). Amarowicz *et al.* (2010) found catechin and epicatechin glucosides, procyanidin dimers, quercetin diglycoside, and *transp*-coumaric acids as the dominant phenolics in green lentils. Madhujith, Amarowicz, and Shahidi (2004) studied the phenolic constituents of bean varieties with different hull colours (white, brown, red, and black) and identified delphinidin, cyanidin, procyanidin, and phenolic acids as the major phenolic compounds present in the bean hulls.

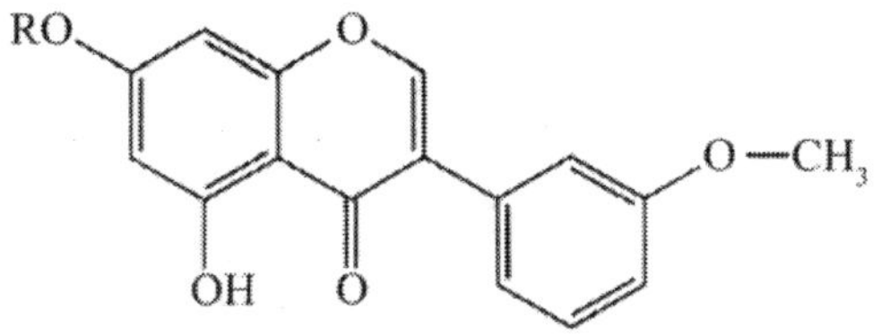

R=H, Mutabilein
R=β-D-glucose, Mutabilin

Figure 8.11: Chemical structures of isoflavonoid derivatives found in *Lupinus mutabilis.*

Bean phenolics exhibited strong protection against *in vitro* 2, 2′-azobis (2-methylpropionamidine) dihydrochloride (AAPH) radicalinduced DNA strand scission. Beans, especially the coloured ones, can be considered an important source of natural antioxidants, hence, a food with potential health benefits (Madhujith and Shahidi, 2005). Siah, Konczak, Agboola,Wood, and Blanchard (2012) found that the crude phenolic extracts obtained from rawand roasted faba beans exhibited potential health-beneficial properties, including potent antioxidant activities (based on both reagent- and cell-based assays), chemopreventative effects (through induction of cancer cell apoptosis), protection against reactive oxygen species, H2O2 and inhibitory effect on angiotensin converting enzyme (ACE),α-glucosidase and lipase (based on *in vitro* methods). Xu and Chang (2009) investigated the effects of boiling and steaming under atmospheric and high pressure on the phenolic components and antioxidant properties of pinto and black beans. In comparison with the original raw beans, both processing methods caused significant ($p < 0.05$) decreases in total phenolic content, total flavonoid content, condensed tannin content, monomeric anthocyanin content, DPPH radical scavenging activity, FRAP and ORAC values in both pinto and black beans. Recently, two anthocyanin compounds, peonidin-3-rutinoside and malvidin-3-*O*-glucosidewere newly identified in the adzuki bean extract (Han *et al.,* 2015).

B. FRUITS AND VEGETABLES

1. Fruits

Fruits are rich in antioxidants that help in lowering the incidence of degenerative diseases such as cancer, arthritis, arteriosclerosis, heart disease, inflammation, brain dysfunction and acceleration of the ageing process (Feskanich *et al.,* 2000; Gordon,

1996; Halliwell, 1996). The most abundant antioxidants in fruits are polyphenols; vitamins A, B, C and E and carotenoids are present to a lesser extent in some fruits. These polyphenols, most of which are flavonoids, are present mainly in the ester and glycoside forms (Lim, Lim, and Tee, 2007). In tropical fruits guava, papaya and star fruit have high antioxidant potential when compared to orange, as measured by scavenging of DPPH radical and iron (III) reducing assays (Lim *et al.,* 2007). Banana, star fruit, water apple, langsat and papaya are though weaker than orange as primary antioxidants, however, they exhibit powerful secondary antioxidant potential as measured by the iron (II) chelating experiment (Lim *et al.,* 2007). García-Alonso, Pascual-Teresa, Santos-Buelga, and Rivas-Gonzalo (2004) reported the highest content of flavanols in plum (*var.* Claudia), apples, custard apple, peach, strawberry and cherry, with concentrations of total flavanols between 70 and 370 mg/100 g of dry weight. On the other hand, the lowest contents were noted in avocado, banana and pear (*var.* blanquilla), with values lower than 4 mg/100 g of dry weight. The total antioxidant activities of 12 fruits and 5 commercial fruit juices were evaluated by Wang, Cao, and Prior (1996) using an automated ORAC assay and a peroxyl radical generator. On the basis of the wet weight of fruits (edible portion), strawberry had the highest ORAC activity followed by plum, orange, red grape, kiwi fruit, pink grapefruit, white grape, banana, apple, tomato, pear, and honeydew melon. On a dry weight basis, strawberry had the highest ORAC followed by plum, orange, pink grapefruit, tomato, kiwi fruit, red grape, white grape, apple, honeydew melon, pear, and banana. Amongst the commercial fruit and vegetable juices, grape juice had the highest antioxidant activity, followed by tomato juice, orange juice, and apple juice. According to Vinson, Su, Zubik, and Bose (2001), cranberry had the highest total phenols, followed by red grape, on a fresh weight basis (Table 8.8). Citrus fruits have a very low phenolic concentration. Contribution of ascorbic acid to the antioxidants in fruits is minor, except for melon, nectarine, orange, white grape and strawberry (Vinson, Teufel, andWu, 2001).The fruit extracts' antioxidant quality (measured by IC_{50}) was better than the vitamin antioxidants and most pure phenols, suggesting a synergism amongst the antioxidants in the extract mixture. The quantity/quality index, PAOXI, is one comprehensive parameters for comparing food antioxidants which is the ratio of phenol concentration (μmol/kg) to the IC50 value (μM) (Vinson and Hontz, 1995). According to PAOXI (Table 8.8), cherry is most effective followed by red grape, blueberry, strawberry, white grape, cranberry, banana and apple. Thus, small fruits such as berries are amongst the best sources of polyphenol antioxidants (Sun, Chu, Wu, and Liu, 2002; Vinson, Teufel, *et al.,* 2001). Cherry had the highest wet weight PAOXI of any fruit or vegetable studied. Finally, it was concluded that fruits had significantly better quantity and quality of phenolic antioxidants than vegetables.

Fruits, especially apples and cranberries, have phenolic antioxidants that can bind to lower density lipoproteins and protect them from oxidation (Vinson, Teufel, *et al.,* 2001).Wolfe et al. (2008) found that pomegranate and berries (wild blueberry, blackberry, raspberry, and blueberry) had the highest cellular antioxidant activity values, whereas banana and melons had the lowest. It is evident that berry fruits are consistently ranked amongst the top sources of total phenolics and antioxidant activity, containing levels up to 4 times greater than other fruits, 10 times higher than vegetables, and 40 times higher than cereals (Halvorsen *et al.,* 2002). Recent review by Alasalvar and Shahidi (2013) reveals that individuals who regularly consume generous amounts of dried fruits (dates, raisins, prunes, figs, acai berries, apples, bananas, black currants, blackberries, cherries, citrus fruits, cranberries, gingers, goji berries, guavas, kiwis, mangoes,mulberries, nectarines, papayas, passion fruits, peaches, pears, pineapples, raspberries, star apples and strawberries, amongst others) have a lower rate of CVD, obesity, various types of cancer, type 2 diabetes, and other chronic diseases.

Table 8.8: Total phenol content and total phenol antioxidant index (PAOXI) of fruits.

Fruit	Total phenols (μmol/g)		Total PAOXI × 10^{-3a}	
	Dry weight	Dry weight	Dry weight	Dry weight
Apple	34.1	6.4	110	20.6
Avocado	12.7	3.5	60.5	16.7
Banana	42.3	11.2	108	28.7
Blueberry	62.0	8.9	273	40.5
Cantaloupe	8.1	0.9	32.4	3.6
Cherry	52.3	10.9	523	109
Cranberry	158.8	22.7	212	31.2
Grape (white)	52.3	6.7	262	33.5
Grape (red)	63.7	13.6	351	50.3
Grape fruit	7.5	0.9	39.5	4.7
Lemon	19.6	2.4	67.6	8.28
Melon (honeydew)	11.4	1.3	36.8	4.2
Nectarine	12.3	1.5	64.7	7.89
Orange	18.9	1.4	55.6	4.1
Peach	27.9	2.4	60.7	5.22
Pear	41.4	6.6	81.2	12.9
Pineapple	11.9	2.3	44.1	8.52
Plum	58.2	7.8	116	15.6
Strawberry	72.3	4.6	603	38.3
Watermelon	19.5	2.2	44.3	5.0

Source: Adapted from Vinson, Su, et al. (2001). [a]Ratio of phenol concentration (μmol/kg) to the IC_{50} value (μM).

a. Bilberries

Bilberry contains the highest amount of anthocyanins amongst different berry varieties, with a ratio of 30:36:13 for cyanidin-, delphinidin-, and malvidin-3-*O*-glycosides (Du, Jerz, and Winterhalter, 2004; Wang, Cao, and Prior, 1997). Bilberry anthocyanosides representing all possible combinations of 5 anthocyanidins (cyanidin, delphinidin, peonidin, petunidin, malvidin) with 3 sugar moieties (3-*O*-arabinosides, 3-*O*glucosides, 3-*O*-galactosides) were separated by HPLC of bilberry extract (Madhavi, Bomser, Smith, and Singletary, 1998; Martinelli, Baj, and Bombardelli, 1986). Du *et al.* (2004) identified delphinidin-3-*O*-sambubioside, and cyanidin-3-*O*-sambubioside in bilberry. Presence of resveratrol in bilberries was reported by Lyons *et al.* (2003). Fresh bilberries were also a source of flavonols.The flavonols content of fresh bilberries was around 41 mg/kg (Jamison, 2003). Presence of caffeic, chlorogenic, *p*-coumaric, ferulic and syringic acids and derivative of *p*-hydroxybenzoic acid, *o*- and *m*-coumaric, gallic, *m*- and *p*-hydroxybenzoic, protocatechuic and vanillic acids in bilberry juice has been reported (Brenneisen and Steingger, 1981). Potent antioxidant properties of *Vaccinium* fruits have been documented (Wang *et al.,* 1997). A commercial extract of *Vaccinium myrtillus* (bilberry), called *V. myrtillus* anthocyanin (VMA), containing largely glycosides of delphinidin and cyanidin (Baj, Bombardelli, Gabetta, and Martinelli, 1983) has been used to treat various microcirculation diseases resulting from capillary fragility and has been used to maintain normal vascular permeability (Wang *et al.,* 1997). A recent study supported the use of bilberry and bilberry extracts in functional foods and food supplements designed for the prevention of chronic diseases associated with oxidative stress (Valentová, Ulrichová, Cvak, and Šimánek, 2007).

b. Blackberries

Blackberries are a rich source of anthocyanins and other polyphenolic antioxidants (Siriwoharn, Wrolstad, Finn, and Pereira, 2004). Extensive studies have been carried out on blackberry anthocyanins, and their identities have been well-characterized as being solely cyanidin-based compounds. Five anthocyanins were detected in blackberries and they were identified as cyanidin 3-glucoside (major anthocyanin), cyanidin 3-rutinoside, malonic acid acylated cyaniding 3-glucoside and xylose-cyanidin derivative (Siriwoharn *et al.,* 2004). Stintzing, Stintzing, Carle, and Wrolstad (2002) isolated cyanidin 3-dioxalyl-glucoside, a novel zwitterionic anthocyanin, from evergreen blackberry. Wada and Ou (2002) also reported the presence of cyanidin 3-(6′-*p*-coumaryl)glucoside in Marion berries and cyanidin 3-arabinoside in Evergreen blackberries, and Dugo, Mondello, Errante, Zappia, and Dugo (2001) reported the presence of cyanidin 3-galactoside, cyaniding 3-arabinoside, pelargonidin 3-glucoside, and malvidin 3-glucoside in a commercial Italian blackberry extract (Siriwoharn et al., 2004). Several phenolic acids have been detected in blackberries, namely gallic, caffeic, ferulic, *p*-coumaric and ellagic acids. Of these, ellagic acid was the major phenolic acid present (Sellappan, Akoh, and Krewer, 2002). Compared with blackberries grown in temperate climates, the tropical highland blackberry (*Rubus adenotrichus*) presents high contents of ellagitannins and lowcontents of anthocyanins (Acosta-Montoya *et al.,* 2010). Elisia, Hu, Popovich, and Kitts (2010) suggested that the anthocyanins and more specifically cyanidin-3-glucoside contribute mainly to the antioxidant ability of blackberry to suppress both peroxyl radicalinduced chemical and intracellular oxidation.

c. Blackcurrants

Blackcurrant (*Ribes nigrum*) berries are widely cultivated for their use in beverages reputed to be excellent for health due to their high content of antioxidant phenolics (Lu and Foo, 2003). Black currants were named as "superfruits" due to the presence of one of the important sources of potential health promoting phytochemicals, with potent immunomodulatory, antimicrobial and anti-inflammatory actions, inhibition of low-density lipoprotein as well as reduction of cardiovascular diseases (Nour, Stampar, Veberic, and Jakopic, 2013). The fruit serves as a rich source of phenolic compounds such as anthocyanins, flavonoids, phenolic acids and proanthocyanidins (Costantino, Albasini, Rastelli, and Benvenuti, 1992; Foo and Porter, 1981; Koeppen and Herrmann, 1977; Shahidi and Naczk, 2004). Flavonoids in blackcurrant, unlike other berries, are dominated by myricetin, followed by quercetin and kaempferol (Häkkinen, Kärenlampi, Heinonen, Mykkänen, and Törrönen, 1999; Mikkonen *et al.*, 2001; Vuorinen, Määttä, and Törrönen, 2000). Eleven delphinidin-, cyanidin-, malvidin-, petunidin-, and peonidin-based anthocyanins were detected in blackcurrant, with the main components being delphinidin-3-*O*-glucoside, delphinidin-3-*O*-rutinoside and cyanidin-3-*O*-rutinose (Borges, Degeneve, Mullen, and Crozier, 2009). Recently, Nour *et al.* (2013) detected and quantified nine individual anthocyanins using HPLC-MS in ethanolic extracts of black currants, i.e. delphinidin 3-glucoside, delphinidin 3-rutinoside, cyanidin 3-glucoside, cyanidin 3-rutinoside, petunidin 3-rutinoside, pelargonidin 3-rutinoside, peonidin 3-rutinoside, petunidin 3-(6-coumaroyl)-glucoside and cyanidin 3-(6-coumaroyl)-glucoside. Blackcurrant seeds are a rich source of potent antioxidants; the seeds also contain high levels of polyunsaturated fatty acids, especially γ-linolenic acid (Traitler, Winter, Richli, and Ingenbleek, 1984), which is stable in the intact seed (Lu and Foo, 2003; Lu, Foo, andWong, 2002; Shahidi and Naczk, 2004). Srivastava *et al.* (2010) reported the antioxidant and anti-inflammatory activities of polyphenolics from Southeastern U.S. range blackberry cultivars. Flavon(ol) glycosides such asmyricetin-3-rutinoside,myricetin-3-glucoside, quercetin-3-rutinoside, quercetin-3-glucoside and kaempferol-3-glucoside have been identified in blackcurrant seeds (Lu and Foo, 2003). Aureusidin, like myricetin, quercetin or kaempferol, could also be a hydrolysis product found in blackcurrant, and its presence was first reported by Lu and Foo (2003). Two novel noncyanogenic nitrile-containing compounds, nigrumin-5-*p*coumarate and nigrumin-5-ferulate were also isolated from seed residues (Lu *et al.*, 2002). Lu et al. (2002) identified a number of phenolic acids and their derivatives in blackcurrant seeds, namely caffeic, ferulic, *p*-coumaric, gallic, protocatechuic and *p*-hydroxybenzoic acids as well as 1-cinnamoyl-β-D-glucoside and 1-*p*-coumaroyl-β-D-glucoside. A recent study on antioxidant activity of berries suggests that blackcurrants had the highest antioxidant capacity in the FRAP assay followed by blueberries, raspberries, and red currants, and the lowest was noted for cranberries (Borges et al., 2009). Furthermore, researchers have provided convincing evidence that these small and softfleshed berry fruits may have an enormous potential for cancer prevention (Bishayee *et al.*, 2011; Gopalan *et al.*, 2012).

d. Blueberries

Several studies have confirmed that blueberries contain high antioxidant activity compared with other fruits (Cao, Sofic, and Prior, 1996; Prior *et al.,* 1998; Vinson, Teufel, *et al.,* 2001; Wang et al., 1996). The high antioxidant capacity of whole blueberries has been highly correlated to their anthocyanin and total phenolic contents (Kalt and Dufour, 1997; Kalt, McDonald, and Donner, 2000; Prior et al., 1998). The anthocyanins in blueberries exist almost exclusively in the skin, whereas phenolics, in general, and antioxidant properties are mostly in the skin (Lee and Wrolstad, 2004). Fifteen anthocyanins were detected in blueberries, many of which exhibited substantial antioxidant activity. Most antioxidant activity was attributed to delphinidin-3-*O*-galactoside, cyanidin-3-*O*-galactoside, delphinidin-3-*O*-arabinoside, petunidin-3-*O*-galactoside, malvidin-3-*O*-galactoside, malvidin-3-*O*-arabinoside, 5-*O*feruloylquinic acid and traces of a quercetin-*O*-diglucoside (Borges et al., 2009). Rabbiteye blueberries also contain a higher level of catechin than southern highbush blueberries (Ballington, Ballinger, and Maness, 1987; Francis, Harborne, and Barker, 1966; Prior, Lazarus, Cao, Muccitelli, and Hammerstone, 2001).The skins of blueberries consist mainly of cinnamic acids and flavonols glycosides, with minor amount of gallic acid and syringic acid. The flesh consists of only cinnamic acids. Chlorogenic acid (5-caffeoylquinic acid) is the main phenolic compound present in the flesh.The seed fraction also has cinnamic acids and flavonols glycosides (Lee and Wrolstad, 2004). Oligomeric B-type procyanidins from dimers to octamers have been identified in blueberry (Gu *et al.,* 2002; Prior *et al.,* 2001). Blueberry processing waste is high in antioxidant activity and total phenols and has the potential to be a good source of natural colourant and nutraceutical (Lee and Wrolstad, 2004). Blueberry leaves were found to serve as a good source of phenolics that possess high antioxidant activity. The antioxidant activity of crude phenolic extracts of blueberry leaves in β-carotene–linoleate model system was comparable to that displayed by BHA. The DPPH radical scavenging effect of crude phenolic extracts, at 100 μg/assay, was over 90% (Naczk, Amarowicz, Zadernowski, Pegg, and Shahidi, 2003). A recent study has shown that phenolic compounds such as chlorogenic acid, quercetin, ellagic acid and quercetin-3-galactoside present in the blueberry extract could be used to control pathogenic microorganisms due to its antimicrobial activity (Shen *et al.,* 2014).

e. Cranberries.

Cranberries have the highest amount of phenolic antioxidants of any commonly consumed fruit (Vinson, Teufel *et al.,* 2001) and this is the world's first government approved use for a fruit to treat urinary tract infection (Vinson, Bose, Proch, Al Kharrat, and Samman, 2008). Cranberry phenolics include simple phenolic acids and flavonoids that include anthocyanins, proanthocyanidins, as well as flavonols (Gregoire, Singh, Vorsa, and Koo, 2007). Anthocyanins of cranberry consist largely of galactoside and arabinoside conjugates of cyanidin and peonidin (Singh, Wilson, Kalk, Cheong, and Vorsa, 2009). Proanthocyanidin isolates contain epicatechin (monomer), dimer (epicatechin-(4β → 8, 2β → O → 7)-epicatechin), and trimer (epicatechin-(4β → 8)-epicatechin-(4β → 8, 2β → O → 7)-epicatechin) (Singh *et al.,*

2009). Isolated cranberry flavonols included myricetin-3-β-galactoside, myricetin-3-α-arabinofuranoside, quercetin-3-β-galactoside, quercetin-3-β-glucoside, quercetin-3-rhamnospyranoside, and quercetin-3-*O*-(6″-*p*-benzoyl)-β-galactoside (Singh *et al.*, 2009; Yan, Murphy, Hammond, Vinson, and Neto, 2002). Only a small percentage of the total flavonol content in cranberry or cranberry juice exist as aglycones such as free myricetin, quercetin, and kaempferol (Haekkinen, Kaerenlampi, Heinonen, Mykkaenen, and Toerroenen, 1999). Yan *et al.* (2002) suggested that compared to a standard dietary antioxidant such as vitamin E, both the flavonol glycosides and cyanidin 3-galactoside from whole cranberry are effective in scavenging free radicals and preventing LDL oxidation, cyanidin 3-galactoside being superior to flavonols present in this regard. Fifteen phenolic acids, namely benzoic, *o*-hydroxybenzoic, cinnamic, *m*-hydroxybenzoic, *p*-hydroxybenzoic, *p*-hydroxyphenylacetic, phthalic, 2,3-dihydroxybenzoic, vanillic, *o*-hydroxycinnamic, 2,4-dihydroxybenzoic, *p*-coumaric, ferulic, caffeic, and sinapic acids were identified in cranberry fruit in the free and bound forms (Zuo,Wang, and Zhan, 2002). Sinapic, caffeic and *p*-coumaric acids were the major bound phenolic acids whilst *p*-coumaric, 2,4-dihydroxybenzoic and vanillic acids were the predominant free phenolic acids (Naczk and Shahidi, 2006; Zuo *et al.*, 2002). Cranberry and cranberry juice phenolics are also known to provide a rich source of antioxidants that can protect LDL from oxidative injury (Wilson, Porcari, and Harbin, 1998; Wilson, Porcari, and Maher, 1999). The order of total polyphenols on a fresh weight basis follows the order of dried cranberries > frozen cranberries > cranberry sauce > jellied sauce. Cranberry powder is not usually considred a food, but it had the most polyphenols of the cranberry products (Vinson et al., 2008). The frozen cranberries, 100% juice, and dried cranberries provide the most antioxidants distantly followed by mixed and 27% cranberry juice and other cranberry products (Vinson *et al.*, 2008).

f. Raspberries

Red raspberries (*Rubus idaeus* L.) are known to possess strong antioxidant capacity, mainly as a result of their high levels of anthocyanins and other phenolic compounds (Kafkas, Özgen, Özog˘ ul, and Türemis,, 2008; Kähkönen, Hopia, and Heinonen, 2001). Raspberries contain a unique phytochemical profile rich in hydrolysable tannins (ellagitannins) and anthocyanins (more specifically cyanidin glycosides) that distinguishes them from other berries and fruits (Rao and Snyder, 2010). The major raspberry anthocyanin (Figure 8.12) is cyanidin-3-sophoroside with smaller quantities of other nthocyanins, including cyanidin-3-(2-glucosylrutinoside), cyanidin-3-glucoside, cyanidin-3-rutinoside, pelargonidin-3-sophoroside, pelargonidin-3(2-glucosylrutinoside), and pelargonidin-3-glucoside (Borges et al., 2009; Bowen-Forbes, Zhang, and Nair, 2010; Mullen, Stewart, et al., 2002; Rommel and Wrolstad, 1993). The two main ellagitannins are sanguiin H-6 and lambertianin C. Raspberries also contain quercetin and kaempferol-based flavonols conjugates and trace levels of (−)-epicatechin and a procyanidin dimer (Mullen, McGinn, et al., 2002). Of these, quercetin 3-glucuronide is the predominant flavonol glycoside in red raspberry juice (Rommel and Wrolstad, 1993). Red and black raspberries are especially high in phenolic compounds compared to other lighter coloured fruits and vegetables (Liu et al., 2002). Black raspberries have much higher levels of anthocyanins (687 mg/100 g of fw) than red raspberries, attributable to their darker colouration, and the majority of these are also cyanidins (Rao and Snyder, 2010). Beekwilder, Hall, and de Vos (2005) published a review on raspberry antioxidants and reported that raspberries had the highest antioxidant capacity followed by strawberries, kiwi, broccoli, leek, apple, and, finally, tomato. Liu et al. (2002) reported that the antioxidant activity of the raspberry was directly related to their total content of phenolics and flavonoids.

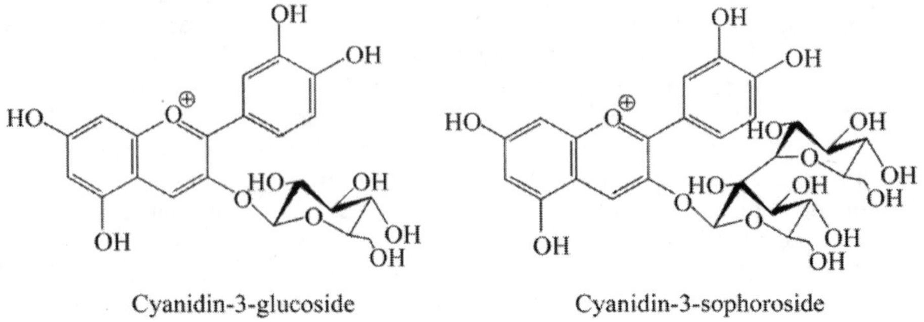

Cyanidin-3-glucoside

Cyanidin-3-sophoroside

Figure 8.12: Chemical structures of raspberry anthocyanins (adapted from Sun, Cao, Bai, Liao, and Hu, 2010).

g. Grapes

Grape phenolics, especially high in the grape skin, are classified into two groups of flavonoids and nonflavonoids. The flavonoids include flavan-3-ols (catechin), flavonols (quercetin) and anthocyanins.The non-flavonoids encompass hydroxybenzoic acids (gallic acid), hydroxycinnamic acids and stilbenes (resveratrol) (Yang, Martinson, and Liu, 2009). In the grape berry, the flavonoids are mainly localized in the skins, such as the anthocyanins and resveratrol (Palomino, Gomez-Serranillos, Slowing, Carretero, and Villar, 2000), whilst flavan-3-ols (catechins and proanthocyanidins) are present both in the skins and in the seeds (Yang, Martinson, *et al.*, 2009). Grape berries and their skins contain phenolic acids such as caftaric acid (*trans*-caffeoyltartaric acid), coutaric acid (*p*coumaryltartaric acid) and *trans*-fertaric acid (Cantos, Espin, and Tomas-Barberan, 2002; Souquet, Labarbe, Le Guerneve, Cheynier, and Moutounet, 2000; Vrhovsek, 1998), flavonols such as quercetin 3-glucuronide, quercetin 3-glucoside, myricetin 3-glucuronide andmyricetin 3-glucuronide (Souquet et al., 2000) as well as flavanonols, such as astilbin (dihydroquercetin 3-rhamnoside) and engeletin (dihydrokaempferol 3-rhamnoside) (Lu and Foo, 1999; Souquet *et al.*, 2000). A number of stilbenes, namely *trans*- and *cis*-resveratrols (3,5,4-trihydroxystilbene), *trans*- and *cis*-piceids (3-*O*-d-glucosides of resveratrol), *trans*- and *cis*-astringins (3-*O*-d-glucosides of 3-hydroxyresveratrol), *trans*and *cis*-resveratrolosides (4-*O*-d-glucosides of resveratrol) and pterostilbene (a dimethylated derivative of stilbene) have been detected in both grape leaves and berries (Jaendet *et al.*, 2002; Naczk and Shahidi, 2006; Wang, Catana, Yang, Roderick, and van Breemen, 2002). Studies have shown that grape seeds are a rich source of monomeric phenolic compounds such as (+)-catechins, (−)-

epicatechin, (+)-gallocatechins, (−)-epigallocatechin, and their dimeric, trimeric, and tetrameric proanthocyanidins (Baoshan Sun and Spranger, 2005; Freitas, Glories, Bourgeois, and Vitry, 1998). Grapes are one of the major dietary sources of anthocyanins, responsible for a wide range of colours of black, red and purple grapes; however, these phenolics are absent in white grapes. In particular, anthocyanins mostly accumulate in the skins, whereas procyanidins are located in the seeds. The anthocyanins in grape skins are predominately the 3-*O*-glucosides of malvidin, cyanidin, delphinidin, peonidin and petunidin. Malvidin, the reddest of all anthocyanins, is the major one in dark red *vinifera* grapes, with higher proportions of cyanidin in red grapes. Cyanidin 3-monoglucoside and delphinidin 3-monoglucoside are the major anthocyanins in Concord grapes (Mazza, 1995; Yang, Martinson, *et al.*, 2009). However, Orak (2007) reported that anthocyanin and phenolic compounds either alone or in combination, are responsible for the antioxidant activity of different grape cultivars. Recently, de Camargo, Regitano-d'Arce, Biasoto, and Shahidi (2014) identified that in grape by products most prominent phenolic compounds are present in the insoluble bound form.

h. Cherries

Sweet cherries and tart cherries are rich sources of flavonoids (Gao and Mazza, 1995). The major anthocyanin pigments in sweet and sour cherries have been identified. Cyanidin-3-rutinoside, cyanidin-3-glucoside, peonidin-3-rutinoside, peonidin-3-glucoside, and pelargonidin-3-rutinoside have been identified in Bing and other sweet cherry cultivars (Chaovanalikit and Wrolstad, 2004; Gao and Mazza, 1995; Lynn and Luh, 1964; Mozetic, Trebse, and Hribar, 2002). Sour cherries are reported to contain cyanidin-3-*O*-glucoside, cyanidin-3-*O*-rutinoside, cyanidin-3-*O*-glucosylrutinoside, cyanidin-3-*O*-sophoroside, pelargonidin-3-*O*-glucoside, peonidin-3-*O*rutinoside and cyanidin-3-*O*-arabinosylrutinoside (Chandra, Nair, and Lezzoni, 1992; Chaovanalikit and Wrolstad, 2004; Gao and Mazza, 1995). Kirakosyan *et al.* (2010) reported that tart cherry produces various kinds of polyphenolics in its fruits that include cyanidin derivatives (mostly cyanidin 3-glucosylrutinoside, cyanidin 3-rutinoside, cyanidin sophoroside), peonidin 3-glucoside; kaempferol, quercetin and isorhamnetin and their derivatives, as well as the alkaloid, melatonin. Several research reports have indicated that sweet cherries are rich in phenolic acids. The most abundant phenolics present are hydroxycinnamic acids, such as caffeic and coumaric acids, with antioxidant and anti-inflammatory properties (Mulabagal, Lang, DeWitt, Dalavoy, and Nair, 2009). Cherries contain higher amounts of anthocyanins and phenolics when compared to several other fruits. It is implied that regular consumption of cherries is beneficial in reducing risk factors for heart disease, diabetes and certain types of cancer (Bobe, Wang, Seeram, Nair, and Bourquin, 2006; Bolleddula, Vareed, Olson, and Nair, 2005; Mulabagal *et al.*, 2009).

i. Apples

Apples are an excellent source of several phenolic compounds and also possess high total antioxidant capacity. Biedrzycka and Amarowicz (2008) reviewed apple polyphenols as antioxidants. Sun *et al.* (2002) found that apples had the highest soluble free phenolics when compared to 10 other commonly consumed fruits (Khanizadeh *et al.*, 2008).There are five major groups of polyphenolic compounds in apple, including hydroxycinnamic acids, flavan-3-ols, anthocyanidins, flavonols, and dihydrochalcones (Tsao, Yang, Sockovie, and Khanizadeh, 2005). The concentration of total phenolic compounds is much greater in the peel of apples than in the flesh and the antioxidant and antiproliferative activities of unpeeled apples were greater than those of peeled apples (Burda, Oleszek, and Lee, 1990; Ju, Yuan, Liu, Zhan, and Wang, 1996; Wolfe, Wu, and Liu, 2003).The flesh of apple contains catechins, procyanidins, phloridzin, phloretin glycosides, caffeic acid and chlorogenic acid; the peel possesses all of these compounds and has additional flavonoids not found in the flesh, such as quercetin glycosides (Wolfe *et al.*, 2003). Huber and Rupasinghe (2009) found specific antioxidant compounds such as quercetin glycosides and cyanidin-3-*O*-galactoside in apple skin extract, which are not found in the flesh of apples. McGhie, Hunt, and Barnett (2005) reported that apple skin contains approximately 46% of the total phenolics in apples. Quercetin and quercetin-3-*O*-glucoside have been shown to exhibit a better antioxidant activity than BHT in bulk fish oil (Huber, Rupasinghe, and Shahidi, 2009). Tsao *et al.* (2005) reported that the major polyphenolic groups present are hydroxycinnamic acids (chlorogenic acid and *p*-coumaroylquinic acid), cyanidin-3-galactoside, flavan-3-ols/procyanidins (mainly catechin, epicatechin, and procyanidins B1 and B2), flavonols (quercetin galactoside, glucoside, rhamnoside, arabinoside, and xyloside), and dihydrochalcones (mainly phloridzin and phloretin-2′-xyloglucoside). No anthocyanins were detected in the flesh (Durkee and Poapst, 1965; Guyot, Marnet, Laraba, Sanoner, and Drilleau, 1998; Sanoner, Guyot, Marnet, Molle, and Drilleau, 1999; Tsao *et al.*, 2005; Van der Sluis, Dekker, Skrede, and Jongen, 2002). Chlorogenic acid was the major hydrocinnamic acid family derivative identified in apple fruit accounting for up to 87% of the total amount (Guyot *et al.,* 1998). Chinnici, Bendini, Gaiani, and Riponi (2004) found that the antioxidant activity of individual phenolic compounds of apple followed the order of quercetin glycosides > procyanidins > chlorogenic > phloridzin. However, Van der Sluis *et al.* (2002) reported that flavan-3-ols were the top contributor, followed by quercetin glycosides, chlorogenic acid, cyanidin-3-galactoside, and phloridzin. Tsao *et al.* (2005) concluded that flavan-3-ols/procyanidins are the most important contributors to the *in vitro* antioxidant activity of apple and that procyanidin B2 and epicatechin are the most important individual antioxidants in apple. Hydroxycinnamic acids may play a significant role in the flesh. Apple pomace is a valuable source of polyphenols relevant for its antioxidant activity, as flavanols and flavonols and the antioxidant activity of apple pomace can be predicted by the contents of phloridzin, procyanidin B2, rutin + isoquercitrin, protocatechuic acid and hyperin (García, Valles, and Lobo, 2009). Recently, Sekhon-Loodu, Warnakulasuriya, Rupasinghe, and Shahidi (2013) found that polyphenols from both dried and frozen apple peel have higher inhibition of lipid oxidation compared to α-tocopherol, butylated hydroxytoluene and crude apple peel extracts.

j. Citrus fruits

In citrus fruits, cinnamic acid derivatives, coumarins and flavonoids (flavonones, flavones and flavonols) are the major groups of phenolic compounds (Horowitz and Gentili, 1977; Manthey and Grohmann, 2001; Naczk and Shahidi, 2006). Phenolic acids are found in the flavedo of citrus fruits (Manthey and Grohmann, 2001; Peleg, Naim, Rouseff, and Zehavi, 1991) in the form of esters, amides and glycosides (Bocco, Cuvelier, Richard, and Berset, 1998; Mouly, Gaydou, Faure, and Estienne, 1997; Rapisarda, Carollo, Fallico, Tomaselli, and Maccarone, 1998). Citrus fruits, citrus fruit extracts and citrus flavonoids exhibit a wide range of promising biological properties including anti-atherogenic, anti-inflammatory and antitumour activity, inhibition of blood clots and strong antioxidant activities (Middleton and Kandaswami, 1994; Ramful, Bahorun, Bourdon, Tarnus, and Aruoma, 2010).Three types of flavonoids occur in citrus fruits: flavanones, flavones and flavonols (Calabrò *et al.*, 2004). Naringin, naringenin 7-neohesperidoside and narirutin, naringenin 7-rutinoside are the major flavanone glycosides found in grapefruit, whilst narirutin and hesperidin, hesperetin 7-rutinoside are present in sweet oranges and naringin, neohesperidin and hesperetin 7-neohesperidoside in sour oranges (Kanes, Tisserat, Berhow, and Vandercook, 1992). On the other hand, hesperidin, narirutin and didymin (isosakuranetin 7-rutinoside) are the predominant flavanone glycosides in navel (Gil-Izquierdo, Gil, and Ferreres, 2002) and blood oranges (Mouly *et al.*, 1997). Polymethoxyflavones and hydroxylated polymethoxyflavones are unique phenolic compounds that exist exclusively in citrus genus, especially in the citrus peels (S. Li *et al.*, 2009). Tangeretin and nobiletin are the most abundant polymethoxyflavones in citrus peels (S. Li *et al.*, 2009). However, Marin and Del Rio (2001) reported that diosmin (4′methoxy-5, 7, 3′-trihydroxyflavone-7-rutinoside) and neodiosmin (4′methoxy-5, 7, 3′-trihydroxyflavone-7-neohepseridoside) are the predominant glycosylated flavones identified in citrus fruits. Nobiletin (5,6,7,8,3′,4′-hexamethoxyflavone) and sinensetin (5,6,7,3′,4′-pentame thoxyflavone) have been identified in orange, whilst tangeretin, 3,5,6,7,8,3′,4′-heptamethoxyflavone, 5,7,8,4′-tetramethoxyflavone, and 5,7,8,3′,4′-pentamethoxyflavone exist in grapefruits (S. Li *et al.*, 2009; Naczk and Shahidi, 2006). Cyanidin-3-glycoside, cyanidin-3-(6″-malonyl)-glycoside and cyanidin 3-rhamnoside are the major anthocyanins in Italian blood oranges (Maccarone, Maccarrone, Perrini, and Rapisarda, 1983; Maccarone, Maccarrone, and Rapisarda, 1985; Maccarone, Rapisarda, Fanella, Arena, and Mondello, 1998) whilst cyanidin-3-glycoside and cyanidin-3-(6″-malonyl)-glycoside are resent in high amounts in Budd blood oranges (Lee, 2002). Recent studies have indicated that the flavedo extracts of citrus fruits represent a significant source of phenolic antioxidants with potential prophylactic properties for the development of functional foods (Ramful *et al.*, 2010). Limonoids are a group of highly oxygenated triterpenoids found in citrus that were also reported as an anticancer agent (Hasegawa, Lam, and Miller, 2000). An extract of multiple varieties of citrus peels containing abundant flavonoids (polymethoxyflavones) effectively suppressed azoxymethane (AOM)-induced colonic tumourigenesis and exhibited anticancer activity, especially for prostate cancer (Lai *et al.*, 2007, 2013; Suzawa, Guo, Pan, Ho, and Li, 2014).

k. Pomegranate

Pomegranates are a rich source of hydrolysable tannins and anthocyanins. Studies have shown that pomegranate fruit contains anthocyanins, ellagitannins, gallotannins, non-coloured flavonoids, and lignans, amongst other (poly) phenolic compounds (Abbas, 2014; Bonzanini, Bruni, Palla, Serlataite, and Caligiani, 2009; Fischer, Carle, and Kammerer, 2011; Mena et al., 2012). Several anthocyanins such as cyanidin 3-glucoside, delphinidin 3-glucoside, cyanidin 3,5-diglucoside, delphinidin 3,5-diglucoside and pelargonidin 3-glucoside have been detected in pomegranate juice (Gil, Tomas-Barberan, Hess-Pierce, Holcroft, and Kader, 2000; Noda, Kaneyuki, Mori, and Packe, 2002). The seed coat of pomegranate contains delphinidin-3-glucoside, delphinidin-3, 5-diglucoside, cyanidin-3-glucoside, cyanidin -3,5-diglucoside, pelargonidin-3-glucoside, and pelargonidin-3,5-diglucoside (Du, Wang, and Francis, 1975). Hydrolysable tannins are the most abundant polyphenols and antioxidant compounds in pomegranates and include gallotannins, ellagitannins and gallagyl esters such as punicalagin and punicalin (Madrigal-Carballo, Rodriguez, Krueger, Dreher, and Reed, 2009). Pomegranate marc (contains about 78% peel and 22% seeds based on wet weight), a byproduct after pomegranate juice processing, is a good raw material for producing natural antioxidants because of its high content of polyphenols (Qu, Pan, and Ma, 2010). Recently, Qin et al. (2013) reported that the antioxidant potential of pomegranate rind powder extract (PRP), pomegranate juice (PJ) and pomegranate seed powder extract in raw ground pork meat stored at 4 °C for 12 days follows the order of BHT > PRP > PJ > PSP > control. Based on several studies, Syed, Chamcheu, Adhami, and Mukhtar (2013) revealed that pomegranate is a potential chemopreventive agent for leukaemia, skin, lung, prostate, colon and breast cancer.Wu, Ma, and Tian (2013) found that pomegranate husk extract, punicalagin and ellagic acid inhibit fatty acid synthase and adipogenesis of 3T3-L1 adipocyte and could have potential effect in the prevention and treatment of obesity.More recently, a study showed that thinning sour–sweet cultivars of pomegranate could be considered as a good source of bioactive compounds (Nuncio-Jáuregui et al., 2015). Pomegranate peel extract has been suggested as the best source of microbial substrates at colonic level in comparison to pulp and juice (Mosele, Macià, Romero, Motilva, and Rubió, 2015).

l. Apricot

Apricot peel and flesh contain four main groups of phenolic compounds, procyanidins, hydroxycinnamic acid derivatives, flavonol, and anthocyanins (Ruiz, Egea, Gil, and Tomas-Barberan, 2005). Chlorogenic and neochlorogenic acids, procyanidins B1, B2, and B4, and some procyanidin trimers, quercetin 3-rutinoside, kaempferol 3-rhamnosyl-hexoside and quercetin 3-acetyl-hexoside, cyanidin 3-rutinoside, and 3-glucoside have been detected in the skin and flesh of the different cultivars of apricot (Han, Liu, Cui, and Zhang, 2013; Ruiz *et al.*, 2005).

m. Dried Fruits: Phytochemicals, Antioxidant Efficacies and Health Benefits

Nutrition plays a major role in the primary and secondary prevention of non-communicable diseases (NCDs). Consumption of fruits and vegetables is one of the essential nutritional recommendations to prevent NCDs. Numerous studies have demonstrated that the intake of 3–5 daily servings of fruits and vegetables would protect against NCDs (He, Nowson, Lucas, and MacGregor, 2007; Lichtenstein et al., 2006; Slavin and Lloyd, 2012). The 2010 Dietary Guidelines for Americans also recommended to make one half of food plates with fruits and vegetables (US Department of Health and Human Services, 2010). Fruits constitute a major part of the human diet. Besides, fruits may be consumed as a part of religious practices and as nutritional therapy in different human traditions around the world (Slavin and Lloyd, 2012; Vayalil, 2012). Studies indicate that the role of fruits together with their nutrients in the prevention of NCDs could be stronger than vegetables (Habauzit and Morand, 2012). This happens because fruits provide essential vitamins, minerals, as well as various phytochemicals that confer significant health benefits other than basic nutrition (Slavin and Lloyd, 2012; US Department of Health and Human Services, 2010). Most of the common fruits are produced on a seasonal basis and hence may not be available in fresh conditions throughout the year. Thus, fresh fruits are processed by various techniques to become dried fruits to prolong their shelf life. Dried fruits are a concentrated form of fresh fruits, albeit with lower moisture content than that of their fresh counterparts since a large proportion of their moisture content has been removed through sun-drying or various modern drying techniques, such as mechanical devices (Alasalvar and Shahidi, 2013a, 2013b). Fruits can be dried whole (e.g., grapes, berries, apricots, and plums), in halves, or as slices (e.g., mangoes, papayas, and kiwis). Dried fruits are important healthy snacks worldwide. They also have the advantage of being easy to store and distribute, available throughout the year, and healthier alternative to salty or sugary snacks. Apples, apricots, dates, figs, peaches, pears, prunes, and raisins are referred to as "conventional'' or "traditional" dried fruits. Meanwhile, some fruits, such as blueberries, cranberries, cherries, strawberries, and mangoes are usually infused with sugar solutions or fruit juice concentrate before drying (Alasalvar and Shahidi, 2013a, 2013b). In terms of production, raisins and currants rank first according to 2014 statistics on global production of commercially important dried fruits with a production of 1,394,566 metric tonnes (MT), followed by dates (756,000 MT), prunes (216,853 MT), cranberries (152,000 MT), figs (135,744 MT), and apricots (88,129 MT). However, statistics on the production of other dried fruits, such as berries, citrus fruits, and cherries are generally scarce (INC, 2014). Although raisins, figs, dates, prunes, and apricots are the most common dried fruits in the marketplace, food stores and local markets offer many more choices such as dried apples, pineapples, berries, mangoes, and papayas, among others. They are rich sources of essential nutrients and health-promoting bioactive compounds. Epidemiological evidence has demonstrated an association between dried fruit consumption and diet quality (Coleman et al., 2008; Keast, O'Neil, and Jones, 2011). Raisins are the most studied among all dried fruits, followed by dates, prunes, figs, apricots, peaches, apples, pears, and other fruits (Alasalvar and Shahidi, 2013a, 2013b). This review discusses the phytochemical compositions, antioxidant efficacies, and potential health-promoting properties of traditional dried fruits as well as cranberries. The potential of several nontropical dried fruits (apples, apricots, figs, peaches, pears, prunes, and raisins) and a selected tropical dried fruit, dates, as functional or healthy foods have been highlighted. This review is intended to stimulate large-scale commercial exploitation and generate interest among researchers in various scientific fields to study dried fruits as functional foods.

PHYTOCHEMICALS AND ANTIOXIDANT EFFICACIES OF DRIED FRUITS

Phytochemicals are non-nutritive, naturally occurring, and biologically active compounds found in the plant kingdom. However, a large percentage of phytochemicals is still unknown and remains to be identified by the scientific community (Quideau, Deffieux, Douat-Casassus, and Pouységu, 2011; Tomás-Barberán and Andrés-Lacueva, 2012). Non-specific methods such as determination of total phenolics and anthocyanins by pH differential, and specific methods such as high performance liquid chromatography (HPLC), ultraviolet-visible spectroscopy (UV-IS), and mass spectrometry (MS) have been used to characterise phytochemicals in dried fruits (Alasalvar and Shahidi, 2013a, 2013b; Vayalil, 2012). Dried fruits provide a wide range of phytochemicals, such as phenolic acids, flavonoids, phytoestrogens, and carotenoids (Alasalvar and Shahidi, 2013a, 2013b). There are a number of review articles reporting the nutritional and functional properties as well as phytochemicals (carotenoids, phytosterols, polyphenols, phenolic acids, flavonoids, anthocyanins, and phytoestrogen) of dates (Al-Farsi and Lee, 2008; Vayalil, 2012). Williamson and Carughi (2010) reviewed the polyphenols and health benefits of raisins. However, literature on phenolic acid, flavonoid, phytoestrogen, and carotenoid profiles of other dried fruits are still scarce. The reported phytochemicals in dried fruits and their corresponding antioxidant activities vary considerably according to the cultivars as well as climatic and agricultural practices (Alasalvar and Shahidi, 2013a, 2013b; Al-Farsi and Lee, 2008; Pellegrini et al., 2006; Vayalil, 2012; Wu et al., 2004).

TOTAL PHENOLICS

Folin–Ciocalteu reagent assay is the common method used to determine the total phenolic content (TPC) of dried fruits. TPCs of dried fruits, expressed as mg of ascorbic acid equivalents (AAE)/100 g dry weight (DW), were reported by Ishiwata, Yamaguchi, Takamura, and Matoba (2004) with a range of 916–2414. Raisins contained the highest TPC (2414 mg AAE/100 g), followed by dried apricots, cranberries, peaches, figs, pears, and prunes (Table 8.9). In another study conducted by Wu et al. (2004), TPC [expressed as mg of gallic acid equivalents (GAE)/g] of dried fruits decreased in the order of prunes > raisins > figs > dates. However, in another study (Vinson, Zubik, Bose, Samman, and Proch, 2005), dates demonstrated the highest TPC [1959 mg catechin equivalents (CE)/100 g fresh weight (FW)], while figs had the lowest TPC (320 mg CE/100 g FW) among six dried fruits (apricots, cranberries, dates, figs, plums, and raisins) studied. A recent study reported the TPC of two Turkish dried fig varieties of *Sarılop* and *Bursa Siyahı* to be 193 and 417 mg GAE/100 g DW, respectively (Kamiloglu and Capanoglu, 2015).

Table 8.9: Phytochemicals in selected dried fruits.

Dried fruit	Total phenolics[a]	Flavonoids[b]	Flavonols[b]	Phytoe-strogens[c]	Isoflavones[c]	Total lignans[c]	Carot-enoids[d]
Apples	916	nr	nr	nr	nr	nr	nr
Apricots	2256	56.8[f]	nr	445	39.8	401	2.2
Cranberries	1819	7.66	4.50	nr	nr	nr	nr
Dates	661[e]	2.63	0.93	330	5.1	324	0.97
Figs	1234	105[f]	nr	nr	nr	nr	0.032
Peaches	1260	nr	nr	nr	nr	nr	2.08
Pears	1196	nr	nr	nr	nr	nr	nr
Prunes	1032	2.58	1.80	184	4.2	178	0.69
Raisins	2414	0.85	0.26	30.2	8.1	22.0	nd

a Data are expressed as mg of AAE/100 g DW and obtained from Ishiwata et al. (2004).

b Data are expressed as mg/100 g edible portion and obtained from Bhagwat, Haytowitz, and Holden (2014).

c Data are expressed as mg/100 g edible portion and obtained from Thompson et al. (2006).

d Data are expressed as mg/100 g FW and obtained from Al-Farsi and Lee (2008).

e Data are expressed as mg of GAE/100 g and obtained from Wu et al. (2004).

f Data are expressed as mg of QE/100 g and obtained from Ouchemoukh, Hachoud, Boudraham, Mokrani, and Louaileche (2012).

AAE, ascorbic acid equivalents; DW, dry weight; FW, fresh weight; GAE, gallic acid equivalents; nd, not detected; nr, not reported; QE, quercetin equivalents.

Flavonoids

The major flavonoids reported in dried fruits are anthocyanidins, dihydrochalcones, flavonols, flavones, and flavan-3-ols. The reported flavonoid contents in selected dried fruits are given in Table 8.10. Apricots contain all the above-mentioned classes of flavonoids (Erdog˘an and Erdemog˘ lu, 2011; Schmitzer et al., 2011). The chemical structures of the representative flavonoids reported in dried fruits are shown in Figure 8.13. Dried fruits contain undetectable amounts of anthocyanins, which are likely degraded to phenolic acids. Dates contain one anthocyanidin (cyanidin) and one flavonol (quercetin) (Harnly et al., 2006). No anthocyanins have been detected in dried pears (Ferreira et al., 2002) and raisins (Chiou et al., 2007; Meng et al., 2011; Williamson and Carughi, 2010). However, a number of anthocyanin (delphinidin-3-glucoside, cyanidin-3-glucoside, petunidin-3-glucoside, pelargonidin-3-glucoside, peonidin-3-glucoside, and malvidin-3-glucoside) and anthocyanin-derived compounds (vitisin A, acetylvitisin A, B-type vitisin of malvidin-3-glucoside, peonidin-3-glucoside, peonidin-3-acetylglucoside, A-type vitisin of malvidin-3-acetylglucoside, and malvidin-3-acetylglucoside) have been detected in raisins from Spanish *Merlot* and *Syrah* varieties (Table 2) (Marquez, Dueñas, Serratosa, and Merida, 2012). These anthocyanin-derived compounds resulted from the cyclo-addition of pyruvic acid (A-type vitisin) and acetaldehyde (B-type vitisin) to anthocyanin molecules due to enzymatic transformations (Marquez et al., 2012).

Dihydrochalcones, such as phloridzin and phloretin, are reported only in dried apples and apricots (Erdog˘an and Erdemog˘ lu, 2011; Joshi, Rupasinghe, and Khanizadeh, 2011). Rutin is the predominant flavonol in prunes (Donovan, Meyer, andWaterhouse, 1998). Dried fruits also contain traces of proanthocyanidins (USDA, 2004; Vallejo, Marín, and Tomás-Barberán, 2012). Proanthocyanidins are detected in plums and grapes, but not in prunes and raisins, suggesting that these compounds are degraded during the drying process (Franke, Custer, Arakaki, and Murphy, 2004; Harnly et al., 2006). Flavonols have been reported in all selected dried fruits, except dried pears and peaches (Ferreira et al., 2002; Rababah, Ereifej, and Howard, 2005; Silva, Shahidi, and Coimbra, 2013; Villarino, Sandín-España, Melgarejo, and De Cal, 2011). In another study conducted by Vallejo et al. (2012), four flavonols (kaempferol rutinoside, quercetin-acetylglucoside, quercetinrutinoside, and quercetin-glucoside) were reported in three dried fig cultivars, among which quercetin-rutinoside was the major one. In a recent study, the total flavonoids, total proanthocyanidins, and total anthocyanin contents of dried figs from *Sarılop* and *Bursa Siyahı* varieties were 14–52 mg GAE/100 g DW, 12–16 mg cyanidin equivalents (CE)/100 g DW, and 0.1–14.5 mg cyanidin-3-glycoside equivalents (C3GE)/100 g DW, respectively (Kamiloglu and Capanoglu, 2015). Specifically, rutin and cyanidin-3-rutinoside were the major flavonol and anthocyanin in both Turkish dried figs varieties. Anthocyanins, such as cyanidin-3-glucoside and cyanidin-3-rutinoside, were reported in Turkish dried figs (Kamiloglu and Capanoglu, 2015), but these anthocyanins were not found in Spanish dried figs (Vallejo et al., 2012). Flavones, such as luteolin, were reported only in dried apricots (Erdog˘an and Erdemog˘lu, 2011), while apigenin was detected in dried figs (Kamiloglu and Capanoglu, 2015).

Table 8.10: Comparison of phenolic compounds in selected dried fruits.

Dried fruit	Types of phenolic compounds		Unit	Content
Apples	Dihydrochalcones	Phloridzin	mg/100 g DW	35.44
		Phloretin		0.32
	Flavan-3-ols	Catechin	mg/100 g DW	0.26
		Epicatechin		3.11
	Flavonols	Cyanidin-3-*O*-galactoside	mg/100 g DW	2.97
		Quercetin-3-*O*-rutinoside		1.71
		Quercetin-3-*O*-galactoside		10.49
		Quercetin-3-*O*-glucoside		5.93
		Quercetin-3-*O*-rhamnoside		9.21
	Phenolic acids	Chlorogenic acid	mg/100 g DW	139
	Flavonols	Quercetin-3-*O*-galactoside	% of total quercetin derivatives	36.2
		Quercetin-3-*O*-glucoside		12.2
		Quercetin-3-*O*-xyloside		6.7
		Quercetin-3-*O*-arabinoside		25.9
		Quercetin-3-*O*-rhamnoside		14.1
		Quercetin		4.9

Dried fruit	Types of phenolic compounds		Unit	Content
Apricots	Flavan-3-ols[a]	Catechin	mg/kg DW	47.3
		Epicatechin		21.5
	Flavonols[a]	Quercetin-3-*O*-glucoside	mg/kg DW	8.14
		Rutin		34.6
	Phenolic acids$_a$	Chlorogenic acid	mg/kg DW	365
		Neochlorogenic acid		221
	Dihydroch-alcones[b]	Phloridzin	mg/kg DW	0.17
	Flavan-3-ols[b]	Catechin	mg/kg DW	3.9
		Epicatechin		20.96
		Epicatechin gallate		13.26
		Epigallocatechin		8.19
		Epigallocatechin gallate		8.21
		Gallocatechin		2955
		Procyanidin B2		5.19
	Flavonols[b]	Myricetin	mg/kg DW	1.29
		Quercetin		0.29
		Rutin		75.01
	Flavone[b]	Luteolin	mg/kg DW	0.43
Cranberries	Flavan-3-ols	Catechin	mg/100 g FW	0.8
		Epicatechin		4.5
		Epigallocatechin gallate		1.9
	Flavonols	Myricetin	mg/100 g FW	16.6
		Quercetin		19.4
	Phenolic acids	Caffeic acid	mg/100 g FW	2.31
		Chlorogenic acid		10.3
		Ferulic acid		2.96
		Gallic acid		14.5
		p-Coumaric acid		25.2
		Protocatechuic acid		51.2
		3-Hydroxybenzoic acid		1.9
		3-Hydroxyphenylpropionic acid		1.53 3.42
		4-Hydroxybenzoic acid		3.21
		4-Hydroxyphenylacetic		
Dates	Anthocyanins	Cyanidin	mg/100 g FW	1.7
	Flavonols	Quercetin	mg/100 g FW	0.9
	Phenolic acids[c]	Caffeic acid	mg/100 g FW	2.52
		Ferulic acid		11.83
		Gallic acid		1.56

Dried fruit	Types of phenolic compounds		Unit	Content
		o-Coumaric acid		2.88
		p-Coumaric acid		5.77
		Protocatechuic acid		4.94
		Syringic acid		2.89
		Vanillic acid		2.26
Figs	Flavan-3-ols[d]	Catechin	mg/100 g FW	1.5–8.67
		Epicatechin		0.6–21.83
	Flavonols[d]	Kaempferol-3-*O*-glucoside	mg/100 g FW	0.39
		Kaempferol-rutinosidee[f]		0.2–2.0
		Luteolin-8-*C*-glucosidee		0.14
		Quercetin-acetylglucosidee		2.6
		Quercetin-rutinosidee		10.2
		Quercetin-3-*O*-glucosidee		2.87
		Quercetin-3-glucosidef		0.2–0.7
		Quercetin-glucosidee		2.5
		Rutine[f]		9.0–15
	Anthocyanins	Cyanidin-3-*O*-rutinosidee	mg/100 g FW	0.19
		Cyanidin-3-glucosidef		0.1
		Cyanidin-3-rutinoside[f]		1.5
	Phenolic acids[g]	Chlorogenic acid[g]	mg/100 g FW	0.8–3.0
		Ellagic acid[g]		0.2
		Ferulic acidg		7.19
		Gallic acid[g]		0.1
		p-Coumaric acid[fg]		1.1–9.89
		Protocatechuic acid[g]		1.96
		Syringic acid[g]		1.44
		Vanillic acid[g]		33.35
Peaches	Anthocyanins	Cyanidin-3-glucoside	mg/100 g FW	5.1
	Phenolic acids	Chlorogenic acid	mg/kg FW	151
		Neochlorogenic acid		85.27
Pears	Flavan-3-ols	Catechin	mg/g DW	2.6
		Epicatechin		15.2
	Procyanidins	Arbutin	mg/g DW	7.5
		Caffeoylquinic acid		6.1
		p-Coumarylmallic acid		0.7
Prunes	Flavan-3-ols	Epicatechin	mg/kg DW	7.2
	Flavonols	Cyanidin 3-rutinoside	mg/kg DW	15
		Rutin		3–89
	Phenolic acids	Caffeic acid	mg/kg DW	1–35h
		Chlorogenic acid		67–562h

Dried fruit	Types of phenolic compounds		Unit	Content
		Neochlorogenic acid		928–3045h
		p-Coumaric acid		2–43h
		Protocatechuic acid		0.5–2.0h
Raisins	Anthocyanins[i]	Cyanidin-3-glucoside	mg/L	1.16–1.56
		Delphinidin-3-glucoside	mg/L	1.45–1.75
		Malvidin-3-glucoside		47.2–74.1
		Pelargonidin-3-glucoside		0.87–1.04
		Peonidin-3- glucoside		20.0–22.7
		Petunidin-3-glucoside		9.34–11.4
		A-type vitisin of malvidin-3-acetylglucoside		0.44–0.53
		B-type vitisin of peonidin-3-glucoside		0.38–0.75
		Vitisin A		
		Vitisin B		0.11–0.12
		Cyanidin-3-acetylglucoside		0.76–0.84
		Delphinidin-3-acetylglucoside		0.67–0.98
		Malvidin-3-acetylglucoside		0.39–0.40
		Malvidin-3-caffeoylglucoside		14.3–16.7
		Peonidin-3-acetylglucoside		1.56–1.73
		Petunidin-3-acetylglucoside		3.81–5.57
		Petunidin-3-caffeoylglucoside		5.17–6.33
		Petunidin-3-coumaroylglucoside		2.86–2.95 2.36–2.54
	Flavonols[j]	Isorhametin 3-*O*-glucoside	mg/100 g FW	1.25
		Kaempferolk		0.06[k]
		Quercetin-3-*O*-rutinoside		0.71
		Quercetin-3-*O*-glucuronide		0.06
		Quercetin-3-*O*-glucoside		2.1
		Quercetin		0.29
	Phenolic acids	Caffeic acid	mg/100 g FW	0.63
		Cinnamic acid		0.16
		Ferulic acid		0.32
		Gallic acid		0.69
		p-Coumaric acid		0.36
		p-Hydroxybenzoic acid		0.23
		p-Hydroxyphenylacetic acid		0.12
		Protocatechuic acid		0.44
		Syringic acid		0.34
		Vanillic acid		1.21
		3,4-Dihydroxyphenylacetic acid		0.1

Dried fruit	Types of phenolic compounds		Unit	Content
	Flavonols[l]	Quercetin	μg/g DW	253
		Rutin		44.76
	Phenolic acids[l]	Gallic acid	μg/g DW	1.59
		Ferulic acid		17.37
		p-Coumaric acid		8.6
		Salicylic acid		61.23
		Syringic acid		17.87
		3,4-Dihydroxybenzoic acid		510.94

a *Cafona* variety from Italy where data are obtained from Madrau et al. (2009).

b Unknown from Turkey where data are obtained from Erdog˘an and Erdemog˘ lu (2011).

c Three varieties from Oman where data are obtained from Al-Farsi et al. (2005).

d Unknown from Spain where data are obtained from Slatnar et al. (2011).

e Autumn free cultivar from Spain where data are obtained from Vallejo et al. (2012).

f *Sarılop* and *Bursa Siyahı* varieties of yellow and purple figs from Turkey where data are obtained from Kamiloglu and Capanoglu (2015).

g Unknown from Turkey where data are obtained from Aegean Exporter's Associations (2009).

h Range (minimum – maximum) where data are obtained from Del Caro et al. (2004); Donovan et al. (1998); Kayano et al. (2002); Kayano et al. (2004).

i *Merlot* and *Syrah* varieties from Spain where data are obtained from Marquez et al. (2012).

j California variety where data are obtained from Williamson and Carughi (2010).

k Data are obtained from Chiou et al. (2007).

l Chinese Thompson seedless where data are obtained from Meng et al. (2011). DW, dry weight; FW, fresh weight.

PHENOLIC ACIDS

The reported phenolic acids determined in selected dried fruits are given in Table 2, while the structures of the representative phenolic acids in selected dried fruits are shown in Figure 8.13. Phenolic acids such as hydroxycinnamic and hydroxybenzoic acids have been reported in selected dried fruits, except in dried pears (Ferreira et al., 2002; Silva, Shahidi, and Coimbra, 2013). In raisins, the most abundant phenolic acids are caftaric and coumaric acids (Williamson and Carughi, 2010). In addition, Californian and Chinese raisins have been reported to contain gallic, vanillic, syringic, ferulic, chlorogenic, 3,4-dihydroxybenzoic, cinnamic, protocatechuic, and phloretic acids (Meng et al., 2011; Parker,Wang, Pazmiño, and Engeseth, 2007;Williamson and Carughi, 2010; Zhao and Hall, 2008). However, Greek raisins contained vanillic, caffeic, gallic, syringic, *p*-coumaric, and protocatechuic acids (Chiou et al., 2007). Raisins of nine grape genotypes from Xinjiang Province of China were studied by Meng et al. (2011), where 3,4-dihydroxybenzoic acid was the most predominant phenolic acid in a majority of raisin samples. Besides, salicylic acid was found in raisins from nine Chinese varieties (Meng et al., 2011), which was different from that reported by Williamson and Carughi (2010), Zhao and Hall (2008), and Parker et al. (2007). Four free phenolic acids (protocatechuic, vanillic,

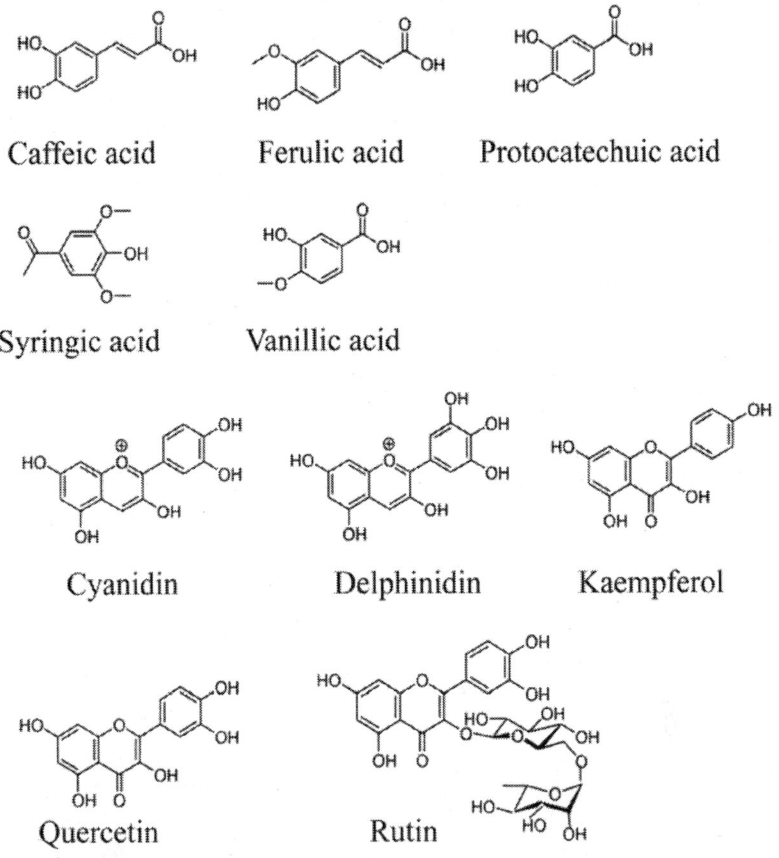

Figure 8.13: Chemical structures of representative phenolic compounds (phenolic acids, flavonols, and anthocyanidins) reported in dried fruits.

syringic, and ferulic) and nine bound phenolic acids [five of which are hydroxylated derivatives of benzoic acid (e.g., gallic, protocatechuic, *p*-hydroxybenzoic, vanillic, and syringic acids) and four of which are cinnamic acid derivatives (e.g., caffeic, *p*-coumaric, ferulic, and *o*-coumaric acids)] have been reported in fresh and sun-dried Omani dates of three native varieties (Al-Farsi, Alasalvar, Morris, Baron, and Shahidi, 2005). Meanwhile, seven phenolic acids (free and bound), four of which being hydroxylated benzoic acid derivatives (gallic, protocatechuic, syringic, and vanillic acids) and three being cinnamic acid derivatives (chlorogenic, ferulic, and *p*-coumaric acids), have been identified inTurkish dried figs. Bound phenolics were most abundant in dried figs where their amounts weremuch higher than those of free phenolics (Aegean Exporter's Associations, 2009; Al-Farsi et al., 2005). However, Vallejo et al. (2012) found only one phenolic acid (chlorogenic acid) in Spanish dried fig cultivars at 1.4–2.0mg/100 g edible portion. Prunes are unique in their combination and overall content of polyphenols (Neveu et al., 2010; Vrhovsek, Rigo, Tonon, and Mattivi, 2004) and are rich in hydroxycinnamic acids, such as neochlorogenic, cryptochlorogenic, and chlorogenic acids (Neveu et al., 2010; Piga, Del Caro, and Corda, 2003). Prunes also contain quinic acid that is metabolised to hippuric acid, which may help alleviate urinary tract infection (Fang, Yu, and Prior, 2002; Kayano, Kikuzaki, Fukutsuka, Mitani, and Nakatani, 2002). Another study reported the presence of different caffeoylquinic acid isomers in prunes (Kayano et al., 2004). Rutin and chlorogenic acid have also been identified as the main phenolic compounds responsible for the *in vitro* anti-cancer (lung, breast, liver, and colon) activities of prunes, as recently shown by Li et al. (2013).

PHYTOESTROGENS

Dietary phytoestrogens present in dried fruits have attracted much interest due to their potential protective effects against various disease conditions such as cancer, cardiovascular disease (CVD), osteoporosis, and menopausal symptoms (Bradford and Awad, 2007; John, Sorokin, and Thompson, 2007).They comprise three major classes: isoflavones, lignans, and coumestan (John et al., 2007). Some dried fruits, such as apricots, currants, dates, prunes, and raisins, contain phytoestrogens [e.g., isoflavones (formononetin, daidzein, genistein, and glycitein), lignans (e.g., matairesinol, lariresinol, pinoresinol, and secoisolariciresinol), and coumestan (coumestrol)]. The structures of the representative phytoestrogens of dried fruits are shown in Figure 8.14. Apricots contain the highest amount of phytoestrogens (445 μg/100 g) among dried fruits, followed by dates (330 μg/100 g), prunes (184 μg/100 g), and raisins (30.2 μg/100 g) (Table 8.9) (Thompson, Boucher, Liu, Cotterchio, and Kreiger, 2006). Dried fruits contain a higher concentration of lignans (ranging from 22.0 to 401 μg/100 g) than isoflavones (ranging from 4.2 to 39.8 μg/100 g) (Table 8.9). Coumestan, expressed as coumestrol, is generally present in low amounts in dried fruits (Thompson et al., 2006). Dried apricots contain daidzein, genistein, and biochanin A in trace amounts (Kuhnle et al., 2009; Liggins et al., 2000). Meanwhile, dried figs contain a relatively low amount of isoflavones (5.97 μg/100 g) compared to other dried fruits, except dried apricots (4.27 μg/100 g) (Liggins et al., 2000). However, detailed quantitative analysis on different classes of phytoestrogens in different forms and varieties of other dried fruits remains unexplored. Thus, more research should be conducted to fill this knowledge gap.

Carotenoids

Five carotenoids, namely, α-carotene, β-carotene, β-cryptoxanthin, lutein, and zeaxanthin, have been reported in some dried fruits. Of these, β-carotene, which acts as provitamin A, is most abundant in dried apricots (2163 μg/100 g), followed by peaches (1074 μg/100 g), and prunes (394 μg/100 g). Lutein + zeaxanthin (559 μg/100 g) and β-cryptoxanthin (444 μg/100 g) are detected only in dried peaches. Dates are the third richest source of carotenoids after dried apricots and peaches (Table 1) (Al-Farsi and Lee, 2008; USDA, 2014). Dates can be considered a moderate source of carotenoids (Vayalil, 2012). No carotenoids have been reported in raisins, while small or trace amounts of carotenoids have been found in dried apples, figs, pears, and prunes (Al-Farsi and Lee, 2008; Delgado-Pelayo, Gallardo-Guerrero, and Hornero-Méndez, 2014; USDA, 2014). The lowlevel of carotenoids in dried fruits may be due to the drying process since carotenoids are sensitive to heat (Namitha andNegi, 2010; Rawson et al., 2011). However, USDA (2014) Nutrient Database reported that drying significantly increased carotenoids concentration (β-carotene, β-cryptoxanthin, and lutein) in dried peaches compared to its fresh counterpart. This happens because of the removal of water that concentrates the phytochemicals.

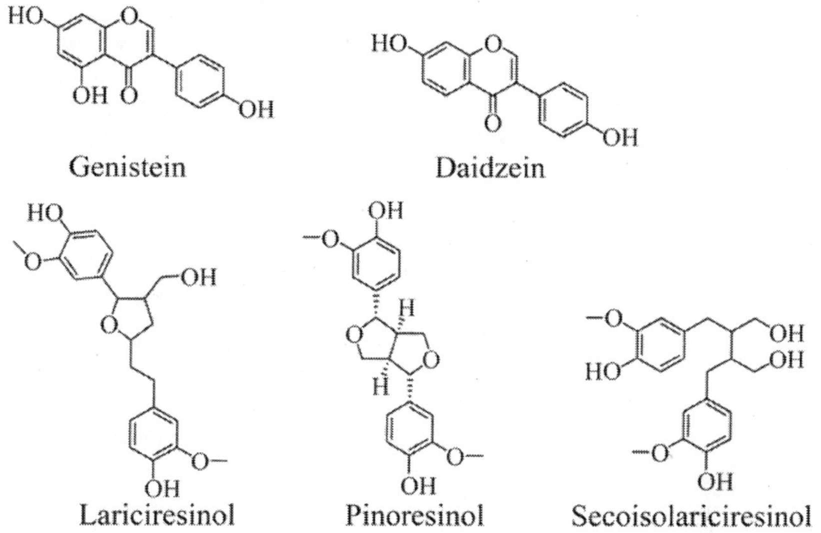

Figure 8.14: Chemical structures of representative phytoestrogens reported in dried fruits.

Antioxidant efficacies of dried fruits

A number of methods have been used to determine the antioxidant activity of selected dried fruits. These include ferric reducing antioxidant power (FRAP), oxygen radical absorbance capacity (ORAC), 2,2′-azino-bis(3-ethylbenzothiazoline-6-sulphonic acid) (ABTS), cupric ion reducing antioxidant capacity (CUPRAC), and 2,2-diphenyl-1-picrylhydrazyl (DPPH) radical scavenging activity. Dried fruits are rich sources of antioxidant polyphenols. The ORAC values for selected dried fruits are given in Table 8.11. Raisins (golden seedless) have the highest ORAC value [10,450 μmol trolox equivalents (TE)/100 g], followed by dried pears, prunes, apples, peaches, figs, and apricots. Interestingly, dates have the lowest ORAC (2387 μmol TE/100 g) and TPC (661 mg GAE/100 g) among nine tested dried fruits (USDA, 2010; Wu et al., 2004). The DPPH radical scavenging activity and the TPC of selected dried fruits (apples, apricots, cranberries, figs, raisins, peaches, pears, and prunes) have been determined by Ishiwata et al. (2004) (Tables 8.9 and 8.11). Apricots contained the highest DPPH radical scavenging activity, followed by prunes and cranberries (Table 8.11). The TPC of dried fruits were highly correlated with DPPH radical scavenging activity (Ishiwata et al., 2004). Pellegrini et al. (2006) determined the total antioxidant activity (using three different *in vitro* assays) of some dried fruits (apricots, figs, prunes, and raisins) in which prunes had the highest value followed by apricots. However, limited information is available on the phenolic profiles and antioxidant components of other dried fruits in the study conducted by Pellegrini et al. (2006). Various studies have reported the bioactive compounds and corresponding antioxidant activities of dried fruits (e.g., peaches and dates) which are always higher than those of their corresponding fresh counterparts (Ishiwata et al., 2004; Rababah et al., 2005; Threlfall, Morris, and Meullenet, 2007; Vinson et al., 2005). This is because antioxidants are concentrated after the dehydration process. Even though there is a loss or change of some phytochemicals during drying, antioxidant activity and TPC of dried fruits remain relatively unaffected during the process, but many of the phenolic compounds remain to be identified (Madrau, Sanguinetti, Del Caro, Fadda, and Piga, 2010).

Table 8.11: Antioxidant activities of selected dried fruits.

Dried fruit	DPPH (mg AAE/ 100 g DW)[a]	ORAC (μmol TE/100 g)[b]	TAC (μmol TE/g)[c]	L-ORACFL (μmol TE/g)[c]	H-ORACFL (μmol TE/g)[c]	FRAP (mmol Fe2+E/kg)[d]
Apples	875	6681	nr	nr	nr	nr
Apricots	3846	3234	nr	nr	nr	36.64
Cranberries	3079	nr	nr	nr	nr	nr
Dates	nr	2387	23.87	0.27	23.6	nr
Figs	1087	3383	33.83	1.83	32.0	14.43
Peaches	1442	4222	67.6[f]	nr	nr	nr
Pears	1301	9496	nr	nr	nr	Nr
Prunes	3114	8578	85.78	1.79	83.99	60.54
Raisins	1346	10450[e]	30.37	0.35	30.02	23.26

a Data are obtained from Ishiwata et al. (2004). b Data are obtained from USDA (2010). c Data are obtained from Wu et al. (2004). d Data are obtained from Pellegrini et al. (2006). e Golden seedless raisins. f Data are obtained from Rababah et al. (2005).

AAE, ascorbic acid equivalents; DPPH, 2,2-diphenyl-1-picrylhydrazyl; DW, dry weight; Fe2+E, Fe2+equivalents; FRAP, ferric-reducing antioxidant power; H-ORACFL, hydrophilic-oxygen radical absorbance capacity; L-ORACFL, lipophilic-oxygen radical absorbance capacity; nr, not reported; ORAC, oxygen radical absorbance capacity; TAC, total antioxidant capacity; TE, trolox equivalents.

The lipophilic and hydrophilic ORAC (L-ORAC and H-ORAC) values for dates, figs, prunes, and raisins are shown in Table 8.11. For those dried fruits, the hydrophilic antioxidants contributed more than 94% to the total antioxidant activity (Wu et al., 2004). Recently, the antioxidant activities of Turkish dried apricots, figs, and raisins have been determined using ABTS, CUPRAC, DPPH, and FRAP assays. The antioxidant activities were in the descending order of raisins > apricots > figs (Table 8.12) (Kamiloglu, Pasli, Ozcelik, and Capanoglu, 2014). Meanwhile, the quality of antioxidants reported as IC50 [expressed in the concentration needed to inhibit *in vitro* oxidation of low-density lipoprotein (LDL) particles by 50%] in dried cranberries, raisins, and prunes were 2.38, 3.45 and 4.38, respectively (Vinson et al., 2005). Antioxidants of dried figs have been reported to protect plasma lipoproteins from subsequent oxidation. In addition, these antioxidants increased the plasma antioxidant capacity up to 4 h after consumption, alleviating the oxidative stress induced after consumption of high fructose corn syrup in a carbonated soft drink (Vinson et al., 2005).

Table 8.12: Changes in TPC and antioxidant activities of selected dried fruits during different stages of *in vitro* gastrointestinal digestion[a]

Dried fruits	Unit	Apricots	Figs	Raisins
TPC		(mg GAE/100 g)		
Initial		211 ± 9.4	123 ± 7.8	1322 ± 59
PG		694 ± 50	575 ± 56	1294 ± 95
IN		133 ± 8.9	85 ± 8.8	104 ± 4.4
OUT		619 ± 15	369 ± 45	1355 ± 129
ABTS		(mg TE/100 g)		
Initial		238 ± 4.4	163 ± 14	2262 ± 186
PG		291± 34	280 ± 9.0	887 ± 119
IN		77 ± 4.3	68 ± 5.3	79 ± 2.0
OUT		380 ± 13	340 ± 64	1071 ± 189
DPPH		(mg TE/100 g)		
Initial		82 ± 3.8	44 ± 4.4	1204 ± 70
PG		281 ± 48	247 ± 25	1818 ± 50
IN		33 ± 2.4	14 ± 1.7	54 ± 2.5
OUT		123 ± 10	81 ± 18	2132 ± 314
CUPRAC		(mg TE/100 g)		
Initial		374 ± 71	152 ± 19	2351 ± 58
PG		627 ± 48	486 ± 42	2274 ± 242
IN		183 ± 35	124 ± 13	214 ± 7.8
OUT		680 ± 67	404 ± 70	2723 ± 243
FRAP		(mg TE/100 g)		
Initial		108 ± 9.7	46 ± 4.4	646 ± 12
PG		218 ± 50	142 ± 9.6	987 ± 69
IN		36 ± 3.5	21 ± 0.5	72 ± 3.7
OUT		219 ± 34	67 ± 0.9	896 ± 131

a Data are obtained with permission from Kamiloglu et al. (2014). ABTS, [2,2′-azino-bis(3-ethylbenzothiazoline-6-sulphonic acid)]; CUPRAC, cupric ion reducing antioxidant capacity; DPPH 2,2-diphenyl-1-picrylhydrazyl; FRAP, ferric-reducing antioxidant power; GAE, gallic acid equivalents; IN, dialysed fraction after intestinal digestion; Initial, as initially determined from sample matrix using 75% aqueousmethanol containing 0.1% formic acid; OUT, non-dialysed fraction after intestinal digestion; PG, compounds remaining after gastric digestion; and TE, trolox equivalents.

Bioaccessibility of bioactive compounds from dried fruits

Dried fruits are popular due to their high nutrient and pharmaceutical values (Alasalvar and Shahidi, 2013a, 2013b). However, limited information is available on the absorption, distribution, metabolism, and excretion of their bioactive compounds in humans, especially about the bioaccessibility and bioavailability of antioxidant polyphenols from dried fruits. Kamiloglu and Capanoglu (2013) have evaluated the total antioxidant activity, total proanthocyanidins, and major phenolic compounds of Turkish dried fig varieties at different phases of simulated gastrointestinal (GI) tract digestion [(after gastric digestion (PG), dialysed fraction after intestinal digestion (IN), and non-dialysed fraction after intestinal digestion (OUT)] using an *in vitro* model. The dialysed fraction after intestinal digestion represents serum bioavailability of dried fruits (Kamiloglu and Capanoglu, 2013).They found that dried figs rendered higher bioavailability of chlorogenic acid compared to fresh figs after intestinal digestion (IN fraction). On the other hand, lower bioavailability of rutin from dried figs after intestinal digestion (IN fraction) was evident. Besides, an increase in the amounts of both cyanidin-3-glucoside and cyanidin-3-rutinoside of dried purple figs was observed after gastric digestion (PG fraction). However, no anthocyanins were detected in dried figs after intestinal digestion (IN fraction), indicating low bioaccessibility of anthocyanins (Kamiloglu and Capanoglu, 2013). The antioxidant activity (determined using ABTS assay) and total proanthocyanidin content of dried figs decreased 83% and increased 140%, respectively, after intestinal digestion (IN fraction). Recently, the impact of GI digestion on the TPC and antioxidant activities of dried apricots, figs, and raisins have been determined, as summarised in Table 8.12 (Kamiloglu et al., 2014). There was an increase in TPC (0.4-4.5-fold) for all samples after the gastric digestion (PG fraction). The antioxidant activities of dried apricots and figs were increased determined using ABTS, CUPRAC, DPPH, and FRAP assays (Table 4). This indicates an increase in the amount of polyphenols after the gastric phase of the *in vitro* digestion process. However, after the pancreatic digestion phase (IN fraction), these antioxidants were degraded by the alkaline pH, giving rise to a significant loss in antioxidant activity after *in vitro* digestion. These results are consistent with a recent study on Turkish dried fig varieties (Kamiloglu and Capanoglu, 2015), demonstrating higher TPC, total flavonoids, and total anthocyanin contents after gastric digestion (PG fraction). However, low bioavailabilities of TPC, total flavonoids, and total anthocyanins were noted after intestinal digestion. Interestingly, total antioxidant activity of dried figs increased after intestinal digestion (Kamiloglu and Capanoglu, 2015). Even though simulated *in vitro* GI tract digestion model cannot directly imitate the human *in vivo* conditions, this model might be helpful for investigating the effect of food matrix and enzymes on polyphenol bioaccessibility. In addition, determination of bioaccessibility by *in vitro* models can be well correlated with the results from human studies and animal models (Kamiloglu and Capanoglu, 2013, 2015; Kamiloglu et al., 2014). More research should be carried out to determine the metabolites of bioactive compounds from dried fruits after urinary excretion coupled together with cellular models such as Caco-2 in order to obtain more qualitatively well-correlated results with human/clinical studies. These data are essential for demonstrating the true biological relevance of these compounds in the context of nutrition and human health (Kamiloglu and Capanoglu, 2015; Kamiloglu et al., 2014).

BENEFICIAL HEALTH EFFECTS OF DRIED FRUITS

Since numerous health-beneficial phytochemicals are found even after processing of fruits, regular intake of dried fruits can exert various health benefits. Dried fruits are essential sources of potassium and dietary fibre with a low amount of fat (0.32–0.93 g/100 g). It has been reported that consuming 40 g (on a per serving basis) of dried fruits supplies 3.8–9.9% of potassium and more than 9% of dietary fibre for recommended dietary allowances (RDA) or adequate intake (AI) for adults (Alasalvar and Shahidi, 2013a, 2013b). High intake of potassium can help in reducing blood pressure (Lichtenstein et al., 2006). High fibre diets are recommended to reduce the risk of developing various NCDs, including type II diabetes, obesity, diverticulitis, colorectal cancer, and CVD (Anderson et al., 2009). Dried fruits are excellent sources of carbohydrates and sugars, such as glucose and fructose (Alasalvar and Shahidi, 2013a, 2013b). Due to their sweetness, dried fruits are expected to exert a high glycaemic index (GI) (70 and above) and insulin response. However, various studies have demonstrated that dried fruits have a low (55 and lower) to moderate (56–69) glycaemic and insulin indices, and hence, good glycaemic and insulin response comparable to those of fresh fruits (Anderson et al., 2011a, 2011b; Esfahani, Lam, and Kendall, 2014; Kim, Hertzler, Byrne, and Mattern, 2008). This could be due to the presence of fibre, phenolic compounds, and tannins that are able to moderate the response. Thus, dried fruits with a low GI may help to reduce the risk of diabetes and are useful for the medical nutritional therapy of hyperglycaemic conditions (Alasalvar and Shahidi, 2013a, 2013b). Dietary phytoestrogens in dried fruits also play beneficial roles in diabetes, bone health, breast cancer, CVD, and metabolic syndrome (Poluzzi et al., 2014). Isoflavones and lignans from dates appear to act through various mechanisms that modulate pancreatic insulin secretion or through antioxidant action. Besides, they may also act *via* estrogen receptormediated mechanisms. Recently, it has been shown that genistein and daidzein from dates play important roles in the regulation of glucose homeostasis (Choi, Jung, Yeo, Kim, and Lee, 2008). Besides, genistein and daidzein also improve plasma triacylglycerol (TAG) and free fatty acid (FFA) concentrations by reducing intestinal cholesterol absorption (John et al., 2007). Hence, it may be presumed that high amounts of phytoesterogens contained in dried fruits may potentially help to maintain normal glucose and lipid metabolism in both healthy populations as well as in obese/diabetic patients (Choi et al., 2008; John et al., 2007).

Various scientific evidence suggest that individuals who consume dried fruits regularly have a lower risk of CVD, obesity, certain types of cancer, type II diabetes, metabolic syndrome, inflammatory bowel disease, and osteoporosis as well as other NCDs (Al-Farsi and Lee, 2008; Anderson and Waters, 2013; Kundu and Surh, 2013; Lever, Cole, Scott, Emery, and Whelan, 2014; Vayalil, 2012). Table 8.13 summarises the health benefits of selected dried fruits and their active components using only well-designed animal or human studies. The health benefits of dried fruits are mainly due to the additive and synergistic combinations of their essential nutrients and phytochemicals (such as anthocyanidins, carotenoids, phytoestrogens, flavan-3-ols, flavones, flavonols, and phenolic acids) related to their antioxidant activities (Liu, 2003; Parker et al., 2007; Rankin, Andreae, Oliver Chen, and O'Keefe, 2008; Vinson et al., 2005). A number of studies demonstrating the health benefits of various dried fruits have been undertaken and these are outlined below.

Table 8.13: Health promoting properties and mechanism of actions demonstrated by selected dried fruits determined in humans and animals.

Health	Dried fruits/ bioactives	Effects Study design	Results/mechanisms
Cardio protective	Cranberries/ phenolics	Protective Randomised, double-blind, and placebo-controlled trial	Improvement in lipid profiles, liver function indices as well as antioxidant defences.
		Artherogenic-induced rats	Improvement in lipid profiles, liver function indices as well as pro-inflammatory biomarkers.
	Raisins	Randomised and controlled intervention trial	Improved lipid profiles and inflammatory biomarkers.
	Apricots/ phenolics	Ischeamia-reperfusion (I/R) rats model	Alleviation of I/R injury in hearts by enhancement of antioxidant defences with reduced lipid peroxidation.
	Prunes/pectin	Apolipoprotein E-deficient mice	Improved arterial atherosclerotic lesion with lowered plasma lipid profiles.
Antipathogenic /antibiotic	Cranberries/ phenolics	Prospective, longitudinal, and followed-up study	Improved antioxidant defence *in vivo*, relief and remission from urological symptoms during regular and frequent urination, and reduced severity and prevalence of urinary problems.
	Cranberry/A-type cranberry proanthocyanidins	Randomised, double blind, controlled, and dose dependent clinical trial	Reduced severity of urinary tract infections with antibiotic effects.
	Apricots	Prospective, longitudinal, and followed-up epidemiological study	Reduced the severity and prevalence of urinary problems.

Health	Dried fruits/ bioactives	Effects Study design	Results/mechanisms
		Randomised and controlled intervention trial	Protection against *Helicobacter pylori*-induced chronic atrophic gastritis, which leads to gastric cancer.
	Prunes	Growing broiler chicks	Beneficial effects in intestinal and immunity.
Anti-diabetic/ hypoglycaemic	Cranberries/ flavonoids	Normal aging rats	Improved antioxidant status, protection and enhancement of pancreatic functions, and maintenance of insulin release in normal aging rats.
	Dates	Randomised and controlled clinical trial	Improved glycaemic and lipid controls in both healthy and diabetic patients.
		Type I diabetic rats	Improved glucose homeostasis in type I diabetes and lipid profiles.
	Raisins (Corinthian)/ *p*hydroxybenzoic acid	Two-armed, randomised, and controlled, prospective intervention trial	Improved diastolic blood pressure and total antioxidant potential in patients with well-controlled type II diabetes.
	Raisins	Randomised and controlled clinical trial	Reduced glycaemia, systolic blood pressure, and cardiovascular risk factors.
	Raisins (Corinthian)	Randomised and controlled clinical trial	Improved glycaemic and insulin emic responses in healthy people and diabetic patients.
	Raisins	Single cross-over clinical trial	Improved glycaemic and insulin response in diabetic patients and college aged students.
	Raisins	Randomised and controlled clinical trial	Improved blood pressure, glycaemic, and insulin response in diabetic patients.
	Raisins	Partial randomised and crossover controlled trial	Improved glycaemic and insulin response in healthy individuals.
	Cranberries	Obese-diabetic rats	Improved insulin emic and glycaemic control with enhanced antioxidant defence.

Health	Dried fruits/ bioactives	Effects Study design	Results/mechanisms
Antioxidative	Apricots, cranberries, dates, figs, plums, and raisins/ phenolics	Randomised and controlled clinical trial	Enhancement in plasma antioxidant capacity, alleviating postprandial oxidative stress after high sugar-drink intake.
	Raisins	Randomised, controlled, and cross-over clinical trial	Protect plasma lipoproteins from oxidation.
			Increased serum antioxidant capacity after postprandial oxidative stress.
		Randomised, placebo-controlled, and crossover clinical trial	Moderate enhancement in serum antioxidant capacity, alleviating pro-inflammatory biomarkers.
	Cranberries	Randomised, double-blind, and placebo controlled trial	Prevention of urinary tract infection as together with oxidative stress *in vivo*.
	Apricots	Methotrexate-induced nephrotoxic rats	Alleviation of nephrotoxicity by up-regulating endogenous antioxidant enzymes, increasing blood antioxidant capacity.
			Prevented apoptosis and glomerulosclerosis in kidney.
Lipid-lowering effects	Peach, apple, and pear/phenolics	Obese-induced rats	Improved plasma and liver lipid profiles as well as antioxidant status (apple).
	Raisins	Randomised, controlled, and cross-over intervention trial	Improved lipid profiles, oxidative defence mechanisms, and colon function.
	Dates	Hyper-choleste-rolaemic rats	Modulate cholesterol absorption or metabolism by improving lipid profiles and body weight.
	Apples	Normal and artherogenic rats	Improved lipid profiles and reduced lipid peroxidation.
	Prunes	Ovariectomy-induced hypercho-lesterolaemic rats	Improved lipid profiles.

Health	Dried fruits/ bioactives	Effects Study design	Results/mechanisms
Bone health/ osteoprotective	Prunes/phenolics	Randomised and controlled clinical trial	Improved biochemical markers of bone turnover and muscular strength in breast cancer survivors.
		C57BL/6J mice	Prevents ovariectomy-induced bone loss while modulating the immune response.
		Randomised, controlled, and clinical trial	Improved bone mineral density in postmenopausal women by partly suppressing the rate of bone turnover.
		Orchidectomised osteopenic rats	Reversed bone loss by improving bone mineral density due to osteoporosis.
		Male osteoporotic rats	Prevents bone loss by increasing bone formation.
		Randomised, controlled, and intervention trial	Improved indices of bone formation in postmenopausal women.
Hepato-protective	Apricots	Ethanol-induced rats	Alleviated ethanol-induced liver damage in rats, enhanced liver antioxidant defences, and prevented liver carcinogenesis.
	Prunes	Randomised, controlled, and clinical trial	Improved levels of liver enzymes for better control of hepatitis.
Appetite and satiety control	Prunes/fibre	Randomised within-subject crossover intervention trial	Better control of hunger, desire, and motivation to eat between snack and meal.
	Raisins	Randomised, controlled, and intervention trial	Altering hormones influencing satiety by reducing hunger and improved lipid profiles.
Metabolic syndrome	Cranberries/ procyanidins	Metabolic syndrome rats model	Improved plasma glucose, lipid profiles, and kidney function associated with metabolic syndrome.
Gastrointestinal function	Prunes	Randomised, controlled, and intervention trial	Managing constipation by improving the frequency of bowel movements, ease on defecation.
Inflammatory bowel disease	Cranberries/ phenolics and fibre	Dextran sulphate sodium induced-colitis rats model	Reduced inflammatory responses and prevented symptoms in colitis.

Improved diet quality and overall health

Epidemiological studies, specifically, the National Health and Nutrition Examination Survey (NHANES) (1999–2004) demonstrate the association of dried fruit consumption and reduced risks of NCDs (Keast et al., 2011). In the diet study, comprising 13,292 participants, dried fruit consumers are defined as those who consume at least one-eighth cup-equivalent of fruit per day. Diet quality was measured using the Healthy Eating Index 2005. Dried fruit consumers had improved MyPyramid food intake, including lower solid fats/alcohol/added sugars intake than non-consumers. The total Healthy Eating Index 2005 score was significantly higher ($p < 0.01$) in consumers (59.3 ± 0.5) than non-consumers (49.4 ± 0.3). The average energy intake of dried fruit consumers was 1038 kJ higher than non-consumers. However, weight, body mass index (BMI), waist circumference, subscapular skinfolds, and risk of overweight were all inversely related to dried fruit consumption. Additionally, diastolic and systolic blood pressures were also inversely associated with dried fruit consumption in this population (Keast et al., 2011). The prevalence and risk for overweight/obesity (56.2 ± 2.3 vs 65.8 ± 0.7; $p < 0.01$; odds ratio, 0.65; and 95% confidence interval, 0.52–0.81) were both lower in dried fruit consumers than non-consumers. In short, dried fruit consumption was associated with improved nutrient intakes (vitamins A, E, and K, magnesium, phosphorus, and potassium), a higher overall diet quality score, and lower body weight/adiposity measures (Keast et al., 2011).

Anti-glaucoma effect

Specific health benefit of dried peach consumption was demonstrated by another cohort study (Coleman et al., 2008). The association between glaucoma and the consumption of specific fruits and vegetables was studied in a cohort of women aged 65 and older, with osteoporotic fractures. The odds ratio for the risk of glaucoma were lowered by 47% in women who consumed at least one serving per week of dried peaches compared to those who consumed less than one serving per month. However, consumption of one, two, or more servings of fresh peaches per week did not have any effect (Coleman et al., 2008). The protective effect of dried peaches may be due to their content of vitamin A or provitamin A carotenoids, especially β-cryptoxanthin, which may become more bioavailable after processing (Coleman et al., 2008; Namitha and Negi, 2010). The authors recommended that randomised controlled trials are needed to determine whether the intake of specific nutrients from dried peaches changes the risk of glaucoma (Coleman et al., 2008).

Anticancer effect

The potential anticancer effects of dried fruits were reported by Mills, Beeson, Phillips, and Fraser (1989) in a cohort study of diet, lifestyle, and prostate cancer risk of approximately 14,000 Seventh-day Adventist men. During a 6-year follow-up period, they found that increased consumption of raisins, dates, and other dried fruits were significantly associated with decreased prostate cancer risk. Moreover, the decreased risk of prostate cancer was not related to exposure to vegetarian lifestyle during childhood, suggesting the potential benefits of dried fruits in the prevention of prostate cancer (Mills et al., 1989).

Lipid-lowering effect

An *in vivo* study was carried out in rats fed on a diet with or without 1% cholesterol and 10% dried peaches, apples, or pears. Dried peach supplementation of a cholesterol-containing diet significantly prevented the increase in plasma and liver lipids, although the values were still higher than those of rats fed cholesterol-free diets (Gorinstein et al., 2002). However, dried peach supplementation had no effect on plasma antioxidant capacity and serum lipid oxidation levels. Meanwhile, dried apples were more effective than dried peaches and pears in alleviating lipid levels and markers of oxidative stress. The authors suggest that this may be due to higher concentrations of phenolic compounds found in dried apples, where the effects on lipid levels may also be dependent on the fibre content, which was high in all dried fruits (Gorinstein et al., 2002).

Antioxidative and anti-inflammation effects

Study of serum antioxidant capacity after consumption of raisins was conducted by Parker et al. (2007). Consumption of raisins for 4 weeks increased serum antioxidant capacity by the second and third week during the study duration. However, antioxidant capacity fell again in the fourth week. The authors postulated that there may be a physiological plateau of 2 or 3 weeks after consistent consumption. At the same time, Cao, Booth, Sadowski, and Prior (1998) reported similar results. It was speculated that the high sugar content of raisins may have affected the antioxidant capacity of blood by inducing postprandial oxidative stress (Parker et al., 2007). A more recent study found that serum antioxidant capacity was modestly increased by daily consumption of raisins, but this did not alter postprandial inflammatory response in these relatively healthy but overweight individuals (Rankin et al., 2008). In another study, feeding sun-dried raisins prior to and during a triathlon to trained athletes significantly lowered the urinary proinflammatory biomarkers compared to feeding of a glucose drink with the same energy content, suggesting the excellent antioxidant capacity of raisins *in vivo* (Spiller, Schultz, Spiller, and Ou, 2002). Meanwhile, the effect of diets containing raisins alone or together with other plant materials with improved blood lipid profiles have also been reported (Bruce, Spiller, and Farquhar, 1997; Bruce, Spiller, Klevay, and Gallagher, 2000). These imply that inclusion of raisins in the daily diet could reduce the risk of CVD by increasing plasma antioxidant capacity, lowering the total and LDL cholesterol concentrations and reducing inflammation (Shahidi and Tan, 2013).

Antibiotic/anti-pathogenic effect

Dried cranberries are particularly effective in reducing reoccurrence and severity of urinary tract infections (UTIs), as already reviewed in the existing literature (Howell, 2007; Howell et al., 2005; Jepson and Craig, 2007; Vasileiou, Katsargyris, Theocharis, and Giaginis, 2013). This approach has attracted increasing attention due to antibiotic resistance which leads to ineffective treatment and persistent reoccurrence of UTIs. In a recent review by Micali et al. (2014), after evaluation of the efficacy/safety ratio of dried cranberry in the prevention of UTIs, the use of dried cranberry in the prevention of recurrent UTIs in young and middle-aged women is strongly recommended. However, evidence of its clinical use among other patients remains to be investigated.

Cardio protection effect

A 1-year randomised cross-over clinical trial was conducted to evaluate the effect of dried apples vs prunes consumption (75 g/day) in reducing CVD risk factors in postmenopausal women (Chai et al., 2012). The authors found that there was no significant difference ($p > 0.05$) between the effects of dried apples and prunes in altering serum levels of cholesterols except total cholesterol after 6 months. However, when within treatment group comparisons are made, consumption of dried apples (about two medium-sized apples) can significantly lower atherogenic cholesterol levels as early as 3 months. Both dried apples and prunes were able to lower serum levels of lipid hydroperoxides and C-reactive protein (CRP). However, serum CRP levels were significantly lower in prunes group compared with the dried apples group at the end of 3 months. The authors conclude that the consumption of dried apples and prunes is beneficial to human health in terms of anti-inflammatory and antioxidative properties.

Bone protection

A recent study compared the efficacies of several dried fruits (prunes, apples, apricots, raisins, and mangoes) in the restoration of bone in an osteopenic ovariectomised (OVX) mouse model (Rendina et al., 2013). As a result, whole body and spine bone mineral density improved in mice consuming prune, apricot, and raisin diets compared to the OVX control mice. However, prunes were the only dried fruits that had an anabolic effect on trabecular bone in the vertebra that prevents bone loss in the tibia. Additionally, restoration of biomechanical properties occurred in conjunction with changes in trabecular bone in spine. Compared to other dried fruits in this study, prunes were unique in their ability to down-regulate osteoclast differentiation coincident with up-regulating osteoblast and glutathione activity (Rendina et al., 2013).The authors suggested that these alterations in bone metabolism and antioxidant status compared to other dried fruits provide insight into prunes' unique effects in bone health. In addition, the bioactive component(s) in prunes remain to be identified as to whether it is a single component or the unique combination of nutrients and phytochemicals in prunes that is responsible for the osteoprotective effects (Rendina et al., 2013).

Anti-diabetic/hypoglycaemic effect

The acute glycaemic and insulin response to dates in diabetic and non-diabetic subjects was carried out by Famuyiwa et al. (1992). Results demonstrated that the well-controlled diabetic patients had significantly lower glucose levels after the consumption of *Sukkari* dates. In addition, stimulation of insulin secretion in healthy volunteers after dates' consumption was 2.7 times less than that of dextrose. The same results were also obtained by Anderson et al. (2011a) using raisins. This shows that dried fruits do not adversely affect the glucose tolerance in healthy individuals. In addition, dried fruits consumption would benefit the control of diabetes in patients

(Anderson et al., 2011a; Famuyiwa et al., 1992). Some recent well-controlled clinical trials have shown the beneficial effects of raisins compared to fresh fruits and processed snacks in improving glycaemic and insulin response in diabetic patients with cardiovascular risk factors (Table 5) (Anderson, Weiter, Christian, Ritchey, and Bays, 2014; Bays,Weiter, and Anderson, 2015; Kanellos et al., 2013, 2014). In summary, well-designed human intervention studies are needed to further validate the health effects of various dried fruits, especially dried peaches, where studies on their potential health benefits are scarce (Table 5). Some limitations from previous studies need to be addressed to demonstrate more affirming results. It has been suggested that future human studies should address subject characteristics, their health status, and other confounding factors before making any reliable conclusions for the health effects of dried fruits (Burton-Freeman, 2010).Various new analytical methods, such as nanotechnology or nutrigenetics, might be promising tools to get a clearer picture about the interactions between diet components from dried fruits and human genetic susceptibility to disease development (Tomás-Barberán and Andrés-Lacueva, 2012).

NOVEL PRODUCT FORMULATIONS AND FUTURE PERSPECTIVES

To the best of our knowledge, limited information is available regarding the functional characteristics and potential applications of various dried fruits and their by-products. Raisins, dried apricots, and dried figs are an indispensable ingredient in breads, cakes, cookies, pies, puddings, tarts, pastries, jams, marmalades, and sugar and confectionery products as a delicious ingredient. They are mixed or added to several cereals and cereal-based products, such as muesli logs, fruit filled cereals, nuggets, yogurts, ice creams, and even in some types of cheeses (Alasalvar, 2013; California Prune Board, 2005; Göncüog˘ lu, Mogol, and Gökmen, 2013; Shahidi and Tan, 2013). Prune puree, obtained from whole prunes by extrusion, in the form of homogenous paste, has a number of applications such as fat replacer for reducing fat content in a series of goods while imparting delicious flavour (California Prune Board, 2005). Besides, the particular composition of prunes also confers them other technological functionalities, such as antioxidant effect in roast beef (Osburn, 2009) and preservative effect in bread, or for inhibiting pathogen growth in extending shelf life in baked goods (Fung and Thompson, 2009). Dried peaches are utilised as pie, tart, and turnover filling, while their powders provide excellent purees, spreads, or glazes, after proper dehydration and preparation (Alvarez-Parrilla, de la Rosa, González-Aguilar, and Ayala-Zavala, 2013). Dried peach by-products have been used as food additives as antioxidants, antimicrobials, colorants, flavourings, and thickeners, where this also applies to the by-products of other dried fruits, such as apricots (Alvarez-Parrilla et al., 2013). More research is needed to enhance the potential functionalities of the by-products of other dried fruits in the future, where these by-products have been shown to contain numerous phyto-chemicals that may be beneficial to human health (Pelentir, Block, Monteiro Fritz, Reginatto, and Amante, 2011).These aspects should be examined to identify other possible value-added uses for human nutrition, in addition to minimising the cost of waste management for the agro-business industry. Utilisation of the by-products as sources for developing functional foods and nutraceuticals may also help to improve the quality of life by preventing the occurrence of diseases and maintaining the wellness of human beings (Galanakis, 2012). One of the major safety concerns with regard to dried fruits is the potential microbial spoilage during storage and the potential health hazard resulting from natural/microbial toxins (Iamanaka, Taniwaki, Menezes, Vicente, and Fungaro, 2005; Iqbal, Asi, and Jinap, 2014; Luttfullah and Hussain, 2011; Masood, Iqbal, Asi, and Malik, 2015). It has been shown that dried fruits can be contaminated with aflatoxins (aflatoxins B1, B2, G1, and G2), ochratoxin-A, kojic acid, and occasionally patulin or zearalenone (Iqbal et al., 2014; Karaca and Nas, 2006; Luttfullah and Hussain, 2011; Masood et al., 2015; Trucksess and Scott, 2008; Van de Perre, Jacxsens, Lachat, El Tahan, and De Meulenaer, 2015). A recent study documented the occurrence of high levels of total aflatoxins (3.28–6.32 μg/kg) and aflatoxin B1 (2.42–4.5 μg/kg) in dried apricots, dates, figs, prunes, and raisins from Pakistan, of which the levels were above the European Union (EU) limits (4 and 2 μg/kg, respectively) (Masood et al., 2015). The same results were also found among five varieties of dates and four date products from Pakistan (Iqbal et al., 2014). The authors have urged the relevant authorities to screen for the contamination of aflatoxins in dried fruits and their products, where strict regulations should be implemented to minimise fungal contamination (Iqbal et al., 2014; Masood et al., 2015). Besides, drying of fresh fruits at high temperatures may generate Maillard reaction products (MRP) due to non-enzymatic browning reactions, which are potentially genotoxic (Rawson et al., 2011). Metabolites of 5-(hydroxymethyl)-2-furfural, an MRP, have been detected in the urine of human volunteers receiving dried prunes or dried prune juices (Prior, Wu, and Gu, 2006). However, Lavelli and Vantaggi (2009) found that Maillard reaction occurs very slowly in dried fruits, most probably due to the low mobility of water and the rate of reaction increases upon a critical water activity level where mobility of water is higher. Hence, the hydroxymethylfurfural content of dehydrated apples was found to be lower than those of apple juice concentrates and apple pomace (Lavelli and Vantaggi, 2009).The difference between the level of MRP detected in dried prunes and dehydrated apples could be due to different samples and different processing methods. Besides, heat-catalysed Maillard reaction, an enzymatic browning reaction, has also been reported to occur in dried fruits (Rawson et al., 2011). To avoid non-enzymatic browning reactions, processing of dried fruits should be carried out at relatively low temperatures. Appropriate processing can prevent browning of dried fruits. For example, pre-treatment of apples with pineapple juice has been shown to reduce the rate and extent of enzymatic browning reaction (Rawson et al., 2011). Therefore, dried fruits should go through routine quality assessment before consumption. Hence, more research is needed to further refine and optimise the drying processes utilised for production of dried fruits. Dried peaches [produced with sulphites (sulphur dioxide or potassium metabisulphite)] are used as preservatives and antioxidants. However, ingested sulphites have been shown to cause several mild to severe and even fatal, adverse effects to the asthmatic population, including broncho-onstriction, urticarial, and anaphylaxis (Alvarez-Parrilla et al., 2013). Therefore, consumption of dried peaches (possibly other dried fruits) containing high levels of sulphites may pose a threat to sensitive individuals. The use of alternative treatments to help preserve the quality of fruits throughout dehydration and storage is recommended

(DiPersio, Kendall, and Sofos, 2004). With the advancement in food processing technologies, a wide variety of dried fruits and their combinations are now available in the shelves of supermarkets, indicating their increased popularity (Vayalil, 2012). There is a great opportunity to develop functional foods from dried fruits evolving and growing at different rates both within and across various countries. However, the development is not spread evenly due to consumer resistance, legislative barriers, ethical variations, cost effectiveness, and technological issues (Sun-Waterhouse, 2011). The future of functional foods based on dried fruits depends on their efficacy in promoting health. Hence, a structured approach for designing and developing dried fruit-based functional finished products should be used to address the technical challenges and their associated solutions during food design, formulation, processing, and storage. In addition, a consumer oriented food product development process should also be used to control the interactions among the targeted bioactive compounds and other food components in dried fruits during processing, handling, and storage. This is the key to ensure that a stable and appealing functional food is produced from dried fruit. In short, controlling the beneficial synergies among food ingredients and food formulation/processing methods has the potential to lead to substantial food innovations for dried fruit based functional foods (Sun-Waterhouse, 2011). The scientific community is expecting more exciting results on the products in the coming years.

Dried fruits, which serve as important healthful snacks worldwide, provide a concentrated form of fresh fruits. They are nutritionally equivalent to fresh fruits in smaller serving sizes, ranging from 30 to 43 g depending on the fruit, in current dietary recommendation in different countries. Daily consumption of dried fruits is recommended in order to gain full benefit of essential nutrients, health-promoting phytochemicals, and antioxidants that they contain, together with their desirable taste and aroma. Recently, much interest in the health benefits of dried fruits has led to many *in vitro* and *in vivo* (animal and human intervention) studies as well as the identification and quantification of various groups of phytochemicals. This information discusses phytochemical compositions, antioxidant efficacies, and potential health benefits of eight traditional dried fruits such as apples, apricots, dates, figs, peaches, pears, prunes, and raisins, together with dried cranberries. Novel product formulations and future perspectives of dried fruits are also discussed. Research findings from the existing literature published within the last 10 years have been compiled and summarised. Dried fruits are nutritionally equivalent to fresh fruits in smaller serving sizes. They have unique combination of taste/aroma, essential nutrients, fibre, and phytochemicals. Dried fruits are important for human health in providing great nourishment and health benefits. More research should be carried out to determine the complete profiles of phytochemicals, such as phenolic acids, flavonoids, phytoestrogens, and carotenoids of other dried fruits in relation to their antioxidant activities or other bioactivities. Considerable opportunities exist in the dried fruit-based functional food products for expansion and innovation. Hence, more sophisticated human intervention studies or clinical trials are needed to validate the health benefits of various dried fruits.

Perceptions of the Health and Convenience Characteristics of Fresh and Dried Fruits

There has been limited research on consumer perception of dried fruits; our results describe and highlight the main differences between consumer perceptions fresh fruits and dried fruits. This study provides a number of interesting findings to improve the product development and marketing of dried fruits.

Fresh fruits compared to dried fruits

As expected, fresh fruits were generally perceived to be healthier than dried fruits. In addition, respondents reported more positive and less negative feelings while eating fresh fruits compared to dried fruits. This finding is consistent with the higher popularity of fresh fruits compared to dried fruits based on the actual consumption of these products. However, for both fresh and dried fruits, the reported feelings were mostly positive. Fresh fruits thus scored higher on the perceptions of health and were associated more with positive feelings. Dried fruits, on the other hand, were perceived as more convenient, especially compared to plums. Thus, dried fruits are generally preferred in situations in which convenience is important, such as when a healthy snack is needed when outside of the home. Dried fruits were also perceived as containing more energy, which suggests opportunities for dried fruits in situations in which a high energy food is desired, such as during sports. In qualitative research, consumers' perception of the healthiness of dried fruits has been ambiguous (Jesionkowska et al., 2007); although dried fruits have not been perceived to be as healthy as fresh fruits, consumers still report positive feelings about dried fruits. Dried fruits were perceived as being a good source of energy, which was also the case for fresh fruits, though to a lesser extent. In addition, while consumers see both fresh fruits and dried fruits as sources of energy, they might be ambivalent about the sugar content. This result suggests that when communicating with consumers, product developers and marketers should communicate about the energy content instead of sugar content.

Health and convenience orientations of consumers

In addition to product characteristics, consumer characteristics should also be considered. With regards to health consciousness, a division between health sacrifice and health preoccupation can be made. The more the respondents reported a willingness to sacrifice for their health, the more positive they were about the health aspects of both fresh and dried fruits and about most of the convenience aspects of both fresh and dried fruits. This group seems to be similar to the health conscious consumers described in literature (e.g., Steptoe et al., 1995). Health reoccupation, on the other hand, was unrelated to most of the convenience perceptions and was associated with lower scores on several perceptions of healthiness. Thus, health preoccupied people seem to be more critical about health and convenience aspects compared to people who sacrifice for their health. Convenience orientation was not consistently correlated with the perceptions of the health and convenience of fresh and dried fruits. Therefore, it might be more difficult to convince this group about the convenience benefits of dried fruits compared to the high health sacrifice group.

LIMITATIONS

This exploratory study in which we considered monthly consumers of fresh fruit and dried fruits may not be seen as representative for the general population, as it is not known yet whether consumers who are less or not familiar with dried fruits share this perception.

When interpreting the results, one should be aware that there are some limitations. First, gathering data by means of a questionnaire represents the intentions of the consumers and may not be representative of how they actually behave in everyday life situations. For example, the perception of fruits or dried fruits in specific situations might be differently perceived than when filling out a questionnaire. Second, although several feelings and associations came forward during the exploratory research, measuring those items by means of a questionnaire was quite difficult. The feelings experienced during consumption in certain circumstances may no longer exist when filling out a questionnaire. Third, although the health consciousness scales provided us with insights, further exploration of these scales might be necessary to better understand the role health consciousness in different countries.

REMARKS AND RECOMMENDATIONS

The perception of fresh fruits differs from the perception of dried fruits. Compared to fresh fruits, the perception of the health characteristics of dried fruits is more complicated; consumers with different orientations have different perceptions that need to be considered when marketers communicate with consumers. Although consumers seem to be aware of the convenience of dried fruits, the health information commonly used and easily recognised by consumers regarding fresh fruits (e.g., fruits are healthy) needs to be adopted for dried fruits. Consumers who sacrifice more for their health seem to be a more relevant target group than those who are preoccupied with their health. This information clearly shows that consumers, except the ones who may score high on measures of health sacrifice, may be difficult to convince about the health and convenience benefits of dried fruits. To identify effective procedures that can induce positive consumer attitudes towards these types of products, a case-specific approach should be considered in which the convenience and health aspects can be mutually strengthened.

B. VEGETABLES

Epidemiological studies have shown that consumption of fruits and vegetables is associated with reduced risk of chronic diseases. Increased consumption of fruits and vegetables containing high levels of phytochemicals has been recommended to prevent chronic diseases related to oxidative stress in the human body (Chu, Sun, Wu, and Liu, 2002). Phenolics in vegetables are present in both free and bound forms. Bound phenolics, mainly in the form of β-glycosides, may survive the human stomach and small intestine digestion and reach the colon intact, where they are released and exert their beneficial effects (Sosulski *et al.*, 1982). A more complete profile of phenolic distributions, including both free and bound phenolics in some vegetables (Figure 8.15) is reported using new and modified methods (Chu *et al.*,

2002). Broccoli contained the highest total phenolic content, followed by spinach, yellow onion, red pepper, carrot, cabbage, potato, lettuce, celery and cucumber. Red pepper had the highest total antioxidant activity, followed by broccoli, carrot, spinach, cabbage, yellow onion, celery, potato, lettuce and cucumber (Chu *et al.*, 2002). Alvarez-Parrilla, de la Rosa, Amarowicz, and Shahidi (2010) found that fresh and processed Jalapeño and Serrano peppers are good sources of phenolics and ascorbic acid, as well as capsaicinoids, and present high antioxidant activity. The phenolics antioxidant index (PAI) was used to evaluate the quality/quantity of phenolic contents of vegetables.With this evaluation, spinach ranked first with the highest PAI (0.979), followed by broccoli, red pepper, carrot, yellow onion, cabbage, potato, celery, lettuce and cucumber (PAI = 0.004) (Table 8.14) (Chu *et al.*, 2002). Antioxidant activity of vegetables is shown in Table 8.15. Chu, Chang, and Hsu (2000) found high amount of flavonoids in both green and purple leaves of sweet potatoes (185.01 and 426.82 mg/kg wet basis, respectively) and the outer leaves of onion (264.03 mg/kg wet basis) (Table 8.15), and also reported that green leaves of sweet potatoes and the outer leaves of onion showed higher reducing power and higher antioxidant activity in a linoleic acid system as compared to cabbage, spinach and potato. Sultana and Anwar (2008) reported that amongst vegetables, kaempferol was the dominant flavonol which was not detected in carrot and garlic. On the basis of ORAC analysis, Ou, Huang, Hampsch-Woodill, Flanagan, and Deemer (2002) reported that green pepper; spinach, purple onion, broccoli, beet, and cauliflower are the leading sources of antioxidants with activities against peroxyl radicals. A Recent study on the antioxidant capacities and total phenolic contents of lipophilic and hydrophilic extracts of 56 commonly consumed vegetables showed that the highest antioxidant capacities and phenolic contents were found in Chinese toon bud, loosestrife, perilla leaf, cowpea, caraway, lotus root, sweet potato leaf, soy bean (green), pepper leaf, ginseng leaf, chives, and broccoli, whilst the values were very low in marrow squash and eggplant (purple) (Deng *et al.*, 2013).

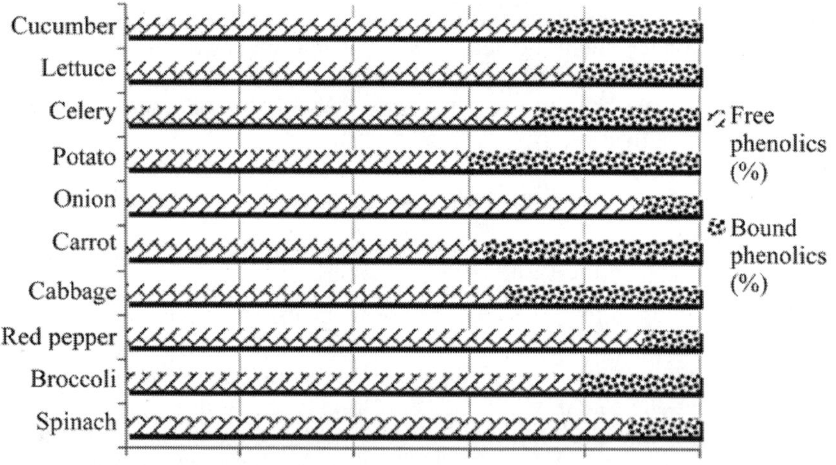

Figure 8.15: Free and bound phenolic content of vegetables (adapted from Chu *et al.*, 2002).

Table 8.14: Antioxidant activity of selected vegetables.

Vegetables	Antioxidant activity[a]	PAI[b]
Spinach	737	0.979
Broccoli	775	0.948
Red pepper	826	0.651
Cabbage	314	0.179
Carrot	750	0.448
Onion	231	0.287
Potato	83	0.030
Celery	90	0.021
Lettuce	49	0.017
Cucumber	21	0.004

Source: Adapted from Chu et al. (2002). [a] μmol of vitamin C eq/100 g.
[b]PAI = free phenolic content/EC_{50} of corrected total antioxidant activity. (EC_{50} of corrected total antioxidant activity = total antioxidant activity - vitamin C antioxidant activity)

Table 8.15: Flavonol and flavone contents (mg/kga) of methanolic extracts of several vegetables after acidic hydrolysis.

Vegetable	Flavonol			Flavone		Total
	Myricetin	Quercetin	Kaempferol	Luteolin	Apigenin	
Sponge gourd	1.29	0.29	-	0.10	0.03	1.71
Sweet potato leaves (green)	38.8	143.78	-	-	185.01	185.01
Sweet potato leaves (purple)	155.87	266.86	-	4.09	–	426.82
Leaf lettuce	0.40	0.40	-	-	0.06	0.63
Chinese kale	0.09	0.65	-	-	0.14	0.90
Cucumber	-	0.01	-	0.09	0.03	0.13
Spinach	0.42	19.60	0.56	-	–	25.02
Water spinach	0.3	1.76	-	0.41	0.05	2.11
Purple cabbage	-	0.24	-	0.16	1.07	1.47
Chinese cabbage	0.14	0.24	0.05	1.18	0.31	1.86
White cabbage	-	0.04	-	0.16	0.92	1.12
Onion (interior)	0.21	26.09	0.56	0.22	0.04	27.11
Onion (outer leaves)	–	258.85	5.18	-	–	264.03
Potato	0.01	0.05	0.50	-	0.02	0.13

Source: Adapted from Chu *et al.* (2000). [a]Wet basis.

a. Brassica vegetables

Brassica vegetables, which include different genus of cabbage (white, red, savoy, swamp, Chinese), broccoli, cauliflower, Brussels sprouts and kale, are consumed all over the world (Podsȩdek, 2007). Broccoli is a source of flavonols and hydroxycinnamoyl derivatives. Price, Casuscelli, Colquhoun, and Rhodes (1998) identified the main flavonol glycosides present in broccoli florets as quercetin and kaempferol 3-*O*-sophoroside. Three minor glucosides of these aglycones were also detected, namely isoquercitrin, kaempferol 3-*O*glucoside and kaempferol diglucoside. The predominant hydroxycinnamoyl acids were identified as 1-sinapoyl-2-feruloylgentiobiose, 1,2-diferuloylgentiobiose, 1,2,2′-trisinapoylgentiobiose, and neochlorogenic acid (Borş *et al.*, 2014; Soengas Fernández, Sotelo Pérez, Velasco Pazos, and Cartea González, 2011; Vallejo, Tomas-Barberan, and Garcia-Viguera, 2003). In addition, 1, 2′-disinapoyl-2-feruloylgentiobiose and 1, 2-disinapoylgentiobiose, 1-sinapoyl-2,2-diferuloyl gentiobiose, isomeric form of 1,2,2′-trisinapoylgentiobiose, and chlorogenic acids were found in broccoli (Fernández-León, Fernández-León, Lozano, Ayuso, and González-Gómez, 2012; Price, Casuscelli, Colquhoun, and Rhodes, 1997; Vallejo *et al.*, 2003). Nielsen, Olsen, and Petersen (1993) showed that cabbage contained a mixture of more than 20 compounds of which seven were identified as 3-*O*-sophoroside-7-*O*-glucosides of kaempferol and quercetin with and without further acylation with hydroxycinnamic acids. In addition, unmodified kaempferol tetraglucosides or their derivatives acylated with either sinapic, ferulic or caffeic acid were found in cabbage leaves (Nielsen, Norbek, and Olsen, 1998). Red pigmentation of red cabbage is caused by anthocyanins. Red cabbage contains more than 15 different anthocyanins which are acylglycosides of cyaniding (Dyrby, Westergaard, and Stapelfeldt, 2001; Podsȩdek, 2007). More recently, Wiczkowski, Topolska, and Honke (2014) identified twenty different anthocyanins with the main structure of cyanidin-3-diglucoside-5-glucosides in red cabbages. Kaempferol and myricetin derivatives were also present in Brassica vegetables, but myricetin was not present in broccoli, white cabbage, purple cabbage, and cauliflower (Podsȩdek, 2007). Apigenin and luteolin as flavones were detected in the hydrolysed extracts of different Brassica vegetables, except for broccoli (Bahorun, Luximon-Ramma, Crozier, and Aruoma, 2004). Chu *et al.* (2000) reported that the levels of flavone were higher than those of flavonol in all tested Brassica vegetables (Table 8.15). Apigenin was the predominant flavone aglycone in these vegetables except Chinese cabbage, where luteolin content was nearly 4-fold higher than that of apigenin. Piironen, Syvaoja, Varo, Salminen, and Koivistoinen (1986) reported that α-tocopherol was the predominant tocopherol in all Brassica vegetables, except in cauliflower, containing predominantly γ-tocopherol. In contrast, Kurilich *et al.* (1999) reported a lower concentration of γ-tocopherol than α-tocopherol in cauliflower (Podsȩdek, 2007). A recent study on raw and cooked broccoli and cauliflower showed that boiling increases the concentrations of β-carotene and vitamin A as well as steaming enhances the concentrations of lutein, zeaxanthin, cryptoxanthin, α-carotene and total carotenoids (dos Reis *et al.*, 2015). Generally, amongst the analysed Brassica vegetables, Brussels sprouts, broccoli, and red cabbage possess the highest antioxidant capacity (Podsȩdek, 2007). Common cabbage had a rather low antioxidant activity. In some cases, however, cabbage and broccoli displayed similar antioxidant activities (Ou *et al.*, 2002; Wu *et al.*, 2004). Brassica vegetables, such as kale, broccoli, cabbage, Brussels sprouts and cauliflower, are characterized by considerable antioxidant activity, due to their high contents of polyphenol compounds and vitamin C. The greatest activity is shown by kale which, in addition to a particularly high content of these compounds, also contains a substantial amount of carotenoids (Sikora, Cies□lik, Leszczyn□ska, Filipiak-Florkiewicz, and Pisulewski, 2009). Cao *et al.* (1996) also reported that kale had the highest antioxidant activity against hydroxyl radicals followed by Brussels sprouts, alfalfa sprouts, beets, spinach and broccoli flowers. Recently, Bacchetti *et al.* (2014) found that supplementation with brassica influenced on serum lipid profile with a significant decrease in total cholesterol, LDL-cholesterol and oxidized LDL. In addition brassica vegetables have been shown to exert a protective effect against many chronic degenerative diseases, including cancer (Herr and Buchler, 2010; Manchali, Murthy, and Patil, 2012; Melega *et al.*, 2013).

b. Bitter melon

Main phenolic constituents in the bitter melon extracts are catechin, gallic acid, gentisic acid, chlorogenic acid and epicatechin (Horax, Hettiarachchy, and Chen, 2010; Kubola and Siriamornpun, 2008). Sathishsekar and Subramanian (2005) found that *in vivo* antioxidant efficacy of bitter melon extracts has highlighted the increased activity of naturally occurring hepatic antioxidant enzymes. Bitter melon possesses antioxidant, anti-tumour, anti-bacterial, anti-inflammatory, anti-viral, anti-obesity, anti-hypertension as well as antihyperglycaemic effects (Ahmad *et al.*, 2012; Chen *et al.*, 2012; Hsu, Fang, Liu, and Chen, 2013; Zhang *et al.*, 2012). More recently, Chang *et al.* (2015) found that the triterpenes 3β,25-dihydroxy-7β-methoxycucurbita-5,23(E)-diene and 3β,7β,25-trihydroxycucurbita-5,23(E)-dien-19-al isolated from bitter melon have insulin-sensitizing and insulin-substitution functions, which are correlated to their effects on inhibiting proteintyrosine phosphatase-1B (PTP-1B) and activating AMP-activated protein kinase (AMPK) inhibitor, respectively.

c. Carrots

Carrots have been ranked 10th in nutritional value amongst 39 fruits and vegetables, and research on carrot health benefits continues (Sharma, Karki, Thakur, and Attri, 2012; Sun, Simon, and Tanumihardjo, 2009).The predominant phenolic acids in carrots are chlorogenic acid, caffeic acid, *p*-hydroxybenzoic acid, ferulic acid, and other cinnamic acid isomers. Chlorogenic acid had the highest content amongst all phenolic acids (Alasalvar, Grigor, Zhang, Quantick, and Shahidi, 2001). Sun, Simon, *et al.* (2009) suggested that purple-yellow or purple-orange carrots had high antioxidant content and capacity. Alasalvar *et al.* (2001) reported that all carrots studied contained mainly hydroxycinnamic acid derivatives, namely 3′-caffeoylquinic acid (neochlorogenic acid), 5′-caffeoylquinic acid (chlorogenic acid), 3′-*p*-coumaroylquinic acid, 3′-feruloyquinic acid, 3′, 4′-dicaffeoylquinic acid, 5′-feruloyquinic acid, 5′-*p*-coumaroylquinic acid, 4′-feruloylquinic acid, 3′, 5′-dicaffeoylquinic acid, 3′, 4′-diferuloylquinic acid and 3′, 5′-diferuloylquinic acid. The presence of dicaffeoylquinic acid, especially in orange carrots may exert a very strong antioxidant activity in the product (Alasalvar *et al.*, 2001; Shahidi and Naczk, 2004). The presence of coumarins, namely 6-methoxymellein and 6-hydroxymellein, was also reported in carrot tissues (Naczk and Shahidi, 2006). These compounds were predominantly accumulated in the periderm tissue of carrot root (Talcott and Howard, 1999). Carrot peels have been reported to be a rich source of phenolic compounds (Kähkönen *et al.*, 1999; Zhang and Hamauzu, 2004). Zhang and Hamauzu (2004) found that phenolic contents in different tissues decreased from peels, phloem to xylem, whilst antioxidants and radical scavenging activities in different tissues decreased in the same order as the phenolic contents. Chantaro, Devahastin, and Chiewchan (2008) suggested that carrot peels could be used as a good raw material to produce a dietary fibre powder with antioxidant potential. Surjadinata and Cisneros-Zevallos (2012) have shown that phenolic profiles are influenced by wounding intensity of carrot tissue and chlorogenic acid and its related isomers are the major phenolics present and synthesized by wounding stress in carrot tissue. In addition, Leja *et al.* (2013) found that red carrots have higher antioxidant activity than orange, yellow and white carrots, and in the season of lower rainfall they accumulated higher amounts of phenolic compounds.

d. Onions

Onions are one of the richest sources of flavonoids in the human diet. Onions possess a high level of antioxidant activity, attributed to their flavonoid constituents, namely quercetin, kaempferol, myricetin, and catechin (Cook and Samman, 1996; Patil, Pike, and Yoo, 1995; Pérez-Gregorio, Regueiro, Simal-Gándara, Rodrigues, and Almeida, 2014). Two major components quercetin monoglucoside and quercetin diglucoside account for 80% of the total flavonoids in onions (Rhodes and Price, 1996). Levels of quercetin glucosides are much higher in onions than in other vegetables (Shahidi and Naczk, 2004). Anthocyanins are only minor components of the flavonoid spectrum in the edible portion of red onion varieties, although the edible bulb of red onions is generally higher in total flavonoids than the bulbs of white or sweet yellow onions due to the presence of anthocyanins (Rhodes and Price, 1996).Yellow onions have been found to contain higher levels of quercetin than red onions, with pink and white onions having the lowest concentrations (Patil *et al.*, 1995). However, Gokce, Kaya, Serce, and Ozgen (2010) suggested that red onions had higher antioxidant activities than yellow and white onions although yellow onions were richest in their phenolic contents. The dominant onion flavonoids are quercetin, quercetin-3-*O*-β-glucoside, quercetin-4′-*O*-β-glucoside, and quercetin-3,4′-di-*O*-β-glucoside and the highest contribution to the antioxidant capacity of onions was provided by quercetin-4′-*O*-β-glucoside (Zielinska, Wiczkowski, and Piskula, 2008). These flavonols are mostly concentrated in the skin. The abaxial epidermis of scales contained a higher level of flavonols than did the mesophyll and approximately 50% of flavonols were detected in the top quarter part of the scales (Naczk and Shahidi, 2006). Onions also contain small quantities of phenolic acids bound to cell walls. Of these, protocatechuic acid was the most abundant phenolic component in the papery scales and was not detected in other tissues. In addition ferulic, *p*-hydroxybenzoic, vanillic and coumaric acids have been found in the papery and fleshy scales (Ng *et al.*, 2000). Anthocyanins, namely peonidin 3-glucoside, cyanidin 3-glucoside and cyaniding 3-arabinoside and their malonylated derivatives, cyaniding 3-laminariobioside and delphinidin and petunidin derivatives (Donner, Gao, and Mazza, 1997), are located in the red onion skin and the outer fleshy layer (Gennaro *et al.*, 2002). Fossen, Andersen, Ovstedal, Pedersen, and Ranknes (1996) reported that 3-(6″-malonyl-3″-glucosylglucoside), 3(3″, 6″dimalonylglucoside), 3-(6″-malonylglucoside), 3-(3″-malonylglucoside), 3-(3″-glucosylglucoside) and 3-glucoside of cyanidin comprise over 95% of total anthocyanins in the whole red onion. Gennaro *et al.* (2002) have demonstrated that cyanidin and delphinidin derivatives constitute over 50% and 30% of total anthocyanins in whole red onion, respectively. Pasteurized 'Recas' paste was chosen to be the most appropriate onion by-product for developing an antioxidant food ingredient amongst all the onion byproducts analysed (Roldán, Sánchez-Moreno, Ancos, and Cano, 2008). It showed several advantages such as a remarkable antioxidant activity, moderately high bioactive composition (total phenols and quercetin), and an excellent antibrowning effect from a technological point of view. Nuutila, Puupponen-Pimiä, Aarni, and Oksman-Caldentey (2003) compared the antioxidant activities of onions and garlic extract and found that onions had clearly higher radical scavenging activities than garlic; red onion is being more active than yellow onion. The skin extracts of yellow and red onion possessed the highest activities. Recently, Albishi, John, Al-Khalifa, and Shahidi (2013a) also reported that pearl, red, yellow and white onion skins contain higher phenolic content (approximately six times) than that of their flesh counterparts as well as exhibited highest antioxidant activity. Furthermore, Wu and Xu (2014) showed the inhibitory effects of onion against α-Glucosidase activity and suggested their use as anti-diabetic agent.

e. Potato

Chlorogenic acid is the predominant phenolic in potato tuber, constituting up to 90% of its total phenolic content (Friedman, 1997). About 50% of the phenolic compounds were located in the peel and the adjoining tissues of potato, whilst the remainder decreased in concentration from outside towards the centre of the tubers (Friedman, 1997; Hasegawa, Johnson, and Gould, 1966). Major phenolics in potato peel are chlorogenic acid, gallic acid, protocatechuic acid, caffeic acid and quercetin (Al-Saikhan, Howard, and Miller, 1995; Al-Weshahy and Rao, 2009; Albishi, John, Al-Khalifa, and Shahidi, 2013b; Mohdaly, Hassanien, Mahmoud, Sarhan, and Smetanska, 2013; Nara, Miyoshi, Honma, and Koga, 2006; Rodriguez de Sotillo, Hadley, and Holm, 1994a, 1994b). Other phenolic compounds in potato include ferulic acid, *p*-coumaric acid as well as small amounts of rutin, quercetin, myricetin, kaempferol, naringenin and other flavonoids (Nara et al., 2006; Reyes, 2005). Purplefleshed potato also contained petunidin- and malvidin-3-rutinoside-5-glycosides acylated with *p*-coumaric and ferulic acid whilst red-fleshed potato had pelargonidin- and peonidin-3-rutinoside-5-glycosides acylated with *p*-coumaric and ferulic acid (Reyes, 2005; Rumbaoa, Cornago, and Geronimo, 2009a). A higher anthocyanidin content and more hydroxylated anthocyanidins (malvidin) can contribute to a higher antioxidant activity of purple-fleshed potatoes (Lachman, Hamouz, Orsak, Pivec, and Dvorak, 2008).Three caffeoylquinic acid isomers, caffeic acid, 3,5-dicaffeoylquinic acid, and N-[2-hydroxy-2-(4-hydroxyphenyl) ethyl] ferulamide were identified as the main phenolic compounds in the peel and flesh of potato (López-Cobo, Gómez-Caravaca, Cerretani, Segura-Carretero, and Fernández-Gutiérrez, 2014). Polyphenolic compounds in potatoes show antioxidative activity in several food systems. Potato peel extract (with petroleum ether), exhibited strong antioxidant activity in soy bean oil during storage which was almost equal to the antioxidant activity of BHA and BHT. However, the level of potato peel extract needed was 8–12 times higher than that of BHA and BHT to control the development of rancidity during storage of cooking oils at elevated temperatures (Rehman, Habib, and Shah, 2004). In related studies, Onyeneho and Hettiaachchy (1993) evaluated the abilities of freeze-dried extracts from six potato peel varieties to prevent soybean oil oxidation and confirmed their strong antioxidant activities.

f. Sweet potato

Oki *et al.* (2002) identified chlorogenic, isochlorogenic and cinnamic acids, and cyanidin and peonidin aglycones, as the phenolic compounds present in sweet potatoes. Philpott, Gould, Markham, Lewthwaite, and Ferguson (2003) found that hydroxycinnamic acid was the major antioxidant component of sweet potatoes. Zhan (1996) found that sweet potato extract had a high antioxidative activity resulting from the presence of phenolic acids, including chlorogenic, isochlorogenic, 4-*O*-caffeoylquinic and neochlorogenic acids. Anthocyanins from sweet potatoes were shown to possess potent antioxidative activity *in vitro* (Oki *et al.*, 2002; Philpott, et al., 2003; Stintzing, Stintzing, Carle, Frei, and Wrolstad, 2002). Another study found that phenols, flavonoids, and anthocyanins of sweet potato flours were positively correlated with the DPPH radical scavenging effects (Huang, Chang, and Shao, 2006). Kano, Takayanagi, Harada, Makino, and Ishikawa (2005) reported that anthocyanin from purple sweet potato had a better radical scavenging activity than that of red cabbage, grape skin, elderberry and purple corn. Furuta, Suda, Nishiba, and Yamakawa (1998) reported that the purple-fleshed sweet potato cultivars had a higher radical-scavenging activity than those with white, yellow or orange flesh. The antioxidant activity of the sweet potato extract was due to a synergistic effect of phenolic components with amino acids (Hayase and Kato, 1984; Shih, Kuo, and Chiang, 2009). Sweet potato extracts showed a better radical scavenging activity, reducing power and oxidation inhibition than α-tocopherol and iron-chelating capacity than EDTA (ethylenediaminetetraacetic acid), which makes sweet potato a viable alternative source for antioxidants (Rumbaoa, Cornago, and Geronimo, 2009b).The leaves of sweet potato serve as the rich source of phenolics. Polyphenolic acids in sweet potato leaves namely 3,5-di-*O*-caffeoylquinic acid > 4,5-di-*O*caffeoylquinic acid > chlorogenic acid (3-*O*-caffeoylquinic acid) > 3,4-di-*O*-caffeoylquinic acid > 3,4,5-tri-*O*-caffeoylquinic acid > caffeic acid. Of these 3, 4, 5-tri-*O*-caffeoylquinic acid and 4,5-di-*O*-caffeoylquinic acid are the predominant phenolics in sweet potato (Islam *et al.*, 2002). In addition, sweet potato leaves also have been shown to possess strong antioxidant activity (Hue, Boyce, and Somasundram, 2012).

g. Beetroot

Ninety-two different phenol-containing plant extracts were recently screened and high antioxidant activity and high total phenolic content were found in berries, especially aronia and crowberry. Beetroot peel was shown to have the second-highest dry weight concentration of total phenols (Kähkönen et al., 1999; Vinson, Yong, Su, and Ligia, 1998). Kujala et al. (2001) isolated a highly unstable phenolic compound from red beetroot peels (*Beta vulgaris*), and proposed its structure to be 5,5′,6,6′-tetrahydroxy-3,3′-biindolyl, a dimer of 5,6-dihydroxyindole. Extra interest in 5,6-dihydroxyindoles has arisen from the recent recognition of their exceptional radical scavenging and photoprotective abilities, which makes them amongst the most effective endogenous antioxidants (Mee, Lee, Baldwin, and Cowley, 2004). Ravichandran *et al.* (2013) showed that different processing conditions affect betalains as well as antioxidant activity of beet.They found that betalains content and antioxidant activity increased during processing with vacuum and with certain microwave treatments. Most varieties of red beetroot contain betanin (5-*O*-*β*-dglucopyranosylbetanidin) as the predominant colouring, and this represents 75 to 90% of the total colourant present (Kujala, Loponen, Klika, and Pihlaja, 2000). Other betacyanins found in different parts and cell cultures of red beetroot are betanidin, isobetanidin, isobetanin (5-*O*-*β*-d-glucopyranosylisobetanidin), prebetanin (betanin 6′-*O*-sulphate), neobetanin (5-*O*-*β*-dglucopyranosylneobetanidin), amaranthin, lampranthin I (4-coumaroylbetanin), and lampranthin II (feruloylbetanin) (Alard, Wray, Grotjahn, Reznik, and Strack, 1985; Bokern *et al.*, 1991).The main known ferulic acid ester *β*-d-fructofuranosyl-*α*-d-(6-*O*-[*E*]-feruloylglucopyranoside) was determined in the peels of beetroot (Kujala *et al.*, 2000). In addition, *p*-hydroxybenzoic, *cis*and *trans*-ferulic, *trans*-coumaric, and vanillic acids, as well as *p*-hydroxybenzaldehyde and vanillin have been detected in beetroot. Of these, ferulic acids are found to be the predominant phenolic acids in red beetroot (Ng, Harvey, Parker, Smith, and Walderon, 1998; Shahidi and Naczk, 2004). Sakihama, Maeda, Hashimoto, Tahara, and Hashidoko (2012) have shown that beetroot betalain could inhibit peroxynitrite-mediated tyrosine nitration and DNA strand cleavage. Beetroot juice induced the level of glutathione S-transferase pi, enzyme involved in active metabolites of 7,12-dimethylbenz[a]anthracene (DMBA) detoxification in the mammary gland (Szaefer, Krajka-Kuz□niak, Ignatowicz, Adamska, and Baer-Dubowska, 2014). Recently, beetroot pomace also has been identified as a potential source of nutraceutical due to its antioxidant, antiradical and hepatoprotective activity (Vulic□*et al.*, 2012, 2014).

h. Asparagus

The antioxidant activity of asparagus was the highest amongst 23 vegetables analysed according to their method of inhibition of low-density lipoprotein oxidation and one of the greatest amongst 43 vegetables tested by the β-carotene bleaching method (Tsushida, Suzuki, and Kurogi, 1994; Vinson et al., 1998). Antioxidants in asparagus can protect human beings against diseases, such as cancer and cardio- and cerebrovascular diseases. The major antioxidant of asparagus is rutin (Sun, Tang, and Powers, 2005; Tsushida *et al.*, 1994). It has been established that flavonoids are the most abundant phenolics in green asparagus (Guillen *et al.*, 2008). Fuentes-Alventosa et al. (2008) confirmed that *triguero* asparagus flavonoids are derivatives of three different aglycones, quercetin being the major flavonol, followed by isorhamnetin and kaempferol. In addition, rutin (quercetin-3-*O*rutinoside), nicotiflorin (kaempferol-3-*O*-rutinoside), narcissi (isorhamnetin-3-*O*-rutinoside), and isorhamnetin-3-*O*-glucoside were identified in asparagus. Rodríguez et al. (2005) suggested that asparagus byproducts should be considered as an excellent source of natural antioxidants. A number of studies have shown antioxidant, anti-inflammatory, anti-proliferative and anti-urease activity of asparagus (Maro *et al.*, 2013; Samad *et al.*, 2014; Shah, Khan, Sattar, Ahmad, and Mirza, 2014).

i. Celery and endive

Studies on celery have identified 7-*O*apiosylglucosides and 7-*O*-glucosides of luteolin, apigenin, and chrysoeriol as well as chlorogenic acid, cinnamic acids, coumarins, and their glycosides in tested products (Galensa and Hermann, 1979; Garg, Gupt, and Sharma, 1980). Lin, Lu, and Harnly (2007) reported the presence of 10 naturally occurring glycosylated flavone malonates in celery, Chinese celery, and celery seeds. Three kaempferol conjugates [kaempferol 3-*O*glucoside, kaempferol 3-*O*-glucuronide, and kaempferol 3-*O*-[(6-*O*-malonyl)glucoside]] were identified in endive varieties and the presence and identity of kaempferol 3-*O*-(6-*O*malonyl) glucoside in endive was shown by DuPont, Mondin, Williamson, and Price (2000) for the first time. Of these, kaempferol 3-*O*-glucuronide comprised over 70% of the total flavonoids present (DuPont et al., 2000). Small quantities of hydroxycinnamoyl derivatives such as dicaffeoyl esters have also been detected in endive extracts (Shahidi and Naczk, 2004).

j. Lettuce

Lettuce is a good source of flavonoids. Five quercetin conjugates [quercetin 3-*O*-galactoside, quercetin 3-*O*glucoside, quercetin 3-*O*-glucuronide, quercetin 3-*O*-(6-*O*malonyl) glucoside, and quercetin 3-*O*-rhamnoside] and luteolin 7-*O*-glucuronide were detected in green-leaf lettuce. In addition, two cyanidin conjugates [cyanidin 3-*O*-glucoside and cyanidin 3-*O*-[(6-*O*-malonyl) glucoside] were also identified in red leaf lettuce varieties (DuPont et al., 2000; Pérez-López, Pinzino, Quartacci, Ranieri, and Sgherri, 2014). Heimler, Isolani, Vignolini, Tombelli, and Romani (2007) identified quercetin, kaempferol, luteolin, apigenin and chrysoeriol in lettuce cultivars. Recently, quercetin and luteolin rhamnosyl-hexosides were found in lettuce (Llorach, Martínez-Sánchez, Tomás-Barberán, Gil, and Ferreres, 2008). Red lettuce varieties contained higher levels of flavonoids than did green lettuce varieties (Crozier, Lean, McDonald, and Black, 1997; DuPont et al., 2000; Naczk and Shahidi, 2006). Red, iceberg and romaine lettuce contain caffeoyltartaric, chlorogenic, dicaffeoyltartaric and 3′,5′-dicaffeoylquinic acids (Cantos, Espin, and Tomas-Barberan, 2001; Ferreres, Gil, Castaner, and Tomas-Barberan, 1997; Llorach et al., 2008; Winter and Herrmann, 1986). Wounding induced the accumulation of phenolic compounds in iceberg and romaine lettuce leaf tissues (Kang and Saltveit, 2002). Altunkaya, Becker, Gökmen, and Skibsted (2009) showed that the degree of lipid oxidation in the liposomes in the presence of α-tocopherol, quercetin and ascorbic acid with lettuce extract was significantly lower when oxidation was initiated by free radicals formed in the water phase compared to initiation in the lipid phase. Antioxidants localized at or near the interface of the liposomes such as quercetin and α-tocopherol acted synergistically with lettuce extract as an antioxidant, whilst the hydrophilic antioxidant ascorbic acid showed no synergism.

k. Spinach

The flavonoids jaceidin 4′-glucuronide; 5, 3′, 4′-trihydroxy-3-methoxy-6:7-methylenedioxyflavone 4′-glucuronide and 5, 4′-dihydroxy-3, 3′-dimethoxy-6: 7-methylenedioxyflavone 4′-glucuronide was identified from *Spinucicr oleraceu* L. leaves (Aritomi and Kawasaki, 1984). Ferreres, Castaner, and Tomas-Barbern (1997) also isolated and identified five novel flavonoids from alcoholic extracts of spinach leaves. The new compounds (Figure 8.16) were identified as spinacetin 3-*O*-*β*-D-glucopyranosyl (1→6)-[*β*-D-apiofuranosyl(1→2)]-*β*-D glucopyranoside, patuletin 3-*O*-*β*-D-(2'feruloylglucopy ranosyl)(1→6)-[*β*-D-apiofuranosyl(1→2)]-*β*-D-glucopyranoside, spinacetin 3-*O*-*β*-D-(2′-p-coumaroylglucopyranosyl)(1→6)-[*β*-Dapiofuranosyl(1→2)]-*β*-D-glucopyranoside, spinacetin 3-*O*-*β*-D-(2′feruloylglucopyranosyl) (1→6)-[*β*-D-apiofuranosyl(1→2)]-*β*-D-glucopyranoside and spinacetin 3-*O*-*β*-D-(2'feruloyl glucopyranosyl)(1→6)-*β*-D-glucopyranoside. In addition, four compounds were identified as glucuronic acid derivatives of flavonoids, three components as *trans* and *cis* isomers of *p*-coumaric acid, and other components as *meso*-tartarate derivatives of *p*-coumaric acid (Bergman, Varshavsky, Gottlieb, and Grossman, 2001).The water soluble antioxidants isolated from spinach and its isolated component, glucorinated flavonoid (GF), exhibited strong antioxidant activities (Bergman, Perelman, Dubinsky, and Grossman, 2003). The occurrence of two major dietary flavonols, namely quercetin and myricetin, was also reported in spinach leaves (Chu et al., 2000). Recently, the presence of three flavonols (quercetin, kaempferol, myricetin) and two flavones (apigenin, luteolin) was reported in the extract of fresh spinach leaves (Dehkharghanian, Adenier, and Vijayalakshmi, 2010). Highest kaempferol levels in spinach amongst the 9 vegetables analysed in the present study are in good agreement with the proclamation that leafy vegetables are a good source of kaempferol (Sultana and Anwar, 2008). Lee, Lee, Lee, Park, and Choe (2002) reported that spinach powder can reduce lipid oxidation in fried products during storage in the dark. Bergman et al. (2001) showed that the antioxidant potential of the purified aromatic polyphenol fractions of spinach was similar to that of EGCG and vitamin E and even superior to that of BHT. Recently, the antioxidants from spinach leaves were combined with selected commercial polyphenolic antioxidants, such as ferulic acid, caffeic acid, and epigallocatechin-3-gallate (EGCG), and found that synergistic activity reduced reactive oxygen species generation. These findings demonstrated the importance of using antioxidant 'cocktails' which may enhance the beneficial effects in alleviating many disease conditions, including cardiovascular disease and cancer (Hait-Darshan, Grossman, Bergman, Deutsch, and Zurgil, 2009).

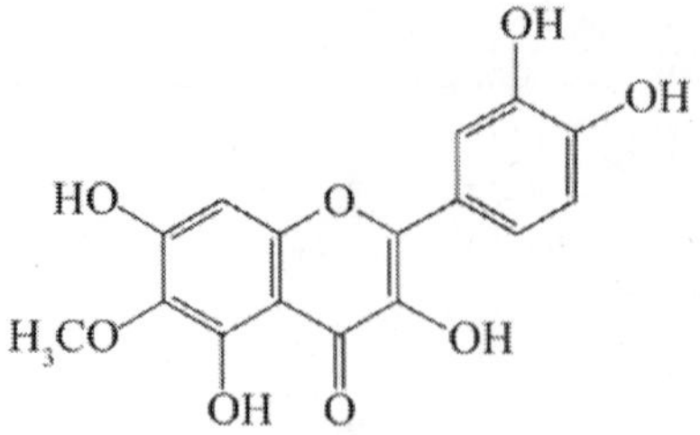

Patuletin 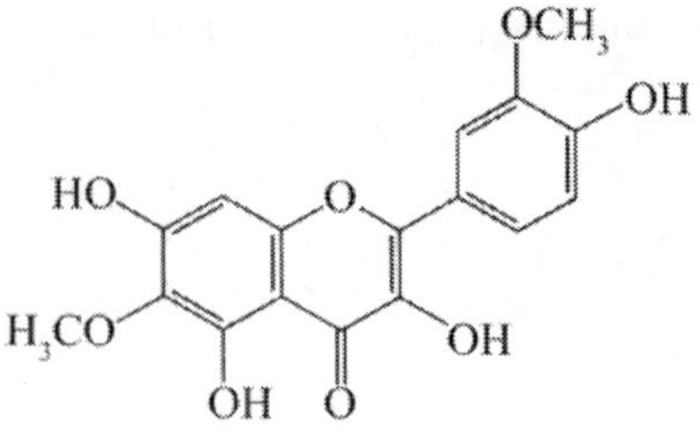 Spinacetin

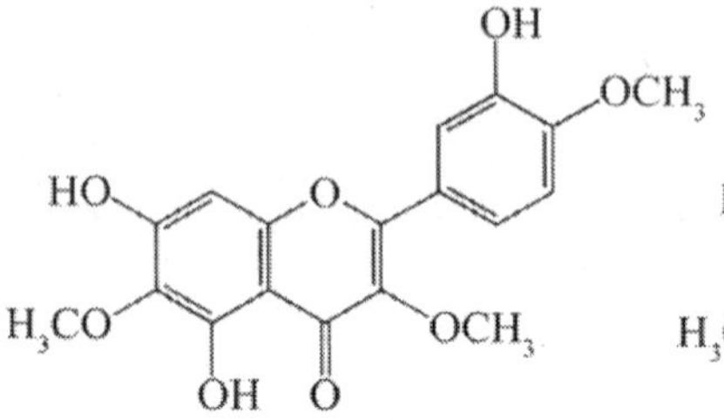

Jaceidin 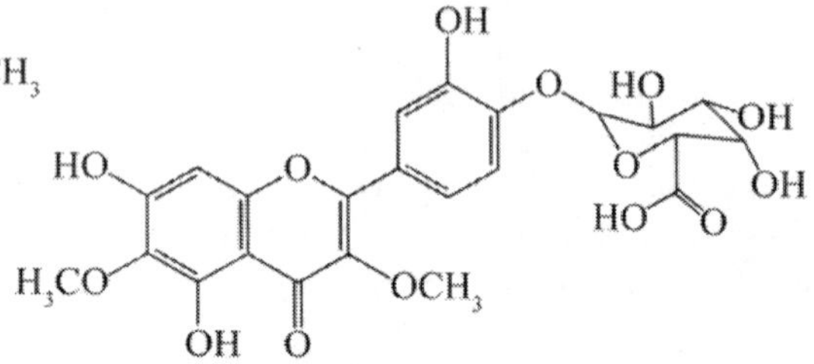 Spinatoside

Figure 8.16: Chemical structures of some phenolic compounds found in spinach.

l. Tomato

Tomato is a reservoir of diverse antioxidant molecules, such as ascorbic acid, carotenoids, flavonoids and phenolic acids (George, Kaur, Khurdiya, and Kapoor, 2004). The presence of flavonoids in tomato is also important in conferring antioxidative health benefits (Takeoka et al., 2001; Vinson et al., 1998). In tomatoes, vitamin E is present mainly within the seeds, and α-tocopherol contribution to the antioxidant properties of tomato is minor (Raffo, Malfa, Fogliano, Maiani, and Quaglia, 2006). Quercetin conjugates are the predominant form of flavonols found in tomatoes, but lower amounts of kaempferol conjugates and traces of free aglycons have also been detected (Crozier *et al.*, 1997). The main quercetin conjugate was identified as rutin (quercetin 3-rhamnosylglucoside) (Stewart et al., 2000). Immature green tomato fruits contain a high level of chlorogenic acid in the pericarp and pulp. The level of chlorogenic acid rapidly declines as the colour of the fruit changes from green to pink and then to red (Buta and Spaulding, 1997; Fleuriet and Macheix, 1981). George *et al.* (2004) suggested that cherry tomatoes, particularly variety 818 cherry, have the highest content of antioxidants (lycopene, ascorbic acid and phenols) and a higher antioxidant activity than larger size tomato cultivars. Along with a diet rich in other plant produce, tomatoes, and their skins and seeds, could play an important role in improving antioxidant intake in the human diet (Toor and Savage, 2005). Recently, Valdez-Morales, Espinosa-Alonso, Espinoza-Torres, Delgado-Vargas and Medina-Godoy (2014) found that the peel and seeds of different tomato types (grape, cherry, bola and saladette type) possess antioxidant and antimutagenic activities.

C. OILSEEDS AND PLANT OILS

Oilseeds constitute a major source of plant antioxidants. Amongst oilseeds of particular interest in recent years are soybean, canola, flax, borage and evening primrose because of the perceived beneficial health effects related to their oil component or their potential by-products. Canada is a major producer and exporter of several oilseeds, including canola and flaxseed. Oilseeds contain a variety of phenolic compounds with varying antioxidant activities (Shahidi, Wanasundara, and Amarowicz, 1995). The type of antioxidant present in each seed is different (Shahidi, 2000b). Total phenolic acids in some oilseeds are shown in Table 8.16.

Table 8.16: Total phenolic acids in some oilseed flours[a]

Phenolic acid	Soybean	Flax	Sesame	Cottonseed[b]	Sunflower	Peanut
p-Hydroxybenzoic	13.9	2.6	Trace	1.1	7.4	2.0
Vanillic	–	Trace	–	–	0.8	–
Syringic	28.9	0.4	–	–	3.6	–
trans-p-Coumaric	9.4	6.1	7.2	4.3	5.6	146.4
trans-Ferulic	15.7	37.6	5.7	4.5	7.2	16.2
trans-Caffeic	6.0	5.3	9.8	0.5	979.1	2.8
trans-Sinapic	–	29.1	–	–	–	8.1
Total	73.6	81.1	22.7	10.4	1003.7	175.5

Source: Adapted from Shahidi and Naczk (2004). [a]Milligram per 100 g of flour. [b]Glanded.

a. Soybean

In soybean oil, the active antioxidant is tocopherol (mainly α-tocopherol) and, to a lesser extent, γ-tocopherol (Schuler, 1990). Soybean flour and other soybean derivatives are good sources of a large variety of antioxidant compounds belonging to the family of isoflavone glycosides and their derivatives (Figure 8.17), as well as phospholipids, tocopherols, amino acids and peptides (Shahidi and Naczk, 2004). It has been reported that 99% of the isoflavones are present as glycosides, of which 64% are genistein, 23% daidzein and 13% glycitein 7-*O*-β-glycoside (Naim, Gestetner, Zilkah, Birk, and Bondi, 1973). Phenolic acids, namely syringic, vanillic, caffeic, ferulic, *p*-coumaric and *p*-hydroxybenzoic acids, possessing antioxidative activity, have also been found in soybeans (Arai, Suzuki, Fujimaki, and Sakrai, 1966; Hammerschmidt and Pratt, 1978; Pratt and Birac, 1979). Amongst polyphenolic antioxidants extracted with methanol or water from dried and fresh soybean, chlorogenic, isochlorogenic, caffeic and ferulic acids were identified, in addition to isoflavones (Pratt and Birac, 1979). Juan and Chou (2009) reported that fermentation enhanced the total phenolic and flavonoid contents as well as the antioxidant activity of black soybean extract due to the liberation of isoflavonoids from their corresponding glycosides by acids formed during the fermentation (Shahidi and Naczk, 2004). The antioxidant activity of soy isoflavones in a β-carotene/linoleate model system in the increasing order was glycetin, diadzein, genistein, quercetin and

6, 7, 4′-trihydroxyisoflavones, respectively. Meanwhile, for the phenolic acids, the increasing order of antioxidant activity was *p*-coumaric, ferulic, chlorogenic and caffeic acids, respectively (Pratt and Birac, 1979; Shahidi and Naczk, 2004).

Compound	R_1	R_2	R_3	R_4
Daidzein	H	H	H	H
Formononetin	CH_3	H	H	H
Genestein	H	H	H	OH
Prunetin	H	CH_3	H	OH
4′,6′,7′-trihydroxyisoflavone	H	H	OH	H
Glycitein (7-*O*-glucoside)	H	Gluc	OCH_3	H

Figure 8.17: Chemical structures of isoflavone and isoflavone glycosides of soybean.

b. Rapeseed and canola

Canola seeds, canola meal and their by-products are rich in phenolics, including caffeic, cinnamic, *p*-coumaric, ferulic, gentisic, *p*-hydroxybenzoic, salicylic, sinapic and syringic acids (Shahidi and Naczk, 2004) with high radical-scavenging activities. Total phenolic content in rapeseed flour is much higher than other oleaginous seeds, exceeding about 28 times the level of phenolics in soybean flour (Table 8.16) (Shahidi and Naczk, 1992, 2004). Phenolic acids and their derivatives, as well as soluble and insoluble tannins, are the predominant phenolic compounds found in canola and rapeseed (Naczk, Amarowicz, Pink, and Shahidi, 2000; Naczk, Amarowicz, Sullivan, and Shahidi, 1998; Naczk, Nichols, Pink, and Sosulski, 1994; Wanasundara and Shahidi, 1994). According to Nowak, Kujawa, Zadernowski, Roczniak, and Kozlowska (1992), amongst oilseeds, rapeseed contains the highest amount of phenolic compounds. The most significant phenolic compounds in rapeseed are sinapic acid derivatives such as sinapine; the choline ester of sinapic acid (constitutes 80% of the total phenolic compounds). Sinapic acid in rapeseed can also exist as a glucosidic ester, glucopyranosyl sinapate (Amarowicz and Shahidi, 1994). Shahidi and Naczk (2004) reported the presence of two flavonoid glucosides in rapeseed meal, namely, 3-(*O*-sinapoyl sophoroside)-7-*O*-glucoside of kaempferol and 3-(*O*-sinapoyl glucoside)-7-*O*-sophoroside of kaempferol. Canolol (4-vinylsyringol) is one of the main phenolic compounds present in processed canola/rapeseed oil from the precursor sinapic acid (Huidrom and Thiyam-Holländer, 2012). Wanasundara and Shahidi (1994) reported that the antioxidant activity of ethanolic extracts of canola meal in canola oil was equivalent to that of TBHQ and stronger than that of BHA, BHT, and BHA/BHT/monoacylglycerol citrate (MGC). The most active component of these extracts was 1-*O*-*β*-D-glucopyranosyl sinapate (Wanasundara, Amarowicz, and Shahidi, 1994). Shahidi, Wanasundara, Amarowicz, and Naczk (1995) demonstrated that addition of canola flour to meat resulted in 73–97% inhibition of lipid oxidation, as determined by the 2-thiobarbituric acid (TBA) test. Use of canola hulls as a source of natural antioxidants was explored by Naczk, Pegg, Zadernowski, and Shahidi (2005). According to Wanasundara *et al.* (1996), the antioxidant activity of different rapeseed phenolic fractions was lower than the activity of crude ethanolic extract due to synergism between different phenolics in a β-carotene–linoleate model system. The most active rapeseed meal phenolic fraction contained several classes of phenolic compounds including phenolic acids, flavones, and flavonols. Canola and rapeseed hulls have been reported to contain up to 6% tannins (Naczk et al., 2000). Therefore, use of hulls, after dehulling, as a source of natural antioxidants may provide a means for their utilization (Amarowicz, Naczk, and Shahidi, 2000).

c. Flaxseed

Flaxseed is a rich source of the lignan secoisolariciresinol diglucoside (Eliasson, Kamal-Eldin, Andersson, and Aman, 2003). Other phenolic compounds have also been found in flaxseed, e.g. the hydroxycinnamic acid derivatives, *p*-coumaric acid-4-*O*-glucoside and ferulic acid-4-*O*-glucoside (Johnsson *et al.*, 2002), and the flavonoid herbacetin diglucoside (Struijs *et al.*, 2007). The ratio of the tocopherol homologues α-, γ-, and δ- in flaxseed hull at 17:61:22 were different from that in the seed (Oomah, Kenaschuk, and Mazza, 1997). Compared to other oilseeds, flaxseed contains low amount of phenolic acids (Amarowicz, Karamac, Wanasundara, and Shahidi, 1997). Besides lignans, flaxseed was reported to contain a number of phenylpropanoids, namely *p*-coumaric, *o*-coumaric, ferulic, *p*-hydroxybenzoic, gentisic, vanillic, and sinapic acids in the free and/or bound forms (Babrowski and Sosulski, 1984; Kozlowska, Zadernowski, and Sosulski, 1983). *Trans*-ferulic and *trans*-sinapic acids are the major phenolic acids and *trans*-caffeic, *p*-coumaric and *p*-hydroxybenzoic are the minor phenolic acids found in dehulled, defatted meal (Dabrowski and Sosulski, 1984). However, Harris and Haggerty (1993) reported that the content of ferulic and chlorogenic acids accounted for 84% of total phenolic acids in methanolic extracts of defatted meal (Shahidi and Naczk, 2004).

d. Sesame seeds

The oil from the sesame seed has a superior oxidative stability when compared with other vegetable oils largely due to the presence of sesamol in sesame seeds (Shahidi and Wanasundara, 1992). The compounds present in sesame cake extract included sesamol, sesamin, sesamolin, sesaminol diglucoside, and sesaminol triglucoside (Suja, Jayalekshmy, and Arumughan, 2004). Both sesamin and sesamolin have the same 2,7-dioxabicyclo-(3,3,0)-octane backbone with two 3,4-methylene dioxyphenyl substituents.These components along with γ-tocopherol confer superior oxidative stability to sesame oil as compared to other sources of vegetable oils (Shahidi, Amarowicz, Abou-Gharbia, and Shehata, 1997). Chemical structures of some phenolic compounds found in sesame seeds are shown in Figure 8.18. Abou-Gharbia, Shehata, and Shahidi (2000) reported that oxidative stability of sesame oil may be due to the presence of unique unsaponifiable constituents namely lignans and tocopherols. These compounds possess strong antioxidant activity and may have the potential of inhibiting the process of ageing in man and in biological systems. Chang, Yen, Huang, and Duh (2002) have shown that sesame hulls possess considerable antioxidant activity, partly due to their high level of phenolic compounds. Shahidi, Liyana-Pathirana, andWall (2006) suggested that defatted sesame extracts and their hulls possess good antioxidant activity. This activitywas higher for black sesame than white sesame and concentrated mainly in the hull fraction. In the linoleic acid system, the brown pigment of black sesame seed showed an equal antioxidant activity to BHA and higher than trolox and α-tocopherol indicated that the brown pigment of sesame seed possessed excellent antioxidant activity (Xu, Chen, and Hu, 2005).

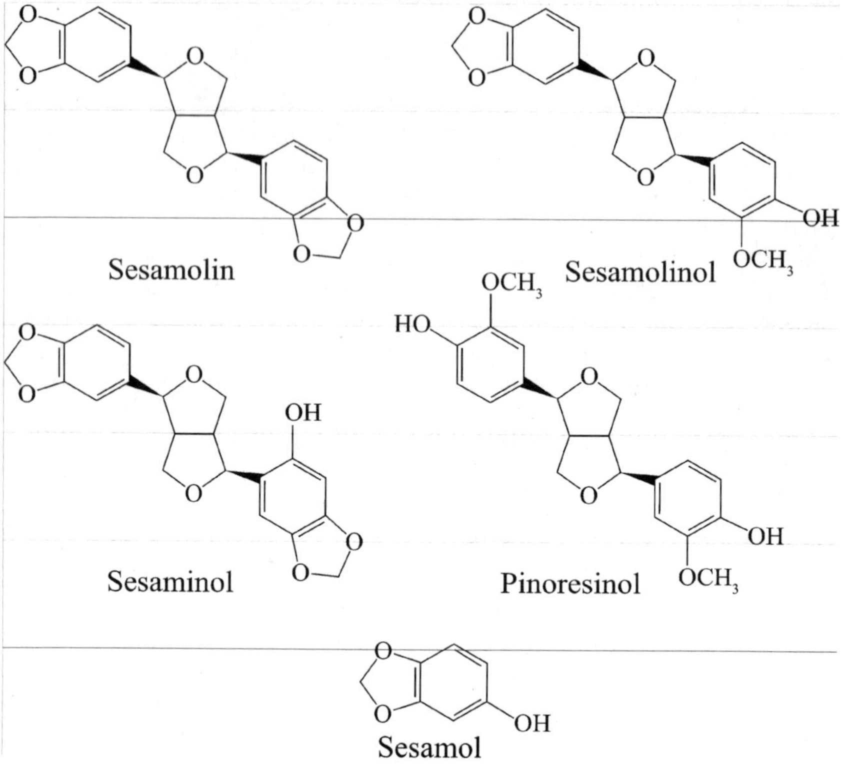

Figure 8.18: Chemical structures of some phenolic compounds found in sesame seeds.

e. Cottonseed

Gossypol is the major phenolic component present in cottonseed kernel in the free and bound forms, a polyphenolic binaphthyl dialdehyde, which is toxic to man and monogastric animals and also act as a male contraceptive (Shahidi and Naczk, 2004). Zhang, Yang, Zhao, Luan, and Ke (2001) identified that glandless cottonseed contains five, kaempferol 3-0-β-D-apiosyl(1→2)-[α-L-rhamnosyl (1→6)]-β-D-glucoside, quercetin-3-0-β-D-apiosyl (1→2)-[α-L-rhamnosyl (1→6)]-β-D-glucoside, quercetin-3-0-β-D-apiosyl (1→2)-α-D-glucoside, rutin and quercetin-3-0-β-D-glucoside.However, Piccinelli *et al.* (2007) identified seven flavonol glycosides, namely quercetin 3-*O*-{β-dapiofuranosyl-(1 → 2)-[α-l-rhamnopyranosyl-(1 → 6)]-β-dglucopyranoside}, kaempferol 3-*O*-{β-d-apiofuranosyl- (1 → 2)-[α-l-rhamn-opyranosyl-(1 → 6)]-β- d-glucopyranoside}, quercetin 3-*O*-[β-d-apiofuranosyl- (1 → 2)-β-d-glucopyranoside], quercetin 3-*O*-β-d-glucopyranoside, kaempferol 3-*O*-[α-lrhamnopyranosyl-(1→6)-β-d- glucopyranoside], quercetin 3-*O*-[α-l-rhamnopyranosyl- (1→6)-β-d-glucopyranoside], kaempferol 3-*O*-α-l-rhamnopyranoside. In addition, gallic acid and 3,4-dihydroxybenzoic acids were identified in whole cottonseed byproducts. According to Rhee, Ziprin, and Calhoun (2001), cottonseed meal may be a potential source of natural antioxidants for cooked meat products as well as other oxidationsusceptible food products. However, they suggested that further research is needed to clearly establish the relationship amongst processing technology, gossypol profile and antioxidant activity of cottonseed meal in food products with different types/amounts of prooxidative constituents.

f. Evening primrose

Evening primrose has earned an important place in the pharmaceutical industry due to the presence of a high amount (9–10%, w/w) of γ-linolenic acid in its seed oil (Gibson, Lines and Neumann, 1992; Shahidi, Amarowicz, He, and Wettasinghe, 1997; Wettasinghe and Shahidi, 1999b). Evening primrose meal contain three major low-molecular-weight phenolic compounds, namely (+)-catechin, (−)-epicatechin, and gallic acid (Wettasinghe, Shahidi, and Amarowicz, 2002). Shahidi (2000b) reported that the stronger antioxidant activity of evening primrose meal may be attributed to its tannin components. More recent studies have suggested that the antioxidant properties of evening primrose may arise from phenolic acids such as gallic, caffeic, *p*-hydroxybenzoic, vanillic, ferulic and salicylic acids, as well as proanthocyanidins and flavanols (Zadernowski, Nowak-Polakowska, and Konopka, 1996), catechin and epicatechin derivatives (Wettasinghe and Shahidi, 2002) or protocatechuic and gallic acids and esters (Peschel, Dieckmann, Sonnenschein, and Plescher, 2007; Schmidt, Niklová, Pokorný, Farkaš, and Sekretár, 2003). Salicylic, *p*-hydroxybenzoic, 2-hydroxy-4-methoxybenzoic, vanillic, *m*- and *p*-coumaric, gallic, ferulic and caffeic acids are found in smaller quantities in evening primrose seeds (Zadernowski, Naczk, and Nowak-Polakowska, 2002).

g. Borage

Borage seeds and leaves have been employed as a source of phenolics/polyphenolics, showing good antioxidant activity and retarding lipid oxidation in various food model systems (Wettasinghe and Shahidi, 1999a, 2000, 2002). Rosmarinic acid, syringic acid and sinapic acid are the major phenolic compounds present in the ethanolic extract of borage meal (Wettasinghe, Shahidi, Amarowicz, and Abou-Zaid, 2001). Although borage oil is rich in polyunsaturated fatty acids, it is highly resistant to oxidation in intact seed tissues due to the presence of tocopherols and several other phenolic compounds in oil-bearing tissues (Wettasinghe and Shahidi, 1999a). Wettasinghe and Shahidi (1999a) suggested that borage extract may be added to bulk oils and meat products in place of synthetic antioxidants in order to retard lipid oxidation. However, the low content of hydrophobic phenolics in the extract might make it less antioxidative in oil-in-water emulsion systems (Wettasinghe and Shahidi, 2000, 2002). Recently, Gómez-Estaca, Giménez, Montero, and Gómez-Guillén (2009) reported that the incorporation of a borage antioxidant extract improved the antioxidant power of the commercial fish gelatin films with only minor modifications of the physico-chemical properties (lower breaking force). The use of polyphenolic extracts from natural sources as the borage extract seems to be a more promising way to improve the antioxidant power of edible films than the use of BHT or α-tocopherol (Gómez-Estaca *et al.*, 2009).

h. Olive and olive oil

Olive oil contains a high amount of natural antioxidants such as tocopherols, carotenoids, sterols and phenolic compounds (Minioti and Georgiou, 2008). Secoiridoids, oleuropein, demethyloleuropein, ligstroside and verbascoside (caffeoylrhamnosylglucoside of hydroxytyrosol) (Figure 8.19) are the main phenolic compounds of olive fruits (Amiot, Fleuriet, and Macheix, 1986; Angerosa, d'Alessandro, Konstantinou, and Di Giacinto, 1995, 1996; Borzillo, Iannota, and Uccella, 2000; Czerwin□ ska, Kiss, and Naruszewicz, 2012; Ucella, 2001). Oleuropein is the major phenolic compound in olive fruit, which can be as much as 14% in dried fruit, whilst hydroxytyrosol is the major phenolic component in olive oil (Amiot *et al.*, 1986).

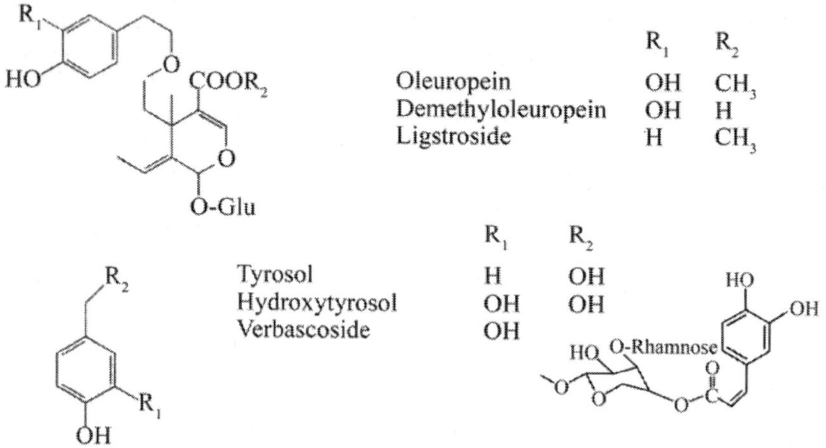

	R_1	R_2
Oleuropein	OH	CH_3
Demethyloleuropein	OH	H
Ligstroside	H	CH_3

	R_1	R_2
Tyrosol	H	OH
Hydroxytyrosol	OH	OH
Verbascoside	OH	

Figure 8.19: Chemical structures of some phenolics found in olive oil.

According to Cabrini *et al.*, (2001), amongst olive phenolics, tyrosol is present at the highest concentration; however, this compound appears to contribute very little, if any, to the stability of olive oil. Other phenolic compounds include hydroxytyrosol, 3, 4-dihydroxyphenylacetic, 4-hydroxyphenylacetic, and 4-hydroxybenzoic acids as well as protocatechuic, vanillic, syringic, and *p*-coumaric acids (Cabrini *et al.*, 2001). However, Owen *et al.* (2000) reported that hydroxytyrosol is one of the main components of virgin olive oil and olive mill waste with a strong antioxidant potential (González-Santiago, Fonollá, and Lopez-Huertas, 2010). Amongst the phenolic compounds found in extra-virgin olive oils, *o*-dihydroxy-phenolics are very potent antioxidants (Boskou, Blekas, and Tsimidou, 2005). New classes of phenolic compounds, hydroxy-isochromans, l-phenyl-6, 7-dihydroxy-isochroman and 1-(3′-methoxy-4′-hydroxy) phenyl-6,7-dihydroxy-isochroman were found in different samples of extra-virgin olive oil (Bianco, Coccioli, Guiso, and Marra, 2002). Rutin and luteolin 7-glucoside are the two main flavonoids in olive fruits. In addition, cyanidin 3-*O*-glucoside and cyaniding 3-*O*-rutinoside are the most abundant anthocyanins in olive fruits (Romani, Mulinacci, Pinelli, Vincieri, and Cimato, 1999). Moreover, Brenes *et al.* (2000) have identified two new phenolic compounds in olive oil, namely pinoresinol and 1-acetoxypinoresinol. Cioffi *et al.* (2010) have reported that phenolic compounds in virgin olive oil are a complex mixture of components that include α- and γ-tocopherols, hydroxytyrosol, tyrosol, phenolic acids (caffeic acid, vanillic acid, syringic acid), lignans (pinoresinol, 1-acetoxypinoresinol) (Montedoro, Baldioli, and Miniati, 1992), and secoiridoids (oleuropein aglycone, oleuropein, demethyloleuropein, ligstroside) (Lavelli and Bondesan, 2005). It has also been reported that one of the well-known phenolic compounds present in newly-pressed extra-virgin olive oil, the dialdehydic form of deacetoxy-ligstroside aglycone, called oleocanthal, is one of the main substances responsible for the bitter taste of olive oil and possesses ibuprofen-like cyclooxygenases (COX-1 and 2) inhibitory activity and responsible for its anti-inflammatory effect (Beuchamp, Keast, Morel, and Lin, 2005). Czerwinska *et al.* (2012) showed that oleacein is a more potent scavenger of hypochlorous acid than oleuropein as well as a stronger inhibitor of reactive oxygen species production and myeloperoxidase release in neutrophils.

D. BEVERAGES

1. Tea

Phenolics in tea are responsible for its antioxidant activity. Polyphenolic compounds and caffeine are major constituents of tea. The pure catechins and phenolic acids found in tea are more powerful than the antioxidant vitamins C and E as well as β-carotene in an *in vitro* lipoprotein oxidation model (Vinson and Dabbagh, 1998). Vinson and Dabbagh (1998) indicated that both catechins and theaflavins contribute to the antioxidant characteristics of tea. Green tea is a gently processed tea where the catechin profile closely resembles that originally present in the leaves at harvest (Astill, Birch, Dacombe, Humphrey, and Martin, 2001). Black tea, unlike green tea, is a more processed product. Fresh tea leaves are fermented in a process by which oxidative enzymes naturally occurring in tea leaves including polyphenol oxidase and peroxidase, oxidize catechin monomers and generate a complex mixture of polyphenol derived products including theaflavins, theasinensins, and other poorly characterized complex oxidation polymers known as thearubigins (Ferruzzi, 2010). Thearubigins have been reported as the most abundant pigment in black tea, which comprises polyphenolic oxidation products (Yassin, Koek, and Kuhnert, 2015), Recently, Yassin *et al.* (2015) reported that oxidation is mainly taking place on the B-ring and the galloyl group, where the oxidized components subsequently undergo oxidative coupling for the formation of theaflavins, theasinensins and polyhydroxylated flavan-3-ols, all precursors for thearubigin formation.The total flavanol level is reduced from 50–53% in green tea to 10% in black tea (Sun, Ho, and Shahidi, 2009). The antioxidant activities of various teas have been shown to descend in the following order: green > oolong > black > pu'erh tea (Yashin, Yashin, and Nemzer, 2011). Catechins are found in particularly high concentration in teas, especially the unfermented green and white varieties (Freeman and Niemeyer, 2008).

Green tea leaves contain relatively large amounts (approximately 30%) of (−)-epicatechin (EC), (−)-epicatechin gallate (ECG), (−)-epigallocatechin (EGC), and (−)-epigallocatechin gallate (EGCG) (Figure 8.20) (Amarowicz and Shahidi, 1996; He and Shahidi, 1997). Theaflavins, which include theaflavin, theaflavin-3-gallate, theaflavin-3′ gallate and theaflavin-3,3′-digallate (Figure 8.21), are key to the characteristic colour and taste of black tea, and account for 2–6% of the solids in brewed black tea (Khan and Mukhtar, 2010). According to Chen, Zhu, Tsang, and Huang (2001), the total content of epicatechin derivatives in Chinese black teas is between 2.4 and 5.1 g/kg; in Chinese oolong tea it ranges from 41.4 to 46.3 g/kg and in Chinese green tea between 80 and 144.4 g/kg. EGCG is the most abundant epicatechin derivative in teas. Besides catechins, flavonol, and their glycosides, anthocyanidin and leucoanthocyanidin, phenolic acids and depsides are also present in fresh tea leaves (Wan, Li, and Zhang, 2009). Hara (1994) has evaluated the antioxidative potency of crude extracts of green tea and individual catechins in lard by the active oxygen method and found that crude tea catechins reduce the formation of peroxides more effectively than α-tocopherol or BHA. Wanasundara and Shahidi (1996) found that tea catechins exhibit an antioxidative activity similar to or better than that of BHA, BHT, and TBHQ.The order of their potency, at 200 ppm, in marine oils was ECG > EGCG > EGC > EC (Shahidi, Wanasundara, He, and Shukla, 1997).

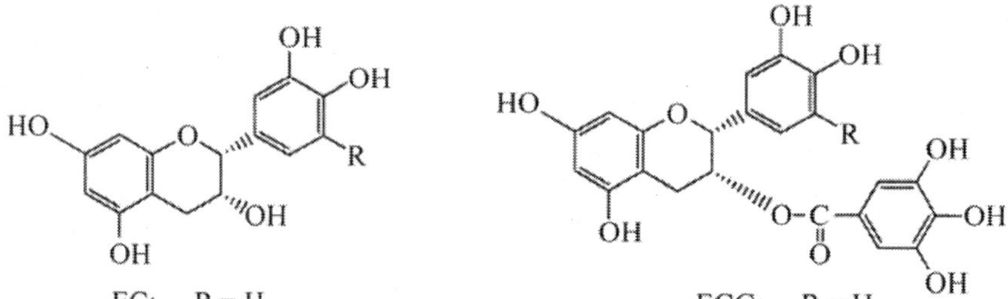

EC: R = H
EGC: R = OH

ECG: R = H
ECGC: R = OH

Figure 8.20: Structures of tea catechins.

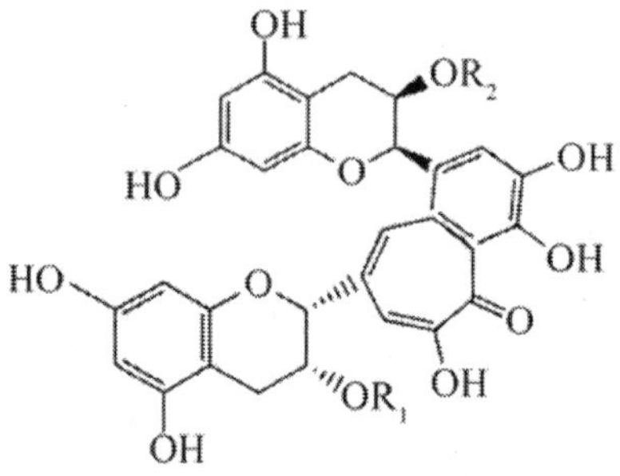

	R1	R2
Theaflavin	H	H
Theaflavin-3-gallate	Galloyl	H
Theaflavin-3'-gallate	H	Galloyl
Theaflavin-3,3'-digallate	Galloyl	Galloyl

Figure 8.21: Structures of major theaflavin in black tea (adapted from Wan *et al.*, 2009).

Several studies have shown that green tea extract were effective in stabilizing meat and fish (Chan, Soh, Tie and Law, 2011; He and Shahidi, 1997; Mustafa, 2013; Senanayake, 2013; Shahidi and Alexander, 1998; Shahidi and Wanasundara, 1992). Zhu *et al.* (2009) reported that EGCG's direct trapping of reactive carbonyl species may also contribute to the significant reduction of acrolein and other aldehydes in the peroxidation of seal blubber oil. Black and green teas were not significantly different in their phenolic contents, and in antioxidant strength as measured by IC50, or in antioxidant potential as measured by the phenolic antioxidant index (PAOXI) (Vinson and Dabbagh, 1998). Vinson and Dabbagh (1998) found that the PAOXI of teas was higher than those of grape juices and wine. Tea has been shown to possess several health benefits, such as inhibition of mutagenesis, protection against metabolic syndrome, antioxidative, anticarcinogenic, anti-inflammatory, antiobesity, anti-hypertensive and hypocholesterolaemic properties (Basu *et al.*, 2013; Ramji, Huang, Shahidi, and Ho, 2009; Suliburska *et al.*, 2012; Yarmolinsky, Gon, and Edwards, 2015). In addition, lyophilized derivatives of green tea have also been shown to exert antioxidant and anticancer activities (Zhong *et al.*, 2012; Zhong and Shahidi, 2011). In recent years, scientists throughout the world have investigated the potential benefits of green tea and its most abundant catechin, EGCG (Karaosmanoglu and Kilmartin, 2015).The potential for use of EGCG in human cancer prevention and treatment seems very promising because EGCG acts against cancer through a number of complementary mechanisms (Khan and Mukhtar, 2010). Recently, Kadasi *et al.* (2014) found that green tea extracts could affect basic ovarian cell functions such as inducing the suppression of proliferation, stimulation of apoptosis, promotion of progestagen and changes in testosterone release.

2. Coffee

Phenolics play an important role in the formation of coffee flavour. From over 800 volatile compounds found in roasted coffee aroma, only 42 have been identified as phenolics (Flament, 1989, 1995; Nijssen, Visscher, Maarse, Willemsens, and Boelens, 1996). Catechol is the predominant volatile phenolic compound found in coffee aroma, followed by 4-ethylguaiacol, 4-ethylcatechol, pyrogallol, quinol and 4-vinylcatechol (Smith, 1985). Green coffee beans may contain at least five major groups of chlorogenic acid isomers (Figure 8.22), namely caffeoylquinic acids, dicaffeoylquinic acids, feruoylquinic acids, *p*-coumaroylquinic acids and caffeoylferuloylquinic acids (Clifford and Ramirez-Martinez, 1991). Chu, Lin, Yu, and Ye (2008) determined catechin, rutin, ferulic acid, o-dihydroxybenzene, chlorogenic acid, caffeic acid, gallic acid and protocatechuic acid in coffee using capillary electrophoresis with amperometric detection (CE-AD). Brezová, Šlebodová, and Staško (2009) reported that antioxidant capacity of instant coffee was approximately 3–4 times higher than that of ground coffee. However, considering the traditional preparation procedures of coffee brews (approximately four times higher amount of ground than instant coffee in the same water volume) the antioxidant capacities of both ground and instant brews were quite comparable. More detailed investigations of two coffee components, namely caffeic acid and caffeine, using moderate oxidants, such as $ABTS^{\bullet +}$, DPPH and TEMPOL (4-hydroxy-2,2,6,6-tetramethylpiperidine *N*-oxyl) radicals, indicated a high antioxidant activity of caffeic acid but no antioxidant action for caffeine (Brezová *et al.*, 2009). However, employing hydroxyl radical, a more reactive oxidant source (generated in the photochemical decomposition of H2O2), both compounds showed a remarkable cavenging activity (Brezová *et al.*, 2009). Recently, Mnatsakanyan *et al.* (2010) reported that the overall antioxidant profile of the "Decaffeinato" coffee sample was similar to "Gold" and "Ristretto" coffee samples which are rich in phenolic compounds. Furthermore, decaffeinated coffee showed similar positive effects on human health.

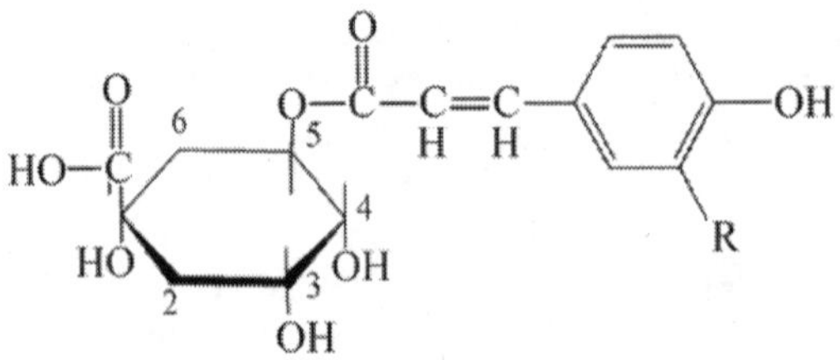

	R
5-p-Coumaroylquinic acid	H
5-Caffeoylquinic acid	OH
5-Feruloylquinic acid	OCH_3

Figure 8.22: Structure of chlorogenic acid and its derivatives.

3. Cocoa

Cocoa bean and its products (cocoa liquor, cocoa powder, and dark chocolate) are rich in phenolic compounds (Othman, Ismail, Ghani, and Adenan, 2007). Indeed cocoa products contain greater antioxidant capacity and higher amounts of flavonoids per serving than either tea or red wine (Lee, Kim, Lee, and Lee, 2003; Steinberg, Bearden, and Keen, 2003). Cocoa beans have a high phenolic content of about 12–18% (dry weight) in unfermented beans (Kim and Keeney, 1984). Dreosti (2000) reported that 60% of the total phenolics in raw cocoa beans are flavan-3-ol monomers (epicatechin and catechin) and procyanidin oligomers (dimer to decamer). Theobromine and caffeine are two major ingredients in chocolate (Vinson, Proch, and Zubik, 1999).The polyphenolic composition of cocoa has been characterized and quantified (Sanchez-Rabaneda *et al.*, 2003; Wollgast and Anklam, 2000; Zumbe, 1998). The compounds identified include epicatechin, gallocatechin and epigallocatechin and other phenolics such as procyanins, anthocyanins, and flavone and flavonol glycosides such as luteolin-7-*O*-glucoside and quercetin-3-*O*-arabinoside. Reports on the polyphenolic content of cocoa products vary greatly in the literature, with values ranging from 3.3 to 65 mg/g in cocoa powder or 1.7 to 36.5 mg/g total polyphenols in dark chocolate (Adamson *et al.*, 1999; Vinson *et al.*, 1999). There are mainly flavan-3-ols (monomeric epicatechin and catechin, as well as their oligomers from dimers to decamers, the procyanidins), with small amounts of anthocyanins (mainly cyanidin glycosides) and flavonols (quercetin glycosides) (Adamson *et al.*, 1999; Hammerstone, Lazarus, Mitchell, Rucker, and Schmitz, 1999). Counet, Callemien, and Collin (2006) identified *trans*-resveratrol and *trans*-piceid in dark chocolate and cocoa liquor extracts and found that chocolate products have higher antioxidant activity than concentrated commercial stilbene extracts. Epicatechin, the major monomeric polyphenol in chocolate (Sanbongi *et al.*, 1998) and an extract of chocolate liquor were both found to stimulate cellular immune response *in vitro* (Sanbongi, Suzuki, and Sakane, 1997). Chocolate has also been reported to be a good source of dietary catechins, second only to green tea (Arts, Hollman, and Kromhout, 1999). According to Saura-Calixto and Goñi (2006), the total polyphenolic content of cocoa was higher than that reported in other foods like cereals, legumes, vegetables, nuts and fruits. Studies have demonstrated that the consumption of cocoa or chocolate reduces the risk of cardiovascular disease (Keen, 2001; Osakabe *et al.*, 1998). Moreover, extracts prepared from cocoa powder and cocoa beans were shown to exhibit antihyperglycaemic effects on streptozotocin-induced diabetic rats (Amin, Faizul, and Azli, 2004; Ruzaidi, Amin, Nawalyah, Hamid, and Faizul, 2005). Recently, Pimentel, Nitzke, Klipel, and de Jong (2010) found that 49 g of dark chocolate (71% cocoa) had the same quantity of flavonoids as that of 196 ml of Tannat wine, which is the daily wine intake recommended to produce health benefits in an adult of 70 kg body weight.

4. Beer

Phenolic compounds identified in beer include phenolic acids, flavonoids, proanthocyanidins, tannins and amino phenolic compounds, all of which have been reported to possess antioxidant and antiradical properties as well as other biological effects (Gorinstein, Caspi, Zemser, and Trakhtenberg, 2000; Montanari, Perretti, Natella, Guidi, and Fantozzi, 1999). Zhao, Chen, Lu, and Zhao (2009) identified nine phenolic compounds, including gallic acid, protocatechuic acid, (+)-catechin, vanillic acid, caffeic acid, syringic acid, (−)-epicatechin, *p*-coumaric acid and ferulic acid in different beer samples.The major free phenolic acids in beers were *m*-, *p*- and *o*-coumaric and ferulic acids (Montanari *et al.*, 1999), whilst vanillic, ferulic and *p*-coumaric acids were the dominant free phenolic acids in Spanish, German and Danish brands (Bartolome, Pen-Neira, and Gomez-Cordoves, 2000). Sinapic, vanillic, chlorogenic, homovanillic, *p*-hydroxybenzoic, 2,6- and 3,5-dihydroxybenzoic, syringic, gallic, protocatechuic and caffeic acids are also identified in beer (Bartolome *et al.*, 2000; Montanari *et al.*, 1999). Moreover, anthocyanins and anthocyanidins (cyanin, cyanidin, pelargonin, pelargonidin and delphinidin), flavones and flavonols such as vitexin, isoquercetin, quercetin, rutin, kaempferol, myricetin and myricitrin have been found in beer at less than 1 mg/l (Drawert, Leupold, and Lessing, 1977). Gorinstein, Zemser, Weisz, Haruenkit, and Trakhtenberg (1998) reported a significant beneficial effect in rats supplemented for 4 weeks with both alcohol-containing and alcohol-free beer. Ghiselli et al. (2000) indicated that beer, which has a moderate antioxidant capacity coupled with low ethanol content, is an alcoholic beverage that is able to improve plasma antioxidant capacity without the negative effects produced by high doses of ethanol. Later, Gasowski *et al.* (2004) found that beer has a positive effect on plasma lipid profile and plasma antioxidant capacity, and increases bile volume and bile acid concentrations mainly in rats fed cholesterol-containing diets.The degree of this positive influence of beer is directly connected to the bioactive compounds (flavonoids) in beer. Beer also contains an array of healthpromoting isoflavonoids (Lapcik, Hill, Hampl, Wahala, and Adlercreutz, 1998). However, despite all the health benefits of alcoholic beverages, it should be noted that excessive alcohol consumption has clear detrimental effects, which cause mortality (Shahidi, 2009).

5. Wine

Wine is a particularly rich source of phenolics, and its moderate intake has been associated with decreased levels of lipid peroxidation and a lower incidence of certain types of cancer (Bianchini and Vainio, 2003; Waffo-Teguo *et al.*, 2001). The moderate consumption of red wine has a relatively higher benefit in the prevention of atherosclerosis and coronary heart disease (CHD) (Faustino, Sobrattee, Edel, and Pierce, 2003; Gambelli and Santaroni, 2004). According to Vinson, Teufel, *et al.* (2001), consumption of 196 ml (~1 glass) of red wine per day is enough to significantly reduce the atherosclerosis risk in a 70 kg adult. *Trans*-esveratrol and some phenolic antioxidants (catechin, epicatechin, quercetin and rutin) were identified in red wine (Anli, Vural, Demiray, and Özkan, 2006). *Trans*- and *cis*-resveratrol have been shown to exist in wine as both the aglycone and the bound glucoside piceid (Vian, Tomao, Gallet, Coulomb, and Lacombe, 2005). The most common flavonoids in wine are flavonols (quercetin, kaempferol, and myricetin), flavan-3-ols (catechin and epicatechin), tannins, and anthocyanins (cyanin). Nonflavonoids comprise stilbenes, hydroxycinnamic acids and benzoic acids (S. Li *et al.*, 2009). Phenolic acids and stilbenes are important non-flavonoid compounds present in grapes and wine. Phenolic acids are present in their free form or as glycosylated derivatives and esters of tartaric, quinic and shikimic acid in both red and white wines (Monagas, Bartolome, and Gomez-Cordoves, 2005). Phenolic compounds in red wine are derived fromthe grape's skin, as well as from grape seeds, grape stems, or grape pulp, all of which are important sources of flavanols that are transferred to the wine during fermentation. On the contrary, white wines are usually made from the free running juice, without the grape mash, having any contact with the grape skins. This is thought to be the main reason for the relatively low polyphenol content and for the lower antioxidant activity of white wine in comparison to red wine (Fuhrman, Volkova, Suraski, and Aviram, 2001; S. Li *et al.*, 2009). Epidemiological studies have shown that moderate wine intake may be beneficial for human health (Mudnic *et al.*, 2010). Amongst many beneficial effects, wine inhibits low density lipoprotein (LDL) oxidation (Frankel, Kanner, German, Parks, and Kinsella, 1993), increases antioxidative capacity in humans (Maxwell, Cruickshank, and Thorpe, 1994) and modulates vascular function by inducing vasodilation through increased production of nitric oxide (NO) (Flesch, Schwarz, and Bohm, 1998). The "French paradox", that is, the low incidence of cardiovascular events despite a diet high in saturated fat, was attributed to the regular drinking of red wine in southern France (Renaud and de Lorgeril, 1992). Recently, Fukui, Choi, and Zhu (2010) have shown the protective effect of resveratrol against oxidative neuronal death based on an *in vitro* model.

E. TREE NUTS

Frequent consumption of nuts has been linked to a lowered risk of cardiovascular disease (Kris-Etherton *et al.*, 1999; Shahidi and John, 2010). Phytochemicals, especially phenolics, in nuts may be considered as the major bioactive compounds for health benefits. In particular, nuts include plant protein, unsaturated fatty acids, dietary fibre, plant sterols, phytochemicals and micronutrients like tocopherols (Kris-

Etherton, Zhao, Binkoski, Coval, and Etherton, 2001). Maguire, O'Sullivan, Galvin, O'Connor, and O'Brien (2004) reported that α-tocopherol was the most dominant tocopherol in almonds, peanuts, hazelnuts and macadamias. However, Kornsteiner, Wagner, and Elmadfa (2006) reported that β- and γ-tocopherols, were the most predominant tocopherols in 7 nut types and the mean amounts in descending order were pistachios > walnuts > pecans > Brazil nuts > pines ≥ peanuts > cashews. Figure 8.23 shows the total phenolic and flavonoid content of tree nuts. Walnuts, pistachios, peanuts and pecans contain the highest total phenolic and flavonoid content amongst all types of nuts (Yang, Liu, and Halim, 2009). Kornsteiner *et al.* (2006) reported that the antioxidants were found to be lower in cashews and pines. Macadamias also contained the lowest level of antioxidative constituents (polyphenols, tocopherols). Nuts are a good source of phytochemicals, including phenolics (tannins, ellagic acid, and curcumin), flavonoids (luteolin, quercetin, myricetin, kaempferol, and resveratrol), isoflavones (genistein and daidzein), terpenes, organosulphuric compounds, and vitamin E (Bravo, 1998; Yang, Martinson, *et al.*, 2009). Miraliakbari and Shahidi (2008) found that minor component extracts of chloroform/methanol extracted pecan and walnut oils possessed the highest antioxidant activities, which is likely due to their high amounts of tocopherols and other antioxidative minor components such as phospholipids and possibly other unidentified phenolic and/or non-phenolic components present in the extracts.

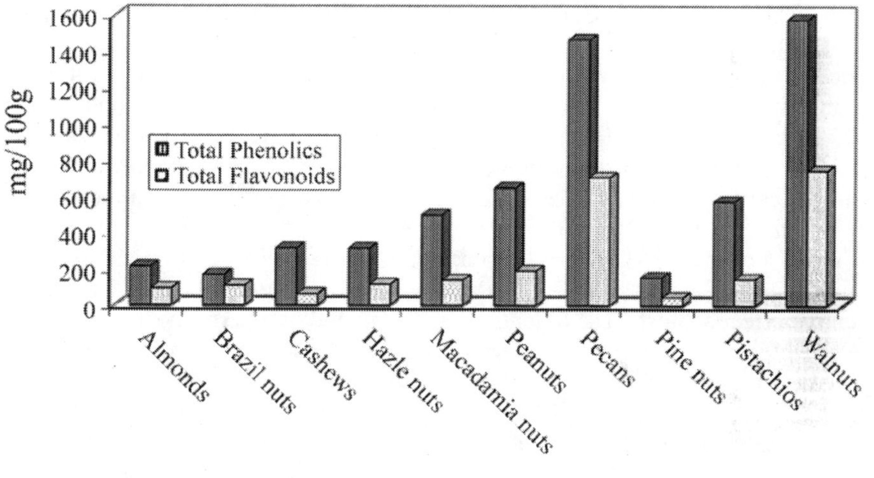

Figure 8.23: Content of total phenolics and flavonoids in tree nuts (mg/100 g) (adapted from Yang, Liu, *et al.*, 2009).

1. Almonds

Extracts of whole almond seed, brown skin, and green shell cover possess potent free radical scavenging capacities (Siriwardhana and Shahidi, 2002).These activities may be related to the presence of flavonoids and other phenolic compounds in nuts (Wijeratne, Abou-Zaid, and Shahidi, 2006). The antiradical activity of almond extracts against free radicals examined follow the order: skin > shell > seed (Siriwardhana and Shahidi, 2002). Almond hulls have been shown to serve as a rich source of triterpenoids, betulinic, urosolic, and oleanolic acids (Takeoka *et al*., 2000) as well as flavonol glycosides and phenolic acids (Sang, Lapsley, Rosen, and Ho, 2002). In addition, Sang, Lapsley, Rosen, *et al*. (2002) isolated catechin, protocatechuic acid, vanillic acid, *p*-hydroxybenzoic acid, and naringenin glucoside, as well as galactoside, glucoside, and rhamnoglucoside of 3′-*O*-methylquercetin and rhamnoglucoside of kaempferol (Esfahlan, Jamei, and Esfahlan, 2010;Wijeratne, Abou-Zaid, *et al*., 2006). Amarowicz, Troszynska, and Shahidi (2005) reported the presence of phenolic compounds such as vanillic, caffeic, *p*-coumaric, ferulic acids, quercetin, kaempferol, isorhamnetin, delphinidin and cyanidin as well as procyanidins B2 and B3 in almond extract using an HPLC method and found that procyanidins B2 and B3 and vanillic acid were the dominant phenolic compounds in almond extract.Wijeratne, Amarowicz, and Shahidi (2006) suggested that almond green shell cover extract, which contained a high phenolic acid content, could serve as a valuable source for phenolic acids. Recently, hydroxybenzoic acids (as protocatechuic, *p*-hydroxybenzoic, chlorogenic, vanillic and *trans*-*p*-coumaric acids), flavan-3-ols (as (+)-catechin and (–)-epicatechin), flavonols (as aglycones; isorhamnetin, quercetin, kaempferol and glycosides; quercetin-3-*O*-rutinoside, quercetin-3-*O*-galactoside, quercetin-3-*O*glucoside, kaempferol-3-*O*-rutinoside, isorhamnetin-3-*O*rutinoside, kaempferol-3-*O*-glucoside, isorhamnetin-3-*O*glucoside) and flavanones (as aglycones; naringenin and eriodictyol and glycosides; eryodictiol-7-*O*-glucoside and naringenin-7-*O*-glucoside) were identified in the almond skin samples (Mandalari *et al*., 2010). Recently, Esfahlan, Jamei, and Esfahlan (2010) reported quercetin, isorhamnetin, quercitrin, kaempferol-3-*O*-rutinoside, isorhamnetin-3-*O*-glucoside, and morin as the major flavonoids in almond extracts. Sfahlan, Mahmoodzadeh, Hasanzadeh, Heidari, and Jamei (2009) found a strong antioxidant activity for methanolic extracts of almond hull and shell.Wijeratne, Abou-Zaid, *et al*. (2006) evaluated antioxidant efficacy of defatted almond whole seed, brown skin, and green shell cover extracts by monitoring inhibition of human low-density lipoprotein (LDL) oxidation, inhibition of DNA scission and metal ion chelation activities and also found that the brown skin of almond exerted the highest preventive effect against LDL oxidation at 10, 50, and 100 ppm levels, compared to those of whole almond and its green shell (leafy) cover. Almonds, when incorporated in the diet, have been reported to reduce colon cancer risk in rats (Davis and Iwahashi, 2001) and increase HDL cholesterol and reduce LDL cholesterol levels in humans (Hyson, Schneeman, and Davis, 2002). Almond appears to be effective in reducing the risk of heart disease and cancer prevention, and consumption of almond is recommended by FDA for better health conditions (Shahidi, Zhong, Wijeratne, and Ho, 2009).

2. Cashew nuts

The major phenolics found in cashew nuts are anacardic acids, cardanols, cardols, tocopherols and other minor phenolic constituents (Shahidi and Tan, 2009).The kernel of cashew nut valued in trade is covered with a thin reddish-brown skin or testa. The testa has been reported to be a good source of hydrolysable tannins (Pillai,

Kedlaya, and Selvarangan, 1963) with catechin and epicatechin as the major polyphenols (Mathew and Parpia, 1970).

The polymeric proanthocyanidins account for 40% of total testa phenolics; other components of testa phenolics include leucocyanidins and leucopelargonidins (Mathew and Parpia, 1970). The presence of gallic, caffeic and quinic acids apart from catechin and leucocyanidin was reported by Kantamoni (1965). γ-Tocopherol is prevalent in cashew, whilst α- and δ-tocopherols are present in lower amounts (Trevisan *et al.*, 2006). Kamath and Rajini (2007) suggested that cashew nut skin, a byproduct of the cashew processing industry, can be used as an economical source of natural antioxidants. Cashew nut shell liquid, a byproduct obtained during the processing of cashew nuts is reported to possess antioxidant activity (Singh, Kale, and Rao, 2004). Recently, Chandrasekara and Shahidi (2011) reported that whole cashew and nuts testa were better sources of antioxidants compared to the kernel as assessed in different food and biological model systems.

3. Hazelnut

The presence of gallic, vanillic, caffeic and ferulic acids in extracts of hazelnut was reported by Alasalvar, Karamac, Amarowicz, and Shahidi (2006). Amongst the identified phenolic acids, *p*-coumaric acid was most abundant in hazelnut kernel, hazelnut green leafy cover, and hazelnut tree leaf, whereas gallic acid was most abundant in hazelnut skin and hazelnut hard shell (Alasalvar *et al.*, 2006, 2009). However, in a previous study, protocatechuic acid was reported to be the predominant phenolic acid in testa (skin) of hazelnut (Senter, Horvat, and Forbus, 1983). Amaral et al. (2005) identified and quantified four phenolic acids, namely, 3-caffeoylquinic acid, 5-caffeoylquinic acid, caffeoyltartaric acid, and *p*-coumaroyltartaric acid, in hazelnut leaves from 10 different cultivars grown in Portugal. Hazelnut contained one anthocyanidin (cyanidin) and four flavan-3-ols [(−)-epicatechin, (−)-epigallocatechin, (−)-epigallocatechin 3-gallate and (+)-catechin]. Cyanidin was the most abundant flavonoid and no flavonones and flavonols were found in hazelnut (Alasalvar, Hoffman, and Shahidi, 2009). Hoffman and Shahidi (2009) found small amounts (8.61 to 68.22 μg/g) of paclitaxel Taxol®) in tombul hazelnut (*Corylus arellana* L.) hard shells, green shell covers and leaves. In addition to paclitaxel, 10-deacetyl baccatin III, baccatin III and cephalomannine were also identified in tombul hazelnut extracts. In addition, hazelnut was found to contain the highest amount of condensed tannins amongst seven tree nuts (hazelnut, almond, cashew, chestnut, pecan, pistachio and walnut) (Alasalvar, Shahidi, Amaral, and Oliveira, 2009). A research carried out on crude extracts obtained from hazelnut by-products supports the hypothesis that hazelnut wastes, especially skin and hard shell, could serve as reliable source of efficient natural antioxidants (Shahidi, Alasalvar, and Liyana-Pathirana, 2007). Recently, Alasalvar, Shahidi, et al. (2009) suggested that hazelnut skin can be considered as a value added byproduct for use as dietary antioxidants.

4. Walnuts and heartnut

Most commonly identified phenolic compounds in walnut are condensed tannins and phenolic acids.Walnut phenolics are found in the highest concentration in the hull fraction (the pellicle surrounding the kernel) (Labuckas, Maestri, Perelló, Martínez, and Lamarque, 2008). A walnut extract containing ellagic acid, gallic acid, and flavonoids was reported to inhibit the oxidation of human plasma and low density lipoproteins (LDL) *in vitro* (Anderson *et al.*, 2001). Luczak, Swiatek, and Zadernowski (1989) identified 10 phenolic acids in the pericarp of walnut: *p*-hydroxybenzoic, vanillic, gentisic, protocatechuic, syringic, *p*-coumaric, gallic, ferulic, caffeic and sinapic acids. Syringic acid was identified as a principle phenolic acid. Subsequently, Anderson *et al.* (2001) reported that ellagitannins were the most abundant phenolics in walnuts. Of these, ellagic acid, valoneic acid dilactone and pedunculagin have been identified in phenolic extracts from shelled walnuts. Walnut leaves are considered as a good source of healthcare compounds, and have been widely used in traditional medicine for the treatment of skin inflammations, hyperhidrosis and ulcers as well as for antidiarrhoeic, anti-helminthic, antiseptic and astringent properties (Carvalho *et al.*, 2010). The compounds 3- and 5-caffeoylquinic acids, 3- and 4-*p*-coumaroylquinic acids, *p*-coumaric acid, quercetin 3-galactoside, quercetin 3-pentoside derivative, quercetin 3-arabinoside, quercetin 3-xyloside and quercetin 3-rhamnoside were identified in different cultivars of walnut leaves grown in Portugal (Pereira *et al.*, 2007). Recently Zhang, Liao, Moore,Wu, andWang (2009) identified seven phenolic compounds, namely, pyrogallol, *p*-hydroxybenzoic acid, vanillic acid, ethyl gallate, protocatechuic acid, gallic acid and 3,4,8,9,10-pentahydroxydibenzo[*b*,*d*]pyran-6-one in English walnut seed by spectroscopic methods.Walnuts contained a number of structurally related naphthoquinones, including 1,4-naphthoquinone, juglone (5-hydroxy-1,4-naphthoquinone), 2-methyl-1,4-naphthoquinone and plumbagin (5-hydroxy-2-methyl-1,4-naphthoquinone) (Muller and Leistner, 1978a, 1978b). Phenolics extracted from different fractions of English walnut showed marked antioxidant activities in different *in vitro* model systems (Samaranayake, John, and Shahidi, 2008).A recent study represents the first evidence that walnut seed, green husk and leaf methanolic extracts possess effective antihaemolytic and human renal cancer cell antiproliferative activities. It is therefore suggested that English walnut may be used as an inexpensive and easily accessible source of effective natural antioxidants and chemopreventive agents (Carvalho *et al.*, 2010). Heartnut (*Juglans ailanthifolia* var. *cordiformis*), a naturally occurring genetic oddity of the more common Japanese walnut, has recently become very popular amongst consumers and nut tree growers in southern Ontario, Canada (Li *et al.*, 2006). Ellagic acid and valoneic acid dilactones are the major polyphenolics in heartnut (Li *et al.*, 2006). γ-Tocopherol was the main tocopherol homologue present in heartnut, followed by δ- and α-tocopherol (Li, Tsao, Yang, Kramer, and Hernandez, 2007). Colaric, Veberic, Solar, Hudina, and Stampar (2005) identified phenolic acids (chlorogenic, caffeic, *p*-coumaric, ferulic, sinapic, ellagic, and syringic acid) as well as syringaldehyde and juglone in ripe fruits of 10 walnut cultivars. Syringic acid, juglone and ellagic acid were found as predominant phenolics. Phenolic compounds from walnut fruits have a positive influence on human health such as a decrease of coronary heart diseases, prevention of several kinds of cancer, and anti-inflammatory and antimutagenic activities (Anderson *et al.*, 2001).

5. Peanuts

Numerous phytochemical compounds, including polyphenolics and tocopherols are present in peanuts with potential antioxidant activity (Hashim, Koehler, and Eitenmiller, 1993; Talcott, Passeretti, Duncan, and Gorbet, 2005). Peanuts are good sources of folic acid and resveratrol (Yang, Martinson, *et al.*, 2009).The predominant phenolic acid present in peanuts with antioxidant potential is *p*-coumaric acid (Talcott, Passeretti, *et al.*, 2005). Fajardo, Waniska, Cuero, and Pettit (1995) demonstrated a stresselicited synthesis of free and bound phenolics in peanuts, with *p*-coumaric and ferulic acids as the major compounds identified. Talcott, Duncan, Pozo-Insfran, and Gorbet (2005) reported that free *p*-coumaric acid, three esterified derivatives of *p*-coumaric, and two esterified derivatives of hydroxybenzoic acid were the predominant polyphenolics present in peanuts. Yu, Ahmedna, and Goktepe (2005) reported that 3 classes of polyphenols were present in peanut skins, including phenolic acids (caffeic acid, chlorogenic acid, ferulic acid, and coumaric acid), flavonoids (catechins and procyanidins), and stilbene (resveratrol). Lou, Yamazaki, Sasaki, Uchida, and Tanaka (1999) identified 6 A-type procyanidins in peanut skin. Identification of phenolic compounds from dry-blanched peanut skins by liquid chromatography-electrospray ionization mass spectrometry was also reported by Ma *et al.* (2014). Recently, de Camargo et al. (2015) identified proanthocyanidins as the major phenolic compounds present in peanut skin. Three compounds possessing potential antioxidant activities in the methanolic extract of peanut shell were identified as 5,7-dihydroxychromone, eriodictyol and luteolin are using the HPLC–DAD–TOF/MS technique (Qiu *et al.*, 2012). Craft, Kosinska, Amarowicz, and Pegg (2010) reported the antioxidant properties of extracts obtained from raw, dry-roasted, and oilroasted peanuts. Potential contributors to antioxidant capacity in peanuts were identified as an ethyl ester of protocatechuic acid isolated from peanut testa (Huang, Yen, Chang, Yen, and Duh, 2003), luteolin in peanut hulls (Duh, Yeh, and Yen, 1992), Maillardderived compounds and protein hydrolysates (Hwang, Shue, and Chang, 2001), tocopherols (Hashim et al., 1993) and resveratrol (Ibern-Gonzalez, Roig-Perez, Lameula-Raventos, and Torre-Boronat, 2000; Sobolev and Cole, 1999) in peanuts and peanut-based products. Research has shown that the consumption of high-oleic acid peanuts has potential health benefits, such as lowering of blood cholesterol levels in hypercholesterolaemic women (O'Byrne, Knauft, and Shireman, 1997). The content of oleic acid in high-oleic acid peanuts is similar to that of olive oil (Cabrini *et al.*, 2001), which is known for its heart-healthy characteristics. However, literature shows that peanut contains multiple allergens (De Jong *et al.*, 1998; Isanga, and Zhang, 2007). A total of 6 protein subunits were recognized by more than 50% of the specific IgE-containing plasma samples of peanut allergic patients and were, therefore, defined as major allergens (De Jong *et al.*, 1998).

6. Pistachio

Phytochemicals identified from pistachios include fatty acids, phytosterols, lutein, resveratrol, and anthocyanins (Phillips, Ruggio, and Ashraf-Khorassani, 2005). Galloyl quinic derivatives were identified in aqueous extracts from leaves and gum of *Pistacia lentiscus* (Romani, Pinelli, Galardi, Mulinacci, and Tattini, 2002), and α-tocopherol was evaluated in lipophilic extracts from the leaves of *P. lentiscus* and *Pistacia therebintus* (Kivcak and Akay, 2005). The polyphenol phytoalexin *trans*-resveratrol was detected in the aqueous extracts from the edible nut of fiveTurkish cultivars of *Pistacia vera* (Tokusoglu, Unal, and Yemis, 2005). Gentile *et al.* (2007) found *trans*-resveratrol, proanthocyanidins, and a remarkable amount of the isoflavones daidzein and genistein (3.68 and 3.40 mg per 100 g of edible nut, respectively) in hydrophilic extract of *P. vera.* Seeram *et al.* (2006) identified quercetin (14.9 μg/g), luteolin (10.0 μg/g), eriodictyol (10.2 μg/g), rutin (1.6 μg/g), naringenin (1.2 μg/g), apigenin (0.2 μg/g), and the anthocyanins, cyanidin-3-galactoside (696 μg/g) and cyanidin-3-glucoside (209 μg/g) in pistachio skin. Recently, Saitta, Giuffrida, Torre, Potortì, and Dugo (2009) identified sixteen different 3-alkylphenols (cardanols) with a saturated, monounsaturated and diunsaturated chain in pistachio kernel. The most abundant cardanols were 3-(8-pentadecenyl)-phenol, 3-(10-pentadecenyl)-phenol, 3-pentadecyl-phenol and 3-(10-eptadecenyl)-phenol. *Trans*-resveratrol and *trans*resveratrol-3-*O*-β-glucoside (*trans*-piceid) were identified in Sicilian pistachio variety (Grippi *et al.*, 2008). Halvorsen *et al.* (2006) ranked the edible pistachio nut amongst the first 50 food products having the highest antioxidant potential. Recent studies show that essential oils and lipophilic extracts of leaves, fruits, gum, and galls of Pistachio may exert various characteristics such as anti-microbial, anti-inflammatory, insecticidal, and anti-nociceptive activities, with various terpenes playing a major role in the observed activities (Alma *et al.*, 2004).

7. Pecan

Pecans are a rich source of γ- and a poor source of α-tocopherol, containing 24.4 and 1.4 mg per 100 g of nut, respectively (Haddad, Jambazian, Karunia, Tanzman, and Sabaté, 2006). Phenolic acids (gallic acid) (Senter, Horvat, and Forbus, 1980), proanthocyanidins (Polles, Hanny, and Harvey, 1981), prodelphinidins (3-*O*-gallates) including epigallocatechin, epicatechin-3-*O*-gallate, and the more common flavan-3-ols, catechin and epicatechin were identified in pecan kernel (Villarreal-Lozoya, Lombardini, and Cisneros-Zevallos, 2007). Miraliakbari and Shahidi (2008) found that chloroform/methanol extracted pecan oil had the highest antioxidant activity and total phenolic content (711 mg/kg gallic acid equivalents) amongst the seven tree nut oils tested. Recent studies have shown that pecan kernels may improve human serum lipid profile and lower low density lipoprotein levels, due to their high monounsaturated fatty acid content (Rajaram, Burke, Connell, Myint, and Sabate, 2001). de la Rosa, Alvarez-Parrilla, and Shahidi (2010) identified five phenolic compounds in Mexican pecan kernels (ellagic, gallic, protocatechuic, and *p*-hydroxybenzoic acids and catechin), and two (ellagic and gallic acids) in shells. In addition they also reported the presence of seven phenolic compounds in kernels, namely, protocatechuic aldehyde, (epi) gallocatechin, one gallic acid–glucose conjugate, three ellagic acid derivatives, and valoneic acid dilactone by means of MS and UV spectral comparison.

8. Chestnut

The principal extract and medicinal constituent of *Aesculus hippocastanum* seeds is aescin, a mixture of triterpenoid saponin glycosides (Sirtori, 2001). Its components include glycosides of protoaescigenin and barringtogenol. A number of other products have been isolated from chestnut seeds, i.e., coumarin derivatives (aesculin, fraxin, scopolin), essential oils (oleic acid, linoleic acid), and tannins (leucocyanidine, proanthocyanidin A2) (Bombardelli, Morazzoni, and Griffini, 1996). The skins of chestnut are rich in tannin. Tannins of chestnut fruit consist mainly of 3,6-digalloylglucose, pyrogallol, and resorcinol (Hwang, Hwang, and Park, 2001). Ogawa *et al.* (2008) reported procyanidin trimers, polymeric proanthocyanidins having a series of heteropolyflavan-3-ols, (+)-catechin, (−)-epicatechin in seed shells of the Japanese horse chestnut. Nine flavonol oligosides of quercetin and kaempferol with glucose, xylose, and rhamnose as sugars were isolated from the seeds of *A. hippocastanum* L (Hubner, Wray, and Nahrstedt, 1999). Kapusta *et al.* (2007) identified a new glycoside, tamarixetin 3-*O*- [β-dglucopyranosyl (1→3)]-*O*-β-d-xylopyranosyl-(1→2)-*O*-β-dglucopyranoside in horse chestnut. Horse chestnut is most often used as a treatment for venous insufficiency. This is a condition associated with varicose veins, when the blood pools in the veins of the legs and causes aching, swelling, and a sense of heaviness (Kapusta *et al.*, 2007). Kimura *et al.* (2006) found that saponins from edible seeds were appreciably effective in inhibiting pancreatic lipase *in vitro* and exhibiting antiobesity effects in mice that had been fed high-fat diets. Kapusta *et al.* (2007) suggested that industrial horse chestnut waste water can be used to obtain quercetin and kaempferol glycosides for cosmetic, nutraceutical, and food supplement industries.

9. Pine nut

Traditionally known as cedar nuts, the seeds of *Pinus sibirica* are considered to be a valuable medicinal raw plant material, being ascribed a wide spectrum of traditional properties. Lantto *et al.* (2009) identified benzoic acid (protocatechuic, syringic and vanillic acids) and cinnamic acid derivatives (*m*-coumaric and (*E*)-cinnamic acids), flavanones (taxifolin, eriodictyol and naringenin) and flavan-3-ols (catechin, epicatechin and epigallocatechin gallate) in Siberian pine. The most abundant component is eriodictyol (383 ± 1.0 mg/100 g) followed by taxifolin (172 ± 3.1 mg/100 g), whilst the least abundant components are *trans*-cinnamic (12.2 ± 1.2 mg/100 g) and *m*-coumaric (trace) acids. Some other components were identified as proanthocyanidins, hydroxylated benzoic and cinnamic acid derivatives, flavan-3-ol and flavonol derivatives. The beneficial properties of *Pinus* species in general primarily related to their anti-inflammatory, antioxidant, antineoplastic and immunomodulatory properties (Guri, Kefalas, and Roussis, 2006; S. Li *et al.*, 2009; Rohdewald, 2002).

10. Macadamia nut

Tocopherols, tocotrienols, and squalene are present in macadamia kernels (Maguire *et al.*, 2004). Macadamia kernels contained low concentrations of α-tocopherol and δ-tocopherol compared to almonds and pecans (Fourie and Basson, 1989), but

Kaijser, Dutta, and Savage (2000) found high concentrations of α-tocotrienol in *Macadamia tetraphylla*, a species grown in Australia (Wall, 2010). Macadamia nuts have generated considerable interest because they are believed to have blood cholesterol lowering properties (Colquhoun, Humphries, Moores, and Somerset, 1996). Quinn and Tang (1996) found catechol, pyrogallol, 3,4,5-trihydroxy phenolic compounds, 2,6-dihydroxybenzoic acid, 2′-hydroxy-4′-methoxyacetophenone, 3′,5′-dimethoxy-4′-hydroxyacetophenone, and 3,5-dimethoxy-4-hydroxycinnamic acid in oil extract of macadamia nut kernels and shells using thin-layer chromatography.

E. HERBS AND SPICES

There is increasing interest both in the industry and in scientific research for spices and aromatic herbs because of their strong antioxidant and antimicrobial properties, which exceed many currently used natural and synthetic antioxidants (Suhaj, 2006). These properties are due to many substances, including some vitamins, flavonoids, terpenoids, carotenoids, phytoestrogens and minerals, amongst others, that render spices and some herbs their antioxidant components as preservative agents in food (Calucci *et al.*, 2003). Important fractions of spices are their volatile oils and oleoresins (Shahidi, Pegg, and Saleemi, 1995). Many herbs and spices, usually used to flavour dishes, are an excellent source of phenolic compounds which have been reported to show good antioxidant activity (Carlsen *et al.*, 2010; Embuscado, 2015; Rice-Evans *et al.*, 1996; Shahidi, Pegg, *et al.*, 1995; Zheng and Wang, 2001). Viegas, Amaro, Ferreira, and Pinho (2012) studied the inhibitory effect of ntioxidantrich marinades containing beer and white wine (with/without alcohol) alone or mixed with herbs commonly used as meat flavouring (garlic, ginger, thyme, rosemary, and red chili pepper) on the formation of heterocyclic aromatic amines (HAs) in panfried beef.The authors found that all selected marinades exhibit a reduction in total HAs formation in pan-fried meat. Shahidi, Pegg, *et al.* (1995) investigated the antioxidant efficacy of ground clove, ginger, oregano, rosemary, sage and thyme in comminuted pork systems. Spices at 200–2000 ppm levels of addition inhibited the formulation of the 2-thiobarbituric acid reactive substances (TBARS) by 12–96% over 21-days of storage at 4°C. The antioxidant efficiency of spices was reported as clove > sage > rosemary = oregano > thyme = ginger. In addition, some herbs such as rosemary and sage are used to produce drugs classified as phytopharmaceuticals, representing a significant part of the world pharmaceutical market (Kähkönen *et al.*, 1999; Madsen and Bertelsen, 1995). A problem which is often encountered when using spices and herbs is that they generally have a distinct odour, taste, and colour, a fact that makes their systematic use in foods (for example, in spreads) rather difficult (Kiokias et al., 2008). Besides, the extraction procedures are quite complicated and cannot easily be made into an economically attractive process. Purification of the extracts and further processing to convert them to additives suitable for foods (Kiokias *et al.*, 2008) are usually needed. Furthermore, plants show seasonal and topographical variations, giving rise to differences in their antioxidant activities, whereas the safety of the extracts is a topic of current discussion (Kiokias, 2006). However, deflavoured spice extracts and oleoresins, such as those of rosemary and sage are commercially available.

1. Rosemary and sage

The greatest level of attention amongst herbs and spices as sources of antioxidants has been focused on rosemary (Berdahl and McKeague, 2015; Erkan, Ayranci, and Ayranci, 2008; Rodríguez-Rojo, Visentin, Maestri, and Cocero, 2012). In earlier studies, sage (*Salvia officinalis* L.) and rosemary were shown to have similar patterns of phenolic compounds and their antioxidant activity was attributed mainly to carnosic acid, carnosol and rosmarinic acid present (Figure 8.24) (Frankel, Huang, Aeschbach, and Prior, 1996; Okamura, Fujimoto, Kuwabara, and Yagi, 1994; Senanayake, 2013; Thorsen and Hildebrandt, 2003; Y. Zhang *et al.*, 2012). These polyphenols also possess important biological activities *in vitro* as anti-tumour, chemopreventive and anti-inflammatory agents (Al-Sereiti, Abu-Amer, and Sen, 1999; Cheung and Tai, 2007; Danilenko *et al.*, 2003; Shuang-sheng and Rong-liang, 2006). It has been proposed that polyphenols of rosemary may greatly increase functionality of food for health and wellness (Shahidi and Naczk, 2004). Antioxidative effect of rosemary ethanolic extract on butter was reported by Zegarska *et al.* (1996, 1998). The main compounds responsible for rosemary's antioxidant properties have been identified as phenolic diterpenes, such as carnosic acid, carnosol, rosmanol, epi- and iso-rosmanol, rosmadial and methyl carnosate (Bandonienė, Murkovic, Pfanhauser, Venskutonis, and Gruzdiéne, 2002; Bandonienė, Venskutonis, Gruzdiene, and Murkovic, 2002; Ibañez *et al.*, 2003; Pizzale, Bortolomeazzi, Vichi, and Conte, 2002). Other compounds, such as rosmarinic acid, caffeic acid and flavonoids, have also been associated with the antioxidant activity of rosemary (del Baño *et al.*, 2003; Suhaj, 2006). Recently, Escriche, Kadar, Juan-Borrás, and Domenech (2014) found kaempferol, chrysin, pinocembrin, caffeic acid and naringenin in rosemary honey. Wang *et al.* (1998) extracted sage with ethanol and identified seven flavonoids, 15 diterpenoids, 17 triterpenes and steroids including rosmarinic acid, 4-*O*-β-dglucopyranosylacetophenone (picein), 6-*O*-(*E*)-feruloyl-(α and β)-glucopyranoside, (+)-1-hydroxypinoresinol-1-β-d-glucoside, homoplantaginin, (−)-isolariciresinol 3α-*O*-β-d-glucopyranoside, luteolin-7-*O*-β-glucopyranoside, 4-hydroxyacetophenone-4-*O*-β-d-apiofuranosyl-(1→6)-*O*-β-d-glucopyranoside, icariside F2, and 2,3-dihydro-2-(4′-hydroxy-3′-methoxyphenyl)-3-(hydroxymethyl)-7-methoxy-5- benzofuranpropanol 4′-*O*-β-glucopyranoside. Recently, Kontogianni *et al.*, (2013) reported that the rosemary extract exhibits more pronounced antioxidant, cytotoxic and immunomodifying activities, due to the presence of betulinic acid and a higher concentration of carnosic acid. Romano, Abadi, Repetto, Vojnov, and Moreno (2009) suggested that the methanolic extract of rosemary synergistically enhances the antiradical efficiency of BHT and the antibacterial activity of BHA. It has also been observed that ascorbic acid (500 mg/kg of ascorbic acid + 200 mg/kg of rosemary extract) enhanced the antioxidant activity of rosemary extract in lard (Chang, Ostric-Matijasevic, Hsieh, and Huang, 1977). Whilst rosemary extracts have been found to be effective antioxidants, they are generally considered to be less effective than BHA and BHT (Berdahl *et al.*, 2010; Sebranek, Sewalt, Robbins, and Houser, 2005). For example, Ahn, Gruen, and Fernando (2002) reported that rosemary was significantly less effective than BHA/BHT for suppression of oxidative changes in cooked ground beef. Several studies have shown that rosemary extracts are effective oxidative stabilizers of various meat and poultry products as well as vegetable and fish oils (Senanayake, 2013; St. Angelo, Crippen, Dupuy, and James, 1990; Ucak, Ozogul, and Durmus, 2011; Wanasundara and Shahidi, 2005). In addition rosemary extracts are effective antioxidants in snack foods, nuts, baked goods and pet foods (Gruiner-Richter, Otto, and Weidner, 2012; Lalas and Dourtoglou, 2003). Rosemary and sage extracts are commercially available as StabilEnhance®, OxyBlock®, Xtrablend®, Fortium®, Herbalox®, Duralox®, Aquarox®, Inolens®, Vivox®, and are manufactured by Naturex, Kemin, Kalsec, Flavex and Vitiva (Embuscado, 2015).

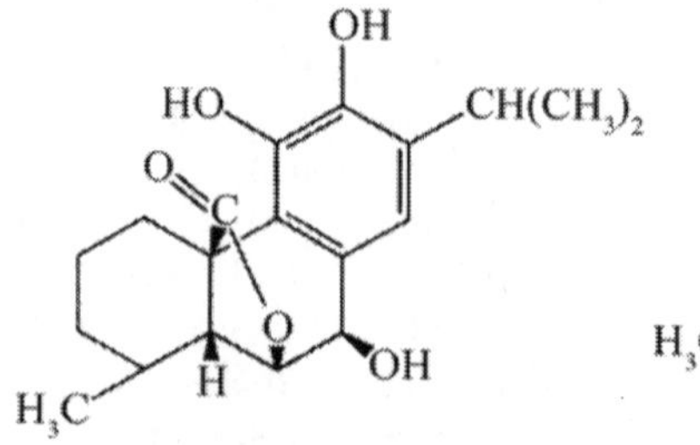

Rosmanol

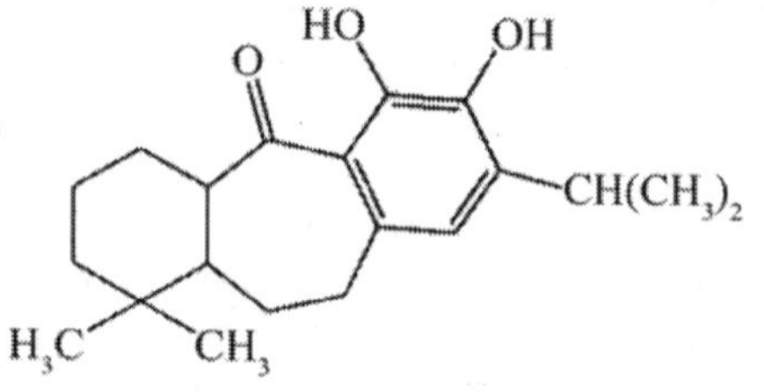

Rosmaridiphenol

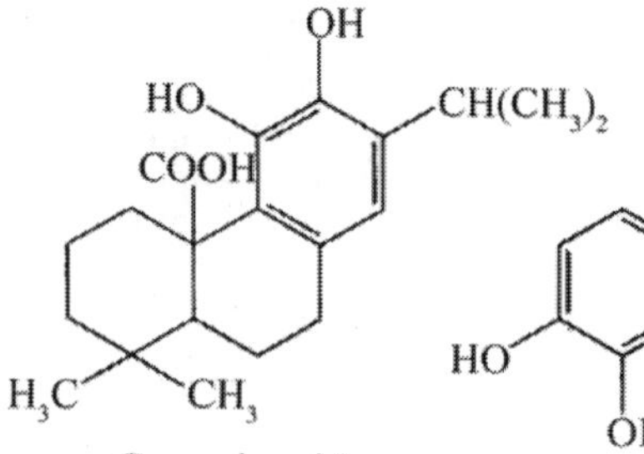

Carnosic acid

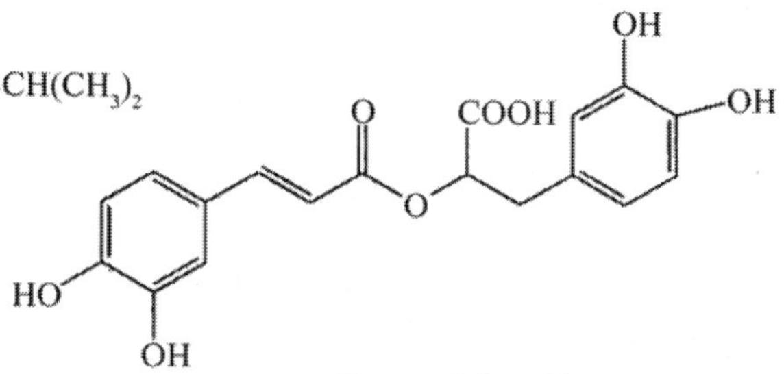

Rosmarinic acid

Figure 8.24: Chemical structures of selected antioxidant compounds in rosemary (Source: Adapted from Shahidi and Naczk, 2004).

2. Clove

Cloves are the dried flower buds of *Syzygium aromaticum* and serve as sources of anti-microbial agents against oral bacteria that are commonly associated with dental caries and periodontal disease (Cai and Wu, 1996). The major aroma constituents of *S. aromaticum* buds are eugenol (2-methoxy-4-allylphenol) and eugenyl acetate (Lee and Shibamoto, 2002). Eugenol is reported to have antifungal activity (Martini, Weidenborner, Adams, and Kunz, 1996) and inhibits malonaldehyde formation from cod liver oil and the formation of hexanal (Lee and Shibamoto, 2001). Sesquiterpenes, in clove may serve as potential anticarcinogenic agents (Zheng, Kenney and Lam, 1992). The main components of clove leaf essential oil are eugenol (76.8%), followed by β-caryophyllene (17.4%), α-humulene (2.1%), and eugenyl acetate (1.2%) (Jirovetz *et al.*, 2006).

3. Ginger

Kikuzaki and Nakatani (1993) reported that chemical constituents like gingerols and shogaols present in ginger exhibit strong antioxidative activity. This has clearly been demonstrated by inhibition of phospholipid peroxidation induced by the $FeCl_3$–ascorbate system (Aeschbach *et al.*, 1994) and its inhibitory effect on xanthine oxidase system (Chang, Chang, Lu, and Chiang, 1994) which is responsible for the generation of reactive oxygen species, such as superoxide anion. Sekiwa, Kubota, and Kobayashi (2000) isolated two glucosides of 6-gingerdiol, namely 1-(4-*O*-*β*-d-glucopyranosyl-3-methoxyphenyl)-3,5-dihydroxydecane and 5-*O*-*β*-d-glucopyranosyl-3-hydroxy-1-(4-hydroxy-3-methoxy phenyl) decane and reported that 5-*O*-*β*-d-glucopyranosyl-3-hydroxy-1-(4-hydroxy-3-methoxy phenyl)decane had a strong antioxidant activity in a linoleic acid model system. Other gingerol-related compounds are formed from phenylalanine via ferulic acid, and diarylheptanoids and terpenoids (Kikuzaki, Tsai, and Nakatani, 1992). Ginger and the components in it are also known to possess such physiological features as antimicrobial, antioxidative, antitumour, and antiplatelet aggregation activities (Conner, 1991; Sekiwa *et al.*, 2000; Yin and Cheng, 1998). The anticancer properties of ginger are attributed to the presence of certain pungent vallinoids, gingerol and paradol, as well as some other constituents like shogaols, zingerone, amongst others (Shukla and Singh, 2007). Nirmala, Prasanna Krishna, and Polasa (2007) reported the *in vivo* antimutagenic effect of ginger in rats and suggested that the antimutagenic and chemopreventive potential of ginger could be due to its antioxidant activity.

4. Black pepper

Agbor, Vinson, Oben, and Ngogang (2006) reported that black pepper is more effective than white pepper in scavenging free radicals and ROS. This property of black pepper is related to its higher polyphenol content. Chemically, peppercorn contains lignans, alkaloids, flavonoids, aromatic compounds, and amides (Jirovetz, Buchbauer, Ngassoum, and Geissler, 2002). Essential oil and oleoresins of black pepper exhibit high antioxidant and radical scavenging activities against various antioxidant assays *in vitro* (Kapoor *et al.*, 2009).The antioxidative action of black pepper, for instance, has been attributed to its content of piperine and piperine isomers, such as chavicine, isopiperine, and isochavicine, and monoterpene (Chipault, Mizuno, and Lundberg, 1956; Milbourne, 1987; Nakatani, Inatani, Ohta, and Nishioka, 1986; Yanishlieva, Marinova, and Porkorny, 2006). However, Kapoor *et al.* (2009) showed that the antioxidant activity of black pepper was due to the presence of β-caryophyllene, limonene, β-pinene, piperine and piperolein in essential oil and oleoresins. Suhaj (2006) reported that according to USDA (2002), ascorbic acid, beta-carotene, camphene, carvacrol, eugenol, gamma-terpinene, lauric acid, linalylacetate, methyleugenol, myrcene, myristic acid, myristicin, palmitic acid, piperine, terpinen-4-ol and ubiquinone were the antioxidants present in black pepper.

5. Turmeric

Curcuminoids [i.e., curcumin (1,7-bis(4-hydroxy-3-methoxyfenil)-1,6-heptadiene-3,5-dione), demethoxy-curcumin, and bisdemethoxy-curcumin] are the major antioxidative compounds of turmeric (Schieffer, 2002). In addition to curcuminoids, other compounds possessing antioxidant capabilities that are present in turmeric include γ-terpinene, ascorbic acid, beta-carotene, beta-sitosterol, caffeic acid, campestrol, camphene, dehydrocurdione, eugenol, *p*-coumaric acid, protocatechuic acid, stigmasterol, syringic acid, turmerin, turmeronol and vanillic acid (Cousins, Adelberg, Chen, and Rieck, 2007; Relajakshmi and Narasimhan, 1996). Curcumin, the principal polyphenolic curcuminoid is reported to have antioxidant, anticarcinogenic, anti-HIV, antibacterial and anti-inflammatory activities (Smitha, Dhananjaya, Dinesha, and Srinivas, 2009). Further studies have revealed that curcumin may suppress cancer development by helping in inhibiting enzymes that lead to tumour production (Surh, 2002) and prevents cancer along with inflammation by inducing production of enzymes used to detoxify electrophyllic species produced in lipid peroxidation (Piper *et al.*, 1998). Verma, Salamone, and Goldin (1997) demonstrated that curcumin inhibited the growth of human breast cancer cells.

6. Oregano

Oregano (*Origanum vulgare* L.), a widely used herb, also commonly used in Western dishes possesses antithrombin, anti-*Helicobacter pylori*, antibiotic, antihyperglycaemia and antioxidation effects (Goun, Gunningham, Solodnikov, Krasnykch, and Miles, 2002; Lin *et al.*, 2008; Stamatis *et al.*, 2003). The compound 4-(3,4-dihydroxybenzoyloxymethyl)phenyl-*O*-β-d-glucopyranoside is a major constituent of oregano and might contribute to its antioxidant activity (Nakatani and Kikuzaki, 1987). Kikuzaki and Nakatani (1993) also isolated five different phenolic compounds from the methanol extract of leaves of oregano, namely, caffeic, protocatechuic, and rosmarinic acids, a phenyl glucoside and a new compound 2-caffeoyloxy-3-[2-(4-hydroxybenzyl)-4,5-dihydroxy]phenylpropionic acid; amongst these rosmarinic acid was present in highest concentration. Rosmarinic acid is a caffeoyl ester and has now been shown to be an important antioxidant and anti-inflammatory compound (Chun, Vattem, Lin, and Shetty, 2005). Antioxidative effect of oregano ethanolic extract on butter was reported by Amarowicz, Carle et al. (2009) and Amarowicz, Zegarska *et al.*, (2009). Phenolic compounds and flavonoids such as luteolin, hispidulin, apigenin, acacetin, diosmetin, herbacetin, quercetin and naringin had also been described in oregano extracts (Cavero *et al.*, 2006; El-Ansari, El-Negoumy, and El-Desoky, 1996; Justesen and Knuthsen, 2001; Pizzale *et al.*, 2002; Zheng andWang, 2001).

7. Fenugreek seed

Fenugreek seeds are a rich source of polyphenols (Kenny, Smyth, Hewage, and Brunton, 2013). Several compounds were identified by HPLC, including apigenin and a number of kaempferol and quercetin glycosides (Chatterjee, Variyar, and Sharma, 2009) as well as the flavonoids; vitexin, tricin, naringenin, quercetin and tricin 7-*O-β-D*-lucopyranoside (Shang et al., 1998). Studies have shown that fenugreek extracts are effective in reducing the total cholesterol and triglyceride levels in patients with coronary artery disease (CAD) (Bordia, Verma, and Srivastava, 1997), alleviating physiological oxidative stress and inflammation in arthritic mice (Sindhu, Ratheesh, Shyni, Nambisan, and Helen, 2012) and protecting against hepatotoxicity caused by the anti-cancer drug adriamycin (ADR) in albino rats (Sakr and Abo-El-Yazid, 2012).

■■■■

9 Effect of processing on antioxidant activity

INTRODUCTION

Interest in the development of policy statements about drought tolerant grains is increasing in several developing countries such as India, China, and some countries of Africa because of water scarcity and increasing populations. In addition, the earmarked funds to scientific research for purposes of improving and increasing their production and utilization as food have also been increased. Millet is one of the most important drought-resistant crops and the 6th cereal crop in terms of world agriculture production. Also, millet has resistance to pests and diseases, short growing season, and productivity under drought conditions, compared to major cereals (Devi *et al.,* 2011). Therefore, millet grains are now receiving specific attention from these developing countries in terms of utilization as food as well as from some developed countries in terms of its good potential in the manufacturing of bioethanol, nutracutical foods and biofilms (Li *et al.*, 2008).

Fruits and vegetables account for a small part of our daily caloric intake; however their benefits to health surpass their caloric contribution. The contributory factors are due to the presence of vitamins and provitamins, such as ascorbic acid, tocopherols and carotenoids and, in addition to that, they are also rich in a wide variety of phenolic substances (Loliger, 1991). Phenolic substances are a category of phytonutrients that exert strong antioxidant properties (Ho, 1992). They can be classified into simple phenols, phenolic acids, hydroxycinnamic acid derivatives and flavonoids. The ability of some of the phenolic substances to act as potent antioxidant components has been reported (Velioglu *et al*., 1998; Kähkönen *et al*., 1999).

AANTIOXIDANT CONTENTS AND ACTIVITIES

The growing public awareness of nutrition and health care research substantiates the potential of phytochemicals such as polyphenols and dietary fiber on their health beneficial properties (Devi *et al.,* 2011). The increased consumption of whole grains and whole grain products has been associated with reduced risk of developing chronic diseases such as cardiovascular disease, type 2 diabetes, some cancers, and all-cause mortality. In addition, whole grains contain unique phytochemicals that complement those in fruits and vegetables when consumed together (Liu, 2007). Polyphenols are the biggest group of phytochemicals that have been found in plant-based foods and have been linked to several health benefits. Therefore, dietary polyphenols have received tremendous attention among nutritionists, food scientists, and consumers due to their roles in human health (Tsao, 2010). It has been reported that soluble- and insoluble-bound phenolic extracts of several varieties of millet (kodo, finger, foxtail, proso, pearl, and little millets) whole grains are rich sources of phenolic compounds and show antioxidant, metal chelating, and reducing powers. However, the potential of whole millets as natural sources of antioxidants depends on the variety used (Chandrasekara and Shahidi, 2010). Much attention has been devoted to investigations of the nutraceutical and antioxidant properties of some major millet varieties, including finger millet, pearl millet, and foxtail millet. It has been reported that foxtail millet contains 47mg polyphenolics/100 g and 3.34 mg tocopherol/100 g (wet basis); however, proso millet contains 29 mg polyphenolics/ 100 g and 2.22 mg tocopherol/100 g (wet basis). In addition, a positive and significant correlation ($R^2 = 0.9973$, $P<0.01$) between polyphenolic content and radical cation scavenging activity was observed (Choi *et al.*, 2007). For finger millet, the high-performance liquid chromatography (HPLC) analysis of polyphenols indicated nearly 30 prominent constituent phenolics, but only about 30% of them could be identified (Chethan and Malleshi, 2007). Furthermore, phenolic acids from the milling fractions of finger millet (whole flour, seed coat, 3%, 5%, and 7%) were also isolated. Acidic methanol extracts from seed coat to whole flour were rich in polyphenol content and were found to be stable up to 48 h at pH 4, 7, and 9 (Viswanath *et al.*, 2009). Currently, over 50 phenolic compounds belonging to several classes, namely, phenolic acids and their derivatives, dehydrodiferulates and dehydrotriferulates, flavan-3-ol monomers and dimers, flavonols, flavones, and flavanonols in 4 phenolics fractions of several whole millet grains (kodo, finger, foxtail, proso, little, and pearl millets) were positively or tentatively identified using HPLC and HPLC-tandem mass spectrometry (MS). However, insoluble bound fraction of kodo millet showed the highest phenolic content as well as antioxidant activity in the *in vitro* test systems employed (Chandrasekara and Shahidim, 2011). Therefore, and based on literature data, millet grains can be used as functional food ingredients and as sources of natural antioxidants.

The antioxidant activity of phenolics and other bioactive components extracted from millet grains and their fractions was evaluated in a number of research studies. It was reported that 3 antioxidative phenolic compounds, 1 serotonin derivative, and 2 flavonoids were isolated from an ethanol extract of Japanese barnyard millet grains. Their structures were established to be *N*-(*p*-coumaroyl) serotonin, luteolin, and tricin. Although the antioxidant activity of luteolin was lower than that of *N*-(*p*coumaroyl) serotonin, it was nearly equal to that of quercetin, whereas the activity of tricin was lower than that of luteolin (Watanabe, 1999). Kodo millet, finger millet, little millet, foxtail millet, barnyard millet, and sorghum *bicolor* grown in India and their white varieties were screened for free radical quenching of 1, 1, diphenyl-2-picrylhydrazyl (DPPH) by electron spin resonance. Methanol extracts of the kodo millet flour showed 70% DPPH quenching in comparison to other millet extracts that showed 15% to 53%. Further, the white varieties of sorghum, finger millet, and foxtail millet showed lower quenching than their colored counterparts, indicating that phenolics in the seed coat could be responsible for the antioxidant activities. However, the content of the phenols and tannin in these grains did not correlate with the antioxidant activities (Hegde and Chandra, 2005). Furthermore, finger millet extracts were found to have a potent radical-scavenging activity that is higher than those of wheat, rice, and other species of millet (Dykes and Rooney, 2006). In addition, the reducing power of finger millet seed coat extract was significantly ($P < 0.05$) higher than that of whole flour extract (Viswanath *et al.,* 2009). Moreover, xylo-oligosaccharides (XOs) mixture of finger millet exhibited relatively higher antioxidant activity than the XOs of rice, wheat, and maize by DPPH and ferric reducing antioxidant power assays (Veenashri and Muralikrishna, 2011). For foxtail millet, methanolic extracts of whole flour and bran-rich fraction exhibited a significantly higher ($P < 0.05$) radical-scavenging activity (44.62% and 51.80%, respectively) using a DPPH model system, and reducing power (0.381 and 0.455, respectively) at 2 mg, than the ethanol and water used for extraction (Suma and Urooj, 2011). On the other hand, 50% ethanol extract from defatted foxtail millet bran was found to be the best-promoting phenolic compound with substantial antioxidant activity (Amadou *et al.*, 2011). In addition, defatted foxtail millet protein hydrolysates also exhibited antioxidant potency (Mohamed *et al.*, 2012). Thus, millet may serve as a natural source of antioxidants in food applications and as a nutraceutical and functional food ingredient in health promotion and disease risk reduction. However, more studies in animal models and with human subjects should be performed to verify their activity and health benefits.

EFFECT OF PROCESSING ON ANTIOXIDANT ACTIVITY OF MILLET GRAINS

The processing methods mentioned above such as malting, decortication, soaking, and cooking were also found to affect the content and activity of the antioxidants in millet grains. In one study, the antioxidant capacity of the fraction containing free phenolic acids was increased (2-fold) after 96 h of malting of finger millet, whereas the antioxidant capacity of the fraction containing bound phenolic acids was decreased (Rao and Muralikrishna, 2002). In another study, a growth-promoting medium was developed to enhance the production of a hydroxyl radical inhibitory water-soluble protein from germinated millet. The single factor test indicated that H_2O_2 plays a key role in the inhibition activity (Li *et al.*, 2007). In addition, to further increase the yield of the hydroxyl radical inhibitory water-soluble protein from stress-germinated millet, the effects of the sprouting conditions (temperature, time, and pH of stress medium) on the hydroxyl radical inhibition were investigated. The optimal conditions were identified as temperature 28°C, culture time 54 h, and stress medium pH 7.5. Under optimum conditions, the highest inhibition (60.38%) was achieved (Li *et al.*, 2008). Also, germination of foxtail millet for 3 d allowed obtaining flour with high DPPH-scavenging activity (Coulibaly and Chen, 2011). Cooking of kodo or finger millet by roasting or boiling resulted in a reduction in antioxidant activity. Fractionation of kodo millet into husk and endosperm also decreased the DPPH quenching activity and the phytochemicals appeared to act synergistically (Hegde and Chandra, 2005). Furthermore, effects of germination, steaming, and roasting on the nutraceutical and antioxidant properties of little millet were also investigated. The results showed that the total phenolic, flavonoid, and tannin contents of processed little millet increased by 21.2, 25.5, and 18.9 mg/100 g, respectively, compared to the native sample (Pradeep and Guha, 2011). However, dehulling and hydrothermal treatments were found to affect the antioxidant potential and phenolic content of pearl millet grains (Chandrasekara *et al.*, 2012). The reduction in antioxidant contents and their activities can be attributed to oxidation and degradation reactions taken place during thermal treatments, such as cooking, boiling, and roasting. However, the reduction happened by dehulling can be attributed to removing of pericarp layer from the grains, which is known rich in polyphenol and antioxidant compounds. Therefore, processing of millet grains, their fractions, and food products must be under optimized conditions to protect their quality and potential health benefits. In addition, the increase in antioxidant contents and their activities by germination can be attributed to conversion of complex components to simpler compounds with higher antioxidant activity by endogenous enzyme during germination.

Buckwheat products obtained after thermal treatment, in extrusion cooking process, still had significantly higher antioxidant capacity when compared to the extruded cereal grains and should be more widely recommended in human nutrition (Henryk Zieliński *et al.*, 2006).

Dehulled and cooked grains of five millet varieties (kodo, finger, proso, foxtail and pearl) were subjected to *in vitro* enzymatic digestion and microbial fermentation under physiological conditions in order to determine the bioaccessibility of their phenolic compounds. Extracts recovered as supernatants from enzymatic digestion and microbial fermentation were employed for the determination of their total phenolic content (TPC) and total flavonoid content (TFC), as well as the scavenging of 2,2-diphenyl-1-picrylhydrazyl (DPPH), peroxyl and hydroxyl radicals. Furthermore, trolox equivalent antioxidant capacity (TEAC), reducing power and ferrous ion chelating activity of the extracts so obtained were evaluated. The DPPH and hydroxyl radical scavenging activities were determined using electron paramagnetic resonance (EPR) spectroscopy. The peroxyl radical activity was measured using an oxygen radical absorbance capacity (ORAC) assay. The TPC ranged from 12.7 to 35.4 and 21.2 to 47.4 lmol ferulic acid equivalents per gram of grain, on a dry weight (dw) basis at the end of intestinal digestion and colonic fermentation, respectively. All five millet varieties exhibited effective antioxidant activity and the order of efficacy differed according to the assay employed. The present study thus demonstrated that phenolic compounds of processed millets were bioaccessible and colonic fermentation released the phenolics bound to the insoluble fibre in the grain (Chandrasekara and Shahidi, 2012).

Millets are generally dehulled and subjected to a hydrothermal treatment before consumption, thus the hulls can be used as a potential source of antioxidants. Several millet grains, namely kodo, finger (Ravi), finger (local), proso, foxtail, little and pearl millet were studied. Antioxidant activities of phenolic extracts obtained from whole grains, as well as their corresponding dehulled and cooked grains and hulls were studied for their total phenolic content (TPC), radical scavenging capacity, and antioxidant activity in a β-carotene/linoleate emulsion. The phenolics present in whole grains were identified and quantified using HPLC and HPLC/MS and results were expressed as total for each of the phenolic groups. The TPC ranged from 2 to 112 μmol ferulic acid equivalents/g defatted meal. All varieties exhibited effective inhibition of 2, 2-diphenyl-1-picrylhydrazyl (DPPH), hydroxyl, peroxyl and superoxide radicals. Dehulling and cooking affected the TPC and radical scavenging and antioxidant activities of the grains, depending on the variety. In general, the antioxidant activity of phenolic extracts was in the order of hull > whole grain > dehulled grain > cooked dehulled grain. With the exception of the two finger millet varieties, hulls of other millet grains had high TPC, thus demonstrating their superior antioxidant activity. Hydroxybenzoic acids, hydroxycinnamic acids and flavonoids in whole grains were identified as contributors to the observed effects. Therefore, dehulling of grain and hydrothermal treatments affect the phenolic content and antioxidant potential of millet grains (Chandrasekara *et al.*, 2012).

Biochemical changes in antioxidant activity during soaking of sorghum

The results shown in Table 9.1 exhibited the changes in antioxidant activity (AO) and antioxidant capacity (AC) during soaking. Data revealed that DPPH and ABTS scavenging activity in raw sorghum three genotypes varied from 21.72 to 27.69% and 25.29% to 31.97% respectively. Shandaweel-6 recorded the highest AO and AC due to its highest content of total phenols and flavonoids contents. Also, AC in raw sorghum varied from 12.82 to 16.39 μmole/g and 13.82 to 17.52 μmole/g, respectively. After soaking DPPH and ABTS scavenging and antioxidant capacity were significantly decreased (Afify *et al.*, 2012).

Table 9.1: Effect of soaking of sorghum grains on AO and AC (mean ± SD).

Treatment	Variety	AC_{ABTS} (μmole/g)	AC_{DPPH} (μmole/g)	AO_{ABTS} (%)	AO_{DPPH} (%)
Raw	Dorado	15.29±0.22	14.33±0.84	28.00±0.42	24.59±1.34
	Shandaweel-6	17.51±1.55	16.39±1.10	31.97±2.82	27.69±1.88
	Giza-15	13.82±1.25	12.82±0.63	25.29±2.28	21.72±1.07
Soaking	Dorado	12.36±1.15	11.66±0.36	23.21±2.15	20.28±0.63
	Shandaweel-6	15.23±0.69	12.99±1.48	28.49±1.29	22.47±2.57
	Giza-15	11.81±0.82	10.70±0.10	22.17±1.54	18.59±0.18
	LSD	1.8462	1.5699	3.4082	2.6648

Effect of heat processing on the antioxidant properties of pearl millet

DPPH is a stable free-radical compound widely used to test the free-radical scavenging ability of various samples (Sakanaka *et al.*, 2005). Antioxidant activity (AOA) of different control pearl millet cultivars varied significantly ($p < 0.05$) and ranged between 31.8 and 46.7% (Siroha *et al.*, 2016). Antioxidant activity (%) after toasting and roasting is shown in Fig. 9.1. Upon toasting, AOA of pearl millet flours increased and this increased by 6.5 to 23.5%, the highest increase was for cv.HHB-

223 whereas lowest was observed for cv.W-445. Dewanto *et al.* (2002) found an increase in AOA after thermal treatment for sweet corn. Pradeep and Guha (2011) reported that the radical scavenging activity of processed millet extracts increased when compared to their native counterpart sample. An increased AOA activity can be attributed to the formation of non-enzymatic browning products especially melanoids at high temperatures which contribute to AOA (Sharma, and Gujral, 2011). Cooking led to a significant decrease in AOA by 9.6 to 37.4%, the highest and the lowest decrease was observed for flours from cv.W-445 and cv.GHB-732, respectively. Chandrasekara *et al.* (2012) reported a decrease in AOA in the millets after cooking. Xu and Chang (2008) also found that after processing different grains, the DPPH radical scavenging was significantly reduced when compared to their counterpart sample. A decrease in AOA after cooking may be due to the activity leached into the boiling water (Chuah *et al.*, 2008). AOA after toasting had higher values than the values after cooking.

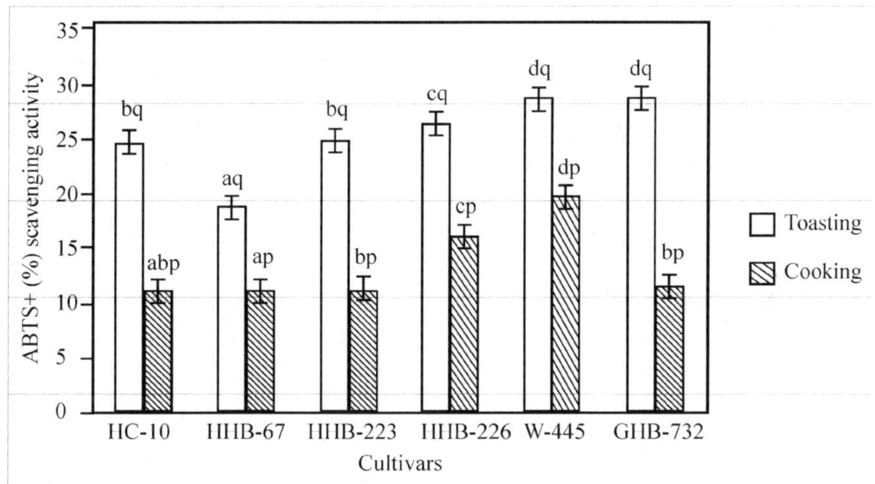

Figure 9.1: Antioxidant activity (%) of flours after toasting and cooking. Superscript (a–e) are significantly ($p < 0.05$) different within cultivars whereas superscript (p and q) are significantly ($p < 0.05$) different between toasting and cooking.

ABTS$^+$ scavenging activity

The antioxidative capacity of test compounds is assessed by measuring their ability to reduce the ABTS radical to its non-radical form. ABTS$^+$ activity of different control pearl millet cultivars varied from 18.3 to 25.0%, with cv.W-445 showing the maximum activity. Salar and Purewal (2016) observed ABTS$^+$ activity from 27.0 to 52.01% for different pearl millet cultivars. ABTS$^+$ (%) scavenging activity of flours after toasting and cooking is shown in Fig. 9.2. After toasting ABTS$^+$ activity of flours increased by 0.3 to 25.5% with cv.GHB-732 showing the maximum effect. Chandrasekara and Shahidi (2012) also reported an increase in AOA values of thermally treated cashew kernels compared to that of its unroasted counterparts. It has been also reported that during thermal treatment Maillard products such as HMF

(5-hydroxymethyl-2-furaldehyde) are formed and contributed to AOA (Sharma and Gujral 2014). Cooking, however, had the reverse effect as compared to toasting and ABTS$^+$ scavenging activity decreased by 22.8 to 54.1%. Most of the phenolic compounds are found to be heat sensitive, which could be the reason for their significant decrease in scavenging activity upon cooking (Gujral *et al.*, 2012). Comparison of both processing methods revealed that toasting resulted in higher ABTS$^+$ scavenging activity than cooking.

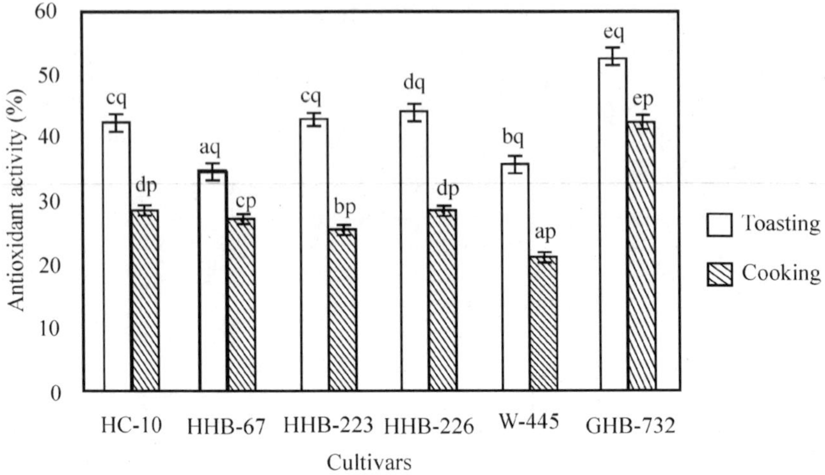

Figure 9.2: ABTS$^+$ (%) scavenging activity of flours after toasting and cooking. Superscript (a–d) are significantly ($p < 0.05$) different within cultivars whereas superscript (p and q) are significantly ($p < 0.05$) different between toasting and cooking.

Pushparaj and Urooj, (2014) reported that the antioxidant activity of pearl millet was influenced both by the processing methods and the cultivars. K and MRB showed low inhibition values of DPPH (about 60%), RPA (about 0.32) and FRAP (about 2.69). Between the cultivars, K exhibited a higher content of antioxidant components reflecting its higher antioxidant capacity. The Bran Rich Fraction showed high antioxidant activity in terms of RPA which was due to the tannin, phytic acid and flavonoid levels. The millet subjected to various heat treatments exhibited higher antioxidant activity (DPPH scavenging activity and RPA) mainly due to its flavonoid content. An inverse relationship was found between radical scavenging activity and yield. It was noteworthy that significantly ($P \leq 0.05$) high radical scavenging activity in heat treated millet had the lowest yield, whilst germinated millet which showed the lowest activity had the highest yield.

Effect of processing on antioxidant activity (AOA) of minor millets

The AOA as determined by the three different methods showed a wide range of values. The DPPH radicals scavenging activity of soaked millets of both the samples was higher which was found to be 51.06% and 52.12% respectively. The increase is due to the lower heat treatment provided to soaked samples after processing. The

FRAP values of roasted sample of foxtail millet and blanched proso millet had highest antioxidant power ranging from 317.5 μmol and 236.8 μmol respectively (Table 9.2). The FRAP values depends on the part of samples used for determination. The blanched millet of both the samples had higher reducing power i.e., 0.426 and 0.418 respectively which determines higher antioxidant activity of these millets. It was observed that ethanol had the higher antioxidant activity when compared with other solvents. The correlation between TPC and AOA among foxtail and proso millet was determined which showed 44.6% significant difference in foxtail millet with respect to DPPH, FRAP assay and reducing power; whereas proso millet showed 56.46% significant difference with respect to AOA. Thus TPC was significantly correlated with AOA as determined by the three methods (Jemima Beryl Mohankumar and Indu Vaishnavi, 2012).

Table 9.2: Antioxidant activity and phenolic content of processed millet flours

Treatment	DPPH (%)		FRAP (μmol of trolox)		Reducing power		Phenolic content (mM of GAE)	
	Foxtail millet	Proso millet	Foxtail millet	Proso millet	Foxtail millet	Proso millet	Foxtail millet	Proso millet
Boiled	31.52	31.04	197.4	115.2	0.151	0.394	4.56	1.44
Blanched	39.02	50.00	214.1	236.8	0.426	0.418	4.16	5.88
Roasted	36.79	42.45	317.5	103.6	0.357	0.357	8.36	4.36
Soaked	51.06	52.12	237.5	207.6	0.341	0.402	9.44	5.40
Germinated	37.81	41.97	258.1	153.8	0.400	0.392	16.92	7.52

α-Tocopherol content of little millet landraces at various stages of processing

Depicts α-tocopherol (mg/100 g) content of different millets at various stages of processing of decorticated grain and bran. The values for α -tocopherol ranged between 1.75 to 5.43 mg/100 g. After milling there was reduction in α -tocopherol content. The sample Mantrodi-77 millet had highest α -tocopherol content (5.43 mg/100 g) followed by Ganjigatti-55 (4.90 mg/100 g) and least in Chadaval-53 (1.75 mg/ 100 g) (Table 9.3). Bran contained higher amount of α - tocopherol compared to decorticated grain. Per cent retention of α -tocopherol varied among the millets. It ranged between 17.07 to 62.85 per cent. The highest retention was observed in Chadaval-53 millet followed by Jekinkatti-82 millet and least in Kamplikoppa-60 millet sample. Youngmin *et al* (2006) also reported that black decorticated grain contained higher amount of vitamin E (1.2 mg/100 g) whereas proso millet contained least or negligible amounts of vitamin E (0.01 mg/100 g).

Table 9.3: α-Tocopherol (mg/100 g) content of different types of little millets at various stages of processing

Millet	Whole grain	Decorticated grain	Bran	% retention in decorticated grain
Chikkayagti-01	2.42	1.47	4.42	60.00
Chikkayagti-07	3.66	1.60	4.80	43.71
Jekinkatti-33	2.48	1.21	3.62	48.79
Chadaval-35	4.26	0.87	2.62	20.42
Chadaval-53	1.75	1.10	3.31	62.85
Ganjigatti-55	4.90	1.18	3.54	24.08
Kamplikoppa-60	4.86	0.83	2.49	17.07
Chadaval-66	3.80	1.10	3.30	28.94
Mantrodi-77	5.44	1.10	3.30	20.22
Jekinkatti-82	2.51	1.53	4.58	60.95
Mean	3.61	0.12	3.60	
SEM±	0.10	0.13	0.13	
CD @ 5%	0.30	0.38	0.38	

Source: Kundgol *et al.* (2014).

Effect of Processing on antioxidant activity of fruits and vegetables

Processing can alter and often damage fruit and vegetable antioxidants. Maceration, heating, and various separation steps can result in oxidation, thermal degradation, leaching, and other events that lead to lower levels of antioxidants in processed food compared with fresh. This is particularly true in the case of vitamin C and phenolic antioxidants. However, in the case of carotenoids, processing can lead to a dissociation of antioxidants from plant matrix components, an increase in carotenoid antioxidants and improved digestive absorption (Shi and le Maguer, 2000). As discussed in an earlier section, the localization of components within plant materials becomes important when tissues such as peels and seeds are separated from other components during processing, as in juice and wine production (Waterhouse and Walzem, 1998). This can reduce the level of certain antioxidant components from processed food products and yield processing byproducts that still contain substantial levels of antioxidant compounds (Waterhouse and Walzem 1998; Skrede *et al.*, 2000). While most studies examine the impact of specific factors on the content of antioxidant phytochemicals, few studies have examined their impact on the actual antioxidant activity of foods. In a survey of commercially available spinach, peas, green beans, and carrots, the highest antioxidant activity was generally found in the fresh, and then frozen produce. The antioxidant activity of canned and jarred products was about equal and was less than the same vegetables that were fresh or frozen. Ascorbate accounted for 47% and 23% of the total antioxidant activity in fresh peas and spinach, respectively (Hunter and Fletcher, 2002). Steam blanching followed by –20°C frozen storage for up to a year resulted in vitamin C declines in carrots, broccoli, and green beans. Most of the loss occurred during blanching and early in the storage period, after which the rate of loss slowed. Ascorbic acid was much more stable after blanching and storage in broccoli and carrots than in green beans, which lost about 50% of their ascorbic acid content (Howard *et al.,* 1999). Additional relatively minor losses of ascorbic acid occurred in these foods during microwave cooking (Howard *et al.,* 1999). Peas lost about 20% of their ascorbate during blanching, and there was no further loss during 21 d of frozen storage (Hunter and Fletcher, 2002). When fresh spinach was boiled in water, approximately 60% of the vitamin C was found in the cooking water and 40% in the cooked tissue (Gil *et al.,* 1999). In a study of various cooking methods of 4 different vegetables (bean sprouts, green beans, nappa cabbage, and spinach), microwave cooking and stir-frying in oil retained more vitamin C than either boiling or stir-frying in water (Masrizal *et al.,* 1997). Because of its water solubility, larger losses of vitamin C occur in processes that use water. Steam blanching is therefore more effective in retaining vitamin C before freezing than hot water blanching. Blanching inactivates enzymes, such as ascorbate oxidase, that are capable of breaking down vitamin C during frozen storage. A recent comprehensive study of the effects of processing on the antioxidants in peas, carrots, spinach, potatoes, and several brassicas found that losses in vitamin C during blanching and frozen storage ranged from about 10% to 40% (Puupponen-Pimiä *et al.,* 2003). Carotenoids appear to be less adversely affected by processing compared with the other major fruit and vegetable antioxidants. In the same study (Puupponen-Pimiä *et al.,* 2003) in which vitamin C content was examined in carrots, spinach, potatoes, and several brassicas, there was essentially no loss in α-carotene and β-carotene after blanching. In fact, some increases were observed that were attributed to the dissociation of carotenoids from plant matrix materials. After blanching and freezing, losses in β-carotene content occurred between 6 and 12 mo of frozen storage. Pumpkin and carrot b-carotene retention was compared after pressure cooking and normal boiling (Gayarthri *et al.*, 2004). Retention was substantially greater in carrots compared with pumpkin. Although pressure cooking helped to retain more b-carotene in carrots, this was not the case with pumpkin, in which more β-carotene was lost during pressure cooking compared with boiling. The effect of processing on tomato lycopene has been well studied partly because of the abundance of processed tomatoes in the Western diet. Compared with other antioxidants, including other carotenoids, lycopene is relatively stable. Lycopene can be lost as a result of thermal degradation and oxidation. During thermal processing, the naturally occurring *trans*-isomer, which is a linear molecule, can be converted to the *cis*-isomer, which is kinked. The degree of isomerization is directly correlated to the degree of thermal treatment (Shi and le Maguer, 2000). Formation of *cis*-lycopene may be of benefit because *cis*-lycopene, although less stable, is believed to be more bioavailable in the body than *trans*-lycopene (Boileau *et al.,* 2002). Lycopene, like other carotenoids, is often higher in processed products because of its dissociation from plant matrix materials. Lycopene is readily degraded by exposure to air, especially in the presence of cationic pro-oxidant metals such as copper. This may be apparent by a loss in redness in products. Lycopene retention can be improved by storage at low temperature, low light, and low water content (Shi and le Maguer, 2000). Phenolic antioxidants are subject to degradation during processing. In the same study in which vitamin C, β-carotene, and α-carotene were examined in carrots, spinach, potatoes, and several brassicas, losses in total phenolic content after blanching and long-term frozen storage ranged from 20% to 30% (Puupponen-Pimiä *et al.,* 2003). In this study, total phenolic content was positively correlated with the total antioxidant capacity of these foods, suggesting that among vitamin C, carotene, and phenolics, phenolics make an important contribution to the antioxidant capacity of these foods. Like vitamin C, phenolic antioxidants are water-soluble and can be leached from fruit and vegetable tissues by processing in water. After fresh spinach was boiled in water, approximately half of the flavonoid content was found in the cooking water and the other half in the cooked tissue (Gil *et al.,* 1999). In processing blueberries into juice, substantial losses of phenolics occurred; the recovery of anthocyanins, procyanidins and chlorogenic acid were 32%, 43%, and 53%, respectively. Heat-labile enzymes in blueberry fruit (for example, polyphenol oxidase) made a large contribution to the loss in anthocyanins. Approximately 20% of the anthocyanins in blueberries were retained in the press cake after juicing (Skrede *et al*., 2000).

EFFECT OF PROCESSING ON ANTIOXIDANT ACTIVITY OF FRUITS

Both refrigerated storage and pasteurization lowered the DPPH antioxidant activity of serviceberry fruit. In the former, the change was not great; however, pasteurisation caused an approximately twofold increase in EC_{50}. On the basis of total anthocyanin content and total polyphenolic content measurements, it would appear that the antioxidant properties of serviceberry are affected mainly by anthocyanins, which are very well preserved in frozen products, even during storage. Higher pasteurisation temperature applied (100°C), usually used for home food processing, allowed slightly better preservation of these compounds during long-term storage (Magdalena Michalczyk and Ryszard Macura, 2010).

Blanching, a mild heat treatment (accomplished with hot water, steam, hot air or microwave) is an important step in the processing of fruits and vegetables. It is done primarily to inactivate potentially harmful enzymes, reduce microbial loads and decrease interstitial gases prior to further processing. Blanching in hot water can provoke large losses of water-soluble vitamins by leaching. The heat treatment in blanching may result in some losses of thermally labile nutrients and bioactive compounds, but the inactivation of oxidative and other degradative enzymes (e.g., lipoxygenase for carotenoids, polyphenol oxidase for flavonoids, ascorbic acid oxidase for vitamin C) will prevent further and greater losses by enzyme-catalyzed degradation during slow processing (as in drying) and storage.

Considerable losses of carotenoids and flavonoids have been observed when fruits and vegetables are processed into juice because substantial amounts of the bioactive compounds are left in the discarded skin and pulp (press cake), these losses often surpassing those of heat treatment.

Thermal processing has long been known to cause significant losses of vitamins, carotenoids and anthocyanins. High temperature accelerates reactions, which would occur more slowly at ambient temperature. Whatever the processing method chosen, retention of these compounds decreases with longer processing time, higher processing temperatures and cutting or maceration of the food.

On the other hand, as shown with carotenoids, mechanical disruption of the food matrix or heat treatment increases bioavailability, presumably by softening or breaking cell walls and membranes, thereby facilitating the release of the health-promoting compounds from the food matrices, the first step in the digestion of food. Processing conditions should therefore be optimized so that bioavailability is increased, but without provoking significant degradation. Bioavailability refers to that portion of the nutrient or bioactive compound consumed that is released from the food matrix, absorbed and used by the body.

In home preparation, retention of vitamins and bioactive compounds decreases in the following order: microwaving > steaming > boiling. For the fat-soluble carotenoids, boiling is less harmful than sautéing. Deep-frying, prolonged cooking, combination of several preparation methods all result in substantial losses. Whatever the cooking or processing method used, retention decreases with longer cooking time, higher cooking temperature and peeling/cutting of the food.

Freezing (especially quick freezing) and frozen storage generally preserve nutrients and bioactive compounds, but slow thawing may be detrimental.

Traditional sun-drying, although the cheapest and most accessible means of food preservation in poor countries, causes considerable destruction of nutrients and bioactive compounds. Drying in a solar dryer, even of simple and inexpensive design, can appreciably reduce losses. Protecting the food from direct sunlight also has a positive effect.

In the last two decades, nonthermal processing technologies (e.g., high pressure processing, high-intensity pulsed electric field processing) have been introduced. These technologies have been reported to inactivate microorganisms and enzymes, without provoking adverse thermal effects on sensory and nutritional properties.

Enzymatic degradation may be a problem in minimally processed foods. Loss of cellular integrity and of cell compartmentalization in fresh-cut fruits and vegetables, due to the cutting operations, destroys the natural protection of health-promoting substances and exposes them to oxidative and hydrolytic degradative enzymes. Modified atmosphere packaging and low temperature storage, used to extend the shelf-life of minimally processed foods, have also been shown to preserve micronutrients and bioactive compounds.

The aim of the present study was to assess the effect of thermal processing and storage period on the antioxidant properties and colour quality of strawberry, sweet and sour cherry low sugar jam (Table 9.4). It was observed that thermal processing of fruits led to statistical significant alterations for all monitored parameters. Most important losses due to thermal processing were recorded for monomeric anthocyanins (92-93% from the value recorded in frozen fruits), followed by vitamin C content (54-78%), FRAP values (30-41%) and total phenolics content (25-43%). Additional, alterations of measured parameters occur during storage. Jams storage for a period of 3 months at 20°C led to a decrease in vitamin C content of 22-33% from the value recorded one day after processing and a loss of anthocyanins content, ranging from 22% in sour cherries jam to 33% in strawberry jam related to the value recorded one day after processing. The FRAP values showed not statistical significant alterations after 3 months of storage (Table 9.5). Also, there was an increase in polymeric colour in the range of 39-63% from the value recorded one day after processing while for the colour density there were no statistically significant alterations (Mariana-atena *et al.*, 2011).

Table 9.4: Alterations of measured parameters as a result of fruits thermal processing.

Samples	Frozen fruits [C]	Jam after a day of processing
Vitamin C (mg·100^{-1} g ds)		
Strawberry	314.43±28.41	69.9±5.43***
Sweet cherry	78.11±6.28	35.65±3.12***
Sour cherry	172.93±14.61	51.12±4.29***
Total phenolics (mM GAE·100^{-1} g ds)		
Strawberry	17.79±1.56	10.24±0.81**
Sweet cherry	19.37±1.64	13.32±1.06**
Sour cherry	21.58±1.79	16.15±1.45*
Total anthocyanins (mg·100-1g ds)		
Strawberry	233.44±20.24	15.80±1.35***
Sweet cherry	303.61±28.27	21.4±1.82***
Sour cherry	547.46±40.11	42.55±3.12***
Antioxidant activity (mM Fe2+·100^{-1} g ds)		
Strawberry	60.22±5.17	35.29±2.85**
Sweet cherry	45.47±3.85	30.17±2.61**
Sour cherry	72.99±6.47	50.60±4.52**

Data are shown as means, relative to control (C) response recorded in the frozen fruits. Statistical differences are indicated as $P<0.05=*$ (significant), $P<0.01=**$ (highly significant) and $P<0.001=***$ (extremely significant).

Table 9.5: The losses of measured parameters as a result of fruit jam storage at 20°C

Losses (%)	Jam in storage after a month			Jam in storage after 3 months		
	Straw berry	Sweet cherry	Sour cherry	Straw berry	Sweet cherry	Sour cherry
Vitamin C	13.97	14.91	7.82	32.69	28.68	22.18
Total phenolics	11.23	9.61	8.98	25.39	21.92	18.2
Monomeric anthocyanins	10.78	14.36	9.56	33.11	27.59	21.57
FRAP	6.83	5.83	4.41	19.1	14.72	10.63

FRAP: Ferric reducing antioxidant power.

TOTAL ANTIOXIDANT ACTIVITY IN SELECTED VEGETABLES

The total antioxidant activity, which reflected the ability of the vegetable extracts to inhibit the bleaching of b-carotene, was measured and compared with that of the control which contained no antioxidant component. The b-carotene bleaching rates of the vegetable extracts are shown in Figs. 9.3 and 9.4. There was a decrease in absorbance values of β-carotene in the absence of vegetable extracts due to the oxidation of β-carotene and linoleic acid. The high absorbance values indicated that vegetable extracts possessed antioxidant activity.

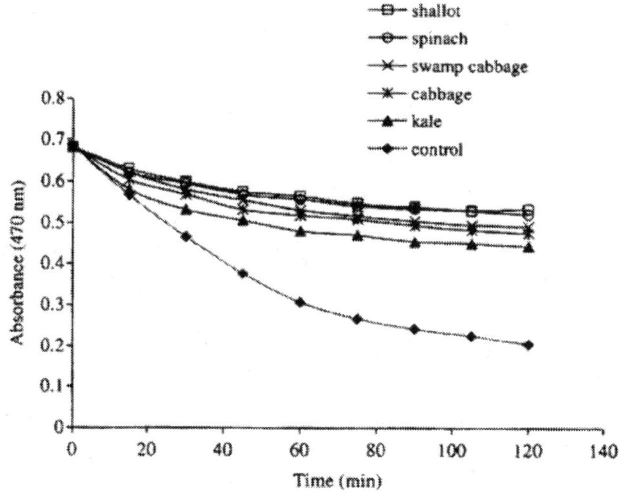

Figure 9.3: Antioxidant activity of fresh vegetable extracts at 1 mg/ml using β–carotene-linoleate system.

The absorbance value of the control was significantly lower (p < 0:05) than the vegetable extracts. All the vegetable extracts showed similar trends with a significant decrease (p < 0:05) of absorbance value until 45 min. This indicated that the samples, whether fresh or thermally treated, had acted as effective antioxidants in the β-carotene-linoleate system, which inhibited the oxidation activity of β-carotene (Figs. 9.3 and 9.4). Among the studied vegetables, shallot exhibited the greatest inhibition of β-carotene bleaching rates, followed by spinach, swamp cabbage, cabbage and kale.

The comparison of mean total antioxidant activity of fresh and thermally treated vegetables is presented in a descending order in Fig. 9.5. Mean antioxidant activity of shallots, spinach, swamp cabbage, cabbage and kale were 69.1±1.1%, 66.4±1.1%, 60.3±1.0%, 59.3±4.1% and 50.2±1.3%, respectively. Results of ANOVA analysis indicated that antioxidant activity of shallots is significantly higher (p < 0:05) than the rest of the fresh vegetables, except spinach. As shown in Fig. 9.5, the mean antioxidant activities of the 1-min boiled vegetables were similar to the fresh vegetable, shallots > spinach > swamp cabbage > cabbage > kale (68.5±2.4%, 61.9±0.6%, 56.5±1.1%, 53.4±0.6% and 45.9±1.3%, respectively).

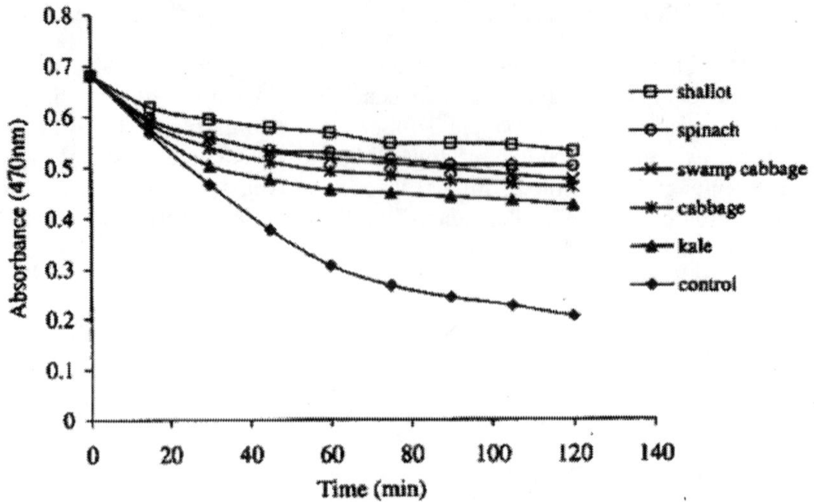

Figure 9.4: Antioxidant activity of thermal treated vegetable extracts at 1 mg/ml using β–carotene-linoleate system.

In a study carried out by Cao *et al.* (1996) on antioxidant activity, using oxygen radical absorbance capacity assay, kale was found to have the highest antioxidant activity, followed by spinach and cabbage. Kurilich *et al.* (1999) reported that kale had higher levels of vitamins (α-carotene, β-carotene, α-tocopherol, γ-tocopherol, and ascorbate) than did cabbage. The differences of the results obtained from this study compared to the previous findings may have been due to the differences in species cultivation of the kale used and differences in the extraction methods. Environmental factors, such as climatic growth conditions, growth, ripening stage, temperature, duration of storage and thermal treatment may have influenced the antioxidant activity (Gazzani, *et al.,* 1998).

Although the amounts of antioxidant vitamins in shallot bulb were lower than other vegetables (Puwastein, *et al.,* 2000), they had the highest antioxidant activity. Therefore, shallots may have some powerful antioxidant compounds other than α-tocopherol, carotenoids and vitamin C. According to Shahidi and Wanasundara (1992), shallots had high flavonoid contents which may have contributed to their high antioxidant activity. A high antioxidant activity of spinach and swamp cabbage may be attributed to the nutrient antioxidants found in these vegetables. According to Tee *et al.* (1996), swamp cabbage and spinach have high contents of carotenoids, ascorbic acid and a-tocopherol and these might have been the main factors that contribute to the high antioxidant activity. The low antioxidant activity of cabbage was in agreement with Cao *et al.* (1996).

Antioxidant activity of vegetable extracts also depends on the type and polarity of the extracting solvent, the isolation procedures and purity of active compounds, as well as the assay techniques and substrate used (Meyer *et al.* 1998). In the study of Tsuda *et al.* (1994), less polar solvents provided slightly more active extracts than mixtures with ethanol or methanol, or methanol alone for tamarind seed coats. The extraction solvent in this study was 70% ethanol (v/v) while 100% acetone was used by Cao *et al.* (1996). This factor may also have affected the results in the different findings.

Thermal treatment has been known to affect antioxidant activity (Gazzani *et al.*, 1998). Antioxidant activity of kale, spinach and swamp cabbage were reduced significantly ($p < 0.05$) after 1 min of thermal treatment. The thermal treatment on cabbage and shallots in this study did not show any significant difference in antioxidant activity (Fig. 9.5). However, increasing boiling time may have affected the antioxidant activity of these vegetables. Yadav and Sehgal (1995) showed that cooking spinach in an open pan for 30 min had reduced the vitamin C content by 90% while the b-carotene content was reduced down to 50%. Microwave cooking of kale for 6 min, in a small amount of water, produced a loss of 15–57% of the xanthophylls and 15% loss of the carotenoids (Micozzi *et al.,* 1990).

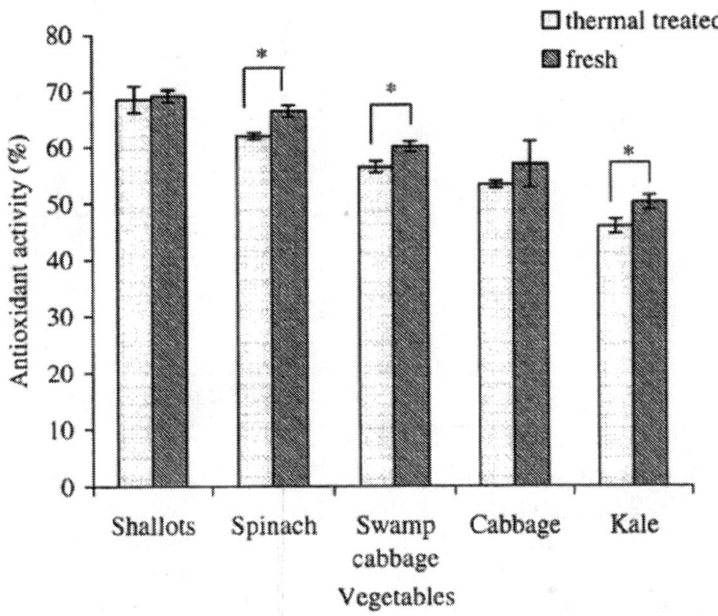

Figure 9.5: Mean total antioxidant activity of vegetable extracts. Asterisk (*) indicates a significant difference at the level $p < 0.05$ between fresh and thermally treated vegetables. Antioxidant activity was measured using a β–carotene-linoleate system. Results are means of three determinations. Values indicate that the coefficient of variation was less than 7%.

Total phenolic content and antioxidant activity in selected vegetables

The total phenolic content of the vegetable extracts is shown in Fig. 9.6. Among all the fresh vegetables, spinach had the highest phenolic content (7167 ± 73 mg/100 g vegetable extract), followed by swamp cabbage (4175 ± 41 mg/100 g), kale (3689 ± 66 mg/100 g), shallots (2528 ± 43 mg/100 g) and cabbage (1107 ± 57 mg/100 g). For thermally treated vegetables, spinach (6168 ± 41 mg/100 g) still had an exceptionally high total phenolic content, followed by kale (3251 ± 123 mg/100 g), swamp cabbage (3095 ± 74 mg/100 g), shallots (2187 ± 49 mg/100 g) and cabbage (886 ± 52 mg/100 g).

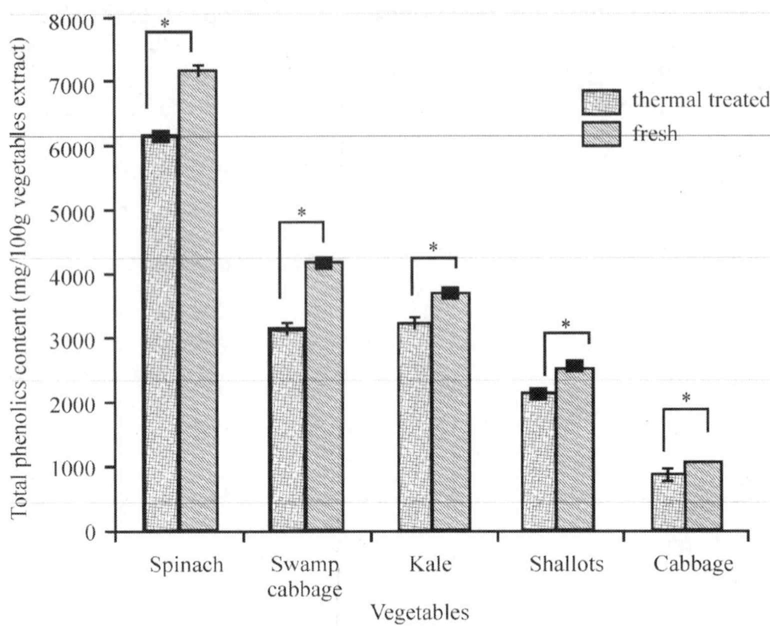

Figure 9.6: Mean total phenolic content of vegetable extracts. Asterisk (*) indicates a significant difference at the level $p < 0.05$ between fresh and thermally treated vegetables. Results are means of three determinations. Values indicate that the coefficient of variation was less than 9%.

Swamp cabbage lost the highest amount of phenolic content (26%), followed by cabbage (20%), spinach (14%), shallots (13%) and kale (12%) after a 1-min blanching in boiling water. ANOVA showed significant differences (p < 0:05) in the total phenolic content between fresh and treated vegetables (Fig. 9.7). The findings indicated that phenolic compounds were very sensitive to heat treatment even in a short period of cooking. Kähkönen *et al.* (1999) found that the total phenolic contents of vegetables were very low compared to fruits. Vinson *et al.* (1998) found that beets had the highest total phenolic content, followed by red onion, broccoli and kidney beans. Meanwhile, Velioglu *et al.* (1998) reported that red onion had a higher total phenolic content than other plant materials.

Several studies have reported on the relationships between phenolic content and antioxidant activity. Some authors found a correlation between the phenolic content and the antioxidant activity, while others found no such relationship. Velioglu *et al.* (1998) reported a strong relationship between total phenolic content and antioxidant activity in selected fruits, vegetables and grain products. No correlation between antioxidant activity and phenolic content was found in the study by Kähkönen *et al.*

(1999) on some plant extracts containing phenolic compounds. In this study, the findings do not show any relationship between antioxidant activity and total phenolic contents (Fig. 9.7). For example, cabbage had the lowest total phenolic content whereas its antioxidant activity was a bit higher than kale that had a higher total phenolic content. Spinach had much higher total phenolic content than other vegetables although its antioxidant activity was lower than was shallots (Fig. 9.7). There is a wide degree of variation between different phenolic compounds in their effectiveness as antioxidant (Robards *et al.* 1999). Flavonoids generally have more hydroxyl groups if compared to ferulic acids. Besides, orthosubstitution with electron-donating alkyl or methoxy groups of flavonoid, increases the stability of the free radical and hence its antioxidant potential (Rice-Evans *et al.* 1995). Therefore, shallots that have high flavonoid content generally have higher antioxidant activity than the other studied vegetables. The findings of this study indicate that each type of vegetable had a different antioxidant activity, contributed by different antioxidant components, such as α-tocopherol, β-carotene, vitamin C, selenium or phenolic compounds. The high antioxidant activity of shallots might be due to its flavonoid contents; for spinach and swamp cabbage the high α-tocopherol, β-carotene and ferulic acid corresponded with the high antioxidant activity.

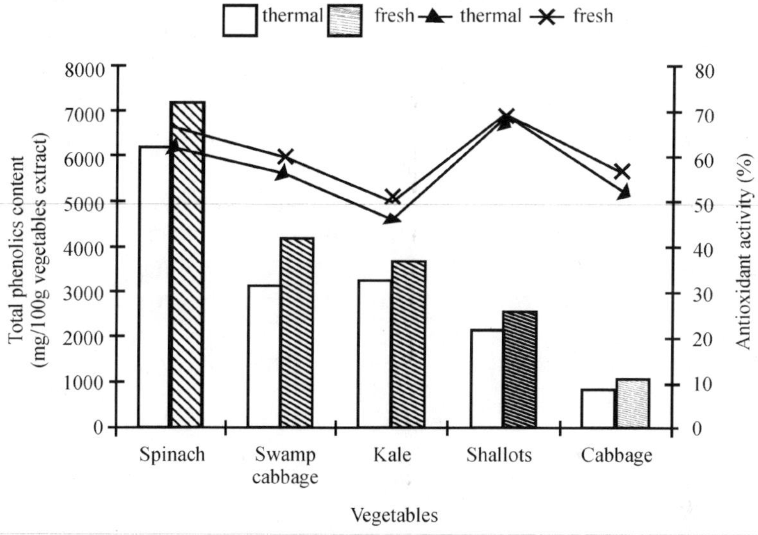

Figure 9.7: Comparison of antioxidant activity and total phenolic content in fresh and thermally treated vegetables.

Effect of processing on antioxidant activity in vegetables

Phenolic compounds and other reductones in vegetables contribute to their antioxidant activities; hence processes that affect the total phenolic content will also affect antioxidant activity. Effects of different heat treatments on the antioxidant activity of some vegetables are shown in Table 9.6.

Table 9.6: Effect of different heat treatments on the antioxidant activities of some vegetables

Vegetable	Treatment	Method of analysis	Effects	Percentage of increase / decrease no change
Broad beans	Steam cooking fresh material	ABTS	Decrease	11
	Steam cooking frozen beans	ABTS	Decrease	10-17
Green asparagus	Convectional cooking	TEAC	Increase	16
Red cabbage	Steam cooking	TEAC	Decrease	5.20
Spinach, *Koma-tosuna, haruna, chingensai* and white cabbage	Mild heating 50ºC (10-30 min.)	Deoxyribose assay	Decrease	0.20
Tomatoes	Semi-drying	TEAC	Decrease	53
Broccoli	Convectional cooking	B-CLAMS	Decrease	4.80
Broccoli	Steam cooking	ORAC	Increase	230

Canning is a common industrial process for preserving vegetables and was shown to cause a more pronounced loss of antioxidant activity on frozen vegetables than fresh vegetables (Murcia et al., 2009). The reduction in antioxidant activity as a result of canning might be causal by the high temperatures used during sterilization and the degradation of phenolic compounds as a result of storage.

Faller and Fialho (2009) evaluated the effect of boiling microwaving and steaming on fresh conventional and organic retail vegetables (potato, carrot, onion, broccoli, and white cabbage). Organic vegetables showed higher sensitivity to heat processing than did conventionally grown vegetables. In contrast significant differences between foods obtained by the two cultivation procedures were not observed by Lima *et al*. (2009). In general, cooking was found to lead to reductions in the antioxidant activity for most vegetables, with small differences between cooking methods applied (Faller and Fialho, 2009). In his studies, Sutana *et al*. (2008) reported that different cooking methods affects the antioxidant properties of vegetables differently, with microwave treatment exhibiting more deleterious effects when compared with those of other treatments.

Normal cooking temperatures detrimentally affects the antiradical and anti-proliferative activities of spinach, *komatsuna, haruna, chingensai*, cabbage and Chinese cabbage. However, mild heating of vegetable juices (50ºC, 10-30 min) preserved the antioxidant activity and cell proliferation inhibition activities (Roy *et al*., 2007). The results showed that foods should be minimally processed so as to preserve the compounds that contribute to their antioxidant and cell proliferation inhibition activities. Hence consumer demand for safe and nutritious food has led to the development of a number of non-thermal food preservation techniques.

Xu and Chang (2008) reported the effects of soaking, boiling and steaming processes on the antioxidant activity in commonly consumed Cool Season Food Legumes (CSFL's), including green pea, yellow pea, chickpea and lentil. As compared to original unprocessed legumes, all processing steps caused significant ($p< 0.05$) decreases in DPPH free radical scavenging activity in all tested CSFL's. All soaking and atmospheric boiling treatments caused significant ($p < 0.05$) decreases in oxygen radical absorbing capacity (ORAC). However, pressure boiling and pressure steaming caused significant ($p<0.05$) increases in ORAC values. Steaming treatments resulted in a greater retention of DPPH and ORAC values in all tested CSFL's as compared to boiling treatments. Steam cooking was also reported to increase the antioxidant activity of broccoli by 230% (Roy *et al.*, 2007). Therefore, steam cooking should be used instead of boiling the vegetables for prolonged periods. When cooking time and temperature combinations should be closely monitored as prolonged exposure of vegetables to cooking conditions may lead to deleterious effects.

A study by Turkmen *et al.* (2005) on pepper, peas and broccoli gave interesting results, were after cooking total antioxidant activity increased or remained unchanged depending on the type of vegetable but not type of cooking. Amin *et al.* (2006) also reported that blanching up to 15 min may affect losses of antioxidant activity, depending on the species of spinach (Amin *et al.*, 2006). The type of vegetable or species differences might be due to the easy of oxidation of phenolic compounds from different plant matrices. Processes, such as freezing plant material prior to cooking were found to reduce antioxidant activities to some greater extent. Processes applied on fresh material reduced the activity against ABTS to a smaller extent (cooking by 11% and freezing by around 20%), while cooking frozen beans caused a further decrease by 10-17% (Wolosiak *et al.*, 2009).

In most studies on effects of heat treatment on vegetables, the antioxidant activity decreased as a result of processing; in contrast Fanasca *et al.* (2009) reported an increase in antioxidant activity. In his study on antioxidant properties of raw and cooked spears of green asparagus cultivars, the cooking process increased the antioxidant activity by 16%. Podsedek *et al.* (2008) reported that steam-cooking is recommended to prevent the major loss of scavenging activity, because under these conditions, the corresponding TEAC (Trolox Equivalent Antioxidant Capacity) values were reduced only by 5-20%. Roy *et al.* (2007) and Xu and Chang (2008) also reported an increase in antioxidant activity as a result of steam cooking.

The effects of different cooking methods (boiling, frying and microwave cooking) on the antioxidant activity of some selected vegetables (peas, carrot, spinach, cabbage, cauliflower, yellow turnip and white turnip) were assessed by measuring the reducing power and percentage inhibition in linoleic acid system (Sultana *et al.*, 2008). The author reported contradicting results found by most researchers as there was a significant ($p<0.05$) increase in reducing power as a result of frying. However, boiling and microwave cooking did not affect reducing power. Inhibition of peroxidation was also increased by boiling and frying, whereas, in contrast it was decreased by microwave cooking.

PHENOLIC COMPOUNDS AND ANTIOXIDANT ACTIVITIES IN PROCESSED (ROASTED) NUTS

Nuts may be consumed as natural (raw) or preferably roasted. The main purpose of roasting is to improve the desirable flavour, colour, crispiness, and the crunchy texture of nuts (Hojjati *et al.*, 2013; Pelvan *et al.,* 2012; Schmitzer *et al.,* 2011). The roasting process involves microstructural and chemical changes, such as the decrease in moisture content, lipid modifications, and changes in colour as well as the formation of compounds responsible for the typical roasted nut flavour, mainly due to Maillard reaction products (Alamprese *et al.,* 2009; Amaral *et al.*, 2006; Saklar *et al.,* 2001). However, microstructural and lipid modifications might lead to an enhanced susceptibility of roasted nuts to lipid oxidation (Alamprese *et al.*, 2009), compared to raw nuts. Besides, changes in the content of metabolites (e.g., antioxidants) may also happen due to roasting (Amaral *et al.*, 2006). Therefore, the roasting process may influence both the formation of health promoting components and those with potentially adverse health effects. To the best of our knowledge, limited studies are available on the effect of roasting on phenolic compounds of nuts. Hence, more research should be carried out to fill this knowledge gap. The impact of processing (roasting and stir-frying) on phenolic compounds of nuts demonstrates a rising trend at lower temperatures or over shorter heating durations, but a declining trend at higher temperatures or over longer heating periods (Bolling *et al.*, 2011; Chandrasekara and Shahidi, 2011; Lin *et al.*, 2016; Pelvan *et al.*, 2012; Schlörmann *et al.*, 2015; Schmitzer *et al.*, 2011; Vinson and Cai, 2012; Yang *et al.*, 2015). However, the impact of roasting on antioxidant activities of nuts and their co-products is inconsistent mainly due to the use of different processing (roasting and stir-frying) duration and temperature, as well as different types of nuts.

Yang *et al.* (2015) have demonstrated that stir-frying pecan at a low temperature over a short time leads to an apparent increase of its extractable chlorogenic acid, gallic acid, and antioxidant activity. However, these values decreased with increases in heating time and temperature. Hence, it is only beneficial to stir-fry pecans at certain conditions during the processing. Pecan phenolics, with the exception of gallic acid, were always the highest in whole unroasted kernel and were significantly reduced after skin removal. Similarly, pecan TPC and antioxidant activity decreased when its skin was removed. Roasting had a significant negative effect on individual phenolics but not on the TPC and antioxidant activity of kernels (Yang *et al.*, 2015). In hazelnut, phloretin-2′-*O*-glucoside and flavan-3-ols (catechin and epicatechin) content decreased significantly when roasted, where significant differences were observed between the raw nut with skin and roasted nut without skin. Thermal treatment had a negative effect on the content of procyanidin dimers and trimers in hazelnut (Schmitzer *et al.*, 2011). Again, significant differences existed between natural and roasted hazelnuts, suggesting that the stability of these compounds is indeed affected by temperature. Moreover, no polyphenols have been detected in roasted hazelnut from which the skin was removed. This shows that polyphenols in hazelnut are mostly present in the skin (Schmitzer *et al.*, 2011). Another study on pistachio nut also reported a significant decrease in several phenolic compounds after raw nuts were roasted. Chlorogenic acid and daidzein were detected only in roasted pistachio, not in natural pistachio (Seeram *et al.*, 2006). Roasting had an ambiguous effect on the content of protocatechuic acid in roasted hazelnut (Monagas *et al.*, 2009). Similarly, roasting had a significant positive effect on the content of gallic acid in roasted hazelnut compared to natural hazelnut. This could be due to degradation of polymerised polyphenols, specifically hydrolysable tannins, and the hydrolysis of other glycosylated flavonoids (Monagas *et al.*, 2009). Other factors may also affect the phenolic acid contents upon roasting, such as the content of phenolic acids in the skin, their thermal decomposition, and their liberation from esters, glucosides, and bound forms (Pelvan *et al.*, 2012). Catechin and ellagic acid are also susceptible to heat, possibly by their highly hydroxylated structure that is vulnerable to redox reactions, causing losses when subjected to heat treatment. Therefore, it seems that roasting reduces the levels of phenolic compounds in most nuts, not only because of the removal of the skin but also due to chemical degradation of many phenolic compounds. Most of them are highly unstable and may be lost during processing, particularly when heat treatment is involved. The thermolability of phenolic compounds depends not only on their structure but also on the food matrix (Jinap *et al.,* 2004).

Antioxidant activities of natural (raw) and roasted nuts determined by various assays and results are explained here. Enhanced roasting conditions led to a significant decrease of the antioxidant activity in hazelnut, macadamia, and walnut, whereas in almond and pistachio the antioxidant activity remained stable or was slightly enhanced. Furthermore, antioxidant activity was restored in hazelnut, almond, and walnut at the most intense roasting conditions (Schlörmann *et al.*, 2015). Similar results were found by Açar *et al.* (2009), who observed a decline of antioxidant activity in several types of nuts by roasting at 150°C for 10–30 min and enhanced activities after 60 min. This effect can be explained by the loss of polyphenols due to thermal treatment which is counteracted by the formation of antioxidant-active compounds due to Maillard reactions as also discussed by Açar *et al.* (2009). Most of the antioxidative compounds of nuts, such as polyphenols, are located in the skin and its removal is accompanied by a loss of antioxidant activity (Arcan and Yemeniciog˘ lu, 2009; Schmitzer *et al.*, 2011).

Vinson and Cai (2012) determined the antioxidant activities of nut and peanut butter polyphenols as measured by the concentration to reduce the oxidation of LDL + VLDL by 50%. The order of decreasing efficacy (increasing IC50) for natural (raw) nuts is walnut > cashew > hazelnut ~ pecan ~ almond ~macademia > pistachio > Brazil nut > peanut. For roasted nuts and peanut butters, the order is almond ~ macademia ~ pecan ~ peanut butter crunchy > peanut butter creamy > peanut > pistachio > hazelnut ~ walnut > Brazil nut > cashew. However, roasting caused a statistically insignificant decline in antioxidant efficacy. The average IC50 for raw nuts was 4.3 ± 2.1 mM and for roasted nuts (not including peanut butters) was 6.9 ± 3.9 mM. Raw walnut had the highest antioxidant activity among all nuts. Besides, the efficacy of nut polyphenols to bind to LDL, thus protecting the LDL from oxidation, was also determined [almond (2.1 mM), peanut (5.6 mM), and walnut (1.0 mM)]. That means walnut-bound polyphenol antioxidants are twice as effective as almond and 5-fold effective compared to peanut in lipoprotein binding (Chang *et al.*, 2016).

Schlörmann *et al.* (2015) have evaluated the effect of roasting (125°C/25 min to 185.5°C/25 min) on antioxidant capacities of almond, hazelnut, macadamia, pistachio, and walnut using TBARS, hydrophilic and lipophilic antioxidant assays. The TBARS levels significantly increased with roasting conditions (both temperature and time).The increase in TBARS levels was relatively low in hazelnut (from 0.94 ± 0.02 in raw up to 1.73 ± 0.1 μg malondialdehyde equivalents (MDAE)/g) and macadamia (from 0.40 ± 0.04 in raw up to 0.69 ± 0.05 μg MDAE/g) compared to pistachio (from 1.41 ± 0.15 in raw up to 3.51 ± 0.36 μg MDAE/g). In contrast, the highest increase of TBARS levels was observed in walnut (from 0.95 ± 0.05 in raw up to 16.3 ± 3.16 μg MDAE/g) and in walnut roasted under extreme conditions (185.5°C/25 min). These results are in agreement with those of Yaacoub *et al.,* (2008) in roasted almond and pistachio. For hydrophilic antioxidant activity, no effect on the antioxidant activity was found in almond and pistachio, whereas antioxidant activity significantly decreased in hazelnut and macadamia after roasting. In contrast, the antioxidant activity significantly increased after roasting (171 °C) in macadamia. In walnut, the hydrophilic antioxidant activity significantly decreased after roasting. In addition, the lipophilic antioxidant activity decreased in pistachio and walnut due to roasting (Schlörmann *et al.*, 2015). According to Xu and Chang (2008), the thermal treatment applied to foods of plant origin by heating or roasting cause's evaporation of intracellular water, triggering chemical reactions that can change the lignocellulosic structure, and promotes protein denaturation, which might result in a greater availability of plant phenolic compounds in the matrix. Enhancement of antioxidant activity might be attributed to the release of some bound antioxidant phenolic compounds acting as free radical scavengers from the cell matrix, and increase phenolic extractability after roasting (Lin *et al.*, 2016; Schmitzer *et al.*, 2011).

DEVELOPMENT OF RISK TOXICITY COMPOUNDS

The safety of nut co-product consumption as a functional food requires consideration. Exposure to irregular food components may pose allergenic or chemical toxicity risks. Thus, the safety of several nut co-products or components has been reported. Doses of up to 2.5 g/kg of almond skin did not pose genotoxicity in Swiss albino mice and up to 10% dietary almond skin was not toxic to Sprague-Dawley rats (Song *et al.*, 2010; Zhang *et al.,* 2011a, b).Topical application of cashew shell liquid/oil had a weak tumour-promoting activity in a skin tumourigenesis model using Swiss albino mice (Banerjee and Rao, 1992; George and Kuttan, 1997). Chronic consumption of 300 mg cashew anacardic acids/kg by albino mice did not pose toxic symptoms, although 600 and 1000 mg/kg increased urea, and lowered haematocrit levels (Carvalho *et al.*, 2011).Walnut coproduct consumption may induce toxicity in dogs, horses, and other animal species (Coleman and Merola, 2016). Occupational exposure to macadamia shell can induce allergic contact dermatitis (Knight and Hausen, 1996). However, the component responsible for this effect is unknown.

Acrylamide, which is classified as probably carcinogenic in humans, is formed during Maillard reaction (Lineback *et al.,* 2012), involving asparagine from protein and reactive carbonyls at temperatures above 120°C (Amrein *et al.*, 2005; Lineback *et al.*, 2012). Acrylamide was detectable in roasted almond and pistachio, ranging from 16–1220 to 14–88 μg/kg, respectively, with increasing concentrations depending on the roasting conditions (Schlörmann *et al.*, 2015). It is noteworthy that the acrylamide levels found in almond did not or only marginally exceeded the recommended level of 1000 μg/kg for potato chips, whereas levels for pistachio clearly do not exceed recommended levels. In another study, Zhang *et al.*, (2011a) demonstrates that acrylamide levels reduced significantly at all roasting time evaluated by controlling the roasting temperature at or below 146°C. Further, stability studies on roasted almond indicated that acrylamide levels decreased by 12.9–68.5% after 3 days of storage at 60°C. This shows that short-term elevated temperature storage may be a good alternative to reduce the acrylamide levels in roasted almond (Zhang *et al.*, 2011b).

SUMMARY AND CONCLUSIONS

There is unprecedented awareness and interest in both the positive and negative impacts of dietary selection in human health. In focusing on approaches to maintain and improve health through dietary means, consumers, public health organizations, and the food industry wish to know more about the content of health-promoting phytochemicals, including antioxidants, in fruits and vegetables. They also wish to know how health-promoting components can be best preserved during storage and processing. Research that has addressed the content and retention of phytochemical antioxidants can help the horticulture and food industries to develop and promote foods based on their antioxidant content.

Some of the major conclusions of this review are summarized in Table 9.7. It is clear that, after food choice, the most effective means of improving the antioxidant content of the diet is by fruit and vegetable varietal selection. The substantial variation that exists in phytochemical antioxidant content means that breeders can select genotypes and develop varieties with high levels of antioxidants. Once high antioxidant varieties are available, new technologies in cultivar identification and traceability can make it possible to offer the consumer a wide selection of health functional varieties and to obtain premium prices for high-antioxidant produce. In addition to genetic factors, environmental factors, before and after harvest, can influence antioxidant content. Among the antioxidant groups discussed, phenolic antioxidants appear to be more responsive to environmental factors such as water availability, light quality, and temperature. These factors undoubtedly contribute to the seasonal variability in fruit and vegetable phenolic content. Phenolics also seem to be more affected by storage factors such as temperature, atmosphere, and light, than either carotenoids or vitamin C. Cultivar selection and manipulation of storage conditions could lead to specific conditions for optimal phenolic antioxidant enhancement. Because vitamin C is the most labile antioxidant and prone to loss during storage and processing, it may serve to indicate the quality of all antioxidant groups in a product. In developing processing technologies that optimize antioxidant retention, it is important to understand more fully the chemical properties of antioxidants, the enzymes that affect their content in plant tissue, and the partitioning of antioxidants among plant tissues. Increasing the consumption of fruit and vegetable antioxidants requires the development of foods with optimal phytochemical antioxidant content without sacrifice to taste or convenience. Continued research into the many avenues presented here will underpin the development of food products with enhanced antioxidant content.

Table 9.7: Summary of some production and processing effects on fruit and vegetable antioxidants

Factor	Ranking
Influence of genotype	Vitamin C = carotenoids = phenolics
Influence of preharvest environmental factors	Phenolics > Vitamin C > carotenoids
Losses during fresh storage	Vitamin C >> carotenoids >> phenolics
Increases during fresh storage	Carotenoids, phenolics
Losses during processing	Vitamin C > phenolics > > carotenoids

Source: Wilhelmina Kalt, 2005.

Cereals and legumes undergo different types of primary processing to enable their further use for product manufacture or cooking. Some of the primary processed products are also in ready-to-eat form such as expanded rice products. Generally processing alters the grain quality. As long as the whole grain is used, all nutrients and phytochemicals are retained; however, abstraction of any part of the grain results in reduced nutrients. The distribution of nutrients and phytochemicals in any grain is not uniform with the outer portion containing more nutrients and fiber contents. Milling has mutual effects on nutritional quality. It results in breakage of cell wall and improves availability of nutrients which are bound in nutrient matrix. On the other side, during milling outer layer of grains which are very rich source of nutrient except starch are separated. Separation of bran/husk decreases nutrients but improves digestibility and/or bioaccessibility. Separation of bran portion by mechanical means such as sieving of flour can also reduce the nutrient content of sieved flour. Processes like soaking and germination reduce the antinutrient content and also increase the availability of nutrients, in particular of minerals. Finally, it can be said that nutritional quality of grains is influenced by pre-processing treatments and processes which retain all parts of whole grains as beneficial for health and consumption of highly refined products should be discouraged.

This chapter presents the recent research carried out to date for purposes of processing and nutritional quality improvement of millet grains and their food products. Based on the results of studies carried out, we can observe that millet grains contain many health-promoting components such as dietary fiber, minerals, vitamins, and phytochemicals that include phenolic compounds, and they are comparable to those of major grains and they also have several potential health benefits. However, novel processing and preparation methods are needed to enhance the bioavailability of the micronutrients and to improve the quality of millet diets. Research is also needed to determine the bioavailability, metabolism, and health contribution of millet grains and their different fractions in humans. Making millet food products that deliver convenience, taste, texture, colour, and shelf-stability at economical cost for poor people is needed. In addition, for promoting utilization of millet grains in urban areas to open new markets for farmers to improve their income, developing highly improved products from millet is needed.

Numerous phenolic compounds (phenolic acids and flavonoids) have been identified and quantified in nuts (natural and roasted) and their co-products (skin, hard shell, and hull/pellet) together with their corresponding antioxidant activities. The compiled results indicated that many of these bioactive compounds remain to be fully identified and characterised in nuts and their co-products (natural and roasted), especially macadamia, Brazil nut, and pine nut. Nuts roasted at suitable temperature and time demonstrated excellent phenolic compositions and antioxidant activities and should be recommended for daily consumption. Therefore, nuts, especially their co-products can be considered as a valuable source of functional foods based on their phenolic compositions and their corresponding antioxidant activities presented in this review article. The phenolic compounds from nuts and their co-products are bioaccessible and bioavailable in humans, with various health benefits, including antioxidant, antiinflammatory, prebiotic, anti-microbial, chemopreventive, and hypocholesterolaemic. More sophisticated human intervention studies or clinical trials are needed to validate the health benefits of nut co-products.

■■■■

10 Future Considerations

There is increasing evidence that consumption of a variety of phenolic compounds present in natural foods may lower the risk of serious health disorders because of their antioxidant activity, amongst other mechanisms. Mode of action of different phenolic compound in specific body organs needs to be found. Due to safety and other limitation surrounding the use of synthetic antioxidants, natural antioxidants obtained from edible sources, by-products and co-products are alternative sources of interest. Further studies on the isolation of phenolic compounds using complementary methods and their effects on antioxidant status in animal models and human subjects are needed to evaluate their potential benefits. In addition it is necessary to further confirm lack of toxicity and bioavailability of such natural phenolic extract. Delivery of isolated phenolics as dietary supplements or functional food ingredients for health promotion and disease risk reduction may also be helpful in improving the efficacy of such materials.

Phenolic compounds are ubiquitous in the plant kingdom. Animals do not synthesize compounds with benzenoid rings from their aliphatic precursors. Phenolics are thus strictly exogenous food components of exclusive plant origins. Of the various complex phenolic compounds that occur in plants, only flavonoid polymers of flavan-3-ols and flavan-3, 4-diols (condensed tannins) have been reported to exhibit antinutritional effects in animal and human subjects (Butler *et al.*, 1984). Populations consuming mainly animal foods may not consume significant amounts of tannins. However, dietary tannins can be a serious problem for people whose diets are mainly vegetarian. Condensed tannins occur in many foods and beverages (Deshpande *et al.*, 1984). Of particular nutritional concern are the tannins and related polyphenolic materials in foods which are major components of human diets or are otherwise consumed intensively. These include the popular betel quid and arecanut powder chewed throughout India and the Asian countries, red wines, fruits and fruit juices, beans, and cereals such as sorghum (Stich *et al.*, 1983; Ribereau-Gapon *et al.*, 1983; Hulse, 1980). A large section of the world population in semiarid tropics depends heavily on sorghum millets and grain legumes as main sources of proteins and calories (Salunkhe *et al.*, 1985a; Salunkhe *et al.*, 1985b). Certain cultivars of sorghums and legumes are known to contain appreciable levels of condensed tannins which drastically lower their nutritional quality. One obvious solution to the tannin problem is to grow cultivars low in tannins. Such low-tannin cultivars that are agronomically high-yielding arc available and indeed predominate in most semiarid tropics. However, such low-tannin cultivars cannot be economically grown in certain areas where the bird-depredation problem is severe (Butler, 1978). Only relatively high-tannin cultivars give dependable yields under such ecological conditions. Hence, the growing of high-tannin cultivars will continue at least in certain parts of the world. In certain grain legumes like dry beans, the phenolics in seeds improve their acceptability over white beans, perhaps due to taste. Besides, tannins are known to function as natural defense agents of plants against insects, fungi, and pre harvest seed gernlination (Salunkhe, 1982). Since tannins are agronomically desirable and nutritionally undesirable to man and livestock, attempts arc needed to develop cultivars that contain appreciable levels of tannins active only towards yield-deteriorating agents, but less toxic to consumers. Recent advances in the chemistry of tannins assay procedures, their occurrence and interactions with dietary components, changes during maturation and ripening, storage, cooking and germination and physiological influence on the function of secretary glands in animals have generated interest in the nutritional significance of dietary tannins. Recent developments in these areas are briefly summarized in this section, and future research needs are addressed.

There is a lack of agreement and uniformity in names given to those compol1ents of food grain that contain biologically active phenolics. The word seed coat is offended used describes the pericarp. The mature ovule wall has been called seed coat, testa, and sub coat under coat, inner integument, outer integument, or nuclear layers (Hulse, 1980). There is a need to have on" common name, perhaps "testa," to describe a grain component that contains tannins further investigations are essential on the genetics of testa color, mode of testa synthesis, biochemical composition of the testa, type of testa layers, and number of genes involved in the synthesis and spreading of phenolic compounds from the testa to the pericarp. The presence of only methanol-extractable tannins in some Group III sorghums and both methanol and acidic methanol-extractable tannins in other Group III cultivars (Butler, 1982) indicates the possible differences in testa structure as well as possibility of involving other genes factors that influence the spreading of tannins from the testa to the pericarp. In legumes most tannins arc located in the seed coat. Further investigations are essential on the chemical composition of the seed coat, location of tannins within the seed coat, nature of tannin association with the seed coat components, and genetics of tannins in grain legumes.

Phenolics in plants include phenolic acids and coumarins, and flavonoid compounds, including anthocyanidins ($C_6 - C_3 - C_6$ structures). Flavonoids, are a large and most diverse range of plant phenolics, include anthocyanin pigments, flavones, flavonols, flavanols, aurones, chalcones, and isoflavones. Of these, only flavanols (flavan-3-ols, flavan-3, 4-diols, flavan-4-ols), and their oligomers and polymers are very common in sorghum and several legumes. These are previously called in literature phenolics, polyphenols, tannin(s), and tannin-like substances. In order to develop proper nomenclature for these compounds, isolation and characterization of various phenolics produced in different genotypes at various stages of maturity need to be made. The available evidence indicates that monomeric phenolics predominate during the early stages of grain development. As the grain matures, monomers get converted to oligomers and finally to polymers at maturity (Bullard, R. W. and Elias, 1980). However, relative proportions of monomeric, oligomeric and polymerized phenolics in mature grain vary widely among the different cultivars of the same crop, like sorghum (Price *et al.*, 1979). The biological activity of the phenolic compounds is largely governed by their chemical forms. Hence, investigations on their structures, mode of biosynthesis, factors that induce plants to produce such compounds, mechanism of polymerization, degree of polymerization, and biological activity of each type of phenolic compounds are essential.

Haslam and Gupta (1978) isolated and characterized the major polyphenol of bird-resistant testa containing sorghum, NK-300, to be a oligomer of flavan-3-01 with five to seven units condensed together having molecular weights of 1700 to 2000. This oligomer, called procyanidin polymer, precipitates proteins. However, recent reports on the degree of polymerization of sorghum tannins indicate that the extent to which monomers will polymerize depends upon the type of cultivar and stage of maturity (Price *et al.*, 1979). The length and composition of the flavan-3-ol oligomers are not directly coded on DNA. The system is polydisperse, like starch. Hence, it is difficult to purify a complete homogenous sample even though it is pure (Butler, 1982). Indirect methods need to be standardized to determine the variations in degree of polymerization of tannins in different cultivars and grain types. Besides, the molecular shape and size are important factors in determining the properties of procyanidin polymers. The sorghum procyanidin has been predicted to possess a right-handed helical structure (Haslam, 1986). However, more investigations are essential on the molecular structure of oligomeric and completely polymerized tannin in order to assess their biological activities. It has been indicated that monomeric and dimeric phenolics are nutritionally harmless; oligomeric phenols with 3 to 10 monomeric residues bind with proteins and are nutritionally harmful, while more completely polymerized phenolics are chemically inert and nutritionally harmless (Bullard and Elias, 1980; Price, 1979). Monomeric phenolics have been tentatively implicated in desirable agronomical properties such as bird repellency (Bullard and Elias, 1980). Thus, from a production and nutritional quality point of view, three groups of phenolic compounds exist in fruits, forages, wines and seeds of millet, sorghums, and legumes. These are monomeric, oligomeric, and polymeric phenolics which differ significantly in their structure, molecular size and shape, chemical properties, and biological activity. They cannot be collectively called tannins or even polyphenols. There is indeed an urgent need to name these compounds on the basis of their structure and properties to bring out the uniformity in reporting experimental results on various aspects of these complex phenolic compounds in plants.

Several methods are available for the analysis of phenolic compounds. However large variations exist in the preparation of samples, particle size, solvent used for extraction solvent to sample ratio time of extraction period elapsed between grinding of sample and extraction of phenolic compounds assay conditions to develop color, and standards used for quantitative calculations (Deshpande *et al.,* 1986). Due to these sources of variations, it has become difficult to make meaningful comparisons of results reported from different laboratories. Since the chemical nature of phenolic compounds is very complex, it is difficult to standardize a simple common method for their determination. Among the available methods, the Folin-Denis Assay (Swain and Hills, 1959) Prussian blue assay (Price and Butler, 1977) modified vanillin-HCI assay (Price *et al.*, 1978) protein-precipitation method (Hagerman and Butler, 1978) and relative degrees of polymerization (Butler, 1982; Butler et al., 1982) are some of the most useful methods. However, each of these methods measure different classes of phenolic compounds and suffer from drawbacks.

From a nutritional quality and agronomical benefits point of view, tannins in sorghums, legumes, forages, wines, and fruits need to be determined as monomeric phenolics, oligomeric phenolics, and large polymeric phenolics. Common assay condition describing extraction procedures, reaction conditions, and standards to be used need to be defined separately for each group of these phenolics to obtain more meaningful and useful results. Although for routine analysis and screening purposes, methods like the Prussian blue or vanillin assay may serve the purpose; more precise and sophisticated techniques need to be standardized for tannin analysis for their chemical characterization. The techniques based on high pressure liquid chromatography on acetylated materials (Glennie *et al.*, 1981) and continuous counter current chromatography (Ito and Bhatnagar, 1981) may be very useful in this context.

Experiments with a variety of laboratory animals. have shown that dietary tannins decrease the growth rate, feed efficiency, egg production by laying hens, digestibilities of dry matter and proteins, net energy, metabolizable energy, and growth promoting value of the diet (Butler *et al.*, 1986). Tannins have been shown to cause leg abnromalities and ultra structural changes in the livers of chicks, and affect the physical or sexual maturity and even death in hamsters, and are also regarded as possible carcinogenic factors. The major anti nutritional effect of dietary tannins is the formation of stable complexes with dietary proteins and perhaps with digestive enzymes, which ultimately results in poor digestion and malabsorption of dietary nutrients. Tannins have also been reported to form complexes with starch and reduce its digestion (Deshpande and Salunkhe, 1982). To develop methods of the elimination of such effects by tannins, it is essential to understand the mechanism of interactions between tannins and proteins, and the conditions that favor the formation of indigestible complexes.

Recent reports on these aspects of dietary tannins indicate that only oligomeric polyphenols with 3 to 10 flavan-3-ol residues form complexes with proteins, while the monomeric, dimeric, and higher polymeric polyphenols are nutritionally harmless (Butler *et al.*, 1986). Hence, developing cultivars in which the monomeric phenolics never polymerize or, if they do, polymerize to the extent where most of the monomers are converted to polymers with more than 10 residues, may be one possible approach to eliminate the tannin problem. Such cultivars need to be developed through genetic selection and hybridization. Suitable chemical treatments of mature grains to induce polymerization during storage can also be thought. The high concentrations of monomeric and short oligomeric phenolics are known to impart an astringent taste to grains (Bullard and Elias, 1980). This may lower acceptability and depress food/feed intake. Hence, an approach to develop cultivars/ chemical treatments in which tannins polymerize almost completely at maturity or during storage may be more useful rather than blocking the formation of oligomers from monomers during maturation. In this context, cultivars of Group II sorghums seem to be promising. Further investigations on the mode of biosynthesis, polymerization and nature of tannin association with seed component(s) in Group II cultivars may be useful in developing desirable plant types in sorghum.

Interactions between oligomeric tannin and protein seem to be extremely complex phenomenon. Sorghum tannin is highly selective in its binding to proteins. *In vitro*-binding specificity implies that tannins interact with specific site(s) of proteins or other macromolecules under particular microenvironments. Tannin-binding salivary proteins share some biochemical characteristics with tannin-associated proteins from sorghum grain. Proteins with high affinities for tannin tend to have open, loose conformations, high molecular weights and high contents of proline and other hydrophobic amino acids. The tannin protein interactions have been reported to be tannin-specific as well as protein-specific (Asquith and Butler, 1986). These reports indicate that oligomeric tannins arc not merely universal protein-binding agents, but different tannins exhibit different affinities for the same protein. Hence, tannins of different food sources need to be isolated, purified, thoroughly characterized, and tested for their binding properties with dietary proteins. Investigations on the association of tannins with proteins have shown that most tannins are associated with proclaims in sorghum (Chavan, *et al.*, 1979) and glutelins in finger millet (Ramachandra *et al.*, 1977). Butler (1982) reported that sorghum tannins have higher affinities for alcohol-soluble hydrophobic-storage proteins than for other seed proteins with more conventional properties. Although more confirmative studies are needed in this respect, it is suggested that developing cultivars that are low in proline and prolamin fractions may be very useful.

Tannins are known to form complexes and inhibit the activities of digestive and hydrolytic enzymes (Deshpande and Salunkhe; 1982). Unfortunately, most of reports on the inhibition of enzyme activities by tannins are based on *in vitro* experiments using synthetic substrates. The results of such experiments do not give a clear picture of tannin-enzyme interactions because most enzymes in the biological system work on their natural substrates. The natural substrates have been shown to be susceptible for tannin binding, while the synthetic substrates are not. In the presence of natural substrates like proteins, tannins have a choice to bind preferably with either the substrate (Mole and Watterman, 1987) or enzyme, or both depending on their relative affinities towards each other. It is rather difficult to determine whether given tannin will preferably bind with an enzyme or protein substrate unless each system is individually studied. Nevertheless, if a system contains a substrate protein that has a low affinity for tannin, the tannin may bind with the enzyme provided other conditions favor such binding and inhibit its activity. It is, therefore, clear that developing cultivars containing proteins with low affinities for tannins may not be sufficient, because tannin in these cultivars will exert its effect by inhibiting the activities of digestive enzymes. The structure and amino acid composition of hydrolytic enzymes of the digestive tract of animals, including man, are difficult to change. Hence, there is a need to develop cultivars that contain either nutritionally harmless tannins or develop a mechanism to inactivate tannins in the biological system before they form complexes with dietary proteins or digestive enzymes. In this context, certain saliva proteins, extremely rich in proline (Meunzer *et al.*, 1979), with very high affinities for tannin seem to be important. Rats fed high-tannin diets showed a dramatic increase in the weight of their parotid glands and synthesis of proline-rich proteins (PRPS) (Mehansho *et al.*, 1983). Tannins selectively bind with these proteins out of a large excess of other proteins with average affinities for tannin. Thus, the parotid gland acts as a guard in inactivating the tannins through an increased secretion of PRPs in response to tannin intake. This can be considered as a natural defense system of animals against tannins. It is, therefore, necessary to further investigate the relationship between tannin intake and stimulation of parotid glands, and the component that induces the secretion of salivary PRPs. This may help in developing lines rich in a component that stimulates the secretion of salivary PRPs in increased amounts so that a complete protection against dietary tannin is obtained.

A wide variety of processing approaches for the elimination of the antinutritional effects of tannins in sorghum and grain legumes have been applied (Deshpande and Salukhe, 1982). These included dehulling and milling, soaking grains in a variety of chemical solutions followed by discarding the soaking medium and drying of grains, spraying of chemicals into ground meals inhibition of water and chemicals into the grain followed by their anaerobic storage, different cooking procedures, germination, freezing drying and storage. In general, methods like cooking, germination, freezing, drying, and storage of grains, although decrease assayable tannins, are not effective in eliminating the antinutritional effects of tannins (Price *et al.*, 1979). During such processing operations, tannins get bound to certain seed components, perhaps proteins, with relatively higher affinities for tannins. Thus, such tannin-bound proteins become nutritionally unavailable. Investigations arc therefore, needed to develop procedures of treating grains before or during cooking, germination, or drying so that tannins do not bind with proteins, but get converted into chemically inert forms. Several grain legumes are consumed as sprouts. When high-tannin seeds are germinated, tannins in the seed coats may form stable complexes with seed proteins and decrease their availability. Further investigations on tannin metabolism during germination are essential. Dehulling of grains appears to be the most simple and economical method for the removal of tannins from grain legumes. For sorghums however, dehulling results in a significant loss of dry matter and seed proteins which may not be economical to compensate the benefits obtained from the removal of tannins. Dehulling will lower the fiber content of diets based on such processed grains, and the method will not be a practical solution unless special dehullers or polishers are designed for sorghums and millets. Although, soaking seeds in dilute solutions of alkali or acids although removes tannins effectively, it results in the alkali destruction of certain amino acids, affects sensory properties of treated materials, and requires additional drying after treatment. The additional cost involved in such processing in relation to benefits obtained from tannin removal needs critical appraisal.

Other approaches for tannin deactivation include supplementing high-tannin diets with extra levels of proteins, nutritionally poor proteins like gelatin, and using suitable tannin binding detergents like PYP and certain methyl donors such as methionine (Deshpande and Salukhe, 1982). All these components are implicated in the deactivation of tannin and/or its metabolites. Although results "with extra protein, gelatin, or PYP are quite impressive, the practical utility of such an approach IS doubtful. High-tannin sorghums, millets, or legumes are principal sources of diets of the economically poor people of Asia and Africa, where both malnutrition and under nutrition are severe nutritional problems. Under such conditions, expenditure on additional dietary proteins, gelatin, PYP, or amino acids for supplementation to deactivate tannins seems to be beyond the reach of common consumers. Although these technologies may be feasible to develop animal feeds in the developed world, their practical applications for human food in the developing world appear to be difficult. Besides, certain problem such as sensory properties of a supplemented diet and toxicological effects of consumption of gelatin or PYP needs investigation. Amino acid supplementation of diets for the deactivation of tannin metabolites in the human diet seems difficult for people of developing countries due to the problems of availability, cost, and lack of educational and technological developments. Methionine and its analogues deactivate only tannin metabolites after their absorption and not oligomeric tannin per se. For animals, methionine requirements vary from species to species (Deshpande and Salukhe, 1982). However, such investigations are warranted to elucidate the mechanism(s) of tannin deactivation.

The deactivation of tannins through oxidation or induced polymerization *in situ* by imbibing grains With aqueous solutions of various alkalies, ammonia, water, or formaldehyde, followed by anaerobic storage of grains for 2 to 3 weeks, appears to be the most simple, practical, and efficient method of tannin deactivation (Deshpande and Salukhe, 1982). High-tannin grains imbibed with aqueous ammonia, water, or formaldehyde have been reported to exhibit improved nutritional quality in feeding trials. Further studies are essential on the mechanism of tannin deactivation in imbibed grains. Conditions that make the polymerization of tannins almost complete need to be standardized to develop the processing technology and suitable machinery for tannin deactivation in intact grains. Such processing treatments seem to be more appropriate than the removal of tannins from grain or meal because tannins arc only nutritionally deactivated but exhibit other desirable qualities in diet, 'ouch as dark color to cooked bean and soups.

Although the precise role of tannins in plants is uncertain (Glennie *et al.*, 1981; Beart *et al.*, 1985) available reports support that tannins offer significant benefits to otherwise vulnerable plants. Tannin-containing plants have been found to exhibit some degree of resistance against attack by birds. Pests, molds and pre-harvest seed germination (Deshpande and Salukhe, 1982). Although tannins may not give complete protection to plants against deteriorative agents, they can be advantageous as components of coordinated plant protection measures. Since the biological effects associated with tannins are genera.lly beneficial in the field and harmful in the diet, investigations are essential to make tannins more beneficial in the field and less toxic in the diet. Chemical forms of tannins that exhibit beneficial effects in the field and during storage need to be precisely ascertained. Available reports on these aspects indicate that monomeric and short oligomeric phenolics are mainly responsible for the resistance against birds, pests, and molds both in grain (Bullard and Elias, 1980; Subrammanian *et al.*, 1983) and leaves (Watterson and Butler, 1983; Jambunathan *et al.*, 1986). However, further investigations are essential to confirm these findings so that strategies can be planned to obtain higher levels of monomeric flavan-3-ols, flavan-3, 4-diols, and flavan-4-ols and other non tannin phenolics in immature grains and leaves to derive the agronomical benefits of tannins.

Several investigators have attempted to resolve the complexity of chemistry of phenolic compounds of sorghums, millets, and grain legumes, their purification and characterization, methods of analyses, interactions with dietary components, antinutrilional effects, and remedies to overcome these effects over the last decade on the basis of recommend at ions by a working group of a symposium held during the 36th annual meeting of the Institute of Food Technologists at St. Louis, MO, in 1979 (Hulse, 1980). However, certain aspects, like interactions of phenolics with starches, lipids, vitamins, and minerals, have not received enough attention. Considerable data are available on sorghum tannins. However, such studies are required on tannins in fruits, forages, and legumes, Reddy *et al.*, (1985) identified the following areas for future research for legume tannins:

1. Identification of low-tannin types in common beans of different colors.
2. Fate of tannins in beans during cooking and other processing.
3. Study of protein-binding and enzyme-inactivation properties of tannins from different bean cultivars and identification of possible sites of chemical reactions.
4. Improved experimental models and analyses to establish relationships between tannin and protein digestibility, and proline quality in beans.
5. Mechanisms of detoxification of bean tannins.
6. Development of simpler processing techniques for improving the nutritional quality of high-tannin bean cultivars.

Tannins will continue to be a component of the human diet and livestock feeds. Therefore, the problem of the dietary effects of tannins must continue to be attacked. Approaches may include breeding livestock having a resistance to the detrimental effects of dietary tannins or modifying tannin-producing crops either by conventional breeding techniques or by genetic engineering to produce tannins less toxic (Deshpande and Salunkhe, 1982). The work has in fact just begun and will take a long time to reach the desired goals.

In the last two decades, the use of high-yielding varieties improved food grain production considerably in developing countries. However, their protection from pests, molds, viruses, and grain-eating birds has still remained a challenging problem. The problem of pests, diseases, and birds is more severe in tropical regions of the world where climatic conditions are favorable for the rapid growth and spread of these agents as compared to temperate regions. The indiscriminate use of chemical pesticides is known to cause the resurgence of pest populations as an outcome of selective pressure and the emergence of a secondary pest outbreak resulting in the elimination of beneficial species. Pesticides are also expensive and contribute to environmental pollution and health hazards. In developing countries, the optimum use of insecticides is not feasible due to their increasing costs and no availability, and the special equipments required for their applications. The agricultural practices in developing countries are labor-intensive as compared to developed nations. Hence, more economical and safe methods to control infestation are needed.

Recently, emphasis has been placed on the optimal use of biological controls and improved agricultural production practices to achieve higher crop yields with minimum disturbances to the ecosystem. Plants possess a pool of resistance factors that can be used to select and further improve resistance varieties. This inbuilt protection has played a major role in suppressing various groups of plant diseases and pests capable of damaging host plants. Various biochemical compounds that impart some degree of inbuilt resistance of plants to various pathogens include phenol-phenolase systems, phytoalexins, metal chelates and antibiotics (Panda, 1979). Recent advances in the physiology of proteinaceous inhibitors in crop plants suggest that inhibitors of trypsin, chymotrypsin and subtilisin may have important roles as protective agents against the invasion of fungi and bacteria (Richardson, 1977; Ryan and Green, 1974; Chavan and Hejgaard, 1981). Phenolic compounds including tannins have been implicated in conferring some degree of resistance in grains and leaves of cereals like sorghum and of several legume seeds against the attacks of fungi, pests, grain-eating birds, and preharvest seed germination (Tipton *et al.*, 1970; Harris and Burns, 1970, 1973; McMillion *et al.*, 1972; Bond, 1976; Mishra *et al.*, 1980; Bullard and Elias, 1980; Dreyer *et al.*, 1981; Chang and Norris, 1983; Kramer et al., 1984; Scalbert and Haslam, 1987).

Sorghum and millets are the staple food crops and provide nourishment for a large segment of the human populations of Asia and Africa (Hulse *et al.*, 1980). The reliance on sorghum, pearl millet and certain grain legumes as staple food crops in these areas of developing countries is due to their ability to give dependable yields under relatively harsh ecological and growing conditions (Doggett, 1970). In some arid regions of Africa, sorghum is the only cereal grain that can be grown with reasonable success. The grains of certain cultivars of sorghum and grain legumes are known to contain an appreciable level of phenolics, including tannins which impart defense properties to seeds against invading pathogens and birds. During the 1960s and early 1970s. the development of bird-resistant inbreeds and hybrids became an integral part of many sorghum research programs (McMillion *et al.*, 1972) Comparatively more literature is available on sorghum tannins as a chemical defense perhaps because sorghum is more severely damaged by birds and certain cultivars of sorghum contain higher levels of tannins which are shown to be resistant to bird attacks as compared to low-tannin cultivars. Natural defenses in these crop' due to tannins is briefly reviewed in this section.

LOSSES DUE TO BIRDS, PESTS AND PATHOGENS

Sorghum millets and grain legumes are infested by a number of pests, fungi and grain viviparous (grain-eating) birds. Among birds, the red-billed weaver *(Quelea quelea)* of Africa is perhaps the most well-known. It has the distinction of being the most numerous and possibly the most destructive bird in the world, comparable only to locusts in the amount of damage it causes. The Quelea population in Africa Crook and Ward, 1968) has been estimated to be in the tune of 10^9 to 10^{10}. Other species in other parts of the world likewise cause significant losses in sorghum. Red-winged black birds, starlings, and sparrows are the major birds damaging crops in North America. In Southern Asia, sparrows, parakeets, crows, mynas, and several others cause havoc to sorghum and millet crops (Rachie, 1970). Eared doves, blackbirds, sparrows, parakeets, and parrots are the major pests creating problems to sorghum growers in several Latin American countries (Bullar and Elias, 1980). Precise reports on the extent of losses in sorghum, millet, and legumes are not available, although losses of up to 100% in sorghum have been reported in Louisiana (Tipton *et al.*, 1970). It is generally considered that birds are better able to perch on the heads of tight-headed hybrids than on the more open-headed hybrids. Major damage to the grain by birds occurs during the milk stage of development. Damage to the matured grains is rather limited.

Sorghum, millets, and grain legumes are infested by a range of pests and microflora during field growth, seed ripening, and postharvest handling and storage (Krishnamurthy *et al.*, 1983; Salunkhe *et al.*, 1985). Reliable estimates of losses due to the activity of pests and diseases both at field and during storage are not available. The extent of losses is influenced by the nature of pests, methods used to control the pests, insect population, storage conditions, crop and variety, and other climatic conditions that favor or disfavor the growth and development of pests and molds. However, a rough estimate of 10% losses in the field due to pests alone is generally agreed (Krishnamurthy *et al.*, 1985). Of the 50% losses that occur during postharvest handling in developing countries (Anon, 1965) a major proportion is contributed by rodents and pests (Parpia, 1970; Salunkhe *et al.*, 1985). Salunkhe *et al.* (1985) recently reviewed the nature and causes of postharvest losses in cereals and legumes, and the suitable biotechnology to minimize them. The development of cultivars with inbuilt resistances through compounds like protease inhibitors and tannins towards damaging birds, pests, and molds can form an integral part of integrated plant protection technology to improve crop productivity.

BIRD RESISTANCE

Bird resistance in sorghum has been attributed to several characteristics including the presence of stiff lemma awans (Boyd, 1965) lax panicles (Voight, 1966; Perumal and Subramanian, 1973) and high-tannin content (Niehaus, 1966; Niehaus and Schmidt, 1970). High-tannin content is the characteristic most often associated with bird resistance (Table 10.1), and many reports support this relationship (Haris, 1969). Several sorghum lines with high-tannin contents have been developed and promoted on the basis of their ability to resist depredation by birds. Many of these resistant sorghums are used extensively in those areas of the world where the risk of such losses is high. Birds are detracted from the grains that possess the astringent taste properties. The repellent characteristics of bird-resistant sorghums are attributed to the group of polyphenols (Joslyn and Goldstein, 1969; Goldstein and Swain, 1963). The astringency of tannins is considered to be the principal means by which high-tannin sorghums offer resistance to bird depredation. Bullard et al. (1980) compared the bird-repellent characteristics of several sorghum cultivars under laboratory conditions. The studies revealed that polyphenols in sorghum are responsible for the taste-repellent properties to wild birds. Of the cultivars tested, seven bird-resistant cultivars contained more than 3% catechin equivalent values and were uniforn1 in polyphenol properties. Subramanian et al. 3H studied the tannin content and grain and agronomic characters of 18 sorghum genotypes with brown per carp colors, 15 of which were known to be bird-resistant. The agronomic characters, such as earhead length, head type, glum color and threshability, varied significantly among the genotypes, indicating that these characters are not likely to be responsible for bird resistance. The condensed tannin contents of these cultivars ranged from 0.13 to 7.22% catechin equivalents. Of the lines tested, 10 cultivars (all Group III type) were analyzed for polyphenols by different methods (Table 10.2). The cultivars IS-2826, -3171, -724, -2880 and -3031 contained condensed tannins in amounts that might be expected to confer bird resistance (Subramanian *et al.* 1983). The other lines however did not contain significant amounts of condensed tannins although they are reported to be bird-resistant. The amount of flavan-4-ols is generally low in sorghum (Watterson and Butler, 1983). However, the average value of flavan-4-ols in the nine bird-resistant cultivars was 19.5 A_{500}. It has been suggested that the presence of relatively higher levels of flavan-4-ols in all these cultivars may be a factor for the bird-resistance property. Bird damage is known to be minimal to mature grains of even low-tannin, white seeded cultivars. The birds attack immature grains more severely than matured grains. Only those cultivars that resist bird depredation during immature stages of grain ripening could be referred to as bird-resistant types. Hence, it is more appropriate to investigate the predominant types of polyphenols in immature grains of proven bird-resistant cultivars, when bird attacks are likely to be severe.

Table 10.1: Seed colour, tannin contents and bird damage in sorghum cultivars

Cultivar	Seed colour	Tannins (%)	Bird damage (%)
Asgrow Double TX	Red	0.2	68
Pioneer 828	Red	0.2	68
ACCO R 1019	Bronze	0.2	73
Dekalb F 64	Bronze	0.2	87
ACCO R 109	Bronze	0.2	70
Dekalb E 57	Bronze	0.2	50
Funk G 522	Bronze	0.2	78
ACCO R 1029	Bronze	0.2	70
Funk BR 79	Brown	5.7	0
GA 615	Brown	5.2	0
Excel Bird Go	Brown	6.0	0
AKS 663	Brown	2.3	0
Dorman BR 100	Brown	4.5	0
ACCO R 1093	Brown	5.2	0
Asgrow Bravia R	Brown	6.9	0
McNair 546	Brown	3.0	0
Dekalb BR 64	Brown	5.3	3
AKS 614	Brown	3.0	3
RS 700	Tan	3.3	7

Table 10.2: Polyphenol analyses of mature dry grains of bird resistant sorghum cultivars.

IS No	Total phenols (μmol/g)	Protein precipitation (mg protein/g)	Vanillin assay (CE/100g)	Proantho-cyanidins (A_{500}/g)	Flavan -4-ols (A_{500}/g)
2649	64	3.28	0.01	0.0	2.9
8765	78	0.37	0.02	1.3	19.7
3153	61	1.94	0.02	1.3	17.3
2827	55	0.0	0.02	1.6	17.1
8748	149	0.0	0.20	9.2	13.4
2826	254	28.30	0.76	26.1	15.1
3171	297	20.80	0.63	23.4	17.3
724	419	47.50	1.14	29.0	22.1
2880	514	60.50	1.89	41.7	20.8
3031	714	74.30	1.43	46.4	32.9

Bird-resistant sorghums contain more tannins and astringency in immature grains (Table 10.3). Both the tannin content and astringency decreases towards maturity. Monomeric and oligomeric phenols predominate in immature grains (Bullard and Elias, 1980). The polyphenol content increases with grain development until the dough-milk stage and then decreases to various degrees in different cultivars as the grains mature. Monomeric and oligomeric phenols, including flavan-4-ols that predominate in immature grains, are responsible for bird resistance (Price *et al.*, 1979). In mature grains, this property may disappear as the polyphenols polymerize.

Table 10.3: Tannic acid and total astringents content of grain sorghum during grain development

	Bird-resistant		Nonbird-resistant			
	Lindsey BR 75		Dekalb 57		Asgrow Ranger B	
Maturity stage	TA (%)	Total astringents (%)	TA (%)	Total astringents (%)	TA (%)	Total astringents (%)
Milk stage	0.68	0.94	0.03	0.20	0.07	0.24
Dough stage	0.50	0.72	0.09	0.25	0.07	0.24
22% moisture	0.44	0.66	0.09	0.24	0.09	0.26
17% moisture	0.57	0.74	0.05	0.18	0.07	0.24
16% moisture	0.53	0.66	0.06	0.19	0.08	0.23

It has been reported that tannins in Group II sorghums are relatively more polymenzed in matured grains and are possibly attached to some grain components so that they are deactivated and exhibit less harmful effects on protein utilization

(Price *et al.*, 1979). It has also been suggested that cultivars of Group II sorghums may be useful in developing the most desirable plant type that will show bird resistance as well as good nutritional quality. The report of Subramanian *et al.* (1983) indicated that similar cultivars can also be identified from Group III sorghums. There are several reports on varieties having resistance to bird depredation in the field and demonstrated good nutritional quality in feeding or digestion trials (Thayer *et al.*, 1957; Damron *et al.*, 1968; Harris et al., 1970; Tipton et al., 1975). Similarly, very little enzyme inhibition activity has been observed in varieties with pigmented testas and/or sizeable levels of tannins (Ayyangar and Krishnaswami, 1941). All Group II varieties have not been found to be bird resistant, indicating the variability of polyphenol contents during the ripening of seeds.

Bird depredation is a complex phenomenon. Bird attacks on growing crops would depend upon the nature of birds, nature, and quantity of alternate food grains available for feeding, bird population and natural defense offered by the grains (Butler, 1982). Grains with repellent properties will also be considered by birds for eating if no other grains are available for feeding and the bird populations are high where competition for available grains is quite high. Polyphenols as a natural defense cannot give complete protection against birds. However, if only bird repellent cultivars are grown over an entire region for several years, it can certainly reduce the population of birds in the region. Polyphenols as a natural defense to birds needs to be viewed as one of the components of plant-protection measures that have to be applied in a coordinated manner to save crops from bird damage. In this context, cultivars that contain higher levels of monomeric phenolics in immature grains, but almost completely polymerized forms in mature grains, hold promise.

Preharvest seed germination

Preharvest seed germination occurs in grain sorghum when mature grains receive rains at the time of harvest and temperatures are conducive for germination. Seed germination in ear heads of standing crops has been observed in fields and experimental plots in the humid Southeastern U.S. for many years. However, reliable estimates of grain losses in the world due to such preharvest seed germination are not available. Brown-seeded hybrids have been reported to show marked degrees of resistance to preharvest seed germination (Harris *et al.*, 1962). High-tannin hybrids were significantly lower in preharvest seed germination than low-tannin hybrids (Table 10.4). Although the tannin content of seeds was apparently the dominant influence inhibiting preharvest seed germination, maturity among the non-brown-seeded hybrids was also involved. Late-maturing, non-brown-seeded hybrids showed a lower preharvest germination index than early hybrids.

Table 10.4: Tannin content and preharvest seed germination in sorghum

Seed colour	No of hybrids	Tannin content	Germination index
Brown	21	10.2	1.3
Non-brown	21	3.6	3.7

Tannins in grain sorghum are present in the testa and pericarp. Tannins in mature grains are in polymerized forms, forming a continuous layer in the testa around the aleurone. This thick layer of testa cells, filled with relatively insoluble tannin polymers, could retard the absorption of water and delay preharvest seed germination. Tannins in grains of Group H sorghums are reported to polymerize almost completely at maturity and bind to some gram component (Price *et al.*, 1979). It is quite likely that this form of tannins would retard the water imbibitions by grain more effectively as compared to grains in which tannins are spread in the pericarp and not fully polymerized. In this context, Group II sorghums may exhibit better resistances to preharvest seed germination than cultivars of Group III. However, such reports are not available. The relative ability of Group II and Group III cultivars to resist preharvest seed germination need to be evaluated in field conditions to elucidate the mechanism of action of tannins in resisting grain sprouting in the field. Because tannins are known to inactivate a variety of enzymes, it was suggested that tannins inhibit preharvest seed germination by inactivating germination enzymes· However, this possibility seems to be remote. When high tannin sorghum is germinated and then milled, the activity of enzymes involved in malting are found to diminish when compared with low-tannin types. However, little inhibition was seen when the grain was left intact. Butler *et al.* (1984) reported that, on germination tannins apparently bind to some component of the seed possibly prolamin proteins in the outer portion of the endosperm. This bound form of tannins probably is much less able to inhibit enzyme activity of the seed. It appears that tannins of mature high-tannin grains impart resistance to preharvest seed germination by acting as a physical barrier for water imbibitions into the endosperm and germ. Further studies on preharvest seed germination in Group II and Group III sorghums may give additional information about the role of tannins in this process.

RESISTANCE TO PESTS AND DISEASES

Although the concept of tannins as chemical defense agents in crop plants has been recently questioned (Beart *et al.,* 1985) several reports support the view that tannins do confer some level of resistance in sorghum and grain legumes against the infestation by pests and molds. Higher tannin content has been reported to offer resistance against anthracnose and greenbugs in sorghum (Mishra *et al.*, 1980; Dreyer *et al.*, 1981). However, the mechanism of such resistance, and the nature of polyphenols that confer resistance and interaction of pests and molds with these polyphenolic compounds, has not been extensively investigated. Most reports on tannins as defense agents in sorghums and legumes are related to field infestation. However, reports on the benefits of a higher level of tannins in grain against stored grain pests and molds are lacking. The attack of insects and molds in the field begins with flowering, grain setting, and grain ripening where the tissues contain ample water, rich in nutrients like sugars, soluble proteins, and amino acids, and tender to cause injury. Mature dry grains are usually not infested by field pests. Molds may grow on mature grains during rainy climate.

Attempts have been made to study the different types of polyphenols in the leaves and grains of several mold-resistant and susceptible cultivars of sorghum.

Jambunathan *et al.* (1986) analyzed sorghum cultivars exhibiting both resistance and susceptibility to grain molds for different polyphenols, including flavan-4-ols and proanthocyanidins. Cultivars with white seed colors were susceptible, while those with brown colors were resistant. However, among the red-colored types, both susceptible and resistant varieties were present. The resistant brown-seed types contained higher levels of total phenols, condensed tannins, protein precipitable phenols, flavan-4-ols, and proanthocyanidins than susceptible red types. The resistant red-seeded types contained two- to three fold higher levels of both methanol and acidic methanol-soluble flavan-4-ols than the susceptible cultivars. The white-seeded susceptible cultivars did not contain any phenolics in appreciable quantities (Table 10.5). These observations indicate that flavan-4-ols may be responsible for grain-mold resistance, rather than condensed tannins in sorghum.

Table 10.5: Polyphenols, flavan-4-ols, and proanthocyanidin contents of sorghum cultivars

Sorghum cultivars				**Total phenols (A_{720}/g)**	**Vanillin assay (A_{500}/g)**	**Protein precipi-tation (A_{510}/g)**	**Flavan -4-ols (A_{550}/g)**	**Proantho-cyanidin (A_{550}/g)**	**Formic acid HCl assay (A_{550}/g)**
Seed color	**Mold resistant/ susceptible**	**No. of culti-vars**	**Extractant**						
White	Susceptible	3	Methanol	3.13	0.0	0.04	0.0	0.0	0.0
			H^+ methanol	3.48	0.0	0.05	0.0	0.0	0.0
Red	Susceptible	2	Methanol	6.65	0.75	0.01	5.25	0.34	15.87
			H^+ methanol	5.37	0.60	0.04	3.28	0.22	10.07
Brown	Resistant	4	Methanol	55.00	3592	3.98	16.56	27.59	29.08
			H^+ methanol	37.95	19.03	3.77	5.29	18.51	9.71
Red	Resistant	3	Methanol	778	1.54	0.05	15.23	1.02	29.34
			H^+ methanol	6.79	1.10	0.06	10.68	0.06	19.42

It appears from available evidences that the monomeric and smaller oligomeric phenolics in leaves and grains of sorghum are mainly responsible for desirable agronomic traits, such as resistance to birds, pests, and molds. Monomeric phenolics are nutritionally not harmful. It therefore, suggests that cultivars higher in monomeric phenolics, such as flavan-3-ols, flavan-3,4-diols, and flavan-4-ols, and low in polymerized tannins need to be developed. Grains that contain higher levels of monomeric phenols, both in immature and mature grains, exhibit desirable properties, i.e., resistance to birds pests and molds in the field, as well as storage and good nutritional quality. Such plant types need to be identified from available germplasm. Butler (1982) suggested that the lines in which tannins never polymerize to anti nutritional forms at maturity need to be developed so that the levels of monomeric and small oligomeric phenolics remain high, even in the mature grains. Approaches of modifying tannin-producing crops, either by conventional breeding techniques or by genetic engineering to produce tannins that are more toxic to birds, pests and molds, and less toxic to humans, would certainly help to improve the production and quality of these food grains.

RISK OF TOXICITY FROM NUTS AND CO-PRODUCTS

The safety of nut co-product consumption as a functional food requires consideration. Exposure to irregular food components may pose allergenic or chemical toxicity risks. Thus, the safety of several nut co-products or components has been reported. Doses of up to 2.5 g/kg of almond skin did not pose genotoxicity in Swiss albino mice and up to 10% dietary almond skin was not toxic to Sprague-Dawley rats (Song et al., 2010; Zhang, Xiang, Cui, Jia, and Li, 2011a).Topical application of cashew shell liquid/oil had a weak tumour-promoting activity in a skin tumourigenesis model using Swiss albino mice (Banerjee and Rao, 1992; George and Kuttan, 1997). Chronic consumption of 300 mg cashew anacardic acids/kg by albino mice did not pose toxic symptoms, although 600 and 1000 mg/kg increased urea, and lowered haematocrit levels (Carvalho et al., 2011).Walnut co-product consumption may induce toxicity in dogs, horses, and other animal species (Coleman and Merola, 2016). Occupational exposure to macadamia shell can induce allergic contact dermatitis (Knight and Hausen, 1996). However, the component responsible for this effect is unknown. Acrylamide, which is classified as probably carcinogenic in humans, is formed during Maillard reaction (Lineback, Coughlin, and Stadler, 2012), involving asparagine from protein and reactive carbonyls at temperatures above 120°C (Amrein *et al.*, 2005; Lineback *et al.*, 2012).

Acrylamide was detectable in roasted almond and pistachio, ranging from 16–1220 to 14–88 μg/kg, respectively, with increasing concentrations depending on the roasting conditions (Schlörmann et al., 2015). It is noteworthy that the acrylamide levels found in almond did not or only marginally exceeded the recommended level of 1000 μg/kg for potato chips, whereas levels for pistachio clearly do not exceed recommended levels. In another study, Zhang, Huang, Xiao, Seiber, and Mitchell (2011b) demonstrates that acrylamide levels reduced significantly at all roasting time evaluated by controlling the roasting temperature at or below 146°C. Further, stability studies on roasted almond indicated that acrylamide levels decreased by 12.9–68.5% after 3 days of storage at 60°C. This shows that short-term elevated temperature storage may be a good alternative to reduce the acrylamide levels in roasted almond (Zhang *et al.*, 2011b).

FUTURE PERSPECTIVES

This book gives an overview of phenolic compounds in cereals, fruits, vegetables and other crops also. Many phenolics found in fruits and vegetables are also detected in cereal grains. However, many of these compounds are unidentified. Therefore, further research is needed to isolate and characterize phenolic components that contribute to health, which is challenging. Many of these compounds are bound to the matrix of the grain, making their extraction difficult. Also, the lack of appropriate standards increases the difficulty of identifying these phenols. The combination of mass spectrometry coupled with liquid chromatography is effective in the isolation, characterization, and identification of those compounds.

Identifying and quantifying cereal phenols will help us select grains with increased levels of these health-promoting compounds. Research is also needed to determine their bioavailability, metabolism, and health contribution in humans. Within cereals, great variation in colors and phenol components occur among genetic materials. Special grains are available and can provide large quantities of potentially health-promoting substances. Often, the cereal varieties commonly grown have been selected for absence of color and bland taste, which means the phenol content, is reduced.

Since phenol profile and quantities depend on the sample's genetics, this information will also help plant breeders produce cereals with high levels of the desired compounds. Developing special varieties of cereals high in phenolics is possible and has been done for sorghum. For example, we know enough about the genetics affecting sorghum phenolics to develop cultivars with high levels of condensed tannins and special anthocyanins. Milling to produce bran concentrates these compounds, which can be used for potential nutraceutical application or as food colorants.

Obtaining phenolic compounds from cereals has several attractive advantages. Compared to fruits and vegetables, cereal grains are dry, easy to store for long periods of time, and possibly easier to process into shelf-stable concentrates. Cereals can provide viable alternatives to diversify sources of healthy components in foods. Obviously, the benefits are highest for whole grain cereal consumption.

Further studies are required to gain a better understanding of the role of lycopene in human health. Processing has traditionally been considered as contributing towards the loss of nutritional quality of foods. There is, however, indication that in the case of lycopene processing might actually enhance its bioavailability and nutritional quality. Although it is generally recognized that heat treatment of tomatoes induces the formation of cis-isomers from the all trans form present in raw tomatoes, specific types of isomers formed and their biological significance is not fully understood. Bioavailability of lycopene is influence not only by its isomeric form but also the presence of other dietary components including dietary fat, other carotenoids, vitamins and minerals. These factors should be evaluated in an attempt to maximize lycopene absorption from the diet. Another area of future research should include elucidation of mechanisms other than antioxidant properties of lycopene in the prevention of chronic diseases. Preliminary data suggests the possibilities of these alternate pathways of lycopene activity. Finally, long term human intervention studies should be carried out with populations at risk for cancer and coronary heart disease. Based on the results from these studies an effective dietary strategy can be developed for the management of chronic diseases in humans that will be based on increased consumption of lycopene containing foods.

Numerous phenolic compounds (phenolic acids and flavonoids) have been identified and quantified in nuts (natural and roasted) and their co-products (skin, hard shell, and hull/pellet) together with their corresponding antioxidant activities. The compiled results indicated that many of these bioactive compounds remain to be fully identified and characterised in nuts and their co-products (natural and roasted), especially macadamia, Brazil nut, and pine nut. Nuts roasted at suitable temperature and time demonstrated excellent phenolic compositions and antioxidant activities and should be recommended for daily consumption. Therefore, nuts, especially their co-products can be considered as a valuable source of functional foods based on their phenolic compositions and their corresponding antioxidant activities presented in this review article. The phenolic compounds from nuts and their co-products are bioaccessible and bioavailable in humans, with various health benefits, including antioxidant, antiinflammatory, prebiotic, anti-microbial, chemopreventive, and hypocholesterolaemic. More sophisticated human intervention studies or clinical trials are needed to validate the health benefits of nut co-products.

■■■■

References

Abbas, M. N. (2014). Pomegranate peels: Source of antioxidants extraction and natural dentifrices preparation. *European Academic Research*, *2*, 3078–3089.

Abdalla, A. E., and Roozen, J. P. (1999). Effect of plant extracts on the oxidative stability of sunfower oil and emulsion. Food Chemistry, 64, 323-329.

Abdullin, I. F., Turova, E. N., Budnikov, G. K., Ziyatdinova, G. K., and Gajsina, G. K. (2002). Electrogenerated bromine – Reagent for determination of antioxidant capacity of juices and extracts. *Zavodskaya Laboratoriya Diagnostika Materialov*, *68*, 12–15.

Abou-Gharbia, H. A., Shehata, A. A. Y. and Shahidi, F. (2000). Effect of processing on oxidative stability and lipid classes of sesame oil. *Food Research International*, *33*, 331–340.

Abram, V., and Donko, M. (1999). Tentative identification of polyphenols in Sempervivum tectorum and assessment of the antimicrobial activity of Sempervivum L. Journal of Agricultural and Food Chemistry, 47, 485-489.

Abuja, P. M., Murkovic, M., and Pfannhauser, W. (1998). Antioxidant and prooxidant activities of Elderberry (*Sambucus nigra*) extract in Low-Density-Lipoprotein oxidation. Journal of Agricultural and Food Chemistry, 46, 4091-4096.

Abushita, A. A., Hebshi, E. A., Daood, H. G., and Biacs, P. A. (1997). Determination of antioxidant vitamins in tomatoes. Food Chemistry, 60, 207-212.

Abuzaytoun, R. and Shahidi, F. (2006). Oxidative stability of algal oils as affected by their minor components. *Journal of Agricultural and Food Chemistry*, *54*, 8253–8260.

Açar, Ö.Ç., Gökmen, V., Pellegrini, N., and Fogliano, V. (2009). Direct evaluation of the total antioxidant capacity of raw and roasted pulses, nuts, and seeds. *European Food Research and Technology*, *229*, 961–969.

Acosta-Montoya, O., Vaillant, F., Cozzano, S., Mertz, C., Pérez, A. M. and Castro, M. V. (2010). Phenolic content and antioxidant capacity of tropical highland blackberry (*Rubus adenotrichus* Schltdl.) during three edible maturity stages. *Food Chemistry*, 119: 1497–1501.

Adams, J. M. (1977). A review of the literature concerning losses in stored cereals and pulses published since 1964. *Trop. Sci.* 19, 1.

Adamson, G. E., Lazarus, S. A., Mitchell, A. E., Prior, R. L., Cao, G., Jacobs, P. H., Kremers, B. G., Hammerstone, J. F., Rucker, R. B., Ritter, K. A. and Schmitz, H. H. (1999). HPLC method for the quantification of procyanidins in cocoa and chocolate samples and correlation to total antioxidant capacity. *Journal of Agricultural and Food Chemistry*, 47: 4184–4188.

Adom, K. K., Sorrells, M. E., and Liu, R. H. (2005). Phytochemicals and antioxidant activity of milled fractions of different wheat varieties. *J. Agric. Food Chem.*, 53:2297-2306.

Adom, K. K. and Liu, R. H. (2002). Antioxidant activity of grains. *Journal of Agricultural and Food Chemistry*, 50: 6182–6187.

Aegean Exporter's Associations. (2009). *Report on food composition and clinical research on seedless raisins and dried figs*. Izmir, Turkey: Aegean Exporter's Associations.

Aeschbach, R., Loliger, J., Scott, B. C., Murcia, A., Butler, J., Halliwell, B. and Aruoma, O. I. (1994). Antioxidant actions of thymol, carvacrol, 6-gingerol, zingerone and hydroxytyrosol. *Food and Chemical Toxicology*, *32*, 31–36.

Afify, AEMR., El-Beltagi, HS., El-Salam, SMA., and Omran, A. A. (2012). Biochemical changes in phenols, flavonoids, tannins, vitamin E, β-carotene and antioxidant activity during soaking of three white sorghum varieties. *Asian Pacific Journal of Tropical Biomedicine*. 203-209.

Afshin, A., Micha, R., Khatibzadeh, S., and Mozaffarian, D. (2014). Consumption of nuts and legumes and risk of incident ischemic heart disease, stroke, and diabetes: A systematic review and meta-analysis. *The American Journal of Clinical Nutrition*, 100, 278–288.

Agbor, G. A., Vinson, J. A., Oben, J. E. and Ngogang, J. Y. (2006). Comparative analysis of the in vitro antioxidant activity of white and black pepper. *Nutrition Research*, 26: 659–663.

Agu, R.C., (2002). A comparison of maize, sorgum and barley as brewing adjuncts. Journal of Institute of Brewing 108, 19-22.

Ahmad, M. S., Fazal, F., Rahman, A., Hadi, S. M. and Parish, J. H. (1992). Activities of flavonoids for the cleavage of DNA in the presence of Cu (II): Correlation with generation of active oxygen species. *Carcinogenesis*, 13: 605–608.

Ahmad, Z., Zamhuri, K. F., Yaacob, A., Siong, C. H., Selvarajah, M., Ismail, A. and Hakim, M. N. (2012). In vitro anti-diabetic activities and chemical analysis of polypeptide-k and oil isolated from seeds of Momordica charantia (bitter gourd). *Molecules*, 17: 9631–9640.

Ahmed, T., Sadia, H., Khalid, A., Batool, S., and Janjua, A. (2010). Report: Prunes and liver function: A clinical trial. *Pakistan Journal of Pharmaceutical Sciences*, *23*, 463–466.

Ahn, J., Gruen, L. U. and Fernando, L. N. (2002). An antioxidant property of natural plant extracts containing polyphenolic compounds in cooked ground beef. *Journal of Food Science*, *67*: 1364–1369.

Akihisa, T., Yasukawa, K., Yamaura, M., Ukiya, M., Kimura, Y., Shimizu, N., and Arai, K. (2000). Triterpene alcohol and sterol ferulates from rice bran and their anti-inflammatory effects. *J. Agric. Food Chem.*, 48:2313–2319.

Akingbala, J.O. (1991).Effect of processing on flavonoids in millet (*Pennisetum americanum*) flour Cereal Chemistry, 68: 180–183.

Alamed, J., Chaiyasit, W., McClements, D. J. and Decker, E. A. (2009). Relationship between free radical scavenging and antioxidant activity in foods. *Journal of Agricultural and Food Chemistry*, 57: 2969–2976.

Al-Amier, H., Mansour, B. M. M., Toaima, N., Korus, R. A., and Shetty, K. (1999). Tissue culture based screening for selection of high biomass and phenolic producing clonal lines of lavender using Pseudomonas and azetidine-2-carboxylate. Journal of Agricultural and Food Chemistry, 47, 2937-2943.

Alamprese, C., Ratti, S., and Rossi, M. (2009). Effects of roasting conditions on hazelnut characteristics in a two-step process. *Journal of Food Engineering*, *95*, 272–279.

Alard, D., Wray, V., Grotjahn, L., Reznik, H. and Strack, D. (1985). Neobetanin: Isolation and identification from *Beta vulgaris*. *Phytochemistry*, 24: 2383–2385.

Alasalvar, C. (2013). Functional characteristics of dried figs. In C. Alasalvar and F. Shahidi (Eds.), *Dried fruits: Phytochemicals and health effects* (pp. 284–299). Oxford: Wiley-Blackwell.

Alasalvar, C., and Bolling, B. W. (2015). Review of nut phytochemicals, fat-soluble bioactives, antioxidant components, and health effects. *The British Journal of Nutrition*, *113*, S68–S78.

Alasalvar, C., and Shahidi, F. (2008). Tree nuts: Composition, phytochemicals, and health effects: An overview. In C. Alasalvar and F. Shahidi (Eds.), *Tree nuts: Composition phytochemicals, and health effects* (pp. 1–10). Boca Raton, FL: CRC Press Taylor and Francis Group.

Alasalvar, C., and Shahidi, F. (2009). Natural antioxidants in tree nuts. *European Journal of Lipid Science and Technology*, *111*, 1056–1062.

Alasalvar, C., and Shahidi, F. (2013a). Composition, phytochemicals and beneficial health effects of dried fruits: an overview. In C. Alasalvar and F. Shahidi (Eds.), *Dried fruits: Phytochemicals and health effects* (pp. 1–18). Oxford: Wiley-Blackwell.

Alasalvar, C., and Shahidi, F. (2013b). Nutritional composition, phytochemicals, and health benefits of dates. In C. Alasalvar and F. Shahidi (Eds.), *Dried fruits: Phytochemicals and health effects* (pp. 428–443). Oxford: Wiley-Blackwell.

Alasalvar, C., Grigor, J. M., Zhang, D., Quantick, P. C. and Shahidi, F. (2001). Comparison of volatiles, phenolics, sugars, antioxidant vitamins, and sensory quality of different colored carrot varieties. *Journal of Agricultural and Food Chemistry*, 49: 1410–1416.

Alasalvar, C., Hoffman, A. M. and Shahidi, S. (2009). Antioxidant activities and phytochemicals in hazelnut (*Corylus avellana* L.) and hazelnut by-products. In C. Alasalvar and F. Shahidi (Eds.), *Tree nuts composition: Phytochemicals and health effects* (pp. 215–235).

Alasalvar, C., Hoffman, A. M., and Shahidi, F. (2008). Antioxidant activities and phytochemicals in hazelnut (*Corylus avellana* L.) and hazelnut by-products. In C. Alasalvar and F. Shahidi (Eds.), *Tree nuts: Composition, phytochemicals and health effects* (pp. 215–235). Boca Raton, FL: CRC Press, Taylor and Francis Group.

Alasalvar, C., Karamac, M., Amarowicz, R. and Shahidi, F. (2006). Antioxidant activities in extracts of defatted hazelnut (*Corylus avellana* L.) and its green shell cover. *Journal of Agricultural and Food Chemistry*, 54: 4826–4832.

Alasalvar, C., Karamac, M., Kosinska, A., Rybarczyk, A., Shahidi, F. and Amarowicz, R. (2009). Antioxidant activity of hazelnut skin phenolics. *Journal of Agricultural and Food Chemistry*, 57: 4645 – 4650.

Alasalvar, C., Shahidi, F., Amaral, J. S. and Oliveira, B. P. P. (2009). Compositional characteristics and health effects of hazelnut (*Corylus avellana* L.): An overview. In C. Alasalvar and F. Shahidi (Eds.), *Tree nuts composition: Phytochemicals and health effects* (pp. 186–214). Boca Raton, FL: CRC Press.

Albishi, T., John, J. A., Al-Khalifa, A. S. and Shahidi, F. (2013b). Phenolic content and antioxidant activities of selected potato varieties and their processing by-products. *Journal of Functional Foods*, 5: 590–600.

Albishi, T., John, J. A., Al-Khalifa, A. S. and Shahidi, F. (2013a). Antioxidative phenolic constituents of skins of onion varieties and their activities. *Journal of Functional Foods*, 5: 1191–1203.

Al-Farsi, M., Alasalvar, C., Morris, A., Baron, M., and Shahidi, F. (2005). Comparison of antioxidant activity, anthocyanins, carotenoids, and phenolics of three native fresh and sundried date (*Phoenix dactylifera* L.) varieties grown in Oman. *Journal of Agricultural and Food Chemistry*, 53: 7592–7599.

Al-Farsi, M., and Lee, C. Y. (2008). Nutritional and functional properties of dates: A review. *Critical Reviews in Food Science and Nutrition*, *48*, 877–887.

Alfawaz, M., Smith, J. S., and Jeon, I. J. (1994). Maillard reaction products as antioxidants in precooked ground beef. Food Chemistry, 51, 311-318.

Alma, M. H., Nitz, S., Kollmannsberger, H., Digrak, M., Efe, F. T. and Yilmaz, N. (2004). Chemical composition and antimicrobial activity of the essential oils from the gum of turkish pistachio (*Pistacia vera* L.). *Journal of Agricultural and Food Chemistry*, 52: 3911–3914.

Alsaif, M. A., Khan, L. K., Alhamdan, A. A., Alorf, S. M., Harfi, S. H., Al-Othman, A. M., and Arif, Z. (2007). Effect of dates and gahwa (Arabian coffee) supplementation on lipids in hypercholesterolemic hamsters. *International Journal of Pharmacology*, *3*, 123–129.

Al-Saikhan, M. S., Howard, L. R. and Miller, J. C., Jr. (1995). Antioxidant activity and total phenolics in different genotypes of potato (*Solanum tuberosum* L. *Journal of Food Science*, 60: 341–347.

Al-Saikhan, M. S., Howard, L. R., and Miller, J. C. (1995). Antioxidant activity and total phenolics in different genotypes of potato (Solanum tuberrosum, L). Journal of Food Science, 60, 341-343.

Al-Sereiti, M. R., Abu-Amer, K. M. and Sen, P. (1999). Pharmacology of rosemary (Rosmarinus officinalis Linn.) and its therapeutic potentials. *Indian Journal of Experimental Biology*, 37: 124–130.

Altunkaya, A., Becker, E. M., Gökmen, V. and Skibsted, L. H. (2009). Antioxidant activity of lettuce extract (*Lactuca sativa*) and synergism with added phenolic antioxidants. *Food Chemistry*, 115: 163–168.

Álvarez, P., Alvarado, C., Matheiu, F., Jiménez, L. and De la Fuente, M. (2006). Diet supplementation for 5 weeks with polyphenol-rich cereals improves several functions of the radox state of mouse leucocytes. *European Journal of Nutrition*, 45(8), 428-438.

Alvarez-Parrilla, E., de la Rosa, L. A., Amarowicz, R. and Shahidi, F. (2010). Antioxidant activity of fresh and processed Jalapeño and Serrano peppers. *Journal of Agricultural and Food Chemistry*, 59: 163–173.

Alvarez-Parrilla, E., de la Rosa, L. A., González-Aguilar, G. A., and Ayala-Zavala, J. F. (2013). Phytochemical composition and health aspects of peach products. In C. Alasalvar and F. Shahidi (Eds.), *Dried fruits: Phytochemicals and health effects* (pp. 309-324). Oxford: Wiley-Blackwell.

Al-Weshahy, A. and Rao, A. V. (2009). Isolation and characterization of functional components from peel samples of six potatoes varieties growing in Ontario. *Food Research International*, 42: 1062–1066.

Amadou I, Amza T, Shi Y. H. and Le, G. W. (2011). Chemical analysis and antioxidant properties of foxtail millet bran extracts. *Songklanakarin J Sci. Technol.* 33(5):509–15.

Amaral, J. S., Casal, S., Seabra, R. M., and Oliveira, B. P. (2006). Effects of roasting on hazelnut lipids. *Journal of Agricultural and Food Chemistry*, *54*, 1315–1321.

Amaral, J. S., Ferreres, F., Andrade, P. B., Valentão, P., Pinheiro, C., Santos, A. and Seabra, R. (2005). Phenolic profile of hazelnut (*Corylus avellana* L.) leaves cultivars grown in Portugal. *Natural Product Research*, 19: 157–163.

Amarovicz, R., and Shahidi, F. (1997). Antioxidant activity of peptide fractions of capelin protein hydrolysates. Food Chemistry, 58, 355-359.

Amarowicz, R. and Pegg, R. B. (2013). Inhibition of proliferation of human carcinoma cell lines by phenolic compounds from a bearberry-leaf crude extract and its fractions. *Journal of Functional Foods*, 5: 660–667.

Amarowicz, R. and Pegg, R. B. (2008). Legumes as a source of natural antioxidants. *European Journal of Lipid Science and Technology*, 110: 865–878.

Amarowicz, R. and Raab, B. (1997). Antioxidative activity of leguminous seed extracts evaluated by chemiluminescence methods. Z. Naturforsch. 5C. 709-712.

Amarowicz, R. and Shahidi, F. (1994). Chromatographic separation of glucopyranosyl sinapate from canola meal. *Journal of the American Oil Chemists' Society*, 7: 551–552.

Amarowicz, R. and Shahidi, F. (1995). Antioxidant activity of green tea catechins in a β-carotene-linoleate model system. J. Food Lipids. 2: 47-56.

Amarowicz, R. and Shahidi, F. (1996). A rapid chromatographic method for separation of individual catechins for green tea. *Food Research International*, 29: 71–76.

Amarowicz, R., Carle, R., Dongowski, G., Durazzo, A., Galensa, R., Kammerer, D., Maiani, G. and Piskula, M. K. (2009). Influence of postharvest processing and storage on the content of phenolic acids and flavonoids in foods. *Molecular Nutrition and Food Research*, 53: 151–S183.

Amarowicz, R., Estrella, I., Hernández, T., Robredo, S., Troszyn□ska, A., Kosin□ska, A. and Pegg, R. B. (2010). Free radical-scavenging capacity, antioxidant activity, and phenolic composition of green lentil (*Lens culinaris*). *Food Chemistry*, 121: 705– 711.

Amarowicz, R., Karamac, M., and Shahidi, F. (1999). Synergistic activity of capelin protein hydrolysates with synthetic antioxidants in a model system. *Journal of Food Lipids*, *6*, 271–275.

Amarowicz, R., Karamac, M., Kmita-Glazewska, H., and Kozlowska, H. (1996b). Antioxidant activity of phenolic fractions of everlasting pea, faba bean and broad bean. J. Food Lipids. 3: 199-211.

Amarowicz, R., Karamac, M.,Wanasundara, J. P. D. Shahidi, F. (1997). Antioxidant activity of hydrophobic phenolic fractions of flaxseed. *Nahrung - Food*, 41: 178–180.

Amarowicz, R., Naczk, M. and Shahidi, F. (2000). Antioxidant activity of crude tannins of canola and rapeseed hulls. *Journal of the American Oil Chemists' Society*, 77: 957–961.

Amarowicz, R., Trosynska, A., Karamac, M., and Kozlowska, H. (1996a). Antioxidative properties of legume seed extracts. In: *Agri-Food Quality an Interdisciplinary Approach*. G. R. Fenwick, C. Hedley, R. L. Richards and S. Khokhar. (Eds.). The Royal Society of Chemistry, Cambridge. p. 376-379.

Amarowicz, R., Troszynska, A. and Shahidi, F. (2005). Antioxidant activity of almond seed extract and its fractions. *Journal of Food Lipids*, 12: 344–358.

Amarowicz, R., Troszynska, A., Barylko-Pikielna, N. and Shahidi, F. (2004). Polyphenolics extracts from legume seeds: Correlations between total antioxidant activity, total phenolics content, tannins content and astringency. *Journal of Food Lipids*, 11: 278–286.

Amarowicz, R., Zegarska, Z., Rafalowski, R., Pegg, R. B., Karamac, M. and Kosinska, A. (2009). Antioxidant activity and free radical-scavenging capacity of ethanolic extracts of thyme, oregano, and marjoram. *European Journal of Lipid Science and Technology*, 111: 1111–1117.

Amatatongchai, M., Laosing, S., Chailapakul, O., and Nacapricha, D. (2012). Simple flow injection for screening of total antioxidant capacity by amperometric detection of DPPH radical on carbon nanotube modified-glassy carbon electrode. *Talanta*, *97*, 267–272.

Amerine, M. A. (1965). Laboratory Procedures for Estimation of Califorma. Davis CA.

Ames, B. N., Shigenaga, M. K. and Gold, L. S. (1993). DNA lesion, inducible DNA repair, and cell division: the three key factors in mutagenesis and carcinogenesis. *Environmental Health Perspectives*, 101(5), 35-44.

Ames, B. N., Shigenaga, M. K. and Hagen, T. M. (1993). Oxidants, antioxidants, and the degenerative diseases of aging. *Proceedings of the National Academy of Science of the United States of America,* 90(17), 7915-7922.

Amin, I., Faizul, H. A. and Azli, R. (2004). Effect of cocoa powder extract on plasma glucose levels in hyperglycaemic rats. *Nutrition and Food Science*, 34: 116–121.

Amin, I., Norazaidah, K. and Hainida, E. (2006). Antioxidant activity and phenolic content of raw and blanched *Amaranthus* species. *Food Chem*. 94: 47-52.

Amiot, M. J., Fleuriet, A. and Macheix, J. J. (1986). Accumulation of oleuropin derivatives during olive maturation. *Journal of Agricultural and Food Chemistry*, 34: 823–826.

Amrein, T. M., Lukac, H., Andres, L., Perren, R., Escher, F., and Amado, R. (2005). Acrylamide in roasted almonds and hazelnuts. *Journal of Agricultural and Food Chemistry*, *53*, 7819–7825.

Andarwulan, N., Fardiaz, D., Wattimena, G. A., and Shetty, K. (1999). Antioxidant activity associated with lipid and phenolic mobilization during seed germination of Pangium edule Reinw. Journal of Agricultural and Food Chemistry, 47, 3158-3163.

Andersen, M. B., Kristensen, M., Manach, C., Pujos-Guillot, E., Poulsen, S. K., Larsen, T. M., Astrup, A., and Dragsted, L. (2014). Discovery and validation of urinary exposure markers for different plant foods by untargeted metabolomics. *Analytical and Bioanalytical Chemistry*, *406*, 1829–1844.

Andersen, M. L., and Skibsted, L. H. (2002). Detection of early events in lipid oxidation by electron spin resonance spectroscopy. *European Journal of Lipid Science and Technology*, *104*, 65–68.

Anderson, J. A., Andersen, K. F., Heimerman, R. A., Larson, M. M., Baker, S. E., Freeman, M. R., Carughi, A., and Wilson, T. (2011a). Glycaemic response of type 2 diabetics to raisins. *FASEB Journal*, *25*, 587.5.

Anderson, J. A., Huth, H. A., Larson, M. M., Colby, A. J., Krieg, E. J., Golbach, L. P., Simon, K. A.,Wasmundt, S. L., Malone, C. J., and Wilson, T. (2011b). Glycaemic and insulin response to raisins, grapes and bananas in college aged students. *FASEB Journal*, *25*, 587.4.

Anderson, J. W., and Waters, A. R. (2013). Raisin consumption by humans: Effects on glycemia and insulinemia and cardiovascular risk factors. *Journal of Food Science*, *78*, A11–A17.

Anderson, J. W., Baird, P., Davis, R. H., Jr., Ferreri, S., Knudtson, M., Koraym, A., Waters, V., and Williams, C. L. (2009). Health benefits of dietary fiber. *Nutrition Reviews*, *67*, 188-205.

Anderson, J. W., Weiter, K. M., Christian, A. L., Ritchey, M. B., and Bays, H. E. (2014). Raisins compared with other snack effects on glycaemia and blood pressure: A randomized, controlled trial. *Postgraduate Medicine*, *126*, 37–43.

Anderson, K. J., Teuber, S. S., Gobeille, A., Cremin, P., Waterhouse, A. L. and Steinberg, F. M. (2001).Walnut polyphenolics inhibit in vitro human plasma and LDL oxidation. *Journal of Nutrition*, 131: 2837–2842.

Anderson, T. J., Grégoire, J., Hegele, R. A., Couture, P., Mancini, G. J., McPherson, R., Francis, G. A., Poirier, P., Lau, D. C., Grover, S., Genest, J., Jr., Carpenter, A. C., Dufour, R., Gupta, M.,Ward, R., Leiter, L. A., Lonn, E., Ng, D. S., Pearson, G. J., Yates, G. M., Stone, J. A., and Ur, E. (2013). 2012 update of the Canadian Cardiovascular Society guidelines for the diagnosis and treatment of dyslipidemia for the prevention of cardiovascular disease in the adult. *The Canadian Journal of Cardiology*, *29*, 151–167.

Andjelkovic M, van Camp J, de Meulenaer B, Depaemelaere G, Socaciu C, Verloo M, Verhe R. (2008) Iron-chelation properties of phenolic acids bearing catechol and galloyl groups. Food Chem 98:23–31

Andre, C., Castanheira, I., Cruz, J. M., Paseiro, P. and Sanches-Silva, A. (2010). Analytical startegies to evaluate antioxidants in food: A review. *Trends in Food Science and technology*, 21: 229–246.

Andreasen, M. F., Christensen, L. P., Meyer, A. S. and Hansen, A. (2000). Content of phenolic acids and ferulic acid dehydrodimers in 17 rye (*Secale cereale* L.) varieties. *Journal of Agricultural and Food Chemistry*, 48: 2837–2842.

Andreasen, M. F., Kroon, P. A., Williamson, G., and Garcia-Conesa, M.-T. (2001). Esterase activity able to hydrolyze dietary antioxidant hydroxycinnamates is distributed along the intestine of mammals. *J. Agric. Food Chem.*, 49:5679-5684.

Andreasen, M. F., Landbo, A. K., Christensen, L. P., Hansen, A. and Meyer, A. S. (2001). Antioxidant effects of phenolic rye (*Secale cereale* L.) extracts, monomeric hydroxycinnamates, and ferulic acid dehydrodimers on human low-density lipoproteins. *Journal of Agricultural and Food Chemistry*, 49: 4090–4096.

Andreu-Navarro, A., Fernandez-Romero, J. M., and Gomez-Hens, A. (2011). Determination of antioxidant additives in foodstuffs by direct measurement of gold nanoparticle formation using resonance light scattering detection. *Analytica Chimica Acta*, *695*, 11–17.

Angerosa, F., and di Giovacchino, L. (1996). Natural antioxidants of virgin olive oil obtained by two and tri-phase centrifugal decanters. Grasas y Aceites, 47, 247-254.

Angerosa, F., d'Alessandro, N., Konstantinou, P. and Di Giacinto, L. (1995). GC-MS evaluation of phenolic compounds in virgin olive oil. *Journal of Agricultural and Food Chemistry*, 43: 1802–1807.

Angerosa, F., d'Alessandro, N., Konstantinou, P. and Di Giacinto, L. (1996). Characterization of phenolic and secoiridoid aglycons present in virgin oil by gas chromatography-chemical ionization mass spectrophotometry. *Journal of Chromatography. A*, 43: 1802–1807.

Anli, E., Vural, N., Demiray, S. and Özkan, M. (2006). Transresveratrol and other phenolic compounds in Turkish red wines with HPLC. *Journal ofWine Research*, 17: 117–125.

Anonymous, (1965). Avoiding losses of food grains after harvesting. *World Crops.* 17, 59.

Anonymous, (2009). Codex General Standard for Food Additives (GSFA) online database. Joint FAO/WHO Codex Alimentarius Commission. Available from <http://www.codexalimentarius.net/gsfaonline/index.html>. Accessed 27.01.2010.

Anson, N. M., Havenaar, R., Bast, A. and Haenen, G. R. M. M. (2010). Antioxidant and anti-inflammatory capacity of bioaccessible compounds from wheat fractions after gastrointestinal digestion. *Journal of cereal Science*, 51(1), 110-114.

Antolovich, M., Prenzler, P. D., Patsalides, E., McDonald, S., and Robards, K. (2002). Methods for testing antioxidant activity. *The Analyst, 127*, 183–198.

AOAC, (1965). Association of Official Agricultural Chemists, Official Methods of Analysis, 10th ed., Association of Official Agricultural Chemists, Washington, D. C.

AOAC. (1980). Association of Official Agricultural Chemists, Official Methods of Analysis, 12th ed., Association of Official Agricultural Chemists, Washington, D.C.

Apak, R., Gorinstein, S., Böhm, V., Schaich, K. M., Özyürek, M., and Güçlü, K. (2013). Methods of measurement and evaluation of natural antioxidant capacity/activity. *Pure and Applied Chemistry, 85*, 957–998.

Apak, R., Güçlü, K., Özyürek, M., and Karademir, S. E. (2004). Novel total antioxidant capacity index for dietary polyphenols, vitamins C and E, using their cupric ion reducing capability in the presence of neocuproine: CUPRAC method. *Journal of Agricultural and Food Chemistry, 52*, 7970–7981.

Aprikian, O., Levrat-Verny, M. A., Besson, C., Busserolles, J., Rémésy, C., and Demigné, C. (2001). Apple favourably affects parameters of cholesterol metabolism and of anti-oxidative protection in cholesterol-fed rats. *Food Chemistry, 75*, 445–452.

Arai, S., Suzuki, H., Fujimaki, M. and Sakrai, Y. (1966). Studies on flavor components in soybeans, II. Phenolic acids in defatted soybean flour. *Agricultural Biology and Chemistry*, 30: 364–369.

Arcan, I., and Yemenicioglu, A. (2009). Antioxidant activity and phenolic content of fresh and dry nuts with or without the seed coat. *Journal of Food Composition and Analysis, 22*, 184–188.

Aritomi, M. and Kawasaki, T. (1984). Three highly oxygenated flavone glucuronides in leaves of *Spinacia oleracea. Phytochemistry*, 23: 2043–2047.

Arjmandi, B. H., Khalil, D. A., Lucas, E. A., Georgis, A., Stoecker, B. J., Hardin, C., Payton, M. E., and Wild, R. A. (2002). Dried plums improve indices of bone formation in postmenopausal women. *Journal of Women's Health and Gender-Based Medicine*, 11, 61–68.

Armstrong, G. S. (1983). A Study of Tannin-protein Interactions. Ph. D. thesis. Purdue University. West Lafayette. IN.

Arnao, M. B. (2000). Some methodological problems in the determination of antioxidant activity using chromogen radicals: A practical case. *Trends in Food Science and Technology, 11*, 419–421.

Arora, A., and Strasburg, G. M. (1997). Development and validation of fuorescence spectroscopic assays to evaluate antioxidant e cacy. Application to metal chelators. Journal of the American Oil Chemists' Society, 74, 1031-1040.

Arts, I. C., Hollman, P. C. and Kromhout, D. (1999). Chocolate as a source of tea flavonoids. *Lancet*, 354: 488.

Aruoma, O. (1999). I Antioxidant actions of plant foods, use of oxidative DNA damage as a tool for studying antioxidant e cacy. Free Radical Research, 30, 419-427.

Aruoma, O. I. (1997). Extracts as antioxidant prophylactic agents. Inform, 8, 1236-1242.

Aruoma, O. I. (1998). Free radicals, oxidative stress and antioxidants in human health and disease. Journal of the American Oil Chemists Society, 75, 199-212.

Aruoma, O. I., Murcia, A., Butler, J., and Halliwell, B. (1993). Evaluation of the antioxidant and prooxidant actions of gallic acid and its derivatives. Journal of Agricultural and Food Chemistry, 41, 1880-1885.

Aruoma, O. I., Spencer, J. P. E., Warren, D., Jenner, P., Butler, J., and Halliwell, B. (1997). Characterization of food antioxidants, illustrated using commercial garlic and ginger preparations. Food Chemistry, 60, 149-156.

Asakawa, T., and Matsushita, S. (1978). Colorimetric determination of peroxide value with potassium iodide-silica gel reagent. *American Oil Chemists' Society*, *55*, 619–620.

Asamarai, A. M., Addis, P. B., Epley, R. J. and Krick, T. P. (1996).Wild rice hull antioxidants. *Journal of Agricultural and Food Chemistry*, 44: 126–130.

Asquith, T. N. and Buller, L. G. (1985). Use of dye-labeled protein as spectrophotometric assay for protein precipitants such as tannin, J. *Chem. Ecol.*, 2, 1535.

Asquith, T. N. and Butler, L. G. (1986). Interactions of condensed tannins with selected proteins. *Phytochemistry.* 25: 1591.

Asquith, T. N. and Butler, L. G. (1986). Interactions of condensed tannins with selected proteins, *Phytochemistry,* 25: 1591.

Astill, C., Birch, M. R., Dacombe, C., Humphrey, P. G. and Martin, P. T. (2001). Factors affecting the caffeine and polyphenol contents of black and green tea infusions. *Journal of Agricultural and Food Chemistry*, 49: 5340–5347.

Athukorala, Y., Lee, K. W., Song, C., Ahn, C. B., Shin, T. S., Cha, Y. J., Shahidi, F., and Jeon, Y. J. (2003). Potential antioxidant activity of marine red alga Grateloupia filicina extracts. *Journal of Food Lipids*, *10*, 251–263.

Atkinson J, Epand RF, Epand RM; Epand; Epand (2008). "Tocopherols and tocotrienols in membranes: a critical review". Free radical biology and medicine 44 (5): 739–64.

Awika, J. M., and Rooney, L. W. (2004). Sorghum phytochemicals and their potential impact on human health. *Phytochem,* 65:1199-1221.

Awika, J. M., Rooney, L. W., Wu, X., Prior, R. L. and Cisneros-Zevallos, L. (2003). Screening methods to measure antioxidant activity of sorghum (Sorghum bicolor) and sorghum products. *Journal of Agricultural and Food Chemistry*, 51: 6657–6662.

Awika, J. M., Rooney, L. W. and Waniska, R. D. (2004a). Anthocyanins from black sorghum and their antioxidant properties. *Food Chemistry*, 90: 293–301.

Awika, J. M., Rooney, L. W. and Waniska, R. D. (2004b). Properties of 3-deoxyanthocyanins from sorghum. *Journal of Agricultural and Food Chemistry*, 52: 4388–4394.

Ayyangar, G. N. R. and Krishnaswami, N. (1941). Studies on the histology and coloration of the pericarp of the sorghum grain. *Proc. Indian Acad. Sci.* 1413. 114.

Azizah, A. H., Nik Ruslawati, N. M., and Swee Tee, T. (1999). Extraction and characterization of antioxidant from cocoa by-products. Food Chemistry, 64, 199-202.

Azzi A, Stocker A; Stocker (2000). "Vitamin E: non-antioxidant roles". Prog Lipid Res 39(3): 231–255.

Babich, H. (1982). Butylated hydroxytoluene: A review. *Environmental Research*, 29: 1–29.

Babrowski, K. J. and Sosulski, F. W. (1984). Composition of free and hydrolyzable phenolic acids in defatted flours of ten oilseeds. *Journal of Agricultural and Food Chemistry*, 32: 128–130.

Bacchetti, T., Tullii, D., Masciangelo, S., Gesuita, R., Skrami, E., Brugè, F., Silvestri, S., Orlando, P., Tiano, L. and Ferretti, G. (2014). Effect of black and red cabbage on plasma carotenoid levels, lipid profile and oxidized low density lipoprotein. *Journal of Functional Foods*, 8: 128–137.

Bahorun, T., Luximon-Ramma, A., Crozier, A. and Aruoma, O. I. (2004). Total phenol, flavonoid, proanthocyanidin and vitamin C levels and antioxidant activities of Mauritian vegetables. *Journal of the Science of Food and Agriculture*, 84:1553–1561.

Bailey, D. T., Dalton, C., Daugherty, F. J., and Tempesta, M. S. (2007). Can a concentrated cranberry extract prevent recurrent urinary tract infections in women? A pilot study. *Phytomedicine: International Journal of Phytotherapy and Phytopharmacology*, *14*, 237–241.

Baj, A., Bombardelli, E., Gabetta, B. and Martinelli, E. M. (1983). Qualitative and quantitative evaluation of *Vaccinium myrtillus* anthocyanins by high-resolution gas chromatography and high-performance liquid chromatography. *Journal of Chromatography*, 279: 365–372.

Balasinska, B., and Troszynska, A. (1998). Total antioxidant activity of evening primrose (Oenothera paradoxa) cake extract measured in vitro by liposome model and murine L1210 cells. Journal of Agricultural and Food Chemistry, 46, 3558-3563.

Balavoigne, P. (1934). Estimation of the tannin content in wine, *Mitt. Geb. Lebensmttlunters. Hyg*, 25, 222.

Ballington, J. R., Ballinger, W. E. and Maness, E. P. (1987). Interspecific differences in the percentage of anthocyanins, aglycons and aglycon-sugars in the fruit of seven species of blueberries. *Journal of American Society for Horticultural Science*, 112: 859–864.

Bandarra, N. M., Campos, R. M., Batista, I., Nunes, M. L., and Empis, J. M. (1999). Antioxidant synergy of α-tocophenol and phospholipids. Journal of the American Oil Chemists' Society, 76, 905-913.

Bandonienė, D., Murkovic, M., Pfanhauser, W., Venskutonis, P. R. and Gruzdiéne, D. (2002). Detection and activity evaluation of radical scavenging compounds by using DPPH free radical and online HPLC–DPPH methods. *European Food Research and Technology*, 214: 143–147.

Bandonienė, D., Venskutonis, P. R., Gruzdiene, D. and Murkovic, M. (2002). Antioxidative activity of sage, savory and borage extracts in rapeseed oil. *European Journal of Lipid Science*, 104: 286–292.

Bandyopadhyay, C., Narayan, V. S., and Variyar, P. S. (1990). Phenolics of green pepper berries (Piper nigrum L.). Journal of Agricultural and Food Chemistry, 38, 1696-1699.

Banerjee, S., and Rao, A. R. (1992). Promoting action of cashew nut shell oil in DMBA-initiated mouse skin tumor-model system. *Cancer Letters*, *62*, 149–152.

Baoshan Sun, M. and Spranger, I. (2005). Review: Quantitative extraction and analysis of grape and wine proanthocyanidins and stilbenes. *Ciência Téc Vitiv*, 20: 59–89.

Barahona, T., Chandia, N. P., Encinas, M. V., Matsuhiro, B., and Zuniga, E. A. (2011). Antioxidant capacity of sulphated polysaccharides from seaweeds. A kinetic approach. *Food Hydrocolloids*, *25*, 529–535.

Barbeau, W. E., and Kinsella, J. E. (1983). Factors affecting the binding of chlorogenic acid to fraction I leaf protein. Journal of Agricultural and Food Chemistry, 31, 993-998.

Bartolome, B., and Gomez-Cordoves, C. (1999). Barley spent grain: release of droxycinnamic acids (ferulic and *p*-coumaric acids) by commercial enzyme preparations. *J. Sci. Food Agric.*, 79:435-439.

Bartolome, B., Pen-Neira, A. and Gomez-Cordoves, C. (2000). Phenolics and related substances in alcohol-free beer. *European Food Research and Technology*, 210: 419–423.

Basu, A., Betts, N. M., Mulugeta, A., Tong, C., Newman, E. and Lyons, T. J. (2013). Green tea supplementation increases glutathione and plasma antioxidant capacity in adults with the metabolic syndrome. *Nutrition Research*, 33: 180–187.

Basu, A., Rhone, M., and Lyons, T.J. (2010). Berries: Emerging impact on cardiovascular health. *Nutrition Reviews*, 68(3): 168–177.

Bate-Smith, C. E., Hughes, J. C. and Swain, T. (1958). After-cooking discoloration in potatoes *Chern. Ind.* 627:

Bate-Smith, E. C. (1954). Leucoanthocyanins. I. Detection and identification of anthocyanidins formed from leucoanthocyanins in plant tissues, *Biochem. J.* 58, 122.

Bate-Smith, E. C. (1973). Haemanalysis of tannins: the concept of relative astringency, *Phytochemistry,* 12, 907.

Bate-Smith, E. C. (1975). Phytochemistry of proanthocynidins. *Phytochemistry,* 14: 1107.

Bate-Smith, E. C. (1977). Astringent tannins of *Acer species. Phytochemistry.* 16, 1421.

Bate-Smith, E. C. and Lerner, N. H. (1954). Leucoanthocyanins II. Systematic distribution of leucoanthocyanins in leaves. *Biochem. J.,* 58, 126.

Bate-Smith, E. C. and Rasper, V. (1969). Tannins of grain sorghum. Luteoforol (Leucoluteolinidin). 3'. 4. 4'. 5.7 pemahydroxyfavan. J. *Food Sci.,* 34, 203.

Bate-Smith, E. C. and Swain, T. (1962). *Comparative Biochemistry* Vol. 3. Mason, H. S. and Flockin, M., Eds., Academic Press. New York.

Bate-Smith, E.C. (1958). The contribution of phenolic substances to quality in plant products. Qualitas Plantarum. 3-4: 440-455.

Baublis, A., Decker, E. A., and Clydesdale, F. M. (2000). Antioxidant effect of aqueous extracts from wheat based ready-to-eat breakfast cereals. Food Chemistry, 68, 1-6.

Bays, H., Weiter, K., and Anderson, J. (2015). A randomized study of raisins versus alternative snacks on glycemic control and other cardiovascular risk factors in patients with type 2 diabetes mellitus. *Physician and Sportsmedicine*, *43*, 37–43.

Beart, J. E., Lilley, T. H., and Haslam, E. (1985). Plant polyphenols secondary metabolism and chemical defense: some observations, *Phytochemistry.* 24: 33.

Becker, E. M., Nissen, L. R., and Skibsted, L. H. (2004). Antioxidant evaluation protocols. Food quality or health effects. *European Food Research and Technology*, *219*, 561–571.

Becker, P. and Martin. J. S. (1982). Protein precipitating capacity of tannins in *Shorea* (Dipterocarpaceae) seedling leaves. J. *Chem. Ecol.* 8, 1353.

Beckman, K. B., and Ames, B. N. (1998). The free radical theory of aging matures. *Physiological Reviews*, *78*, 547–581.

Beekwilder, J., Hall, R. D. and de Vos, C. H. (2005). Identification and dietary relevance of antioxidants from raspberry. *Biofactors (Oxford, England)*, 23: 197–205.

Bell EF (1987). "History of vitamin E in infant nutrition". American Journal of Clinical Nutrition 46 (1 Suppl): 183–186.

Bellido, G. G. and Beta, T. (2009). Anthocyanin composition and oxygen radical scavenging capacity (ORAC) of milled and pearled purple, black, and common barley. *Journal of Agricultural and Food Chemistry*, 57(3), 1022-1028.

Benavente-Garcia, O., Castillo, J., Marin, F. A., Ortuño, A. and Del Río, J. A. (1997). Uses and properties of citrus flavonoids. *Journal of Agricultural and Food Chemistry*, 45: 4505–4515.

Bener, M.,O″ zyürek, M., Güçlü, K., and Apak, R. (2010). Development of a low-cost optical sensor for cupric reducing antioxidant capacity measurement of food extracts. *Analytical Chemistry*, *82*, 4252–4258.

Bener, M.,O″ zyürek, M., Güçlü, K., and Apak, R. (2013). Novel optical fiber reflectometric CUPRAC sensor for total antioxidant capacity measurement of food extracts and biological samples. *Journal of Agricultural and Food Chemistry*, *61*, 8381–8388.

Beninger, C. W., Hosfield, G. L. and Bassett, M. J. (1999). Flavonol composition of three genotypes of dry bean (*Phaseolus vulgaris* L.) differing in seed coat color. *Journal of the American Society for Horticultural Science*, 124: 514–518.

Beninger, C. W., Hosfield, G. L. and Nair, M. G. (1998). Flavonol glycosides from the seed coat of new Manteca-type dry bean (*Phaseolus vulgaris* L.). *Journal of Agricultural and Food Chemistry*, 46: 2906–2910.

Beninger, C. W. and Hosfield, G. L. (2003). Antioxidant activity extracts, condensed tannin fractions, and pure flavonoids from *Phaseolus vulgaris* L. seed coat color genotypes. *Journal of Agricultural and Food Chemistry*, 51: 7879–7883.

Benvegnu, D. M., Barcelos, R. C. S., Boufleur, N., Reckziegel, P., Pase, C. S., Muller, L. G., Martins, N. M. B., Vareli, C., and Burger, M. E. (2010). Protective effects of a by-product of the pecan nut industry (*Carya illinoensis*) on the toxicity induced by cyclophosphamide in rats. *Journal of Environmental Pathology, Toxicology and Oncology*, *29*, 185–197.

Benvegnu, D. M., Barcelos, R. C. S., Roversi, K., Boufleur, N., Pase, C. S., Trevizol, F., Segat, H. J., Dias, V. T., Dolci, G. S., Antoniazzi, C. T. D., Reckziegel, P., Lima, F., de Lima, L. A. R., de Carvalho, L. M., Da Silva Junior, V. A., and Burger, M. E. (2013). Aqueous extract of pecan nut shell [*Carya illinoensis* (Wangenh.) K. Koch] exerts protection against damage induced by cyclophosphamide in rat testis. *Journal of Environmental Pathology, Toxicology and Oncology*, *32*, 329–341.

Benzie, I. F. F., and Strain, J. J. (1999). Ferric reducing/antioxidant power assay: Direct measure of total antioxidant activity of biological fluids and modified version for simultaneous measurement of total antioxidant power and ascorbic acid concentration. *Methods in Enzymology*, *299*, 15–27.

Benzie, I. F. F., and Szeto, Y. T. (1999). Total antioxidant capacity of teas by the ferric reducing/antioxidant power assay. Journal of Agricultural and Food Chemistry, 47, 633-636.

Berdahl, D. R. and McKeague, J. (2015). Rosemary and sage extracts as antioxidants for food preservation. In F. Shahidi (Ed.), *Handbook of antioxidants for food preservation* (pp. 177–217). Cambridge, UK: Woodhead Publishing.

Berdahl, D., Nahas, R. I. and Barren, J. P. (2010). Synthetic and natural additives in food stabilization: Current applications and future research. In E. A. Decker, R. J. Eliasand D. J.

Bergman, C. J., and Xu, Z. (2003). Genotype and environment effects on tocopherol, tocotrienol, and g-oryzanol contents of Southern U.S. rice. *Cereal Chem.*, 80:446-449.

Bergman, M., Perelman, A., Dubinsky, Z. and Grossman, S. (2003). Scavenging of reactive oxygen species by a novel glucurinated flavonoid antioxidant isolated and purified from spinach. *Phytochemistry*, 62: 753–762.

Bergman, M., Varshavsky, L., Gottlieb, H. E. and Grossman, S. (2001). The antioxidant activity of aqueous spinach extract: Chemical identification of active fractions. *Phytochemistry*, 58: 143–152.

Berker, K. I., Demirata, B., and Apak, R. (2012). Determination of total antioxidant capacity of lipophilic and hydrophilic antioxidants in the same solution by using ferric-ferricyanide assay. *Food Analytical Methods*, *5*, 1150–1158.

Berker, K. I., Gueclue, K., Tor, I., Demirata, B., and Apak, R. (2010). Total antioxidant capacity assay using optimized ferricyanide/Prussian blue method. *Food Analytical Methods*, *3*, 154–168.

Berker, K. I., Olgun, F. A. O., Ozyurt, D., Demirata, B., and Apak, R. (2013). Modified Folin-Ciocalteu antioxidant capacity assay for measuring lipophilic antioxidants. *Journal of Agricultural and Food Chemistry*, *61*, 4783–4791.

Berryman, C. E., Grieger, J. A.,West, S. G., Chen, C. Y., Blumberg, J. B., Rothblat, G. H., Sankaranarayanan, S., and Kris-Etherton, P. M. (2013). Acute consumption of walnuts and walnut components differentially affect postprandial lipemia, endothelial function, oxidative stress, and cholesterol efflux in humans with mild hypercholesterolemia. *The Journal of Nutrition*, *143*, 788–794.

Bersuder, P., Hole, M., and Smith, G. (1998). Antioxidants from a heated histidine-glucose mode system. I, Investigation of the antioxidant role of histidine and isolation of antioxidants by High-Performance Liquid Chromatography. Journal of Agricultural and Food Chemistry, 75, 181-187.

Bertholet, R., Kusy, A., Rivier, V. and Colarow, L. (1998). Antioxidant composition and process for the preparation thereof. US Patent US5714094.

Beta, T., Nam, S., Dexter, J. E., and Sapirstein, H. D. (2005). Phenolic content and ntioxidant activity of pearled wheat and roller-milled fractions. *Cereal Chem.*, 82:390-393.

Beuchamp, G. K., Keast, R. S., Morel, D. and Lin, J. (2005). Enzymes in an inflammation pathway are inhibited by oleocanthal, a component of olive oil. *Nature*, 437: 45–46.

Bezzi, S., Loupassaki, S., Petrakis, C., Kefalas, P., and Calokerinos, A. (2008). Evaluation of peroxide value of olive oil and antioxidant activity by luminol chemiluminescence. *Talanta*, *77*, 642–646.

Bhagwat, S., Haytowitz, D. B., and Holden, J. M. (2014). *USDA Database for the flavonoid content of selected foods, release 3.1.* <http://www.ars.usda.gov/nutrientdata/flav> Accessed 25.03.15.

Bhanja Dey, T. and Kuhad, R. C. (2014). Upgrading the antioxidant potential of cereal by their fungal fermentation under solid-state cultivation condition. *Letters in Applied Microbiology*, 59(5), 493-499.

Bhanja, T., Kumari, A., and Banerjee, R. (2009) Enrichment of phenolic and free radical scavenging property of wheat koji prepared with two filamentous fungi. *Bioresource Technology*, 100(1), 2861-2866.

Bhanja, T., Rout, S., Banerjee, R. and Bhattacharyya B. C. (2008). Studies on the performance of a new bioreactor for improving antioxidant of rice. *LWT-Food Science and Technology*, 41(8), 1459-1465.

Bianchini, F. and Vainio, H. (2003).Wine and resveratrol: Mechanisms of cancer prevention? *European Journal of Cancer Prevention*, 12: 417–425.

Bianco, A., Coccioli, F., Guiso, M. and Marra, C. (2002). The occurrence in olive oil of a new class of phenolic compounds: Hydroxyl-isochromans. *Food Chemistry*, 77: 405–411.

Biedrzycka, D. and Amarowicz, R. (2008). Diet and health: Apple polyphenols as antioxidants. *Food Reviews International*, 24: 235–251.

Bieri JG, Evarts RP; Evarts (1974). "γ-Tocopherol: metabolism, biological activity and significance in human vitamin E nutrition". American Journal of Clinical Nutrition 27 (9): 980–986.

Bindu, V. and Krishnaveni, B. (2013). Effect of supplementation of oats porridge on lipid profile of the hypertension subjects. *Current Biotica*, 6: 480–488.

Biparva, P., Ehsani, M. and Hadjmohammadi, M. R. (2012). Dispersive liquid–liquid microextraction using extraction solvents lighter than water combined with high performance liquid chromatography for determination of synthetic antioxidants in fruit juice samples. *Journal of Food Composition and Analysis*, 27: 87–94.

Bishayee, A., Mbimba, T., Thoppil, R. J., Háznagy-Radnai, E., Sipos, P., Darvesh, A. S., Folkesson, H. G. and Hohmann, J. (2011). Anthocyanin-rich black currant (*Ribes nigrum* L.) extract affords chemoprevention against diethylnitrosamine-induced hepatocellular carcinogenesis in rats. *The Journal of Nutritional Biochemistry*, 22: 1035–1046.

Bland, J. S. (1995). Oxidants and antioxidants in clinical medicine: Past, present and future potential. *Journal of Nutritional and Environmental Medicine*, 5(3), 255-280.

Blasco, A. J., Rogerio, M. C., Gonzalez, M. C., and Escarpa, A. (2005). Electrochemical index as a screening method to determine total polyphenolics in foods: A proposal. *Analytica Chimica Acta, 539*, 237–244.

Blekas, G., and Boskou, D. (1998). Antioxidative activity of 3,4-dihydroxtphenil acetic acid and a-tocopherol on the triglyceride matrix of olive oil. Effect of acidity. Grasas y Aceites, 49, 34-37.

Blekas, G., Tsimidu, M. and Bosku, D. (1995). Contribution of tocopherol to olive oil stability. *Food Chemistry*, 52: 289–294.

Blois, M. S. (1958). Antioxidant determination by use of a stable freeradical. *Nature*, *181*, 1199–1200.

Bobe, G., Wang, B., Seeram, N. P., Nair, M. G. and Bourquin, L. D. (2006). Dietary anthocyanin-rich tart cherry extract inhibits intestinal tumorigenesis in APC (Min) mice fed sub optional levels of sulindac. *Journal of Agricultural and Food Chemistry*, 54: 9322–9328.

Bocco, A., Cuvelier, M.-E., Richard, H. and Berset, C. (1998). Antioxidant activity and phenolic composition of citrus peel and seed extracts. *Journal of Agricultural and Food Chemistry*, 46: 2123–2129.

Boileau, T. W., Boileau, A. C. and Erdman, J. W. (2002). Bioavailability of all-trans and all-cis isomers of lycopene. *Exp. Biol. Med.* 227:914–9.

Bokern, M., Heuer, S., Wray, V.,Witte, L., Macek, T., Vanek, T. and Strack, D. (1991). Ferulic acid conjugates and betacyanins from cell cultures of *Beta vulgaris*. *Phytochemistry*, 30: 3261–3265.

Bolleddula, J., Vareed, S. K., Olson, L. K. and Nair, M. G. (2005). Insulin secretion by bioactive anthocyanins and anthocyanidins present in fruits. *Journal of Agricultural and Food Chemistry*, 53: 28–31.

Bolling, B. W., Chen, C. Y. O., McKay, D. L., and Blumberg, J. B. (2011). Tree nut phytochemicals: Composition, antioxidant capacity, bioactivity, impact factors. A systematic review of almonds, Brazils, cashews, hazelnuts, macadamias, pecans, pine nuts, pistachios and walnuts. *Nutrition Research Reviews*, *24*, 244–275.

Bolling, B. W., McKay, D. L., and Blumberg, J. B. (2010b). The phytochemical composition and antioxidant actions of tree nuts. *Asia Pacific Journal of Clinical Nutrition*, *19*, 117–123.

Bombardelli, E., Morazzoni, P. and Griffini, A. (1996). *Aesculus hippocastanum* L. *Fitoterapia*, 67: 483–511.

Bonanni, A., Campanella, L., Gatta, T., Gregori, E., and Tomassetti, M. (2007). Evaluation of the antioxidant and prooxidant properties of several commercial dry spices by different analytical methods. *Food Chemistry*, *102*, 751–758.

Bond, D. A. (1976). *In vitro* digestibility of the testa in tannin free field beans *(Vida faba* L.), *J. Agric. Sci. Camb.* 86, 561.

Bonilla, F., Mayen, M., Merida, J., and Medina, M. (1999). Extraction of phenolic compounds from red grape marc for use as food lipid antioxidants. Food Chemistry, 66, 209-215.

Bonzanini, F., Bruni, R., Palla, G., Serlataite, N. and Caligiani, A. (2009). Identification and distribution of lignans in *Punica granatum* L. fruit endocarp, pulp, seeds, wood knots and commercial juices by GC–MS. *Food Chemistry*, 117: 745–749.

Booth, A. N., Emerson, O. H., Jones, F. T. and Deeds, F. (1957). Urinary metabolites of caffeic and chlorogenic acids. *The Journal of Biological Chemistry*, 229: 51–59.

Bordia, A., Verma, S. K. and Srivastava, K. C. (1997). Effect of ginger (*Zingiber officinale* Rosc.) and fenugreek (*Trigonella foenumgraecum* L.) on blood lipids, blood sugar and platelet aggregation in patients with coronary artery disease. *Prostaglandins, Leukotrienes and Essential Fatty Acids*, 56: 379–384.

Borges, G., Degeneve, A., Mullen,W. and Crozier, A. (2009). Identification of flavonoid and phenolic antioxidants in black currants, blueberries, raspberries, red currants, and cranberries. *Journal of Agricultural and Food Chemistry*, 58: 3901–3909.

Bors,, M. D., Socaci, S., Tofana˘, M., Mures,̇an, V., Pop, A. V., Nagy, M. and Vlaic, R. (2014). Determination of total phenolics, flavonoids and antioxidant capacity of methanolic extracts of some Brassica seeds. *Bulletin of University of Agricultural Sciences and Veterinary Medicine Cluj-Napoca. Food Science and Technology*, 71: 205–206.

Borzillo, A., Iannota, N. and Uccella, N. (2000). Oinotria table olives: Quality evaluation during ripening and processing by biomolecular components. *European Food Research and Technology*, 212: 113–121.

Boskou, D., Blekas, G. and Tsimidou, M. (2005). Phenolic compounds in olive oil and olives. *Current Topics in Nutraceutical Research*, 3: 125–136.

Boudet, A. M. (2007). Evolution and current status of research in phenolic compounds. *Phytochemistry*, 68: 2722–2735.

Bourn E, L.C., and Rive-Evans, C. (1998). Bioavailability of ferulic acid. Biochemical and Biophysical Research Communication, 253: 222–227.

Bourne, L., Paganga, G., Baxter, D., Hughes, P. and Rice-Evans, C., (2000). Absorption of ferulic acid from low-alcohol beer. Free Radical Research 32, 273-280.

Bowen-Forbes, C. S., Zhang, Y. and Nair, M. G. (2010). Anthocyanin content, antioxidant, anti-inflammatory and anticancer properties of blackberry and raspberry fruits. *Journal of Food Composition and Analysis*, 23: 554–560.

Boyd. F. T. Green. V. E., Jr. and Chapman, H. L., Jr. (1965). Plantation Pride: A Bird and Disease Resistant Feed Sorghum for South Florida, Circ. 5·167. Agriculture Experiment Station, University of Florida. Gainesville. FL. 10.

Bradford, P. G., and Awad, A. B. (2007). Phytosterols as anticancer compounds. *Molecular Nutrition and Food Research*, *51*, 161–170.

Brainina, K. Z., Varzakova, D. P., and Gerasimova, E. L. (2012). A chronoamperometric method for determining total antioxidant activity. *Journal of Analytical Chemistry*, *67*, 364–369.

Brajdes, C. and Vizireanu, C. (2012). Sprouted buckwheat an important vegetable source of antioxidants. *Annals of the University Dunarea de Jos of Galati-Fascicle VI: Food Technology*, 36: 53–60.

Brand-Williams, W., Cuvelier, M. E., and Berset, C. (1995). Use of a free radical method to evaluate antioxidant activity. Lebensmittel Wissenschaftund Technologie, 28, 25-30.

Bravo, L. (1998). Polyphenols: Chemistry, dietary sources, metabolism, and nutritional significance. *Nutrition Reviews*, 56: 317–333.

Brcanovic, J. M., Pavlovic, A. N., Mitic, S. S., Stojanovic, G. S., Manojlovic, D. D., Kalicanin, B. M., and Veljkovic, J. N. (2013). Cyclic voltammetric determination of antioxidant capacity of cocoa powder, dark chocolate and milk chocolate samples: Correlation with spectrophotometric assays and individual phenolic compounds. *Food Technology and Biotechnology*, *51*,460–470.

Brenes, M., Hidalgo, F. J., Garcia, A., Rios, J. J., Garcia, P., Zamora, R. and Garrido, A. (2000). Pinoresinol and 1-acetoxypinoresinol, two new phenolic compounds identified in olive oil. *Journal of the American Oil Chemists' Society*, 77: 715–720.

Brenneisen, V. R. and Steingger, E. (1981). Analysis of polyphenols from *Vaccinium myrtillus* L. (Ericaceae) fruit. *Pharmaceutica Acta Helvetiae*, 56: 180–185.

Bressani, R., Elias, L. G., Wolzak. A. Hagerman. A. E. and Butler. L. G. (1983). Tannin in common bean: methods of analysis and effects on protein quality. *J. Food Sci.* 48, 1000.

Brezová, V., Šlebodová, A. and Staško, A. (2009). Coffee as a source of antioxidants: An EPR study. *Food Chemistry*, 114: 859–868.

Brigelius-Flohé R, Davies KJ; Davies (2007). "Is vitamin E an antioxidant, a regulator of signal transduction and gene expression, or a 'junk' food? Comments on the two accompanying papers: "Molecular mechanism of alpha-tocopherol action" by A. Azzi and "Vitamin E, antioxidant and nothing more" by M. Traber and J. Atkinson". Free radical biology and medicine 43 (1): 2–3.

Brigelius-Flohé R, Traber MG; Traber (1999). "Vitamin E: function and metabolism". FASEB J. 13 (10): 1145–55.

Broadhurst, R. B. and Jones, W. J. (1978). Analysis of condensed tannins using acidified vanillin, *J. Sri. Food Agric.* 29, 788.

Brown, J. E., Khodr, H., Hider, R. C. and Rice-Evans, C. A. (1998). Structural dependence of flavonoid interactions with Cu2+ions: Implications for their antioxidant properties. *Biochemistry Journal*, 330: 1173–1178.

Brown, S. A. (1964). Lignin and tannin biosynthesis. in *Biochemistry of Phenolic Compounds,* Harbome, J. G., Ed. Academic Press, New York, 361.

Bruce, B., Spiller, G. A., and Farquhar, J. W. (1997). Effects of a plantbased diet rich in whole grains, sun-dried raisins and nuts on serum lipoproteins. *Vegetarian Nutrition: An International Journal*, *1*, 58–63.

Bruce, B., Spiller, G. A., Klevay, L. M., and Gallagher, S. K. (2000). A diet high in whole and unrefined foods favourably alters lipids, antioxidant defences, and colon function. *Journal of the American College of Nutrition*, *19*, 61–67.

Bruun-Jensen, L., Skovgaard, I. M., Madsen, E. A., and Skibsted, L. H. (1996). The antioxidative activity of RRR-a-tocopherol vs RRR-dtocopherol in combination with ascorbyl palmitate in cooked minced turkey. Food Chemistry, 56, 347-354.

Bruun-Jensen, L., Skovgaard, I. M., Madsen, E. A., Skibsted, L. H., and Bertelsen, G. (1996). The combined effect of tocopherols, L-ascarbyl palmitate and L-ascorbic acid on the development of warmed-over favour in cooked, minced turkey. Food Chemistry, 55, 41-47.

Bu, S. Y., Lucas, E. A., Franklin, M., Marlow, D., Brackett, D. J., Boldrin, E. A., Devareddy, L., Arjmandi, B. H., and Smith, B. J. (2007). Comparison of dried plum supplementation and intermittent PTH in restoring bone in osteopenic orchidectomized rats. *Osteoporosis International*, *18*, 931–942.

Bub, A., Watzl, B., Heeb, D., Rechkemmer, G., and Briviba, K. (2001). Malvidin-3-glucoside bioavailability in humans after ingestion of red wine dealcoholized red wine and red grape juice. *Eur J Nutr.* 40(3): 113:120.

Buck, D. F. (1984). Food antioxidants – Applications and uses insnack foods. *Cereal FoodsWorld*, 29: 301–303.

Budini, R., Bub, A., Watzl, B., Heeb, D., Rechkemmer, G., and Briviba, K. (2001). Bulk oil and oil-in-water emulsion. *Journal of Agricultural and Food and Chemistry*, 44: 131-135.

Budini, R., Tonelli, D. and Girotti, S. (1980). Analysis of total phenols using the Prussian blue. Method. *J Agric. Food Chem.* 28, 1236.

Bullard, R. W. and Elias, B. D. (1980). Sorghum polyphenols and bird resistance, in *Polyphenols in Cereals and Legumes.* Hulse. J. H. Ed., International Development Research Council, Ottawa. Canada. 43.

Bullard, R. W., Garrison, M. V., Kilburn, S. R., and York, J. 0. (1980). Laboratory comparisons of polyphenols and their repellent characteristics in bird-resistant sorghum grains, *J. Agric. Food Chem. 28,* 1006.

Bullard, R. W., York, J. O. and Kilburn. S. R. (1981). Polyphenolic changes in ripening bird-resistant sorghums. J. *Agric. Food Chem.* 29, 973.

Buller, L. G., Riedl, D. J., Lebryk, D. G., and Blytt, H. J. (1984). Interaction of proteins with sorghum tannin: mechanism, specificity and significance, J. *Am. Oil Chmn. Soc.,* 61, 916.

Burda, S., Oleszek, W. and Lee, C. Y. (1990). Phenolic compounds and their changes in apples during maturation and cold storage. *Journal of Agricultural and Food Chemistry*, 38: 945–48.

Burns, J., Yokota, T., Ashihara, H., Lean, M. E. J. and Crozier, A. (2002). Plant foods and herbal sources of resveratrol. *Journal of Agricultural and Food Chemistry*, 50: 3337–3340.

Burns, R. E. (1963). Methods of tannin analysis for forage crop evaluation, Ga. *Agric. Exp. Sm. Bull.* M.S., 32.

Burns, R. E. (1971). Methods for estimation of tannin in grain sorghum, *Agron. J.,* 63, 511.

Burton, G. W. and Ingold, K. U. (1981). Autoxidation of biological molecules. 1. Antioxidant activity of vitamin E and related chain-breaking phenolic antioxidants *in vitro*. *Journal of the American Chemical Society*, 103: 6472–6477.

Burton-Freeman, B. (2010). Postprandial metabolic events and fruit-derived phenolics: A review of the science. *British Journal of Nutrition*, *104*, S1–S14.

Buta, J. G. and Spaulding, D. W. (1997). Endogenous levels of phenolics in tomato fruit during growth and maturation. *Journal of Plant Growth Regulation*, 16: 43–46.

Butler, L. G. (1978). Tannins in sorghum grain: problems, solutions and opportunities. *Proc. 33rd Annu. Corn Sorghum Res. Con.* 33: 190.

Butler, L. G. (1982). Polyphenols and their effects on sorghum quality, in *Proc. Int. Symp. Sorghum Grain Qual.,* International Crops Research Institute for the Semi-Arid Tropics, Patancheru, India, 294.

Butler, L. G. (1982). Relative degree of polymerization of sorghum tannin during seed development and maturation, *J. Agric. Food Chem.,* 30: 1090.

Butler, L. G., Price, M. L. and Brotherton, J. E. (1982). Vanillin assay for proanthocyanidins (Condensed tannins): modification of the solvent for estimation of the degree of polymerization, *J. Agric. Food Chem.,* 30, 1087.

Butler, L. G., Riedl, D. J., Lebryk, D. G., and Blytt, H. J. (1984). Interactions of proteins with sorghum tannin: mechanism, specificity and significance. J. *Am. Oil Chem. Soc.,* 61: 916.

Butler, L. G., Rogier, J. C., Mehansho, H., and Carlson, D. M. (1986). Dietary effects of tannins, in *Planl Flavonoids in Biology and Medicine: Biochemical, Pharmacological and Structure -Activity Relationships,* Alan R. Liss, New York. 141.

Butler, L. G., Rogier, J. C., Mehansho, H., and Carlson, D. M. (1986). Dietary effects of tannins. In *Plane Flavonoids in Biology and Medicine: Biochemical. Pharmacological and Structure Activity Relationships,* Alan R. Liss, New York, 141.

Butsat, S. and Siriamornpun, S. (2010). Antioxidant capacities and phenolic compounds of the husk, bran and endosperm of Thai rice. *Food Chemistry*, 119: 606–613.

Cabrini, L., Barzanti, V., Cipollone, M., Fiorentini, D., Grossi, G., Tolomelli, B., Zambonin, L. and Landi, L. (2001). Antioxidants and total peroxyl radical-trapping ability of olive and seed oils. *Journal of Agricultural and Food Chemistry*, 49: 6026–6032.

CadahõÂa, E., Conde, E., FernaÂ ndez de SimoÂ n, B., and GarcõÂa-Vallejo, M. C. (1998). Changes in tannic composition of reproduction cork Quercus suber througout industrial processing. Journal of Agricultural and Food Chemistry, 46, 2332-2336.

Cai, L. and Wu, C. D. (1996). Compounds from Syzygium aromaticum possessing growth inhibitory activity against oral pathogens. *Journal of Natural Products*, 59: 987–990.

Cai, S., Gao, F., Zhang, X., Wang, O., Wu, W., Zhu, S., Zhang, D., Zhou, F. and Ji, B. (2014). Evaluation of γ-aminobutyric acid, phytate and antioxidant activity of tempeh-like fermented oats (*Avena sativa* L.) prepared with different filamentous fungi. *Journal of Food Science and Technology*, 51(10), 2544-2551.

Cai, S., Wang, O., Wu., W., Zhu, S., Zhou, F., Ji, B., Gao, F., Zhang, D., Liu, J. and Cheng, Q. (2012). Comparative study of the effects of solid-state fermentation with three filamentous fungi on the total phenolics content, flavonoids, antioxidant activities of subfractions from oats (*Avena sativa* L.). *Journal of Agricultural and Food Chemistry*, 60(1), 507-513.

Caimari, A., Puiggros, F., Suarez, M., Crescenti, A., Laos, S., Ruiz, J. A., Alonso, V., Moragas, J., Del Bas, J. M., and Arola, L. (2015). The intake of a hazelnut skin extract improves the plasma lipid profile and reduces the lithocholic/deoxycholic bile acid faecal ratio, a risk factor for colon cancer, in hamsters fed a high-fat diet. *Food Chemistry*, *167*, 138–144.

Calabrò, M. L., Galtieri, V., Cutroneo, P., Tommasini, S., Ficarra, P. and Ficarra, R. (2004). Study of the extraction procedure by experimental design and validation of a LC method for determination of flavonoids in Citrus bergamia juice. *Journal of Pharmaceutical and Biomedical Analysis*, 35: 349–363.

California Prune Board. (2005). *California dried plums offer bakers innovative solutions to control and reduce fat, calories, sugar and carbohydrates while extending shelf life.* Pleasanton, CA: California Prune Board.

Calucci, L., Pinzono, C., Zandomeneghi, M., Capocchi, A., Ghiringhelli, S., Saviozzi, F., Tozzi, S. and Galleschi, L. (2003). Effects of gamma-irradiation on the free radical and antioxidant contents in nine aromatic herbs and spices. *Journal of Agricultural and Food Chemistry*, 51: 927–934.

Camire, M. E., and Dougherty, M. P. (1998). Added phenolic compounds enhance stability in extruded corn. Journal of Food Science, 63, 516-518.

Campos-Vega, R., Loarca-Piña, G. and Oomah, B. D. (2010). Minor components of pulses and their potential impact on human health. *Food Research International*, 43: 461–482.

Canizares-Macias, M. P., Garcia-Mesa, J. A., and Luque de Castro, M. D. (2004). Determination of the oxidative stability of olive oil, using focused-microwave energy to accelerate the oxidation process. *Analytical and Bioanalytical Chemistry, 378*, 479–483.

Cano, A., Acosta, M., and BanÄ oÂ n, M. (1998). EstimacioÂ n de la actividad antioxidante de distintas bebidas comerciales. Alimentaria, 9, 73-76.

Cantos, E., Espin, J. C. and Tomas-Barberan, F. A. (2001). Effect of wounding on phenolic enzymes in six minimally processed lettuce cultivars upon storage. *Journal of Agricultural and Food Chemistry*, 49: 322–330.

Cantos, E., Espin, J. C. and Tomas-Barberan, F. A. (2002). Varietal differences among polyphenols of seven table grape cultivars studied by LC-DAD-MS-MS. *Journal of Agricultural and Food Chemistry*, 50: 5691–5696.

Cao, G., (1998). Oxygen reactive capacity (ORAC) and ferric reducing antioxidant power (FRAP) assays: A comparative study. *Journal of Agricultural and Food Chemistry*, 50: 3122–3128.

Cao, G., Booth, S. L., Sadowski, J. A., and Prior, R. L. (1998). Increases in human plasma antioxidant capacity after consumption of controlled diets high in fruit and vegetables. *American Journal of Clinical Nutrition, 68*, 1081–1087.

Cao, G., Sofic, E. and Prior, R. L. (1996). Antioxidant capacity of tea and common vegetables. *Journal of Agricultural and Food Chemistry*, 44: 3426–3431.

Cao, G., Sofic, E. and Prior, R. L. (1997). Antioxidant and prooxidant behavior of flavonoids: Structure-activity relationships. *Free Radical Biology and Medicine*, 22: 749–760.

Caponio, F., Allogio, V., and Gomes, T. (1999). Phenolic compounds of virgin olive oil, infuence of paste preparation techniques. Food Chemistry, 64, 203-209.

Carbonneau, M. A., LeÂger, C. I., Descomps, B., Michel, F., and Monnier, L. (1998). Improvement in the antioxidant status of plasma and low-density lipoprotein in subjects receiving a red wine phenolics mixture. Journal of the American Oil Chemists' Society, 75, 235-240.

Carlsen, M. H., Halvorsen, B. L., Holte, K., Bohn, S. K., Dragland, S., Sampson, L. and Blomhoff, R. (2010). The total antioxidant content of more than 3100 foods, beverages, spices, herbs and supplements used worldwide. *Nutrition Journal*, 9: 1–11.

Carrol, K. K., Kurowska, E. M. and Guthrie, N. (1999). Use of citrus limonoids and favonoids as well as tocotrienols for the treatment of cancer. Int. Patent WO9915167.

Carvalho, A. L., Annoni, R., Silva, P. R., Borelli, P., Fock, R. A., Trevisan, M. T., and Mauad, T. (2011). Acute, subacute toxicity and mutagenic effects of anacardic acids from cashew (*Anacardium occidentale* Linn.) in mice. *Journal of Ethnopharmacology*, 135, 730–736.

Carvalho, M., Ferreira, P. J., Mendes, V. S., Silva, R., Pereira, J. A., Jerónimo, C. and Silva, B. M. (2010). Human cancer cell antiproliferative and antioxidant activities of *Juglans regia* L. *Food and Chemical Toxicology*, 48: 441–447.

Cavalcanti, R.N., Santos, D.T., and Meireles M.A.A. (2011). Non-thermal stabilization mechanisms of anthocyanins in model and food systems-An overview. *Food Research International*. 44(2): 499–509.

Cavero, S., García-Risco, M. R., Marín, F. R., Jaime, L., Santoyo, S., Señoráns, F. J., Reglero, G. and Ibañez, E. (2006). Supercritical fluid extraction of antioxidant compounds from oregano: Chemical and functional characterization via LC–MS and in vitro assays. *The Journal of Supercritical Fluids*, 38: 62–69.

Cerda, B., Tomás-Barberán, F. A., and Espin, J. C. (2005). Metabolisms of antioxidant and chemopreventive ellagitannins from strawberries, raspberries, walnuts, and oak-aged wine in humans: Identification of biomarkers and individual variability. *Journal of Agricultural and Food Chemistry*, *53*, 227–235.

Cerretani, L., Giuliani, A., Maggio, R. M., Bendini, A., Gallina Toschi, T., and Cichelli, A. (2010). Rapid FTIR determination of water, phenolics and antioxidant activity of olive oil. *European Journal of Lipid Science and Technology*, *112*, 1150–1157.

Cerretani, L., Lerma-Garcia, M. J., Herrero-Martinez, J. M., Gallina-Toschi, T. and Simo-Alfonso, E. F. (2010). Determination of tocopherols and tocotrienols in vegetable oils by nanoliquid chromatography with ultraviolet–visible detection using a silica monolithic column. *Journal of Agricultural and Food Chemistry*, 58: 757–761.

Cerrutti, P., Alzamora, S. M., and Vidales, S. L. (1997). Vanillin as an antimicrobial for producing shelf-stable strawberry puree. Journal of Food Science, 62, 608-610.

Cesa, S. (2004). Malondialdehyde contents in infant milk formulas. *Journal of Agricultural and Food Chemistry*, *52*, 2119–2122.

Cevallos-Casals, B. A. and Cisneros-Zevallos, L. (2010). Impact of germination on phenolic content and antioxidant activity of 13 edible seed species. *Food Chemistry*, 119: 1485–1490.

Chai, S. C., Hooshmand, S., Saadat, R. L., Payton, M. E., Brummel-Smith, K., and Arjmandi, B. H. (2012). Daily apple versus dried plum: Impact on cardiovascular disease risk factors in postmenopausal women. *Journal of the Academy of Nutrition and Dietetics*, *112*, 1158–1168.

Chambers, L. J., Jimbin, M. and McDonald, R. (1988). Antioxidant composition. EP0267630.

Chambers, S. J., Lambert, N., Plumb, G. W., and Williamson, G. (1996). Evaluation of the antioxidant properties of a methanolic extract from `Juice Plus fruit' and `Juice Plus vegetable' (dietary supplements). Food Chemistry, 57, 271-274.

Chan, E. W., Soh, E. Y., Tie, P. P. and Law, Y. P. (2011). Antioxidant and antibacterial properties of green, black, and herbal teas of Camellia sinensis. *Pharmacognosy Research*, 3: 266–271.

Chandra, A., Nair, M. G. and Lezzoni, A. (1992). Evaluation and characterization of the anthocyanin pigments in tart cherries (*Prunus cerasus* L.). *Journal of Agricultural and Food Chemistry*, 40: 967–969.

Chandrasekara, A., and Shahidi, F. (2011b). Inhibitory activities of soluble and bound millet seed phenolics on free radicals and reactive oxygen species. Journal of Agricultural and Food Chemistry, 59: 428–436.

Chandrasekara, A., and Shahidi, F. (2011c). Bioactivities and antiradical properties of millet grains and hulls. Journal of Agricultural and Food Chemistry, 59: 563–9571.

Chandrasekara, A., and Shahidi, F. (2012). Bioaccessibility and antioxidant potential of millet grain phenolics as affected by simulated in vitro digestion and microbial fermentation. Journal of Function al Foods, 4:226–237.

Chandrasekara, A., Naczk, M., and Shahidi, F. (2012a). Effect of processing on the antioxidant activity of millet grains. Food Chemistry, 133: 1–9.

Chandrasekara, A. and Shahidi, F. (2010). Content of insoluble bound phenolics in millets and their contribution to antioxidant capacity. *Journal of Agricultural and Food Chemistry*, 58: 6706–6714.

Chandrasekara, N., and Shahidi, F. (2011). Effect of roasting on phenolic content and antioxidant activities of whole cashew nuts, kernels, and testa. *Journal of Agricultural and Food Chemistry*, *59*, 5006–5014.

Chandrasekara, N. and Shahidi, F. (2011a). Antioxidative potential of cashew phenolics in food and biological model systems as affected by roasting. *Food Chemistry*, 129: 1388–1396.

Chandrasekara A and Shahidi F. (2011). Antioxidant phenolics of millet control lipid peroxidation in human ldl cholesterol and food systems. *J Am Oil Chem. Soci.* DOI: 10.1007/s11746-011-1918-5. Available from Springer [http://www.springerlink.com]. Posted August 20, 2012.

Chan-Eam, S., Teerasong, S., Damwan, K., Nacapricha, D., and Chaisuksant, R. (2011). Sequential injection analysis with electrochemical detection as a tool for economic and rapid evaluation of total antioxidant capacity. *Talanta*, *84*, 1350–1354.

Chang, C. I., Chou, C. H., Liao, M. H., Chen, T. M., Cheng, C. H., Anggriani, R., Tsat, T. P., Tseng, H. I. and Cheng, H. L. (2015). Bitter melon triterpenes work as insulin sensitizers and insulin substitutes in insulin-resistant cells. *Journal of Functional Foods*, 13: 214–224.

Chang, J., Rimando, A., Pallas, M., Camins, A., Porquet, D., Reeves, J., Shukitt-Hale, B., Smith, M. A., Joseph, J. A. and Casadesus, G. (2012). Low-dose pterostilbene, but not resveratrol, is a potent neuromodulator in aging and Alzheimer's disease. *Neurobiology of Aging*, 33: 2062–2071.

Chang, L. W., Yen, W. J., Huang, S. C. and Duh, P. D. (2002). Antioxidant activity of sesame coat. *Food Chemistry*, 78: 347–354.

Chang, S. K., Alasalvar, C., and Shahidi, F. (2016). Review on dried fruits: Phytochemicals, antioxidant efficacies and health benefits. *Journal of Functional Foods*, *21*, 113–132.

Chang, S. K., Cesarettin Alasalvar, C., BradleyW. Bolling, B. W. and Shahidi F. (2016). Nuts and their co-products: The impact of processing (roasting) on phenolics, bioavailability, and health benefits – A comprehensive review. *Journal of Functional Foods.* 26: 88–122.

Chang, S. S., Ostric-Matijasevic, B., Hsieh, O. A. L. and Huang, C. L. (1977). Natural antioxidants from rosemary and sage. *Journal of Food Science*, 42: 1102–1106.

Chang, W. S., Chang, Y. H., Lu, F. J. and Chiang, H. C. (1994). Inhibitory effects of phenolics on xanthine oxidase. *Anticancer Research*, 14: 501–506.

Chantaro, P., Devahastin, S. and Chiewchan, N. (2008). Production of antioxidant high dietary fiber powder from carrot peels. *LWT - Food Science and Technology*, 41: 1987–1994.

Chaovanalikit, A. and Wrolstad, R. E. (2004). Anthocyanin and polyphenolic composition of fresh and processed cherries. *Journal of Food Science*, 69: 73–83.

Chapman, A. C. (1908). Hop tannin, J. Inst. *Brew. London,* 13, 644.

Chapon, L. Chollot, B., and Urian, E. (1961). Peroxidase hazes of beers, basic phenomenon. *Eur. Brew. Cony. Proc. Congr.,* 319.

Charalampidis, P.S.; Veltsistas, P.; Karkabounas, S.; Evangelou, A. (2009). Blue CrO_5 assay: A novel spectrophotometric method for the evaluation of the antioxidant and oxidant capacity of various biological substances. *Eur. J. Med. Chem. 44*, 4162-4168.

Chatterjee, S., Variyar, P. S. and Sharma, A. (2009). Stability of lipid constituents in the radiation processed fenugreek seeds and turmeric: Role of phenolic antioxidants. *Journal of Agricultural and Food Chemistry*, 57: 9226–9233.

Chavan, J. K. and Hejgaard, J. (1981). Detection and partial characterization of subtilisin inhibitors in legume seeds by isoelectric focusing, J. *Sci. Food Agric.*, 32, 857.

Chavan, J. K., Ghonsikar, C. P., Kadam, S. S., and Salunkhe, D. K. (1979). Protein, tannin and starch changes in developing seeds of low and high lannin cuhivars of sorghum. *J. Food Biochem.* 3: 13.

Che Man, Y. B., and Tan, C. P. (1999). Effects of natural and synthetic antioxidants on changes in refined, bleached, and deodorized palm olein during deep-fat frying of potato chips. Journal of the American Oil Chemists' Society, 76, 331-339.

Chen, B. H., and Tang, Y. C. (1998). Processing and stability of carotenoid powder from carrot pulp waste. Journal of Agricultural and Food Chemistry, 46, 2312-2318.

Chen, C. Y., and Blumberg, J. B. (2008). Phytochemical composition of nuts. *Asia Pacific Journal of Clinical Nutrition, 17*, 329–332.

Chen, C. Y., Milbury, P. E., Collins, F. W. and Blumberg, J. B. (2007). Antioxidant activity in human after acute consumption of an enriched mixture from oats. *The Journal of Nutrition,* 136(6), 1375-1382.

Chen, C. Y., Milbury, P. E., Kwak, H. K., Collins, F. W., Samuel, P. and Blumberg, J. B. (2004). Avenanthramides and phenolic acids from oats are bioavailable and act synergistically with vitamin C to enhance hamster and human LDL resistance to oxidation. *The Journal of Nutrition*, 134: 1459–1466.

Chen, C. Y., Milbury, P. E., Lapsley, K., and Blumberg, J. B. (2005). Flavonoids from almond skins are bioavailable and act synergistically with vitamins C and E to enhance hamster and human LDL resistance to oxidation. *The Journal of Nutrition*, *135*, 1366–1373.

Chen, P. H., Chen, G. C., Yang, M. F., Hsieh, C. H., Chuang, S. H., Yang, H. L., Kuo, Y. H., Chyuan, J. H. and Chao, P. M. (2012). Bitter melon seed oil-attenuated body fat accumulation in diet-induced obese mice is associated with cAMP-dependent protein kinase activation and cell death in white adipose tissue. *The Journal of Nutrition*, 142: 1197–1204.

Chen, X., and Ahn, D. U. (1998). Antioxidant activities of six natural phenolics against lipid oxidation induced by Fe2+ or ultraviolet light. Journal of the American Oil Chemists' Society, 75, 1717-1721.

Chen, Z. Y., Chan, P. T., Kwan, K. Y., and Zhang, A. (1997). Reassessment of the antioxidant activity of conjugated linoleic acids. Journal of the American Oil Chemists' Society, 74, 749-753.

Chen, Z. Y., Zhu, Q. Y., Tsang, D. and Huang, Y. (2001). Degradation of green tea catechins in tea drinks. *Journal of Agricultural and Food Chemistry*, 49: 477–482.

Chen, Z.-Y., Wang, L.-Y., Chan, P. T., Zhang, Z., Chung, H. Y., and Liang, C. (1998). Antioxidative activity of green tea catechin extract compared with that of rosemary extract. Journal of the American Oil Chemists' Society, 75, 1141-1145.

Cheronis, S. D. and Entrikin, J. B. (1957). Semimicro-qualitative Organic Analysis. 2nd ed., Inter Science. New York.

Chethan S. and Malleshi, N. G. (2007). Finger millet polyphenols: optimization of extraction and the effect of pH on their stability. *Food Chem.* 105(2):862–70.

Cheung, S. and Tai, J. (2007). Anti-proliferative and antioxidant properties of rosemary *Rosmarinus officinalis*. *Oncology Reports*, 17: 1525–1531.

Chevillard, G., Nouhi, Z., Anna, D., Paquet, M. and Blank, V. (2010). Nrf3-deficient mice are not protected against acute lung and adipose tissue damages induced by butylated hydroxytoluene. *FEBS Letter*, 584: 923–928.

Chiang, A. N., Wu, H. L., Yeh, H., Chu, C. S., Lin, H. C. and Lee, W. C. (2006). Antioxidant effect of black rice extract through the induction of superoxide dismutase and catalase activities. *Lipids*, 41(8), 797-803.

Chiang, H. and Norris, D. M. (1983). Phenolic and tannin content as related to anatomical parameters of soybean resistance to agromyzid bean flies, J. *Agric. Food Chem.* 31, 726.

Chimi, H., Cillard, J., Cillard, P., and Rahmani, M. (1991). Peroxyl and hydroxyl radical scavenging activity of some natural phenolic antioxidants. Journal of the American Oil Chemists' Society, 68, 307-311.

Chinnici, F., Bendini, A., Gaiani, A. and Riponi, C. (2004). Radical scavenging activities of peels and pulps from cv. Golden Delicious apples as related to their phenolic composition. *Journal of Agricultural and Food Chemistry*, 52: 4684–4689.

Chiou, A., Karathanos, V. T., Mylona, A., Salta, F. N., Preventi, F., and Andrikopoulos, N. K. (2007). Currants (*Vitis vinifera* L.) content of simple phenolics and antioxidant activity. *Food Chemistry*, *102*, 516–522.

Chipault, J. R., Mizuno, G. R. and Lundberg,W. O. (1956). The antioxidant properties of natural spices. *Journal of Food Technology*, 10: 207–210.

Chitarrini, G., Nobili, C., Pinzari, F., Antonini, A., De Rossi, P., Del Fiore, A., Procacci, S., Tolaini, V., Scala, V., Scarpari, M. and Reverberi, M. (2014). Buckwheat achenes antioxidant profile modulates Aspergillus flavus growth and aflatoxin production. *International Journal of Food Microbiology*, 189: 1–10.

Cho, S. H., Choi, Y., and Ha, T. Y. (2000). In vitro and in vivo effects of prosomillet, buckwheat and sorghum on cholesterol metabolism. *J. Fed. Am. Soc. Exp. Biol.,* 14: A249.

Choi, M. S., Jung, U. J., Yeo, J. K. M. J., Kim, M. J., and Lee, M. K. (2008). Genistein and daidzein prevent diabetes onset by elevating insulin level and altering hepatic gluconeogenic and lipogenic enzyme activities in non-obese diabetic (NOD) mice. *Diabetes/Metabolism Research and Reviews*, *24*, 74–81.

Choi, Y. M., Jeong, H. S., and Lee, J. S. (2006). Antioxidant activity of methanolic extracts from some grains consumed in Korea. *Food Chem.,* 103:130–138.

Chopin, J. and Bouillant, M. L. (1975). C-Glycosylflavones, In *The Flavonoids,* Harbome, J. B., Mabry, T. J., and Mabry, H., Eds. Chapman and Hall, London, 632.

Christensen, K. D. (1974). *Improving the Plant Proteins by Nuclear Techniques,* International Atomic Energy Agency, Vienna.

Christodouleas, D., Fotakis, C., Papadopoulos, K., Yannakopoulou, E., and Calokerinos, A. C. (2009). Development and validation of a chemiluminogenic method for the evaluation of antioxidant activity of hydrophilic and hydrophobic antioxidants. *Analytica Chimica Acta*, *652*, 295–302.

Chuah, A. M., Lee, Y., Yamaguchi, T., Takamura, H., Yin, I. and Matoba, T. (2008). Effect of cooking on the antioxidant properties of coloured peppers. *Food Chem.* 111: 20–28.

Chu, Q., Lin, M., Yu, X. and Ye, J. (2008). Study on extraction efficiency of natural antioxidant in coffee by capillary electrophoresis with amperometric detection. *European Food Research and Technology*, 226: 1373–1378.

Chu, Y. F., Sun, J.,Wu, X. and Liu, R. H. (2002). Antioxidant and antiproliferative activities of common vegetables. *Journal of Agricultural and Food Chemistry*, 50: 6910–6916.

Chu, Y. F.,Wise, M. L., Gulvady, A. A., Chang, T., Kendra, D. F., van Klinken, B. J. W., Shi, Y. and O'Shea, M. (2013). In vitro antioxidant capacity and anti-inflammatory activity of seven common oats. *Food Chemistry*, 139: 426–431.

Chu, Y. H., Chang, C. L. and Hsu, H. F. (2000). Flavonoid content of several vegetables and their antioxidant activity. *Journal of the Science of Food and Agriculture*, 80: 561–566.

Chu, Y.-H., and Hsu, H.-F. (1999). Effects of antioxidants on peanut oil stability. Food Chemistry, 66, 29-34.

Chuda, Y., Ono, H., Ohnishi-Kameyama, M., Nagata, T., and Tsushida, T. (1996). Structural identification of two antioxidant quinic acid derivatives from garland (Chrysanthemum coronarium L). Journal of Agricultural and Food Chemistry, 44, 2037-2039.

Chun, S. S., Vattem, D. A., Lin, Y. T. and Shetty, K. (2005). Phenolic antioxidants from clonal oregano (*Origanum vulgare*) with antimicrobial activity against *Helicobacter pylori*. *Process Biochemistry*, 40: 809–816.

Chung, H. S., Chang, L. C., Lee, S. K., Shamon, L. A., van Breemen, R. B., Mehta, R. G., Farnsworth, N. R., Pezzuto, J. M., and Douglas, K. A. (1999). Flavonoids constituents of Chorizanthe diffusa with potential cancer chemopreventive activity. Journal of Agricultural and Food Chemistry, 47, 36-41.

Chung, J., Lee, Y., and Choe, E. (2006). Effects of sesame oil addition to soybean oil during frying on the lipid oxidative stability and antioxidants contents of the fried products during storage in the dark. *Journal of Food Science*, *71*, C222–C226.

Cilliers, J. J. L., and Singleton, V. L. (1990). Antioxidative phenolic ring opening under alkaline conditions as a model for natural polyphenols in food. Journal of Agricultural and Food Chemistry, 38, 1797-1798.

Cilliers, J. J. L., and Singleton, V. L. (1991). Characterization of the products of nonenzymic autoxidative phenolic reactions in a caffeic acid model system. Journal of Agricultural and Food Chemistry, 31, 1298-1303.

Cioffi, G., Pesca, M. S., Caprariis, P. D., Braca, A., Severino, L. and Tommasi, N. D. (2010). Phenolic compounds in olive oil and olive pomace from Cilento (Campania, Italy) and their antioxidant activity. *Food Chemistry*, 121: 105–111.

Clarke, I. D., Frey, R., and Hyland, H. L. (1939). Seasonal variation in tannin content of *Lespedeza sericea, J. Agric. Res.* 58, 131.

Clifford, M. N. (1999). Chlorogenic acids and other cinnamates –Nature, occurrence, and dietary burden. *Journal of Science and Food Agriculture*, 79: 362–372.

Clifford, M. N. and Ramirez-Martinez, J. R. (1991). Phenols and caffeine in wet-processed coffee beans and coffee pulp. *Food Chemistry*, 40: 35–42.

Clifford, M.N. (2000). Anthocyanins: nature, occurrence and dietary burden. *Journal of the Science of Food and Agriculture*. 80(7): 1063–1072.

Colaric, M., Veberic, R., Solar, A., Hudina, M. and Stampar, F. (2005). Phenolic acids, syringaldehyde, and juglone in fruits of different cultivars of *Juglans regia* L. *Journal of Agricultural and Food Chemistry*, 53: 6390–6396.

Coleman, A. E., and Merola, V. (2016). Clinical signs associated with ingestion of black walnut tree (*Juglans nigra*) wood, nuts, and hulls in dogs: 93 cases (2001–2012). *Journal of the American Veterinary Medical Association*, *248*, 195–200.

Coleman, A. L., Stone, K. L., Kodjebacheva, G., Yu, F., Pedula, K. L., Ensrud, K. E., Cauley, J. A., Hochberg, M. C., Topouzis, F., Badala, F., and Mangione, C. M. (2008). Glaucoma risk and the consumption of fruits and vegetables among older women in the study of osteoporotic fractures. *American Journal of Ophthalmology*, *145*, 1081–1089.

Collomb, M., and Spahni, M. (1996). Review of the methods for determining lipid oxidation products, in particular lipids from milk products. *Schweizerische Milchwirtschaftliche Forschung*, *25*, 3–24.

Colquhoun, D. M., Humphries, J. A., Moores, D. and Somerset, S. M. (1996). Effects of macadamia enriched diet on serum lipids and lipoproteins compared to low fat diet. *Food Australia*, 48: 216–221.

Conde, E., CadahõÂa, E., GarcõÂa-Vallejo, M., and FernaÂ ndez de SimoÂ n, B. (1998). Polyphenolic composition of Quercus suber cork from different spanish provenances. Journal of Agricultural and Food Chemistry, 44, 3166-3171.

Conner, M. J. (1991). Modulation of tumor promotion in mouse skin by the food additive citral. *Cancer Letters*, 56: 25–28.

Cook, N. C. and Samman, S. (1996). Flavonoids – Chemistry, metabolism, cardioprotective effects and dietary sources. *Journal of Nutritional Biochemistry*, 7: 66–76.

Copeland, E. L., Clifford, M. N., and Williams, C. M. (1998). Preparation of (ÿ) epigallocatechin gallate from commercial green tea by caffein precipitation and solvent partition. Food Chemistry, 61, 81-87.

Costantino, L., Albasini, A., Rastelli, G. and Benvenuti, S. (1992). Activity of polyphenolic extracts as scavengers of superoxide radicals and inhibitors of xanthine oxidase. *Planta Medica*, 58: 342–344.

Cotroneo, M. S., Wang, J., Fritz,W. A., Eltoum, I. E. and Lamartiniere, C. A. (2002). Genistein action in the prepubertal mammary gland in a chemoprevention model. *Carcinogenesis*, 23: 1467–1474.

Counet, C., Callemien, D. and Collin, S. (2006). Chocolate and cocoa:New sources of *trans*-resveratrol and *trans*-piceid. *Food Chemistry*, 98: 649–657.

Coulibaly A, and Chen J. (2011). Evaluation of energetic compounds, antioxidant capacity, some vitamins and minerals, phytase and amylase activity during germination of foxtail millet. *Am J Food Technol*. 6(1):40–51.

Cousins, M., Adelberg, J., Chen, F. and Rieck, J. (2007). Antioxidant capacity of fresh and dried rhizomes from four clones of turmeric (*Curcuma longa* L.) grown *in vitro*. *Industrial Crops and Products*, 25: 129–135.

Craft, B. D., Kosinska, A., Amarowicz, R. and Pegg, R. B. (2010). Antioxidant properties of extracts obtained from raw, dryroasted, and oil-roasted US peanuts of commercial importance. *Plant Foods for Human Nutrition*, 65: 311–318.

Craft, C.C. (1961). Polyphenolic compounds in Elberta peaches during storage and ripening. *Proc. Am. Soc. Hort. Sci*. 78: 119-125.

Crook, J. H. and Ward, P. (1968). The quelea problem in Africa, in *The Problems of Birds and Pests*. Murton, R. K. and Wright, E. N., Eds., Academic Press. London. 211.

Crozier, A., Lean, M. E. J., McDonald, M. S. and Black, C. (1997). Quantitative analysis of the flavonoid content of commercial tomatoes, onions, lettuce and celery. *Journal of Agricultural and Food Chemistry*, 45: 590–595.

Cruz, J. M., DomõÂnguez, J. M., DomõÂnguez, H., and ParajoÂ, J. C. (1999). Solvent extraction of hemicellulosic wood hydrolysates, a procedure useful for obtaining both detoxified fermentation media and polyphenols with antioxidant activity. Food Chemistry, 67, 147-153.

Cukelj, N., Novotni, D. and Curic, D. (2010). Antioxidant properties of whole grain cereal. *Croatian Journal of Food Technology, Biotechnology and Nutrition*, 5(1-2), 18-23.

Cumby, N., Zhong, Y., Naczk, M., and Shahidi, F. (2008). Antioxidant activity and water-holding capacity of canola protein hydrolysates. *Food Chemistry*, *109*, 144–148.

Cummings. D. P. and Axtell, J. D. (1973). Inheritance and Improvement of Protein Quality and Content in Sorghum. Research Progress Repon. Department of Agronomy. Purdue University, West Lafayette. IN. 112.

Cuvelier, M. E., Richard, H., and Berset, C. (1992). Comparison of the antioxidative activity of some acid-phenols: structure-activity relationship. Bioscience, Biotechnology and Biochemistry, 56, 324-325.

Cuvelier, M. E., Richard, H., and Berset, C. (1996). Antioxidative activity and phenolic composition of pilot-plant and commercial extracts of sage and rosemary. Journal of the American Oil Chemists' Society, 73, 645-652.

Czerniwski, M. (1958). Tannins. I. Stability and determination of tannins, *Chem. Anal. Warsaw,* 3, 37.

Czerwin□ska, M., Kiss, A. K. and Naruszewicz, M. (2012). A comparison of antioxidant activities of oleuropein and its dialdehydic derivative from olive oil, oleacein. *Food Chemistry*, 131: 940–947.

Dabrowski, K. and Sosulski, F. (1984). Composition of free and hydrolyzable phenolic acids in defatted flours of ten oilseeds. *Journal of Agricultural and Food Chemistry*, 32: 128–130.

Daiber, K. H. (1975). Enzyme inhibition by polyphenols of sorghum grain and malt. J. *Sci. Food Agric. 26,* 1399.

Daiber, K. H. (1975). Enzyme inhibition by polyphenols of sorghum grain and mall, J. *Sci. Food Agric.* 26. 1399.

Dalton, T. P., Shertzer, H. G., and Puga, A. (1999). Regulation of gene expression by reactive oxygen. *Annual Review of Pharmacology and Toxicology*, *39*, 67–101.

Damron, B. L., Price, G. M., and Harms, R. H. (1968). Evaluation of various bird-resistant and non-resistant varieties of grain sorghum for use in broiler diets, *Poult. Sci.,* 47, 1648.

Dangles O, Dufour C (2008) Flavonoid-protein binding processes and their potential impact on human health. In: Daayf F, Lattanzio V (Eds) Recent advances on polyphenol research, vol 1. Wiley-Blackwell, Oxford.

Danilenko, M., Wang, Q.,Wang, X., Levy, J., Sharoni, Y. and Studzinski, G. P. (2003). Carnosic acid potentiates the antioxidant and prodifferentiation effects of 1α 25-dihydroxyvitamin D3 in leukemia cells but does not promote elevation of basal levels of intracellular calcium. *Cancer Research*, 63: 1325–1332.

Das S, Lekli I, Das M, Szabo G, Varadi J, Juhasz B, Bak I, Nesaretam K, Tosaki A, Powell SR, Das DK; Lekli; Das; Szabo; Varadi; Juhasz; Bak; Nesaretam; Tosaki; Powell; Das (2008). "Cardioprotection with palm oil tocotrienols: comparison of different isomers".American journal of physiology. Heart and circulatory physiology 294 (2): H970–8.

Das, D. K. (1994). Naturally occurring flavonoids: Structure, chemistry, and high-performance liquid chromatography methods for separation and characterization. *Methods in Enzymology*, 234: 410–420.

Das, N. P., and Pereira, T. A. (1990). Effects of favonoids on thermal autoxidation of palm oil structure-activity relationships. Journal of the American Oil Chemists' Society, 67, 255-258.

Devi, P. B, Vijayabharathi, R., Sathyabama, S., Malleshi, N. G. and Priyadarisini, V. B. (2011). Health benefits of finger millet (*Eleusine coracana* L.) polyphenols and dietary fiber: a review. *J Food Sci Technol* DOI: 10.1007/s13197-011-0584-9. Available from Springer [http://www.springerlink.com]. Posted November 22, 2011.

Davies, K. J. (2000). Oxidative stress, antioxidant defenses, and damage removal, repair, and replacement systems. *IUBMB Life*, *50*, 279–289.

Davies, K. J. A. (1995). Oxidative stress: The paradox of aerobic life. In C. Rice-Evans, B. Halliwell, and G. G. Lunt (Eds.), *Free radical and oxidative stress: Environments, drugs and food additives* (pp. 1–31). London, UK: Portland Press Ltd.

Davis, A. B. and Hoseney, R. C. (1979). Grain sorghum condensed tannins. I. Isolation, estimation and selective adsorption by starch. *Cereal Chem.* 56, 310.

Davis, A. B. and Hoseney, R. C. (1979a). Grain sorghum condensed tannins. II. Pre harvest changes. *Cereal Chem.,* 56, 314.

Davis, A.B. and Hoseney, R.C. (1979b). Grain sorghum condensed tannins. I. Isolation, estimation and selective adsorption by starch. *Cereal Chem.* 56: 310-314.

Davis, K. R. (1982). Effects of processing on composition and *Tetrahymena* relative nutritive value of green and yellow peas, lentils and white pea beans, *Cereal Chem,* 58, 454.

Davis, P. A. and Iwahashi, C. K. (2001). Whole almonds and almond fractions reduce aberrant crypt foci in a rat model of colon carcinogenesis. *Cancer Letters*, 165: 27–33.

Dawes, H. W., and Keene, J. B. (1999). Phenolic composition of kiwi-fruit juice. Journal of Agricultural and Food Chemistry, 47, 2398-2403.

Dawidowicz, A. L., and Olszowy, M. (2010). Influence of some experimental variables and matrix components in the determination of antioxidant properties by β-carotene bleaching assay: Experiments with BHT used as standard antioxidant. *European Food Research and Technology*, *231*, 835–840.

Dawidowicz, A. L.,Wianowska, D., and Olszowy, M. (2012). On practical problems in estimation of antioxidant activity of compounds by DPPH method (problems in estimation of antioxidant activity). *Food Chemistry*, *131*, 1037–1043.

De Araujo, T. A., Barbosa, A. M. J., Viana, L. H. and Ferreira, V. S. (2011). Electroanalytical determination of TBHQ, a synthetic antioxidant, in soybean biodiesel samples. *Fuel*, 90.

de Camargo, A. C., Regitano-d'Arce, M. A. B., Biasoto, A. C. T. and Shahidi, F. (2014). Low molecular weight phenolics of grape juice and winemaking by-products: Antioxidant activities and inhibition of oxidation of human low-density lipoprotein cholesterol and DNA strand breakage. *Journal of Agricultural and Food Chemistry*, 62: 12159–12171.

de Camargo, A. C., Regitano-d'Arce, M. A. B., Gallo, C. R. and Shahidi, F. (2015). Gamma-irradiation induced changes in microbiological status, phenolic profile and antioxidant activity of peanut skin. *Journal of Functional Foods*, 12: 129–143.

De Jong, E. C., Van Zijverden, M., Spanhaak, S., Koppelman, S. J., Pellegrom, H. and Penninks, A. H. (1998). Identification and partial characterization of multiple major allergens in peanut proteins. *Clinical and Experimental Allergy*, 28: 743–751.

De la Presa-Owens, S., Lopez-Sabater, M. C., and Rivero-Urgell, M. (1995). Shelf-life prediction of an infant formula using an accelerated stability test (Rancimat). *Journal of Agricultural and Food Chemistry*, *43*, 2879–2882.

de la Rosa, L. A., Alvarez-Parrilla, E. and Shahidi, F. (2010). Phenolic compounds and antioxidant activity of kernels and shells of Mexican pecan (*Carya illinoinensis*). *Journal of Agricultural and Food Chemistry*, 59: 152–162.

De la Torre Boronat, M. C., and LoÂ pez Tamames, E. (1997). El papel de los antioxidantes. Alimentaria, 6, 19-27.

de las Heras, A., Schoch, A., Gibis, M., and Fischer, A. (2003). Comparison of methods for determining malondialdehyde in dry sausage by HPLC and the classic TBA test. *European Food Research and Technology, 217*, 180–184.

de Lima, A. C., Soares, D. J., da Silva, L. M., de Figueiredo, R. W., de Sousa, P. H., and de Abreu Menezes, E. (2014). *In vitro* bioaccessibility of copper, iron, zinc and antioxidant compounds of whole cashew apple juice and cashew apple fibre (*Anacardium occidentale* L.) following simulated gastrointestinal digestion. *Food Chemistry, 161*, 142–147.

de Lucas, A., Martinez de la Ossa, E., Rincón, J., Blanco, M. A. and Gracia, I. (2002). Supercritical fluid extraction of tocopherol concentrates from olive tree leaves. *The Journal of Supercritical Fluids*, 22: 221–228.

Debre, B., Conquy, S., Amsallem-Ouazanal, D., Bruel, P., Rahhali,N., and Charles, J. (2010). PIH35 impact of preventive treatment with titrated dry cranberry extract in a patient suffering from recurrent cystitis. *Value in Health, 13*, A186.

Decker, E. A.,Warner, K., Richards, M. P., and Shahidi, F. (2005). Measuring antioxidant effectiveness in Foods. *Journal of Agricultural and Food Chemistry, 53*, 4303–4310.

Deeds, F. and Couch, J. F. (1948). Rutin in green asparagus. *Food Res.,* 13: 378.

Dehkharghanian, M., Adenier, H. and Vijayalakshmi, M. A. (2010). Study of flavonoids in aqueous spinach extract using positive electrospray ionisation tandem quadrupole mass spectrometry. *Food Chemistry*, 121: 863–870.

del Baño, M. J., Lorente, J., Castillo, J., Benavente-Garcia, O., del Rio, J. A., Ortuno, A., Quirin, K.-W. and Gerard, D. (2003). Phenolic diterpenes, flavones, and rosmarinic acid distribution during the development of leaves, flowers, stems, and roots of Rosmarinus officinalis. Antioxidant activity. *Journal of Agricultural and Food Chemistry*, 51: 4247–4253.

Del Caro, A., Piga, A., Pinna, I., Fenu, P. M., and Agabbio, M. (2004). Effect of drying conditions and storage period on polyphenolic content, antioxidant capacity, and ascorbic acid of prunes. *Journal of Agricultural and Food Chemistry, 52*, 4780-4784.

Delgado-Pelayo, R., Gallardo-Guerrero, L., and Hornero-Méndez, D. (2014). Chlorophyll and carotenoid pigments in the peel and flesh of commercial apple fruit varieties. *Food Research International, 65*, 272–281.

Delmonte, P., Perry, J. and Rader, J. I. (2006). Determination of isoflavones in dietary supplements containing soy, Red Clover and kudzu: Extraction followed by basic or acid hydrolysis. *Journal of Chromatography. A*, 1107: 59–69.

Deng, G. F., Xu, X. R., Guo, Y. J., Xia, E. Q., Li, S., Wu, S., Chen, F., Ling, W. H. and Li, H. B. (2012). Determination of antioxidant property and their lipophilic and hydrophilic phenolic components in cereal grains. *Journal of Functional Foods*, 4(4), 906-914.

Deng, G.-F., Lin, X., Xu, X.-R., Gao, L.-L., Xie, J.-F. and Li, H.-B. (2013). Antioxidant capacities and total phenolic contents of 56 vegetables. *Journal of Functional Foods*, 5: 260–266.

Derdelinckx, G., (2008). Polyphenols in wort and beer: state of the art in 2008: where and why? Cerevisia Belgian Journal of Brewing and Biotechnology 33, 174-187.

Deschner, E. E., Ruperto, J., Wong, G. and Newmark, H. L. (1991). Quercetin and rutin as inhibitors of azoxymethanol-induced colonic neoplasia. *Carcinogenesis*, 12: 1193–1196.

Deshpande, S. S. (1985). Investigations on Dry Beans *(Phaseolus vulgaris* L.): Microstructure. Processing and Antinutrients. Ph.D. thesis. University of Illinois, Urbana-Champaign.

Deshpande, S. S. and Cheryan, M. (1985). Evaluation of vanillin assay for tannin analysis of dry beans. *J. Food Sci.* 50, 905.

Deshpande, S. S. and Cheryan, M. (1987). Determination of phenolic compounds of dry beans using vanillin redox and precipitation assays. *J. Food Sci.* 52, 332.

Deshpande, S. S. and Salunkhe, D. K. (1982). Interactions of tannic acid and catechin with legume starches, *J. Food Sci.,* 47: 2050.

Deshpande, S. S., Cheryan, M., and Salunkhe, D. K. (1986). Tannin analyses of food products. *CRC Crit. Rev. Food Sci. Nutr.* 24: 401.

Deshpande, S. S., Sathe, S. K., and Salunkhe, D. K. (1984). Chemistry and safety of plant polyphenols, In *Nutritional and Toxicological Aspects of Food Safety.* Friedman, M. Ed. Plenum Press, New York; Adv. *Exp. Med. Biol.* 177, 457.

Deshpande, S. S., Sathe, S. K., and Salunkhe, D. K. (1984). Chemistry and safety of plant polypenols. *Adv. Exp. Med. Bioi.,* 177: 457.

Deshpande, S. S., Sathe, S. K., Salunkhe, D. K., and Cornforth, D. P. (1982). Effects of dehulling on phytic acid, polyphenols and enzyme inhibitors of dry beans *(Phaseolus vulgaris* L.). *J. Food Sci.* 47, 1846.

Deshpande, S.S., Sathe, S.K., Salunkhe, D.K., and Cornforth, D.P. (1982). Effects of dehulling on phytic acid, polyphenols and enzyme inhibitors of dry beans (*Phaseolus vulgaris* L.). J. Food Sci. 47: 1846-1850.

Desobry, S. A., Netto, F. M., and Labuza, T. P. (1997). Comparison of spray-drying, drum-drying and freeze-drying for b-carotene encapsulation and preservation. Journal of Food Science, 62, 1158-1162.

Devaraj S, Hugou I, Jialal I; Hugou; Jialal (2001). "-Tocopherol decreases CD36 expression in human monocyte-derived macrophages". J Lipid Res 42 (4): 521–527.PMID 11290823.

Dewanto, V., Wu, X., and Liu, R. H. (2002). Processed sweet corn has higher antioxidant activity. *J. Agric. Food Chem.*, 50:4959-4964.

Diack, M., and Saska, M. (1994). Separation of vitamin E and γ-oryzanols from rice bran by normal-phase chromatography. *J. Am. Oil Chem. Soc.*, 71:1211-1217.

Diehl, B. W. L. (1998). Multinuclear high-resolution nuclear magnetic resonance spectroscopy. In R. J. Hamilton (Ed.), *Lipid analysis in oils and fats* (pp. 87–135). London: Blackie Academic and Professional.

Diemair, W. and Polaster, A. (1967). Uber Gerbsloffeim Rotwein. II. Eigenschaften und Isolierung, Z. *Lebensm. Unlers Forsch.* 134, 345.

Dietrych-Szostak, D. and Oleszek,W. (1999). Effect of processing on the flavonoid content in buckwheat (*Fagopyrum esculentum* Möench) grain. *Journal of Agricultural and Food Chemistry*, 47: 4384–4387.

Dikinson, D., Knight, M., and Rees, D. I. (1957). Varieties of broad beans suitable for canning, *Chern. Ind.,* 1503.

Dimberg, L. H., Theander, O., and Lingnert, H. (1993). Avenanthramidess - A group of phenolic antioxidants in oats. *Cereal Chem.*, 70:637-641.

Dini, I., Schettino, O. and Dini, A. (1998). Studies on the constituents of *Lupinus mutabilis* (Fabaceae). Isolation and characterization of two new isoflavonoid derivatives. *Journal of Agricultural and Food Chemistry*, 46: 5089–5092.

Dionisi, F., Prodolliet, J. and Tagliaferri, E. (1995). Assessment of olive oil adulteration by reversed-phase high-performance liquid chromatography/amperometric detection of tocopherols and tocotrienols. *Journal of the American Oil Chemists' Society*, 72: 1505–1511.

DiPersio, P. A., Kendall, P. A., and Sofos, J. N. (2004). Inactivation of *Listeria monocytogenes* during frying and storage of peach slices treated with acidic or sodium metabisulphite solutions. *Food Microbiology*, *21*, 641–648.

Dobarganes, M. C., and Velasco, J. (2002). Analysis of lipid hydroperoxides. *European Journal of Lipid Science and Technology*, *104*, 420–428.

Doggett, H. (1970). *Sorghum.* Longmans Green and Co., London.

Dolatabadi, J. E. N. and Kashanian, S. (2010). A review on DNA interaction with synthetic phenolic food additives. *Food Research International*, 43: 1223–1230.

Doleschall, F., Kemeny, Z., Recseg, K., and Kovari, K. (2002). A new analytical method to monitor lipid peroxidation during bleaching. *European Journal of Lipid Science and Technology*, *104*, 14–18.

Dominguez-Avila, J. A., Alvarez-Parrilla, E., Lopez-Diaz, J. A., Maldonado-Mendoza, I. E., Gomez-Garcia Mdel, C., and de la Rosa, L. A. (2015). The pecan nut (*Carya illinoinensis*) and its oil and polyphenolic fractions differentially modulate lipid metabolism and the antioxidant enzyme activities in rats fed high-fat diets. *Food Chemistry*, *168*, 529–537.

Donath, H. (1962). Discoloration phenomena in sterile Cauliflower packs. *Nahrang.* 6: 470.

Dong, J., Ma, K., van de Voort, F. R., and Ismail, A. A. (1997). Stoichiometric determination of hydroperoxides in oils by Fourier transform near infrared spectroscopy. *Journal of AOAC International*, *80*, 345–352.

Donner, H., Gao, L. and Mazza, G. (1997). Separation and characterization of simple and malonylated anthocyanins in red onions, *Allium cepa* L. *Food Research International*, 30: 637–643.

Donovan, J. L., Meyer, A. S., and Waterhouse, A. L. (1998). Phenolic composition and antioxidant activity of prunes and prune juice (*Prunus domestica*). *Journal of Agricultural and Food Chemistry*, *46*, 1247–1252.

Đor_evic, T. M., Šiler-Marinkovic, S. S. and Dimitrijevic-Brankovic, S. I. (2010). Effect of fermentation on antioxidant properties of some cereals and pseudo cereals. *Food Chemistry*, 119(3), 957-963.

Dordevic, T. M., Siler-Marinkovic, S. S. and Dimitrijevic-Brankovic, S. I. (2010). Effect of fermentation on antioxidant properties of some cereals and pseudo cereals. *Food Chemistry*, 119: 957–963.

dos Reis, L. C. R., de Oliveira, V. R., Hagen, M. E. K., Jablonski, A., Flôres, S. H. and de Oliveira Rios, A. (2015). Carotenoids, flavonoids, chlorophylls, phenolic compounds and antioxidant activity in fresh and cooked broccoli (*Brassica oleracea* var. Avenger) and cauliflower (*Brassica oleracea* var. Alphina F1). *LWT-Food Science and Technology*, 63: 177–183.

Dowd P, Zheng ZB; Zheng (1995). "On the mechanism of the anticlotting action of vitamin E quinone". Proc Natl Acad Sci U S A. 92 (18): 8171–8175.

Drawert, F., Leupold, G. and Lessing, V. (1977). Gaschromatographiische analyse von phenolischen verbindugen in bier. *Brauwissenschaft*, 30: 13–18.

Dreosti, I. E. (2000). Antioxidant polyphenols in tea, cocoa and wine. *Nutrition (Burbank, Los Angeles County, Calif.)*, 16: 7–8.

Dreyer, D. L., Reese, J. C., and Jones, K. C. (1981). Aphid feeding deterrents in sorghum: bioassay, isolation and characterization, J. *Chem. Ecol.*, 7, 273.

Du, C. T., Wang, P. L. and Francis, F. J. (1975). Anthocyanins of pomegranate, Punica granatum. *Journal of Food Science*, 40: 417–418.

Du, Q., Jerz, G. and Winterhalter, P. (2004). Isolation of two anthocyanin sambubiosides from bilberry (*Vaccinium myrtillus*) by high-speed counter-current chromatography. *Journal of Chromatography. A*, 1045: 59–63.

Duenas, M., Hernandez, T. and Estrella, I. (2002). Phenolic composition of the cotyledon and the seed coat of lentils (*Lens culinaris* L.). *European Food Research and Technology*, 215: 478–483.

Duenas, M., Sun, B., Hernandez, T., Estrella, I. and Spranger, M. I. (2003). Proanthocyanidin composition in the seed coat of lentils (*Lens culinaris* L.). *Journal of Agricultural and Food Chemistry*, 51: 7999–8004.

Dugo, P., Mondello, L., Errante, G., Zappia, G. and Dugo, G. (2001). Identification of anthocyanins in berries by narrow-bore highperformance liquid chromatography with electrospray ionization detection. *Journal of Agricultural and Food Chemistry*, 49: 3987–3992.

Duh, P. D., Yeh, D. B. and Yen, G. C. (1992). Extraction and identification of an antioxidative component of peanut hulls. *Journal of the American Oil Chemists' Society*, 69: 814–818.

Duh, P.-D. (1998). Antioxidant activity of Burdock (Arctium lappa LinneÂ), its scavenging effect on free-radical and active oxygen. Journal of the American Oil Chemists' Society, 75, 455-461.

Duh, P.-D. (1999). Antioxidant activity of water extract of four Harng Jyur (Chrysanthemum morifolium Ramat) varieties in soybean oil emulsion. Food Chemistry, 66, 471-476.

Duh, P.-D., and Yen, G.-C. (1995). Changes in antioxidant activity and components of methanolic extracts of peanut hulls irradiated with ultraviolet light. Food Chemistry, 54, 27-131.

Duh, P.-D., and Yen, G.-C. (1997a). Antioxidant e cacy of methanolic extracts of peanut hulls in soybean and peanut oils. Journal of the American Oil Chemists' Society, 74, 745-748.

Duh, P.-D., and Yen, G.-C. (1997b). Antioxidative activity of three herbal water extracts. Food Chemistry, 60, 639-645.

Duh, P.-D., Yen, W. J., Du, P.-C., and Yen, G.-C. (1997). Antioxidant activity of mung bean hulls. Journal of the American Oil Chemists' Society, 74, 1059-1063.

DuPont, M. S., Mondin, Z.,Williamson, G. and Price, K. R. (2000). Effect of variety, processing, and storage on the flavonoids glycoside content and composition of lettuce and endive. *Journal of Agricultural and Food Chemistry*, 48: 3957–3964.

Durge, A.V., Sarkar, S., and Singhal, R.S. (2013). Stability of anthocyanins as pre-extrusion coloring of rice extrudates. *Food Research International*. 50: 641-646.

Durkee, A. B. and Poapst, P. A. (1965). Phenolic constituents in core tissues and ripe seeds of Mcintosh apples, J. *Agric. Food Chern,* 13: 137.

Durkee, A. B. and Poapst, P. A. (1965). Ripe seed of McIntosh apples. *Journal of Agricultural and Food Chemistry*, 13: 137–139.

Durmaz, G. (2012). Freeze-dried ABTS+ method: A ready-to-use radical powder to assess antioxidant capacity of vegetable oils. *Food Chemistry*, *133*, 1658–1663.

Duthie GG, Gardner PT, Kyle A. M. (2003) Plants polyphenols: are they the new magic bullet? Proc Nutr Sci 62:599–603.

Duthie, D. W. (1928). The determination of tannins in cacao kernel, *Analyst.* 63, 27.

Duthie, S. J., Ma, A., Ross, M. A. and Collins, A. R. (1996). Antioxidant supplementation decreases oxidative DNA damage in human lymphocytes. *Cancer Research*, 56(6), 1291-1295.

Dykes, L., and Rooney, L.W. (2006). Sorghum and millet phenols and antioxidants. *J. Cereal Sci.* 44: 236-243.

Dykes, L., and Rooney, L.W. (2007). Phenolic compounds in cereal grains and their health benefits. *Cereal Food World,* 52:105-111.

Dykes, L., Rooney, L. W., Waniska, R. D., and Rooney, W. L. (2005). Phenolic compounds and antioxidant activity of sorghum grains of varying genotypes. *J. Agric. Food Chem.,* 53:6813-6818.

Dykes, L. and Rooney, L. W. (2006). Sorghum and millet phenols and antioxidants. *Journal of Cereal Science*, 44: 236–251.

Dyrby, M., Westergaard, N. and Stapelfeldt, H. (2001). Light and heat sensitivity of red cabbage extract in soft drink model systems. *Food Chemistry*, 72: 431–437.

Dziezak, J. D. (1986). Preservatives: Antioxidants. *Food Technology*, 40: 94–102.

Earp, C. G., Akingbala, J. O., Ring, S. H., and Rooney, L. W. (1981). Evaluation of several methods to determine tannins in sorghums with varying kernel characteristics. *Cereal Chem.* 58. 234.

Edge, M. S., Jones, J. M., and Marquart, L. (2005). A new life for whole grains. *J. Am. Diet. Assoc.,* 105:1856-1860.

El-Ansari, M. A., El-Negoumy, S. L. and El-Desoky, S. K. (1996). Flavonoid constituents of *Origanum syriacum* L. (Labiatae). Proceedings of Polyphenols Communications (p. 85). Bordeaux.

Eler, G. J., Peralta, R. M. and Bracht, A. (2009). The action of n-propyl gallate on gluconeogenesis and oxygen uptake in the rat liver. *Chemico-Biological Interactions*, 181: 390–399.

Eliasson, C., Kamal-Eldin, A., Andersson, R. and Aman, P. (2003). High-performance liquid chromatographic analysis of secoisolariciresinol diglucoside and hydroxycinnamic acid glucosides in flaxseed by alkaline extraction. *Journal of Chromatography. A*, 1012: 151–159.

Elisia, I., Hu, C., Popovich, D. G. and Kitts, D. D. (2010). Antioxidant assessment of an anthocyanin-enriched blackberry extract. *Food Chemistry*, 101: 1052–1058.

El-Magoli, S. B., Karel, M., and Yong, S. (1980). Acceleration of lipid oxidation by volatile products of hydroperoxide decomposition. *Journal of Food Biochemistry*, *3*, 111–124.

Embuscado, M. E. (2015). Herbs and spices as antioxidants for food preservation. In F. Shahidi (Ed.), *Handbook of antioxidants for food preservation* (pp. 251–283). Cambridge, UK: Woodhead Publishing Ltd.

Emmons, C. L., Peterson, D. M., and Paul, G. L. (1999). Antioxidant capacity of oat (*Avena sativa* L.) extracts. 2. In vitro antioxidant activity and contents of phenolic and tocolantioxidants. *J. Agric. Food Chem., 47*, 4894-4898.

Emmons, C. L., Peterson, D. M. and Paul, G. L. (1999). Antioxidant capacity of oat (*Avena sativa* L.) extracts. 2. In vitro antioxidant activity and contents of phenolic and tocol antioxidants. *Journal of Agricultural and Food Chemistry*, 47: 4894–4898.

Engelmann, M. D., Hutcheson, R. and Cheng, I. F. (2005). Stability of ferric complexes with 3-hydroxyflavone (flavonol), 5,7-dihydroxyflavone (Chrysin), and 3',4'-dixydroxyflavone. *Journal of Agricultural and Food Chemistry*, 53: 2953–2960.

Enomoto, S., Yanaoka, K., Utsunomiya, H., Niwa, T., Inada, K., Deguchi, H., Ueda, K., Mukoubayashi, C., Inoue, I., Maekita, T., Nakazawa, K., Iguchi, M., Arii, K., Tamai, H., Yoshimura, N., Fujishiro, M., Oka, M., and Ichinose, M. (2010). Inhibitory effects of Japanese apricot (*Prunus mume* Siebold et Zucc.; Ume) on *Helicobacter pylori*-related chronic gastritis. *European Journal of Clinical Nutrition, 64*, 714-719.

Erdog˘an, S., and Erdemog˘ lu, S. (2011). Evaluation of polyphenol contents in differently processed apricots using accelerated solvent extraction followed by high-performance liquid chromatography-diode array detector. *International Journal of Food Sciences and Nutrition, 62*, 729–739.

Erkan, N., Ayranci, G. and Ayranci, E. (2008). Antioxidant activities of rosemary (*Rosmarinus officinalis* L.) extract blackseed (*Nigella sativa* L.) essential oil, carnosic acid, rosmarinic acid and sesamol. *Food Chemistry*, 110: 76–82.

Escarpa, A., and González, M. C. (2001). An overview of analytical chemistry of phenolic compounds in foods. *Critical Reviews in Analytical Chemistry, 31*, 57–139.

Escriche, I., Kadar, M., Juan-Borrás, M. and Domenech, E. (2014). Suitability of antioxidant capacity, flavonoids and phenolic acids for floral authentication of honey. Impact of industrial thermal treatment. *Food Chemistry*, 142: 135–143.

Esfahani, A., Lam, J., and Kendall, C. W. (2014). Acute effects of raisin consumption on glucose and insulin responses in healthy individuals. *Journal of Nutritional Science, 3*, 1–6.

Esfahlan, A. J., Jamei, R. and Esfahlan, R. J. (2010). The importance of almond (*Prunus amygdalus* L.) and its by-products. *Food Chemistry*, 120: 349–360.

Eskin, N. A. M., Hoehn, E., and Frenkel, C. (1978). A simple and rapid quantitative method for total phenols, J. *Agric. Food Chem.* 26, 973.

Estrada-MunÄ oz, R., Boyle, E. A. E., and Mardsen, J. L. (1998). Liquid smoke effects on Escherichia coli O157, H7, and its antioxidant properties in beef products. Journal of Food Science, 63, 150-153.

Evershed, R. P., Spooner, N., Prescott, M. C., and Goad, L. J. (1988). Isolation and characterization of intact steryl ferulates from seeds. *J. Chromatogr.*, 440:23-25.

Ewald, C., Fjelkner-Modig, S., Johansson, K., SjoÈ holm, I., and A Ê kesson, B. (1999). Effect of processing on major favonoids in processed onions, green beans, and peas. Food Chemistry, 64, 231-235.

Eymard, S., and Genot, C. (2003). A modified xylenol orange method to evaluate formation of lipid hydroperoxides during storage and processing of small pelagic fish. *European Journal of Lipid Science and Technology, 105*, 497–501.

Fajardo, J. E., Waniska, R. D., Cuero, R. G. and Pettit, R. E. (1995). Phenolic compounds in peanut seeds: Enhanced elicitation by chitosan and effects on growth and aflatoxin B1 production by *Aspergillus flavus*. *Food Biotechnology*, 9: 59–78.

Famuyiwa, O. O., El-Hazmi, M. A. F., Al-Jasser, S. J., Sulimani, R. A., Jayakumar, R. V., Al-Nuaim, A. A., and Mekki, M. O. (1992). A comparison of acute glycaemic and insulin response to dates (*Phoenix dactylifera*) and oral dextrose in diabetic and nondiabetic subjects. *Saudi Medical Journal, 13*, 397–402.

Fang, N. B., Yu, S. G., and Prior, R. L. (2002). LC/MS/MS characterization of phenolic constituents in dried plums. *Journal of Agricultural and Food Chemistry, 50*, 3579–3585.

Fantozzi, P., Montanari, L., Mancini, F., Gasbarrini, A., Addolorato, G., Simoncini, M., Cardini, M., Ghiselli, A. and Scaccini, C., (1998). In vitro antioxidant capacity from wort to beer. Lebensmittel Wissenschaft Und Technologie 31, 221-227.

FAO (1992). *Maize in human nutrition. FAO Food and Nutrition Sereies, No. 25*. Rome (Italy).

Faller, A. L. K. and Fialho, E. (2009). The antioxidant capacity and polyphenol content of organic and conventional retail vegetables after domestic cooking. *Food Res. Int.* 42: 210-215.

Fanasca, S., Rouphael, Y., Venneria, E., Azzini, E., Durazzo, A. and Maiani, G. (2009). Antioxidant properties of raw and cooked spears of green asparagus cultivars. *Int. J. Food Sci. Technol.* 44: 1017-1023.

Farahpour, M. R., Mirzakhani, N., Doostmohammadi, J., and Ebrahimzadeh, M. (2015). Hydroethanolic *Pistacia atlantica* hulls extract improved wound healing process: Evidence for mast cells infiltration, angiogenesis and RNA stability. *International Journal of Surgery*, *17*, 88–98.

Farajian, P., Katsagani, M., and Zampelas, A. (2010). Short-term effects of a snack including dried prunes on energy intake and satiety in normal-weight individuals. *Eating Behaviors*, *11*,201–203.

Fardet, A. (2010). New hypotheses for the health-protective mechanisms of whole-grain cereals: what is beyond fibre? *Nutrition Research Reviews*, 23 (1), 65-134.

Fardet, A., Canlet, C., Gottardi, G., Lyan, B., Llorach, R., Rémésy, C., Mazur, A., Paris, A. and Scalbert, A. (2007). Whole grain and refined wheat flours show distinct metabolic profiles in rats as assessed by a 1H NMR-based metabonomic approach. *Journal of Nutrition*, 137(4), 923-929.

Fardet, A., Edmond, R. and Remesy, C. (2008). Is in vitro antioxidant potential of whole-grain cereals and cereal products well reflected in vivo? *Journal of Cereal Science*, 48(2), 258-276.

Farombi, E. O., and Britton, G. (1999). Antioxidant activity of palm oil carotenes in organic solution, effects of structure and chemical reactivity. Food Chemistry, 64, 315-321.

Farvin, K. H. S., Andersen, L. L., Nielsen, H. H., Jacobsen, C., Jakobsen, G., Johansson, I., and Jessen, F. (2014). Antioxidant activity of Cod (Gadus morhua) protein hydrolysates: In vitro assays and evaluation in 5% fish oil-in-water emulsion. *Food Chemistry*, *149*, 326–334.

Fassoula, E., Economou, A., and Calokerinos, A. (2011). Development and validation of a sequential-injection method with chemiluminescence detection for the high throughput assay of the total antioxidant capacity of wines. *Talanta*, *85*, 1412–1418.

Fausch, H, Kunding, W. and Keukom, H. (1963). Ferulic acid as a component of a glycoprotein from wheat flour, *Nature,* 199: 287.

Faustino, R. S., Sobrattee, S., Edel, A. L. and Pierce, G. N. (2003). Comparative analysis of the phenolic content of selected Chilean, Canadian and American merlot red wines. *Molecular and Cellular Biochemistry*, 249: 11–19.

Faustman, C., Liebler, D. C., McClure, T. D. and Sun, Q. (1999). α,β-Unsaturated aldehydes accelerate oxymyoglobin oxidation. *Journal of Agricultural and Food Chemistry*, 47: 3140–3144.

Feen y, P. P. and Bostock, H. (1968). Seasonal changes in the tannin content of oak leaves, *Phytochemistry,* 7. 871.

Fenwick, G.R., Curl, C.L., Pearson, A.W., and Butler, A.W. (1984). The treatment of rapeseed meal and its effect on the chemical composition and egg tainting potential. J. Sci. Food Agric. 35: 757-761.

FernaÂ ndez de SimoÂ n, B., CadahõÂa, E., Conde, E., and GarcõÂa-Vallejo, M. C. (1996). Low molecular weight phenolic compounds in Spanish oakwoods. Journal of Agricultural and Food Chemistry, 44, 1507-1511.

Fernandez, M. S., Kapran, I., Souley, S., Abdou, M., Maiga, I. H., Acharya, C. B., Hambilin, M. T. and Kresovich, S. (2009). Collection and characterization of yellow endosperm sorghums fromWest Africa for biofortification. *Genetic Resources and Crop Evolution*, 56: 991–1000.

Fernández-León, M. F., Fernández-León, A. M., Lozano, M., Ayuso, M. C. and González-Gómez, D. (2012). Identification, quantification and comparison of the principal bioactive compounds and external quality parameters of two broccoli cultivars. *Journal of Functional Foods*, 4: 465–473.

Ferreira, D., Guyot, S., Marnet, N., Delgadillo, I., Renard, C. M., and Coimbra, M. A. (2002). Composition of phenolic compounds in a Portuguese pear (*Pyrus communis* L. var. S. Bartolomeu) and changes after sun-drying. *Journal of Agricultural and Food Chemistry*, *50*, 4537–4544.

Ferreira, D., Kamara, B. I., Brandt, E. V., and Joubert, E. (1998). Phenolic compounds from Cyclopia intermedia (Honeybush tea)1. Journal of Agricultural and Food Chemistry, 46, 3406-3410.

Ferreira, D. and Li, X.-C. (2000). Oligomeric proanthocyanidins: Naturally occurring O-heterocycles. *Natural Products Report*, 17: 193–212.

Ferreres, F., Castaner, M. and Tomas-Barbern, A. (1997). Acylated flavonol glycosides from spinach leave (*Spinacia oleraceas*). *Phytochemistry*, 45: 1701–1705.

Ferreres, F., Gil, M. I., Castaner, M. and Tomas-Barberan, F. A. (1997). Phenolic metabolites in red pigmented lettuce (*Lactuca sativa*). Changes with minimal processing and cold storage. *Journal of Agricultural and Food Chemistry*, 45: 4249–4254.

Ferruzzi, M. G. (2010). The influence of beverage composition on delivery of phenolic compounds from coffee and tea. *Physiology and Behavior*, 100: 33–41.

Ferudenberg, K. and Weinges, K. (1962). Catechins and flavonoid tannins. In: *The Chemistry of Flavonoid Compounds*. T. A. Geissman (Ed.), p. 197-211. Pergamon Press, Oxford.

Feskanich, D., Ziegler, R. G., Michaud, D. S., Giovannucci, E. L., Speizer, F. E.,Willett,W. C. and Coldits, G. A. (2000). Prospective study of fruit and vegetable consumption and risk of lung cancer among men and women. *Journal of the National Cancer Institute*, 92: 1812–823.

Fieser, L. F. and Fieser, M. (1956). Organic Chemistry. Reinhold, New York.

Fischer, U. A., Carle, R. and Kammerer, D. R. (2011). Identification and quantification of phenolic compounds from pomegranate (*Punica granatum* L.) peel, mesocarp, aril and differently produced juices by HPLC–DAD–ESI/MSn. *Food Chemistry*, 127: 807–821.

Flament, I. (1989). Coffee, cocoa and tea. *Food Reviews International*, 5: 317–414.

Flament, I. (1995). Coffee, cocoa and tea. In H. Maarse (Ed.), *Volatile compounds in foods and beverages* (pp. 617–669). New York: Marcel Dekker.

Flesch, M., Schwarz, A. and Bohm, M. (1998). Effects of red and white wine on endothelium-dependent vasorelaxation of rat aorta and human coronary arteries. *American Journal of Physiology, Heart and Circulatory Physiology*, 275: H1183–H1190.

Fletcher, A. C. and Porter, L. J. (1977). Plant proanthocyanidins. III. Conformational and configurational studies of natural procyanidins. *J. Chem. Sac. Perkin Trans. J.* 14: 1628.

Fleuriet, A. and Macheix, J. J. (1981). Quinyl esters and glucose derivatives of hydroxycinnamic acids during growth and ripening of tomato fruit. *Phytochemistry*, 20: 667–671.

Floyd, R. A., and Hensley, K. (2002). Oxidative stress in brain aging. Implications for therapeutics of neurodegenerative diseases. *Neurobiology of Aging*, *23*, 795–807.

Fogliano, V., Verde, V., Randazzo, G., and Ritiene, A. (1999). Method for measuring antioxidant activity and its application to monitoring the antioxidant capacity of wines. Journal of Agricultural and Food Chemistry, 47, 1035-1040.

Folin, O. and Ciocalteau, V. (1927). On tyrosine and tryptophan determination in proteins. *J. Biol. Chem., 73.* 627-650.

Folin, O. and Denis, W. (1912). A new colorimetric method of the determination of vanillin in vanilla extracts. *J. Ind. Eng. Chem.* 4, 680.

Folin, O. and Denis, W. (1912). On phosphotungstic-phosphomolybdic compounds as color reagents. *J. Biol .Chem.* 12, 239.

Folin, O. and Denis, W. (1912). Tyrosine in proteins determined by a new colorimetric method. J. *Biol. Chem.* 12, 245.

Foo, L. Y. and Porter, L. J. (1981). The structure of tannins of some edible fruits. *Journal of the Science of Food and Agriculture*, 32: 711–716.

Food and Agricullural Organizalion, (1985). *Produclion Yearbook, Food and Agricullural Organizalion. Rome, Italy.*

Forsyth, W. G. C. and Quesnel, V. C. (1957a). Caeao-polyphenolic substances. IV. The anthocyanin pigments. *Biochem. J.* 65, 177.

Forsyth, W. G. C. and Quesnel, V. C. (1957b). Intermediates in the enzymic oxidation of catechol. *Bioche11l. Biophys. Acta,* 65, 177.

Forsyth, W. G. C., Quesnel, V. C., and Roberts, J. B. (1960). Diphenylenediomide-2, 3-quinone: an intermediate in the enzymic oxidation of calechol, *Biochem. Biophys. Acta,* 37. 322.

Fossen, T., Andersen, O. M., Ovstedal, D. O., Pedersen, A. T. and Ranknes, A. (1996). Characteristic anthocyanin pattern from onions and other Allium spp. *Journal of Food Science*, 61: 703–706.

Foti, M., Piattelli, M., Baratta, M. T., and Ruberto, G. (1996). Flavonoids, coumarins, and cinnamic acids as antioxidants in a micellar system structure-activity relationship. Journal of Agricultural and Food Chemistry, 44, 497-501.

Fourie, P. C. and Basson, D. S. (1989). Changes in the tocopherol content of almond, pecan and macadamia kernels during storage. *Journal of the American Oil Chemists' Society*, 66: 1113–1115.

Fraga, C. G., Motchnik, P. A., Shigenaga, M. K., Helbock, H. J., Jacob, R. A. and Ames, B. N. (1991). Ascorbic acid protects against endogenous oxidative DNA damage in human sperm. *Proceedings of the National Academy of Science of the United States of America,* 88(24), 11003-11006.

Francis, F. J., Harborne, J. B. and Barker, W. G. (1966). Anthocyanins in the lowbush blueberry, *Vaccinium angustifolium*. *Journal of Food Science*, 31: 583–587.

Franke, A. A., Custer, L. J., Arakaki, C., and Murphy, S. P. (2004). Vitamin C and flavonoid levels of fruits and vegetables consumed in Hawaii. *Journal of Food Composition and Analysis*, *17*, 1–35.

Franke, A. A., Halm, B. M., Kakazu, K. and Li, X. (2010). Metabolism, bioavailability, and analysis of dietary isoflavones. In C. G. Fraga (Ed.), *Plant phenolics and human health* (pp. 165–183). Hoboken, NJ: JohnWiley and Sons, Inc.

Frankel, E. N. (1996). Antioxidants in lipid foods and their impact on food quality. Food Chemistry, 57, 51-55.

Frankel, E. N., Huang, S.-W., and Aeschbach, R. (1997). Antioxidant activity of green teas in different lipid systems. Journal of the American Oil Chemists' Society, 74, 1309-1315.

Frankel, E. N., Kanner, J., German, J. B., Parks, E. and Kinsella, J. E. (1993). Inhibition of oxidation of human low-density lipoprotein by phenolic substances in red wine. *Lancet*, 341: 454–457.

Frankel, E. N. and Finley, J. W. (2008). How to standardize the multiplicity of methods to evaluate natural antioxidants. *Journal of Agricultural and Food Chemistry*, 56: 4901–4908.

Franklin, M., Bu, S. Y., Lerner, M. R., Lancaster, E. A., Bellmer, D., Marlow, D., Lightfoot, S. A., Arjmandi, B. H., Brackett, D. J., Lucas, E. A., and Smith, B. J. (2006). Dried plum prevents bone loss in a male osteoporosis model *via* IGF-I and the RANK pathway. *Bone*, *39*, 1331–1342.

Freeman, J. D. and Niemeyer, E. D. (2008). Quantification of tea flavonoids by high performance liquid chromatography. *Journal of Chemical Education*, 85: 951–953.

Freitas, V. A. P., Glories, Y., Bourgeois, G. and Vitry, C. (1998). Characterisation of oligomeric and polymeric procyanidins from grape seeds by liquid secondary ion mass spectrometry. *Phytochemistry*, 49: 1435–1441.

Frejnagel, S., Zdunczyk, Z., Amarowicz, R., and Juskiewicz, J. (1994). Effect of faba bean phenolic compounds on nutrients absorption in the rat gut. In: *Bioactive substances in food of plant origin,* Vol. 1. H. Kozlowska, J. Fornal and Z. Zdunczyk (Eds.), p. 192-197. Centre for Agrotechnology and Veterinary Sciences, Polish Academy of Sciences, Olsztyn, Poland.

Freudenberg, K. (1920). Die Chemic der Nawrlicherl Gerbsloffe, Springer-Verlag, Berlin.

Freudenberg, K. and Weinges, K. (1962). Acid catalyzed autocondensation of hydroxyOavans: condensed proan-thoeyanidins, *Tetrahedroll Lett.* 1073.

Freudenberg, K. and Weinges, K. (1962). The Chemistry of Flavonoid Compounds. Geissman. T. A., Ed., Pergamon Press. Oxford.

Friedman, M. (1997). Chemistry, biochemistry and dietary role of potato polyphenols. A review. *Journal of Agricultural and Food Chemistry*, 45: 1523–1540.

Friedman, M. (1997). Chemistry, biochemistry, and dietary role of potato polyphenols. Journal of Agricultural and Food Chemistry, 45, 1523-1540.

Friedrich, W. and Galensa, R., (2002). Identification of a new flavanol glucoside from barley (Hordeum vulgare L.) and malt. European Food Research and Technology 214, 388-393.

Friend, J., (1981. Plant phenolics, lignifications and plant diseases. *Prog. Phylochon.* 7: 197.

Fuentes-Alventosa, J. M., Jaramillo, S., Rodriguez-Gutierrez, G., Cermeo, P., Espejo, J. A., Jimenez-Araujo, A., Guillen-Bejarano, R., Fernandez-Bolanos, J. and Rodriguez-Arcos, R. (2008). Flavonoid profile of green asparagus genotypes. *Journal of Agricultural and Food Chemistry*, 56: 6977–6984.

Fuhrman, B., Volkova, N., Suraski, A. and Aviram, M. (2001). White wine with red wine-like properties: Increased extraction of grape skin polyphenols improves the antioxidant capacity of the derived white wine. *Journal of Agricultural and Food Chemistry*, 49: 3164–3168.

Fukuda, T., Ito, H., and Yoshida, T. (2004). Effect of the walnut polyphenol fraction on oxidative stress in type 2 diabetes mice. *Biofactors (Oxford, England)*, *21*, 251–253.

Fukui, M., Choi, H. J. and Zhu, B. T. (2010). Mechanism for the protective effect of resveratrol against oxidative stressinduced neuronal death. *Free Radical Biology and Medicine*, 49: 800–813.

Fung, D. Y. C., and Thompson, L. K. (2009). "Natural" suppression of the growth of foodborne pathogens in meat products. *International Review of Food Science and Technology*, *2*, 80–81.

Furuta, S., Nishiba, Y., and Suda, I. (1997). Fluorometric assay for screening antioxidative activity of vegetables. Journal of Food Science, 62, 526-528.

Furuta, S., Suda, I., Nishiba, Y. and Yamakawa, O. (1998). High tertbutylperoxyl radical scavenging activities of sweet potato cultivars with purple flesh. *Food Science and Technology-International Tokyo*, 4: 33–35.

Gabrielska, J., Oszmianski, J., and Lamer-Zarawska, E. (1997). Protective effect of plant favonoids on the oxidation of lecithin liposomes. Pharmazie, 52, 2-3.

Galanakis, C. M. (2012). Recovery of high added-value components from food wastes: Conventional, emerging technologies and commercialized applications. *Trends in Food Science and Technology*, *26*, 68–87.

Galensa, R. and Hermann, K. (1979). Flavone glycosides of leaves and tubers of celery (*Apium graveolens* L.). 9. Flavonols and flavones of vegetables. *Zeitschrift für Lebensmitteluntersuchung und - Forschung A*, 169: 170–172.

Gallaher, C. M., and Gallaher, D. D. (2009). Dried plums (prunes) reduce atherosclerosis lesion area in apolipoprotein E-deficient mice. *British Journal of Nutrition*, *101*, 233–239.

Gambelli, L. and Santaroni, G. P. (2004). Polyphenols content in some Italian red wines of different geographical origins. *Journal of Food Composition and Analysis*, 17: 613–618.

Gamel, T. H. and Abdel-Aal, E-S. M. (2012). Phenolic acids and antioxidant properties of barley wholegrain and pearling fractions. *Agricultural and Food Science*, 21(2), 118-131.

Ganthavorn, C., and Hughes, J. S. (1997). Inhibition of soybean oil oxidation by extracts of dry beans (*Phaseolus vulgaris*). Journal of the American Oil Chemists' Society, 74, 1025-1030.

Gao, L. and Mazza, G. (1995). Characterization, quantitation and distribution of anthocyanins and colorless phenolics in sweet cherries. *Journal of Agricultural and Food Chemistry*, 43: 343–346.

García, Y. D., Valles, B. S. and Lobo, A. P. (2009). Phenolic and antioxidant composition of by-products from the cider industry: Apple pomace. *Food Chemistry*, 117: 731–738.

García-Alonso, M., Pascual-Teresa, S., Santos-Buelga, C. and Rivas-Gonzalo, J. C. (2004). Evaluation of the antioxidant properties of fruits. *Food Chemistry*, 84: 13–18.

Garcia-Aloy, M., Llorach, R., Urpi-Sarda, M., Tulipani, S., Estruch, R., Martinez-Gonzalez, M. A., Corella, D., Fito, M., Ros, E., Salas-Salvado, J., and Andres-Lacueva, C. (2014). Novel multimetabolite prediction of walnut consumption by a urinary biomarker model in a free-living population: The PREDIMED study. *Journal of Proteome Research*, *13*, 3476–3483.

Garcia-Conesa, M.T., Puumb, G.W., Kroon, P.A.,Walace, G., and Williamson, G. (1997). Antioxidant properties of ferulic acid dimmers. Redox Report, 3, 239–244.

GarcõÂa, E., Filisetti, T. M. C. C., Udaeta, J. E. M., and Lajolo, F. (1998). Hard-to-cook beans (*Phaseolus vulgaris*), involvement of phenolic compounds and pectates. Journal of Agricultural and Food Chemistry, 46, 2110-2116.

GarcõÂa-Conesa, M. T., Wilson, P. D., Plumb, G. W., Ralph, J., and Williamson, G. (1999). Antioxidant properties of 4,40-dihydroxy-3,30-dimethoxy-beta, beta0-bicinnamic acid (8-8-diferulic acid, noncyclic form). Journal of the Science of Food and Agriculture, 79, 379-384.

Gardner, P. T., McPhail, D. B., and Duthie, G. G. (1998). Electron spin resonance spectroscopic assessment of the antioxidant potential of teas in aqueous and organic media. *Journal of the Science of Food and Agriculture*, *76*, 257–262.

Garg, S. K., Gupt, S. R. and Sharma, N. D. (1980). Coumarins from *Apium graveolens* seed. *Phytochemistry*, 18: 1580–1581.

Garrido, I., Monagas, M., Gómez-Cordovés, C., and Bartolomé, B. (2008). Polyphenols and antioxidant properties of almond skins: Influence of industrial processing. *Journal of Food Science*, *73*, C106–C115.

Garrido, I., Urpi-Sarda, M., Monagas, M., Gomez-Cordoves, C., Martin-Alvarez, P. J., Llorach, R., Bartolome, B., and Andres-Lacueva, C. (2010). Targeted analysis of conjugated and microbial-derived phenolic metabolites in human urine after consumption of an almond skin phenolic extract. *The Journal of Nutrition*, *140*, 1799–1807.

Gasowski, B., Leontowicz, M., Leontowicz, H., Katrich, E., Lojek, A., Ciz, M., Trakhtenberg, S. and Gorinstein, S. (2004). The influence of beer with different antioxidant potential on plasma lipids, plasma antioxidant capacity, and bile excretion of rats fed cholesterol-containing and cholesterol-free diets. *The Journal of Nutritional Biochemistry*, 15: 527–533.

Gayathri, G.N., Platel, K., Prakash, J. and Srinivasan, K. (2004). Influence of antioxidant spices on the retention of b-carotene in vegetables during domestic cooking processes. *Food Chem*. 84:35–43.

Gazzani, G., Papetti, A., Massolini, G., and Daglia, M. (1998). Antiand prooxidant activity of water soluble components of some common diet vegetables and the effect of thermal treatment. Journal of Agricultural and Food Chemistry, 46, 4118-4122.

Gebicki, J. M., and Guille, J. (1989). Spectrophotometric and high performance chromatographic assays of hydroperoxides by the iodometric technique. *Analytical Biochemistry*, *176*, 360–364.

Geissman, T. A. (1955). Modern Methods of Plant Analysis. Vol. 3. Peach, K. and Tracey, V. Eds. Springer Verlag. Berlin.

Gelmez, N., K□ncal, N. S. and Yener, M. E. (2009). Optimization of supercritical carbon dioxide extraction of antioxidants from roasted wheat germ based on yield, total phenolic and tocopherol contents, and antioxidant activities of the extracts. *The Journal of Supercritical Fluids*, 48: 217–224.

Gennaro, L., Leonardi, C., Esposito, F., Salucci, M., Maiani, G., Quaglia, G. and Fogliano, V. (2002). Flavonoid and carbohydrate contents in Tropea red onions: Effects of homelike peeling and storage. *Journal of Agricultural and Food Chemistry*, 50: 1904–1910.

Genot, C., Kansci, G., and Laroche, M. (1994). Measurement of phospholipid oxidation in model membranes by determination of oxygen consumption with a semi-automic polarographic method. *Sciences Des Aliments*, *14*, 673–682.

Gentile, C., Tesoriere, L., Butera, D., Fazzari, M., Monastero, M., Allegra, M. and Livrea, M. A. (2007). Antioxidant activity of sicilian pistachio (*Pistacia vera* L. var. *Bronte*) nut extract and its bioactive components. *Journal of Agricultural and Food Chemistry*, 55: 643–648.

George, B., Kaur, C., Khurdiya, D. S. and Kapoor, H. C. (2004). Antioxidants in tomato (*Lycopersicum esculentum*) as a function of genotype. *Food Chemistry*, 84: 45–51.

George, J., and Kuttan, R. (1997). Mutagenic, carcinogenic and cocarcinogenic activity of cashew nut shell liquid. *Cancer Letters*, *112*, 11–16.

Gester, H. (1993). Anticarcinogenic effect of common carotenoids. *International Journal of Vitamin Nutrition Research*, 63: 93–121.

Ghiselli, A., Nardini, M., Baldi, A., and Scaccini, C. (1998). Antioxidant activity of different phenolic fractions separated from Italian red wine. Journal of Agricultural and Food Chemistry, 46, 361-367.

Ghiselli, A., Natella, F., Guidi, A., Montanari, L., Fantozzi, P. and Scaccini, C. (2000). Beer increases plasma antioxidant capacity in humans. *The Journal of Nutritional Biochemistry*, 11: 76–80.

Ghosh, D. and Konishi, T. (2007). Anthocyanins and anthocyanin-rich extracts: role in diabetes and eye function. *Asia Pacific Journal of Clinical Nutrition*. 16(2): 200-208.

Giao, M. S., Leitao, I., Pereira, A., Borges, A. B., Guedes, C. J., Fernandes, J. C., Belo, L., Santos-Silva, A., Hogg, T. A., Pintado, M. E., and Malcata, F. X. (2010). Plant aqueous extracts: Antioxidant capacity via haemolysis and bacteriophage P22 protection. *Food Control*, *21*, 633–638.

Gibson, R. A., Lines, D. R. and Neumann, M. A. (1992). Gamma linolenic acid (GLA) content of encapsulated evening primrose products. *Lipids*, 27: 82–84.

Gil, M. I., Ferreres, F., and TomaÂ s-BarberaÂ n, F. A. (1999). Effect of postharvest storage and processing on the antioxidant constituents (Flavonoids and vitamin C) of fresh-cut spinach. Journal of Agricultural and Food Chemistry, 47, 2213-2217.

Gil, M. I., Holcroft, D. M., and Kader, A. A. (1997). Changes in strawberry anthocyanins and other polyphenols in response to carbon dioxide treatment. Journal of Agricultural and Food Chemistry, 45, 1662-1667.

Gil, M. I., Tomas-Barberan, F. A., Hess-Pierce, B., Holcroft, D. M. and Kader, A. A. (2000). Antioxidant activity of pomegranate juice and its relationship with phenolic composition and processing. *Journal of Agricultural and Food Chemistry*, 48: 4581– 4589.

Gil-Izquierdo, A., Gil, M. I. and Ferreres, F. (2002). Effect of processing techniques at industrial scale on orange juice antioxidant and beneficial health compounds. *Journal of Agricultural and Food Chemistry*, 50: 5107–5114.

Giokas, D. L., Vlessidis, A. G., and Evmiridis, N. P. (2007). On-line selective detection of antioxidants free-radical scavenging activity based on Co(II)/EDTA-induced luminol chemiluminescence by flow injection analysis. *Analytica Chimica Acta*, *589*, 59–65.

Gioti, E. M., Fiamegos, Y. C., Skalkos, D. C., and Stalikas, C. D. (2007). Improved method for the in vitro assessment of antioxidant activity of plant extracts by headspace solid-phase microextraction and gas chromatography-electron capture detection. *Journal of Chromatography. A*, *1152*, 150–155.

Giusti, M.M., and Wrolstad, R.E. (2003). Acylated anthocyanins from edible sources and their applications in food systems. *Biochemical Engineering Journal*. 14(3): 217–225.

Glennie, C. W., Kaluza, W. A., and van Niekerk, P. J. (1981). High performance liquid chromatography of procyanidins in developing sorghum grain, *J. Agric. Food Chem.* 29: 965.

Gliwa, J., Gunenc, A., Ames, N., Willmore, W.G., and Hosseinian, F. S. (2011). Antioxidant activity of alkylresorcinols from rye bran and their protective effects on cell viability of PC-12 AC cells. *J. Agric. Food Chem.*, 59:11473-11482.

Godoy-Navajas, J., Aguilar-Caballos, M. P., and Gomez-Hens, A. (2011). Long-wavelength fluorimetric determination of food antioxidant capacity using Nile blue as reagent. *Journal of Agricultural and Food Chemistry*, *59*, 2235–2240.

Gokce, A. F., Kaya, C., Serce, S. and Ozgen, M. (2010). Effect of scale color on the antioxidant capacity of onions. *Scientia Horticulturae*, 123: 431–435.

Goldstein, J. L, and Swain, T. (1963). Changes in tannins in ripening fruits. *Phytochemistry*. 2, 371.

Goldstein, J. L. and Swain, T. (1965). The inhibition of enzymes of tannins, *Phytochemistry*, 4, 185.

Gómez-Estaca, J., Giménez, B., Montero, P. and Gómez-Guillén, M. C. (2009). Incorporation of antioxidant borage extract into edible films based on sole skin gelatin or a commercial fish gelatin. *Journal of Food Engineering*, 92: 78–85.

Göncüog˘ lu, N., Mogol, B. A., and Gökmen, V. (2013). Phytochemicals and health benefits of dried apricots. In C. Alasalvar and F. Shahidi (Eds.), *Dried fruits: Phytochemicals and health effects* (pp. 226–242). Oxford: Wiley-Blackwell.

Gong, L., Jin, C., Wu, X. and Zhang Y. (2013). Relationship between total antioxidant capacities of cereal measured before and after the *in vitro* digestion. *International Journal of Food Science and Nutrition*, 64(7), 850-856.

Gong, Y., Plander, S., Xu, H., Simandi, B., and Gao, Y. (2011). Supercritical CO2 extraction of oleoresin from marigold (*Tagetes erecta* L.) flowers and determination of its antioxidant components with online HPLC-ABTS+ assay. *Journal of the Science of Food and Agriculture*, *91*, 2875–2881.

González-Santiago, M., Fonollá, J. and Lopez-Huertas, E. (2010). Human absorption of a supplement containing purified hydroxytyrosol, a natural antioxidant from olive oil, and evidence for its transient association with low-density lipoproteins. *Pharmacological Research*, 61: 364–370.

González-Sarrías, A., Giménez-Bastida, J. A., García-Conesa, M. T., Gómez-Sánchez, M. B., García-Talavera, N. V., Gil-Izquierdo, A., Sálchez-Álvarez, C., Fontana-Compiano, L. O., Morga-Egea, J. P., Pastor-Quirante, F. A., Martínez-Díaz, F., Tomás-Barberán, F. A., and Espín, J. C. (2010). Occurrence of urolithins, gut microbiota ellagic acid metabolites and proliferation markers expression response in the human prostate gland upon consumption of walnuts and pomegranate juice. *Molecular Nutrition and Food Research*, *54*, 311–322.

Gopalan, A., Reuben, S. C., Ahmed, S., Darvesh, A. S., Hohmann, J. and Bishayee, A. (2012). The health benefits of black currants. *Food and Function*, 3: 795–809.

Gorden, M. H. (1990). The mechanism of antioxidant action *in vitro*. In B. J. F. Hudson (Ed.), *Food antioxidants* (pp. 1–18). Amsterdam, Netherlands: Elsevier.

Gordon, M. (2001). Measuring antioxidant activity. In J. Pokorny, N. Yanishlieva, and M. Gordon (Eds.), *Antioxidants in food: Practical applications* (pp. 71–84). Cambridge, England: Woodhead Publishing Ltd.

Gordon, M. H. (1996). Dietary antioxidants in disease prevention. *Natural Product Reports*, 13: 265–273.

Gorinstein, S., Caspi, A., Libman, I., Leontowicz, H., Leontowicz, M., Tashma, Z., Katrich, E., Jastrzebski, Z. and Trakhtenberg, S., (2007). Bioactivity of beer and its influence on human metabolism. International Journal of Food Sciences and Nutrition 58, 94-107.

Gorinstein, S., Caspi, A., Zemeser, M. and Trrakhtenberg, S., (2000). Comparative contents of some phenolics in beer, red and white wines. Nutrition Research 20, 131-139.

Gorinstein, S., Martin-Belloso, O., Lojek, A., Cˇ íž, M., Soliva-Fortuny, R., Park, Y. S., Caspi, A., Libman, I., and Trakhtenberg, S.(2002). Comparative content of some phytochemicals in Spanish apples, peaches, and pears. *Journal of the Science of Food and Agriculture*, *82*, 1166–1170.

Gorinstein, S., Zemser, M., Weisz, M., Haruenkit, R. and Trakhtenberg, S. (1998). The influence of dry matter of different alcoholic beverages on lipids, proteins, and antioxidant activity in serum of rats. *Journal of Nutritional Biochemistry*, 9: 131–135.

Gorjanovic, S.Z., Novarkovic, M., Potkonjak, N.I., Leskosek-Chkalovic, I. and Suznjevic, D.Z., (2010). Application of a novel antioxidative assay in beer analysis abd brewing process monitoring. Journal of Agricultural and Food Chemistry 58, 744-751.

Goto, T. Hoshino, T. and Ohba, M. (1976). Studies on anthocyanin complexes. I. Stabilization effect of neutral salts on anthocyanins: flavylium salts, anhydrobases, and genuine anthocyanins, *Agric. Biol. Chem., 40,* 1593.

Gottstein, T. and Grosch,W. (1990). Model study of different antioxidant properties of α- and γ-tocopherol in fats. *Fat Science Technology*, 92: 139–144.

Goulart, L. A., Teixeira, A. R. L., Ramalho, D. A., Terezo, A. J. and Castilho, M. (2014). Development of an analytical method for the determination of tert-butylhydroquinone in soybean biodiesel. *Fuel*, 115: 126–131.

Goun, E., Gunningham, G., Solodnikov, S., Krasnykch, O. and Miles, H. (2002). Antithrombin activity of some constituents from *Origanum vulgare*. *Fitoterapia*, 73: 692–694.

Goupy, P., Hugues, M., Boivin, P. and Amiot, M.J., (1999). Antioxidant composition and activity of barley (Hordeum vulgare) and malt extracts and of isolated phenolic compounds. *Journal of the Science of Food and Agriculture* 79, 1625-1634.

Goupy, P., Hugues, M., Boivin, P. and Amiot, M. J. (1999). Antioxidant composition and activity of barley (Hordeum vulgare) and malt extracts and of isolated phenolic compounds. *Journal of the Science of Food and Agriculture*, 79: 1625–1634.

Grampp, G.; Landgraf, S.; Wesierskia, T.; Jankowska, B.; Kalisz, E.; Sabou, D.M.; Mladenova, B. (2002). Kinetics of the formation of the blue complex CrO (O-2)(2) formed by dichromate and H_2O_2 in acid solutions. A stopped-flow investigation using rapid-scan UV-VIS detection. *Mon. Chem. 133*, 1363-1372.

Gregoire, S., Singh, A. P., Vorsa, N. and Koo, H. (2007). Influence of cranberry phenolics on glucan synthesis by glucosyltransferases and *Streptococcus mutans* acidogenicity. *Journal of Applied Microbiology*, 103: 1960–1968.

Grice, H. C. (1986). Safety evaluation of butylated hydrotoluene (BHT) in the liver, lung and gastrointestinal tract. *Food and Chemical Toxicology*, 24: 1127–1130.

Griffiths, D. W. and Jones, D. H. (1977). Cellulase inhibition by tannins in the testa of field beans *(Vicia foba)* J. *Agric. Food Chem.,* 28, 983.

Grippi, F., Crosta, L., Aiello, G., Tolomeo, M., Oliveri, M., Gebbia, N. and Curione, A. (2008). Determination of stilbenes in Sicilian pistachio by high-performance liquid chromatographic diode array (HPLC-DAD/FLD) and evaluation of eventually mycotoxin contamination. *Food Chemistry*, 107: 483–487.

Grisebach, H. (1967). Biosynthetic patterns in Microorganisms and Higher Plana, John Wiley and Sons. New York.

Grosso, G., Yang, J., Marventano, S., Micek, A., Galvano, F., and Kales, S. N. (2015). Nut consumption on all-cause, cardiovascular, and cancer mortality risk: A systematic review and metaanalysis of epidemiologic studies. *The American Journal of Clinical Nutrition, 101*, 783–793.

Gruiner-Richter, S., Otto, F. and Weidner, E. (2012). Impregnation of oil containing fruits. *Journal of Supercritical Fluids*, 66: 321–327.

Gu, L., Kelm, M. A., Hammerstone, J. F., Beecher, G., Holden, J., Haytowitz, D. and Prior, R. L. (2003). Screening of foods containing proanthocyanidins and their structural characterization using LC–MS/MS and thiolytic degradation. *Journal of Agricultural and Food Chemistry*, 51: 7513–7521.

Gu, L., Kelm, M., Hammerstone, J. F., Beecher, G., Cunningham, D., Vannozzi, S. and Prior, R. L. (2002). Fractionation of polymeric procyanidins from lowbush blueberry and quantification of procyanidins in selected foods with an optimized normalphase

Guadagni, D. G., Sorber, D. B. and Wilber, J. S. (1949). Enzymic oxidation of phenolic compounds of frozen peaches, *Food Technol.* 33. 359.

Guajardo-Flores, D., Cardenas-Hinojosa, A. P., Dykers, L., McDonough, C. M. and Rooney, I. W. (2006). Comparison of total phenol, antioxidant activity and tannin content in different grains. AACCI Annual Meeting Abstracts. Published online at http://www.aacci.org/meeting/2006/abstracts/p-231.htm.

Gujral, H. S., Angurala, M., Sharma, P. and Singh, J. (2012). Phenolic content and antioxidant activity of germinated and cooked pulses. *Int. J. Food Prop.* 14: 1366–1374.

Guclu, K., Kibrislioglu, G., Ozyurek, M., and Apak, R. (2014). Development of a fluorescent probe for measurement of peroxyl radical scavenging activity in biological samples. *Journal of Agricultural and Food Chemistry*, *62*, 1839–1845.

GuÈ ntensperger, B., HaÈ mmerli-Meier, D. E., and Escher, F. E. (1998). Rosemary extract and precooking effects on lipid oxidation in heat sterilized meat. Journal of Food Science, 63, 955-957.

Guesnel, V. C. (1968). Fractionation and properties of the polymeric leucocyanidin of the seeds of *Theobroma* cacao, *Phytochemistry,* 7: 1583.

GuilleÂn, M. D., and Ibargoitia, M. L. (1998). New components with potential antioxidant and organoleptic properties, detected for the first time in liquid smoke favouring preparations. Journal of Agricultural and Food Chemistry, 46, 1276-1285.

GuilleÂn, M. D., and Manzanos, M. J. (1999). Extractable components of the aerial parts of Salvia lavandulifolia and composition of the liquid smoke favoring obtained from them. Journal of Agricultural and Food Chemistry, 47, 3016-3027.

Guillen, R., Rodriguez, R., Jaramillo, S., Rodriguez, G., Espejo, J. A., Fernandez-Bolanos, J., Heredia, A. and Jimenez, A. (2008). Antioxidants from asparagus spears: Flavonoids. *Acta Horticulturae*, 776: 247–254.

Guillen-Sans, R., and Guzman-Chozas, M. (1998). The thiobarbituric acid (TBA) reaction in foods: A review. *Critical Reviews in Food Science and Nutrition*, *38*, 315–330.

Guillot, F. L., Malnoeè, A. and Stadler, R. H. (1996). Antioxidant properties of novel tetraoxygenated phenylindan isomers formed during thermal decomposition of caffeic acid. Journal of Agricultural and Food Chemistry, 44, 2503-2510.

Gulaboski, R., Mirceski, V., and Mitrev, S. (2013). Development of a rapid and simple voltammetric method to determine total antioxidative capacity of edible oils. *Food Chemistry*, *138*, 116–121.

Gülçin, I. (2010). Antioxidant properties of resveratrol: A structure–activity insight. *Innovative Food Science and Emerging Technologies*, 11: 210–218.

Gulcin, I.; Bursal, E.; Sehitoglu, M.H.; Bilsel, M.; Goren, A.C. (2010). Polyphenol contents and antioxidant activity of lyophilized aqueous extract of propolis from Erzurum, Turkey. *Food Chem. Toxicol. 48*, 2227-2238.

Guo, H., Ling, W., Wang, Q., Liu, C., Hu, Y., Xia, M., Feng, X. and Xia, X. (2007). Effect of Anthocyanin-rich extract from black rice (*Oryza sativa* L. *indica*) on hyperlipidemia and insulin resistance in fructose-fed rats. *Plant Foods for Human Nutrition*, 62(1), 1-6.

Guo, X. D.,Wu, C. S., Ma, Y. J., Parry, J., Xu, Y. Y., Liu, H. and Wang, M. (2012). Comparison of milling fractions of tartary buckwheat for their phenolics and antioxidant properties. *Food Research International*, 49: 53–59.

Gupta, R. K. and Haslam, E. (1980). Vegetable tannins - structure and biosynthesis. In *Polyphenols En Cereals and Legumes,* Hulse, J. H., Ed. International Development Research Centre, Ottawa, 16.

Gupta, R. K. and Haslam, E. (1980). Vegetable tannins: structure and biosynthesis, in *Polyphenols in Cereals and Legumes.* Hulse. J. H. Ed. International Development Research Center. Ottawa. Canada. 15.

Gupta, R. K. and Haslam, E. (1978). Plant proanthocyanidins. Part 5. Sorghum polyphenols. *Journal of the Chemical Society, Perkin Transactions I*, 4: 892–896.

Gupta, R.K. and Haslam, E. (1980). Vegetable tannins-structure and biosynthesis. In: *Polyphenols in Cereals and Legumes*. J. H. Hulse (Ed.), p. 15-24. IDRC, Ottawa.

Guri, A., Kefalas, P. and Roussis, V. (2006). Antioxidant potential of six pine species. *Phytotherapy Research*, 20: 263–266.

Guyot, S., Marnet, N., Laraba, D., Sanoner, P. and Drilleau, J. F. (1998). Reversed-phase HPLC following thiolysis for quantitative estimation and characterization of the four main classes of phenolic compounds in different tissue zones of a French cider apple variety (Malus domestica Var. Kermerrien). *Journal of Agricultural and Food Chemistry*, 46: 1698–1705.

Gyamfi, M. A. and Aniya, Y. (2002). Antioxidant properties of Thonningianin A, isolated from the African medicinal herb, *Thonningia sanguinea. Biochemical Pharmacology*, 63: 1725–1737.

Ha, T. Y., Cho, I. J., and Lee, S. H. (1998). Screening of HMG-CoA reductase inhibitory activity of ethanol and methanol extracts from cereals and legumes. *Korean J. Food Sci. Technol.*, 30:224–229.

Habauzit, V., and Morand, C. (2012). Evidence for a protective effect of polyphenols-containing foods on cardiovascular health: An update for clinicians. *Therapeutics Advances in Chronic Disease*, *3*, 87–106.

Haddad, E. H., Gaban-Chong, N., Oda, K., and Sabate, J. (2014). Effect of a walnut meal on postprandial oxidative stress and antioxidants in healthy individuals. *Nutrition Journal*, *13*, 4.

Haddad, E., Jambazian, P., Karunia, M., Tanzman, J. and Sabaté, J. (2006). A pecan-enriched diet increases γ-tocopherol/cholesterol and decreases thiobarbituric acid reactive substances in plasma of adults. *Nutrition Research*, 26: 397–402.

Hadolin, M., Skerget, M., Knez, Z. and Bauman, D. (2001). High pressure extraction of vitamin E-rich oil from *Silybum marianum*. *Food Chemistry*, 74: 355–364.

Haekkinen, S. H., Kaerenlampi, S. O., Heinonen, I. M., Mykkaenen, H. M. and Toerroenen, A. R. (1999). Content of the flavonols quercetin, myricetin, and kaempferol in 25 edible berries. *Journal of Agricultural and Food Chemistry*, 47: 2274–2279.

Hag, M.E.E., Tinay, A.H.E., and Yousif, N.E. (2002). Effect of fermentation and dehulling on starch, total polyphenols, phytic acid content and in vitro protein digestibility of pear l millet. Food Chemistry, 77:193–196.

Hagerman, A. E. and Butler, L. G. (1978). Protein precipitation method for the quantitative determination of tannins, *J. Agric. Food Chem.* 26: 809.

Hagerman, A. E. and Butler, L. G. (1980). Condensed tannin purification and characterization of tannin, associated proteins. J. *Agric. Chem.* 28, 98.

Hagerman, A. E. and Butler, L. G. (1980). Determination of protein in tannin-protein precipitafe. *J. Agric. Food Chem.*, 28, 944.

Hagerman, A. E., Riedl, K. M., Jones, G. A., Sovik, K. N., Ritchard, N. T., Hartzfeld, P. W., and Riechel, T. L. (1998). High molecular weight plant polyphenolics (tannins) as biological antioxidants. Journal of Agricultural and Food Chemistry, 46, 1887-1892.

Hagerman, A.E. and Butler, L.G. (1978). Protein precipitation method for the quantitative determination of tannins. J. Agric. Food Chem. 26: 809-812.

Hagerman, A.E. and Butler, L.G. (1994). Assay of condensed tannins or flavonoid oligomers and related flavonoids in plants. Meth. Enz. 234: 429-437.

Hagerman, A.E., Zhao, Y., and Johnson, S. (1997). Methods for determination of condensed and hydrolyzable tannins. In: *Antinutroents and Phytochemicals in Food*. F. Shahidi (Ed.), p. 209-222. American Chemical Society, Washington. DC.

Hahlbrock, K. and Grisebach, H. (1975). Biosynthesis of flavonoids, in *The Flavonoids*. Harbome, J. B., Mabry. T. J. and Mabry, H., Eds., Chapman and Hall. London, 866.

Hahn, D. H., Faubion, J. M. and Rooney, L. W. (1983). Sorghum phenolic acids, their high performance liquid chromatography separation and their relation to fungal resistance. *Cereal Chemistry*, 60: 255–259.

Hahn, D. H., Rooney, L. W., and Earp, C. F. (1984). Tannins and phenols of sorghum. *Cereal Foods World*, 29, 776.

Hait-Darshan, R., Grossman, S., Bergman, M., Deutsch, M. and Zurgil, N. (2009). Synergistic activity between a spinachderived natural antioxidant (NAO) and commercial antioxidants in a variety of oxidation systems. *Food Research International*, 42: 246–253.

Hakala, P., Lampi, A. M., Ollilainen, V., Werner, U., Murkovic, K., Wahala, K., Karkola, S. and Piironen, V. (2002). Steryl phenolic acid esters in cereals and their milling fractions. *Journal of Agricultural and Food Chemistry*, 50: 5300–5307.

Häkkinen, S. H., Kärenlampi, S. O., Heinonen, I. M., Mykkänen, H. M. and Törrönen, A. R. (1999). Content of the flavonols quercetin, myricetin, and kaempferol in 25 edible berries. *Journal of Agricultural and Food Chemistry*, 47: 2274–2279.

Hall, C. A., Cupett, S. L., and Dussault, P. (1998). Hydrogen-donating mechanism of rosmariquinone, an antioxidant found in rosemary. Journal of the American Oil Chemists' Society, 75, 1147-1154.

Hall, D. W. (1970). Handling and Storage of Food Grains in Tropical and Subtropical Areas. No. 90. Food and Agriculture Organization. Geneva. 1.

Halliwell, B. (1996). Antioxidants in human health and disease. *Annual Review of Nutrition*, 16: 33–50.

Halliwell, B., Aeschbacht, R., Loligert, J. and Aruoma, O. I. (1995). The characterization of antioxidant. *Food and Chemical Toxicology*, 33: 601–617.

Halliwell, B., and Gutteridge, J. M. C. (1999). *Free radicals in biology and medicine*. NewYork, NY: Oxford University Press.

Halliwell, B., Rafter, J. and Jenner, A. (2005). Health promotion by flavonoids, tocopherols, tocotrienols, and other phenols: Direct of indirect effects? Antioxidant or not? *American Journal of Clinical Nutrition*, 81: 268S–276S.

Halvorsen, B. L., Carlsen, M. H., Phillips, K. M., Bohn, S. K., Holte, K., Jacobs, D. R., Jr. and Blomhoff, R. (2006). Content of redoxactive compounds (ie, antioxidants) in foods consumed in the United States. *American Journal of Clinical Nutrition*, 84: 95–135.

Halvorsen, B. L., Holte, K., Myhrstad, M. C. W., Barikmo, I., Hvattum, E., Remberg, S. F.,Wold, A. B., Haffner, K., Baugerad, H., Andersen, L. F., Moskaug, O., Jacobs, D. R., Jr. and Blomhoff, R. A. (2002). Systematic screening of total antioxidants in dietary plants. *Journal of Nutrition*, 132: 461–471.

Hamam, F. and Shahidi, F. (2006). Acidolysis reactions lead to esterification of endogenous tocopherols and compromised oxidative stability of modified oils. *Journal of Agricultural and Food Chemistry*, 54: 7319–7323.

Hamama, A. A., and Nawar, W. W. (1991). Thermal decomposition of some phenolic antioxidants. Journal of Agricultural and Food Chemistry, 39, 1063-1069.

Hammerschmidt, P. A. and Pratt, D. E. (1978). Phenolic antioxidants of dried soybeans. *Journal of Food Science*, 43: 556–559.

Hammerstone, J. F., Lazarus, S. A., Mitchell, A. E., Rucker, R. and Schmitz, H. H. (1999). Identification of procyanidins in cocoa (*Theobroma cacao*) and chocolate using high-performance liquid chromatography/mass spectrometry. *Journal of Agricultural and Food Chemistry*, 47: 490–496.

Hammond, M. D., and Skonberg, D. I. (2012). Antioxidant properties of chitosan coatings on frozen Atlantic salmon fillet portions. *Journal of Aquatic Food Product Technology*, *21*, 351–361.

Han, D., Kim, S. J., Kim, S. H., and Kim, D. M. (1998). Repeated regeneration of degraded red beet juice pigments in the presence of antioxidants. Journal of Food Science, 63, 69-72.

Han, D., Yi, O. S., and Shin, H. K. (1990). Antioxidative effect of ascorbic acid solubilized in oils via reversed micelles. Journal of Food Science, 55, 247-249.

Han, D., Yi, O.-S. and Shin, H. K. (1991). Solubilization of vitamin C in fish oil and synergistic effect with vitamin E in retarding oxidation. *Journal of American Oil Chemists' Society*, 68: 740–743.

Han, K. H., Kitano-Okada, T., Seo, J. M., Kim, S. J., Sasaki, K., Shimada, K. I. and Fukushima, M. (2015). Characterisation of anthocyanins and proanthocyanidins of adzuki bean extracts and their antioxidant activity. *Journal of Functional Foods*, 14: 692–701.

Han, R. M., Tian, Y. X., Liu, Y., Chen, C. H., Ai, X. C., Zhang, J. P. and Skibsted, L. H. (2009). Comparison of flavonoids and isoflavonoids as antioxidants. *Journal of Agricultural and Food Chemistry*, 57: 3780–3785.

Han, Y. H. and Park, W. H. (2009). Propyl gallate inhibits the growth of HeLa cells via regulating intracellular GSH level. *Food and Chemical Toxicology*, 47: 2531–2538.

Han, Z. P., Liu, R. L., Cui, H. Y. and Zhang, Z. Q. (2013). Microwaveassisted extraction and LC/MS analysis of phenolic antioxidants in sweet apricot (*Prunus armeniaca L.*) kernel skins. *Journal of Liquid Chromatography and Related Technologies*, 36: 2182–2195.

Hanasaki, Y., Ogawa, S. and Fukui, S. (1994). The correlation between active oxygen scavenging and antioxidative effects of flavonoids. *Free Radical Biology and Medicine*, 16: 845–850.

Hancock, J. T., Desikan, R. and Neill, S. J. (2001). Role of reactive oxygen species in cell signalling pathways. *Biochemical Society Transactions*, 29(Pt2), 345-350.

Hara, H., Ohta, M., Ohta, K., Kuno, S. and Adachi, T. (2003). Increase of antioxidative potential by *tert*-butylhydroquinone protects against cell death associated with 6-hydroxydopamineinduced oxidative stress in neuroblastoma SH-SY5Y cells. *Molecular Brain Research*, 119: 125–131.

Hara, S., and Totani, Y. (1988). A highly sensitive method for the micro-determination of lipid hydroperoxides by potentiometry. *Journal of the American Oil Chemists' Society*, *65*, 1948–1950.

Hara, Y. (1994). Prophylactic functions of tea polyphenols. *phytochemicals for cancer prevention. II. Tea, spices and herbs* (pp. 34–50). ACS Symposium Series 547. Washington, DC: American Chemical Society.

Harbone, J.B. (1965). Plant phenols. XIV. Characterization of flavonoid glycosides by acidic and enzymic hydrolyses. Phytochemistry. 4: 107-120.

Harborne, J. B. (1965). Flavonoid pigments. In: Plant Biochemistry, Bonner. J. and Varner, E. V., Eds., Academic Press, New York, 618.

Harborne, J. B. (1965). Plant phenols. XIV. Characterization of flavonoid glycosides by acidic and enzymic hydrolyses. *Phytochemistry,* 4, 107.

Harborne, J. B. (1967). *Comparative Biochemistry of Flavonoids,* Academic Press, New York.

Harborne, J. B. and Simmonds, N. W. (1964). The natural distribution of the phenolic aglycones, in *Biochemistry of Phenolic Compounds,* Harborne. J. B., Ed. Academic Press, New York.

Harborne, J. B. and Williams, C. A. (2000). Advances in flavonoids research since 1992 (2000). *Phytochemistry*, 55: 481–504.

Harder, H., Tetens, I., Let, M. B. and Meyer, A. S. (2004). Rye bran bread intake evaluates urinary excretion of ferulic acid in human, but does not affect the susceptibility of LDL to oxidation *ex vivo. European Journal of Nutrition,* 43(4), 230-236.

Harhorne, J. B., Mabry, T. J. and Mabry, H. (1975). Eds. The Flavonoids. Harhorne, J. B. Mabry. T. J. and Mabry. H. Eds. Chapman and Hall. London.

Harnly, J. M., Doherty, R. F., Beecher, G. R., Holden, J. M., Haytowitz, D. B., Bhagwat, S., and Gebhardt, S. (2006). Flavonoid content of US fruits, vegetables, and nuts. *Journal of Agricultural and Food Chemistry*, *54*, 9966–9977.

Harris, H. B. (1969). Bird resistance in grain sorghum, *Proc. 24th Annu. Corn Sorghum Res. Conf.,* 24, 113.

Harris, H. B. and Burns, R. E. (1970). Influence of tannin contents on preharvest seed gcrminatitll1 in sorghum. *Agron. J.* 62: 835.

Harris, H. B. and Burns, R. E. (1973). Relationship between tannin content of sorghum and preharvest seed molding, *Agron. J.*, 65, 957.

Harris, H. B., Cummings, D. G., and Burns, R. E. (1970). Tannin content and digestibility of sorghum grain as influenced by bagging. *Agron. J.* 62. 633.

Harris, R. K. and Haggerty, W. J. (1993). Assays for potentially carcinogenic phytochemicals in flaxseed. *Cereal Foods World*, 38: 147–151.

Harris. H. B., Johnson, B. J. and Stacy, S. V. (1962). Georgia 609 - a weather resistant grain sorghum, *Univ.* G. *Agric. Exp. Stn. Leafl. N. S.* 29.

Hasegawa, D., Johnson, R. M. and Gould, W. A. (1966). Effect of cold storage on chlorogenic acid content of potatoes. *Journal of Agricultural Food Chemistry*, 14: 165–169.

Hasegawa, S., Lam, L. K. T. and Miller, E. G. (2000). Citrus limonoids: Biochemistry and possible importance to human nutrition. In F. Shahidi and C.-T. Ho (Eds.), *Phytochemicals and phytopharmaceuticals* (pp. 79–94). Champaign, IL: AOCS Press.

Hashim, I., Koehler, P. E. and Eitenmiller, R. R. (1993). Tocopherols in runner and Virginia peanut cultivars at various maturity stages. *Journal of the American Oil Chemists' Society*, 70: 633–635.

Haslam, E. (1966). *Chemistry of Vegetable Tannins,* Academic Press, New York.

Haslam, E. (1977). Symmetry and promiscuity in procyanidin biochemistry, *Phytochemistry,* 16: 1625.

Haslam, E. and Gupta, R.K. (1978). Plant proanthocyanidins. V. Sorghum polyphenols. J. Chem. Soc. Perkin. 1: 892-897.

Haslam. E., Opie, C.T., and Porter, L.J. (1977). Procyanidin metabolism, a hypothesis. *Phytochemistry.* 16: 99-102.

Hatcher, D. W., and Kruger, J. E. (1997). Simple phenolic acids in flours prepared from Canadian wheat: relationship to ash content, color and polyphenol oxidase activity. *Cereal Chem.,* 74:337–343.

Hathaway, D. E. and Seakins, J. W. T. (1958). The influence of tannins on the degradation of pectin by pectinase enzymes. *Biochem. J.* 70, 158.

Hathway, D. E. (1958). Oak-bark tannins. *Biochem. J.* 70: 34.

Hathway, D. E. (1969). Chromatography and Electrophoresis Techniques. Smith, I. ed. Vol. 1. Heinemann. London.

Hathway, D. E. and Seakins, J. W. T. (1957). Autoxidation of Polyphcnols. III. Autoxidation in neutral aqueous solution of flavans related to catechin. *J. Chem. Soc.* p. 1562.

Hathway, D.E. and Seakins, J.W.T. (1957). Enzymic oxidation of catechin to a polymer structurally related to some phlobatannins. *Biochem. J.* 67: 239-245.

Hathway, O. E. (1958). Autoxidation of polyphenols. IV. Oxidative degradation of the catechin autoxidative polymer. J. *Chem. Soc.,* p. 1520.

Hatier HB, Gould KS (2008) Anthocyanin function in vegetative organs. In: Gould K, Davies K, Winefield C (Eds) Anthocyanins—biosynthesis, functions, and applications. Springer, New York.

Hattori, M., Yamaji-Tsukamoto, K., Kimagai, H., Feng, Y., and Takahashi, K. (1998). Antioxidative activity of soluble elastin peptides. Journal of Agricultural and Food Chemistry, 46, 2167-2170.

Hayase, F. and Kato, H. (1984). Antioxidative components of sweet potato. *Journal of Nutritional Science and Vitaminology*, 30: 37–46.

Hayashi, S., Yamaguchi, K., Ito, M., and Yonezawa, H. (1991). The development of barley foods based on a new milling approach. In Proceedings of Cereal International, Brisbane, Australia (pp. 357–359).

Haysmans, A. A. C. (1970). Storage of food grains - problems and prospects. *Bull. Grain Technol.,* 8, 92.

He, F. J., Nowson, C. A., Lucas, M., and MacGregor, G. A. (2007). Increased consumption of fruit and vegetables is related to a reduced risk of coronary heart disease: Meta-analysis of cohort studies. *Journal of Human Hypertension, 21*, 717–728.

He, Y. and Shahidi, F. (1997). Antioxidant activity of green tea and its catechins in a fish meat model system. *Journal of Agricultural and Food Chemistry*, 45: 4262–4266.

Heftmann, F. (1961). Chromatography Reinhold. New York.

Hegde, P. S., Anitha, B., and Chandra,T. S.(2005). In vivo effect of whole grain flour of finger millet (*Eleucine coracana*) and kodo millet (*Paspalum scrobiculatum*) on rat dermal wound healing. Indian Journal of Experimental Biology, 43: 254–258.

Hegde, P. S., Rajasekaran, N. S. and Chandra, T. S. (2005). Effects of the antioxidant properties of millet species on oxidative stress and glycemic status in alloxan-induced rats. *Nutrition Research*, 25: 1109–1120.

Hegerman, A. E., Riedl, K. M., Jones, G., Sovik, K. N., Rechard, N. T., Hartzfeld, P. W., and Reichel, T. L. (1998). High molecular weight plant polyphenolics (tannins) as biological antioxidants. *Journal of Agricultural and Food Chemistry, 46*, 1887–1892.

Heim, K. E, Tagliafeno A. R. and Bobilya, D. J. (2002) Flavonoid antioxidants: chemistry, metabolism and structure-activity relationships. *Journal of Nutrition and Biochemistry*, 13: 572–584.

Heimler, D., Isolani, L., Vignolini, P., Tombelli, S. and Romani, A. (2007). Polyphenol content and antioxidant activity some species of freshly consumed salads. *Journal of Agricultural and Food Chemistry*, 55: 1724–1729.

Heinonen, M., Haila, K., Lampi, A.-M., and Pironen, V. (1997). Inhibition of oxidation in 10% oil-in-water emulsions by β-carotene with α- and γ- tocopherols. Journal of the American Oil Chemists' Society, 74, 1047-1051.

Heinonen, M., Lehtonen, P. J., and Hopia, A. L. (1998). Antioxidant activity of berry and fruit wines and liquors. Journal of Agricultural and Food Chemistry, 46, 25-31.

Heinonen, M., Meyer, A. S., and Frankel, E. N. (1998). Antioxidant activity of berry phenolics on human low-density-lipoprotein and liposome oxidation. *Journal of Agricultural and Food Chemistry*, 46, 107-4112.

Hempel, J. and Bohm, H. (1996). Quality and quantity of prevailing flavonoid glycosides of yellow and green French beans (*Phaseolus vulgaris* L.). *Journal of Agricultural and Food Chemistry*, 44: 2114–2116.

Henderson, D. E., Slickman, A. M., and Henderson, S. K. (1999). Quatitative HPLC determination of the antioxidant activity of capsaicin on the formation of lipid hydroperoxides of linoleic acid, a comparative study against BHT and melatonin. Journal of Agricultural and Food Chemistry, 47, 2563-2570.

HernaÂ ndez-PeÂrez, M., HernaÂ ndez, T., GoÂ mez-CordoveÂ s, C., Estrella, I., and Rabanal, R. M. (1996). Phenolic composition of the ``MocaÂ n" (Visnea mocanera L.F.). Journal of Agricultural and Food Chemistry, 44, 3512-3515.

Hernanz, D., Nunez, V., Sacho, A. I., Faulds, C. B., Williamson, G., Bartolome, B., and Gomez-Cordoves, C. (2001). Hydroxycinnamic acids and ferulic acid dehydrodimers in barley and processed barley. *J. Agric. Food Chem.*, 49:4884–4888.

Hernanz, D., Nuñez, V., Sancho, A.I., Faulds, C.B., Williamson, G., Bartolomé, B. and Gómez-Cordovés, C., (2001). Hydroxycinnamic acids and ferulic acid dehydrodimers in barley and processed barley. Journal of Agricultural and Food Chemistry 49, 4884-4888.

Henryk Zielin□ski, Anna Michalska, Mariusz K. Piskuła and Halina Kozłowska (2006). Antioxidants in thermally treated buckwheat groats. *Mol. Nutr. Food Res.* 50: 824 – 832.

Herr, I. and Buchler, M. W. (2010). Dietary constituents of broccoli and other cruciferous vegetables: Implications for prevention and therapy of cancer. *Cancer Treatment Reviews*, 36: 377–383.

Herrera E, Barbas C; Barbas (2001). "Vitamin E: action, metabolism and perspectives". Journal of Physiology and Biochemistry 57 (2): 43–56.

Herrmann, K. (1989). Occurrence and content of hydroxycinnamic and hydroxybenzoic acid compounds in foods. *CRC Critical Review in Food Science and Nutrition*, 28: 315–347.

Hertog, M. G. L., Hollman, P. C. H., and Katan, M. B. (1992). Content of potentially anticarcinogenic favonoids of 28 vegetables and 9 fruits commonly consumed in the Netherlands. Journal of Agricultural and Food Chemistry, 40, 2379-2383.

Hettiarachchy, N. S. and Kalapathy, U. (2000). Food additives. In G. L. Christen and J. S. Smith (Eds.), *Food chemistry: Principles and applications* (pp. 261–273).West Sacramento, CA: Science Technology System.

Hicks, M., and Gebicki, J. M. (1979). Spectrophotometric method for the determination of lipid hydroperoxides. *Analytical Biochemistry*, *99*, 249–253.

Hincks, M.J. and Stanley, D.W. (1986). Multiple mechanism of bean hardening. *J. Food Technol*. 21: 731-750.

Hirawan, R., Diehl-Jones, W. and Beta, T. (2011). Comparative evaluation of the antioxidant potential of infant cereals produced from purple wheat and red rice grains and LC-MS analysis of their anthocyanins. *Journal of Agricultural and Food Chemistry,* 59(23), 12330-12341.

Ho, C.-T. (1992). Phenolic compounds in food: An overview. In M.-T. Huang, C.-T. Ho, and C. Y. Lee (Eds.), ACS symposium series: 507. Phenolic compounds in food and their effects on health II: Antioxidants and cancer prevention (pp. 2–7). Washington, DC: American Chemical Society.

Ho, E., Galougahi, K. K., Bhindi, R. and Figtree, G. A. (2013). Biological markers of oxidative stress: Application to cardiovascular research and practice. *Redox Biology*, 1(1), 483-491.

Hoff, J. E. and Singldon, K. I. (1977). A method for determination of tannins in foods by means of immobilized protein.-J. *Food Sci.,* 42, 1566.

Hoffman, A. and Shahidi, F. (2009). Paclitaxel and other taxanes in hazelnut. *Journal of Functional Foods*, 1: 33–37.

Hojjati, M., Calín-Sánchez, Á., Razavi, S. H., and Carbonell-Barrachina, Á.A. (2013). Effect of roasting on colour and volatile composition of pistachios (*Pistacia vera* L.). *International Journal of Food Science and Technology*, *48*, 437–443.

Holasova, M., Fiedlerova, V., Smrcinova, H., Orsak, M., Lachman, J. and Vavreinova, S. (2002). Buckwheat – The source of antioxidant activity in functional foods. *Food Chemistry*, 35: 207–211.

Hollman, P. C. H., Hertog, M. G. L., and Katan, M. B. (1996). Analysis and health effects of favonoids. Food Chemistry, 57, 43-46.

Holtekjolen, A. K., Baevere, A. B., Rodbotten, M., Berg, H., and Knutsen, S. H. (2008). Antioxidant properties and sensory profiles of breads containing barley flour. *Food Chem.,* 110:414–421.

Hooshmand, S., Chai, S. C., Saadat, R. L., Payton, M. E., Brummel-Smith, K., and Arjmandi, B. H. (2011). Comparative effects of dried plum and dried apple on bone in postmenopausal women. *British Journal of Nutrition*, *106*, 923–930.

Hopia, A., and Heinonen, M. (1999). Antioxidant activity of favonol aglycones and their glycosides in methyl linoleate. Journal of the American Oil Chemists' Society, 76, 139-144.

Horax, R., Hettiarachchy, N. and Chen, P. (2010). Extraction, quantification and antioxidant activities of phenolics from harvested at three maturity stages (immature, mature and ripe). *Journal of Agricultural and Food Chemistry*, 58: 4428–4433.

Horowitz, R. M. (1964). Relation between the taste and structure of some phenolic glycosides, in *Biochemistry of Phenolic Compounds,* Harhorne, 1. B., Ed., Academic Press, New York, 54.

Horowitz, R. M. and Gentili, B. (1964). Structure of vitexin and isovitexin, *Chem. Ind.* 498.

Horowitz, R. M. and Gentili, B. (1977). Flavonoid constituents of citrus. In S. Nagy, P. E. Shawand M. K. Veldhuis (Eds.), *Citrus science and technology* (Vol. 1, pp. 397–426). Westport, CT: AVI Publishing.

Horowitz, R.M. (1964). Relation between the taste and structure of some phenolic glycosides. In: *Biochemistry of Phenolic Compounds*. J. B. Harborne (Ed.), p. 545- 556. Academic Press, New York, NY.

Hoshino, T., Ujihara, K., and Shikata, S. (1979). Bird Damage and Resistance in Grain Sorghum. *Bull. Chugoku Narl. Ag. Exp. Station.* (Fukuyama. Japan). A26. 25.

Hosny, M., and Rosazza, J. P. N. (1999). Novel isofavone, cinnamic acid, and triterpenoid glycosides in soybean molasses. Journal of Natural Products, 62, 853-858.

Hosseinian, F. S., Li, W. D. and Beta, T. (2008). Measurement of anthocyanins and other phytochemical in purple wheat. *Food Chemistry*, 109(4):916-924.

Hou,W. C., Lin, R. D., Cheng, K. T., Hung, Y. T., Cho, C. H., Chen, C. H., Hwang, S. Y. and Lee, M. H. (2003). Free radical-scavenging activity of Taiwanese native plants. *Phytomedicine: International Journal of Phytotherapy and Phytopharmacology*, 10: 170–175.

Houlihan, C. M., Ho, C.-T., and Chang, S. S. (1984). Elucidation of the chemical structure of a novel antioxidant, rosmaridiphenol, isolated from rosemary. Journal of the American Oil Chemists' Society, 61, 1018-1036.

Howard, L. A., Wong, A. D., Perry, A. K. and Klein, B. P. (1999). β-carotene and ascorbic acid retention in fresh and processed vegetables. *J Food Sci.* 64:929–36.

Howell, A. B. (2007). Bioactive compounds in cranberries and their role in prevention of urinary tract infections. *Molecular Nutrition and Food Research*, *51*, 732–737.

Howell, A. B., Reed, J. D., Krueger, C. G.,Winterbottom, R., Cunningham, D. G., and Leahy, M. (2005). A-type cranberry proanthocyanidins and uropathogenic bacterial antiadhesion activity. *Phytochemistry*, *66*, 2281–2291.

HPLC-DAD-MS (1998). Fluorescent detection method. *Journal of the Science of Food and Agriculture*, 50: 4852–4860.

Hsieh, R. H., Lien, L. M., Lin, S. H., Chen, C. W., Cheng, H. J. and Cheng, H. H. (2005). Alleviation of oxidative damage in multiple tissues in rats with streptozotocin-induced diabetes by rice bran oil supplementation. *Annual of the New York Academy Science*, 1042, 365-371.

Hsu, C. L., Fang, S. C., Liu, C. W. and Chen, Y. F. (2013). Inhibitory effects of new varieties of bitter melon on lipopolysaccharidestimulated inflammatory response in RAW 264.7 cells. *Journal of Functional Foods*, 5: 1829–1837.

Htock. H. J., Durram. E. L. and Zweig G. (1958). Manual of Paper Chromatography and plant Electrophoresis. 2nd ed. Academic Press. New York.

Huang, D., Boxin, O. and Prior, R. L. (2005). The chemistry behind antioxidant capacity assays. *Journal of Agricultural and Food Chemistry*, 53: 1841–1856.

Huang, D., Ou, B., Hampsch-Woodill, M., Flanagan, J. A., and Deemer, E. K. (2002). Development and validation of oxygen radical absorbance capacity assay for lipophilic antioxidants using randomly methylated β-cyclodextrin as the solubility enhancer. *Journal of Agricultural and Food Chemistry, 50*, 1815–1821.

Huang, S. C., Yen, G.-C., Chang, L.-W., Yen,W.-J. and Duh, P.-D. (2003). Identification of an antioxidant, ethyl protocatechuate, in peanut seed testa. *Journal of Agricultural and Food Chemistry*, 51: 2380–2383.

Huang, Y. C., Chang, Y. H. and Shao, Y. Y. (2006). Effect of genotype and treatment on the antioxidant activity of sweet potato in Taiwan. *Food Chemistry*, 98: 529–538.

Huang, Y. W., Liu, Y., Dushenkov, S., Ho, C. T. and Huang, M. T. (2009). Anti-obesity effects of epigallocatechin-3-gallate, orange peel extract, black tea extract, caffeine and their combinations in a mouse model. *Journal of Functional Foods*, 1: 304–310.

Huang, Z., Wang, B., Williams, P., and Pace, D. (2009). Identification of anthocyanins in muscadine grapes with HPLC-ESI-MS. *Food Science and Technology*. 42(4): 819–824.

Hubbard, B. P., Gomes, A. P., Dai, H., Li, J., Case, A. W., Considine, T. and Sinclair, D. A. (2013). Evidence for a common mechanism of SIRT1 regulation by allosteric activators. *Science*, 339: 1216–1219.

Huber, G. M., Rupasinghe, H. V. and Shahidi, F. (2009). Inhibition of oxidation of omega-3 polyunsaturated fatty acids and fish oil by quercetin glycosides. *Food Chemistry*, 117: 290–295.

Huber, G. M. and Rupasinghe, H. P. V. (2009). Phenolic profiles and antioxidant properties of apple skin extracts. *Journal of Food Science*, 74: C693–C700.

Hubner, G.,Wray, V. and Nahrstedt, A. (1999). Flavonol oligosaccharides from the seeds of *Aesculus hippocastanum*. *Planta Medica*, 65: 636–642.

Hudson, B. J. F., and Lewis, J. I. (1983). Polyhydroxy favonoid antioxidants for edible oils Structural criteria for activity. Food Chemistry, 47, 47-55.

Hudson, B. J. F. and Mahgoub, S. E. O. (1981). Synergism between phospholipids and naturally occurring antioxidants in leaf lipids. *Journal of Science and Food Agriculture*, 32: 208–210.

Hudthagosol, C., Haddad, E. H., McCarthy, K.,Wang, P., Oda, K., and Sabate, J. (2011). Pecans acutely increase plasma postprandial antioxidant capacity and catechins and decrease LDL oxidation in humans. *The Journal of Nutrition, 141*, 56–62.

Hue, S. M., Boyce, A. N. and Somasundram, C. (2012). Antioxidant activity, phenolic and flavonoid contents in the leaves of different varieties of sweet potato ('ipomoea batatas'). *Australian Journal of Crop Science*, 6: 375–380.

Huidrom, D. and Thiyam-Holländer, U. (2012). Canolol as a promising nutraceutical. In U. Thiyam-Holländer, N. A. M. Eskinand B. Matthäus (Eds.), *Canola and rapeseed: Production, processing, food quality, and nutrition* (pp. 337–346). Boca Raton, FL: CRC Press.

Hulse, J. H. (1984). Ed., Polyphenols in cereals *and legumes,* International Development Research Centre Ottawa. Canada.

Hulse, J. H., Ed. (1980). *Polyphenols in Cereals and Legumes,* International Research Development Council, Ottawa, Canada, 1980.

Hulse, J. H., Laing, E. M., and Pearson, O. E. (1980). *Sorghum and the Millets: Their Composition and Nutritive Value.* Academic Press, London.

Hung, P. V., Hatcher, D. W., and Barker W. (2011). Phenolic acid composition of sprouted wheats by ultra-performance liquid chromatography (UPLC) and their antioxidant activities. *Food Chem.,* 126:1896-1901.

Hung, P. V., Maeda, T., Kiyatake, K., and Morita, N. (2009). Total phenolic compounds and antioxidant capacity of wheat graded flours by polishing method. *Food Res. Int.,* 42:185-190.

Hunter J. and Fletcher, J. M. (2002). The antioxidant activity of fresh, frozen, jarred and canned vegetables. *Innov Food Sci. Emerg Technol.* 3:399-406.

Hurrell, R. F. and Finot, P. A. (1985). Effects of food processing on protein digestibility and amino acid availability. In J. W. Finley and D. T. Hopkins, Digestibility and Amino Acid Availability in Cereals and Oilseeds (Chapter 8). Minnesota: Am. Assoc. Cereal Chemists.

Hurtado, I., CalduÂ , P., Gonzalo, A., Ramon, J. M., MõÂnguez, S., and Fiol, C. (1997). Antioxidative capacity of wine on human LDL oxidation in vitro, effect of skin contact in winemaking of white wine. Journal of Agricultural and Food Chemistry, 45, 1283-1289.

Hwang, J. Y., Hwang, I. K. and Park, J. B. (2001). Analysis of physicochemical factors related to the automatic pellicle removal in Korean chestnut (*Castanea crenata*). *Journal of Agricultural and Food Chemistry*, 49: 6045–6049.

Hwang, J. Y., Shue, Y. S. and Chang, H. M. (2001). Antioxidative activity of roasted and defatted peanut kernels. *Food Research International*, 34: 639–647.

Hyson, D., Schneeman, B. O. and Davis, P. A. (2002). Almonds and almond oil have similar effects on plasma lipids and LDL oxidation in healthy men and women. *Journal of Nutrition*, 132: 703–707.

Iamanaka, B. T., Taniwaki, M. H., Menezes, H. C., Vicente, E., and Fungaro, M. H. P. (2005). Incidence of toxigenic fungi and ochratoxin A in dried fruits sold in Brazil. *Food Additives and Contaminants*, *22*, 1258–1263.

IARC (1987) IARC Monographs on the evaluation of the carcinogenic risk of chemicals to humans. Overall evaluation of carcinogenicity: An updating of IARC monograph, International Agency for Research on Cancer, Lyon, France. pp. 59.

IbaÂnÄ ez, E., Oca, A., de Murga, G., LoÂ pez-SebastiaÂ n, S., Tabera, J., and Reglero, G. (1999). Supercritical fuid extraction and fractionation of different preprocessed rosemary plants. Journal of Agricultural and Food Chemistry, 47, 1400-1404.

Ibañez, E., Kubatova, A., Senorans, F. J., Cavero, S., Reglero, U. and Hawthorne, S. B. (2003). Subcritical water extraction of antioxidant compounds from rosemary plants. *Journal of Agricultural and Food Chemistry*, 51: 375–382.

Ibern-Gonzalez, M., Roig-Perez, S., Lameula-Raventos, R. M. and Torre-Boronat, M. C. (2000). Resveratrol and piceid levels in natural and blended peanut butters. *Journal of Agricultural and Food Chemistry*, 48: 6352–6354.

Iglesias, J., Pazos, M., Torres, J. L. and Medina, I. (2012). Antioxidant mechanism of grape procyanidins in muscle tissues: Redox interactions with endogenous ascorbic acid and α-tocopherol. *Food Chemistry*, 134: 1767–1774.

Ikken, Y., Morales, P., MartõÂnez, A., MarõÂn, M. L., Haza, A. I., and Cambero, M. I. (1999). Antimutagenic effect of fruit and vegetable ethanolic extracts against N-nitrosamines evaluated by the Ames test. Journal of Agricultural and Food Chemistry, 47, 3257-3264.

INC. (2014).*World dried fruits production.* Reus, Spain: International Nut and Dried Fruit Council Foundation. Iqbal, S. Z., Asi, M. R., and Jinap, S. (2014). Aflatoxins in dates and dates products. *Food Control, 43,* 163–166.

Innawong, B., Mallikarjunan, P., Irudayaraj, J., and Marcy, J. E. (2004). The determination of frying oil quality using Fourier transform infrared attenuated total reflectance. *Lebensmittel Wissenschaft Und Technologie, 37,* 23–28.

Iqbal, S., Bhanger, M. I. and Anwar, F. (2005). Antioxidant properties and components of some commercially available varieties of rice bran in Pakistan. *Food Chemistry,* 93:265–272.

Irakli, M. N., Samanidou, V. F., Biliaderis, C. G. and Papadoyannis, I. N. (2012). Development and validation of an HPLC-method for determination of free and bound phenolic acids in cereals after solid-phase extraction. *Food Chemistry,* 134: 1624–1632.

Isanga, J. and Zhang, G.-N. (2007). Biologically active components and nutraceuticals in peanuts and related products: review. *Food Reviews International,* 23: 123–140.

Ishiwata, K., Yamaguchi, T., Takamura, H., and Matoba, T. (2004). DPPH radical-scavenging activity and polyphenol content in dried fruits. *Food Science and Technology Research, 10,* 152-156.

Isik, E., Sahin, S., and Demir, C. (2013). Development of a new chromium reducing antioxidant capacity (CHROMAC) assay for plants and fruits. *Talanta, 111,* 119–124.

Islam, M. S., Yoshimoto, M., Yahara, S., Okuno, S., Ishiguro, K. and Yamakawa, O. (2002). Identification and characterization of foliar polyphenolic composition in sweetpotato (*Ipomoea batatas* L.) genotypes. *Journal of Agricultural and Food Chemistry,* 50: 3718–3722.

Ito, A., Shamon, L. A., Yu, B., Mata-Greenwood, E., Lee, S. K., van Breemen, R. B., Mehta, R. G., Farnsworth, N. R., Fong, H. H. S., Pezzuto, J. M., and Kinghorn, A. D. (1998). Antimutagenic constituents of Casimiroa edulis with potential cancer chemopreventive activity. Journal of Agricultural and Food Chemistry, 46, 3509-3516.

Ito, N., Hirose, M., Fukushima, S., Tsuda, H., Shirai, T. and Talematsu, M. (1986). Studies on antioxidants: Their carcinogenic and modifying effects on chemical carcinogenesis. *Food Chemical Toxicology,* 24: 1071–1082.

Ito, Y. and Bhatnagar, R. (1981). Preparative counter current chromatography with a rotating coil assembly, *J. Chromarogr.,* 207: 171.

Iverson, F. (1995). Phenolic antioxidants: Health protection branch studies on butylated hydroxyanisole. *Cancer Letters,* 93: 49–54.

Iverson, F. (1999). *In vivo* studies on butylated hydroxyanisole. *Food and Chemical Toxicology,* 37: 993–997.

Jacobi, H., Eicke, B. and Witte, I. (1998). DNA strand break induction and enhanced cytotoxicity of propyl gallate in the presence of copper(II). *Free Radical Biology and Medicine,* 24: 972–978.

Jacobi, H., Hinrichsen, M. L.,Web. D. and Witte, I. (1999). Induction of lipid peroxidation in human fibroblasts by the antioxidant propyl gallate in combination with copper (II). *Toxicology Letters,* 110: 183–190.

Jacques, D. and Haslam, E. (1974). Biosynthesis of plant proanlhocyanidins. *J. Chem. Sac. D.* 6: 231.

Jacques, D. and Haslam, E. (1974). Plant proanthocyanidins. I. Proanthocyandin A-2 and its derivatives. 1 *Chem. Soc. Perkin Trans. I,* 11: 2663.

Jacques, D., Opic, C. T., Potter, L. J. and Haslam, E. (1977). Plant proanthocyanidin. IV. Biosynthesis of proanthocyanins and observations on the metabolism of cyanidin in plants, J. *Chon. Soc. Perkin Trans.* 1, 14: 1637.

Jadhav, S. J., Lutz, S. E., Ghorpade, V. M., and Salunkhe, D. K. (1998). Barley: chemistry and value-added processing. *Crit. Rev. Food Sci. Nutr.,* 38:123–171.

Jaendet, P., Douillet-Breuil, A. C., Bessis, R., Debord, S., Sbaghi, M. and Adrian, M. (2002). Phytoalexins from the Vitaceae: Biosynthesis, phytoalexin gene expression in transgenic plants, antifungal activity, and metabolism. *Journal of Agricultural and Food Chemistry*, 50: 2731–2741.

Jaganath, I. B. and Crozier, A. (2010). Dietary flavonoids and phenolic compounds. In C. G. Fraga (Ed.), *Plant phenolics and human health* (pp. 1–49). Hoboken, NJ: JohnWiley and Sons, Inc.

Jagtap, U.B.; Panaskar, S.N.; Bapat, V.A. (2010). Evaluation of Antioxidant Capacity and Phenol Content in Jackfruit (Artocarpus heterophyllus Lam.) Fruit Pulp. *Plant Food Hum. Nutr. 65*, 99-104.

Jambunathan, R., Butler, L. G., Bandopadhyay, R., and Mughogho, L. K. (1986). Polyphenol concentrations in grain, leaf and callus tissues of mold-susceptibile and mold-resistant sorghum cultivars, *J. Agric. Food Chem.,* 34: 425.

Jamison, J. R. (2003). Bilberry fruit (*Vaccinium myrtillus*). In J. R. Jamison (Ed.), *Clinical guide to nutrition and dietary supplements in disease management* (pp. 449–451). London, UK: Churchill Livingstone.

Jang, I., Kang, S., and Ko, Y. (2013). Influence of plum (*Prunus mume* Siebold and Zucc.) products on growth performance, intestinal function and immunity in broiler chicks. *Journal of Poultry Science*, *50*, 28–36.

Jang, S. and Xu, Z. (2009). Lipophilic and hydrophilic antioxidants and their antioxidant activity in purple rice bran. *Journal of Agricultural and Food Chemistry*, 57(3), 858-862.

Jardine, D., Antolovich, M., Prenzler, P. D., and Robards, K. (2002). Liquid chromatography-mass spectrometry (LC-MS) investigation of the thiobarbituric acid reactive substances (TBARS) reaction. *Journal of Agricultural and Food Chemistry*, *50*, 1720–1724.

Jariwalla, R. J. (2001). Rice-bran products: Phytonutrients with potential applications in preventive and clinical medicine. *Drugs Exp. Clin. Res.*, 27:17-26.

Jaswir, I., and Che Man, Y. B. (1999). Use optimisation of natural antioxidants in refined bleached, deodorized palm olein during repeated deep-fat frying using response surface methodology. Journal of the American Oil Chemists' Society, 76, 341-348.

JECFA (1996). *Toxicological evaluation of certain food additives and contaminants in food* (pp. 3–86). Who Food Additives Series, No. 35. Geneva, Switzerland: Joint FAO/WHO Expert committee on Food Additives.

Jemima Beryl Mohankumar and Indu Vaishnavi R. (2012). Nutrient and antioxidant analysis of raw and processed minor millets. *Elixir Food Science*. 52A: 11279-11282.

Jensen, H. R. (1985). Cacao tannin and its determination, *Analyst,* 53, 365.

Jensen, J. S.,Wertz, C. F. and O'Neill, V. A. (2010). Preformulation stability of *trans*-resveratrol and *trans*-resveratrol glucoside (piceid). *Journal of Agricultural and Food Chemistry*, 58: 1685–1690.

Jepson, R. G., and Craig, J. C. (2007). A systematic review of the evidence for cranberries and blueberries in UTI prevention. *Molecular Nutrition and Food Research*, *51*, 738–745.

Jerkovic, I.; Marijanovic, Z. (2010). Oak (*Quercus frainetto* Ten.) Honeydew Honey-Approach to Screening of Volatile Organic Composition and Antioxidant Capacity (DPPH and FRAP Assay). *Molecules 15*, 3744-3756.

Jiang, Z. Y.,Woollard, A. C. S., and Wolff, S. P. (1991). Lipid hydroperoxide measurement by oxidation of Fe2+ in the presence of xylenol orange. Comparison with the TBA assay and an iodometric method. *Lipids*, *26*, 853–856.

Jinap, S., Jamilah, B., and Nazamid, S. (2004). Effect of polyphenol concentration on pyrazine formation during cocoa liquor roasting. *Food Chemistry*, *85*, 73–80.

Jirovetz, L., Buchbauer, G., Ngassoum, M. B. and Geissler, M. (2002). Aroma compounds analysis of *Piper nigrum* and *Piper guineense* essential oils from Cameroon using solid-phase microextraction-gas chromatography, solid-phase microextraction-gas chromatography-mass spectrometry and olfactometry. *Journal of Chromatography. A*, 976: 265–275.

Jirovetz, L., Buchbauer, G., Stoilova, I., Stoyanova, A., Krastanov, A. and Schmidt, E. (2006). Chemical composition and antioxidant properties of clove leaf essential oil. *Journal of Agricultural and Food Chemistry*, 54: 6303–6307.

John, S., Sorokin, A. V., and Thompson, P. D. (2007). Phytosterols and vascular disease. *Current Opinion in Lipidology*, *18*, 35-40.

Johnsson, P., Peerlkamp, N., Kamal-Eldin, A., Andersson, R. E., Andersson, R., Lundgren, L. N. and Aman, P. (2002). Polymeric fractions containing phenol glucosides in flaxseed. *Food Chemistry*, 76: 207–212.

Jones, W.T., Broadhurst, R.B., and Lyttleton, J.W. (1976). The condensed tannins of pasture legume species. Phytochemistry. 15: 1407-1409.

Jood, S., Kapoor, A. C., and Singh, R. (1995). Polyphenol and phytic acid contents of cereal grains as affected by insect infestation. Journal of Agricultural and Food Chemistry, 43, 435-438.

Joshi, A. P. K., Rupasinghe, H. P. V., and Khanizadeh, S. (2011). Impact of drying processes on bioactive phenolics, vitamin C, and antioxidant capacity of red-fleshed apple slices. *Journal of Food Processing and Preservation*, *35*, 453–457.

Joslyn, M. A. (1950). *Methods in Food Analysis,* Academic Press, New York. 471.

Joslyn, M. A. and Goldstein, J. L. (1969). Astringency of fruit and fruit products in relation to phenolic content. *Adv. Food Res.,* 13. 179.

Ju, Z., Yuan, Y., Liu, C., Zhan, S. and Wang, M. (1996). Relationships among simple phenol, flavonoid and anthocyanin in apple fruit peel at harvest and scald susceptibility. *Postharvest Biology and Technology*, 8: 83–93.

Juan, M. Y. and Chou, C. C. (2009). Enhancement of antioxidant activity, total phenolic and flavonoid content of black soybeans by solid state fermentation with *Bacillus subtilis* BCRC 14715. *Food Microbiology*, 27: 586–591.

Julkunen-Tiito, R. (1985). Phenolic constituents in the leaves of northern willows, methods for the analysis of certain phenolics. Journal of Agricultural and Food Chemistry, 33, 213-217.

Jung, J. C., Lim, E., Lee, Y., Kang, J. M., Kim, H., Jang, S., Oh, S. and Jung, M. (2009). Synthesis of novel trans-stilbene derivatives and evaluation of their potent antioxidant and neuroprotective effects. *European Journal of Medicinal Chemistry*, 44: 3166–3174.

Jung, M. Y., Kim, J. P., and Kim, S. Y. (1999). Methanolic extract of Coptis japonica Makino reduces photosensitised oxidation of oils. Food Chemistry, 67, 261-268.

Justesen, U. and Knuthsen, P. (2001). Composition of flavonoids in fresh herbs and calculation of flavonoid intake by use of herbs in traditional Danish dishes. *Food Chemistry*, 73: 245–250.

Kadam, S.S., Kute, L.S., Lawande, K.M., and Salunkhe, D.K. (1982). Changes in chemical composition of winged bean during seed development. J. Food Sci. 47: 2051-2053, 2057.

Kadasi, A., Maruniakova, N., Kolesarova, A., Stochmalova, A., Alexa, R., Grossmann, R. and Sirotkin, A. V. (2014). The effect of green tea extracts on steroidogenesis, proliferation and apoptosis in ovarian granulosa cells. *Endocrine Abstracts*, 34: 348.

KaÈhkoÈ nen, M. P., Hopia, A. I., Vuorela, H. J., Rauha, J.-P., Pihlaja, K., Kujala, T. S., and Heinonen, M. (1999). Antioxidant activity of plant extracts containing phenolic compounds. Journal of Agricultural and Food Chemistry, 47, 3954-3962.

Kafkas, E., Özgen, M., Özogˇ ul, Y. and Türemis,, N. (2008). Phytochemical and fatty acid profile of selected red raspberry cultivars: A comparative study. *Journal of Food Quality*, 31: 67–74.

Kago, T. and Terao, J. (1995). Phospholipids increase radical scavenging activity of vitamin E in a bulk oil model system. *Journal of Agricultural and Food Chemistry*, 43: 1450–1454.

Kähkönen, M. P., Hopia, A. I. and Heinonen, M. (2001). Berry phenolics and their antioxidant activity. *Journal of Agriculture and Food Chemistry*, 49: 4076–4082.

Kähkönen, M. P., Hopia, A. L., Vuorela, H. J., Rauha, J. P., Pihlaja, K., Kujala, T. S. and Heinonen, M. (1999). Antioxidant activity of plant extracts containing phenolic compounds. *Journal of Agricultural and Food Chemistry*, 47: 3954–3962.

Kaijser, A., Dutta, P. and Savage, G. (2000). Oxidative stability and lipid composition of macadamia nuts grown in New Zealand. *Food Chemistry*, 71: 67–70.

Kalt, W., Forney, C. F., Martin, A., and Prior, R. L. (1999). Antioxidant capacity, vitamin C. Phenolics, and anthocyanins after fresh storage of small fruits. Journal of Agricultural and Food Chemistry, 47, 4638-4644.

Kalt,W., McDonald, J. E. and Donner, H. (2000). Anthocyanins, phenolics, and antioxidant capacity of processed lowbush blueberry products. *Journal of Food Science*, 65: 390–393.

Kalt,W. and Dufour, D. (1997). Health functionality of blueberries. *Horticultural Technology*, 7: 216–221.

Kamal-Eldin, A. and Appelqvist, L. A. (1996). The chemistry and antioxidant properties of tocopherols and tocotrienols. *Lipids*, 31: 671–701.

Kamal-Eldin, A. and Budilarto, E. (2015). Tocopherols and tocotrienols as antioxidants for food preservation. In F. Shahidi (Ed.), *Handbook of antioxidants for food preservation* (pp. 141–159). Cambridge, UK: Woodhead Publishing Ltd.

Kamath, V. and Rajini, P. S. (2007). The efficacy of cashew nut (*Anacardium occidentale* L.) skin extract as a free radical scavenger. *Food Chemistry*, 103: 428–433.

Kamiloglu, S., and Capanoglu, E. (2013). Investigating the *in vitro* bioaccessibility of polyphenols in fresh and sun-dried figs (*Ficus carica* L.). *International Journal of Food Science and Technology*, *48*, 2621–2629.

Kamiloglu, S., and Capanoglu, E. (2015). Polyphenol content in figs (*Ficus carica* L.): Effect of sun-drying. *International Journal of Food Properties*, *18*, 521–535.

Kamiloglu, S., Pasli, A. A., Ozcelik, B., and Capanoglu, E. (2014). Evaluating the *in vitro* bioaccessibility of phenolics and antioxidant activity during consumption of dried fruits with nuts. *LWT-Food Science and Technology*, *56*, 284–289.

Kamiloglu, S., Pasli, A. A., Ozcelik, B., and Capanoglu, E. (2014). Evaluating the *in vitro* bioaccessibility of phenolics and antioxidant activity during consumption of dried fruits with nuts. *LWT - Food Science and Technology*, *56*, 284–289.

Kanatt, S. R., Paul, P., D'Souza, S. F., and Thomas, P. (1998). Lipid peroxidation in chicken meat during chilled storage as affected by antioxidants combined with low-dose gamma irradiation. Journal of Food Science, 63, 198-200.

Kancheva, V., Slavova-Kazakova, A., Fabbri, D., Dettori, M. A., Delogu, G., Janiak, M. and Amarowicz, R. (2014). Protective effects of equimolar mixtures of monomer and dimer of dehydrozingerone with α-tocopherol and/or ascorbyl palmitate during bulk lipid autoxidation. *Food Chemistry*, 157: 265–274.

Kanellos, P. T., Kaliora, A. C., Liaskos, C., Tentolouris, N. K., Perrea, D., and Karathanos, V. T. (2013). A study of glycaemic response to Corinthian raisins in healthy subjects and in type 2 diabetes mellitus patients. *Plant Foods for Human Nutrition*, *68*, 145–148.

Kanellos, P. T., Kaliora, A. C., Tentolouris, N. K., Argiana, V., Perrea, D., Kalogeropoulos, N., Kountouri, A. M., and Karathanos, V. T. (2014). A pilot, randomized controlled trial to examine the health outcomes of raisin consumption in patients with diabetes. *Nutrition (Burbank, Los Angeles County, Calif.)*, *30*, 358–364.

Kanes, K., Tisserat, B., Berhow, M. and Vandercook, C. (1992). Phenolic composition of various tissues in Rutaceae species. *Phytochemistry*, 31: 967–974.

Kang, H. M. and Saltveit, M. E. (2002). Antioxidant capacity of lettuce leaf tissue increases after wounding. *Journal of Agricultural and Food Chemistry*, 50: 7536–7541.

Kano, M., Takayanagi, T., Harada, K., Makino, K. and Ishikawa, F. (2005). Antioxidant activity of anthocyanins from purple sweet potato, *Ipomoea batatas* cultivar Ayamurasaki. *Bioscience, Biotechnology, and Biochemistry*, 69: 979–988.

Kansci, G., Genot, C., Meynier, A., and Gandemer, G. (1997). The antioxidant activity of carnosine and its consequences on the volatile profiles of liposomes during iron/ascorbate induced phospholipid oxidation. Food Chemistry, 60, 165-175.

Kantamoni, S. (1965). Isolation and identification of the constituents of cashew nut husk. *Leather Science*, 12: 396–398.

Kapadin, G. J., Subha Rao, G., and Morton, J. F. (1983). Herbal tea consumption and esophageal cancer. In *Carcinogens and Mutagens in the Environments. Vol. /II, Naturally Occurring Compounds: Epidemiology and Distribution,* CRC Press, Boca Raton, FL. 3.

Kapoor, I. P. S., Singh, B., Singh, G., Heluani, C. S. D., Lampasona, M. P. D. and Catalan, C. A. N. (2009). Chemistry and *in vitro* antioxidant activity of volatile oil and oleoresins of black pepper (*Piper nigrum*). *Journal of Agricultural and Food Chemistry*, 57: 5358–5364.

Kapusta, I., Janda, B., Szajwaj, B., Stochmal, A., Piacente, S., Pizza, C., Franceschi, F., Franz, C. and Oleszek,W. (2007). Flavonoids in horse chestnut (*Aesculus hippocastanum*) seeds and powdered waste water by-products. *Journal of Agricultural and Food Chemistry*, 55: 8485–8490.

Karaca, H., and Nas, S. (2006). Aflatoxins, patulin, and ergosterol contents of dried figs in Turkey. *Food Additives and Contaminants*, *23*, 502–508.

Karadag, A., Ozcelik, B., and Saner, S. (2009). Review of methods to determine antioxidant capacities. *Food Analytical Methods*, *2*, 41–60.

Karakaya, S. (2004). Bioavailability of phenolic compounds. *Critical Reviews in Food Science and Nutrition*, 44: 453–464.

Karakaya, S., and Nehir, S. (1999). Quercetin, luteolin, apigenin and kaempferol contents of some foods. Food Chemistry, 66, 289-292.

Karamac, M., Amarowicz, R., Weinder, S., Abe, S. and Shahidi, F. (2002). Antioxidant activity of rye caryopses and embryos extracts. *Czech Journal of Food Science*, 20: 209–214.

Karamac, M., and Amarowicz, R. (1997). Antioxidant activity of BHA, BHT and TBHQ examined with Miller's test. Grasas y Aceites, 48, 83-86.

Karaosmanoglu, H. and Kilmartin, P. A. (2015). Tea extracts as antioxidants for food preservation. In F. Shahidi (Ed.), *Handbook of antioxidants for food preservation* (pp. 219–230). Cambridge, UK: Woodhead Publishing Ltd.

Karawita, R., Siriwardhana, N., Lee, K. W., Heo, M. S., Yeo, I. K., Lee, Y. D., and Jeon, Y. J. (2005). Reactive oxygen species scavenging, metal chelation, reducing power and lipid peroxidation inhibition properties of different solvent fractions from *Hizikia fusiformis*. *European Food Research and Technology*, *220*, 363–371.

Kashanian, S. and Dolatabadi, J. E. N. (2009). DNA binding studies of 2-*tert*-butylhydroquinone (TBHQ) food additive. *Food Chemistry*, 116: 743–747.

Kaur, C. and Kapoor, H.C., 2002. Anti-oxidant activity and total phenolic content of some Asian vegetables. International Journal of Food Science and Technology 37, 153-161.

Kawaii, S., Tomono, Y., Katase, E., Ogawa, K., and Yano, M. (1999). HL-60 Differentiating activity and favonoid content of the readily extractable fraction prepared from Citrus juices. Journal of Agricultural and Food Chemistry, 47, 128-135.

Kayano, S. I., Kikuzaki, H., Fukutsuka, N., Mitani, T., and Nakatani, N. (2002). Antioxidant activity of prune (*Prunus domestica* L.) constituents and a new synergist. *Journal of Agricultural and Food Chemistry*, *50*, 3708–3712.

Kayano, S. I., Kikuzaki, H., Ikami, T., Suzuki, T., Mitani, T., and Nakatani, N. (2004). A new bipyrrole and some phenolic constituents in prunes (*Prunus domestica* L.) and their oxygen radical absorbance capacity (ORAC). *Bioscience, Biotechnology, and Biochemistry*, *68*, 942–944.

Keast, D. R., O'Neil, C. E., and Jones, J. M. (2011). Dried fruit consumption is associated with improved diet quality and reduced obesity in US adults: National Health and Nutrition Examination Survey, 1999-2004. *Nutrition Research*, *31*, 460–467.

Kedare, S. B., and Singh, R. P. (2011). Genesis and development of DPPH method of antioxidant assay. *Journal of Food Science and Technology*, *48*, 412–422.

Keen, C. L. (2001). Chocolate: Food as medicine/medicine as food. *Journal of the American College of Nutrition*, 20: 436S–439S.

Kelly, S. A. M., Summerbell, C. D, Brynes, A., Whittaker, V. and Frost, G. (2007). Whole-grain cereals for coronary heart disease. *The Cochrane Database of Systematic Reviews*, 18(2), CD005051.

Kendall, C. W., Josse, A. R., Esfahani, A., and Jenkins, D. J. (2010). Nuts, metabolic syndrome and diabetes. *The British Journal of Nutrition*, *104*, 465–473.

Kenny, O., Smyth, T. J., Hewage, C. M. and Brunton, N. P. (2013). Antioxidant properties and quantitative UPLC-MS analysis of phenolic compounds from extracts of fenugreek (*Trigonella foenum-graecum*) seeds and bitter melon (*Momordica charantia*) fruit. *Food Chemistry*, 141: 4295–4302.

Keyrouz, R., Abasq, M. L., le Bourvellec, C., Blanc, N., Audibert, L., ArGall, E., and Hauchard, D. (2011). Total phenolic contents, radical scavenging and cyclic voltammetry of seaweeds from Brittany. *Food Chemistry*, *126*, 831–836.

Khan, M. A., and Shahidi, F. (2000). Oxidative stability of stripped and non-stripped borage and evening primrose oils and their emulsions in water. *Journal of the American Oil Chemists' Society*, *77*, 963–968.

Khan, M. R. and Shahidi, F. (2001). Effects of natural and synthetic antioxidants on the oxidative stability of borage and evening primrose triacylglycerols. *Food Chemistry*, 75: 431–437.

Khan, N. and Mukhtar, H. (2010). Green tea catechins: Anticancer effects and molecular targets. In C. G. Fraga (Ed.), *Plant phenolics and human health* (pp. 1-49). Hoboken, NJ: JohnWiley and Sons, Inc.

Khanal, R. C., Rogers, T. J., Wilkes, S. E., Howard, L. R., and Prior, R. L. (2010). Effects of dietary consumption of cranberry powder on metabolic parameters in growing rats fed high fructose diets. *Food and Function*, *1*, 116–123.

Khanbabaee K, Van Ree T (2001) Tannins: classification and definition. Nat Prod Rep 18:641–649.

Khanbabaee, K. and van Ree, T. (2001). Tannins: Classification and definition. *Natural Product Report*, 18: 641–649.

Khanizadeh, S., Tsao, R., Rekika, D., Yang, R., Charles, M. T. and Rupasinghe, H. P. V. (2008). Polyphenol composition and total antioxidant capacity of selected apple genotypes for processing. *Journal of Food Composition and Analysis*, 21: 396–401.

Khanna S, Roy S, Slivka A, Craft TK, Chaki S, Rink C, Notestine MA, DeVries AC, Parinandi NL, Sen CK; Roy; Slivka; Craft; Chaki; Rink; Notestine; Devries; Parinandi; Sen (2005). "Neuroprotective Properties of the Natural Vitamin E α-Tocotrienol". Stroke 36(10): 2258–64.

Khanna, S. K., Viswanathan, P. N., Krishnan, P. S., and Sanwal, G. G. (1968). Extraction of total phenolics in the presence of reducing agents, *Phytochemistry.* 7, 1513.

Kikugawa, K., Kunugi, A. and Kurechi, T. (1990). Chemistry and implication of degradation of phenolic antioxidants. In B. J. F. Hudson (Ed.), *Food antioxidants* (pp. 65–98). Amsterdam: Elsevier.

Kikuzaki, H., and Nakatani, N. (1993). Antioxidant effects of some ginger constituents. Journal of Food Science, 58, 1407-1410.

Kikuzaki, H., Tsai, S. M. and Nakatani, N. (1992). Gingerdiol related compounds from the rhizomes of *Zingiber officinale*. *Phytochemistry*, 43: 273–277.

Kikuzaki, H. and Nakatani, N. (1993). Antioxidant effect of some ginger constituents. *Journal of Food Science*, 58: 1407–1410.

Kil, H. Y., Seong, E. S., Ghimire, B. K., Chung, I. M., Kwon, S. S., Goh, E. J., Heo, K., Kim, M. J., Lim, J. D., Lee, D. and Yu, C. Y. (2009). Antioxidant and antimicrobial activities of crude sorghum extract. *Food Chemistry*, 115: 1234–1239.

Kim, H. and Keeney, P. G. (1984). (-)-Epicatechin content in fermented and unfermented cocoa beans. *Journal of Agricultural and Food Chemistry*, 47: 3693–3701.

Kim, J. H., Kang, N. J., Lee, B. K., Lee, K. W. and Lee, H. J. (2008). Gallic acid, a metabolite of the antioxidant propyl gallate, inhibits gap junctional intercellular communication via phosphorylation of connexin 43 and extracellular-signalregulated kinase 1/2 in rat liver epithelial cells. *Mutation Research/Fundamental and Molecular Mechanisms of Mutagenesis*, 638: 175–183.

Kim, J. Y., Kim, J. H., Lee, D. H., Kim, S. H. and Lee, S. S. (2008). Meal replacement with mixed rice is more effective than white rice in weight control, while improving antioxidant enzyme activity in obese women. *Nutrition Research*, 28(2), 66-71.

Kim, K.-H., Tsao, R., Yang, R., and Cui, S. W. (2006). Phenolic acid profiles and antioxidant activities of wheat bran extracts and the effects of hydrolysis conditions. *Food Chem.*, 95:466–473.

Kim, M. J., Chung, J. Y., Kim, J. H., and Kwak, H. K. (2013). Effects of cranberry powder on biomarkers of oxidative stress and glucose control in db/db mice. *Nutrition Research and Practice*, *7*, 430–438.

Kim, M. J., Ohn, J., Kim, J. H., and Kwak, H. K. (2011). Effects of freeze-dried cranberry powder on serum lipids and inflammatory markers in lipopolysaccharide treated rats fed an atherogenic diet. *Nutrition Research and Practice*, *5*, 404–411.

Kim, M., Rhee, S. K., and Cheigh, H. S. (1995). Effect of tocopherols and carotene on the oxidation of purified pinenut oil in the model system. *Journal of the Korean Society of Food Science and Nutrition*, *24*, 60–66.

Kim, M.-J., Hyun, J.-N., Kim, J.-A., Park, J.-C., Kim, M.-Y., Kim, J.-G., Lee, S.-J., Chun, S.-C., and Chung, I.-M. (2007). Relationship between phenolic compounds, anthocyanins content and antioxidant activity in colored barley germplasm. *J. Agric. Food Chem.*, 55:4802-4809.

Kim, S. Y., Kim, J. H., Kim, S. K., Oh, M. J., and Jung, M. Y. (1994). Antioxidant activity of selected oriental herb extracts. Journal of the American Oil Chemists' Society, 71, 633-640.

Kim, T. S., Yeo, J. D., Kim, J. Y., Kim, M. J., and Lee, J. H. (2013). Determination of the degree of oxidation in highly-oxidised lipids using profile changes of fatty acids. *Food Chemistry*, *138*, 1792–1799.

Kim, Y., Hertzler, S. R., Byrne, H. K., and Mattern, C. O. (2008). Raisins are a low to moderate glycaemic index food with a correspondingly low insulin index. *Nutrition Research*, *28*, 304–308.

Kimura, H., Ogawa, S., Jisaka, M., Kimura, Y., Katsube, T. and Yokota, K. (2006). Identification of novel saponins from edible seeds of Japanese horse chestnut (*Aesculus turbinata* BLUME) after treatment with wooden ashes and their nutraceutical activity. *Journal of Pharmaceutical and Biomedical Analysis*, 41: 1657–1665.

King, H. G. C. and Pruden, G. (1970). Lower limit-, of molecular weights of compounds excluded from Sephadex G-25 eluted with aqueous acetone mixture. Application of the results to the separation of the components of tannic acid. J. *Chromatogram,* 52, 285.

Kiokias, S. (2006). Botanicals as ingredients of functional foods. *Greek Journal of Food and Drinks*, 296: 76–79.

Kiokias, S., Varzakas, T. and Oreopoulou, V. (2008). In vitro activity of vitamins, flavanoids, and natural phenolic antioxidants against the oxidative deterioration of oil-based systems. *Critical Reviews in Food Science and Nutrition*, 48: 78–93.

Kirakosyan, A., Seymour, E. M., Noon, K. R., Llanes, D. E. U., Kaufman, P. B.,Warber, S. L. and Bolling, S. F. (2010). Interactions of antioxidants isolated from tart cherry (*Prunus cerasus*) fruits. *Food Chemistry*, 122: 78–83.

Kiritsakis, A., Kanavouras, A., and Kiritsakis, K. (2002). Chemical analysis, quality control and packaging issues of olive oil. *European Journal of Lipid Science and Technology*, *104*, 628–638.

Kirkpatrick, D. C. and Lauer, B. H. (1986). Intake of phenolic antioxidants from foods in Canada. *Food and Chemical Toxicology*, 24: 1035–1037.

Kitagaki, H., and Tsugawa, M. (1999). 1,1-Diphenil-2-picrylhydrazyl radical (DPPH) scavenging ability of sake during storage. Journal of Bioscience and Bioengineering, 87, 328-332.

Kitta, K., Hagiwara, Y. and Shibamoto, T. (1992). Antioxidative activity of an isoflavonoid, 2''-*O*-glycosylisovitexin isolated from green barley leaves. *Journal of Agricultural and Food Chemistry*, 40: 1843–1845.

Kivcak, B. and Akay, S. (2005). Quantitative determination of α-tocopherol in *Pistacia lentiscus, Pistacia lentiscus* var. chia, and *Pistacia terebinthus* by TLC densitometry and colorimetry. *Fitoterapia*, 76: 62–66.

Klaus, S., Pultz, S., Thone-Reineke, C. and Wolfram, S. (2005). Epigallocatechin gallate attenuates diet-induced obesity in mice by decreasing energy absorption and increasing fat oxidation. *International Journal of Obesity*, 29: 615–623.

Kljak, K. and Grbeša, D. (2015). Carotenoids content and antioxidant activity of hexane extracts from selected Croatian corn hybrids. *Food Chemistry*, 167, 402-408.

Knight, T. E., and Hausen, B. M. (1996). Dermatitis in a nutshell: Occupational exposure to *Macadamia integrifolia. Journal of the American Academy of Dermatology, 35*, 482–484.

Kobayashi-Hattori, K., Mogi, A., Matsumoto, Y. and Takita, T. (2005). Effect of caffeine on the body fat and lipid metabolism of rats fed on a high-fat diet. *Bioscience, Biotechnology, and Biochemistry*, 69: 2219–2223.

Kochar, J., Gaziano, J. M., and Djoussé, L. (2010). Nut consumption and risk of type II diabetes in the Physicians' Health Study. *European Journal of Clinical Nutrition*, *64*, 75–79.

Koeppen, B. H. (1964). Structure and interrelation of orient in and homorientin, Z. *Naturforsch.* 19b, 173.

Koeppen, B. H. and Herrmann, K. (1977). Flavonoid glycosides and hydroxy-cinnamic acid esters of blackcurrants (*Ribes nigrum*). Phenolics of fruits 9. *Zeitschrift für Lebensmittel-Untersuchung und-Forschung*, 164: 263–268.

Koga, T., Moro, K., Nakamori, K., Yamakoshi, J., Hosoyama, H., Kataoka, S., and Ariga, T. (1999). Increase of antioxidative potential of rat plasma by oral administration of proanthocyanidin-rich extract from grape seeds. Journal of Agricultural and Food Chemistry, 47, 1892-1897.

Kohyama, N., Nagata, T., Fujimoto, S.-I. and Sekiya, K. (1997). Inhibition of arachidonate lipoxygenase activities by 2-(3,4-dihydroxyphenyl)ethanol, a phenolic compound from olives. *Bioscience, Biotechnology, and Biochemistry*, 61: 347–350.

Koide, T., Kamei, H., Hashimoto, Y., Kojima, T., and Hasegawa, M. (1996). Antitumor effect of hydrolyzed anthocyanin from grape rinds and red rice. *Cancer Biotherapy and Radiopharmaceuticals*. 11: 273-277.

Kong, J. M., Chia, L. S., Goh, N. K., Chia, T. F. and Brouillard, R. (2003). Analysis and biological activities of anthocyanins. *Phytochemistry*, 64: 923–933.

Kong, S. and Lee, J. (2010). Antioxidants in milling fractions of black rice cultivars. *Food Chemistry*, 120: 278–281.

Kontogianni, V. G., Tomic, G., Nikolic, I., Nerantzaki, A. A., Sayyad, N., Stosic-Grujicic, S., Stojanovic, I., Gerothanassis, L. P. and Tzakos, A. G. (2013). Phytochemical profile of *Rosmarinus officinalis* and *Salvia officinalis* extracts and correlation to their antioxidant and anti-proliferative activity. *Food Chemistry*, 136: 120–129.

Kornsteiner, M., Wagner, K. H. and Elmadfa, I. (2006). Tocopherols and total phenolics in 10 different nut types. *Food Chemistry*, 98: 381–387.

Korycinska, M., Czelna, K., Jaromin, A., and Kozubek, A. (2009). Antioxidant activity of rye bran alkylresorcinols and extracts from whole-grain cereal products. *Food Chem.*, 116:1013-1018.

Koshihara, Y., Neichi, T., Murota, S. I., Lao, A. N., Fujimoto, Y. and Tatsuno, T. (1984). Caffeic acid is a selective inhibitor for leukotriene biosynthesis. *Biochimica Biophysica Acta*, 792: 92–97.

Kozlowska, H., Zadernowski, R. and Sosulski, F. W. (1983). Phenolic acids in oilseed flour. *Die Nahrung*, 27: 449–453.

Kozubek, A., and Tyman, J. H. P. (1999). Resorcinolic lipids, the natural nonisoprenoid nonisoprenoid phenolic amphiphiles and their biological activity. *Chem. Rev.*, 99:1-25.

Kramer, R. P., Hindorf, H., Zha, H. C., Kallage, J., and Zilliken, F. (1984). Antifungal activity of soybean and chickpea isoflavones and their reduced derivatives, *Phytochemistry,* 23, 2203.

Kris-Etherton, P. M., Yu-Poth, S., Sabaté, J., Ratcliffe, H. E., Zhao, G. and Etherton, T. D. (1999). Nuts and their bioactive constituents: Effects on serum lipids and other factors that affect disease risk. *American Journal of Clinical Nutrition*, 70: 504S–511S.

Kris-Etherton, P. M., Zhao, G., Binkoski, A. E., Coval, S. M. and Etherton, T. D. (2001). The effects of nuts on coronary heart disease risk. *Nutrition Reviews*, 59: 103–111.

Krishnamurthy Rao, B. H. and Murthy, K. S. R. K. (1983). Eds., *Crop Losses due 10 bisect Pests.* Entomological Society of India. Hyderabad. India.

Kristinova, V., Mozuraityte, R., Storrø, I. and Rustad, T. (2009). Antioxidant activity of phenolic acids in lipid oxidation catalyzed by different prooxidants. *Journal of Agricultural and Food Chemistry*, 57: 10377–10385.

Kruglov, A. G., Nikiforova, A. B., Shatalin, Y. V., Shubina, V. V., Fisyuk, A. S., and Akatov, V. S. (2010). Sulfur-containing compounds quench 3,7- dihydro-2-methyl-6-(4-methoxyphenyl) imidazole [1,2-a] pyrazine-3-one chemiluminescence: Discrimination between true antioxidants and quenchers using xanthine oxidase. *Analytical Biochemistry*, *406*, 230–232.

Kruidenier, L., and Verspaget, H. W. (2012). Review article: Oxidative stress as a pathogenic factor in inflammatory bowel disease –Radicals or ridiculous? *Alimentary Pharmacology and Therapeutics*, *16*, 1997–2015.

Krygier, K., Sosulski, F., and Hodge, L. (1982a). Free, esterified, and insoluble bound phenolic acids. 1. Extraction and purification procedure. *J. Agric. Food Chem.*, 30:330-334.

Krygier, K., Sosulski, F., and Hogge, L. (1982b). Free, esterified and insolublebound phenolic acids. 2. Composition of phenolic acids in rapeseed flour and hulls. *J. Agric. Food Chem.*, 30:334-336.

Kubicka, E., JcÎedrychowski, L., and Amarowicz, R. (1999). Effect of phenolic compounds extracted from sunfower seeds on native lipoxygenase activity. Grasas y Aceites, 50, 3206-3209.

Kubola, J. and Siriamornpun, S. (2008). Phenolic contents and antioxidant activities of bitter gourd (Momordica charantia L.) leaf stem and fruit extracts *in vitro. Food Chemistry*, 110: 881–890.

Kuhnau, J. (1976). The flavonoids. A Class of semiessential food components: their role in human nutrition *World Rev. Nutr. Diet.* 24: 117.

Kuhnle, G. G. C., Dell'Aquila, C., Aspinall, S. M., Runswick, S. A., Joosen, A. M. C. P., Mulligan, A. A., and Bingham, S. A. (2009). Phytoestrogen content of fruits and vegetables commonly consumed in the UK based on LC-MS and 13C-labelled standards. *Food Chemistry*, *116*, 542–554.

Kuiper, G. G. J. M., Carlsson, B., Grandien, K., Enmark, E., Häggblad, J., Nilsson, S. and Gustafsson, J. A. (1997). Comparison of the ligand binding specificity and transcript tissue distribution of estrogen receptors α and β. *Endocrinology*, 138: 863–870.

Kujala, T. S., Loponen, J. M., Klika, K. D. and Pihlaja, K. (2000). Phenolics and betacyanins in red beetroot (*Beta vulgaris*) root: Distribution and effect of cold storage on the content of total phenolics and three individual compounds. *Journal of Agricultural and Food Chemistry*, 48: 5338–5342.

Kujala, T., Klika, K., Ovcharenko, V., Loponen, J., Vienola, M. and Pihlaja, K. (2001). 5,5′,6,6′-Tetrahydroxy.3,3′-biindolyl from beetroot (Beta vulgaris) peel extract. *Zeitschrift fur Naturforschung - Section C Journal of Biosciences*, 56: 714–718.

Kumar, R. and Singh, M. (1984). Tannins: their adverse role in ruminant nutrition. J. Agric. Food Chem. 32: 447-453.

Kundgol, N. G., Kasturiba, B., Kk Math and Kamatar, M. Y. (2014). Effect of preliminary processing on antiradical properties of little millet landraces. *International Journal of Farm Sci.* 4(2): 148-154.

Kundu, J. K., and Surh, Y.-J. (2013). Cancer chemopreventive effects of selected dried fruits. In C. Alasalvar and F. Shahidi (Eds.), *Dried fruits: Phytochemicals and health effects* (pp. 19–51). Oxford: Wiley-Blackwell.

Kuo, J. M., Yeh, D. B., and Pan, B. S. (1999). Rapid photometric assay evaluating antioxidant activity in edible plant material. Journal of Agricultural and Food Chemistry, 47, 3206-3209.

Kurilich, A. C., and Juvik, J. A. (1999). Quantification of carotenoid and tocopherol antioxidants in Zea mays. Journal of Agricultural and Food Chemistry, 47, 1948-1955.

Kurilich, A. C., Tsau, G. J., Brown, A., Howard, L., Klein, B. P., Jeffery, E. H., Kushad, M.,Wallig, M. A. and Juvik, J. A. (1999). Carotene, tocopherol, and ascorbate contents in subspecies of Brassica oleracea. *Journal of Agricultural and Food Chemistry*, 47: 1576–1581.

Kurilich, A. C. and Juvik, J. A. (1999). Quantification of carotenoids and tocopherol antioxidants in *Zea mays*. *Journal of Agricultural and Food Chemistry*, 47: 1948–1955.

Kwak, C. S., Lim, S. J., Kim, S. A., Park, S. C., and Lee, M. S. (2004). Antioxidative and antimutagenic effects of Korean buckwheat, sorghum, millet and job's tears. *J. Korean Soc. Food Sci. Nutr,* 33:921–929.

Labuckas, D. O., Maestri, D. M., Perelló, M., Martínez, M. L. and Lamarque, A. L. (2008). Phenolics from walnut (*Juglans regia* L.) kernels: Antioxidant activity and interactions with proteins. *Food Chemistry*, 107: 607–612.

Lachman, J., Hamouz, K., Orsak, M., Pivec, V. and Dvorak, P. (2008). The influence of flesh colour and growing locality on polyphenolic content and antioxidant activity in potatoes. *Scientia Horticulturae*, 117: 109–114.

Ladikos, D., and Lougovois, V. (1990). Lipid oxidation in muscle foods: A review. *Food Chemistry*, *35*, 295–314.

Lai, C. S., Li, S., Miyauchi, Y., Suzawa, M., Ho, C. T. and Pan, M. H. (2013). Potent anti-cancer effects of citrus peel flavonoids in human prostate xenograft tumors. *Food and Function*, 4: 944–949.

Lai, C.-S., Li, S., Chai, C.-Y., Lo, C.-Y., Ho, C.-T.,Wang, Y.-J. and Pan, M.-H. (2007). Inhibitory effect of citrus 5-hydroxy-3,6,7,8,3. 4′-hexamethoxyflavone on 12-*O* tetradecanoylphorbol 13-acetate-induced skin inflammation and tumor promotion in mice. *Carcinogenesis*, 28: 2581–2588.

Lai, P., Li, K. Y., Lu, S. and Chen, H. H. (2009). Phytochemicals and antioxidant properties of solvent extracts from *Japonica* rice bran. *Food Chemistry*, 117: 538–544.

Lainas, K., Alasalvar, C., and Bolling, B. W. (2016). Effects of roasting on proanthocyanidin contents of Turkish Tombul hazelnut and its skin. *Journal of Functional Foods*, *23*, 647–653.

Lakshmi Kumari, P., and Sumathi,S.(2002).Effect of consumption of finger millet on hyperglycemia in non-insulin dependent diabetes mellitus (NIDDM) subjects. Plant Foods for Human Nutrition, 57: 205–213.

Lalas, S. and Dourtoglou, V. (2003). Use of rosemary extract in preventing oxidation during deep-fat frying of potato chips. *Journal of the American Oil Chemists' Society*, 80: 579–583.

Landete, J. M. (2012). Updated knowledge about polyphenols: Functions, bioavailability, metabolism, and health. *Critical Reviews in Food Science and Nutrition*, *52*, 936–948.

Langcake, P. and Pryce, R. J. (1976). The production of resveratrol by *Vitis vinifera* and other members of the Vitaceae as a response to infection or injury. *Physiological Plant Pathology*, 9: 77–86.

Lantto, T. A., Dorman, H. J. D., Shikov, A. N., Pozharitskaya, O. N., Makarov, V. G., Tikhonov, V. P., Hiltunen, R. and Raasmaja, A. (2009). Chemical composition, antioxidative activity and cell viability effects of a Siberian pine (*Pinus sibirica* Du Tour) extract. *Food Chemistry*, 112: 936–943.

Lapcik, O., Hill, M., Hampl, R.,Wahala, K. and Adlercreutz, H. (1998). Identification of isoflavanoids in beer. *Steroids*, 63: 14–20.

Lapidot, T., Harel, S., Akiri, B., Granit, R., and Kranner, J. (1999). pH-Dependent forms of red-wine anthocyanins as antioxidants. Journal of Agricultural and Food Chemistry, 47, 67-70.

Larrauri, J. A., RupeÂ rez, P., and Saura-Calixto, F. (1997). Effect of drying temperature on the stability of polyphenols and antioxidant activity of red grape pomace peels. Journal of Agricultural and Food Chemistry, 44, 1390-1393.

Larrauri, J. A., SaÂnchez-Moreno, C., and Saura-Calixto, F. (1998). Effect of temperature on the free radical scavenging capacity of extracts from red and white grape pomace peels. Journal of Agricultural and Food Chemistry, 46, 2694-2697.

Larrauri, J. A., SaÂnchez-Moreno, C., RupeÂ rez, P., and Saura-Calixto, F. (1999). Free radical scavenging capacity in the aging of selected red spanish wines. Journal of Agricultural and Food Chemistry, 47, 1603-1606.

Larrosa, M., Tomás-Barberán, F. A. and Espín, J. C. (2003). Grape polyphenol resveratrol and the related molecule 4-hydroxystilbene induce growth inhibition, apoptosis, S-phase arrest, and upregulation of cyclins A, E, and B1 in human SK-Mel-28 melanoma cells. *Journal of Agricultural and Food Chemistry*, 51: 4576–4584.

Lattanzio V, Kroon P, Quideau S, Treutter D (2008) Plant phenolics-secondary metabolites with diverse functions. In: Daayf F, Lattanzio V (Eds) Recent advances in polyphenol research, vol 1. Wiley-Blackwell, Oxford.

Laurena, A. C. Van Den, T., and Mendoza. E. M. T. (1984). Effects of condensed tannins on the *in vitro* protein digestibility of cowpeas *(Vigna unguiculata* L. Walp), *J. Agric. Food Chem,* 32. 1045.

Lavelli, V., and Vantaggi, C. (2009). Rate of antioxidant degradation and color variations in dehydrated apples as related to water activity. *Journal of Agricultural and Food Chemistry*, *57*, 4733–4738.

Lavelli, V. and Bondesan, L. (2005). Secoiridoids, tocopherols, and antioxidant activity of monovarietal extravirgin olive oils extracted form destoned fruits. *Journal of Agricultural and Food Chemistry*, 53: 1102–1107.

Lea, C. H. and Ward, R. J. (1959). Relative antioxidant activities of the seven tocopherols. *Journal of the Science of Food and Agriculture*, 10: 537–548.

Leavy, O. (2014). Inflammation: Regulating ROS: Nature Review. *Nature Reviews Immunology*, 14, 357.

Leclercq, C., Arcella, D. and Turrini, A. (2000). Estimates of the theoretical maximum daily intake of erythorbic acid, gallates, butylated hydroxyanisole (BHA) and butylated hydroxytoluene (BHT) in Italy: A stepwise approach. *Food and Chemical Toxicology*, 38: 1075–1084.

Lee, B. J., and Hendricks, D. G. (1997). Antioxidant effects of L-carnosine on liposomes and beef homogenates. Journal of Food Science, 62, 931-934.

Lee, B. J., Hendricks, D. G., and Cornforth, D. P. (1998). Antioxidant effects of carnosine and phytic acid in a model beef system. Journal of Food Science, 63, 394-398.

Lee, D. M. (1994). α-Phenyl *N-tert*-butylnitrone as antioxidant for low density lipoproteins. *Methods in Enzymology, 234*, 513–523.

Lee, H. J., Ju, J., Paul, S., So, J. Y., DeCastro, A., Smolarek, A., Lee, M. J., Yang, C. S., Newmark, H. L. and Suh, N. (2009). Mixed tocopherols prevent mammary tumorigenesis by inhibiting estrogen action and activating PPAR-γ. *Clinical Cancer Research*, 15: 4242–4249.

Lee, H. S. (2002). Characterization of major anthocyanins and the color of red-fleshed Budd blood orange (*Citrus sinensis*). *Journal of Agricultural and Food Chemistry*, 50: 1243–1246.

Lee, J. M., Kim, D. H., Chang, P. S., and Lee, J. H. (2007). Headspacesolid phase microextraction (HS-SPME) analysis of oxidized volatiles from free fatty acids (FFA) and application for measuring hydrogen donating antioxidant activity. *Food Chemistry*, *105*, 414–420..

Lee, J., Lee, S., Lee, H., Park, K. and Choe, E. (2002). Spinach (*Spinacia oleracea*) powder as a natural food-grade antioxidant in deepfat-fried products. *Journal of Agricultural and Food Chemistry*, 50: 5664–5669.

Lee, J. and Wrolstad, R. E. (2004). Extraction of anthocyanins and polyphenolics from blueberry processing waste. *Journal of Food Science*, 69: 564–573.

Lee, K. and Shibamoto, T. (2001). Antioxidant property of aroma extract isolated from clove buds. *Food Chemistry*, 50: 4947–4952.

Lee, K. G. and Shibamoto, T. (2002). Antioxidant property of aroma extract isolated from clove buds [*Syzygium aromaticum* (L.) Merr. Et Perry]. *Food Chemistry*, 74: 443–448.

Lee, K. W., Kim, Y. J., Lee, H. J. and Lee, C. Y. (2003). Cocoa has more phenolic phytochemicals and a higher antioxidant capacity than teas and red wine. *Journal of Agricultural and Food Chemistry*, 51: 7292–7295.

Lee, S.H., Chung, I. M., Cha, Y. S., and Park, Y. (2010). Millet consumption decreased serum concentration of triglyceride and C-reactive protein but not oxidative status in hyperlipidemic rats. Nutrition Resear ch, 30: 290–296.

Lee, W. K. and Lee, H. J. (2006). The roles of polyphenols in cancer chemoprevention. *Biofactors (Oxford, England)*, 26: 105–121.

Lehtinen, P., and Laakso, S. (1998). Effect of extraction conditions on the recovery and potency of antioxidants in oat fiber. Journal of Agricultural and Food Chemistry, 46, 4842-4845.

Lehtinen, P. and Laakso, S. (1997). Antioxidative-like effect of different cereals and cereal fractions in aqueous suspension. *Journal of Agricultural and Food Chemistry*, 45: 4606–4611.

Leitao, C., Marchioni, E., Bergaentzle, M., Zhao, M., Didierjean, L., Taidi, B. and Ennahar, S., (2011). Effects of processing steps on the phenolic content and antioxidant activity of beer. Journal of Agricultural and Food Chemistry 59, 1249-1255.

Leja, M., Kamin□ska, I., Kramer, M., Maksylewicz-Kaul, A., Kammerer, D., Carle, R. and Baranski, R. (2013). The content of phenolic compounds and radical scavenging activity varies with carrot origin and root color. *Plant Foods for Human Nutrition*, 68: 163–170.

Leontowicz, H., Leontowicz, M., Gorinstein, S., Martin-Belloso, O., and Trakhtenberg, S. (2007). Apple peels and pulp as a source of bioactive compounds and their influence digestibility and lipid profile in normal and atherogenic rats-in English. *MedycynaWeterynaryjna*, *63*, 1434–1436.

Leopoldini M, Russo N, Toscani M (2011). The molecular basis of working mechanism of natural polyphenolic antioxidants. Food Chem 125:288–306.

Leung, J., Fenton, T. W., and Clandinin, D. R. (1981). Phenolic components of sunfower four. *Journal of Food Science*, 46, 1386-1393.

Leung, J., Fenton, T.W., Mueller, M.M., and Clandinin, D.R. (1979). Condensed tannins of rapeseed meal. J. Food Sci. 44: 1313-1316.

Lever, E., Cole, J., Scott, S. M., Emery, P. W., and Whelan, K. (2014). Systematic review: The effect of prunes on gastrointestinal function. *Alimentary Pharmacology and Therapeutics*, *40*(7), 750–758.

Levis, S., Strickman-Stein, N., Doerge, D. R. and Krischer, J. (2010). Design and baseline characteristics of The Soy Phytoestrogens As Replacement Estrogen (SPARE) Study – A clinical trial of the effects of soy isoflavones in menopausal women. *Contemporary Clinical Trials*, 31: 293–302.

Lewak, S. (1968). Determination of the degree of polymerization of leucoantheyanidins, *Phytochemistry,* 7,665.

Lewak, S. (1968). Determination of the degree of polymerization of leucoanthocyanidins, *Phytochemistry,* 7.655.

Li AS, Bandy B, Tsang SS, Davison AJ (2000) DNA-breaking versus DNA-protecting activity of four phenolic compounds in vitro. Free Radic Res 33:551–566.

Li, F., Li, S., Li, H. B., Deng, G. F., Ling,W. H.,Wu, S., Xu, X. R., and Chen, F. (2013). Antiproliferative activity of peels, pulps and seeds of 61 fruits. *Journal of Functional Foods*, *5*, 1298–1309.

Li, H.,Wang, X., Li, Y., Li, P. and Wang, H. (2009). Polyphenolic compounds and antioxidant properties of selected China wines. *Food Chemistry*, 112: 454–460.

Li J, Chen Z, Guan X, Liu J, Zhang M, and Xu B. (2008). Optimization of germination conditions to enhance hydroxyl radical inhibition by water soluble protein from stress millet. *J Cereal Sci.* 48:619–24.

Li J, Chen Z, Yao H. and Xu, Y. (2007). Optimization of stress medium enhance hydroxyl radical inhibition by water-soluble protein from germinated millet. *LWT-Food Sci. Technol.* 40:1630–6.

Li, L., Tsao, R., Yang, R., Kramer, J. K. G. and Hernandez, M. (2007). Fatty acid profiles, tocopherol contents, and antioxidant activities of heartnut (*Juglans ailanthifolia* Var. *cordiformis*) and Persian walnut (*Juglans regia* L.). *Journal of Agricultural and Food Chemistry*, 55: 1164–1169.

Li, L., Tsao, R., Yang, R., Liu, C. M., Zhu, H. H. and Young, J. C. (2006). Polyphenolic profiles and antioxidant activities of heartnut (*Juglans ailanthifolia* var. *cordiformis*) and Persian walnut (*Juglans regia* L.). *Journal of Agricultural and Food Chemistry*, 54: 8033–8040.

Li, S. J., Seymour, A. J., King, A. J., and Morrissey, M. T. (1998). Color stability and lipid oxidation of rockfish as affected by antioxidant from shrimp shell waste. Journal of Food Science, 63, 438-441.

Li, S., Pan, M. H., Lo, C. Y., Tan, D.,Wang, Y., Shahidi, F. and Ho, C.-T. (2009). Chemistry and health effects of polymethoxyflavones and hydroxylated polymethoxyflavones. *Journal of Functional Foods*, 1: 2–12.

Li, W., Wei, C., White, P. J., and Beta, T. (2007). High-amylose corn exhibits better antioxidant activity than typical and waxy genotypes. *J. Agric. Food Chem.*, 55:291-298.

Lichtenstein, A. H., Appel, L. J., Brands, M., Carnethon, M., Daniels, S., Franch, H. A., Franklin, B., Kris-Etherton, P., Harris, W. S., Howard, B., Karanja, N., Lefevre, M., Rudel, L., Sacks, F., Van, H. L.,Winston, M., and Wylie-Rosett, J. (2006). Diet and lifestyle recommendations revision 2006: A scientific statement from the American Heart Association Nutrition Committee. *Circulation*, *114*, 82–96.

Liggins, J., Bluck, L. J. C., Runswick, S., Atkinson, C., Coward,W. A., and Bingham, S. A. (2000). Daidzein and genistein content of fruits and nuts. *Journal of Nutritional Biochemistry*, *11*, 326–331.

Lim, Y. Y., Lim, T. T. and Tee, J. J. (2007). Antioxidant properties of several tropical fruits: A comparative study. *Food Chemistry*, 103: 1003–1008.

Lima, G. P. P., Lopes, T. D. V. C., Rossetto, M. R. M. and Vianello, F. (2009). Nutritional composition, phenolic compounds, nitrate content in eatable vegetables obtained by conventional and certified organic grown culture subject to thermal treatment. *Int. J. Food Sci. Technol.* 44: 1118-1124.

Lin, J. T., Liu, S. C., Hu, C. C., Shyu, Y. S., Hsu, C. Y., and Yang, D. J. (2016). Effects of roasting temperature and duration on fatty acid composition, phenolic composition, Maillard reaction degree and antioxidant attribute of almond (*Prunus dulcis*) kernel. *Food Chemistry*, *190*, 520–528.

Lin, J.-K., Lin, C.-H., Liang, Y.-C., Lin-Shiau, S.-Y., and Juan, I-M. (1998). Survey of catechins, gallic acid, and methylxantines in green, oolong, pu-erh and black teas. Journal of Agricultural and Food Chemistry, 46, 3635-3642.

Lin, L. Y., Liu, H. M., Yu, Y. W., Lin, S. D. and Mau, J. L. (2009). Quality and antioxidant property of buckwheat enhanced wheat bread. *Food Chemistry*, 112: 987–991.

Lin, L. Z., Lu, S. and Harnly, J. M. (2007). Detection and quantification of glycosylated flavonoid malonates in celery, Chinese celery and celery seed by LC-DAD-ESI/MS. *Journal of Agricultural and Food Chemistry*, 55: 1321–1326.

Lin, M.-Y., and Yen, C.-L. (1999). Antioxidative ability of lactic acid bacteria. Journal of Agricultural and Food Chemistry, 47, 1460-1466.

Lin, S. P., Tsai, S. Y., Lin, Y. L., Kuo, S. C., Hou, Y. C. and Chao, P. D. L. (2008). Biotransformation and pharmacokinetics of 4-(3,4-dihydroxybenzoyl-oxymethyl)phenyl-*O*-β-d-glucopyranoside, an antioxidant isolated from *Origanum vulgare*. *Journal of Agricultural and Food Chemistry*, 56: 2852–2856.

Lin, Y.-L., Juan, I.-M., Liang, Y.-C., and Lin, J.-K. (1996). Composition of polyphenols in fresh tea leaves and associations of their oxygen-radical-absorbing capacity with antiproliferative actions in fibroblast cells. Journal of Agricultural and Food Chemistry, 44, 1387-1394.

Lineback, D. R., Coughlin, J. R., and Stadler, R. H. (2012). Acrylamide in foods: A review of the science and future considerations. *Annual Review of Food Science and Technology, 3*, 15–35.

Ling, W. H., Wang, L, L. and Ma, J. (2002). Supplementation of the black rice outer layer fraction of rabbits decreases atherosclerosis plaque formation and increases antioxidant status. *The Journal of Nutrition*, 132(1), 20-26.

Lingnert, H., and Waller, G. R. (1983). Stability of antioxidants formed from histidine and glucose by the Maillard reaction. Journal of Agricultural and Food Chemistry, 31, 27-30.

Linow, F. and Mieth, G. (1976). The fat-stabilizing properties of phosphatides. III. The synergistic action of selected phosphatides. *Die Nahrung*, 20: 19–24.

List, G. R., Evans, C. D., Kwolek,W. F.,Warner, K., and Boundy, B. K. (1974). Oxidation and quality of soybean oil: A preliminary study of the anisidine test. *Journal of the American Oil Chemists' Society*, *51*, 17–21.

Litwinienko, G., Daniluk, A., and Kasprzycka-Guttman, T. (1999). A differential scanning calorimetry study on the oxidation of C12-C18 saturated fatty acids and their esters. *Journal of the American Oil Chemists' Society*, *76*, 655–657.

Liu, C. L., Chen, Y. S., Yang, J. H. and Chiang, B. H. (2007). Antioxidant activity of tartary (*Fagopyrum tataricum* (L.) Gaertn.) and common (*Fagopyrum esculentum* Moench) buckwheat sprouts. *Journal of Agricultural and Food Chemistry*, 56: 173–178.

Liu, M., Li, X. Q.,Weber, C., Lee, C. Y., Brown, J. and Liu, R. H. (2002). Antioxidant and antiproliferative activities of raspberries. *Journal of Agricultural and Food Chemistry*, 50: 926–2930.

Liu, Q. and Yao, H., (2007). Antioxidant activities of barley seeds extracts. Food Chemistry 102, 732-737.

Liu, Q., Qiu, Y. and Beta, T. (2010). Comparison of antioxidant activities of different colored wheat grains and analysis of phenolic compounds. *Journal of Agricultural and Food Chemistry,* 58(16), 9235-9241.

Liu, R. H. (2003). Health benefits of fruit and vegetables are from additive and synergistic combinations of phytochemicals. *American Journal of Clinical Nutrition*, *78*, 517S–520S.

Liu, R. H. (2004). Potential synergy of phytochemicals in cancer prevention: Mechanism of action. *The Journal of Nutrition*, 134: 3479S–3485S.

Liu, R. H. (2007). Whole grain phytochemicals and health. *J Cereal Sci*. 46:207–19.

Liu, S., Manson, J. E., Lee, I. M., Cole, S. R., Hennekens, C. H., Willett,W. C. and Buring, J. E. (2000). Fruit and vegetable intake and risk of cardiovascular heart disease: The women's health study. *American Journal of Clinical Nutrition*, 72: 922–928.

Liu, S., Manson, J. E., Stampfer, M. J., Hu, F. B., Giovannucci, E., Colditz, G. A., Hennekens, C. H. and Willett, W.C. (2000). A prospective study of whole-grain intake and risk of type 2 diabetes mellitus in US women. *American Journal of Public Health*, 90(9), 1409-1415.

Liu, S., Yang, N., Hou, Z. H., Yao, Y., Lü, L., Zhou, X. R. and Ren, G. X. (2011). Antioxidant effect of oat avenanthramide on human serum. *Agricultural Science in China*, 10(8), 1301-1305.

Liu, Y., Blumberg, J. B., and Chen, C. Y. O. (2014a). Quantification and bioaccessibility of California pistachio bioactives. *Journal of Agricultural and Food Chemistry*, *62*, 1550–1556.

Liu, Z., Lin, X., Huang, G., Zhang,W., Rao, P., and Ni, L. (2014b). Prebiotic effects of almonds and almond skins on intestinal microbiota in healthy adult humans. *Anaerobe*, *26*, 1–6.

Liyana-Pathirana, C. M., and Shahidi, F. (2006). Importance of insoluble-bound phenolics to antioxidant properties of wheat. *J. Agric. Food Chem.*, 54:1256-1264.

Liyana-Pathirana, C. M. and Shahidi, F. (2005). Antioxidant activity of commercial soft and hard wheat (*Triticum aestivum* L.) as affected by gastric pH conditions. *Journal of Agricultural and Food Chemistry*, 53: 2433–2440.

Liyana-Pathirana, C. M. and Shahidi, F. (2006). Antioxidant properties of commercial soft and hard winter wheats (Triticum aestivum L.) and their milling fractions. *Journal of the Science of Food and Agriculture*, 86: 477–485.

Liyana-Pathirana, C. M. and Shahidi, F. (2007a). The antioxidant potential of milling fractions from breadwheat and durum. *Journal of Cereal Science*, 45: 238–247.

Liyana-Pathirana, C. M. and Shahidi, F. (2007b). Antioxidant and free radical scavenging activities of whole wheat and milling fractions. *Food Chemistry*, 101: 1151–1157.

Liyana-Pathirana, C., Dexter, J., and Shahidi, F. (2006). Antioxidant properties of wheat as affected by pearling. *Journal of Agricultural and Food Chemistry*, *54*, 6177–6184.

Liyana-Pathirana, C., Dexter, J. and Shahidi, F. (2006). Antioxidant properties of wheat as affected by pearling. *Journal of Agricultural and Food Chemistry*, 54: 6177–6184.

Llorach, R., Martínez-Sánchez, A., Tomás-Barberán, F. A., Gil, M. I. and Ferreres, F. (2008). Characterisation of polyphenols and antioxidant properties of five lettuce varieties and escarole. *Food Chemistry*, 108: 1028–1038.

Lloyd, B. J., Siebenmorgen, T. J. and Beers, K. W. (2000). Effect of commercial processing on antioxidants in rice bran. *Cereal Chemistry*, 77(5), 551-555.

Lloyd-Jones, D. M., Hong, Y., Labarthe, D., Mozaffarian, D., Appel, L. J., Van Horn, L., Greenlund, K., Daniels, S., Nichol, G., Tomaselli, G. F., Arnett, D. K., Fonarow, G. C., Ho, P. M., Lauer, M. S., Masoudi, F. A., Robertson, R. M., Roger, V., Schwamm, L. H., Sorlie, P., Yancy, C. W., and Rosamond,W. D. (2010). Defining and setting national goals for cardiovascular health promotion and disease reduction: The American Heart Association's Strategic Impact Goal through 2020 and beyond. *Circulation*, *121*, 586–613.

Loliger, J. (1991). The use of antioxidants in foods. In O. I. Aruoma and B. Halliwell (Eds.), Free radicals and food additives (pp. 121–150). London: Taylor Francis.

Long-Ze, L., Harnly, J. M., Pastor-Corrales, M. S. and Luthria, D. L. (2008). The polyphenolic profiles of common bean (*Phaseolus vulgaris* L.). *Food Chemistry*, 107: 399–410.

López-Cobo, A., Gómez-Caravaca, A. M., Cerretani, L., Segura-Carretero, A. and Fernández-Gutiérrez, A. (2014). Distribution of phenolic compounds and other polar compounds in the tuber of Solanum tuberosum L. by HPLC-DAD-q-TOF and study of their antioxidant activity. *Journal of Food Composition and Analysis*, 36: 1–11.

Lou, H. X., Yamazaki, Y., Sasaki, T., Uchida, M. and Tanaka, H. (1999). A-type procyanidins from peanut skin. *Phytochemistry*, 51: 297–308.

Lowry, O. H., Rosebrough, N. J., Farr, A. L. and Randall, R. J. (1951). Protein measurement with the Folin phenol reagent. *J. Biol. Chem.* 193.165.

Lu, J., Zhao, H., Chen, J., Fan, W., Dong, J., Kong, W., Sun, J., Cao, Y. and Cai, G., (2007). Evolution of phenolic compounds and antioxidant activity during malting. Journal of Agricultural and Food Chemistry 55, 10994-11001.

Lu, Y., and Foo, L. Y. (1997). Identification and quantification of mayor polyphenols in apple pomace. Food Chemistry, 59, 187-194.

Lu, Y., and Foo, L. Y. (1999). The polyphenol constituents of grape pomace. Food Chemistry, 65, 1-8.

Lu, Y., and Foo, L. Y. (2000). Antioxidant and radical scavenging activities of polyphenols from apple pomace. Food Chemistry, 68, 81-85.

Lu, Y., Foo, L. Y. and Wong, H. (2002). Nigrumin-5-*p*-coumarate and nigrumin-5-ferulate, two unusual nitrile-containing metabolites from blackcurrant (*Ribes nigrum*) seed. *Phytochemistry*, *59:* 465–468.

Lu, Y. and Foo, L. Y. (1999). The polyphenol constituents of grape pomace. *Food Chemistry*, 65: 1–8.

Lu, Y. and Foo, L. Y. (2003). Polyphenolic constituents of blackcurrant seed residue. *Food Chemistry*, 80: 71–76.

Lucas, E. A., Juma, S., Stoecker, B. J., and Arjmandi, B. H. (2000). Prune suppresses ovariectomy-induced hypercholesterolemia in rats. *Journal of Nutritional Biochemistry*, *11*, 255–259.

Lucas-Abellan, C., Mercader-Ros, M. T., Zafrilla, M. P., Gabaldon, J. A., and Nunez-Delicado, E. (2011). Comparative study of different methods to measure antioxidant activity of resveratrol in the presence of cyclodextrins. *Food and Chemical Toxicology*, *49*, 1255–1260.

Luczak, S., Swiatek, L. and Zadernowski, R. (1989). Phenolic acids in leaves and pericarpium of walnut *Juglans regia,* L. *Acta Poloniae Pharmaceutica*, 46: 494–498.

Luh, B. S., Leonard, S. J., and Patel, D. S. (1960). Pink discoloration in canned Bartlett pears, *Food Technol.,* 14: 53.

Lui, R. H. (2007). Whole grain phytochemicals and health. *Journal of Cereal Science*, 46(3), 732-737.

Luo, A., Peng, P., Fei,W., Yang, L. and Fan, Y. (2014). Isolation and antioxidant activity (in vitro and in vivo) of the flavonoid from Tartarian-buckwheat. *Journal of Scientific and Innovative Research*, 3: 168–172.

Luo, C., Zhang, Y., Ding, Y., Shan, Z., Chen, S., Yu, M., Hu, F. B., and Liu, L. (2014). Nut consumption and risk of type 2 diabetes, cardiovascular disease, and all-cause mortality: A systematic review and meta-analysis. *The American Journal of Clinical Nutrition*, *100*, 256–269.

Luthria, D. L. and Liu, K. (2013). Localization of phenolic acids and antioxidant activity in sorghum kernels. *Journal of Functional Foods*, 5: 1751–1760.

Luthria, D. L. and Pastor-Corrales, M. A. (2006). Phenolic acids content of fifteen dry edible bean (*Phaseolus vulgaris* L.) varieties. *Journal of Food Composition and Analysis*, 19: 205–211.

Luttfullah, G., and Hussain, A. (2011). Studies on contamination level of aflatoxins in some dried fruits and nuts of Pakistan. *Food Control*, *22*, 426–429.

Lyford, S. J., Smart. W. W. G., and Bell, T. A. (1967). Inhibition of rumen cellulose digestion by extracts of *Sericeq Lejpedea. J. Am. Sci.*, 26, 632.

Lynn, D. Y. C. and Luh, B. S. (1964). Anthocyanin pigments in Bing cherries. *Journal of Food Science*, 29: 735–743.

Lyons, M. M., Yu, C., Toma, R. B., Cho, S. Y., Reiboldt,W., Lee, J. and van Breemen, R. B. (2003). Resveratrol in raw and baked blueberries and bilberries. *Journal of Agricultural and Food Chemistry*, 51: 5867–5870.

Lyras, L., Cairns, N. J., Jenner, A., and Jenner, P. (1997). An assessment of oxidative damage to proteins, lipids, and DNA in brain from patients with Alzheimer's disease. *Neurochemistry*, *68*, 2061–2069.

Ma, Y. and Bliss, F.A. (1978). Tannin content and inheritance in common bean. Crop. Sci. 18: 201-204.

Ma, Y., Kosinska-Cagnazzo, A., Kerr,W. L., Amarowicz, R., Swanson, R. B. and Pegg, R. B. (2014). Separation and characterization of phenolic compounds from dry-blanched peanut skins by liquid chromatography-electrospray ionization mass spectrometry. *Journal of Chromatography. A*, 1356: 64–81.

Mabarouk, A. F. and Dugan, L. R. (1961). Kinetic investigation into glucose, fructose and sucrose activated oxidation of methyl linoleate emulsion. *Journal of the American Oil Chemists' Society*, 53: 692–695.

Mabbayad, B. B. and Tipton, K. W. (1975). Tannin concentration and *in vitro* dry matter disappearance of seeds of bird-resistant sorghum hybrids. *Philipp. Agric.* 51. 1.

Mabry, T.J., Markham, K.R., and Thomas, M.B. (1970). The Systematic Identification of Flavonoids. p. 35-61. Springer, New York.

Maccarone, E., Maccarrone, A., Perrini, G. and Rapisarda, P. (1983). Anthocyanins of the Moro orange juice. *Annali Di Chimica*, 73: 533–539.

Maccarone, E., Maccarrone, A. and Rapisarda, P. (1985). Acylated anthocyanins from oranges. *Annali Di Chimica*, 75: 79–86.

Maccarone, E., Rapisarda, P., Fanella, F., Arena, E. and Mondello, L. (1998). Cyanidin-3-(6″-malonyl)-β-glucoside. One of the major anthocyanins in blood orange juice. *Italian Journal of Food Science*, 10: 367–372.

Madhavi, D. L., Bomser, J., Smith, M. A. L. and Singletary, K. (1998). Isolation of bioactive constituents from *Vaccinium myrtillus* (bilberry) fruits and cell cultures. *Plant Science*, 131: 95–103.

Madhujith, T., Amarowicz, R. and Shahidi, F. (2004). Phenolic antioxidants in beans and their effects on inhibition of radical-induced DNA damage. *Journal of the American Oil Chemists' Society*, 81: 691–696.

Madhujith, T., and Shahidi, F. (2007). Antioxidative and antiproliferative properties of selected barley (Hordeum vulgarae L.) cultivars and their potential for inhibition of low-density lipoprotein (LDL) cholesterol oxidation. *J. Agric. Food Chem.*, 55:5018–5024.

Madhujith, T., and Shahidi, F. (2009). Antioxidant potential of barley as affected by alkaline hydrolysis and release of insoluble-bound phenolics. *Food Chem.*, 117:615-620.

Madhujith, T., Izydorczyk, M. and Shahidi, F. (2006). Antioxidant properties of pearled barley fractions. *Journal of Agricultural and Food Chemistry, 54:* 3283–3289.

Madhujith, T., Naczk, M. and Shahidi, F. (2006). Antioxidant activity of common beans (*Phaseolus vulgaris* L). *Journal of Food Lipids*, 11: 220–233.

Madhujith, T. and Shahidi, F. (2005). Antioxidant potential of pea beans (*Phaseolus vulgaris* L.). *Journal of Food Science*, 70: S85–S90.

Madhujith, T. and Shahidi, F. (2006). Optimization of the extraction of antioxidative constituents of six barley cultivars and their antioxidant properties. *Journal of Agricultural and Food Chemistry*, 54: 8048–8057.

Madhujith, T. and Shahidi, F. (2007a). Antioxidative and antiproliferative properties of selected barley (*Hordeum vulgarae* L.) cultivars and their potential for inhibition of lowdensity lipoprotein (LDL) cholesterol oxidation. *Journal of Agricultural and Food Chemistry*, 55: 5018–5024.

Madhujith, T. and Shahidi, F. (2007b). Antioxidant potential of barley as affected by alkaline hydrolysis and release of insoluble-bound phenolics. *Food Chemistry*, 117: 615–620.

Madhujith, T. and Shahidi, F. (2009). Antioxidant potential of barley as affected by alkaline hydrolysis and release of insolublebound phenolics. *Food Chemistry*, 117: 615–620.

Madrau, M. A., Piscopo, A., Sanguinetti, A. M., Del Caro, A., Poiana, M., Romeo, F. V., and Piga, A. (2009). Effect of drying temperature on polyphenolic content and antioxidant activity of apricots. *European Food Research and Technology, 228*, 441–448.

Madrau, M. A., Sanguinetti, A., Del Caro, A., Fadda, C., and Piga, A. (2010). Contribution of melanoidins to the antioxidant activity of prunes. *Journal of Food Quality, 33*, 155–170.

Madrigal-Carballo, S., Rodriguez, G., Krueger, C. G., Dreher, M. and Reed, J. D. (2009). Pomegranate (*Punica granatum*) supplements: Authenticity, antioxidant and polyphenol composition. *Journal of Functional Foods*, 1: 324–329.

Madsen, H. L., Sùrensen, B., Skibsted, L. H., and Bertelsen, G. (1998). The antioxidative activity of summer savory (*Satureja hortensis* L.) and rosemary (Rosmarinus o cinalis L.) in dressing stored exposed to light or in darkness. Food Chemistry, 63, 73-180.

Madsen, H. L. and Bertelsen, G. (1995). Spices as antioxidants. *Trends in Food Science and Technology*, 6: 271–277.

Maeda, T. (2001). Studies on the baking property of polished-graded wheat flours. Doctoral thesis, Osaka Prefecture University, Sakai, Japan.

Maeda-Yamamoto, M., Kawahara, H., Tahara, N., Tsuji, K., Hara, Y., and Isemura, M. (1999). Effects of tea polyphenols on the invasion and matrix metalloproteinases activities of human fibrosarcoma HT1080 cells. Journal of Agricultural and Food Chemistry, 47, 2350-2354.

Magalhaes, L. M., Segundo, M. A., Reis, S., and Lima, J. L. F. C. (2008). Methodological aspects about in vitro evaluation of antioxidant properties. *Analytica Chimica Acta, 613*, 1–19.

Magdalena Michalczyk and Ryszard Macura, (2010). Effect of processing and storage on the antioxidant activity of frozen and pasteurized shadblow serviceberry (*Amelanchier canadensis*). *International Journal of Food Properties*, 13:1225–1233.

Maguire, L. S., O'Sullivan, S. M., Galvin, K., O'Connor, T. P., and O'Brien, N. M. (2004). Fatty acid profile, tocopherol, squalene and phytosterol content of walnuts, almonds, peanuts, hazelnuts and the macadamia nut. *International Journal of Food Sciences and Nutrition, 55*, 171–178.

Maguire, L. S., O'Sullivan, S. M., Galvin, K., O'Connor, T. P. and O'Brien, N. M. (2004). Fatty acid profile, tocopherol, squalene and phytosterol content of walnuts, almonds, peanuts, hazelnuts and the macadamia nut. *International Journal of Food Sciences and Nutrition*, 55: 171–178.

Maillard, M. N., and Berset, C. (1995). Evolution of antioxidant activity during kilning, role of insoluble bound phenolic acids of barley and malt. Journal of Agricultural and Food Chemistry, 43, 1789-1793.

Maillard, M. N., Soum, M.-H., Boivin, P., and Berset, C. (1996). Antioxidant activity of barley and malt, relationship with phenolic content. Lebensmittel Wissenshatf und Technologie, 29, 238-244.

Majumder, S. K, and Parpia, H. A. B. (1966). Possible losses of food grains in India *Vijnan Karme* 18, 5.

Makahleh, A., Saad, B. and Bari, M. F. (2015). Synthetic phenolics as antioxidants for food preservation. In F. Shahidi (Ed.), *Handbook of antioxidants for food preservation* (pp.51–78). Cambridge, UK: Woodhead Publishing Ltd.

Malan, R. R. and Roux, D. G. (1975). Flavonoids and tannins of *Acacia* species. *Phytochemistry*. 14: 1835.

Mallet, J. F., Cerrati, C., Ucciani, E., Gamisans, J., and Gruber, M. (1994). Antioxidant activity of plant leaves in relation to their alpha-tocopherol content. Food Chemistry, 49, 61-65.

Malunga, L. N. and Beta, T. (2015). Antioxidant capacity of water-extractable arabinoxylan from commercial barley, wheat, and wheat fractions. *Cereal Chemistry*, 92(1), 29-36.

Manach, C., Gary, W., Morand, C., Scalbert, A., and Remesy, C. (2005). Bioavailability and bioefficacy of polyphenols in humans. I. Review of 97 bioavailability studies. *American Journal of Clinical Nutrition*. 81: 230-242.

Manach, C., Mazur, A. and Scalbert, A. (2005). Polyphenols and prevention of cardiovascular diseases. *Current Opinions in Lipidology*, 16: 77–84.

Manach, C., Scalbert, A., Morand, C., Rémésy, C., and Jiménez, L. (2004). Polyphenols: Food sources and bioavailability. *The American Journal of Clinical Nutrition*, *79*, 727–747.

Manandhar, P. P., Nagao, A. and Yamazaki, M. (1986). Determination of content of linseed oil in edible soybean oil. *Journal of Japan Oil Chemists' Society*, 35: 725–732.

Manchali, S., Murthy, K. C. and Patil, B. S. (2012). Crucial facts about health benefits of popular cruciferous vegetables. *Journal of Functional Foods*, 4: 94–106.

Mandalari, G. (2012). Potential health benefits of almond skin. *Journal of Bioprocess and Biotechnology*, *2*, e110.

Mandalari, G., Arcoraci, T., Martorana, M., Bisignano, C., Rizza, L., Bonina, F. P., Trombetta, D., and Tomaino, A. (2013). Antioxidant and photoprotective effects of blanch water, a by-product of the almond processing industry. *Molecules: A Journal of Synthetic Chemistry and Natural Product Chemistry. [electronic resource]*, *18*, 12426–12440.

Mandalari, G., Bisignano, C., D'Arrigo, M., Ginestra, G., Arena, A., Tomaino, A., and Wickham, M. S. J. (2010a). Antimicrobial potential of polyphenols extracted from almond skins. *Letters in Applied Microbiology*, *51*, 83–89.

Mandalari, G., Bisignano, C., Genovese, T., Mazzon, E.,Wickham, M. S., Paterniti, I., and Cuzzocrea, S. (2011a). Natural almond skin reduced oxidative stress and inflammation in an experimental model of inflammatory bowel disease. *International Immunopharmacology*, *11*, 915–924.

Mandalari, G., Faulks, R. M., Bisignano, C.,Waldron, K. W., Narbad, A., and Wickham, M. S. (2010b). *In vitro* evaluation of the prebiotic properties of almond skins (*Amygdalus communis* L.), *FEMS Microbiology Letters*, *304*, 116–122.

Mandalari, G., Genovese, T., Bisignano, C., Mazzon, E.,Wickham, M. S., Di Paola, R., Bisignano, G., and Cuzzocrea, S. (2011b). Neuroprotective effects of almond skins in experimental spinal cord injury. *Clinical Nutrition: Official Journal of the European Society of Parenteral and Enteral Nutrition*, *30*, 221–233.

Mandalari, G., Tomaino, A., Arcoraci, T., Martorana, M., Lo Turco, V., Cacciola, F., Rich, G. T., Bisignano, C., Saija, A., Dugo, P., Cross, K. L., Parker, M. L.,Waldron, K. W. and Wickham, M. S. J. (2010). Characterization of polyphenols, lipids and dietary fibre from almond skins (*Amygdalus communis* L.). *Journal of Food Composition and Analysis*, 23: 166–177.

Mandalari, G., Tomaino, A., Rich, G. T., Lo Curto, R., Arcoraci, T., Martorana, M., Bisignano, C., Saija, A., Parker, M. L., and Waldron, K. W. (2010d). Polyphenol and nutrient release from skin of almonds during simulated human digestion. *Food Chemistry*, *122*, 1083–1088.

Manthey, J. A. and Grohmann, K. (2001). Phenols in citrus peel byproducts. Concentrations of hydroxycinnamates and polymethoxylated flavones in citrus peel molasses. *Journal of Agricultural and Food Chemistry*, 49: 3268–3273.

Maranville, F. L. and Goldschmid, O. (1954). Ultraviolet absorption spectra as a measure of phenolic hydroxyl group content in polyphenolic tannin like materials. *Anal. Chem.*, 26, 1523.

Marco, G. J. (1968). A rapid method for evaluation of antioxidants. Journal of the American Oil Chemists' Society, 45, 594-598.

Mariana-atena, P., Diana M., Diana D., Constantin M., Diana R., Iosif G. (2011). Processing and storage impact on the antioxidant properties and colour quality of some low sugar fruit jams. *Romanian Biotechnological Letters*. 16 (5): 6504-6512.

Marin, F. R. and Del Rio, J. A. (2001). Selection of hybrids and edible Citrus species with a high content in the diosmin functional compound. modulating effect of plant growth regulators on contents. *Journal of Agricultural and Food Chemistry*, 49: 3356–3362.

Marinou, K. A., Georgopoulou, K., Agrogiannis, G., Karatzas, T., Iliopoulos, D., Papalois, A., Chatziioannou, A., Magiatis, P., Halabalaki, M., Tsantila, N., Skaltsounis, L. A., Patsouris, E. and Dontas, I. A. (2010). Differential effect of *Pistacia vera* extracts on experimental atherosclerosis in the rabbit animal model: An experimental study. *Lipids in Health and Disease*, *9*, 73.

Marinova, E. M., and Yanishlieva, N. VI. (1996). Antioxidative activity of phenolic acids on triacylgycerols and fatty acid methyl esters from olive oil. Food Chemistry, 56, 139-145.

Marinova, E. M., and Yanishlieva, N.VI. (1997). Antioxidative activity of extracts from selected species of the family Lamiaceae in sunfower oil. Food Chemistry, 58, 245-248.

Marinova, E. M., Yanishlieva, N.VI., and Kostova, I. N. (1994). Antioxidative action of the ethanolic extract and some hydroxycoumarins of Fraxinus ornus bark. Food Chemistry, 51, 125-132.

Markham, K, R. (1975). Isolation techniques for flavonoids, in *The Flavonoids,* Harbome, J. B., Mabry, T. J., and Mabry, H., Eds., Chapman and Hall, London.

Markus, F., Daood, H. G., KapitaÂ ny, J., and Biacs, P. A. (1999). Change in the carotenoid and antioxidant content of spice red pepper (Paprika) as a function of ripening and some technological factors. Journal of Agricultural and Food Chemistry, 47, 100-107.

Maro, D. A., Pacifico, S., Fiorentino, A., Galasso, S., Gallicchio, M., Guida, V., Severino, V., Monaco, P. and Parente, A. (2013). Raviscanina wild asparagus (Asparagus acutifolius L.): A nutritionally valuable crop with antioxidant and antiproliferative properties. *Food Research International*, 53: 180–188.

Marquardt, R. R., Campbell, L. D. and Ward, T. (1975). Studies with chicks on the growth depressing factor(s) in faba beans, *J. Nutri.* 106, 275.

Marques, F. Z., Markus, M. A. and Morris, B. J. (2009). Resveratrol: Cellular actions of a potent natural chemical that confers a diversity of health benefits. *The International Journal of Biochemistry and Cell Biology*, 41: 2125–2128.

Marquez, A., Dueñas, M., Serratosa, M. P., and Merida, J. (2012). Formation of vitisins and anthocyanin–flavanol adducts during red grape drying. *Journal of Agricultural and Food Chemistry*, *60*, 6866–6874.

Masrizal, M. A., Giraud, D. W. and Driskell, J. A. (1997). Retention of vitamin C, iron and β-carotene in vegetables prepared using different cooking methods. *J Food Qual.* 20:403–18.

Martinelli, E. M., Baj, A. and Bombardelli, E. (1986). Computer-aided evaluation of liquid-chromatographic profiles for anthocyanins in *Vaccinium myrtillus* fruits. *Analytica Chimica Acta*, 191: 275–281.

Martínez-Tomé, M., Murcia, M. A., Frega, N., Ruggieri, S., Jiménez, A. M., Roses, F. and Parras, P. (2004). Evaluation of antioxidant capacity of cereal brans. *Journal of Agricultural and Food Chemistry*, 52: 4690–4699.

Martini, H.,Weidenborner, M., Adams, S. and Kunz, B. (1996). Eugenol and carvacrol: The main fungicidal compounds in clove and savory. *Italian Journal of Food Science*, 8: 63–67.

Martin-Tanguy, J., Guillaume, J., and Kossa, A. (1977). Condensed tannins in horse bean seeds: chemical structure and apparent effects on poultry. J. Sci. Food Agric. 28: 757-761.

Martin-Tanguy, J., Guillaume, J. and Kossa, A. (1977). Condensed tannins in horse bean seeds: Chemical structure and apparent effects on poultry. *Journal of the Science of Food and Agriculture*, 28: 757–765.

Martin-Tanguy, J., Vermovel, M., Lenoble, M., Martin, C. and Gallet, M. (1976). Sorghum seed tannins: importance in nutritional nitrogen use in growing rats, *Ann. Biol. Anim. Biochem. Biophys.* 16, 879.

Martorana, M., Arcoraci, T., Rizza, L., Cristani, M., Bonina, F. P., Saija, A., Trombetta, D., and Tomaino, A. (2013). *In vitro* antioxidant and *in vivo* photoprotective effect of pistachio (*Pistacia vera* L., variety Bronte) seed and skin extracts. *Fitoterapia*, *85*, 41–48.

Masisi, K., Beta, T., Moghadasian, M.H., Antioxidant properties of diverse cereal grains: a review on *in vitro* and *in vivo* studies, *Food Chemistry* (2015), doi: http://dx.doi.org/ 10.1016/j. oodchem. 2015.09.021

Masisi, K., Diehl-Jones, W. L., Gordon, J., Chapman, D., Moghadasian, M. H. and Beta, T. (2015). Carotenoids of aleurone, germ, and endosperm fractions of barley, corn and wheat differentially inhibit oxidative stress. *Journal of Agricultural and Food Chemistry*, DOI: 10.1021/jf5058606.

Masood, M., Iqbal, S. Z., Asi, M. R., and Malik, N. (2015). Natural occurrence of aflatoxins in dry fruits and edible nuts. *Food Control*, *55*, 62–65.

Masquelier, J., Vitte, G. and Ortega, M. (1959). Dosage colorimetrique des leucoanthocyanncs dans les yins rouqes. *Bult. Trav. Soc. Pharm. Bordeaux,* 98, 145.

Masuda, T., Yonemori, S., Oyama, Y., Takeda, Y., Tanaka, T., Andoh, T., Shinohara, A., and Nakata, M. (1999). Evaluation of the antioxidant activity of environmental plants, activity of the leaf extracts from seashore plants. Journal of the Agricultural and Food Chemistry, 47, 1749-1754.

Matera, R., Gabbanini, S., Nicola, G. R., Iori, R., Petrillo, G. and Valgimigli, L. (2012). Identification and analysis of isothiocyanates and new acylated anthocyanins in the juice of Raphanus sativus cv. Sango sprouts. *Food Chemistry.* 133 (2): 563–572.

Mathew, A. G. and Parpia, H. A. B. (1971). Food browning as a polyphenol reaction, *Adv. Food Re.* 19:75.

Mathew, A. G. and Parpia, H. A. B. (1970). Polyphenols of cashew skin. *Journal of Food Science*, 35: 140–143.

Matsufuji, H., Nakamura, H., Chino, M., and Takeda, M. (1998). Antioxidant activity of capsanthin and the fatty acid esters in paprika (*Capsicum annuum*). Journal of Agricultural and Food Chemistry, 46, 3468-3472.

Matsumoto, H., Inaba, H., Kishi, M., Tominaga, S., Hirayama, M., and Tsuda, T. (2001). Orally administered delphinidin 3-rutinoside and cyanidin 3-rutinoside are directly absorbed in rats and humans and appear in the blood as the intact forms. Journal of Agricultural and *Food Chemistry.* 49: 1546-1551.

Mattila, P., Pihlava, J. M. and Hellstrom, J. (2005). Contents of phenolic acids, alkyl- and alkenylresorcinols, and avenanthramides in commercial grain products. *Journal of Agricultural and Food Chemistry*, 53: 8290–8295.

Mattila, P. and Hellström, J. (2007). Phenolic acids in potatoes, vegetables, and some of their products. *Journal of Food Composition and Analysis*, 20: 152–160.

Matuschek, E.,Towo, E. and Savanberg, U. (2001). Oxidation of polyphenols in phytate-reduced high-tannin cereals: Effect on different phenolic groups and on in vitro accessible iron. Journal of Agricultural and Food Chemistry, 49: 5630–5638.

Maxson, E. D. and Rooney, L. W. (1972). Evaluation of methods for tannin analysis in sorghum grain, *Cereal Chem.,* 49, 719.

Maxson, E. D., Rooney, L. W., Lewis, R. W., Clark, L. E., and Johnson, J. W. (1973). The relationship between tannin content, enzyme inhibition, rat performance and characteristics of sorghum grain. *Nut. Rep. Int.* 8, 145.

Maxwell, S., Cruickshank, A. and Thorpe, G. (1994). Red wine and antioxidant activity in serum. *Lancet*, 344: 193–194.

Mazur,W. M., Duke, J. A.,Wahala, K., Rasku, S. and Adlercreutz, H. (1998). Isoflavonoids and lignans in legumes: Nutritional and health aspects in humans. *Journal of Nutritional Biochemistry*, 9: 193–200.

Mazza, G. (1995). Anthocyanins in grapes and grape products. *Critical Reviews in Food Science and Nutrition*, 35: 341–371.

McClements (Eds.), *Oxidation in foods and beverages and antioxidant applications* (pp. 272–320). Oxford, UK: Woodhead Publishing.

McClements, D. J. and Decker, E. A. (2007). Lipids. In S. Damodaran, K. L. Parkinand O. R. Fennema (Eds.), *Fennema's food chemistry* (pp. 156–212). NewYork: CRC Press.

McCormack, D. and McFadden, D. (2012). Pterostilbene and cancer: Current review. *Journal of Surgical Research*, 173: e53–e61.

McDonald, M., Mila, I., and Scalbert, A. (1996). Precipitation of metal ions by plant polyphenols, optimal conditions and origin of precipitation. Journal of Agricultural and Food Chemistry, 44, 599-606.

McGhie T.K., and Walton, M.C. (2007). The bioavailability and absorption of anthocyanins: towards a better understanding. *Mol Nutr Food Res.* 51(6): 702–713.

McGhie, T. K., Hunt, M. and Barnett, L. E. (2005). Cultivar and growing region determine the antioxidant polyphenolic concentration and composition of apples grown in New Zealand. *Journal of Agricultural and Food Chemistry*, 53: 3065–3070.

McLeod, M. N. (1974). Plant tannins: their role in forage quality, *Nutr. Abstr. Rev.*, 44, 803.

McMillan, W. W., Wiseman, B. R., Burns, R. E., Harris, H. B., and Greene, G. L. (1972). Bird resistance in diverse germplasm of sorghum, *Agron.* J. 64, 821.

McPhail, D. B., Gardner, P. T., Duthie, G. G., Steele, G. M., and Reid, K. (1999). Assessment of the antioxidant potential of scotch whiskeys by electron spin resonance spectroscopy, relationship to hydroxyl-containing aromatic components. Journal of Agricultural and Food Chemistry, 47, 1937-1941.

MeÂndez, E., Sanhueza, J., Nieto, S., Speeisky, H., and Valenzuela, A. (1998). Fatty acid composition, extraction, fractionation, and stabilization of Bullfrog (Rana catesbeiana) oil. Journal of Agricultural and Food Chemistry, 75, 67-71.

Medina, I., Sacchi, R., Giudicianni, I., and Aubourg, S. (1998). Oxidation in fish lipids during thermal stress as studied by ^{13}C nuclear magnetic resonance spectroscopy. *Journal of the American Oil Chemists' Society, 75*, 147–154.

Mee, S. P. H., Lee, V., Baldwin, J. E. and Cowley, A. (2004). Total synthesis of 5,5′,6,6′-tetrahydroxy-3,3′-biindolyl, the proposed structure of a potent antioxidant found in beetroot (*Beta vulgaris*). *Tetrahedron*, 60: 3695–3712.

Mehansho, H. T Hagerman, A., Clements, S. T Butler, L. G., Rogier. J. C. and Carlson. D. M. (1983). Modulation of proline-rich protein biosynthesis in rat parotid glands by sorghums with high tannin levels. *Proc. Natl. Acad. Sci. U.S.A.*, 80: 3948.

Melega, S., Canistro, D., De Nicola, G. R., Lazzeri, L., Sapone, A. and Paolini, M. (2013). Protective effect of Tuscan black cabbage sprout extract against serum lipid increase and perturbations of liver antioxidant and detoxifying enzymes in rats fed a high-fat diet. *British Journal of Nutrition*, 110: 988–997.

Melton, S. L. (1983). Methodology for following lipid oxidation in muscle foods. *Food Technology, 37*, 105–111.

Mena, P., Calani, L., Dall'Asta, C., Galaverna, G., García-Viguera, C., Bruni, R., Crozier, A. and Del Rio, D. (2012). Rapid and comprehensive evaluation of (Poly) phenolic compounds in pomegranate (*Punica granatum* L.) juice by UHPLC–MS. *Molecules*, 17: 14821–14840.

Meng, J., Fang, Y., Zhang, A., Chen, S., Xu, T., Ren, Z., Han, G., Liu, J., Li, H., Zhang, Z., and Wang, H. (2011). Phenolic content and antioxidant capacity of Chinese raisins produced in Xinjiang Province. *Food Research International, 44*, 2830–2836.

Meunzer, J., Bildstein, C., Gleason, M., and Carlson, D. M. (1979). Purification of proline-rich proteins from parotid glands of isoproterenol·trealed rats, *J. Bioi. Chem* 254: 5623.

Meyer, A. S., Heinonen, M., and Frankel, E. N. (1998). Antioxidant interactions of catechin, cyanidin, caffeic acid, quercetin, and ellagic acid on human LDL oxidation. Food Chemistry, 61, 71-75.

Meyer, A. S., Jepsen, S. M., and Sùrensen, N. S. (1998). Enzymatic release of antioxidants for human low-density lipoprotein from grape pomace. Journal of Agricultural and Food Chemistry, 46, 2439-2446.

Meyer, K. A., Kushi, L. H., Jacobs, D. R. Jr., Slavin, J., Sellers, T. A. and Folsom, A. R. (2000) Carbohydrates, dietary fiber, and incident type 2 diabetes in older women. *The American Journal of Clinical Nutrition*, 71(4), 921-930.

Micali, S., Isgro, G., Bianchi, G., Miceli, N., Calapai, G., and Navarra, M. (2014). Cranberry and recurrent cystitis: More than marketing? *Critical Reviews in Food Science and Nutrition, 54*, 1063–1075.

Micozzi, M. S., Beecher, G. R., Taylor, P. R., and Khachik, F. (1990). Carotenoid analysis of raw and cooked foods associated with a lower risk for cancer. *Journal of the National Cancer Institute.* 82: 282–285.

Middleton, E., Jr., Kandaswami, C. and Theoharides, T. C. (2000). The effects of plant flavonoids on mammalian cells: Implications for inflammation, heart disease, and cancer. *Pharmacological Reviews*, 52: 673–751.

Middleton, E., Jr. and Kandaswami, C. (1992). Effects of flavonoids on immune and inflammatory cell functions. *Biochemical Pharmacology*, 43: 1167–1179.

Middleton, E., Jr. and Kandaswami, C. (1994). Potential healthpromoting properties of Citrus flavonoids. *Food Technology*, 11: 115–119.

Mikkonen, T. P., Määttä, K. R., Hukkanen, A. T., Kokko, H. I., Törrönen, A. R., Kärenlampi, S. O. and Karjainen, R. O. (2001). Flavonol content varies among blackcurrant cultivars. *Journal of Agricultural and Food Chemistry*, 49: 3274–3277.

Mikyska, A., Hrabak, M., Haskova, D. and Srogl, J., (2002). The role of malt and hop polyphenols in beer. Quality, flavour and haze stability. Journal of Institute of Brewing 108, 78-85.

Milardovic, S., Kerekovic, I., and Rumenjak, V. (2007). A flow injection biamperometric method for determination of total antioxidant capacity of alcoholic beverages using bienzymatically produced ABTS. *Food Chemistry, 105*, 1688–1694.

Milbourne, K. (1987). The role of natural antioxidants in lipid oxidation. *Food Science and Technology Today*, 1: 171–172.

MilicÏ , B. L. J., Djilas, S. M., and CÏ anadanovicÂ-Brunet, J. M. (1998). Antioxidative activity of phenolic compounds on the metal-ion breakdown of lipid peroxidation system. Food Chemistry, 62, 443-447.

Miller, C. J., Dunn, E. V., and Hashim, I. B. (2003). The glycaemic index of dates and date/ yoghurt mixed meals. Are dates 'the candy that grows on trees? *European Journal of Clinical Nutrition, 57*, 427–430.

Miller, G. J. and Quakenbush, F. W. (1957). A comparison of alkylated phenols as antioxidants for lard. *Journal of the American Oil Chemists' Society*, 34: 249–250.

Miller, H. E. (1971). A simplified method for the evaluation of antioxidants. Journal of the American Oil Chemists' Society, 45, 91.

Miller, N. J., Rice-Evans, C. A., Davies, M. J., Gopinathan, V. and Milner, A. (1993). A novel method for measuring antioxidant capacity and its application to monitoring the antioxidant status in premature neonates. *Clinical Science*, 84: 407–412.

Mills, P. K., Beeson,W. L., Phillips, R. L., and Fraser, G. E. (1989). Cohort study of diet, lifestyle, and prostate cancer in Adventist men. *Cancer, 64*, 598–604.

Min, B., and Ahn, D. U. (2005). Mechanism of lipid peroxidation in meat and meat products — A review. *Food Science and Biotechnology, 14*, 152–163.

Minioti, K. S., and Georgiou, C. A. (2008). High throughput flow injection bioluminometric method for olive oil antioxidant capacity. *Food Chemistry, 109*, 455–461.

Minioti, K. S. and Georgiou, C. A. (2008). High throughput flow injection bioluminometric method for olive oil antioxidant capacity. *Food Chemistry*, 109: 455–461.

Mira, L., Fernandez, M. T., Santos, M., Rocha, R., Florencio, M. H. and Jennings, K. R. (2002). Interactions of flavonoids with iron and copper ions: A mechanism for their antioxidant activity. *Free Radical Research*, 36: 1199–1208.

Miraliakbari, H. and Shahidi, F. (2008). Antioxidant activity of minor components of tree nut oils. *Food Chemistry*, 111: 421–427.

Miranda, C. L., Reed, R. L., Cheeke, P. R. and Buhler, D. R. (1981). Protective effects of butylated hydroxyanisole against the acute toxicity of monocrotaline in mice. *Toxicology and Applied Pharmacology*, 59: 424–430.

Mishra, A., Siradhana, B. S., and Shivpuri, A. (1980). Phenols in relation to resistance of sorghum to anthracnose. *Philipp. Agric*. 63. 71.

Mitaru, B. N. Reichert, R. D. and Blair, R. (1983). Improvement of the nutritive value of high tannin sorghums for broiler chickens by high moisture storage (reconstitution). *Poult. Sci.* 62, 2065.

Miura, K., Kikuzaki, H., and Nakatani, N. (2002). Antioxidant activity of chemical components from sage (*Salvia officinalis* L.) and thyme (*Thymus vulgaris* L.) measured by the oil stability index method. *Journal of Agricultural and Food Chemistry*, *50*, 1845–1851.

Miyake, T., and Shibamoto, T. (1998). Inhibition of malonaldehyde and acetaldehyde formation from blood plasma oxidation by naturally occurring antioxidants. Journal of Agricultural and Food Chemistry, 46, 3694-3697.

Miyake, Y., Murakami, A., Sugiyama, Y., Isobe, M., Koshimizu, K., and Ohigashi, H. (1999). Identification of coumarins from lemon fruit (*Citrus limon*) as inhibitors of in vitro tumor promotion and superoxide and nitric oxide generation. Journal of Agricultural and Food Chemistry, 47, 3151-3157.

Miyake, Y., Sasaki, S., Ohya, Y., Miyamoto, S., Matsunaga, I., Yoshida, T., Hirota, Y. and Oda, H. (2005). Soy, isoflavones, and prevalence of allergic rhinitis in Japanese women: The Osaka Maternal and Child Health Study. *Journal of Allergy and Clinical Immunology*, 115: 1176–1183.

Mnatsakanyan, M., Goodie, T. A., Conlan, X. A., Francis, P. S., McDermott, G. P., Barnett, N. W., Shock, D., Gritti, F., Guiochon, G. and Shalliker, R. A. (2010). High performance liquid chromatography with two simultaneous on-line antioxidant assays: Evaluation and comparison of espresso coffees. *Talanta*, 81: 837–842.

Mohamed, T. K., Issoufou, A. and Zhou, H. (2012). Antioxidant activity of fractionated foxtail millet protein hydrolysate. *Intl Food Res. J.* 19(1):207–13.

Mohdaly, A. A. A., Hassanien, M. F. R., Mahmoud, A., Sarhan, M. A. and Smetanska, I. (2013). Phenolics extracted from potato, sugar beet, and sesame processing by-products. *International Journal of Food Properties*, 16: 1148–1168.

Mohsen, S. M. and Ammar, A. S. M. (2009). Total phenolic contents and antioxidant activity of corn tassel extracts. *Food Chemistry*, 112: 595–598.

Mole, S. and Watterman, P. G. (1987). Tannic acid and proteolylic enzymes: enzyme inhibition or substrate deprivation. *Phyrochemistry,* 26: 99.

Molina-Diaz, A., Ortega-Carmona, I., and Pascual-Reguera, M. I. (1998). Indirect spectrophotometric determination of ascorbic acid with ferrozine by flow-injection analysis. *Talanta*, *47*, 531–536.

Monagas, M., Bartolome, B. and Gomez-Cordoves, C. (2005). Updated knowledge about the presence of phenolic compounds in wine. *Critical Reviews in Food Science and Nutrition*, 45: 85–118.

Monagas, M., Garrido, I., Lebrón-Aguilar, R., Gómez-Cordovés, M. C., Rybarczyk, A., Amarowicz, R., and Bartolomé, B. (2009). Comparative flavan-3-ol profile and antioxidant capacity of roasted peanut, hazelnut, and almond skins. *Journal of Agricultural and Food Chemistry*, *57*, 10590–10599.

Monedero, L., Olalla, M., MartõÂn-Lagos, F., LoÂ pez, H., and LoÂ pez, M. C. (1999). Application of chemometric techniques in obtaining macerates with phenolic compound content similar to that of wines from the Jerez-Sherry region subjected to oxidative aging. Journal of Agricultural and Food Chemistry, 47, 1836-1844.

Montanari, L., Perretti, G., Natella, F., Guidi, A. and Fantozzi, P. (1999). Organic and phenolic acids in beer. *Lebensmittel-Wissenschaft Und-Technoogie*, 32: 535–539.

Montedoro, G. S. M., Baldioli, M. and Miniati, E. (1992). Simple and hydrolyzable phenolic compounds in virgin olive oil. 1. Their extraction, separation, quantitative and semiquantitative evaluation by HPLC. *Journal of Agricultural and Food Chemistry*, 40: 1571–1576.

Moon, J. H., and Terao, J. (1998). Antioxidant activity of caffeic acid and dihydrocaffeic acid in lard and human low-density lypoprotein. Journal of Agricultural and Food Chemistry, 46, 5062-5065.

Moore, J., Cheng, Z., Hao, J., Guo, G., Liu, J., Lin, C. J. and Yu, L. (2007). Effects of solid-state yeast treatment on the antioxidant properties and protein and fiber composition of common hard wheat bran. *Journal of Agricultural and Food Chemistry*, 55(25), 10173-10182.

Moore, J., Hao, Z., Zhou, K., Luther, M., Costa, J. and Yu, L. L. (2005). Carotenoid, tocopherol, phenolic acid, and antioxidant properties of Maryland-grown soft wheat. *Journal of Agricultural and Food Chemistry*, 53: 6649–6657.

Morais, T. C., Pinto, N. B., Carvalho, K. M., Rios, J. B., Ricardo, N. M., Trevisan, M. T., Rao, V. S., and Santos, F. A. (2010). Protective effect of anacardic acids from cashew (*Anacardium occidentale*) on ethanol-induced gastric damage in mice. *Chemico-Biological Interactions*, *183*, 264–269.

Morton, J. F. (1978). Economic botany in epidemiology, *Eco". Bot.* 32: 111.

Mosele, J. I., Macià, A., Romero, M. P., Motilva, M. J. and Rubió, L. (2015). Application of in vitro gastrointestinal digestion and colonic fermentation models to pomegranate products (juice, pulp and peel extract) to study the stability and catabolism of phenolic compounds. *Journal of Functional Foods*, 14: 529–540.

Moskovitz, J., Yim, K. A., and Choke, P. B. (2002). Free radicals and disease. *Archives of Biochemistry and Biophysics.* 397: 354–359.

Motilva MJ, Serra A, Macia, A (2013) Analysis of food polyphenols by ultra-high-performance liquid chromatography coupled to mass spectrometry: an overview. J Chromatogr A 1292:66–82.

Mouly, P. P., Gaydou, E. M., Faure, R. and Estienne, J. M. (1997). Blood orange juice authentication using cinnamic acid derivatives.Variety differentiations associated with flavanone glycoside content. *Journal of Agricultural and Food Chemistry*, 45: 373–377.

Moure, A., Franco, D., Sineiro, J., DomõÂnguez, H., NuÂnÄ ez, M. J. and Lema, J. M. (1999). Antioxidant properties of Gevuina avellana hulls. 8th Mediterranean Congress on Chemical Engineering. Barcelona. 10-12 November.

Moure, A., Franco, D., Sineiro, J., DomõÂnguez, H., NuÂnÄ ez, M. J. and Lema, J. M. (2000). Antioxidant activity of extracts from Gevuina avellana and Rosa rubiginosa deffated seeds. Acepted for publication in Food Research International.

Mozetic, B., Trebse, P. and Hribar, J. (2002). Determination and quantitation of anthocyanins and hydroxycinnamic acids in different cultivars of sweet cherries (*Prunu savium* L.) from Nova Gorica Region (Slovenia). *Food Technology and Biotechnology*, 40: 207–212.

Muanza, D., Robert, R. and Sparks, W. (1998). Antioxidant derived from lentil and its preparation and uses. US Patent. US5762936.

Mudnic, I., Modun, D., Rastija, V., Vukovic, J., Brizic, I., Katalinic, V., Kozina, B., Medic-Saric, M. and Boban, M. (2010). Antioxidative and vasodilatory effects of phenolic acids in wine. *Food Chemistry*, 119: 1205–1210.

Mulabagal, V., Lang, G. A., DeWitt, D. L., Dalavoy, S. S. and Nair, M. G. (2009). Anthocyanin content, lipid peroxidation and cyclooxygenase enzyme inhibitory activities of sweet and sour cherries. *Journal of Agricultural and Food Chemistry*, 57: 1239–1246.

Mullen, W., Stewart, A. J., Lean, M. E. J., Gardner, P., Duthie, G. G. and Crozier, A. (2002). Effect of freezing and storage on the phenolics, ellagitannins, flavonoids, and antioxidant capacity of red raspberries. *Journal of Agricultural and Food Chemistry*, 50: 5197–5201.

Mullen,W., McGinn, J., Lean, M. E. J., MacLean, M. R., Gardner, P., Duthie, G. G., Yokota, T. and Crozier, A. (2002). Ellagitannins, flavonoids and other phenolics: Their contribution to the antioxidant capacity and vasorelaxation properties of red raspberries. *Journal of Agricultural and Food Chemistry*, 50: 5191–5196.

Muller DP (2010). "Vitamin E and neurological function. Review". Mol. Nutr. Food Res 54(5): 710–718.

Muller, C. H. (1969. The "Co"-in coevolution. *Science,* 164. 197.

Mùller, J. K.S, Madsen, H. L., Aaltonen, T., and Skibsted, L. H. (1999). Dittany (Origanum dictamnus) as a source of water-extractable antioxidants. Food Chemistry, 64, 215-219.

Muller, L. G., Pase, C. S., Reckziegel, P., Barcelos, R. C., Boufleur, N., Prado, A. C., Fett, R., Block, J. M., Pavanato, M. A., Bauermann, L. F., da Rocha, J. B., and Burger, M. E. (2013). Hepatoprotective effects of pecan nut shells on ethanol-induced liver damage. *Experimental and Toxicologic Pathology*, *65*, 165–171.

Mùller, R. E., Stapelfeldt, H., and Skibsted, L. H. (1998). Thiol reactivity in pressure-unfolded b-lactoglobulin. Antioxidative properties and thermal refolding. Journal of Agricultural and Food Chemistry, 46, 425-430.

Muller,W. U. and Leistner, E. (1978a). Metabolic relation between naphthalene derivatives in *Juglans regia*. *Phytochemistry*, 17: 1735–1738.

Muller,W. U. and Leistner, E. (1978b). Aglycones and glycosides of oxygenated naphthalenes and glycosyltransferase from *Juglans* (spp). *Phytochemistry*, 17: 1739–1742.

Munne-Bosch, S. and Alegre, L. (2002). The function of tocopherols and tocotrienols in plants. *Critical Reviews in Plant Sciences*, 21: 31–57.

Murcia, M., Jimenez, A. M., and Martinez-Tome, M. (2009). Vegetables antioxidant losses during industrial processing and refrigerated storage. *Food Res. Int*. 42: 1046-1052.

Musa, K. H., Abdullah, A., Kuswandi, B., and Hidayat, M. A. (2013). A novel high throughput method based on the DPPH dry reagent array for determination of antioxidant activity. *Food Chemistry*, *141*, 4102–4106.

Musialik, M., Kuzmicz, R., Pawłowski, T. S. and Litwinienko, G. (2009). Acidity of hydroxyl groups: An overlooked influence on antiradical properties of flavonoids. *Journal of Organic Chemistry*, 74: 2699–2709.

Mustafa, F. A. (2013). Effects of green tea extract on color and lipid oxidation in ground beef meat. *Journal of Tikrit University for Agriculture Sciences*, 13: 351–354.

Naczk, M., Amarowicz, R., Pink, D. and Shahidi, F. (2000). Insoluble condensed tannins of canola and rapeseed hulls. *Journal of Agricultural and Food Chemistry*, 48: 1758–1762.

Naczk, M., Amarowicz, R., Sullivan, A. and Shahidi, F. (1998). Current research developments on polyphenols of rapeseed/canola: A review. *Food Chemistry*, 62: 489–502.

Naczk, M., Amarowicz, R., Zadernowski, R., Pegg, R. and Shahidi, F. (2003). Antioxidant activity of crude phenolic extracts from wild blueberry leaves. *Polish Journal of Food and Nutrition*, 53: 170S–173S.

Naczk, M., Nichols, T., Pink, D. and Sosulski, F. (1994). Condensed tannins in canola hulls. *Journal of Agricultural and Food Chemistry*, 42: 2196–2200.

Naczk, M., Oickle, D., Pink, D., and Shahidi, F. (1996). Protein precipitating capacity of crude canola tannins, effect of pH, tannin, and protein concentrations. Journal of Agricultural and Food Chemistry, 44, 2144-2148.

Naczk, M., Pegg, R. B., Zadernowski, R. and Shahidi, F. (2005). Radical scavenging activity of canola hull phenolics. *Journal of the American Oil Chemists' Society*, 82: 255–260.

Naczk, M., Shahidi, F., and Sullivan, A. (1992a). Recovery of rapeseed tannins by various solvent systems. Food Chem. 45: 51-54.

Naczk, M., Wanasundhara, J., and Shahidi, F. (1992b). Facile spectrophotometric quantification method of sinapic acid in hexane-extracted and methanol-ammonia- water treated mustard and rapeseed meals. J. Agric. Food Chem. 40: 444-448.

Naczk, M. and Shahidi, F. (2004). Extraction and analysis of phenolics in food. *Journal of Chromatography. A*, 1054: 95–111.

Naczk, M. and Shahidi, F. (2006). Phenolics in cereals, fruits and vegetables: Occurrence, extraction and analysis. *Journal of Pharmaceutical and Biomedical Analysis*, 41: 1523–1542.

Naim, M., Gestetner, B., Zilkah, S., Birk, Y. and Bondi, A. (1973). A new isoflavone from soybean. *Phytochemistry*, 12: 169–170.

Nakagami, T., Nanaumi-Tamura, N., Toyomura, K., Nakamura, T., and Shigehisa, T. (1995). Dietary favonoids as potential natural biological response modifiers affecting the autoimmune system. Journal of Food Science, 60, 653-656.

Nakagawa, K., Ninomiya, M., Okubo, T., Aoi, N., Juneja, L. R., Kim, M., Yamanaka, K., and Miyazawa, T. (1999). Tea catechin supplementation increases antioxidant capacity and prevents phospholipid hydroperoxidation in plasma of humans. Journal of Agricultural and Food Chemistry, 47, 3967-3973.

Nakagawa, Y., Nakajima, K., Tayama, S. and Moldeus, P. (1995). Metabolism and cytotoxicity of propyl gallate in isolated rat hepatocytes: Effects of a thiol reductant and an esterase inhibitor. *Molecular Pharmacology*, 47: 1021–1027.

Nakamura, S., Ogawa, M., Nakai, S., Kato, A., and Kitts, D. D. (1998). Antioxidant activity of a Maillard-type phosvitin-galactomannan conjugate with emulsifying properties and heat stability. Journal of Agricultural and Food Chemistry, 46, 3958-3963.

Nakashima, H., Murakami, T., Yamamoto, N., Sakagami, H., Tanuma, S. I., Hatano, T., Yoshida, T. and Okuda, T. (1992). Inhibition of human immunodeficiency viral replication by tannins and related compounds. *Antiviral Research*, 18: 91–103.

Nakasugi, T., and Komai, K. (1998). Antimutagens in the brazilian folk medicinal plant Carqueja (Baccharis trimera Less.). Journal of Agricultural and Food Chemistry, 46, 2560-2564.

Nakatani, N., Inatani, R., Ohta, H. and Nishioka, A. (1986). Chemical constituents of peppers and application to food preservation: Naturally occurring antioxidative compounds. *Environmental Health Perspectives*, 67: 135–142.

Nakatani, N. and Kikuzaki, H. A. (1987). New antioxidative glucoside isolated from oregano (*Origanum vulgare* L.). *Agricultural and Biological Chemistry*, 51: 2727–2732.

Nakayama, R., Tamura, Y., Kikuzaki, H., and Nakatani, N. (1999). Antioxidant effect of the constituents of Susabinori (Porphyra yezoensis). Journal of the American Oil Chemists' Society, 76, 649-653.

Namitha, K., and Negi, P. S. (2010). Chemistry and biotechnology of carotenoids. *Critical Reviews in Food Science and Nutrition, 50*, 728–760.

Nanditha, B. and Prabhasankar, P. (2009). Antioxidants in bakery products: A review. *Critical Reviews in Food Science and Nutrition*, 49: 1–27.

Nara, K., Miyoshi, T., Honma, T. and Koga, H. (2006). Antioxidative activity of bound-form phenolics in potato peel. *Bioscience, Biotechnology, and Biochemistry*, 70: 1489–1491.

National Institute of Health (2009). "Vitamin E fact sheet".

Naurato, N., Wong, P., Lu, Y., Wroblewski, K., and Bennick, A. (1999). Interaction of tannin with human salivary histatins. Journal of Agricultural and Food Chemistry, 47, 2229-2234.

Nawar,W. W. (1996). Lipids. In R. O. Fennema (Ed.), *Food chemistry* (pp. 225–320).

Ndolo, V. U. and Beta, T. (2013). Distribution of carotenoids in endosperm, germ, and aleurone fractions of cereal grain kernels. *Food Chemistry,* 139(1-4), 663-671.

Ndolo, V. U. and Beta, T. (2014). Comparative studies on composition and distribution of phenolic acids in cereal grain botanical fractions. *Cereal Chemistry*, 91: 522–530.

Neish, A. C. (1960). Biosynthetic pathways of aromatic compounds. *Ann. Rev. Plant Physiol.*11:55.

Neish, A. C. (1965). Coumarins, phenylpronanes and lignin, in *Plant Biochemistry.* Bonner, J. and Varner, E. V., Eds. Academic Press, New York, 581.

Nelson, J. S., Stephenson, E. L., Burgos, A., Floyd, J. and York, J. O. (1975). Effect of tannin content and dry matter digestion on energy utilization and average amino acid availability on hybrid sorghum grains, *Poult. Sci.,* 54, 1620.

Neudorffer D, Bonnefont-Rousselot A, Legrand M, Fleury B, Largeron M (2004). 4-Hydroxycinnamic ethyl ester derivatives and related dehydrodimers: relationship between oxidation potential and protective effects against oxidation of low-density lipoproteins. J Agric Food Chem 52:2084–2091.

Neveu, V., Perez-Jiménez, J., Vos, F., Crespy, V., du Chaffaut, L., Mennen, L., Knox, C., Eisner, R., Cruz, J.,Wishart, D., and Scalbert, A. (2010). Phenol-Explorer: An online comprehensive database on polyphenol contents in foods. *Database (Oxford), 2010*, bap024. doi:10.1093/database/bap024.

NewYork, NY: Marcel Dekker, Inc. Nikokavoura, A., Christodouleas, D., Yannakopoulou, E., Papadopoulos, K., and Calokerinos, A. C. (2011). Evaluation of antioxidant activity of hydrophilic and lipophilic compounds in edible oils by a novel fluorimetric method. *Talanta, 84*, 874–880.

Ng, A., Harvey, A. J., Parker, M. L., Smith, A. C. and Walderon, K. W. (1998). Effect of oxidative coupling on the thermal stability of texture and cell wall chemistry of beet root (*Beta vulgaris*). *Journal of Agricultural and Food Chemistry*, 46: 3365–3370.

Ng, A., Parker, M. L., Parr, A. J., Saunders, P. K., Smith, A. C. and Waldron, K. W. (2000). Physicochemical characteristics of onion (*Allium cepa* L.) tissues. *Journal of Agricultural and Food Chemistry*, 48: 5612–5617.

Nichenametla, S.N., Taruscio, T.G., Barney, D.L., and Exon, J.H. (2006). A review of the effects and mechanisms of polyphenolics in cancer. *Critical Reviews in Food Science and Nutrition*. 46(2): 161–183.

Niehaus, M. H. (1966). Bird resistant grain sorghum, *Ohio Rep.* 51, 70.

Niehaus, M. H. and Schmidt, W. H. (1970). Bird Resistant Grain Sorghum: a New Crop for Ohio, Res. Circ. 182, Ohio Agriculture Research Development Center, Worthington, OH, 19.

Nielsen, J. K., Norbek, R. and Olsen, C. E. (1998). Kaempferol tetraglucosides from cabbage leaves. *Phytochemistry*, 34: 539–544.

Nielsen, J. K., Olsen, C. E. and Petersen, M. K. (1993). Acylated flavonol glycosides from cabbage leaves. *Phytochemistry*, 49: 2171–2176.

Nieman, D. C., Scherr, J., Luo, B., Meaney, M. P., Dréau, D., Sha,W., Dew, D. A., and Pappan, K. L. (2014). Influence of pistachios on performance and exercise-induced inflammation, oxidative stress, immune dysfunction, and metabolite shifts in cyclists: A randomized, crossover trial. *PLoS ONE*, *9*, e113725.

Nieto, S., Garrido, A., Sanhueza, J., Loyola, L. A., Morales, G., Leighton, F., and Valenzuela, A. (1993). Flavonoids as stabilizers of fish oil, an alternative to synthetic antioxidants. Journal of the American Oil Chemists' Society, 70, 773-778.

Nijssen, L. M., Visscher, C. A., Maarse, H.,Willemsens, L. C. and Boelens, M. H. (1996). *Volatile compounds in food. Qualitative and quantitative data* (7th ed., pp. 72.1–72.23). Zeist, Netherlands: TNO Nutrition and Food Research Institute.

Niki, E., Tsuchiya, J., Tanimura, R. and Kamiya, Y. (1982). Regeneration of vitamin E from α-chromanoxyl radical by glutathione and vitamin C. *Chemistry Letters*, 11: 789–792.

Nirmala, K., Prasanna Krishna, T. and Polasa, K. (2007). Protective effect of ginger against Benzo(a)pyrene induced DNA damage. *International Journal of Cancer Research*, 3: 13–24.

Niwa, K. and Motoyama, S. (1991). Antioxidant composition of natural products and producing method thereof. US Patent US5009891.

Niwa, T., Doi, U., Kato, Y. and Osawa, T. (2001). Antioxidative properties of phenolic antioxidants isolated from corn steep liquor. *Journal of Agricultural and Food Chemistry*, 49: 177–182.

Nkhili, E., and Brat, P. (2011). Reexamination of the ORAC assay: Effect of metal ions. *Analytical and Bioanalytical Chemistry*, *400*, 1451–1458.

Noda, Y., Kaneyuki, T., Mori, A. and Packe, L. (2002). Antioxidant activities of pomegranate fruit extract and its anthocyanidins: Delphinidin, cyanidin and pelargonidin. *Journal of Agricultural and Food Chemistry*, 50: 166–171.

Noguchi, Y., Fukuda, K., Matsushima, A., Haishi, D., Hiroto, M., Kodera, Y., Nishimura, H., and Inada, Y. (1999). Inhibition of Dfprotease associated with allergic diseases by polyphenol. Journal of Agricultural and Food Chemistry, 47, 2969-2972.

Nogueira, V. and Hay, N. (2013). Molecular pathways: Reactive oxygen species homeostasis in cancer cells and implication for cancer therapy. *Clinical Cancer Research*, 19(16), 4309-4314.

Nordkvist, E., Salomonsson, A.-C. and Aman, P. (1984). Distribution of insoluble-bound phenolic acids in barley grain. *Journal of the Science and Food Agriculture*, 35: 657–661.

Nour, V., Stampar, F., Veberic, R. and Jakopic, J. (2013). Anthocyanins profile, total phenolics and antioxidant activity of black currant ethanolic extracts as influenced by genotype and ethanol concentration. *Food Chemistry*, 141: 961–966.

Nowak, H., Kujawa, K., Zadernowski, R., Roczniak, B. and Kozlowska, H. (1992). Antioxidative and antibactericidal properties of phenolic compounds in rapeseeds. *Fat Science and Technology*, 94: 149–152.

Nuengchamnong, N., de Jong, C. F., Bruyneel, B., Niessen,W. M. A., Irth, H., and Ingkanina, K. (2005). HPLC coupled on-line to ESI-MS and a DPPH-based assay for the rapid identification of antioxidants in Butea superba. *Phytochemical Analysis*, *16*, 422–428.

Nuncio-Jáuregui, N., Nowicka, P., Munera-Picazo, S., Hernández, F., Carbonell-Barrachina, Á. A. and Wojdyło, A. (2015). Identification and quantification of major derivatives of ellagic acid and antioxidant properties of thinning and ripe Spanish pomegranates. *Journal of Functional Foods*, 12: 354–364.

Nuutila, A. M., Puupponen-Pimiä, R., Aarni, M. and Oksman-Caldentey, K. M. (2003). Comparison of antioxidant activities of onion and garlic extracts by inhibition of lipid peroxidation and radical scavenging activity. *Food Chemistry*, 81: 485–493.

Nyam, K. L., Tan, C. P., Karim, R., Lai, O. M., Long, K. and Man, Y. B. C. (2010). Extraction of tocopherol-enriched oils from Kalahari melon and roselle seeds by supercritical fluid extraction (SFECO2). *Food Chemistry*, 119: 1278–1283.

Nyström, L., Mäkinen, M., Lampi, A. M. and Piironen, V. (2005). Antioxidant activity of steryl ferulate extracts from rye and wheat bran. *Journal of Agricultural and Food Chemistry*, 53: 2503–2510.

O″zyürek, M., Bektas¸og˘g˘ lu, B., Güçlü, K., Güngör, N., and Apak, R. (2008). Simultaneous total antioxidant capacity assay of lipophilic and hydrophilic antioxidants in the same acetone–water solution containing 2% methyl-b-cyclodextrin using the cupric reducing antioxidant capacity (CUPRAC) method. *Analytica Chimica Acta*, *630*, 28–39.

O″zyürek, M., Güngör, N., Baki, S., Güçlü, K., and Apak, R. (2012). Development of a silver nanoparticle-based method for the antioxidant capacity measurement of polyphenols. *Analytical Chemistry*, *84*, 8052–8059.

O'Byrne, D. J., Knauft, D. A. and Shireman, R. B. (1997). Low fatmonounsaturated rich diets containing high-oleic peanuts improve serum lipoprotein profiles. *Lipids*, 32: 687–695.

Oak, M. H., El Bedoui, J. and Schini-Kerth, V. B. (2005). Antiangiogenic properties of natural polyphenols from red wine and green tea. *Journal of Nutritional Biochemistry*, 16: 1–8.

Oberthur, E. E., Nicholson, R. L., and Butler, L. G. (1983). Presence of phenolic materials including condensed tannin in sorghum callus, 1. *Agric. Food Chem.* 31. 660.

Oezcan, M. (1999). Antioxidant activity of rosemary (Rosmarinus ocinalis L.) extracts on natural olive and sesame oils. Grasas y Aceites, 50, 355-358.

Ogata, M., Hoshi, M., Shimotohno, K., Urano, S., and Endo, T. (1997). Antioxidant activity of magnolol, honokiol and related phenolic compounds. Journal of the American Oil Chemists' Society, 74, 557-568.

Ogawa, S., Kimura, H., Niimi, A., Katsube, T., Jisaka, M. and Yokota, K. (2008). Fractionation and structural characterization of polyphenolic antioxidants from seed shells of Japanese horse chestnut (*Aesculus turbinata* BLUME). *Journal of Agricultural and Food Chemistry*, 56: 12046–12051.

Ohta, T., Yamasaki, S., Egashira, Y., and Sanada, H. (1994). Antioxidative activity of corn bran hemicellulose fragments. Journal of Agricultural and Food Chemistry, 42, 653-656.

Okada, Y., and Okada, M. (1998). Scavenging effect of water soluble proteins in broad beans on free radicals and active oxygen species. Journal of Agricultural and Food Chemistry, 46, 401-406.

Okamura, N., Fujimoto, Y., Kuwabara, S. and Yagi, A. (1994). High performance liquid chromatographic determination of carnosic acid and carnosol in *Rosmarinus officinalis* and *Salvia officinalis*. *Journal of Chromatography. A*, 679: 381–386.

Okarter, N., Liu, C. S., Sorrells, M. E. and Liu, R. H. (2010). Phytochemical content and antioxidant activity of six diverse varieties of whole wheat. *Food Chemistry*, 119: 249–257.

Oki, T., Masuda, M., Furuta, S., Nishiba, Y., Terahara, N. and Suda, I. (2002). Involvement of anthocyanins and other phenolic compounds in radical-scavenging activity of purple-fleshed sweet potato cultivars. *Journal of Food Science*, 67: 1752–1756.

Okubo, T., Yokoyama, Y., Kano, K. and Kano, I. (2003). Cell death induced by the phenolic antioxidant *tert*-butylhydroquinone and its metabolite *tert*-butylquinone in human monocytic leukemia U937 cells. *Food and Chemical Toxicology*, 41: 679–688.

Olcott, H. S. and Van Der Ven, J. (1968). Comparison of antioxidant activities of tocol and its methyl derivatives. *Lipids*, 3: 331–334.

Oliveira, I., Sousa, A., Valentão, P., Andrade, P. B., Ferreira, I. C., Ferreres, F., Bento, A., Seabra, R., Estevinho, L., and Pereira, J. A. (2007). Hazel (*Corylus avellana* L.) leaves as source of antimicrobial and antioxidative compounds. *Food Chemistry*, *105*, 1018–1025.

Olthof, M. R., Hollman, P. C. H. and Katan, M. (2001). Chlorogenic acid and caffeic acid are absorbed in humans. *Human Nutrition and Metabolism*, 131: 66–71.

Omata, Y., Yoshida, Y., and Niki, E. (2010). Assessment of the antioxidant capacity of natural fruit extracts by inhibition of probe decay and plasma lipid peroxidation. *Bioscience, Biotechnology, and Biochemistry*, *74*, 531–535.

Onyeneho, S. N., and Hettiarachchy, N. S. (1992). Antioxidant activity of durum wheat bran. Journal of Agricultural and Food Chemistry, 40, 1496-1500.

Onyeneho, S.-N. and Hettiaachchy, N. S. (1993). Antioxidant activity, fatty acid and phenolic acid composition of potato peels. *Journal of the Science of Food Agriculture*, 62: 345–350.

Oomah, B. D., Cardador-Martinez, A. and Loarca-Pina, G. (2005). Phenolics and antioxidative activities in common beans (Phaseolus vulgaris L). *Journal of the Science of Food and Agriculture*, 85: 935–942.

Oomah, B. D., Caspar, F., Malcolmson, L. J. and Bellido, A.-S. (2011). Phenolics and antioxidant activity of lentil and pea hulls. *Food Research International*, 44: 436–441.

Oomah, B. D., Kenaschuk, E. O., and Mazza, G. (1995). Phenolic acids in faxseed. Journal of Agricultural and Food Chemistry, 43, 2016-2019.

Oomah, B. D., Kenaschuk, E. O. and Mazza, G. (1997). Tocopherols in flaxseed. *Journal of Agricultural and Food Chemistry*, 45: 2076–2080.

Oomah, B. D., Luc, G., Leprelle, C., Drover, J. C. G., Harrison, J. E. and Olson, M. (2011). Phenolics, phytic acid, and phytase in Canadian-grown low-tannin faba bean (*Vicia faba* L.) genotypes. *Journal of Agricultural and Food Chemistry*, 59: 3763–3771.

Opoku, A. R., Ohenhen, S. O., and Ejiofor, N. (1981).Nutrient composition of millet (*Pennisetum typhoides*) grains and malts. Journal of Agricultural and Food Chemistry, 29: 1247–1248.

Orak, H. H. (2007). Total antioxidant activities, phenolics, anthocyanins, polyphenoloxidase activities of selected red grape cultivars and their correlations. *Scientia Horticulturae*, 111: 235–241.

Orhan, I., Kupeli, E., Aslan, M., Kartal, M., and Yesilada, E. (2006). Bioassay-guided evaluation of anti-inflammatory and antinociceptive activities of pistachio, *Pistacia vera* L. *Journal of Ethnopharmacology*, *105*, 235–240.

Ortiz, R., Antilen, M., Speisky, H., Aliaga, M. E., Lopez-Alarcon, C., and Baugh, S. (2012). Application of a microplate-based ORACpyrogallol red assay for the estimation of an antioxidant capacity: First action 2012.03. *Journal of AOAC International*, *9*, 1558–1561.

Osakabe, N., Yamagishi, M., Sanbogi, C., Natsume, M., Takizawa, T. and Osawa, T. (1998). The antioxidative substances in cacao liquor. *Journal of Nutritional Science and Vitaminology*, 53: 4290–4302.

Osawa, T. K., Katsuzaki, H., Hagiwara, Y., Hagiwara, H. and Shibamoto, T. (1992). A novel antioxidant isolated from young green barley leaves. *Journal of Agricultural and Food Chemistry*, 40: 1135–1138.

Osawa, T., Narasimhan, R., Kawakishi, S., Namdci, M. and Tashiro, T. (1985). Antioxidative defense systems in rice hull against damage caused by oxygen radicals. *Agricultural and Biological Chemistry*, 49: 3085–3087.

Osburn,W. N. (2009). *Evaluation of dried plum powder in meat products destined for convenience and foodservice outlets*. Pleasanton, CA: California Prune Board.

Osuna GarcõÂa, J. A., Wall, M. M., and Waddell, C. A. (1997). Natural antioxidants for preventing color loss in stored paprika. Journal of Food Science, 62, 1017-1021.

Osuna GarcõÂa, J. A., Wall, M. M., and Waddell, C. A. (1998). Endogenous levels of tocopherols and ascorbic acid during fruit ripening of new Mexican-type chile (Capsicum annuum L.) cultivars. Journal of Agricultural and Food Chemistry, 46, 5093-5096.

Oswalt, D. L. (1975). Estimating the Biological Effects of Tannins in Grain Sorghum. Proc. Int. Workshop. Mayaguez, Puerto Rico.

Othman, A., Ismail, A., Ghani, N. A. and Adenan, I. (2007). Antioxidant capacity and phenolic content of cocoa beans. *Food Chemistry*, 100: 1523–1530.

Ou, B.X.; Huang, D.J.; Hampsch-Woodill, M.; Flanagan, J.A.; Deemer, E.K. (2002). Analysis of antioxidant activities of common vegetables employing oxygen radical absorbance capacity (ORAC) and ferric reducing antioxidant power (FRAP) assays: A comparative study. *J. Agric. Food Chem. 50*, 3122-3128.

Ouchemoukh, S., Hachoud, S., Boudraham, H., Mokrani, A., and Louaileche, H. (2012). Antioxidant activities of some dried fruits consumed in Algeria. *LWT-Food Science and Technology*, *49*, 329–332.

Owades, J. L., Rubin, G., and Brenner, M. 'V. (1958). Determination of tannins in beer and brewing materials by ultraviolet spectroscopy. *Am. Soc. Brew. Chem. Proc.*, 6, 66.

Owades, J. L., Rubin, G., and Brenner, M. W. (1958). Determination of food tannins by ultraviolet spectrophotometry, J. *Agric. Food Chem.* 6, 44.

Owen, R. W., Mier, W., Giacosa, A., Hull,W. E., Spiegelhalder, B. and Bartsch, H. (2000). Phenolic compounds and squalene in olive oils: The concentration and antioxidant potential of total phenols, simple phenols, secoiridoids, lignans and squalene. *Food and Chemical Toxicology*, 38: 647–659.

Ozyurt, D., Demirata, B., and Apak, R. (2010). Modified cerium(IV)- based antioxidant capacity (CERAC) assay with selectivity over citric acid and simple sugars. *Journal of Food Composition and Analysis*, *23*, 282–288.

Packer L, Weber SU, Rimbach G; Weber; Rimbach (2001). "Molecular aspects of α-tocotrienol antioxidant action and cell signalling". Journal of Nutrition 131 (2): 369S–73S.PMID 11160563.

Packer, L. (1995). Nutrition and biochemistry of the lipophilic antioxidants, vitamin E and carotenoids. In: Nutrition, Lipids, Health, and Disease. pp. 8-35. Ong, A. S. H., Niki, E. and Packer, L., Eds., American Oil Chemists' Society, Champaign, IL.

Padmarajaiah, N., Anandamurthy, S., Narayana, A., Anantharaman, S., Krishnegowda, A., and Honnur, K. (2012). Quantification of antioxidants in medicinal plants and foodstuffs using Ce(IV) with indigo carmine as chromogenic probe. *Food Analytical Methods*, *5*, 909–919.

Palikova, I., Vostalova, J., Zdarilova, A., Svobodova, A., Kosina, P., Vecera, R., Stejskal, D., Proskova, J., Hrbac, J., Bednar, P., Maier, V., Cernochova, D., Simanek, V., and Ulrichova, J. (2010). Longterm effects of three commercial cranberry products on the antioxidative status in rats: A pilot study. *Journal of Agricultural and Food Chemistry*, *58*, 1672–1678.

Palomino, O., Gomez-Serranillos, M. P., Slowing, K., Carretero, E. and Villar, A. (2000). Study of polyphenols in grape berries by reversed-phase high-performance liquid chromatography. *Journal of Chromatography. A*, 870: 449–451.

Panda, N. (1979). *Principles of Host-Plant Resistance to Insect Pests,* Hindustan Publishing Corp., Delhi, Inida.

Papadopoulos, G., and Boskou, D. (1991). Antioxidant effect of natural phenols on olive oil. Journal of the American Oil Chemists' Society, 68, 669-671.

Papadopoulos, K., Triantis, T., Yannakopoulou, E., Nikokavoura, A., and Dimotikali, D. (2003). Comparative studies on the antioxidant activity of aqueous extracts of olive oils and seed oils using chemiluminescence. *Analytica Chimica Acta*, *494*, 41–47.

Papastergiadis, A., Mubiru, E., van Langenhove, H., and de Meulenaer, B. (2012). Malondialdehyde measurement in oxidized foods: Evaluation of the spectrophotometric thiobarbituric acid reactive substances (TBARS) test in various foods. *Journal of Agricultural and Food Chemistry*, *60*, 9589–9594.

Parejo, L.; Codina, C.; Petrakis, C.; Kefalas, P. (2000). Evaluation of scavenging activity assessed by Co(II)/EDTA-induced luminol chemiluminescence and DPPH center dot (2,2-diphenyl-1-picrylhydrazyl) free radical assay. *J. Pharmacol. Toxicol. Methods 44*, 507-512.

Parker, T. L.,Wang, X. H., Pazmiño, J., and Engeseth, N. J. (2007). Antioxidant capacity and phenolic content of grapes, sundried raisins, and golden raisins and their effect on *ex vivo* serum antioxidant capacity. *Journal of Agricultural and Food Chemistry*, *55*, 8472–8477.

Parkhurst, R. M., Skinner,W. A. and Strum, P. A. (1968). The effect of various concentrations of tocopherols and tocopherol mixtures on the oxidative stabilities of a sample of lard. *Journal of the American oil Chemists' Society*, 45: 641–642.

Parlakpinar, H., Olmez, E., Acet, A., Ozturk, F., Tasdemir, S., Ates, B., Gul, M., and Otlu, A. (2009). Beneficial effects of apricotfeeding on myocardial ischemia-reperfusion injury in rats. *Food and Chemical Toxicology*, *47*, 802–808.

Parpia, H. A. B. (1970). Conservation and technological production of food, *Indian Food Packer,* 24, 1.

Pasalar, M., and Lankarani, K. B. (2015). Letter: Prunes for the treatment of constipation. *Alimentary Pharmacology and Therapeutics, 41*(2), 234.

Pasalar, M., Lankarani, K. B., Mehrabani, D., Tolide-ie, H. R., and Nasri, M. (2013). The effect of *Descureania sophia* L. and *Prunus domestica* L. in prevention of constipation among Iranian Hajj pilgrims, Saudi Arabia. *Research Journal of Pharmaceutical, Biological and Chemical Sciences, 4*, 1195–1204.

Pascoe, H.M., Ames, J.M. and Chandra, S., (2003). Critical stages of the brewing process for changes in antioxidant activity and levels of phenolic compounds in Ale. Journal of American Society of Brewing Chemistry 61, 203-209.

Patil, B. S., Pike, L. M. and Yoo, K. S. (1995). Variation in quercetin content in different colored onions. *Journal of the American Society for Horticultural Science*, 120: 909–913.

Pazanatto, C., Malta, L. G., Pastore, G. M. and Netto, F. M. (2013). Antioxidant capacity of amaranth products: effects on thermal and enzymatic treatments. *Food Science and Technology (Campinas)*, 33(3), 485-493.

PekicÂ , B., KovacÏ , V., Alonso, E., and Revilla, E. (1998). Study of the extraction of proanthocyanidins from grape seeds. Food Chemistry, 61, 201-206.

Pekkarinen, S. S., Heinonen, I. M. and Hopia, A. I. (1999). Flavonoids quercetin,myricetin, kaempferol and (+)-catechin as antioxidants in methyl linoleate. *Journal of the Science of Food and Agriculture*, 79: 499–506.

Pekkarinen, S. S., StoÃ ckmann, H., Schwarz, K., Heinonen, M., and Hopia, A. I. (1999). Antioxidant activity and partitioning of phenolic acids in bulk and emulsified methyl linoleate. Journal of Agricultural and Food Chemistry, 47, 3036-3043.

Peleg, H., Naim, M., Rouseff, R. L. and Zehavi, U. (1991). Distribution of bound and free phenolic acids in oranges (*Citrus sinensis*) and grapefruits (*Citrus paradisi*). *Journal of the Science of Food and Agriculture*, 60: 417–426.

Pelentir, N., Block, J., Monteiro Fritz, A. R., Reginatto, V., and Amante, E. R. (2011). Production and chemical characterization of peach (*Prunus persica*) kernel flour. *Journal of Food Process Engineering, 34*, 1253–1265.

Pellegrini, N., Serafini, M., Salvatore, S., Del Rio, D., Bianchi, M., and Brighenti, F. (2006). Total antioxidant capacity of spices, dried fruits, nuts, pulses, cereals, and sweets consumed in Italy assessed by three different *in vitro* assays. *Molecular Nutrition and Food Research, 50*, 1030–1038.

Pelvan, E., Alasalvar, C., and Uzman, S. (2012). Effects of roasting on the antioxidant status and phenolic profiles of commercial Turkish hazelnut varieties (*Corylus avellana* L.). *Journal of Agricultural and Food Chemistry, 60*, 1218–1223.

Peng, X., Nie, Y.,Wu, J., Huang, Q., and Cheng, Y. (2015). Juglone prevents metabolic endotoxemia-induced hepatitis and neuroinflammation *via* suppressing TLR4/NF-kappaB signaling pathway in high-fat diet rats. *Biochemical and Biophysical Research Communications, 462*, 245–250.

Pereira, J. A., Oliveira, I., Sousa, A., Valentão, P., Andrade, P. B., Ferreira, I. C. F. R., Ferreres, F., Bento, A., Seabra, R. and Estevinho, L. (2007).Walnut (*Juglans regia* L.) leaves: Phenolic compounds, antibacterial activity and antioxidant potential of different cultivars. *Food and Chemical Toxicology*, 45: 2287–2295.

Pérez-Gregorio, M. R., Regueiro, J., Simal-Gándara, J., Rodrigues, A. S. and Almeida, D. P. F. (2014). Increasing the added-value of onions as a source of antioxidant flavonoids: A critical review. *Critical Reviews in Food Science and Nutrition*, 54: 1050–1062.

Pérez-Jiménez, J., Neveu, V., Vos, F., and Scalbert, A. (2010). Identification of the 100 richest dietary sources of polyphenols: An application of the Phenol-Explorer database. *European Journal of Clinical Nutrition, 64*, S112–S120.

Pérez-López, U., Pinzino, C., Quartacci, M. F., Ranieri, A. and Sgherri, C. (2014). Phenolic composition and related antioxidant properties in differently colored lettuces: A study by electron paramagnetic resonance (EPR) kinetics. *Journal of Agricultural and Food Chemistry*, 62: 12001–12007.

Perkins, E. G. (1992). Effect of lipid oxidation on oil and food quality in deep frying. In A. J. St. Angelo (Ed.), *Lipid oxidation in food* (pp. 310–321).Washington, DC: American Chemical Society.

Perron NR, Brumaghin JL (2009) A review of the antioxidant mechanisms of polyphenol com- pounds related to iron binding. Cell Biochem Biophys 53:75–100.

Perron, N. R., Hodges, J. N., Jenkins, M., and Brumaghim, J. L. (2008). Predicting how polyphenol antioxidants prevent DNA damage by binding to iron. *Inorganic Chemistry, 47*, 6153–6161.

Perumal. R. S. and Subramanian, T. R. (1973). Studies on panicle characters associated with bird resistance in sorghum, *Madras Agric. J.,* 60, 256.

Peschel,W., Dieckmann,W., Sonnenschein, M. and Plescher, A. (2007). High antioxidant potential of pressing residues from evening primrose in comparison to other oilseed cakes and plant antioxidants. *Industrial Crops and Products*, 25: 44–54.

Peterson, D. M. (1995). Oat tocols: Concentration and stability in oat products and distribution within the kernel. *Cereal Chemistry*, 72: 21–24.

Peterson, D. M. (2001). Oat antioxidants. *Journal of Cereal Science*, 33: 115–129.

Peterson, D. M., Hahn, M. J. and Emmons, C. L. (2002). Oat avenanthramides exhibit antioxidant activities in vitro. *Food Chemistry*, 79: 473–478.

Petroni, A., Blasevich, M., Papini, N., Salami, M., Sala, A. and Galli, C. (1997). Inhibition of leukocyte leukotriene B4 production by an olive oil-derived phenol identified by mass-spectrometry. *Thrombosis Research*, 87: 315–322.

Petroni, A., Blasevich, M., Salami, M., Servili, M., Montedoro, G. F. and Galli, C. (1994). A phenolic antioxidant extracted from olive oil inhibits platelet aggregation and arachidonic acid metabolism *in vitro.World Review of Nutrition and Dietetics*, 75: 169–172.

Phillips, K. M., Ruggio, D. M. and Ashraf-Khorassani, M. (2005). Phytosterol composition of nuts and seeds commonly consumed in the United States. *Journal of Agricultural and Food Chemistry*, 53: 9436–9445.

Philpott, M., Gould, K. S., Markham, K. R., Lewthwaite, S. L. and Ferguson, L. R. (2003). Enhanced coloration reveals high antioxidant potential in new sweet potato cultivars. *Journal of the Science of Food and Agriculture*, 83: 1076–1082.

Piccinelli, A. L., Veneziano, A., Passi, S., Simone, F. D. and Rastrelli, L. (2007). Flavonol glycosides from whole cottonseed byproduct. *Food Chemistry*, 100: 344–349.

Pietta, P., Simonetti, P., and Mauri, P. (1998). Antioxidant activity of selected medicinal plants. Journal of Agricultural and Food Chemistry, 46, 4487-4490.

Piga, A., Del Caro, A., and Corda, G. (2003). From plums to prunes: Influence of drying parameters on polyphenols and antioxidant activity. *Journal of Agricultural and Food Chemistry, 51*, 3675–3681.

Pigman, W. Anderson, E., Fisher, R., Buchanan, M. A., and Browning, B. L. (1953). Color precursors in spruce and western hemlock woods and inner barks, *Tappi,* 36, 4.

Piironen, V., Syvaoja, E. L., Varo, P., Salminen, K. and Koivistoinen, P. (1986). Tocopherols and tocotrienols in Finnish foods: Vegetables, fruits, and berries. *Journal of the Agricultural and Food Chemistry*, 34: 742–746.

Pikul, J. (1992). Lipid oxidation and warmed-over flavour formation in heated and stored meat. Part I. *Gosp Miêsna*, *7*, 20–23, (in Polish).

Pillai, M. K. S., Kedlaya, K. J. and Selvarangan, R. (1963). Cashew seed skin as a tanning material. *Leather Science*, 10: 317–318.

Pimentel, F. A., Nitzke, J. A., Klipel, C. B. and de Jong, E. V. (2010). Chocolate and red wine – A comparison between flavonoids content. *Food Chemistry*, 120: 109–112.

Piper, J. T., Singhal, S. S., Salameh, M. S., Torman, R. T., Awasthi, Y. C. and Awasthi, S. (1998). Mechanisms of anticarcinogenic properties of curcumin: The effect of curcumin on glutathione linked detoxification enzymes in rat liver. *International Journal of Biochemistry and Cell Biology*, 30: 445–456.

Pischetsrieder, M., Rinaldi, F., Gross, U., and Severin, T. (1998). Assessment of the antioxidative and prooxidative activities of two aminoreductones formed during the Maillard reaction, effects on the oxidation of b-carotene, Na-acetylhistidine, and cis-alkenes. Journal of Agricultural and Food Chemistry, 46, 2945-2950.

Pizzale, L., Bortolomeazzi, R., Vichi, S. and Conte, L. (2002). Antioxidant activity of sage and oregano extracts related to their phenolic compound content. *Journal of the Science of Food and Agriculture*, 82: 1645–1651.

Plank, D. W., Szpylka, J., Sapirstein, H.,Woollard, D., Zapf, C. M., Lee, V., Chen, C. Y., Liu, R. H., Tsao, R., Düsterloh, A., and Baugh, S. (2012). Determination of antioxidant activity in foods and beverages by reaction with 2,2′-diphenyl-1-picrylhydrazyl (DPPH): Collaborative study First Action 2012.04. *Journal of AOAC International*, *95*, 1562–1569.

Plumb, G. W., GarcõÂa, C. M. T., Kroon, P. A., Rhodes, M., Ridley, S., and Williamson, G. (1999). Metabolism of chlorogenic acid by human plasma, liver, intestine and gut fora. Journal of the Science of Food and Agriculture, 79, 390-392.

Plumb, G. W., Price, K. R., and Williamson, G. (1999a). Antioxidant properties of favonol glycosides from green beans. Redox Report, 4, 123-127.

Plumb, G. W., Price, K. R., and Williamson, G. (1999b). Antioxidant properties of favonol glycosides from tea. Redox Report, 4, 13-16.

Podsedek, A. (2007). Natural antioxidants and antioxidant capacity of Brassica vegetables: A review. *LWT - Food Science and Technology*, 40: 1–11.

Podsedek, A., Sosnowska, D., Redzynia, M. and Koziolkiewicz, M. (2008). Effect of domestic cooking on the red cabbage hydrophilic antioxidants. *Int. J. Food Sci. Technol.* 43: 1770-1777.

Polles, S. G., Hanny, B. W. and Harvey, A. J. (1981). Condensed tannins in kernels of 31 pecan [*Carya illinoensis* (Wangenh.) K. Koch] cultivars. *Journal of Agricultural and Food Chemistry*, 29: 196–197.

Poluzzi, E., Piccinni, C., Raschi, E., Rampa, A., Recanatini, M., and De Ponti, F. (2014). Phytoestrogens in postmenopause: The state of the art from a chemical, pharmacological and regulatory perspective. *Current Medicinal Chemistry*, *21*, 417–436.

Pomeranz, Y. (1988). Chemical composition of kernel structures. In: Wheat: Chemistry and Technology (Vol. 1). pp. 97-158. Pomeranz, Y., Ed., American Association of Cereal Chemists, St. Paul, MN.

Ponginebbi, L., Nawar, W. W., and Chinachoti, P. (1999). Oxidation of Linoleic acid in emulsions, effect of substrate, emulsifier, and sugar concentration. Journal of the American Oil Chemists' Society, 76, 131-138.

Popp, R. and Schimmer, O. (1991). Induction of sister-chromatid exchanges (SCE), polyploidy, and micronuclei by plant flavonoids in human lymphocyte cultures. A comparative study of 19 flavonoids. *Mutation Research*, 246: 205–213.

Porter, L. J. and Woodruff, J. (1984). Haemanalysis: the relative astringency of proanthocyanidin polymers, *Phytochemistry,* 23, 1255.

Porter,W. L. (1993). Paradoxical behavior of antioxidants in food and biological systems. In G. M. Williams (Ed.), *Antioxidants: Chemical, physiological, nutritional and toxicological aspects* (pp. 93–122). Princeton, NJ: Princeton Scientific.

Povilaityte, V., Cuvelier, M. E., and Berset, C. (2001). Antioxidant properties of Moldavian dragonhead (Dracocephalum moldavica L.). *Journal of Food Lipids*, *8*, 45–64.

Pradeep, S. R. and Guha, G. (2011). Effect of processing methods on the nutraceutical and antioxidant properties of little millet (*Panicum sumatrense*) extracts. *Food Chem.* 126:1643–1647.

Prajapati, M. R., Patel, V., Parekh, T. and Subhash, R. (2013). Effect of in bio-processing on antioxidant activity of selected cereals. *Asian Journal of Plant Science and Research,* 3(2), 66-72.

Pratt, D. E. and Birac, P. (1979). Source of antioxidant activity of soybean and soy products. *Journal of Food Science*, 44: 1720–1722.

Preedy, V.R., (2009). Beer in Health and Disease Prevention. Academic Press Publisher, New York, NY, USA. 843-855.

Price, K. R., Casuscelli, F., Colquhoun, I. J. and Rhodes, M. J. C. (1997). Hydroxycinnamic acid esters from broccoli florets. *Phytochemistry*, 45: 1683–1687.

Price, K. R., Casuscelli, F., Colquhoun, I. J. and Rhodes, M. J. C. (1998). Composition and content of flavonol glycosides in broccoli florets (*Brassica olearacea*) and their fate during cooking. *Journal of the Science of Food and Agriculture*, 77: 468–472.

Price, M. L. and Butler, L. G. (1972). Rapid visual estimation and Spectrophotometric determination of tannin content of sorghum grain, *J. Agric. Food Chem.,* 25, 1268.

Price, M. L. and Butler, L. G. (1977). Rapid visual estimation and Spectrophotometric determination of tannin content of sorghum grain, *J. Agric. Food Chem.* 25: 1268.

Price, M. L. and Butler, L. G. (1980). Tannin and Nutrition, Agric. Exp. Stn. Bull. No. 272, Prudue University, West Lafayelte, IN. 37.

Price, M. L. Van Scoyoc, S. and Butler, L. G. (1978). A critical evaluation of the vanillin reaction as an assay for tannin in sorghum grain, *J. Agric. Food Chem.* 26. 1214.

Price, M. L., Stromberg, A. M., and Butler, L. G. (1979). Tannin content as a function of grain maturity and drying conditions in several varieties of *Sorghum bicolor* (L.) Moench, *J. Agric. Food Chem* 27: 1270.

Price, M.L. and Butler, L.G. (1977). Rapid visual estimation an spectrophotometric determination of tannin content of sorghum grain. J. Agric. Food Chem. 25: 1268- 1273.

Price, R. K., Wallace, J. M. W., Hamill, L. L., Keaveney, E. M., Strain, J. J., Parker, M. J. and Welch, R. W. (2012). Evaluation of the effect of wheat aleurone-rich foods on markers of antioxidant status, inflammation and endothelial function in apparently healthy men and women. *British Journal of Nutrition*, 108(9), 1644-1651.

Price, R. K., Welch, R. W., Lee-Manion, A. M., Bradbury, I. and Strain, J. J. (2008). Total phenolics and antioxidant potential in plasma and urine of humans after consumption of wheat bran. *Cereal Chemistry*, 85(2), 152-157.

Price, W.E. and Spiro, M. (1985). Kinetics and equilibria of tea infusion: the aflavin and caffein concentrations and partition constants in several whole teas and sieved fractions. J. Sci. Food Agric. 36: 1303-1308.

Price, W.E. and Spitzer, J.C. (1993). Variations in the amounts of individual flavonols in a range of green teas. Food Chem. 47: 271-276.

Pridham, J. B. (1960). Phenolics of Plants in Health and Diseases, Pergamon Press, New York.

Pridham. J. B. (1964). Methods in Polyphenols Chemistry, Pergamon Press, Oxford.

Prior, R. L., Cao, G., Martin, A., Sofic, E., McEwen, J., O'Brien, C., Lischner, N., Ehlenfeldt, M., Kalt, W., Krewer, G. and Mainland, C. M. (1998). Antioxidant capacity as influenced by total phenolic and anthocyanin content, maturity, and variety of *Vaccinium* species. *Journal of Agricultural and Food Chemistry*, 46: 2686–2693.

Prior, R. L., Hoang, H., Gu, L.,Wu, X., Bacchiocca, M., Howard, L., Hampsch-Woodill, M., Huang, D., Ou, B., and Jacob, R. (2003). Assays for hydrophilic and lipophilic antioxidant capacity (oxygen radical absorbance capacity (ORACFL) of plasma and other biological and food samples. *Journal of Agricultural and Food Chemistry, 51*, 3273–3279.

Prior, R. L., Lazarus, S. A., Cao, G., Muccitelli, H. and Hammerstone, J. F. (2001). Identification of procyanidins and anthocyanins in blueberries and cranberries (*Vaccinium* spp.) using highperformance liquid chromatography/mass spectrometry. *Journal of Agricultural and Food Chemistry*, 49: 1270–1276.

Prior, R. L., Rogers, T. R., Khanal, R. C., Wilkes, S. E.,Wu, X., and Howard, L. R. (2010). Urinary excretion of phenolic acids in rats fed cranberry. *Journal of Agricultural and Food Chemistry, 58*, 3940–3949.

Prior, R. L., Wu, X., and Gu, L. (2006). Identification and urinary excretion of metabolites of 5-(hydroxymethyl)-2-furfural in human subjects following consumption of dried plums or dried plum juice. *Journal of Agricultural and Food Chemistry, 54*, 3744–3749.

Prior, R. L., Wu, X. and Schaich, K. (2005). Standardized methods for the determination of antioxidant capacity and phenolics in foods and dietary supplements. *Journal of Agricultural and Food Chemistry*, 53: 4290–4302.

Pro, M. J. (1952). Reports on spectrophotometric determination of tannin in wines and whiskies, J. *Assoc. Off. Agric. Chem.* 37, 255.

Pro, M. J. (1955). Report on spectrophotomelric determination of tannin in wines and spirits, *J. Assoc. Off. Agric. Chem.* 38, 757.

Proceso de extraccioÂ n con fuidos supercrõÂticos para la produccioÂ n de antioxidantes naturales y antioxidante obtenido mediante dicho proceso. Patente espanÄ ola 2 128 996).

Pryzbylski, R., Lee, Y. C., and Eskin, N. A. M. (1998). Antioxidant and radical-scavenging activities of buckwheat seed components. Journal of the American Oil Chemists' Society, 75, 1595-1601.

Puglisi, M. J., Mutungi, G., Brun, P. J., McGrane, M. M., Labonte, C., Volek, J. S., and Fernandez, M. L. (2009). Raisins and walking alter appetite hormones and plasma lipids by modifications in lipoprotein metabolism and up-regulation of the lowdensity lipoprotein receptor. *Metabolism: Clinical and Experimental, 58*, 120–128.

Puglisi, M. J., Vaishnav, U., Shrestha, S., Torres-Gonzalez, M., Wood, R. J., Volek, J. S., and Fernandez, M. L. (2008). Raisins and additional walking have distinct effects on plasma lipids and inflammatory cytokines. *Lipids in Health and Disease*, *7*, 1–9.

Pulido, R., Bravo, L., and Saura-Calixto, F. (2000). Antioxidant activity of dietary polyphenols as determined by a modified ferric reducing antioxidant power assay. *Journal of Agricultural and Food Chemistry*, *48*, 3396–3402.

Pushparaj, F. S. and Urooj, A. (2014). Antioxidant Activity in Two Pearl Millet (*Pennisetum typhoideum*) Cultivars as Influenced by Processing. *Antioxidants.* 3: 55-66.

Puupponen-Pimiä, R., Nohynek, L., Meier, C., Kahkonen, M., Heinonen, M., Hopia, A. and Oksman-Caldentey, K.-M. (2001). Antimicrobial properties of phenolic compounds from berries. *Journal of Applied Microbiology*, 90: 494–507.

Puupponen-Pimiä, R., Häkkinen, S. T., Aarni, M., Suortti, T., Lampi, A. M., Eurola, M., Piironen, V., Nuutila, A. M. and Oksman-Caldentey, K. M. (2003). Blanching and long-term freezing affect various bioactive compounds of vegetables in different ways. *J Sci. Food Agric*. 83:1389–1402.

Puwastein, P., Burlingame, B., Raroengwichit, M., and Sungpuag, P. (2000). ASEAN food composition tables. ASEAN network of food data system (ASEANFOODS). Thailand: Institute of Nutrition, Mahidol University.

Qamar,W., and Sultana, S. (2011). Polyphenols from *Juglans regia* L. (walnut) kernel modulate cigarette smoke extract induced acute inflammation, oxidative stress and lung injury in Wistar rats. *Human and Experimental Toxicology*, *30*, 499–506.

Qin, Y. Y., Zhang, Z. H., Li, L., Xiong,W., Shi, J. Y., Zhao, T. R. and Fan, J. (2013). Antioxidant effect of pomegranate rind powder extract, pomegranate juice, and pomegranate seed powder extract as antioxidants in raw ground pork meat. *Food Science and Biotechnology*, 22: 1063–1069.

Qingming, Y., Xianhui, P., Weibao, K., Hong, Y., Yidan, S., Li, Z., Yanan, Z., Lan, D. and Guoan, L. (2010). Antioxidant activities of malt extract from barley (*Hordeum vulgare* L.) toward oxidative stress *in vitro* and *in vivo*. *Food Chemistry*, 118(1), 84-89.

Qiu, J., Chen, L., Zhu, Q., Wang, D., Wang, W., Sun, X. and Du, F. (2012). Screening natural antioxidants in peanut shell using DPPH–HPLC–DAD–TOF/MS methods. *Food Chemistry*, 135: 2366–2371.

Qiu, Y., Liu, Q. and Beta, T. (2010). Antioxidant activity of commercial wild rice and analysis of soluble and insoluble phenolic acids. *Food Chemistry*, 121(1), 140-147.

Qiu, Y., Liu, Q. and Beta, T. (2009). Antioxidant activity of commercial wild rice and identification of flavonoids compounds in active fractions. *Journal of Agricultural and Food Chemistry*, 57: 7543–7551.

Qiu, Y., Liu, Q. and Beta, T. (2010). Antioxidant properties of commercial wild rice and analysis of soluble and insoluble phenolic acids. *Food Chemistry*, 121: 140–147.

Qu, W., Pan, Z. and Ma, H. (2010). Extraction modeling and activities of antioxidants from pomegranate marc. *Journal of Food Engineering*, 99: 16–23.

Quaranta, M., Nugroho Prasetyo, E., Koren, K., Nyanhongo, G. S., Murkovic, M., Klimant, I., and Guebitz, G. M. (2013). Enzymebased online monitoring and measurement of antioxidant activity using an optical oxygen sensor coupled to an HPLC system. *Analytical and Bioanalytical Chemistry*, *405*, 2371–2377.

Quesnel, V. C. (1968). Fractionation and properties of the polymeric leucocyanidin of the seeds of *Theobramll* cacao. *Phytochemistry,* 7:1583.

Quideau, S., Deffieux, D., Douat-Casassus, C., and Pouységu, L. (2011). Plant polyphenols: Chemical properties, biological activities, and synthesis. *Angewandte Chemie International Edition*, *50*, 586–621.

Quinn, L. A. and Tang, H. H. (1996). Antioxidant properties of phenolic compounds in macadamia nuts. *Journal of the American Oil Chemists' Society*, 73: 1585–1587.

Qureshi, A. A., Mo, H., Packer, L., and Peterson, D. M. (2000). Isolation and identification of novel tocotrienols from rice bran with hypocholesterolemic, antioxidant and antitumor properties. *J. Agric. Food Chem.*, 48:3130–3140.

Rababah, T. M., Ereifej, K. I., and Howard, L. (2005). Effect of ascorbic acid and dehydration on concentrations of total phenolics, antioxidant capacity, anthocyanins, and color in fruits. *Journal of Agricultural and Food Chemistry*, *53*, 4444-4447.

Rachie, K. O. (1970). Sorghum in Asia, in *Sorghum Production and Utilization,* Wall, J. S. and Ross, W. M., Eds. AVI Publishing. Westport, CT. 328.

Raffo, A., Malfa, G. L., Fogliano, V., Maiani, G. and Quaglia, G. (2006). Seasonal variations in antioxidant components of cherry tomatoes (*Lycopersicon esculentum* cv. Naomi F1). *Journal of Food Composition and Analysis*, 19: 11–19.

Ragaee, S., Abdel-Aal, E. M. and Noaman, M. (2006). Antioxidant activity and nutrient composition of selected cereals for food use. *Food Chemistry*, 98: 32–38.

Ragubeer, N., Beukes, D. R., and Limson, J. L. (2010). Critical assessment of voltammetry for rapid screening of antioxidants in marine algae. *Food Chemistry*, *121*, 227–232.

Rahman, A., Fazal, F., Greensill, J., Ainley, K., Parish, J. H. and Hadi, S. M. (1992). Strand scission in DNA induced by dietary flavonoids: Role of Cu (I) and oxygen free radicals and biological consequence of scission. *Molecular and Cellular Biochemistry*, 111: 3–9.

Rajaram, S., Burke, K., Connell, B., Myint, T. and Sabate, J. (2001). A monounsaturated fatty acid-rich pecan-enriched diet favorably alters the serum lipid profile of healthy men and women. *Journal of Nutrition*, 131: 2275–2279.

Rajase karan, N.S.,Nith ya , M.,Rose,C., and Chandra, T. S. (2004). The effect of finger millet feeding on the early responses during the process of wound healing in diabetic rats. Biochimica Biophysica Acta, 1689: 190–201.

Ramachandra, G., Virupaksha, T. K., and Shadaksharaswamy, M. (1977). Relalionship between tannin levels and *in vitro* protein digestibility in finger millet *(Eleusifle eoracana Gaertn), J. Agrie. Food Chern,* 25: 1101.

Ramamurthy, M. S., Maiti, B., Thomas, P., and Nair, M. (1992). High-performance liquid chromatography determination of phenolic acids in potato tubers (Solanum tuberosum) during wound healing. Journal of Agricultural and Food Chemistry, 40, 569-572.

Ramarathnam, N., Osawa, T., Namiki, M. and Kawakishi, S. (1989). Chemical studies of novel rice hull antioxidants. 2. Identification of isovitexin, a C-glycosyl flavonoid. *Journal of Agricultural and Food Chemistry*, 37: 316–319.

Ramezanzadeh, F., Rao, R. M., Windhauser, M., Priniawiwatkul, W., and Marshall, E. (1999). Prevention of oxidative rancidity in rice bran during storage. Journal of Agricultural and Food Chemistry, 47, 2997-3000.

Ramful, D., Bahorun, T., Bourdon, E., Tarnus, E. and Aruoma, O. I. (2010). Bioactive phenolics and antioxidant propensity of flavedo extracts of Mauritian citrus fruits: Potential prophylactic ingredients for functional foods application. *Toxicology*, 278: 75–87.

Ramji, D., Huang, M. T., Shahidi, F. and Ho, C. T. (2009). Effect of tea and tea constituents on inflammation. In C. T. Ho, J. K. Linand F. Shahidi (Eds.), *Tea and tea products:*

Chemistry and healthpromoting properties (pp. 177–189). NewYork: CRC Press, Taylor and Francis Group.

Rankin, J. W., Andreae, M. C., Oliver Chen, C. Y., and O'Keefe, S. F. (2008). Effect of raisin consumption on oxidative stress and inflammation in obesity. *Diabetes, Obesity and Metabolism*, *10*, 1086–1096.

Rao, A. V. and Snyder, D. M. (2010). Raspberries and human health: A review. *Journal of Agricultural and Food Chemistry*, 58: 3871–3883.

Rao, B. S. N. and Prabhavati, T. (1982). Tannin content of foods commonly consumed in India and its innuence on ionizable iron, J. *Sci. Food Agric.* 33: 89.

Rao, M.V.S.S.T.S. , and Muralikrishna, G. (2002). Evaluation of the antioxidant properties of free and bound phenolic acids from native and malted finger millet (Ragi, *Elucine coracana* Indaf- 15). Journal of Agricultural and Food Chemistry, 50: 889–892.

Rapisarda, P., Carollo, G., Fallico, B., Tomaselli, F. and Maccarone, E. (1998). Hydroxycinnamic acids as markers of Italian blood orange juices. *Journal of Agricultural and Food Chemistry*, 46: 464–470.

Rapisarda, P., Tomaino, A., Lo Cascio, R., Bonina, F., De Pasquale, A., and Saija, A. (1999). Antioxidant effectiveness as infuenced by phenollic content of orange juices. Journal of Agricultural and Food Chemistry, 47, 4718-4723.

Raudonis, R., Bumblauskiene, L., Jakstas, V., Pukalskas, A., and Janulis, V. (2010). Optimization and validation of post-column assay for screening of radical scavengers in herbal raw materials and herbal preparations. *Journal of Chromatography. A*, *1217*, 7690–7698.

Ravichandran, K., Saw, N. M. M. T., Mohdaly, A. A., Gabr, A. M., Kastell, A., Riedel, H. and Smetanska, I. (2013). Impact of processing of red beet on betalain content and antioxidant activity. *Food Research International*, 50: 670–675.

Rawson, A., Patras, A., Tiwari, B. K., Noci, F., Koutchma, T., and Brunton, N. (2011). Effect of thermal and non-thermal processing technologies on the bioactive content of exotic fruits and their products: Review of recent advances. *Food Research International*, *44*, 1875–1887.

Rayalam, S., Yang, J.-Y., Ambati, S., Della-Fera, M. A. and Baile, C. A. (2008). Resveratrol induces apoptosis and inhibits adipogenesis in 3T3-L1 adipocytes. *Phytotherapy Research*, 22: 1367–1371.

Raynal, J., Moutounet, M., and Souquet, J. M. (1989). Intervention ofphenolic compounds in plum technology. 1. Changes during drying. Journal of Agricultural and Food Chemistry, 37, 1046-1050.

Reboul E, Richelle M, Perrot E, Desmoulins-Malezet C, Pirisi V, Borel P; Richelle; Perrot; Desmoulins-Malezet; Pirisi; Borel (2006). "Bioaccessibility of carotenoids and vitamin E from their main dietary sources". Journal of Agricultural and Food Chemistry 54(23): 8749–8755.

Rechner, A. R., Kuhnle, G., Bremner, P., Hubbard, G. P., Moore, K. P. and Rice-Evans, C. A. (2002). The metabolic fate of dietary polyphenols in humans. *Free Radical Biology and Medicine*, 33: 220–235.

Reckziegel, P., Boufleur, N., Barcelos, R. C., Benvegnu, D. M., Pase, C. S., Muller, L. G., Teixeira, A. M., Zanella, R., Prado, A. C., Fett, R., Block, J. M., and Burger, M. E. (2011). Oxidative stress and anxiety-like symptoms related to withdrawal of passive cigarette smoke in mice: Beneficial effects of pecan nut shells extract, a by-product of the nut industry. *Ecotoxicology and Environmental Safety*, *74*, 1770–1778.

Reddy, N. R., Pierson, M. D., Sathe, S. K., and Salunkhe, D. K. (1985). Dry bean tannins: a review of nutritional implications, *J. Am. Oil Chem. Soc.,* 62: 541.

Reddy, N.R., Pierson, M.D., Sathe, S.K., and Salunkhe, D.K. (1985). Dry bean tannins: a review of nutritional implications. J. Am. Oil Chem. Soc. 62: 541- 553.

Reed, U. J. S., Horrath, P. J., Allen, M. S., and Van Soest, P. J. (1985). Gravimetric determination of soluble phenolics including tannins from leaves by precipitation with trivalent ytterbium, *J. Sci. Food Agric., 36,* 255.

Reglero, R. G. J., Tabera, G. J. J., IbaÂnÄ ez, E. M. E., LoÂ pez-SebastiaÂ n, L. S., Ramos, M. E., Ballester, S. L. and Bueno, M. J. M. (1999).

Rehman, Z. U., Habib, F. and Shah,W. H. (2004). Utilization of potato peels extract as a natural antioxidant in soy bean oil. *Food Chemistry*, 85: 215–220.

Relajakshmi, D. and Narasimhan, S. (1996). Food antioxidants: Sources and methods of evaluation. In D. L. Madhavi (Ed.), *Food antioxidants technological* (pp. 65–157). NewYork: Marcel Dekker Inc.

Ren, S. C. and Sun, J. T. (2014). Changes in phenolic content, phenylalanine ammonia-lyase (PAL) activity, and antioxidant capacity of two buckwheat sprouts in relation to germination. *Journal of Functional Foods*, 7: 298–304.

Renaud, S. and de Lorgeril, M. (1992).Wine alcohol, platelets and the French paradox for coronary heart disease. *Lancet*, 339: 1523–1526.

Rendina, E., Hembree, K. D., Davis, M. R., Marlow, D., Clarke, S. L., Halloran, B. P., Lucas, E. A., and Smith, B. J. (2013). Dried plum's unique capacity to reverse bone loss and alter bone metabolism in postmenopausal osteoporosis model. *PLoS ONE, 8*, e60569.

Rendina, E., Lim, Y. F., Marlow, D., Wang, Y., Clarke, S. L., Kuvibidila, S., Lucas, E. A., and Smith, B. J. (2012). Dietary supplementation with dried plum prevents ovariectomyinduced bone loss while modulating the immune response in C57BL/6J mice. *Journal of Nutritional Biochemistry*, *23*, 60–68.

Renger, A. and Steinhart, H. (2000). Ferulic acid dehydrodimers as structural elements in cereal dietary fiber. *European Food Research Technology*, 211: 422–428.

Reyes, L. F. (2005). Antioxidant capacity, anthocyanins and total phenolics in purple- and red-fleshed potato (*Solanum tuberosum* L.) genotypes. *American Journal of Potato Research*, 82: 271–277.

Reyes, L. F., and Zevallos, L. C. (2007). Degradation kinetics and color of anthocyanins in aqueous extracts of purple and red-flesh potatoes (Solanum tuberosum L.). *Food Chemistry*. 100 (3): 885–894.

Reynolds, H. and Vaile, J. E. (1942). Chemical composition of Arkansas-grown American grapes, *Ark. Agric. Exp. Sin. Bull.,* 420. 1.

Rhee, K. S., Ziprin, Y. A. and Calhoun, M. C. (2001). Antioxidative effects of cottonseed meals as evaluated in cooked meat. *Meat Science*, 58: 117–123.

Rhodes, M. J. C. and Price, K. R. (1996). Analytical problems in the study of flavonoid compounds in onions. *Food Chemistry*, 57: 113–117.

Rhodes, M.J.C. (1985). The physiochemical significance of plant phenolic compounds. In: *Annual Proceedings of the Phytochemical Society of Europe*. C. F. Van Sumere and P. J. Lea (Eds.), p. 99-119. Clarendon Press, Oxford.

Ribereau, Gayon, P. (1972). Plant Phenolics. Oliver and Boyd. Edinburgh. U.K.

Ribereau-Gapon. P. Pontallier, P. and Gluies, Y. (1983). Some interpretation of colour changes in young red wines during their conservation. *J.* Sci. *Food Agri.* 34: 505.

Ribereau-Gayon, P. (1972). *Plant phenolics*. p. 1-254 Olive and Boyd. Edinburgh, London.

Ribereau-Gayon, P. and Peyand, E. (1958). Analyse et Controle des Vins. Beranger, Paris.

Ribereau-Gayon, P. and Stonestret, E. (1966). Dosage des tannins due vin rouge et delennination de deur structure, *Chem. Anal.,* 48, 188.

Rice-Evans, C. A., Miller, N. J., Bolwell, P. G., Bramley, P. M. and Pridham, J. B. (1995). The relative antioxidant activities of plant-derived polyphenolic flavonoids. *Free Radical Research*, 22: 375–383.

Rice-Evans, C. A., Miller, N. J. and Paganga, G. (1996). Structureantioxidant activity relationships of flavonoids and phenolic acids. *Free Radical Biology and Medicine*, 20: 933–956.

Richardson, M. (1977). The proteinase inhibitors of plant and microorganisms. *Phytochemistry,* 16, 159.

Riedl, K. M., Carando, S., Alessio, H. M., McCarthy, M. and Hagerman, A. E. (2002). Antioxidant activity of tannins and tannin-protein complexes: Assessment *in vitro* and *in vivo*. In M. Morello, F. Shahadiand C. T. Ho (Eds.), *Free radicals in foods: Chemistry, nutrition and health* (pp. 188–200). ACS Symposium Series.Washington, DC: American Chemical Society.

Rimbach, G., Boesch-Saadatmandi, C., Frank, J., Fuchs, D.,Wenzel, U., Daniel, H., Hall,W. L. and Weinberg, P. D. (2008). Dietary isoflavones in the prevention of cardiovascular disease – A molecular perspective. *Food and Chemical Toxicology*, 46: 1308–1319.

Riuz, A., Ayora-Canada, M. J., and Lendl, B. (2001). A rapid method for peroxide value determination in edible oils based on flow analysis with Fourier transform infrared spectroscopic detection. *The Analyst*, *126*, 242–246.

Robards, K., Prenzler, P.D, Tucker, G., Swatsitang, P., and Glover, W. (1999). Phenolic compounds and their role in oxidative processes in fruits. Food Chemistry, 66, 401-436.

Robbins, R. J. (2003). Phenolic acids in foods: An overview of analytical methodology. *Journal of Agricultural and Food Chemistry*, 51: 2866–2887.

Roch, P., Dreyer, A., Lacan, D., Baccou, J. C. and Ginoux, J. P. (1998). Cucumis melo protein extract with antioxidant activity and process for preparing it, cosmetic composition of food composition containing such an extract. US Patent US5747043.

Rodil, R., Quintana, J. B., Basaglia, G., Pietrogrande, M. C. and Cela, R. (2010). Determination of synthetic phenolic antioxidants and their metabolites in water samples by downscaled solid-phase extraction, silylation and gas chromatography–mass spectrometry. *Journal of Chromatography. A*, 1217: 6428–6435.

Rodriguez de Sotillo, D., Hadley, M. and Holm, E. T. (1994a). Phenolics in aqueous potato peel extract: Extraction, identification and degradation. *Journal of Food Science*, 59: 649–651.

Rodriguez de Sotillo, D., Hadley, M. and Holm, E. T. (1994b). Potato peel waste: Stability and antioxidant activity of a freeze-dried extract. *Journal of Food Science*, 59: 1031–1033.

Rodríguez, R., Jaramillo, S., Rodríguez, G., Espejo, J. A., Guillén, R., Fernández-Bolaños, J., Heredia, A. and Jiménez, A. (2005). Antioxidant activity of ethanolic extracts from several asparagus cultivars. *Journal of Agricultural and Food Chemistry*, 53: 5212–5217.

Rodríguez-Rojo, S., Visentin, A., Maestri, D. and Cocero, M. J. (2012). Assisted extraction of rosemary antioxidants with green solvents. *Journal of Food Engineering*, 109: 98–103.

RodrõÂguez de Sotillo, D., Hadley, M., and Holm, E. T. (1994a). Phenolics in aqueous potato peel extract, extraction, identification and degradation. Journal of Food Science, 59, 649-651.

RodrõÂguez de Sotillo, D., Hadley, M., and Holm, E. T. (1994b). Potato peels waste, stability and antioxidant activity of a freeze-dried extract. Journal of Food Science, 59, 1031-1033.

RodrõÂguez de Sotillo, D., Hadley, M., and Wolf-Hall, C. (1998). Potato peel extracts a nonmutagenic antioxidant with potential antimicrobial activity. Journal of Food Science, 63, 907-910.

Roedig-Penman, A., and Gordon, M. H. (1998). Antioxidant properties of myricetin and quercetin in oil and emulsions. Journal of the American Oil Chemists' Society, 75, 169-179.

Rogers, E. J., Rice, S. M., Nicolosi, R. J., Carpenter, D. R., McClelland, C. A., and Romanczyk, L. J. (1993). Identification and quantitation of γ-oryzanol components and simultaneous assessment of tocols in rice bran oil. *J. Am. Oil Chem. Soc.*, 70:301-307.

Roginsky, V., and Liss, E. A. (2005). Review of methods to determine chain-breaking antioxidant activity in food. *Food Chemistry, 92*, 235–254.

Rohdewald, P. (2002). A review of the French maritime pine bark extract (Pycnogenol®), a herbal medication with a diverse clinical pharmacology. *International Journal of Clinical Pharmacology and Therapeutics*, 40: 158–168.

Roldán, E., Sánchez-Moreno, C., Ancos, B. D. and Cano, M. P. (2008). Characterization of onion (*Allium cepa* L.) by-products as food ingredients with antioxidant and antibrowning properties. *Food Chemistry*, 108: 907–916.

Rollman, B., Vancraenenbroeck, R. and Lonti, R. (1966). *Arch. 1m. Physiol. Biochem.,* 74, 714.

Romani, A., Mulinacci, N., Pinelli, P., Vincieri, F. F., and Cimato, A. (1999). Polyphenolic content in five Tuscany cultivars of Olea europaea L. Journal of Agricultural and Food Chemistry, 47, 964-967.

Romani, A., Mulinacci, N., Pinelli, P., Vincieri, F. F. and Cimato, A. (1999). Polyphenolic content in five Tuscany cultivars of *Olea europaea* L. *Journal of Agricultural and Food Chemistry*, 47: 964–967.

Romani, A., Pinelli, P., Galardi, C., Mulinacci, N. and Tattini, M. (2002). Identification and quantification of galloyl derivatives, flavonoid glycosides, and anthocyanins in leaves of *Pistacia lentiscus* L. *Phytochemical Analysis*, 13: 79–86.

Romani, A., Vignolini, P., Galardi, C., Mulinacci, N., Benedettelli, S. and Heimler, D. (2004). Germplasm characterization of Zolfino landraces (*Phaseolus vulgaris* L.) by flavonoid content. *Journal of Agricultural and Food Chemistry*, 52: 3838–3842.

Romano, C. S., Abadi, K., Repetto, V., Vojnov, A. A. and Moreno, S. (2009). Synergistic antioxidant and antibacterial activity of rosemary plus butylated derivatives. *Food Chemistry*, 115: 456–461.

Rommel, A. and Wrolstad, R. (1993). Influence of acid and base hydrolysis on the phenolic composition of red raspberry juice. *Journal of Agricultural and Food Chemistry*, 41: 1237–1241.

Rondini, L., Peyrat-Maillard, M.-N, Fromentin, G., Durand, P., Tome, D., Prost, M., and Berset, C. (2004). Bound ferulic acid from bran is more bioavailable than the free compound in rat. Journal of Agricultural and Food Chemistry, 52: 4338–4343.

Rong, N., Ausman, L.M., and Nicolosi, R. J. (1997). Oryzanol decreases cholesterol absorption and aortic fatty streaks in hamsters. *Lipids,* 32:303–309.

Rooney, L. W., and Waniska, R. D. (2000). Sorghum food and industrial utilization. In: Sorghum: Origin, History, Technology, and Production. pp. 689-729. Smith, C. W., and Frederiksen, R. A., Eds., John Wiley and Sons, New York.

Ros, E. (2010). Health benefits of nut consumption. *Nutrients, 2*, 652–682.

Ros, E. (2015). Nuts and CVD. *The British Journal of Nutrition, 113*,S111–S120.

Roseman, A. S., Livingston, G. E., and Esselen, W. B. (1957). Non-enzymatic discoloration of green bean puree. II. Studies on the thermal discoloration of acetone-extracted tissues. *Food Res.,* 22: 542.

Rosenblatt, M. and Peluso, J. V. (1941). Determination of tannins by photocolorimeter, *J. Assoc. Off. Agric. Chem.* 24, 170.

Ross, A. B., Shepherd, M. J., Schupphaus, M., Sinclair, V., Alfaro, B., Kamal-Eldin, A., and Aman, P. (2003). Alkylresorcinols in cereals and cereal products. *J. Agric. Food Chem.,* 51:4111-4118.

Rossi, J. A, and Singleton, V. L. (1966). Flavor effects and adsorptive properties of purified fractions of grape seed phenols, *Am. J. Enol. Vitic,* 17, 240.

Rossi, J. A. and Singleton, V. L. (1966). Contribution of grape phenols 10 oxygen absorption and brewing of wines, *Am. J. Enol. Vitic,* 17, 231.

Rossi, J.A.Jr. and Singleton, V.L. (1966a). Contributions of grape phenols to oxygen absorption and crowning of wines. Am. J. Enol. Viticul. 17: 231-239.

Rossi, J.A.Jr. and Singleton, V.L. 1966b. Flavour effects and absorptive properties of purified fractions of grape-seed phenols. Am. J. Enol.Viticul. 17: 240-248.

Roux, D. G. and Maihs, A. E. (1960). Selective spray reagents for the identification and estimation of flavonoid compounds associated with condensed tannins. J. *Chromatography.* 4, 65.

Roux, D. G. and Paulus, E. (1961). Condensed tannins. II. Isolation of a condensed tannin from black-wattle heartwood and synthesis of (±) -7:3':4' trihydroxyflavan-4-ol, *Biochem. J*: 80: 476.

Roux, O. G. (1970). Some recent advances in the identification of I proanthocyanins and the chemistry of condensed tannins. *Nature.* 180, 973.

Roy, M. K., Takenaka, M., Isobe, S. and Tsushida, T. (2007). Antioxidant potential, antiproliferative activities and phenolic content in water-soluble fractions of some commonly consumed vegetables: Effects of thermal treatment. *Food Chem.* 103: 106-114.

Rubin, L. J., and Shahidi, F. (1988). Lipid oxidation and the flavor of meat products. Proc. 34th International Congress in Meat Science and Technology, ICMST, Brisbane, pp 295–301.

Rufer, C. E. and Kulling, S. E. (2006). Antioxidant activity of isoflavones and their major metabolites using different in vitro assays. *Journal of Agricultural and Food Chemistry,* 54: 2926–2931.

Ruiz, D., Egea, J., Gil, M. I. and Tomas-Barberan, F. A. (2005). Characterization and quantitation of phenolic compounds in new apricot (*Prunus armeniaca* L.) varieties. *Journal of Agricultural and Food Chemistry*, 53: 9544–9552.

Ruiz, J. A., PeÂrez-Vendrell, A. M., and Esteve-GarcõÂa, E. (1999). Effect of b-carotene and vitamin E on oxidative stability in leg meat of broilers fed different supplemental fats. Journal of Agricultural and Food Chemistry, 47, 448-454.

Rumbaoa, R. G. O., Cornago, D. F. and Geronimo, I. M. (2009a). Phenolic content and antioxidant capacity of Philippine potato (*Solanum tuberosum*) tubers. *Journal of Food Composition and Analysis*, 22: 546–550.

Rumbaoa, R. G. O., Cornago, D. F. and Geronimo, I. M. (2009b). Phenolic content and antioxidant capacity of Philippine sweet potato (*Ipomoea batatas*) varieties. *Food Chemistry*, 113: 1133–1138.

Russell, A. E., Shuttleworth, S. G. and Williamswynn, D. A. (1968). Further studies on the mechanism of vegetable tannage. II. Effect of urea extraction on hydrothermal stability of leathers tanned with a range of organic tanning agents, *J. Soc. Leother Trades Chem.* 52, 220. Russell, A. E. (1935). The natural tannins. *Chem. Rev.* 17, 155.

Ruzaidi, A., Amin, I., Nawalyah, A. G., Hamid, M. and Faizul, H. A. (2005). The effect of Malaysian cocoa extract on glucose levels and lipid profiles in diabetic rats. *Journal of Ethnopharmacology*, 98: 55–60.

Ryan, C. A. and Green, T. R. (1974). Proteinase inhibitors in natural plant protection, in *Metabolism and Regulation of Secondary Plant Products.* Runeckles, V. C. and Conn, E. E., Eds. Academic Press. New York, 123.

Ryan, E., Galvin, K., O'Connor, T. P., Maguire, A. R., and O'Brien, N. M. (2006). Fatty acid profile, tocopherol, squalene and phytosterol content of Brazil, pecan, pine, pistachio and cashew nuts. *International Journal of Food Sciences and Nutrition, 57*, 219–228.

Saad, B., Sing, Y. Y., Nawi, M. A., Hashim, N. H., Ali, A. S. M., Saleh, M. I., Sulaiman, S. F., Talib, K. M. and Ahmad, K. (2007). Determination of synthetic phenolic antioxidants in food items using reversed-phase HPLC. *Food Chemistry*, 105: 389–394.

Sabir, M. A., Sosulski, F. W., and Finlayson, A. J. (1974). Chlorogenic acid-protein interactions in sunfower. Journal of Agricultural and Food Chemistry, 22, 575-578.

Sabliov, C. M., Fronczek, C., Astete, C. E., Khachaturyan, M., Khachatryan, L. and Leonardi, C. (2009). Effects of temperature and UV light on degradation of α-tocopherol in free and dissolved form. *Journal of the American Oil Chemists' Society*, 86: 895–902.

Sahu, S. C. and Gray, G. C. (1993). Interactions of flavonoids, trace metals, and oxygen: Nuclear DNA damage and lipid peroxidation induced by myricetin. *Cancer Letters*, 70: 73–79.

Sahu, S. C. and Gray, G. C. (1994). Kaempferol-induced nuclear DNA damage and lipid peroxidation. *Cancer Letters*, 85: 159–164.

Saint-Cricq de Gaulejac, N., Provost, C., and Vivas, N. (1999). Comparative study of polyphenol scavenging activities assessed by different methods. Journal of Agricultural and Food Chemistry, 47, 425-431.

Saito, H., and Ishihara, K. (1997). Antioxidant activity and active sites of phospholipids as antioxidants. Journal of the American Oil Chemists' Society, 74, 1531-1536.

Saito, M., Hosoyama, H., Ariga, T., Kataoka, S., and Yamaji, N. (1998). Antiulcer activity of grape seed extract and procyanidins. Journal of Agricultural and Food Chemistry, 46, 1460-1464.

Saitta, M., Giuffrida, D., Torre, G. L. L., Potortì, A. G. P. and Dugo, G. (2009). Characterisation of alkylphenols in pistachio (*Pistacia vera* L.) kernels. *Food Chemistry*, 117: 451–455.

Sajilata, M. G., Bajaj, P. R., and Singhai, R. S. (2008). Tea polyphenols as nutraceuticals. *Comprehensive Reviews in Food Science and Food Safety*, *7*, 229–254.

Sakanaka, S., Tachibana, Y. and Okada, Y. (2005). Preparation and antioxidant properties of extracts of Japanese persimmon leaf tea (*kakinoha cha*). *Food Chem.* 89: 569–575.

Sakihama, Y., Maeda, M., Hashimoto, M., Tahara, S. and Hashidoko,Y. (2012). Beetroot betalain inhibits peroxynitrite-mediated tyrosine nitration and DNA strand cleavage. *Free Radical Research*, 46: 93–99.

Saklar, S., Katnas, S., and Ungan, S. (2001). Determination of optimum hazelnut roasting conditions. *International Journal of Food Science and Technology*, *36*, 271–281.

Sakr, S. B. and Abo-El-Yazid, S. M. (2012). Effect of fenugreek seed extract on adriamycin-induced hepatoxicity and oxidative stress in albino rats. *Toxicology and Industrial Health*, 28: 876–885.

Salah, N., Miller, N. J., Paganga, G., Tijburg, L., Bolwell, G. P. and Rice-Evans, C. A. (1995). Polyphenolic flavonols as scavengers of aqueous phase radicals and as chain-breaking antioxidants. *Archives of Biochemistry and Biophysics*, 322: 339–346.

Salami, M., Galli, C., De Angelis, L. and Visioli, F. (1995). Formation of F2-isoprostanes in oxidized low-density lipoprotein: Inhibitory effect of hydroxytyrosol. *Pharmacological Research*, 31: 275–279.

Salar, R. K. and Purewal, S. S. (2016). Phenolic content, antioxidant potential and DNA damage protection of pearl millet (*Pennisetum glaucum*) cultivars of north Indian region. *Food Meas. Charact.* doi:10.1007/s11694-016-9379-z

Saleh, M. M., Hashem, F. A. E.-M., and Glombitza, K. W. (1998). Study of Citrus taitensis and radical scavenger activity of the avonoids isolated. Food Chemistry, 63, 397-400.

Salinas-Moreno, Y., Rojas-Herrera, L., Sosa-Montes, E. and Pérez-Herrera, P. (2005). Anthocyanin composition in black bean (*Phaseolus vulgaris* L.) varieties grown in México. *Agrociencia*, 39: 385–394.

Salovaara, H., and Autio, K. (2001). Rye and triticale. In: Cereals and cereal products. pp. 391-410. Dendy, D. A. V., and Dobraszczyk, B. J., Eds., Aspen Publishers, Maryland.

Salunkhe, D. K. Kadam, S. S., and Chavan. J. K. (1985). Postharvest Biochemistry of Food Legumes. CRC, Press. Boca Raton. FL.

Salunkhe, D. K., Chavan, J. K., and Kadam, S. S. (1985a). Postharvest Biotechnology of Cereals, CRC Press, Boca Raton, Fl. 215.

Salunkhe, D. K., jadhav, S. J., Kadam, S. S., and Chavan, J. K. (1982). Chemical, biochemical and biological significance of polyphenols in cereals and legumes, *Crit. Rev. Food* Sri. *Nutr.* 17: 277.

Salunkhe, D. K., Kadam, S. S., and Chavan, J. K., (1985b). *Postharvest Biotechnology of Food Legumes.* CRC Press, Boca Raton, FL, 1985.

Salunkhe, D.K., Sathe, S.K., and Reddy, N.R. (1982). In: Chemistry and Biochemistry of Legumes. S. K. Arora (Ed.), p. 58-82. Oxford and IBH, New Delhi.

Samad, N. B., Debnath, T., Hasnat, A., Pervin, M., Kim, D. H., Jo, J. E., Park, S. R. and Lim, B. O. (2014). Phenolic contents, antioxidant and anti-inflammatory activities of Asparagus cochinchinensis (Loureiro) Merrill. *Journal of Food Biochemistry*, 38: 83–91.

Samaranayake, A. G. P., John, J. A. and Shahidi, F. (2008). Antioxidant activity of English walnut (*Juglans regia* L.). *Journal of Food Lipids*, 15: 384–397.

Samaras, T.S., Camburn, P.A., Chandra, S.X., Gordon, M.H. and Ames, J.M., (2005). Antioxidant properties of kilned and roasted malts. Journal of Agricultural and Food Chemistry 53, 8068-8074.

Sanbongi, C., Osakabe, N., Natsume, M., Takizawa, T., Gomi, S. and Osawa, T. (1998). Antioxidative polyphenols isolated from *Theobroma cacao. Journal of Agricultural and Food Chemistry*, 48: 454–457.

Sanbongi, C., Suzuki, N. and Sakane, T. (1997). Polyphenols in chocolate, which have antioxidant activity, modulate immune functions in humans in vitro. *Cellular Immunology*, 177: 129–136.

Sánchez-Moreno, C. (2002). Methods used to evaluate the free radical scavenging activity in foods and biological systems. *Food Science and Technology International*, 8(3), 121-137.

Sanchez-Rabaneda, F., Jauregui, O., Casals, I., Andres-Lacueva, C., Izquierdo-Pulido, M. and Lamuela-Raventos, R. M. (2003). Liquid chromatographic/electrospray ionization tandem mass spectrometric study of the phenolic composition of cocoa *Theobroma cocao. Journal of Mass Spectrometry*, 38: 35–42.

Sang, S., Lapsley, K., Jeong,W. S., Lachance, P. A., Ho, C. T. and Rosen, R. T. J. (2002). Antioxidative phenolic compounds isolated from almond skins (*Prunus amygdalus* Batsch). *Journal of Agricultural and Food Chemistry*, 50: 2459–2463.

Sang, S., Lapsley, K., Rosen, R. T. and Ho, C.-T. (2002). New prenylated benzoic acid and other constituents from almond hulls (Prunus amygdalus Batsch). *Journal of Agricultural and Food Chemistry*, 50: 607–609.

Sanghani, A., Kandra, T., Bavadekar, S. and Vansal, S. (2013). Pterostilbene inhibits VEGF production in human breast cancer [MDA-MB-231] cells. *The FASEB Journal, 27*(Meeting Abstracts Suppl.), 1104–1106.

Sanoner, P., Guyot, S., Marnet, N., Molle, D. and Drilleau, J. F. (1999). Polyphenol profiles of French cider apple varieties (*Malus domestica* sp). *Journal of Agricultural and Food Chemistry*, 47: 4847–4853.

Sariahmetoglu, M., Wheatley, R. A., Cakici, I., Kanzik, I., and Townshend, A. (2003). Flow injection analysis for monitoring antioxidant effects on luminol chemiluminescence of reactive oxygen species. *Analytical Letters, 36*, 749–765.

Sarkar, S. K. and Haworth, R. E. (1976). Specificity of the vanillin test for flavonols, *J. Agric. Food Chem.*, 24.317.

Sarma, B. K., Singh, D. P., Mehta, S., Singh, H. B. and Singh, U. P. (2002). Plant growth-promoting rhizobacteria-elicited alterations in phenolic profile of chickpea (*Cicer arietinum*) infected by Sclerotium rolfsii. *Journal of Phytopathology*, 150: 277–282.

Sarni-Machado, P., Cheynier, V., and Montounet, M. (1999). Interactions of grape seed tannins with salivary proteins. Journal of Agricultural and Food Chemistry, 47, 42-47.

Sathishsekar, D. and Subramanian, S. (2005). Antioxidant properties of Momordica charantia (bitter gourd) seeds on streptozotocin induced diabetic rats. *Asia Pacific Journal of Clinical Nutrition*, 14: 153–158.

Sato, M., Ramarathnam, N., Suzuki, Y., Ohkubo, T., Takeuchi, M., and Ochi, H. (1996). Varietal differences in the phenolic content and superoxide radical scavenging potential of wines from different sources. Journal of Agricultural and Food Chemistry, 44, 37-41.

SatueÂ-Gracia, M. T., AndreÂs-Lacueva, C., Lamuela-RaventoÂ s, R. M., and Frankel, E. N. (1999). Spanish sparkling wines (cavas) as inhibitors of in vitro human low-density lipoprotein oxidation. Journal of Agricultural and Food Chemistry, 47, 2198-2202.

Satue-Gracia, M. T., Heinonen, M., and Frankel, E. N. (1997). Anthocyanins as antioxidants on human low-density lipoprotein and lecithin-liposome systems. *Journal of Agricultural and Food Chemistry*, *45*, 3362–3367.

Saucier, C. T., and Waterhouse, A. L. (1999). Synergetic activity of catechin and other antioxidants. Journal of Agricultural and Food Chemistry, 47, 4491-4494.

Saura-Calixto, F. (1998). Antioxidant dietary fiber product, a new concept and a potential food ingredient. Journal of Agricultural and Food Chemistry, 46, 4303-4306.

Saura-Calixto, F. and Goñi, I. (2006). Antioxidant capacity of the Spanish Mediterranean diet. *Food Chemistry*, 94: 442–447.

Sawa, T., Nakao, M., Akaike, T., Ono, K., and Maeda, H. (1999). Alkylperoxyl radical-scavenging activity of various favonoids and other phenolic compounds, implications for the anti-tumor-promoter effect of vegetables. Journal of Agricultural and Food Chemistry, 47, 397-402.

Scalbert, A. (1991). Antimicrobial properties of tannins. *Phytochemistry*, *30*, 3875–3883.

Scalbert, A. and Haslam, E. (1987). Polyphenols and chemical defense of the leaves of Quercus robur, *Phytochemistry*. 26, 3191.

Scalbert, A. and Williamson, G. (2000). Dietary intake and bioavailability of polyphenols. *Journal of Nutrition*, 130: 2073S–2085S.

Schaich, K. M. (2005). Lipid oxidation: Theoretical aspects. In F. Shahidi (Ed.), *Bailey's industrial oil and fat products* (6th ed., Vol. 1, pp. 269–356). Hoboken, NJ: JohnWiley and Sons, Inc.

Schanderl, S. H. (1970). Tannins and related phenolics. In *Methods in Food Analysis,* Joslyn, M. A., Ed., Academic Press, New York, 701.

Schanderl, S. H. (1970). Tannins and related phenolics. In Methods in Food Analysis. Joslyn. M. A. Ed., Academic Press, New York, 701.

Schatzkin, A., Mouw, T., Park, Y., Subar, A. F., Kipnis, V., Hollenbeck, A., Leitzmann, M. F. and Thompson, F. E. (2007). Dietary fiber and whole-grain consumption in relation to colorectal cancer in the NIH-AARP Diet and Health Study. *The American Journal of Clinical Nutrition*, 85(5), 1353-1360.

Schieffer, G. W. (2002). Pressurized liquid extraction of curcuminoids and curcuminoid degradation products from turmeric (*Curcuma longa*) with subsequent HPLC assays. *Journal of Liquid Chromatography and Related Technologies*, 25: 3033–3044.

Schlörmann,W., Birringer, M., Böhm, V., Löber, K., Jahreis, G., Lorkowski, S., Muller, A. K., Schone, F., and Glei, M. (2015). Influence of roasting conditions on health-related compounds in different nuts. *Food Chemistry*, *180*, 77–85.

Schmidt, S., Niklová, I., Pokorný, J., Farkaš, P. and Sekretár, S. (2003). Antioxidant activity of evening primrose phenolics in sunflower and rapeseed oils. *European Journal of Lipid Science and Technology*, 105: 427–435.

Schmitzer, V., Slatnar, A., Mikulic-Petkovsek, M., Veberic, R., Krska, B., and Stampar, F. (2011). Comparative study of primary and secondary metabolites in apricot (*Prunus armeniaca* L.) cultivars. *Journal of the Science of Food and Agriculture*, *91*, 860-866.

Schmitzer, V., Slatnar, A., Veberic, R., Stampar, F., and Solar, A. (2011). Roasting affects phenolic composition and antioxidative activity of hazelnuts (*Corylus avellana* L.). *Journal of Food Science*, *76*, S14–S19.

Schneider C (2005). "Chemistry and biology of vitamin E". Mol Nutr Food Res 49 (1): 7–30.

Scholz, S., and Williamson, G. (2007). Interactions affecting the bioavailability of dietary polyphenols *in vivo*. *International Journal for Vitamin and Nutrition Research*, *77*, 224–235.

Schuler, P. (1990). Natural antioxidants exploited commercially. In B. J. F. Hudson (Ed.), *Food antioxidants* (pp. 99–170). London, UK: Elsevier.

Schultz, J. C., Baldwin, I. T., and Nothnagle, P. J. (1981). Hemoglobin as a binding substrate in the quantitative analyses of plant tannins, J. *Agric. Food Chem.*, 29, 823.

Schulze, B., Hubbermann, E. M., and Schwarz, K. (2014). Stability of quercetin derivatives in vacuum impregnated apple slices after drying (microwave vacuum drying, air drying, and freeze drying) and storage. *LWT-Food Science and Technology*, *57*, 426-433.

Schwartz, H., Ollilainen, V., Piironen, V. and Lampi, A. M. (2008). Tocopherol, tocotrienol and plant sterol contents of vegetable oils and industrial fats. *Journal of Food Composition and Analysis*, 21: 152–161.

Schwarz, K., Bertelsen, G., Nissen, L. R., Gardner, P. T., Heinonen, M. I., Hopia, A., Huynh-Ba, T., Lambelet, P., McPhail, D., Skibsted, L. H., and Tijburg, L. (2001). Investigation of plant extracts for the protection of processed foods against lipid oxidation. Comparison of antioxidant assays based on radical scavenging, lipid oxidation and analysis of the principal antioxidant compounds. *European Food Research and Technology*, *212*, 319–328.

Scott, G. (1997). *Antioxidants: In science, technology, medicine and nutrition* (pp. 36–79). Chichester, UK: Albion Publishing.

Scott, R. D. (1922). Tannin in whiskey. J. *Am. Pharm. Assoc.* Sri. *Ed.* II, 1017.

Scott, S. L., Chen, W., Bakac, A. and Espenson, J. H. (1993). Spectroscopic parameters, electrode potentials, acid ionization constants, and electron exchange rates of the 2, 20-azinobis (3-ethylbenzothia-zoline-6-sulfonate) radicals and ions. *Journal of Physics and Chemistry*, 97: 6710–6714.

Sebranek, J. G., Sewalt, V. J. H., Robbins, K. L. and Houser, T. A. (2005). Comparison of a natural rosemary extract and BHA/BHT for relative antioxidant effectiveness in pork sausage. *Meat Science*, 69: 289–296.

Sedman, J., van de Voort, F. R., and Ismail, A. A. (1997). Application of Fourier transform infrared spectroscopy in edible-oil analysis. In R. E. McDonald and M. M. Mossoba (Eds.), *New techniques and applications in lipid analysis* (pp. 283–324). Champaign, IL: AOCS Press.

Seeram, N. P., Zhang, Y., Henning, S. M., Lee, R., Niu, Y., Lin, G. and Heber, D. (2006). Pistachio skin phenolics are destroyed by bleaching resulting in reduced antioxidative capacities. *Journal of Agricultural and Food Chemistry*, 54: 7036–7040.

Seeram, N. P., Zhang, Y., Henning, S. M., Lee, R., Niu, Y., Lin, G., and Heber, D. (2006). Pistachio skin phenolics are destroyed by bleaching resulting in reduced antioxidative capacities. *Journal of Agricultural and Food Chemistry*, *54*, 7036–7040.

Seguin, A. (1980). Ann. Chim. Paris. 20: 15, 1976. (Cited from Hulse, J. H., Ed., *Polyphenols* ill *Cereals and Legumes,* International Development Research Centre, Ottawa, Canada, 1980).

Seikel, M. (1964). Isolation and identification of phenolic compounds in biological material in Biochemistry of Phenolic Compounds Harborne, J. B. ed., Academic Press. London.

Sekhon-Loodu, S., Warnakulasuriya, S. N., Rupasinghe, H. V. and Shahidi, F. (2013). Antioxidant ability of fractionated apple peel phenolics to inhibit fish oil oxidation. *Food Chemistry*, 140: 189–196.

Sekiwa, Y., Kubota, K. and Kobayashi, A. (2000). Isolation of novel glucosides related to gingerdiol from ginger and their antioxidative activities. *Journal of Agricultural and Food Chemistry*, 48: 373–377.

Sellappan, S., Akoh, C. C. and Krewer, G. (2002). Phenolic compounds and antioxidant capacity of Georgia-grown blueberries and blackberries. *Journal of Agricultural and Food Chemistry*, 50: 2432–2438.

Sen CK, Khanna S, Roy S; Khanna; Roy (2006). "Tocotrienols: Vitamin E beyond Tocopherols". Life Sciences 78 (18): 2088–98.

Sen, A., Bergvinson, D., Miller, S. S., Atkinson, J., Fulcher, R. G. and Arnason, J. T. (1994). Distribution and microchemical detection of phenolic acids, flavonoids, and phenolic acid amides in maize kernels. *Journal of Agricultural and Food Chemistry*, 42: 1879–1883.

Sen, C. K., Khanna, S., Roy, S. and Khanna, Roy (2004). "Tocotrienol: the natural vitamin E to defend the nervous system?" Annals of the New York Academy of Sciences 1031: 127–42.

Senanayake, S. N. (2013). Green tea extract: Chemistry, antioxidant properties and food applications – A review. *Journal of Functional Foods*, 5: 1529–1541.

Senanayake, S. P. J., and Shahidi, F. (1999). Oxidative deterioration of borage and evening primrose oils as assessed by NMR spectroscopy. *Journal of Food Lipids*, *6*, 195–203.

Sengupta, K., Alluri, K., Golakoti, T., Gottumukkala, G., Raavi, J., Kotchrlakota, L., Sigalan, S., Dey, D., Ghosh, S., and Chatterjee, A. (2011). A randomized, double blind, controlled, dose dependent clinical trial to evaluate the efficacy of a proanthocyanidin standardized whole cranberry (*Vaccinium macrocarpon*) powder on infections of the urinary tract. *Current Bioactive Compounds*, *7*, 39–46.

Senter, S. D., Horvat, R. J. and Forbus,W. R. (1980). Relation between phenolic acid content and stability of pecans in accelerated storage. *Journal of Food Science*, 45: 1380–1382.

Senter, S. D., Horvat, R. J. and Forbus,W. R. (1983). Comparative GLC-MS analysis of phenolic acids of selected tree nuts. *Journal of Food Science*, 48: 798–799.

Seppanen, C. M., Song, Q. and Csallany, A. S. (2010). The antioxidant functions of tocopherol and tocotrienol homologues in oils, fats, and food systems. *Journal of the American Oil Chemists' Society*, 87: 469–481.

Serpen, A., Gökmen, V., Pellegrini, N. and Fogliano, V. (2008). Direct measurement of the total antioxidant capacity of cereal products. *Journal of Cereal Science*, 48: 816–820.

Serra, A., Macia, A., Romero, M. P., Angles, N., Morello, J. R., and Motilva, M. J. (2011). Distribution of procyanidins and their metabolites in rat plasma and tissues after an acute intake of hazelnut extract. *Food and Function*, *2*, 562–568.

Serra, A., Macia, A., Rubio, L., Angles, N., Ortega, N., Morello, J. R., Romero, M. P., and Motilva, M. J. (2013). Distribution of procyanidins and their metabolites in rat plasma and tissues in relation to ingestion of procyanidin-enriched or procyanidin-rich cocoa creams. *European Journal of Nutrition*, *52*, 1029–1038.

Serra, B. J., and Ventura, C. F. (1997). Evaluation of bitterness and astringency of polyphenolic compounds in cocoa powder. Journal of Agricultural and Food Chemistry, 60, 365-370.

Serrano, J., Puupponen-Pimi, R., Dauer, A., Aura, A. M. and Saura-Calixto, F. (2009). Tannins: Current knowledge of food sources, intake, bioavailability and biological effects. *Molecular Nutrition and Food Research*, 53: S310–S329.

Sesso HD, Buring JE, Christen WG, Kurth T, Belanger C, MacFadyen J, Bubes V, Manson JE, Glynn RJ, Gaziano JM; Buring; Christen; Kurth; Belanger; MacFadyen; Bubes;

Manson; Glynn; Gaziano (2008). "Vitamins E and C in the Prevention of Cardiovascular Disease in Men: The Physicians' Health Study II Randomized Trial". JAMA: the Journal of the American Medical Association 300 (18): 2123–33.

Seymour, T. A., Li, S. J., and Morrissey, M. T. (1996). Characterization of a natural antioxidant from shrimp shell waste. Journal of Agricultural and Food Chemistry, 44, 682-685.

Sfahlan, A. J., Mahmoodzadeh, A., Hasanzadeh, A., Heidari, R. and Jamei, R. (2009). Antioxidants and antiradicals in almond hull and shell (*Amygdalus communis* L.) as a function of genotype. *Food Chemistry*, 115: 529–533.

Shah, N. A., Khan, M. R., Sattar, S., Ahmad, B. and Mirza, B. (2014). HPLC-DAD analysis, antioxidant potential and anti-urease activity of Asparagus gracilis collected from District Islamabad. *BMC Complementary and Alternative Medicine*, 14: 347.

Shahidi, F. (1998). Indicators for evaluation of lipid oxidation and off-flavor development in food. *Developments in Food Science*, 40: 55–68.

Shahidi, F. (2000a). Antioxidants in food and food antioxidants. *Die Nahrung*, *44*, 158–163.

Shahidi, F. (2000b). Antioxidant factors in plant foods and selected oilseeds. *Biofactors (Oxford, England)*, 13: 179–185.

Shahidi, F. (2004). Quality characteristics of edible oils. *Advances in Experimental Medicine and Biology*, 542: 239–249.

Shahidi, F. (2009). Nutraceuticals and functional foods: Whole *versus* processed foods. *Trends in food Science and Technology*, 20: 376–387.

Shahidi, F. (2015). Antioxidants: Principles and applications. In F. Shahidi (Ed.), *Handbook of antioxidants for food preservation* (pp. 1–14). Cambridge, UK: Woodhead Publishing Ltd.

Shahidi, F. and Naczk, M. (2009). Phenolic compounds in grains. In *Food Phenolics: Sources, chemistry, effect, application* (pp. 3-39); Technomic publishing Company Inc.

Shahidi, F., Alasalvar, C., and Liyana-Pathirana, C. M. (2007). Antioxidant phytochemicals in hazelnut (*Corylus avellana* L.) and its by-products. *Journal of Agricultural and Food Chemistry*, *55*, 1212–1220.

Shahidi, F., Amarowicz, R., Abou-Gharbia, H. A. and Shehata, A. A. Y. (1997). Endogenous antioxidants and stability of sesame oil as affected by processing and storage. *Journal of the American Oil Chemists' Society*, 74: 143–148.

Shahidi, F., Amarowicz, R., He, Y. and Wettasinghe, M. (1997). Antioxidant activity of phenolic extracts of evening primrose (*Oenothera biennis*): A preliminary study. *Journal of Food Lipids*, 4: 75–86.

Shahidi, F., and Amarowicz, R. (1996). Antioxidant activity of protein hydrolysates from aquatic species. *Journal of the American Oil Chemists' Society*, *73*, 1197–1199.

Shahidi, F., and Ambigaipalan, P. (2015). Phenolics and polyphenolics in foods, beverages and spices: Antioxidant activity and health effects – A review. *Journal of Functional Foods*, *18*, 820–897.

Shahidi, F., and Han, X. Q. (1993). Encapsulation of food ingredients. *Critical Reviews in Food Science and Nutrition*, *33*, 501.

Shahidi, F., and Ho, C.-T. (2005). Phenolics in food and natural health products: An overview. In F. Shahidi and C.-T. Ho (Eds.), *Phenolic compounds in foods and natural products* (pp. 1–8). ACS Symposium Series 909.Washington, DC: ACS.

Shahidi, F., and Naczk, M. (1995). Food phenolics. Sources, chemistry, effects and applications. Lancaster, USA: Technomic Pub. Co. Shahidi, F., Amarovicz, R., Abou-Gharbia, H. A., and Shehata, A. A. Y. (1997). Endogenous antioxidants and stability of sesame oil as affected by processing and storage. Journal of the American Oil Chemists' Society, 74, 143-147.

Shahidi, F., and Tan, Z. (2013). Raisins: Processing, phytochemicals, and health benefits. In C. Alasalvar and F. Shahidi (Eds.), *Dried fruits: Phytochemicals and health effects* (pp. 372–392). Oxford: Wiley-Blackwell.

Shahidi, F., and Wanasundara, U. N. (2002). Methods for measuring oxidative rancidity in fats and oils. In C. C. Akoh and D. B. Min (Eds.), *Food lipids: Chemistry, nutrition and biotechnology* (pp. 465–487). NewYork, NY: Marcel Dekker Inc.

Shahidi, F., and Zhong, Y. (2005). Lipid oxidation: Measurement methods. In F. Shahidi (Ed.), *Bailey's industrial oil and fat products* (6th ed., Vol. 1, pp. 357–386). Hoboken, NJ: JohnWiley and Sons, Inc.

Shahidi, F., and Zhong, Y. (2007). Measurement of antioxidant activity in food and biological systems. In F. Shahidi and C. T. Ho (Eds.), *Antioxidant measurement and applications* (pp. 36–66). ACS symposium series 956.Washington, DC: American Chemical Society.

Shahidi, F., and Zhong, Y. (2010). Lipid oxidation and improving the oxidative stability. *Chemical Society Reviews, 39*, 4067–4079.

Shahidi, F., and Zhong, Y. (2011). Revisiting the polar paradox theory: A critical overview. *Journal of Agricultural and Food Chemistry, 59*, 3499–3504.

Shahidi, F., Chavan, U. D., Naczk, M. and Amarowicz, R. (2001). Nutrient distribution and phenolic antioxidants in airclassified fractions of beach pea (*Lathyrus maritimus* L.). *Journal of Agricultural and Food Chemistry*, 49: 926–933.

Shahidi, F., Desilva, C., and Amarowicz, R. (2003). Antioxidant activity of extracts of defatted seeds of niger (*Guizotia abyssinica*). *Journal of the American Oil Chemists' Society, 80*, 443–450.

Shahidi, F., Liyana-Pathirana, C. M. and Wall, D. S. (2006). Antioxidant activity of white and black sesame seeds and their hull fractions. *Food Chemistry*, 99: 473–483.

Shahidi, F., McDonald, J., Chandrasekara, A. and Zhong, Y. (2008). Phytochemicals of foods, beverages and fruit vinegars: Chemistry and health effects. *Asia Pacific Journal of Clinical Nutrition*, 17: 380–382.

Shahidi, F., Pegg, R. B. and Saleemi, Z. O. (1995). Stabilization of meat lipids with ground spices. *Journal of Food Lipids*, 2: 145–153.

Shahidi, F., Wanasundara, U. N., He, Y. and Shukla, V. K. S. (1997). Marine lipids and their stabilization with green tea catechins. In F. Shahidi and K. R. Cadwallader (Eds.), *Flavor and lipid chemistry of seafoods* (pp. 186–197). ACS Symposium Series 674. Washington, DC: American Chemical Society.

Shahidi, F., Wanasundara, U. N. and Amarowicz, R. (1994). Natural antioxidants from low-pungency mustard flour. *Food Research International*, 27: 489–493.

Shahidi, F., Wanasundara, U., Amarowicz, R. and Naczk, M. (1995). Natural antioxidants from canola meal. In Proceedings of 8th Rapeseed International Congress (pp. 873–875). Saskatoon, SK, Canada.

Shahidi, F., Wanasundara, U. and Amarowicz, R. (1995). Isolation and partial characterization of oilseed phenolics and evaluation of their antioxidant activity. *Developments in Food Science*, 37: 1087–1099.

Shahidi, F., Wanasundara, U.N., and Amarowicz, R. (1994). Natural antioxidants from low-pungency mustard flour. Food Res. Int. 27: 489-493.

Shahidi, F., Zhong, Y., and Chandrasekara, G. A. (2012). Antioxidants and human health. In L. Yu, R. Tsao, and F. Shahidi (Eds.), *Cereals and pulses: Nutraceutical properties and health benefits* (1st ed., pp. 273–308). Chichester, UK: JohnWiley and Sons, Inc.

Shahidi, F., Zhong, Y., Wijeratne, S. S. K. and Ho, C.-T. (2009). Almond and almond products: Nutraceutical components and health effects. In C. Alasalvar and F. Shahidi (Eds.), *Tree nuts composition: Phytochemicals and health effects* (pp. 127–141). Boca Raton, FL: CRC Press.

Shahidi, F., Zhong, Y.,Wijeratne, S. S. K., and Ho, C.-T. (2008). Almond and almond products: Nutraceutical components and health effects. In C. Alasalvar and F. Shahidi (Eds.), *Tree nuts: Composition, phytochemicals and health effects* (pp. 127–141). Boca Raton, FL: CRC Press Taylor and Francis Group.

Shahidi, F. and Alexander, D. M. (1998). Green tea catechins as inhibitors of oxidation of meat lipids. *Journal of Food Lipids*, 5: 125–133.

Shahidi, F. and Chandrasekara, A. (2010). Hydroxycinnamates and their in vitro and in vivo antioxidant activities. *Phytochemistry Review*, 9: 147–170.

Shahidi, F. and Chandrasekara, A. (2013). Millet grain phenolics and their role in disease risk reduction and health promotion: A review. *Journal of Functional Foods*, 5: 570–581.

Shahidi, F. and Ho, C. T. (2007). Antioxidant measurement and applications: An overview. In F. Shahidi and C. T. Ho (Eds.), *Antioxidant measurement and applications* (pp. 2–7). ACS Symposium series 956.Washington, DC: American Chemical Society.

Shahidi, F. and John, J. A. (2010). Oxidation and protection of nuts and nut oils. In E. A. Decker, R. J. Eliasand D. J. McClements (Eds.), *Oxidation in foods and beverages and antioxidant applications* (Vol. 2, pp. 274–305). Management in Different Industry Sectors. Cambridge, UK: Woodhead Publishing Ltd.

Shahidi, F. and Naczk, M. (1992). An overview of the phenolics of canola and rapeseed: Chemical, sensory and nutritional significance. *Journal of the American Oil Chemists' Society*, 69: 917–924.

Shahidi, F. and Naczk, M. (2004). *Phenolics in food and nutraceuticals* (pp. 1–558). Boca Raton, FL: CRC Press.

Shahidi, F. and Tan, Z. (2009). Bioactive compounds from cashew nut and its coproducts. In C. Alasalvar and F. Shahidi (Eds.), *Tree nuts composition: Phytochemicals and health effects* (pp. 157–184). Boca Raton, FL: CRC Press.

Shahidi, F. and Wanasundara, J. P. D. (1992). Phenolic antioxidants. *Critical Reviews in Food Science and Nutrition*, 32: 67–103.

Shahidi, F. and Wanasundara, U. (1995). Effect of natural antioxidants on the stability of canola oil. *Developments in Food Science*, 37: 469–479.

Shahidi, F. and Zhong, Y. (2005). Lipid oxidation: Measurement methods. In F. Shahidi (Ed.), *Bailey's industrial oil and fat products* (Vol. 1, pp. 491–512). Hoboken, NJ: JohnWiley and Sons Ltd.

Shahidi, F. and Zhong, Y. (2007). Measurement of antioxidant activity in food and biological systems. In F. Shahidi and C. T. Ho (Eds.), *Antioxidant measurement and applications* (pp. 36–66). ACS Symposium series 956.Washington, DC: American Chemical Society.

Shahidi, F. and Zhong, Y. (2015). Measurement of antioxidant activity. *Journal of Functional Foods*, doi:10.1016/j.jff.2015.01.047.

Shamanthaka, S. M. C., and Sastry, R. N. M. S. (1990). Binding of chlorogenic acid by the isolated polyphenol-free 11S protein of sunfower (Helianthus annuus) seed. Journal of Agricultural and Food Chemistry, 38, 2103-2110.

Shang, M., Cai, S., Han, J., Li, J., Zhao, Y., Namba, T., Kadota, S., Tezuka, Y. and Fan,W. (1998). Studies on flavonoids from fenugreek (*Trigonaella foenum graecum* L.). *Zhongguo Zhong Yao Za Zhi*, 23: 614–639.

Shao, Y., Xu, F., Sun, X., Bao, J. and Beta, T. (2014). Identification and quantification of phenolic acids and anthocyanins as antioxidants in bran, embryo and endosperm of white, red and black rice kernel (*Oryza sativa* L.). *Journal of Cereal Science*, 59(2), 211-218.

Shao, Y., Xu, F., Sun, X., Bao, J. and Beta, T. (2014). Phenolic acids, anthocyanins, and antioxidant capacity in rice (*Oryza sativa* L.) grains at four stages of development after flowering. *Food Chemistry*, 143:90-96.

Sharma, K. D., Karki, S., Thakur, N. S. and Attri, S. (2012). Chemical composition, functional properties and processing of carrot – A review. *Journal of Food Science and Technology*, 49: 22–32.

Sharma, P. and Gujral, H. S. (2014). Antioxidant potential of wheat flour chapattis as affected by incorporating barley flour. *LWT-Food Sci. Technol.* 56: 118–123.

Sharma, M. and Buettner, G. R. (1993). Interaction of vitamin C and vitamin E during the free radical stress in plasma: An ESR study. *Free Radical Biology and Medicine*, 14: 649–653.

Sharma, P., and Gujral, H. S. (2010). Antioxidant and polyphenol oxidase activity of germinated barley and its milling fractions. *Food Chem.*, 120:673-678.

Sharp, R.N., Sharp, C.O., and Kattan, A.A. (1978). Tannin content of sorghum grain by UV spectroscopy. Cereal Chem. 55: 117-118.

Sheabar, F. Z., and Neeman, I. (1988). Separation and concentration of natural antioxidants from the rape of olives. Journal of the American Oil Chemists' Society, 65, 990-993.

Shen, X., Sun, X., Xie, Q., Liu, H., Zhao, Y., Pan, Y., Hwang, C.-A. and Wu, V. C. H. (2014). Antimicrobial effect of blueberry (*Vaccinium corymbosum* L.) extracts against the growth of Listeria monocytogenes and Salmonella enteritidis. *Food Control*, 35: 159–165.

Sherman, L., Luh, B. S. and Hinreincr, E. (1953). Flavor evaluation of canned cling peaches. *Food Technol.* 7, 480.

Sherwin, E. R. (1990). Antioxidants. In A. L. Branen, P. M. Davidson and S. Salminen, Food antioxidants. New York: Marcel Dekker Inc.

Sherwin, E. R. (1990). Antioxidants. In L. Branen (Ed.), *Food additives* (pp. 139–193). NewYork: Marcel Dekker. Singleton, V. L., Orthofer, R., and Lamuela-Raventos, R. M. (1999). Analysis of total phenols and other oxidation substrates and antioxidants by means of Folin–Ciocalteu reagent. *Methods in Enzymology*, *299*, 152–178.

Shetty, K. (2004). Review. Role of proline-linked pentose phosphate pathway in biosynthesis of plant phenolics for functional food and environmental applications: A review. *Process Biochemistry*, 39: 789–803.

Shi J, and le Maguer M. (2000). Lycopene in tomatoes: Chemical and physical properties affected by food processing. *Crit Rev Food Sci Nutr* 40:1–42.

Shih, M. C., Kuo, C. C. and Chiang,W. (2009). Effects of drying and extrusion on color, chemical composition, antioxidant activities and mitogenic response of spleen lymphocytes of sweet potatoes. *Food Chemistry*, 117: 114–121.

Shimoda, H., Tanaka, J., Kikuchi, M., Fukuda, T., Ito, H., Hatano, T., and Yoshida, T. (2008).Walnut polyphenols prevent liver damage induced by carbon tetrachloride and Dgalactosamine: Hepatoprotective hydrolyzable tannins in the kernel pellicles of walnut. *Journal of Agricultural and Food Chemistry*, *56*, 4444–4449.

Shimoda, H., Tanaka, J., Kikuchi, M., Fukuda, T., Ito, H., Hatano, T., and Yoshida, T. (2009). Effect of polyphenol-rich extract from walnut on diet-induced hypertriglyceridemia in mice *via* enhancement of fatty acid oxidation in the liver. *Journal of Agricultural and Food Chemistry*, *57*, 1786–1792.

Shimoni, E., Ampel, M., ZaÈ hner, H., and Neeman, I. (1998). Antioxidant properties of Desferrioxamine E, a new hydroxamate antioxidant. Journal of the American Oil Chemists' Society, 75, 1453-1455.

Shobana, S., and Malleshi, N. G. (2007). Preparation and functional properties of decorticated finger millet (*Eleusine coracana*). Journal of Food Engineering, 79: 529–538.

Shobana, S., Sreeama, Y.N., and Malleshi, N. G. (2009). Composition and enzyme inhibitory properties of finger millet (*Eleusine coracana* L.) seed coat phenolics: Mode of inhibition of α-glucosidase and pancreatic amylase. Food Chemistry, 115: 1268–1273.

Shuang-sheng, H. and Rong-liang, Z. (2006). Rosmarinic acid inhibits angiogenesis and its mechanism of action in vitro. *Cancer Letters*, 239: 271–280.

Shukla, Y. and Singh, M. (2007). Cancer preventive properties of ginger: A brief review. *Food Chemistry*, 45: 683–690.

Siah, S. D., Konczak, I., Agboola, S.,Wood, J. A. and Blanchard, C. L. (2012). *In vitro* investigations of the potential health benefits of Australian-grown faba beans (*Vicia faba* L.): Chemopreventive capacity and inhibitory effects on the angiotensin-converting enzyme, α-glucosidase and lipase. *British Journal of Nutrition*, 108: S123–S134.

Siddalingappa, B., Benson, H. A., Brown, D. H., Batty, K. T. and Chen, Y. (2015). Stabilization of resveratrol in blood circulation by conjugation to mPEG and mPEG-PLA polymers: Investigation of conjugate linker and polymer composition on stability, metabolism, antioxidant activity and pharmacokinetic profile. *PLoS ONE*, 10: e0118824.

Siebert, K. J. (1999). Effect of protein-polyphenol interactions on beverage haze, stabilization, and analysis. Journal of Agricultural and Food Chemistry, 47, 353-362.

Siebert, K. J., Carrasco, A., and Lynn, P. Y. (1996). Formation of protein-polyphenol haze in beverages. Journal of Agricultural and Food Chemistry, 44, 1997-2005.

Siebert, K. J., Troukhanova, N. V., and Lynn, P. Y. (1996). Nature of polyphenol-protein interactions. Journal of Agricultural and Food Chemistry, 44, 80-85.

Sikora, E., Cies□lik, E., Leszczyn□ska, T., Filipiak-Florkiewicz, A. and Pisulewski, P. M. (2009). The antioxidant activity of selected cruciferous vegetables subjected to aquathermal processing. *Food Chemistry*, 107: 55–59.

Silva, L., Shahidi, F., and Coimbra, M. A. (2013). Dried pears: Phytochemicals and potential health effects. In C. Alasalvar and F. Shahidi (Eds.), *Dried fruits: Phytochemicals and health effects* (pp. 325–356). Oxford: Wiley-Blackwell.

Silveira, C., Oliveira, F., Dos Santos, M. L., de Freitas, T., Imparato, J. C., and Magalhaes, A. C. (2014). Anacardic acid from Brazilian cashew nut trees reduces dentine erosion. *Caries Research*, *48*,549–556.

Simonavice, E., Liu, P. Y., Ilich, J. Z., Kim, J. S., Arjmandi, B., and Panton, L. B. (2013). The effects of a 6-month resistance training and dried plum consumption intervention on strength, body composition, blood markers of bone turnover, and inflammation in breast cancer survivors 1. *Applied Physiology, Nutrition, and Metabolism*, *39*, 730–739.

Simonetti, P., Pietta, P., and Testolin, G. (1997). Polyphenol content and total antioxidant potential of selected Italian wines. Journal of Agricultural and Food Chemistry, 45, 1152-1155.

Sindhu, G., Ratheesh, M., Shyni, G. L., Nambisan, B. and Helen, A. (2012). Antiinflammatory and antioxidative effects of mucilage of Trigonella foenum graecum (Fenugreek) on adjuvant induced arthritic rats. *International Immunopharmacology*, 12: 205–211.

Singh, A. P., Wilson, T., Kalk, A. J., Cheong, J. and Vorsa, N. (2009). Isolation of specific cranberry flavonoids for biological activity assessment. *Food Chemistry*, 116: 963–968.

Singh, B., Kale, R. K. and Rao, A. R. (2004). Modulation of antioxidant potential in liver of mice by kernel oil of cashew nut (*Anacardium occidentale* L.) and its lack of tumor promoting ability in DMBA induced skin papillomagenesis. *Indian Journal of Experimental Biology*, 42: 372–377.

Singh, H. P., Ravindranath, S. D., and Singh, C. (1999). Analysis of tea shoots catechins, spectrophotometric quantitation and selective visualization on two-dimensional paper chromatograms using diazotized sulfanilamide. Journal of Agricultural and Food Chemistry, 47, 1041-1045.

Singlcton. V. L. (1981). Naturally occurring food toxicants: phenolic substances of plant origin Common in foods. *Adv. Food Res.* 27: 149.

Singleton, V. L. (1964). Application of charcoal in wine clarification, *Wines Vines.* 45, 29.

Singleton, V. L. (1967a). Fining-phenolic relationships, Wines *Vines,* 48. 23.

Singleton, V. L. (1967b). Adsorption of natural phenols from beer and wine, *Tech. Q. Master Brew. Assoc. Am.,* 4, 245.

Singleton, V. L. and Draper, D. E. (1962). Adsorbent and wines. I. Selection of activated charcoals for treatment of wine, *Am.* J. *Enol. Vitic.* 13, 114.

Singleton, V. L. and Esau, P. (1969). Phenolic Substances in Grapes and Wine and Their Significance, Academic Press, New York, 1969.

Singleton, V. L. and Kratzer, F. H. (1973). Plant phenolics, in *Toxicants Occurring Naturally in Foods,* National Academy of Sciences. Washington, D.C.

Singleton, V. L. and Kratzer, F. H. (1973). Plant phenolics, in *Toxicants Occurring Naturally in Foods,* National Academy of Sciences. Washington, D.C., 309.

Singleton, V. L. and Rossi, J. A. (1965). Colorimetry of total phenolics with phosphomolybdic phosphotungstic acid reagent. *Am. J. Enol. Vitic.* 16. 144.

Singleton, V.L. (1981). Naturally occuring food toxicants: Phenolic substances of plant origion common in foods. Adv. Food Res. 27: 149-154.

Sireerat, I., and Schulte, A. (2012). Automated electrochemical free radical scavenger screening in dietary samples. *Analytical Chemistry*, *84*, 6767–6774.

Siriwardhana, S. S. K. W. and Shahidi, F. (2002). Antiradical activity of extracts of almond and its by-products. *Journal of the American Oil Chemists' Society*, 79: 903–908.

Siriwoharn, T., Wrolstad, R. E., Finn, C. E. and Pereira, C. B. (2004). Influence of cultivar, maturity, and sampling on blackberry (*Rubus* L. hybrids) anthocyanins, polyphenolics, and antioxidant properties. *Journal of Agricultural and Food Chemistry*, 52: 8021–8030.

Siroha, A. K., Sandhu, K. S. and Kaur, M. (2016). Physicochemical, functional and antioxidant properties of flour from pearl millet varieties grown in India. *Food Meas.* 10: 311–318.

Sirtori, C. R. (2001). Aescin: Pharmacology, pharmacokinetics and therapeutic profile. *Pharmacological Research*, 44: 183–193.

Skrede, G, Wrolstad, R. E. and Durst, R. W. (2000). Changes in anthocyanins and polyphenolics during juice processing of highbush blueberries (*Vaccinium corymbosum* L.). *J Food Sci.* 65:357–64.

Slatnar, A., Klancar, U., Stampar, F., and Veberic, R. (2011). Effect of drying of figs (*Ficus carica* L.) on the contents of sugars, organic acids, and phenolic compounds. *Journal of Agricultural and Food Chemistry*, *59*, 11696–11702.

Slatnar, A., Mikulic-Petkovsek, M., Stampar, F., Veberic, R., and Solar, A. (2014). eHPLC-MSn identification and quantification of phenolic compounds in hazelnut kernels, oil and bagasse pellets. *Food Research International*, *64*, 783–789.

Slatnar, A., Mikulic-Petkovsek, M., Stampar, F., Veberic, R., and Solar, A. (2015). Identification and quantification of phenolic compounds in kernels, oil and bagasse pellets of common walnut (*Juglans regia* L.). *Food Research International*, *67*, 255–263.

Slavin, J. (2003). Why whole grains are protective, biological mechanisms. *The Proceedings of Nutrition Society*, 62(1), 129-134.

Slavin, J. L., and Lloyd, B. (2012). Health benefits of fruits and vegetables. *Advances in Nutrition*, *3*, 506–516.

Slavin, J. L., Jacobs, D., Marquart, L., and Wiemer, K. (2001). Grain processing and nutrition. Critical Reviews in Biotechnology, 21: 49–66.

Smith, A. W. (1985). Introduction. In R. J. Clark and R. Macrae (Eds.), *Coffee* (Vol. 1, pp. 1–41). Chemistry. London: Elsevier Applied Science.

Smith, B. D. (1998). The emergence of agriculture. Freeman Press, New York. Sosulski, F., Krzysztof, K., and Lawrence, H. (1982). Free, esterified, and insoluble-bound phenolic acids. 3. Composition of phenolic acids in cereal and potato flours. *J. Agric. Food Chem.*, 30:337-340.

Smitha, S., Dhananjaya, B. L., Dinesha, R. and Srinivas, L. (2009). Purification and characterization of a ~34 kDa antioxidant protein (β-turmerin) from turmeric (*Curcuma longa*) waste grits. *Biochimie*, 91: 1156–1162.

Smulikowska, S., Pastuszewska, B., Swiech, E., Ochtabinska, A., Mieczkowska, A., Nguyen, V. C. and Buraczewska, L. (2001). Tannin content affects negatively nutritive value of pea for monogastrics. *Journal of Animal and Feed Sciences*, 10: 511–523.

Snell, F. O. and Snell, C. T'. (1953). *Colorimetric Methods of Analysis.* Vol. 3, 3rd ed. Van Nostrand. Princeton. NI. 104.

Snell. F. *D.,* Snell, C.' T., and Snell, C. A. (1961). Colorimetric Methods of Analysis. Vol. 3A. 3rd ed. Van Nostrand. Princeton. NJ. 77.

Sobha, R. I. and Andallu, B. (2013). Oxidative stress and cancer: role of anti-carcinogenic herbs and species. *American Journal of Phytomedicine and Clinical Therapeutics,* 3:351-369.

Sobolev, V. S. and Cole, R. J. (1999). *trans*-Resveratrol content in commercial peanuts and peanut products. *Journal of Agricultural and Food Chemistry*, 47: 1435–1439.

Söderholm, P. P., Alfthan, G., Tikkanen, M. J. and Adlercreutz, H. (2012). Rye bread intake improves oxidation resistance of LDL in healthy humans. *Atherosclerosis*, 221(2), 583-586.

Soengas Fernández, M. D. P., Sotelo Pérez, T., Velasco Pazos, P. and Cartea González, M. E. (2011). Antioxidant properties of Brassica vegetables. *Functional Plant Science and Biotechnology*, *5:* 43–55.

Soleas, G. J., Diamandis, E. P. and Goldberg, D. M. (1997). Resveratrol: A molecule whose time has come and gone? *Clinical Biochemistry*, 30: 91–113.

Song, Y.,Wang,W., Cui,W., Zhang, X., Zhang,W., Xiang, Q., Liu, Z., Li, N., and Jia, X. (2010). A subchronic oral toxicity study of almond skins in rats. *Food and Chemical Toxicology: An International Journal Published for the British Industrial Biological Research Association*, *48*, 373–376.

Song,W., Derito, C. M., Liu, K. M., He, X., Dong, M., and Liu, R. H. (2010). Cellular antioxidant activity of common vegetables. *Journal of Agricultural and Food Chemistry*, *58*, 6621–6629.

Sosulski, F. W. (1979). Organoleptic and nutritional effects of phenolic compounds on oilseed protein products, a review. Journal of the American Oil Chemists' Society, 56, 711-715.

Sosulski, F., Krygier, K. and Hogge, L. (1982). Free, esterified, and insoluble-bound phenolic acids. 3. Composition of phenolic acids in cereal and potato flours. *Journal of Agricultural and Food Chemistry*, 30: 337–340.

Sosulski, F., Krzysztof, K., and Lawrence, H. (1982). Free, esterified, and insoluble-bound phenolic acids. 3. Composition of phenolic acids in cereal and potato flours. *J. Agric. Food Chem.*, 30:337-340.

Sosulski, F. and Dabrowski, K. (1984). Composition of free and hydrolyzable phenolic acids in the flours and hulls of ten legume species. *Journal of Agricultural and Food Chemistry*, 32: 131–133.

Sosulski, F.W. and Dabrowski, K.J. (1984). Composition of free hydrolyzable phenolic acids in flours and hull of ten leguminious species. J. Agric. Food Chem. 32: 131-133.

Souquet, J. M., Labarbe, B., Le Guerneve, C., Cheynier, V. and Moutounet, M. (2000). Phenolic composition of grape stems. *Journal of Agricultural and Food Chemistry*, 48: 1076–1080.

Spanos, G. A., and Wrolstad, R. E. (1990). Infuence of processing and storage on the phenolic composition of Thompson seedless grape juice. Journal of Agricultural and Food Chemistry, 38, 1565-1571.

Spanos, G. A., and Wrolstad, R. E. (1992). Phenolics of apple, pear, and white grape juices and their changes with processing and storage, a review. Journal of Agricultural and Food Chemistry, 40, 1478-1487.

Spiller, G. A., Schultz, L., Spiller, M., and Ou, B. (2002). Sun-dried raisins help prevent oxidative DNA damage during intense athletic activity. *Journal of the American College of Nutrition*, *21*, 487.

Sripriya, G., Chandrasekaran, K., Urty , V.S., and Chandra, T. S. (1996). ESR spectroscopic studies on free radical quenching act ion of finger millet (*Elusine coracana*). Food Chemistry, 57: 537–540.

Srivastava, A., Greenspan, P., Hartle, D. K., Hargrove, J. L., Amarowicz, R. and Pegg, R. B. (2010). Antioxidant and anti-inflammatory activities of polyphenolics from Southeastern U.S. range blackberry cultivars. *Journal of Agricultural and Food Chemistry*, 58: 6102–6109.

St.Angelo, A., Crippen, K. L., Dupuy, H. P. and James, C., Jr. (1990). Chemical and sensory studies of antioxidant treated beef. *Food Science*, 55: 1501–1505.

Stamatis, G., Kyriazopoulos, P., Golegou, S., Basayiannis, A., Skaltsas, S. and Skaltsa, H. (2003). *In vitro* anti-Helicobacter pylori activity of Greek herbal medicines. *Journal of Ethnopharmacology*, 88: 175–179.

Stanton,W. R. and Francis, B. J. (1966). Ecological significance of anthocyanin in the seed coats of the *Phaseoleae*. *Nature*, 211: 970–971.

Stein, K., Borowicki, A., Scharlau, D. and Glei, M. (2010). Fermented wheat aleurone induces enzymes involved in detoxification of carcinogen and in antioxidative defence in human colon cells. *The British Journal of Nutrition*, 104(8), 1101-1111.

Steinberg, F. M., Bearden, M. M. and Keen, C. L. (2003). Cocoa and chocolate flavonoids: Implications for cardiovascular health. *Journal of the American Dietetic Association, 103:* 215–223.

Stepanyan, V., Arnous, A., Petrakis, C., Kefalas, P., and Calokerinos, A. (2005). Chemiluminescent evaluation of peroxide value in olive oil. *Talanta, 65*, 1056–1058.

Stewart, A. J., Bozonnet, S., Mullen,W., Jenkins, G. I., Lean, M. E. J. and Crozier, A. (2000). Occurrence of flavonols in tomatoes and tomato-based products. *Journal of Agricultural and Food Chemistry*, 48: 2663–2669.

Stich. H. F., Bohm, B. A. Chatterjee. K. and Saito. J. I. (1983). The role of saliva-borne mutagens and carcinogens in the etiology of oral and esophageal carcinomas of octal nut and tobacco chewers. In *Carcinogens and Mutagens in the Evirollmt, Vol. III.Naturally Occurring Compounds: Epidemiology and Distribution.* Stich. H. F., Ed. CRC Prco. Boca Raton. FL. 43.

Stintzing, F. C., Stintzing, A. S., Carle, R., Frei, B. and Wrolstad, R. E. (2002). Color and antioxidant properties of cyanidin-based anthocyanin pigments. *Journal of Agricultural and Food Chemistry*, 50: 6172–6181.

Stintzing, F. C., Stintzing, A. S., Carle, R. and Wrolstad, R. E. (2002). A novel zwitterionic anthocyanin from Evergreen blackberry (*Rubus laciniatus*Willd.). *Journal of Agricultural and Food Chemistry*, 50: 396–399.

Stintzing, F.C., and Carle, R. (2004). Functional properties of anthocyanins and betalains in plants, food, and in human nutrition. *Trends in Food Science and Technology*. 15(1): 19–38.

Stivala, L. A., Savio, M., Carafoli, F., Perucca, P., Bianchi, L., Maga, G., Forti, L., Pagnoni, U. M., Albini, A., Prosperi, E. and Vannini, V. (2001). Specific structural determinants are responsible for the antioxidant activity and the cell cycle effects of resveratrol. *Journal of Biological Chemistry*, 276: 22586–22594.

Stodolak, B., Starzynska, A., Czyszczon, M., and Zyla, K. (2007). The effect of phytic acid on oxidative stability of raw and cooked meat. *Food Chemistry, 101*, 1041–1045.

Stone, N. J., Robinson, J. G., Lichtenstein, A. H., Merz, C. N. B., Blum, C. B., Eckel, R. H., Goldberg, A. C., Gordon, D., Levy, D., Lloyd-Jones, D. M., McBride, P., Schwartz, J. S., Shero, S. T., Smith, S. C., Jr.,Watson, K., and Wilson, P. W. F. (2014). 2013 ACC/AHA guideline on the treatment of blood cholesterol to reduce atherosclerotic cardiovascular risk in adults: A report of the American College of Cardiology/American Heart Association Task Force on Practice Guidelines. *Circulation, 129*, S46–S48.

Struijs, K., Vincken, J. P., Verhoef, R., van Oostveen-van Casteren, W. H. M., Voragen, A. G. J. and Gruppen, H. (2007). The flavonoids herbacetin diglucoside as a constituent of the lignin macromolecule from flaxseed hulls. *Phytochemistry*, 68: 1227–1235.

Strumeyer, D. H. and Malin, M. J. (1975). Condensed tannins in grain sorghum: isolation, fractionation and characterization, *J. Agric. Food Chem.*, 23,909.

Subba Rao, M. V. and Muralikrishna, G. (2002). Evaluation of the antioxidant properties of free and bound phenolic acids from native and malted finger millet (Ragi, *Eleusine coracana* Indaf-15). *Journal of Agricultural and Food Chemistry*, 50: 889–892.

Subramanian, V, Butler, L. G., Jambunathan, R., and Prasada Rao, K. E. (1983). Some agronomic and biochemical characters of brown sorghums and their possible role in bird resistance, J. *Agric. Food Chem.* 31. 1303.

Subramanian, V., Butler, L. G., Jambunathan, R., and Prasad Rao, K. E. (1983). Some agronorllic and biochemical characters of brown sorghums and their (X) possible role in bird resistance, *J. Agric. Food Chem.* 31: 1363.

Sugamura, K. and Keaney, J. F., Jr. (2011). Reactive species in cardiovascular disease. Free Radical Biology and Medicine, 51(5), 978-992.

Suhaj, M. (2006). Spice antioxidants isolation and their antiradical activity: A review. *Journal of Food Composition and Analysis*, 19: 531–537.

Suja, K. P., Jayalekshmy, A. and Arumughan, C. (2004). Free radical scavenging behavior of antioxidant compounds of sesame (*Sesamum indicum* L.) in DPPH• system. *Journal of Agricultural and Food Chemistry*, 52: 912–915.

Suliburska, J., Bogdanski, P., Szulinska, M., Stepien, M., Pupek-Musialik, D. and Jablecka, A. (2012). Effects of green tea supplementation on elements, total antioxidants, lipids, and glucose values in the serum of obese patients. *Biological Trace Element Research*, 149: 315–322.

Sultana, B., Anwar, F. and Przybylski, R. (2007). Antioxidant potential of corncob extracts for stabilization of corn oil subjected to microwave heating. *Food Chemistry*, 104: 997–1005.

Sultana, B. and Anwar, F. (2008). Flavonols (kaempeferol, quercetin, myricetin) contents of selected fruits, vegetables and medicinal plants. *Food Chemistry*, 108: 879–884.

Sultana, B., Anwar, F. and Iqbal, S. (2008). Effect of different cooking methods on the antioxidant activity of some vegetables from Pakistan. *Int. J. Food Sci. Technol.* 43: 560-567.

Suma, P. F. and Urooj, A. (2011). Nutrients, antinutrients and bioaccessible mineral content (invitro) of pearl millet as influenced by milling. *J Food Sci. Technol.* DOI: 10.1007/s13197-011-0541-7. Available from Springer [http://www.springerlink.com]. Posted October 1, 2011.

Sun, J., Cao, X., Bai, W., Liao, X. and Hu, X. (2010). Comparative analyses of copigmentation of cyanidin 3-glucoside and cyanidin 3-sophoroside from red raspberry fruits. *Food Chemistry*, 120: 1131–1137.

Sun, J., Chu, Y. F., Wu, X. and Liu, R. H. (2002). Antioxidant and antiproliferative activities of common fruits. *Journal of Agricultural and Food Chemistry*, 50: 7449–7454.

Sun, T., Ho, C.-T. and Shahidi, F. (2009). Bioavailability and metabolism of tea catechins in human subjects. In C. T. Ho, J. K. Linand F. Shahidi (Eds.), *Tea and tea products: Chemistry and health-promoting properties* (pp. 111–129). NewYork: CRC Press, Taylor and Francis Group.

Sun, T., Simon, P. W. and Tanumihardjo, S. A. (2009). Antioxidant phytochemicals and antioxidant capacity of biofortified carrots (*Daucus carota* L.) of various colors. *Journal of Agricultural and Food Chemistry*, 57: 4142–4147.

Sun, T., Tang, J. and Powers, J. R. (2005). Effect of pectolytic enzyme preparations on the phenolic composition and antioxidant activity of asparagus juice. *Journal of Agricultural and Food Chemistry*, 53: 42–48.

Sun, T. and Powers, J. R. (2007). Antioxidants and antioxidant activities of vegetables. In F. Shahidi and C. T. Ho (Eds.), *Antioxidant measurements and applications* (pp. 160–183). ACS Symposium Series 956.Washington, DC: American Chemical Society.

Sun-Waterhouse, D. (2011). The development of fruit-based functional foods targeting the health and wellness market: A review. *International Journal of Food Science and Technology*, *46*, 899–920.

Surendrian, G., Goh, C., Le, K., Zhao, Z., Askarian, F., Othman, R., Nicholson, T., Moghadasian, P., Wang, Y. J., Aliani, M., Shen, G., Beta, T. and Moghadasian, M. H. (2013). Wild rice (*Zizania paplustris* L.) prevents atherogenesis in LDL receptor knockout mice. *Atherosclerosis*, 230(2), 284-292.

Surh, Y.-J. (2002). Anti-tumor promoting potential of selected spice ingredients with antioxidative and anti-inflammatory activities: A short review. *Food and Chemical Toxicology*, 40: 1091–1097.

Surjadinata, B. B. and Cisneros-Zevallos, L. (2012). Biosynthesis of phenolic anti-oxidants in carrot tissue increases with wounding intensity. *Food Chemistry*, 134: 615–624.

Suzawa, M., Guo, L., Pan, M. H., Ho, C. T. and Li, S. (2014). In vivo anti-carcinogenic property of a formulated citrus peel extract. *Functional Foods in Health and Disease*, 4: 120–129.

Swain, T. (1965). The tannins in *Plant Biochemistry.* Bonner, J. and Varner, E. Y., Eds., Academic Press. New York, 552.

Swain, T. (1976). Flavonoids. In: Chemistry and Biochemistry of Plant Pigments, Vol. 2. Goodwin. T. W. Ed. 166.

Swain, T. and Bate-Smith, E.C. (1962). Flavonoid compounds. In: *Comparative Biochemistry.* H. S. Mason, and A. M. Florkin. (Eds.), Vol. 3, p. 755-809. Academic Press. New York, NY.

Swain, T. and Hillis, W. E. (1959). Phenolic constituents of *Pruflus domes/iea.* I. Quantitative analysis of phenolic ·constituents. *J. Sci. Food Agric.,* 10: 63.

Syed, D. N., Chamcheu, J. C., Adhami, V. M. and Mukhtar, H. (2013). Pomegranate extracts and cancer prevention: Molecular and cellular activities. *Anti-cancer Agents in Medicinal Chemistry*, 13: 1149.

Szaefer, H., Krajka-Kuz niak, V., Ignatowicz, E., Adamska, T. and Baer-Dubowska,W. (2014). Evaluation of the effect of beetroot juice on DMBA-induced damage in liver and mammary gland of female Sprague–Dawley rats. *Phytotherapy Research*, 28: 55–61.

Tai, A., Sawano, T., Yazama, F., and Ito, H. (2011). Evaluation of antioxidant activity of vanillin by using multiple antioxidant assays. *Biochimica et Biophysica Acta – General Subjects*, *1810*, 170–177.

Takebayashi, J., Iwahashi, N., Ishimi, Y., and Tai, A. (2012). Development of a simple 96-well plate method for evaluation of antioxidant activity based on the oxidative haemolysis inhibition assay (OxHLIA). *Food Chemistry*, *134*, 606–610.

Takeoka, G. R., Dao, L., Flessa, S., Gillespie, D. M., Jewell, W. T., Heupner, B., Bertow, D. and Ebeler, S. E. (2001). Processing effects on lycopene content and antioxidant activity of tomatoes. *Journal of Agricultural and Food Chemistry*, 49: 3713–3717.

Takeoka, G., Dao, L., Teranishi, R., Wong, R., Flessa, S., Harden, L. and Edwards, R. (2000). Identification of three triterpenoids in almond hulls. *Journal of Agricultural and Food Chemistry*, 48: 3437–3439.

Talcott, S. T., and Howard, L. R. (1999). Phenolic autooxidation is responsible for color degradation in processed carrot puree. Journal of Agricultural and Food Chemistry, 47, 2109-2115.

Talcott, S. T., Duncan, C. E., Pozo-Insfran, D. D. and Gorbet, D. W. (2005). Polyphenolic and antioxidant changes during storage of normal, mid, and high oleic acid peanuts. *Food Chemistry*, 89: 77–84.

Talcott, S. T., Passeretti, S., Duncan, C. E. and Gorbet, D. W. (2005). Polyphenolic content and sensory properties of normal and high oleic acid peanuts. *Food Chemistry*, 90: 379–388.

Talcott, S. T. and Howard, L. R. (1999). Determination and distribution of 6-methoxymellein in fresh and processed carrot puree by a rapid spectrophotometric assay. *Journal of Agricultural Food Chemistry*, 47: 3237–3242.

Tamura, H., and Yamagami, A. (1994). Antioxidative activity of monoacylated anthocyanins isolated from Muscat Bailey A grape. *J. Agric. Food Chem.* 42: 1612–1615.

Tan, C. P., and Che Man, Y. B. (1999). Differential scanning calorimetric analysis for monitoring the oxidation of heated oils. *Food Chemistry*, *67*, 177–184.

Tan, C. P., and Che Man, Y. B. (2002). Recent developments in differential scanning calorimetry for assessing oxidative deterioration of vegetable oils. *Trends in Food Science and Technology*, *13*, 312–318.

Tan, C. P., Che-Man, Y. B., Selamat, J., and Yusoff, M. S. A. (2002). Comparative studies of oxidative stability of edible oils by differential scanning calorimetry and oxidative stability index methods. *Food Chemistry*, *76*, 385–389.

Tanabe, M., Kanda, T. and Yanagida, A. (1995). Fruit polyphenol; process for production thereof; and antioxidant, hypotensive agent, antimutagenic agent, antiallergenic agent and anticariogenic agent, each comprising said polyphenol. EP0657169.

Tappel, A. (1997). Vitamin E as a biological lipid antioxidant. Inform, 8, 392-396.

Tarladgis, B. G.,Watts, B. M., Younathan, M. T., and Dugan, L. (1960). A distillation method for the quantitative determination of malonaldehyde in rancid foods. *Journal of the American Oil Chemists' Society*, *37*, 44–48.

Tasioula-Margari, M. and Okogeri, O. (2001). Simultaneous determination of phenolic compounds and tocopherols in virgin olive oil using HPLC and UV detection. *Food Chemistry*, 74: 377–383.

Taylor, J. R., Belton, P. S., Beta, T. and Duodu, K. G. (2014). Increasing the utilisation of sorghum, millets and pseudocereals: Developments in the science of their phenolic phytochemicals, biofortification and protein functionality. *Journal of Cereal Science*, 59: 257–275.

Tebib, K., Rouanet, J. M., and BesancËon, P. A. (1994). Effect of grape seed tannins on the activity of some rat intestinal enzyme activities. Enzyme Protein, 48, 51-60.

Tee, E. S., Lim, C. L., Chong, Y. H., and Khor, S. C. (1996). A study of the biological utilization of carotenoids of carrot and swap cabbage in rats. *Food Chemistry*. 56: 21–32.

Temple, N. J. (2000). Antioxidant and disease: More question than answers. *Nutrition Research*, 20(3), 449-459.

Terao, J., Piskula, M. and Yao, Q. (1994). Protective effect of epicatechin gallate, and quercetin on lipid peroxidation in phospholipid bilayers. *Archives of Biochemistry and Biophysics*, 308: 278–284.

Thayer, R. H., Sieglinger, J. B., and Heller, V. G. (1957). Oklahoma grain sorghum for growing chicks. *Okla. Agric. Exp.* Stn. *Bull.* BA87.

Thompson, L. U. (1994). Antioxidant and hormone-mediated health benefits of whole grains. *Critical Reviews in Food Science and Nutrition*, 34(5-6), 473-497.

Thompson, L. U., Boucher, B. A., Liu, Z., Cotterchio, M., and Kreiger, N. (2006). Phytoestrogen content of foods consumed in Canada, including isoflavones, lignans, and coumestan. *Nutrition and Cancer*, *54*, 184–201.

Thompson, R. S., Jacques, D., Haslam, E. and Tonner, R. J. M. (1972). Plant proanthocyanidins. I. The isolation, structure and distribution in nature of plant procyanidins. 1. *Chem. Soc. Perkin Trans. I. II.* 1387.

Thomson, R. H. (1964). Biochemistry of Phenolic Compounds, Harbome, J. B., Ed., Academic Press. London.

Thorsen, M. A. and Hildebrandt, K. S. (2003). Quantitative determination of phenolic diterpenes in rosemary extracts. Aspects of accurate quantification. *Journal of Chromatography. A*, 995: 119–125.

Threlfall, R., Morris, J., and Meullenet, J. F. (2007). Product development and nutraceu-tical analysis to enhance the value of dried fruit. *Journal of Food Quality*, *30*, 552–566.

Tian, X., and Schaich, K. M. (2013). Effects of molecular structure on kinetics and dynamics of the trolox equivalent antioxidant capacity assay with $ABTS^{+\bullet}$. *Journal of Agricultural and Food Chemistry*, *61*, 5511–5519.

Tipton, K. W., Floyd, E. H. Marshall, J. C., and McDevill, J. B. (1970). Resistance of certain grain sorghum hybrids 10 bird damage in Louisiana. *Agron.* J. 62, 211.

Tipton, K. W., Mabbayad, B. B. Marshall, L. G., Robb, J. L., and Sloane, L. W. (1975). Tannin Content of Brown Seeded Grain Sorghum as Influenced by Year, Location and Hybrid, Proc. 9th Biennial Grain Sorghum Res. Util. Conf. Lubbock, TX.

Tobinaga, S., Hashimoto, M., Utsunomiya, I., Taguchi, K., Nakamura, M., and Tsunematsu, T. (2012). Chronic administration of cardanol (ginkgol) extracted from *Ginkgo biloba* leaves and cashew nutshell liquid improves working memory-related learning in rats. *Biological and Pharmaceutical Bulletin*, *35*, 127–129.

Tokusoglu, O., Unal, M. K. and Yemis, F. (2005). Determination of the phytoalexin resveratrol (3,5,4'-trihydroxystilbene) in peanuts and pistachios by high-performance liquid chromatographic diode array (HPLC-DAD) and gas chromatography-mass spectrometry (GC-MS). *Journal of Agricultural and Food Chemistry*, 53: 5003–5009.

Tomás-Barberán, F. A., and Andrés-Lacueva, C. (2012). Polyphenols and health: Current state and progress. *Journal of Agricultural and Food Chemistry*, *60*, 8773–8775.

Toor, R. K. and Savage, G. P. (2005). Antioxidant activity in different fractions of tomatoes. *Food Research International*, 38: 487-494.

Topinka, J., Binkova, B., Sram, R. J., and Erin, A. N. (1989). The influence of R-tocopherol and pyritinol on oxidative DNA damage and lipid peroxidation in human lymphocytes. *Mut. Res.*, 225:131-136.

Torres, A. M., Mau-Lastovicka, T., and Rezaaiyan, R. (1987). Total phenolics and high-performance liquid chromatography of phenolic acids of avocado. Journal of Agricultural and Food Chemistry, 35, 921-925.

Torres, P., Poveda, A., Jimenez-Barbero, J., Ballesteros, A. and Plou, F. J. (2010). Regioselective lipase-catalyzed synthesis of 3-*O*-acyl derivatives of resveratrol and study of their antioxidant properties. *Journal of Agricultural and Food Chemistry*, 58: 807–813.

Torronen, R., and Maatta, K. (2002). Bioactive substances and health benefits of strawberries. *Acta Horticulturae.* 567: 797–803.

Towo , E.E.,Svanber G, U., and Ndossi, G. D. (2003).Effect of gr ain pre-treatment on different extracta ble phenolic groups in cereals and legumes commonly consumed in Tanzania. Journal of the Science of Food and Agriculture, 83: 980–986.

Traber MG, Atkinson J; Atkinson (2007). "Vitamin E, Antioxidant and Nothing More". Free radical biology and medicine 43 (1): 4–15.

Traber MG, Packer L; Packer, L (1995). "Vitamin E: beyond antioxidant function". American Journal of Clinical Nutrition 62 (6): 1501S–1509S.

Traber MG, Sies H; Sies, H (1996). "Vitamin E in humans: demand and delivery". Annual review of nutrition 16: 321–47.

Traber MG, Stevens JF; Stevens (2011). "Free Radical Biology and Medicine – Vitamins C and E: Beneficial effects from a mechanistic perspective". Free Radical Biology and Medicine 51 (5): 1000–13.

Traber, M.G. (1997). In Ross, A. Catherine. Modern Nutrition in Health and Disease (11 Ed.). Philadelphia, PA: Lippincott Williams and Wilkins. pp. 293–294.

Traber, MG (1998). "The biological activity of vitamin E". The Linus Pauling Institute. Retrieved 6 March 2011.

Traber, MG. "Chapter 15: vitamin E". In Bowman BA and Russell RM. Current Knowledge in Nutrition I (9 Ed.). Washington DC, USA: ILSI. ISBN 978-1-57881-199-1.

Traitler, H., Winter, H., Richli, U. and Ingenbleek, Y. (1984). Characterization of γ-linolenic acid in *Ribes* seed. *Lipids*, 19: 923–928.

Tremblay, A., Masson, E., Leduc, S., Houde, A. and Despres, J.P. (1988). Caffeine reduces spontaneous energy intake in men but not in women. *Nutrition Research*, 8: 553–558.

Tresserra-Rimbau, A., Medina-Remón, A., Pérez-Jiménez, J., Martínez-González, M. A., Covas, M. I., Corella, D., Salas-Salvadó, J., Gómez-Gracia, E., Lapetra, J., Arós, F., Fiol, M., Ros, E., Serra-Majem, L., Pintó, X., Muñoz, M. A., Saez, G. T., Ruiz-Gutiérrez, V.,Warnberg, J., Estruch, R., and Lamuela-Raventós, R. M. (2013). Dietary intake and major food sources of polyphenols in a Spanish population at high cardiovascular risk: The PREDIMED study. *Nutrition, Metabolism, and Cardiovascular Diseases*, *23*, 953–959.

Trevisan, G., Rossato, M. F., Hoffmeister, C., Muller, L. G., Pase, C., Cordova, M. M., Rosa, F., Tonello, R., Hausen, B. S., Boligon, A. A., Moresco, R. N., Athayde, M. L., Burguer, M. E., Santos, A. R., and Ferreira, J. (2014). Antinociceptive and antiedematogenic effect of pecan (*Carya illinoensis*) nut shell extract in mice: A possible beneficial use for a by-product of the nut industry. *Journal of Basic and Clinical Physiology and Pharmacology*, *25*, 401– 410.

Trevisan, M. T. S., Pfundstein, B., Haubner, R., Würtele, G., Spiegelhalder, B., Bartsch, H. and Owen, R. W. (2006). Characterization of alkyl phenols in cashew (*Anacardium occidentale*) products and assay of their antioxidant capacity. *Food and Chemical Toxicology*, 44: 188–197.

Troszyn□ska, A., Estrella, I., López-Amóres, M. L. and Hernández, T. (2002). Antioxidant activity of pea (*Pisum sativum* L.) seed coat acetone extract. *Lebensmittel-Wissenschaft und-Technologie*, 35: 158–164.

Trucksess, M. W., and Scott, P. M. (2008). Mycotoxins in botanicals and dried fruits: A review. *Food Additives and Contaminants. Part A: Chemistry, Analysis, Control, Exposure and Risk Assessments*, *25*, 181–192.

Truong, V. L., Bak, M. J., Jun, M., Kong, A. N., Ho, C.-T., and Jeong,W. S. (2014). Antioxidant defense and hepatoprotection by procyanidins from almond (*Prunus amygdalus*) skins. *Journal of Agricultural and Food Chemistry*, *62*, 8668–8678.

Tsaliki, E., Lagouri, V., and Doxastakis, G. (1999). Evaluation of the antioxidant activity of lupin seed four and derivatives (*Lupinus albus* ssp. Graecus). Food Chemistry, 65, 71-75.

Tsao, R., Yang, R., Sockovie, E. and Khanizadeh, S. (2005). Which polyphenolic compounds contribute to the total antioxidant activities of apple? *Journal of Agricultural and Food Chemistry*, 53: 4989–4995.

Tsao R. (2010). Chemistry and biochemistry of dietary polyphenols. *Nutrients* 2:123–146.

Tsuda, T., Horio, F., Uchida, K., Aoki, H., and Osawa, T. (2003). Dietary cyanidin 3-O-α-Dglucose-rich purple corn color prevents obesity and ameliorates hyperglycemia in mice. *J. Nutr.*, 133:2125-2130.

Tsuda, T., Ohshima, K., Kawakishi, S., and Osawa, T. (1994). Antioxidant pigments isolated from the seeds of *Phaseouls vulgaris*. Journal of Agricultural and Food Chemistry, 42, 248-251.

Tsuda, T., Osawa, T., Nakayama, T., Kawakishi, S., and Ohshima, K. (1993). Antioxidant activity of pea bean (*Phaseolus vulgaris* L.) extract. J. Am. Oil Chem. Soc. 70: 909- 913.

Tsuda, T., Shiga, K., Ohshima, K., Kawakishi, S., and Osawa, T. (1996). Inhibition of lipid peroxidation and the active oxygen radical scavenging effect of anthocyanin pigments isolated from phaseolus vulgaris L. *Biochemical Pharmacology.* 52 (7): 1033-1039.

Tsuda, T., Watanabe, M., Ohshima, K., Yamamoto, A., Kawakishi, S., and Osawa, T. (1994). Antioxidative components isolated from the seed of tamarind (*Tamarindus indica* L.). Journal of Agricultural and Food Chemistry, 42, 2671-2674.

Tsushida, T., Suzuki, M. and Kurogi, M. (1994). Evaluation of antioxidant activity of vegetable extracts and determination of some active compounds. *Journal of the Japanese Society for Food Science and Technology*, 41: 611–618.

Tubaro, F., Micossi, E., and Ursini, F. (1996). The antioxidant capacity of complex mixtures by kinetic anaÂ lisis of crocin bleaching inhbition. Journal of the American Oil Chemists' Society, 73, 173-179.

Tuck, K. L. and Hayball, P. J. (2002). Major phenolic compounds in olive oil: Metabolism and health effects. *The Journal of Nutritional Biochemistry*, 13: 636–644.

Tufan, A. N., Celik, S. E.,O″ zyürek, M., Güçlü, K., and Apak, R. (2013). Direct measurement of total antioxidant capacity of cereals: QUENCHER-CUPRAC method. *Talanta*, *108*, 136–142.

Tulipani, S., Llorach, R., Jauregui, O., Lopez-Uriarte, P., Garcia-Aloy, M., Bullo, M., Salas-Salvado, J., and Andres-Lacueva, C. (2011). Metabolomics unveils urinary changes in subjects with metabolic syndrome following 12-week nut consumption. *Journal of Proteome Research*, *10*, 5047–5058.

Tulipani, S., Urpi-Sarda, M., Garc a-Villalba, R., Rabassa, M., López-Uriarte, P., Bulló, M., Jáuregui, O., Tomás-Barberán, F., Salas-Salvadó, J., Esp n, J. C., and Andrés-Lacueva, C. (2012). Urolithins are the main urinary microbial-derived phenolic metabolites discriminating a moderate consumption of nuts in free-living subjects with diagnosed metabolic syndrome. *Journal of Agricultural and Food Chemistry*, *60*, 8930–8940.

Tung, Y. T., Wu, J. H., Kuo, Y. H. and Chang, S. T. (2007). Antioxidant activities of natural phenolic compounds from *Acacia confuse* bark. *Bioresource Technology*, 98: 1120–1123.

Turkmen, N., Sari, F. and Velioglu, S. (2005). The effect of cooking methods on total phenolics and antioxidant activity of selected green vegetables. *Food Chem.* 93: 713-718.

Tzeng, S. H., Ko, W. C., Ko, F. N. and Teng, C. M. (1991). Inhibition of platelet aggregation by some flavonoids. *Thrombosis Research*, 64: 91–100.

Ucak, I., Ozogul, Y. and Durmus, M. (2011). The effects of rosemary extract combination with vacuum packing on the quality changes of Atlantic mackerel fish burgers. *International Journal of Food Science and Technology*, 46: 1157–1163.

Ucella, N. (2001). Olive biophenols: Novel ethnic and technological approach. *Trends in Food Science and Technology*, 11: 357–363.

Upis, B. V., Lyalikova, R. Y., and Chernichuk, L. L. (1964). Spectrophotometricheskii method opredeleniya dubilnykh i krasyashchikh veshchestv vinogradnom sulse i vine, *Tr. Mold. Nauchno Issled Inst, Pisheh. Promsti.* 4, 109.

Urpi-Sarda, M., Garrido, I., Monagas, M., Gómez-Cordovés, C., Medina-Remón, A., Andres-Lacueva, C., and Bartolomé, B. (2009). Profile of plasma and urine metabolites after the intake of almond [*Prunus dulcis* (Mill.) D.A.Webb] polyphenols in humans. *Journal of Agricultural and Food Chemistry*, *57*, 10134–10142.

US Department of Health and Human Services, and US Department of Agriculture. (2015). *2015–2020 dietary guidelines for Americans* (8th ed.). <http://health.gov/ dietaryguidelines /2015/ guidelines/> Accessed 13.04.16.

US Department of Health and Human Services. (2010). *Dietary guidelines for Americans* (7th ed., p. 2010).Washington, DC: US Department of Health and Human Services.

US Food and Drug Administration. (2003). *Qualified health claims: Letter of enforcement discretion – Nuts and coronary heart disease.* Rockville, MD: US Food and Drug Administration.

USDA (2002). *USDA-Iowa State University database on the isoflavone content of foods.*Washington, DC: Agricultural Research Service, U.S. Department of Agriculture.

USDA. (2004). *USDA database for the proanthocyanidin content of selected foods.* http://www.ars.usda.gov/nutrientdata> Accessed 30.03.15.

USDA. (2010). Oxygen Radical Absorbance Capacity (ORAC) of Selected Foods, Release 2. <http://www.ars.usda.gov/News/ docs.htm?docid=15866>.

USDA. (2010). USDA database for the oxygen radical absorbance capacity (ORAC) of selected foods, release 2.0. <http://www.ars.usda.gov/nutrientdata> Accessed 25.09.14.

USDA. (2014). *USDA national nutrient database for standard reference, release 27.* <http://www.ars.usda.gov/ba/bhnrc/ndl> Accessed 25.09.14.

USDA. (2015a). USDA database for the flavonoid content of selected foods, release 3.2. <http://www.ars.usda.gov/ nutrientdata> Accessed 05.04.16.

USDA. (2015b). USDA database for the proanthocyanidin content of selected foods, release 2. <http://www.ars.usda.gov/ nutrientdata> Accessed 05.04.16.

Vadivel, V., Kunyanga, C. N., and Biesalski, H. K. (2012). Health benefits of nut consumption with special reference to body weight control. *Nutrition*, *28*, 1089–1097.

Valdez-Morales, M., Espinosa-Alonso, L. G., Espinoza-Torres, L. C., Delgado-Vargas, F. and Medina-Godoy, S. (2014). Phenolic content and antioxidant and antimutagenic activities in tomato peel, seeds, and byproducts. *Journal of Agricultural and Food Chemistry*, 62: 5281–5289.

Valentová, K., Stejskal, D., Bednár, P., Vostálová, J., Cíhalík, C., Vecˇerová, R., Koukalová, D., Kolár, M., Reichenbach, R., Sknˇ ourˇil, L., Ulrichová, J., and Šimánek, V. (2007). Biosafety, antioxidant status, and metabolites in urine after consumption of dried cranberry juice in healthy women: A pilot double-blind placebo-controlled trial. *Journal of Agricultural and Food Chemistry*, *55*, 3217–3224.

Valentová, K., Ulrichová, J., Cvak, L. and Šimánek, V. (2007). Cytoprotective effect of a bilberry extract against oxidative damage of rat hepatocytes. *Food Chemistry*, 101: 912–917.

Vallejo, F., Marín, J. G., and Tomás-Barberán, F. A. (2012). Phenolic compound content of fresh and dried figs (*Ficus carica* L.). *Food Chemistry*, *130*, 485–492.

Vallejo, F., Tomas-Barberan, F. and Garcia-Viguera, C. (2003). Health-promoting compounds in broccoli as influenced by refrigerated transport and retail sale period. *Journal of Agricultural and Food Chemistry*, 51: 3029–3034.

Van Buren, J. (1970). Fruit phenolics, in *The Biochemistry of Fruits and Their Products,* Vol. I, Hulme, A. C., Ed. Academic Press. New York.

Van Buren, J., de Vos, L., and Pilnik, W. (1976). Polyphenols in golden delicious apple juice in relation to method of preparation. Journal of Agricultural and Food Chemistry, 24, 448-451.

Van de Perre, E., Jacxsens, L., Lachat, C., El Tahan, F., and De Meulenaer, B. (2015). Impact of maximum levels in European legislation on exposure of mycotoxins in dried products: Case of aflatoxin B1 and ochratoxin A in nuts and dried fruits. *Food and Chemical Toxicology*, *75*, 112–117.

van den Berg, R., Haenen, G. R. M. M., van den Berg, H. and Bas, A. (1999). Applicability of an improved trolox equivalent antioxidant capacity (TEAC) assay for evaluation of antioxidant capacity measurements of mixtures. *Food Chemistry*, 66: 511–517.

Van der Sluis, A. A., Dekker, M., Skrede, G. and Jongen,W. M. F. (2002). Activity and concentration of polyphenolic antioxidants in apple juice. 1. Effect of existing production methods. *Journal of Agricultural and Food Chemistry*, 50: 7211–7219.

Vardi, N., Parlakpinar, H., Ates, B., Cetin, A., and Otlu, A. (2013). The protective effects of *Prunus armeniaca* L. (apricot) against methotrexate-induced oxidative damage and apoptosis in rat kidney. *Journal of Physiology and Biochemistry*, *69*, 371–381.

Vargas-Arispuro, I., Sanz, B. I., MartõÂnez-TeÂ llez, M. A., and PrimoYuÂ fera, E. (1998). Actividad antioxidante de compuestos aislados del residuo no-volaÂ til del aceite esencial de naranja. Grasas y Aceites, 49, 159-164.

Vasileiou, I., Katsargyris, A., Theocharis, S., and Giaginis, C. (2013). Current clinical status on the preventive effects of cranberry consumption against urinary tract infections. *Nutrition Research*, *33*, 595–607.

Vayalil, P. K. (2012). Date fruits (*Phoenix dactylifera* Linn): An emerging medicinal food. *Critical Reviews in Food Science and Nutrition*, *52*, 249–271.

Veenashri, B. R. and Muralikrishna, G. F. (2011). In vitro anti-oxidant activity of xylo-oligosaccharides derived from cereal and millet brans – a comparative study. *Food Chem* 126(3):1475–81.

Velasco, J., and Dobarganes, C. (2002). Oxidative stability of virgin olive oil. *European Journal of Lipid Science and Technology*, *104*, 661–676.

Velasco, J., Anderson, M. L., and Skibsted, L. H. (2004). Evaluation of oxidative stability of vegetable oils by monitoring the tendency to radical formation. A comparison of electron spin resonance spectroscopy with the Rancimat method and differential scanning calorimetry. *Food Chemistry*, *85*, 623–632.

Velioglu, Y. S., Mazza, G., Gao, L., and Oomah, B. D. (1998). Antioxidant activity and total phenolics in selected fruits, vegetables, and grain products. *Journal of Agricultural and Food Chemistry*, 46, 4113–4117.

Venkatamuru, K., Patel, J. D., and Rao, S. (1983). Fractionation of wood phenolic and their use in brandy. Journal of Food Science and Technology, 20, 16-18.

Vercellotti, J. R., St. Angelo, A. J., and Spanier, A. M. (1992). Lipid oxidation in foods: An overview. In A. J. St. Angelo (Ed.), *Lipid oxidation in food* (pp. 1–11).Washington, DC: American Chemical Society.

Verhagen, H., Schilderman, P. A. E. L. and Kleinjans, J. C. S. (1991). Butylated hydroxyanisole in perspective. *Chemico-Biological Interactions*, 80: 109–134.

Verma, B., Hucl, P. and Chibbar, R. N. (2008). Phenolic content and antioxidant properties of bran in 51 wheat cultivars. *Cereal Chemistry*, 85(4), 544-549.

Verma, S. P., Salamone, E. and Goldin, B. (1997). Curcumin and genistein, plant natural products, show synergistic inhibitory effects on the growth of human breast cancer MCF-7 cells induced by estrogenic pesticides. *Biochemical and Biophysical Research Communications*, 233: 692–696.

Vian, A., Tomao, M., Gallet, S., Coulomb, P. O. and Lacombe, J. M. (2005). Simple and rapid method for *cis*- and *trans*-resveratrol and pieced isomers determination in wine by HPLC using Chromolith columns. *Journal of Chromatography. A*, 1085: 224–229.

Viegas, O., Amaro, L. F., Ferreira, I. M. and Pinho, O. (2012). Inhibitory effect of antioxidant-rich marinades on the formation of heterocyclic aromatic amines in pan-fried beef. *Journal of Agricultural and Food Chemistry*, 60: 6235–6240.

Vilegas, W., Sanomimiya, M., Rastrelli, L., and Pizza, C. (1999). Isolation and structure elucidation of two new favonoid glycosides from the infusion of Maytenus aquifolium leaves. Evaluation of the antiulcer activity of the infusion. Journal of Agricultural and Food Chemistry, 47, 403-406.

Villacorta L, Graça-Souza AV, Ricciarelli R, Zingg JM, Azzi A; Graça-Souza; Ricciarelli; Zingg; Azzi (2003). "α-Tocopherol induces expression of connective tissue growth factor and antagonizes tumor necrosis factor-α-mediated downregulation in human smooth muscle cells". Circ. Res. 92 (1): 104–110.

Villarino, M., Sandín-España, P., Melgarejo, P., and De Cal, A. (2011). High chlorogenic and neochlorogenic acid levels in immature peaches reduce *Monilinia laxa* infection by interfering with fungal melanin biosynthesis. *Journal of Agricultural and Food Chemistry*, *59*, 3205–3213.

Villarreal-Lozoya, J. E., Lombardini, L. and Cisneros-Zevallos, L. (2007). Phytochemical constituents and antioxidant capacity of different pecan [*Carya illinoinensis* (Wangenh.) K. Koch] cultivars. *Food Chemistry*, 102: 1241–1249.

Villiams, A. H. (1966). Dihydrochalcones, in *Comparative Phytochemistry,* Swain, *T.,* Ed., Academic Press, New York.

Vinson, J. A., and Cai, Y. (2012). Nuts, especially walnuts, have both antioxidant quantity and efficacy and exhibit significant potential health benefits. *Food and Function*, *3*, 134–140.

Vinson, J. A., Bose, P., Proch, J., Al Kharrat, H. and Samman, N. (2008). Cranberries and cranberry products: Powerful in vitro, ex vivo, and in vivo sources of antioxidants. *Journal of Agriculture and Food Chemistry*, 56: 5884–5891.

Vinson, J. A., Dabbagh, Y. A., Serry, M. M., and Jang, J. (1995). Plant favonols, are powerful antioxidants using an in vitro oxidation model for heart disease. Journal of Agricultural and Food Chemistry, 43, 2800-2802.

Vinson, J. A., Hang, J., Yang, J., Dabbagh, Y., Liang, X., Serry, M., Proch, J., and Cai, S. (1999). Vitamins and especially favonoids in common beverages are powerful in vitro antioxidants which enrich lower density lipoproteins and increase their oxidative resistance after ex vivo spiking in human plasma. Journal of Agricultural and Food Chemistry, 47, 2502-2504.

Vinson, J. A., Hao, Y., Su, X., and Zubik, L. (1998). Phenol antioxidant quantity and quality in food, vegetables. Journal of Agricultural and Food Chemistry, 46, 3630-3634.

Vinson, J. A., Proch, J. and Zubik, L. (1999). Phenol antioxidant quantity and quality in foods: Cocoa, dark chocolate, and milk chocolate. *Journal of Agricultural and Food Chemistry*, 47: 4821–4824.

Vinson, J. A., Su, X., Zubik, L. and Bose, P. (2001). Phenol antioxidant quantity and quality in foods: Fruits. *Journal of Agricultural and Food Chemistry*, 49: 5315–5321.

Vinson, J. A., Teufel, K. and Wu, N. (2001). Red wine, dealcoholized red wine, and especially grape juice, inhibit atherosclerosis in a hamster model. *Atherosclerosis*, 156: 67–72.

Vinson, J. A., Yong, H., Su, X. and Ligia, Z. (1998). Phenol antioxidant quantity and quality in foods: Vegetables. *Journal of Agricultural Food Chemistry*, 46: 3630–3634.

Vinson, J. A., Zubik, L., Bose, P., Samman, N., and Proch, J. (2005). Dried fruits: Excellent *in vitro* and *in vivo* antioxidants. *Journal of the American College of Nutrition, 24*, 44–50.

Vinson, J. A. and Dabbagh, Y. A. (1998). Tea phenols: Antioxidant effectiveness of teas, tea components, tea fractions and their binding with lipoproteins. *Nutrition Research*, 18: 1067–1075.

Vinson, J. A. and Hontz, B. A. (1995). Phenol antioxidant index: Comparative antioxidant effectiveness of red and white wines. *Journal of Agricultural and Food Chemistry*, 43: 401–403.

Visioli, F., Romani, A., Mulinacci, N., Zarini, S., Conte, D., Vincieri, F. F., and Galli, C. (1999). Antioxidant and other biological activities of olive mill waste waters. Journal of Agricultural and Food Chemistry, 47, 3397-3401.

Viswanath, V., Urooj, A., and Malleshi, N. G. (2009). Evaluation of antioxidant and antimicrobial properties of finger millet polyphenols (*Eleusine coracana*). *Food Chem.*, 114:340-346.

Viswanath, V., Urooj, A. and Malleshi, N. G. (2009). Evaluation of antioxidant and antimicrobial properties of finger millet polyphenols (*Eleusine coracana*). *Food Chemistry*, 114: 340–346.

Vitamin E. (2015). Health Professional Fact Sheet". dietary-supplements.info.nih.gov. Retrieved 5 February 2015.

Voight. R. L. (1966). Bird-tolerant sorghums boost takehome yields. *Prog. Agric. Ariz.* 18, 30.

von Gadow, A., Joubert, E., and Hansmann, C. F. (1997a). Effect of extraction and additional heating on the antioxidant activity of rooibos tea (*Aspalathus linearis*) extracts. Journal of Agricultural and Food Chemistry, 45, 1370-1374.

von Gadow, A., Joubert, E., and Hansmann, C. F. (1997b). Comparisonof the antioxidant activity of rooibos tea (*Aspalathus linearis*) with green, oolong and black tea. Food Chemistry, 60, 73-77.

von Gadow, A., Joubert, E., and Hansmann, C. F. (1997c). Comparison of the antioxidant activity of aspalathin with that of other plant phenols of rooibos tea (*Aspalathus linearis*), a-tocopherol, BHT and BHA. Journal of Agricultural and Food Chemistry, 45, 632-638.

Votruba, M.; Stopka, P.; Hroudova, J.; Vesely, K.; Nejedlova, L. (1999). A simple method for quantitative estimation of free radicals in serum. *Klin. Biochem. Met. 7*, 96-101.

Vrhovsek, U. (1998). Extraction of hydroxycinnamoyltartaric acids from berries of different grape varieties. *Journal of Agricultural and Food Chemistry*, 46: 4203–4208.

Vrhovsek, U., Rigo, A., Tonon, D., and Mattivi, F. (2004). Quantitation of polyphenols in different apple varieties. *Journal of Agricultural and Food Chemistry*, *52*, 6532–6538.

Vuataz, L., Brandenberger, H., and Egli, R. H. (1959). Plant phenols. I. Separation of the tea leaf polyphenols by cellulose column chromatography. J. *Chromatogr*, 2, 173.

Vulic, J., C anadanovic-Brunet, J.,C etkovic, G., Tumbas, V., Djilas, S.,C etojevic-Simin, D. andC anadanovic, V. (2012). Antioxidant and cell growth activities of beet root pomace extracts. *Journal of Functional Foods*, 4: 670–678.

Vulic', J. J.,C ebovic, T. N.,C anadanovic-Brunet, J. M.,C etkovic, G. S.,C anadanovic, V. M., Djilas, S. M. and Šaponjac, V. T. T. (2014). In vivo and in vitro antioxidant effects of beetroot pomace extracts. *Journal of Functional Foods*, 6: 168–175.

Vuorela, S., Salminen, H., Makela, M., Kivikari, R., Karonen, M., and Heinonen, M. (2005). Effect of plant phenolics on protein and lipid oxidation in cooked pork meat patties. *Journal of Agricultural and Food Chemistry, 53*, 8492–8497.

Vuorinen, H., Määttä, K. and Törrönen, R. (2000). Content of the flavonoids myricetin, quercetin and kaempferol in Finnish berry wines. *Journal of Agricultural and Food Chemistry*, 48: 2675–2680.

Wada, L. and Ou, B. (2002). Antioxidant activity and phenolic content of Oregon caneberries. *Journal of Agricultural and Food Chemistry*, 50: 3495–3500.

Waffo-Teguo, P., Hawthorne, M. E., Cuendet, M., Merillon, J. M., Kinghorn, A. D., Pezzuto, J. M. and Mehta, R. G. (2001). Potential cancer-chemopreventive activities of wine stilbenoids and flavans extracted from grape (*Vitis vinifera*) cell cultures. *Nutrition and Cancer*, 40: 173–179.

Wagner, J. R, Hu, C. and Ames, B. N. (1992). Endogenous oxidative damage of deoxycytidine in DNA. *Proceedings of the National Academy of Science of the United States of America*, 89(8), 3380-3384.

Wakabayashi, Y. (1999). Effect of red wine comsumption on lowdensity lipoprotein oxidation and artherosclerosis in aorta and coronary artery in Watanabe heritable hyperlipidemic rabbits. Journal of Agricultural and Food Chemistry, 42, 1285-1290.

Walker, J. R. L. (1975). *The Biology of Plont Phetlolics,* Edward Arnold, London.

Wall, M. M. (2010). Functional lipid characteristics, oxidative stability, and antioxidant activity of macadamia nut (*Macadamia integrifolia)* cultivars. *Food Chemistry*, 121: 1103–1108.

Wallace, T. C. (2011). Anthocyanins in Cardiovascular Disease. *Advances in Nutrition.* 2:1-7.

Wan, X., Li, D. and Zhang, Z. (2009). Antioxidant properties and mechanisms of tea polyphenols. In C. T. Ho, J. K. Linand F. Shahidi (Eds.), *Tea and tea products: Chemistry and healthpromoting properties* (pp. 131–159). NewYork: CRC Press, Taylor and Francis Group.

Wanasundara, P. K. J. P. D. and Shahidi, F. (2005). Science, technology and applications. In F. Shahidi (Ed.), *Bailey's industrial oil and fat products* (pp. 431–489). Hoboken, NJ: John Wiley and Sons, Inc.

Wanasundara, U. N., Amarovicz, R., and Shahidi, F. (1994). Isolation and identification of an antioxidative component in canola. Journal of Agricultural and Food Chemistry, 42, 1285-1290.

Wanasundara, U. N., Amarowicz, R. and Shahidi, F. (1996). Partial characterization of natural antioxidants in canola meal. *Food Research International*, 28: 525–530.

Wanasundara, U. N., and Shahidi, F. (1998). Antioxidant and pro-oxidant activity of green tea extracts in marine oils. Food Chemistry, 63, 335-342.

Wanasundara, U. N., Shahidi, F., and Jablonski, C. R. (1995). Comparison of standard and NMR methodologies for assessment of oxidative stability of canola and soybean oils. *Food Chemistry*, *52*, 249–253.

Wanasundara, U., Amarowicz, R. and Shahidi, F. (1994). Isolation and identification of an antioxidative component in canola meal. *Journal of Agricultural and Food Chemistry*, 42: 1285–1290.

Wanasundara, U. and Shahidi, F. (1994). Canola extracts alternative natural antioxidants for canola oil. *Journal of the American Oil Chemists' Society*, 71: 817–822.

Wanasundara, U. and Shahidi, F. (1996). Stabilization of seal blubber and menhaden oils with green tea catechins. *Journal of the American Oil Chemists' Society*, 73: 1183–1190.

Wanasundara, U.N., Amarowicz, R., and Shahidi, F. (1996). Partial characterization of natural antioxidants in canola meal. Food Res. Int. 28: 525-530.

Wang X, Quinn PJ; Quinn (1999). "Vitamin E and its function in membranes". Progress in Lipid Research 38 (4): 309–36.

Wang, H., Cao, G. and Prior, R. L. (1996). Total antioxidant capacity of fruits. *Journal of Agricultural and Food Chemistry*, 44: 701–705.

Wang, H., Cao, G. and Prior, R. L. (1997). Oxygen radical absorbing capacity of anthocyanins. *Journal of Agricultural and Food Chemistry*, 45: 304–309.

Wang, H., Nair, M. G., Strasburg, G. M., Booren, A. M., and Gray, J. I. (1999). Novel antioxidant compounds from tart cherries (*Prunus cerasus*). Journal of Natural Products, 62, 86-88.

Wang, H., Nair, M. G., Strasburg, G. M., Chang, Y.-C., Booren, A. M., and Gray, J. I. (1999). Antioxidant polyphenols from tart cherries (*Prunus cerasus*). Journal of Agricultural and Food Chemistry, 47, 840-844.

Wang, H., Nair, M. G., Strasburg, G. M., Chang, Y.-C., Booren, A. M., Gray, J. I., and DeWitt, D. L. (1999). Antioxidant and antiinfamatory activities of anthocyanins and their aglycon, cyanidin, from tart cherries. Journal of Natural Products, 62, 294-296.

Wang, J., Sun, B., Cao, Y. nd Wang, C. (2010). Wheat bran feruloyl oligosaccharides enhance the antioxidant activity of rat. *Food Chemistry*, 123(2), 472-476.

Wang, J., Zhou, N., Zhu, Z., Huang, J., and Li, G. (2007). Detection of flavonoids and assay for their antioxidant activity based on enlargement of gold nanoparticles. *Analytical and Bioanalytical Chemistry*, *388*, 1199–1205.

Wang, M., Li, J., Rangarajan, M., Shao, Y., LaVoie, E. J., Huang, T. C. and Ho, C.-T. (1998). Antioxidative phenolic compounds from sage (*Salvia officinalis*). *Journal of Agricultural and Food Chemistry*, 46: 4869–4873.

Wang, M., Li, J., Rangarajan, M., Shao, Y., LaVoie, E. J., Huang, T.-C., and Ho, C.-T. (1998). Antioxidative phenolic compounds from sage (Salvia o cinalis). Journal of Agricultural and Food Chemistry, 46, 4869-4873.

Wang, M., Shao, Y., Li, J., Zhu, N., Rangarajan, M., LaVoie, E. J., and Ho, C.-T. (1999). Antioxidative phenolic glycosides from sage (*Salvia officinalis*). Journal of Natural Products, 62, 454-456.

Wang, Q., Han, P., Zhang, M., Xia, M., Zhu, H., Ma, J., Hou, M., Tang, Z. and Ling, W. (2007). Supplementation of black rice pigment fraction improves antioxidant and anti-inflammatory status in patients with coronary heart diseases. *Asia Pacific Journal of Clinical Nutrition*, 16(1), 295-301.

Wang, S.-L., Yieh, T.-C., and Shih, I.-L. (1999). Production of antifungal compounds by Pseudomonas aeruginosa K-187 using shrimp and crab shell powder as carbon source. Enzyme and Microbial Technology, 25, 142-148.

Wang, Y., Catana, F., Yang, Y., Roderick, R. and van Breemen, R. B. (2002). An LC-MS method for analyzing total resveratrol in grape juice, cranberry juice, and in wine. *Journal of Agricultural and Food Chemistry*, 50: 431–435.

Waniska, R. D., Poe, J. H., and Bandyopadhyay, R. (1989). Effects of growth conditions on grain molding and phenols in sorghum caryopsis. *J. Cereal Sci.,* 10:217–225.

Wardhani, D. H., Fucinos, P., Vazquez, J. A., and Pandiella, S. S. (2013). Inhibition kinetics of lipid oxidation of model foods by using antioxidant extract of fermented soybeans. *Food Chemistry*, *139*, 837–844.

Watanabe, M. (1998). Catechins as antioxidants from buckwheat (Fagopyrum esculentum MoÈ ench) groats. Journal of Agricultural and Food Chemistry, 46, 839-845.

Watanabe, M. (1999). Antioxidative phenolic compounds from Japanese barnyard millet (*Echinochloa utilis*) grains. *Journal of Agricultural and Food Chemistry*, 47: 4500–4505.

Watanabe, M., Ohshita, Y. and Tsushida, T. (1997). Antioxidant compounds from buckwheat (*Fagopyrum esculentum* Möench) hulls. *Journal of Agricultural and Food Chemistry*, 45: 1039–1044.

Waterhouse, A. L. and Walzem, R. L. (1998). Nutrition of grape phenolics. In: Watkins TR, editor. Flavonoids in health and disease. Washington, D.C.: American Chem. Soc. p 359–85.

Wattenberg, L. W. (1986). Protective effects of 2(3)-tert-butyl 4-hydroxanisole on chemical carcinogenesis. *Food and Chemical Toxicology*, *24*, 1099–1102.

Watterson, J.J. and Butler, L. G. (1983). Occurrence of a unusualleucoanthocyanidin and absence of proan-thocyanidins in sorghum leaves, *J. Agric. FoodChem.,* 31: 41.

Webster, J. E. and Cross, F. B. (1944). Chemical composition of grape juices: a varietal study. *Proc. Okla. Acad. Sci.,* 24.92.

Wefers H, Sies H; Sies (1988). "The protection of ascorbate and glutathione against microsomal lipid peroxidation is dependent on Vitamin E". European Journal of Biochemistry 174 (2): 353–357.

Weinberg, Z. G., Akiri, B., Potoyevski, E., and Kanner, J. (1999). Enhancement of polyphenol recovery from rosemary (*Rosmarinus ocinalis*) and sage (*Salvia ocinalis*) by enzyme-assisted ensiling (ENLAC). Journal of Agricultural and Food Chemistry, 47, 2959-2962.

Weinges, K. (1968). Mechanism of the formation and the constitution of flavonoid tannin, *J. Polym. Sci. Part* C. 16: 3625.

Weinges, K., Bahr, W., Ebert, W., Goritz, K. and Marx, H. D. (1969). Konstitution, entseehung und bedeutung der flavonoid-gerostoffe, *Fortschr Chem. Org. Natraw,* 27: 158.

Weinges, K., Bahr, W., Ebert, W., Goritz, K., and Marx, H.D. (1969). Konstitution, entstehung and bedeutung der flavonoid-gerbstofe. Forstschr Chem. Org. Natrustoffe. 27: 158-160.

Wen, L., Wrolstad, R. E., and Hsu, V. L. (1999). Characterization of sinapyl derivatives in pineapple (*Ananas comosus*) and sage (*Salvia ocinalis*) by enzyme-assisted ensiling (ENLAC). Journal of Agricultural and Food Chemistry, 47, 2959-2962.

Wettasinghe, M., and Shahidi, F. (1999). Evening primrose meal, a source of natural antioxidants and scavenger of hydrogen peroxide and oxygen-derived free radicals. Journal of Agricultural and Food Chemistry, 47, 1801-1812.

Wettasinghe, M., and Shahidi, F. (2002). Iron (II) chelation activity of extracts of borage and evening primrose meals. *Food Research International*, *35*, 65–71.

Wettasinghe, M., Shahidi, F., Amarowicz, R. and Abou-Zaid, M. M. (2001). Phenolic acids in defatted seeds of borage (*Borago officinalis* L.). *Food Chemistry*, 75: 49–56.

Wettasinghe, M., Shahidi, F. and Amarowicz, R. (2002). Identification and quantification of low molecular weight phenolic antioxidants in seeds of evening primrose (*Oenothera biennis* L.). *Journal of Agricultural and Food Chemistry*, 50: 1267–1271.

Wettasinghe, M. and Shahidi, F. (1999a). Antioxidant and free radical-scavenging properties of ethanolic extracts of defatted borage (*Borago officinalis* L.) seeds. *Food Chemistry*, 67: 399–414.

Wettasinghe, M. and Shahidi, F. (1999b). Evening primrose meal: A source of natural antioxidants and scavenger of hydrogen peroxide and oxygen-derived free radicals. *Journal of Agricultural and Food Chemistry*, 47: 1801–1812.

Wettasinghe, M. and Shahidi, F. (2000). Scavenging of reactiveoxygen species and DPPH free radicals by extracts of borage and evening primrose meals. *Food Chemistry*, 70: 17–26.

Wettasinghe, M. and Shahidi, F. (2002). Iron (II) chelation activity of extracts of borage and evening primrose meals. *Food Research International*, 35: 65–71.

White, P. J. (1995). Conjugated diene, anisidine value and carbonyl value analyses. In K. Warner and N. A. M. Eskin (Eds.), *Methods to assess quality and stability of oils and fat-containing foods* (pp.159–178). Champaign, IL: AOCS Press.

White, T. (1957). Tannins - their occurrence and significance, *J. Sci. Food Agric.* 8: 377.

Whitney, Ellie; Sharon Rady Rolfes (2011). Peggy Williams, ed. Understanding Nutrition (Twelfth Ed.). California: Wadsworth, Cengage Learning. ISBN 0-538-73465-5.

Wichi, H. P. (1988). Enhanced tumor development by butylated hydroxyanisole (BHA) from the perspective of effect on forestomach and oesophageal squamous epithelium. *Food and Chemical Toxicology*, 26: 717–723.

Wiczkowski,W., Topolska, J. and Honke, J. (2014). Anthocyanins profile and antioxidant capacity of red cabbages are influenced by genotype and vegetation period. *Journal of Functional Foods*, 7: 201–211.

Wijeratne, S. S. K., Abou-Zaid, M. M. and Shahidi, F. (2006). Antioxidant polyphenols in almond and its coproducts. *Journal of Agricultural and Food Chemistry*, 54: 312–318.

Wijeratne, S. S. K., Amarowicz, R. and Shahidi, F. (2006). Antioxidant activity of almonds and their by-products in food model systems. *Journal of the American Oil Chemists' Society*, 83: 223–230.

Wijeratne, S. S., Abou-Zaid, M. M., and Shahidi, F. (2006). Antioxidant polyphenols in almond and its coproducts. *Journal of Agricultural and Food Chemistry*, *54*, 312–318.

Wijewickreme, A. N., and Kitts, D. D. (1998). Oxidative reactions of model Maillard reaction products and a-tocopherol in a four-lipid mixture. Journal of Food Scence, 63, 466-471.

Wijewickreme, A. N., Krejpcio, Z., and Kitts, D. D. (1999). Hydroxyl scavenging activity of glucose, fructose, and ribose-lysine model Maillard products. Journal of Food Science, 64, 457-461.

Wilhelmina kalt (2005). Effects of Production and Processing Factors on Major Fruit and Vegetable Antioxidants. Journal of food science. 70 (1): 11-19.

Willet, W. C. (1994). Diet and health: What should we eat? *Science*, 254(5158), 532-537.

Williams, G. M., Iatropoulos, M. J. and Whysner, J. (1999). Safety assessment of butylated hydroxyanisole and butylated hydroxytoluene as antioxidant food additives. *Food and Chemical Toxicology*, 37: 1027–1038.

Williamson, G., and Carughi, A. (2010). Polyphenol content and health benefits of raisins. *Nutrition Research, 30*, 511-519.

Wilson, T., Porcari, J. P. and Harbin, D. (1998). Cranberry extract inhibits low density lipoprotein oxidation. *Life Sciences*, 62: 381–386.

Wilson, T., Porcari, J. P. and Maher, M. A. (1999). Cranberry juice inhibits metal- and non-metal initiated oxidation of lowdensity lipoprotein. *Journal of Nutraceuticals, Functional and Medical Foods*, 2: 5–14.

Winrow, V. R.,Winyard, P. G., Morris, C. J., and Blake, D. R. (1993). Free radicals in inflammation: Second messengers and mediators of tissue destruction. *British Medical Bulletin, 49*, 506–522.

Winter, M. and Herrmann, K. (1986). Esters and glucosides of hydrocinnamic acids in vegetables. *Journal of Agricultural and Food Chemistry*, 34: 616–620.

Wolfe, K. L., and Liu, R. H. (2007). Cellular antioxidant activity (CAA) assay for assessing antioxidants, foods, and dietary supplements. *Journal of Agricultural and Food Chemistry, 55*, 8896–8907.

Wolfe, K. L., Kang, X., He, X., Dong, M., Zhang, Q., and Liu, R. H. (2008). Cellular antioxidant activity of common fruits. *Journal of Agricultural and Food Chemistry, 56*, 8418–8426.

Wolfe, K. L., Kang, X., He, X., Dong, M., Zhang, Q. and Liu, R. H. (2008). Cellular antioxidant activity of common fruits. *Journal of Agricultural and Food Chemistry*, 56: 8418–8426.

Wolfe, K., Wu, X. and Liu, R. H. (2003). Antioxidant activity of apple peels. *Journal of Agricultural and Food Chemistry*, 51: 609–614.

Wollgast, J. and Anklam, E. (2000). Review on polyphenols in *Theobroma cacao*: Changes in composition during the manufacture of chocolate and methodology for identification and quantification. *Food Research International, 33:* 423–447.

Wolosiak, R., Worobiej, E., Piecyk, M., Druzynska, B., Nowak, D. and Lewicki, P. P. (2009). Activities of amine and phenolic antioxidants and their changes in broad beans (*Vicia faba*) after freezing and steam cooking. *Int. J. Food Sci. Technol.* 45: 29-37.

Wong, E. (1976). Biosynthesis of flavonoids, in *Chemistry and Biochemistry of Pla1l1 Pigments.* Vol. I, Goodwin. T., Ed. 464.

Woodhead. S., Padgham, D. E., and Bernays, E. A. (1980). Insect feeding on different sorghum cultivars in relation to cyanide and phenolic acid content. *Ann. Appl. Biol.,* 95: 151.

Wright JS, Johnson ER, DiLabio GA (2001) Predicting the activity of phenolic antioxidants: theoretical method, analysis of substituent effects, and application to major families of antioxidants. J Am Chem Soc 123:1173–1183.

Wu, D., Ma, X. and Tian, W. (2013). Pomegranate husk extract, punicalagin and ellagic acid inhibit fatty acid synthase and adipogenesis of 3T3-L1 adipocyte. *Journal of Functional Foods*, 5: 633–641.

Wu, H. and Xu, B. (2014). Inhibitory effects of onion against α-glucosidase Activity and its correlation with phenolic antioxidants. *International Journal of Food Properties*, 17: 599–609.

Wu, J. M.,Wang, Z. R., Hsieh, T. C., Bruder, J. L., Zou, J. G. and Huang, Y. Z. (2001). Mechanism of cardioprotection by resveratrol, a phenolic antioxidant present in red wine (review). *International Journal of Molecular Medicine*, 8: 3–17.

Wu, J. W., Lee, M-H., Ho, C-T., and Chang, S. S. (1982). Elucidation of the chemical structures of natural antioxidants isolated from rosemary. Journal of the American Oil Chemists' Society, 59, 339-345.

Wu, K., Zhang, W., Addis, P. B., Epley, R. J., Salih, A. M., and Lehrfeld, J. (1994). Antioxidant properties of wild rice. Journal of Agricultural and Food Chemistry, 42, 34-37.

Wu, X., Beecher, G. R., Holden, J. M., Haytowitz, D. B., Gebhardt, S. E. and Prior, R. L. (2004). Lipophilic and hydrophilic antioxidant capacities of common foods in the United States. *Journal of Agricultural and Food Chemistry*, 52: 4026–4037.

Wu, X., Beecher, G. R., Holden, J. M., Haytowitz, D. B., Gebhardt, S. E., and Prior, R. L. (2004). Lipophilic and hydrophilic antioxidant capacities of common foods in the United States. *Journal of Agricultural and Food Chemistry*, *52*, 4026–4037.

Wulff, H. (1997). Natural antioxidant composition. WO9739632.

Xiao, X., Kim, J., Sun, Q., Kim, D., Park, C. S., Lu, T. S., and Park, Y. (2015). Preventive effects of cranberry products on experimental colitis induced by dextran sulphate sodium in mice. *Food Chemistry*, *167*, 438–446.

Xie, J., and Schaich, K. M. (2014). Re-evaluation of the 2,2-diphenyl-1-picrylhydrazyl free radical (DPPH) assay for antioxidant activity. *Journal of Agricultural and Food Chemistry*, *62*, 4251–4260.

Xing, Y., and White, P. (1997). Identification and function of antioxidants from oat groats and hulls. Journal of the American Oil Chemists' Society, 74, 303-307.

Xu, B. J. and Chang, S. K. C. (2007). A comparative study on phenolic profiles and antioxidant activities of legumes as affected by extraction solvents. *Journal of Food Science*, 72: S159–S166.

Xu, B., and Chang, S. K. C. (2008). Total phenolics, phenolic acids, isoflavones, and anthocyanins and antioxidant properties of yellow and black soybeans as affected by thermal processing. *Journal of Agricultural and Food Chemistry*, *56*, 7165–7175.

Xu, B. and Chang, S. K. C. (2008). Effect of soaking, boiling and steaming on total phenolic content and antioxidant activities of cool season food legumes. *Food Chem*. 110: 1-13.

Xu, B. and Chang, S. K. C. (2009). Total phenolic, phenolic acid, anthocyanin, flavan-3-ol, and flavonol profiles and antioxidant properties of pinto and black beans (*Phaseolus vulgaris* L.) as affected by thermal processing. *Journal of Agricultural and Food Chemistry*, 57: 4754–4764.

Xu, G., Ye, X., Liu, D., Ma, Y. and Chen, J. (2008). Composition and distribution of phenolic acids in Ponkan (*Citrus poonensis* Hort. ex Tanaka) and Huyou (*Citrus paradisi* Macf. Changshanhuyou) during maturity. *Journal of Food Composition and Analysis*, 21: 382–389.

Xu, J., Chen, S. and Hu, Q. (2005). Antioxidant activity of brown pigment and extracts from black sesame seed (*Sesamum indicum* L.). *Food Chemistry*, 91: 79–83.

Xu, X., and Hua, N., and Godber, S. (2001). Antioxidant activity of tocopherols, tocotrienols, and γ-oryzanol components from rice bran against cholesterol oxidation accelerated by 2,2'-azobis-(2-methylpropionamidine) dihydrochloride. *J. Agric. Food Chem.*, 49:2077-2081.

Xu, Z., and Godber, J. S. (1999). Purification and identification of components of γ-oryzanol in rice bran oil. *J. Agric. Food Chem.*, 47:2724–2728.

Yaacoub, R., Saliba, R., Nsouli, B., Khalaf, G., and Birlouez-Aragon, I. (2008). Formation of lipid oxidation and isomerization products during processing of nuts and sesame seeds. *Journal of Agricultural and Food Chemistry*. 56: 7082–7090.

Yadav, S. K., and Sehgal, S. (1995). Effect of home processing on ascorbic acid and b-carotene content of spinach and amaranth leaves. *Plant Foods for Human Nutrition*. 47: 125–131.

Yamaguchi, F., Yoshimura, Y., Nakazawa, H., and Ariga, T. (1999). Free radical scavenging activity of grape seed extract and antioxidants by electron spin resonance spectrometry in an H_2O_2/NaOH/DMSO system. Journal of Agricultural and Food Chemistry, 47, 2544-2548.

Yamagushi, N. and Fujimaki, M. (1974). Browning reaction products from reducing sugars and amino acids. XV. Comparison of antioxidative activity of melanoidin with that of tocopherol homologues and the synergistic effect of melanoidin with tocopherols. *Japan Journal of Food Science and Technology*, 21: 13–18.

Yan, X., Murphy, B. T., Hammond, G. B., Vinson, J. A. and Neto, C. C. (2002). Antioxidant activities and antitumor screening of extracts from cranberry fruit (*Vaccinium macrocarpon*). *Journal of Agricultural and Food Chemistry*, 50: 5844–5849.

Yan, X., Suzuki, M., Ohnishi-Kameyama, M., Sada, Y., Nakanishi, T., and Nagata, T. (1999). Extraction and identification of antioxidants in the roots of yacon (*Smallanthus sonchifolius*). Journal of Agricultural and Food Chemistry, 47, 4711-4713.

Yang, C. S., Landau, J. M., Huang, M. T. and Newmark, H. L. (2001). Inhibition of carcinogenesis by dietary polyphenolic compounds. *Annual Review of Nutrition*, 21: 381–406.

Yang, C. S. and Wang, Z. Y. (1993). Effect of tea polyphenols and EGCG on nasopharyngeal carcinoma cell proliferation and the mechanisms involved. *Journal of National Cancer Institute*, 85: 1038–1049.

Yang, J., Liu, R. H. and Halim, L. (2009). Antioxidant and antiproliferative activities of common edible nut seeds. *LWT -Food Science and Technology*, 42: 1–8.

Yang, J., Martinson, T. E. and Liu, R. H. (2009). Phytochemical profiles and antioxidant activities of wine grapes. *Food Chemistry*, 116: 332–339.

Yang, J., Zhou, F., Xiong, L., Mao, S., Hu, Y., and Lu, B. (2015). Comparison of phenolic compounds, tocopherols, phytosterols and antioxidant potential in Zhejiang pecan (*Carya cathayensis*) at different stir-frying steps. *LWT – Food Science and Technology*, *62*, 541–548.

Yang, Z. and Zhai, W. (2010). Identification and antioxidant activity of anthocyanins extracted from the seed and cob of purple corn (*Zea mays* L.). *Innovative Food Science and Emerging Technologies*, 11: 169–176.

Yanishlieva, N. V., and Marinova, E. M. (2001). Stabilisation of edible oils with natural antioxidants. *European Journal of Lipid Science and Technology*, *103*, 752–767.

Yanishlieva, N. VI., and Marinova, E. M. (1995). Effects of antioxidants on the stability of triacylglycerols and methyl esters of fatty acids of sunfower oil. Food Chemistry, 54, 377-382.

Yanishlieva, N. VI., Marinova, E. M., Gordon, M. H., and Raneva, V. G. (1999). Antioxidant activity and mechanism of action of thymol and carvacrol in two lipid systems. Food Chemistry, 64, 59-66.

Yanishlieva, N., Marinova, E. and Porkorny, J. (2006). Natural antioxidants from herbs and spices. *European Journal of Lipid Science and Technology*, 108: 776–793.

Yanishlieva-Maslarova, N. V. (2001). Inhibiting oxidation. In J. Pokorny, N. V. Yanishlieva, and M. Gordon (Eds.), *Antioxidants in food. Practical applications* (pp. 22–70). Cambridge, UK: Woodhead Publishing Ltd.

Yao, L. H., Jiang, Y. M., Shi, J., Tomas-Barberan, F. A., Datta, N., Singanusong, R. and Chen, S. S. (2004). Flavonoids in food and their health benefits. *Plant Foods for Human Nutrition*, 59: 113–122.

Yarmolinsky, J., Gon, G. and Edwards, P. (2015). Effect of tea on blood pressure for secondary prevention of cardiovascular disease: A systematic review and meta-analysis of randomized controlled trials. *Nutrition Reviews*, 73: 236–246.

Yashin, A., Yashin, Y. and Nemzer, B. (2011). Determination of antioxidant activity in tea extracts, and their total antioxidant content. *American Journal of Biomedical Sciences*, *3:* 322–335.

Yassin, G. H., Koek, J. H. and Kuhnert, N. (2015). Model systembased mechanistic studies of black tea thearubigin formation. *Food Chemistry*, 180: 272–279.

Yen, G.-C., and Duh, P. D. (1994). Scavenging effect of methanolic extracts of peanut hulls on free-radical and active-oxygen species. Journal of Agricultural and Food Chemistry, 42, 629-632.

Yen, G.-C., and Duh, P.D. (1995). Antioxidant activity of methanolic extracts of peanut hulls from various cultivars. Journal of the American Oil Chemists' Society, 72, 1065-1067.

Yen, G.C., and Hsieh, C.L. (1998). Antioxidant activity of extracts from Du-zhong (Eucommia ulmoides) toward various lipid peroxidation models in vitro. Journal of Agricultural and Food Chemistry, 46, 3952-3957.

Yen, G.C., and Wu, J.Y. (1999). Antioxidant and radical scavenging properties of extracts from Ganoderma tsugae. Food Chemistry, 65, 375-379.

Yen, G.C., Chen, H.Y., and Peng, H.H. (1997). Antioxidant and prooxidant effects of various tea extracts. Journal of Agricultural and Food Chemistry, 45, 30-34.

Yen, G.C., Duh, P.D., and Tsai, C.L. (1993). Relationship between antioxidant activity and maturity of peanut hulls. Journal of Agricultural and Food Chemistry, 41, 67-70.

Yeo, J. and Shahidi, F. (2015). A critical evaluation of changes in the ratio of insoluble-bound to soluble phenolics on antioxidant activity of lentils during germination. *Journal of Agricultural and Food Chemistry*, 63: 379–381.

Yi, O. S., Han, D., and Shin, H. K. (1991). Synergistic antioxidative effects of tocopherol and ascorbic acid in fish oil/lecithin/water system. Journal of the American Oil Chemists' Society, 68, 881-883.

Yildiz, G.,Wehling, R. L., and Cuppett, S. L. (2003). Comparison of four analytical methods for the determination of peroxide value in oxidized soybean oils. *Journal of the American Oil Chemists' Society*, *80*, 103–107.

Yin, M. C. and Cheng,W. S. (1998). Inhibition of *Aspergillus niger* and *Aspergillus flavus* by some herbs and spices. *Journal of Food Protection*, 61: 123–125.

Yokozawa, T., Dong, E., Nakagawa, T., Kashiwagi, H., Nakagawa, H., Takeuchi, S., and Chung, H. Y. (1998). In vitro and in vivo studies on the radical-scavenging activity of tea. Journal of Agricultural and Food Chemistry, 46, 2143-2150.

Yoshida, T., Nakata, F. and Okuda, T. (1999). Tannins and related polyphenols of melastomataceous plants. VIII. Nobotanins L, M and N, trimeric hydrolyzable tannins from *Tibouchina semidecandra. Chemical and Pharmaceutical Bulletin-Tokyo*, 47: 824–827.

Young, I. S. and Woodside, J.V. (2001). Antioxidant in health and disease. *Journal of Clinical Pathology*, 54(3), 176-186.

Youssef, A. M. (1998). Extractability, fractionation and nutritional value of low and high tannin sorghum proteins. Food Chemistry, 63, 325-329.

Yu, J., Ahmedna, M. and Goktepe, I. (2005). Effects of processing methods and extraction solvents on concentration and antioxidant activity of peanut skin phenolics. *Food Chemistry*, 90: 199–206.

Yu, J., Vasanthan, T. and Temelli, F. (2001). Analysis of phenolic acids in barley by high performance liquid chromatography. *Journal of Agricultural and Food Chemistry*, 49: 4352–4358.

Yu, L., Nanguet, A. L. and Beta, T. (2013). Comparison of antioxidant properties of refined and whole wheat flour and bread. *Antioxidant,* 2(4), 370-383.

Yu, S., Nehus, Z. T., Badger, T. M., and Fang, N. (2007). Quantification of vitamin E and γ-oryzanol components in rice germ and bran. *J. Agric. Food Chem.,* 55:7308-7313.

Yurt, B., and Celik, I. (2011). Hepatoprotective effect and antioxidant role of sun, sulphited-dried apricot (*Prunus armeniaca* L.), and its kernel against ethanol-induced oxidative stress in rats. *Food and Chemical Toxicology*, *49*, 508–513.

Zadernowski, R., Naczk, M. and Nowak-Polakowska, H. (2002). Phenolic acids of borage (*Borago officinalis* L.) and evening primrose (*Oenothera biennis* L.). *Journal of the American Oil Chemists' Society*, 79: 335–338.

Zadernowski, R., Nowak-Polakowska, H. and Konopka, I. (1996). Effect of heating on antioxidative activity of rapeseed and evening primrose extracts. *Polish Journal of Food and Nutrition Sciences*, 46: 13–21.

Zandi, P., and Gordon, M. H. (1999). Antioxidant activity of extracts from old tea leaves. Food Chemistry, 64, 285-288.

Zargoosh, K., Ghayeb, Y., Aeineh, N., and Qandalee, M. (2014). Evaluation of antioxidant capacity of hydrophilic and hydrophobic antioxidants using peroxyoxalate chemiluminescence reaction of the novel furandicarboxylate derivative. *Food Analytical Methods*, *7*, 283–290.

Zegarska, Z., Amarowicz, R., Karamac, M. and Rafalowski, R. (1996). Antioxidative effect of rosemary ethanolic extract on butter. *Milchwissenschaft*, 51: 195–198.

Zegarska, Z., Rafalowski, R., Amarowicz, R., Karamac, M. and Shahidi, F. (1998). Stabilization of butter with deodorized rosemary extract. *Zeitschrift für Lebensmittel -Untersuchung und -Forschung*, 206: 99–102.

Zhan, P. X. (1996). Antioxidative activity of extracts from potato and sweet potato. *Food and Fermentation Industries*, 2: 30–33.

Zhang, D., Chu, L., Liu, Y., Wang, A., Ji, B., Wu,W., Zhou, F., Wei, Y., Cheng, Q., Cai, S., Xie, L., and Jia, G. (2011). Analysis of the antioxidant capacities of flavonoids under different spectrophotometric assays using cyclic voltammetry and density functional theory. *Journal of Agricultural and Food Chemistry*, *59*, 10277–10285.

Zhang, D. and Hamauzu, Y. (2004). Phenolic compounds and their antioxidant properties in different tissues of carrots. *Food Agriculture and Environment*, 2: 95–100.

Zhang, G., Huang, G. W., Xiao, L., Seiber, J., and Mitchell, A. E. (2011b). Acrylamide formation in almonds (*Prunus dulcis*): Influences of roasting time and temperature, precursors, varietal selection, and storage. *Journal of Agricultural and Food Chemistry*, *59*, 8225–8232.

Zhang, H. Y. (1999). Theoretical methods used in elucidating activity differences of phenolic antioxidants. Journal of the American Oil Chemists' Society, 76, 745-748.

Zhang, H., Cao, P., Agellon, L. B. and Zhai, C. K. (2009). Wild rice (*Zizania latifolia* (Griseb) Turcz) improves the serum lipid profile and antioxidant status of rats fed with high fat/cholesterol diet. *The British Journal of Nutrition*, 102(12), 1723-1727.

Zhang, J., Huang, Y., Kikuchi, T., Tokuda, H., Suzuki, N., Inafuku, K., Miura, M., Motohashi, S., Suzuki, T. and Akihisa, T. (2012). Cucurbitane triterpenoids from the leaves of *Momordica charantia*, and their cancer chemopreventive effects and cytotoxicities. *Chemistry and Biodiversity*, 9: 428–440.

Zhang, Q.-J., Yang, M., Zhao, Y.-M., Luan, X.-H. and Ke, Y.-G. (2001). Isolation and structure identification of flavonols glycosides from glandless cotton seeds. *Yaoxue Xuebao*, 36: 827–831.

Zhang, W. M., Zhang, R. F., Zhang, F. X. and Liu, R. H. (2010). Phenolic profiles and antioxidant activity of black rice bran of different commercially available varieties. *Journal of Agricultural and Food Chemistry,* 58(13), 7580-7587.

Zhang, X., Xiang, Q., Cui,W., Jia, X., and Li, N. (2011a). Evaluation of the *in vitro* and *in vivo* genotoxicity of almond skins. *Biomedical and Environmental Sciences*, *24*, 415–421.

Zhang, Y., Li, Q., Xing, H., Lu, X., Zhao, L., Qu, K., and Bi, K. (2013). Evaluation of antioxidant activity of ten compounds in different tea samples by means of an on-line HPLC-DPPH assay. *Food Research International*, *53*, 847–856.

Zhang, Y., Smuts, J. P., Dodbiba, E., Rangarajan, R., Lang, J. C. and Armstrong, D. W. (2012). Degradation study of carnosic acid, carnosol, rosmarinic acid, and rosemary extract (*Rosmarinus officinalis* L.) assessed using HPLC. *Journal of Agricultural and Food Chemistry*, 60: 9305–9314.

Zhang, Y., Vareed, S.K., and Nair, M.G. (2005). Human tumor cell growth inhibition by ontoxic anthocyanidins, the pigments in fruits and vegetables. *Life Science*. 76: 1465-1472.

Zhang, Z., Liao, L., Moore, J., Wu, T. and Wang, Z. (2009). Antioxidant phenolic compounds from walnut kernels (*Juglans regia* L.). *Food Chemistry*, 113: 160–165.

Zhao, B., and Hall, C. A., III (2008). Composition and antioxidant activity of raisin extract obtained from various solvents. *Food Chemistry*, *108*, 511–518.

Zhao, H., Chen, W., Lu, J. and Zhao, M. (2009). Phenolic profiles and antioxidant activities of commercial beers. *Food Chemistry*, 119: 1150–1158.

Zhao, H., Fan, W., Dong, J., Lu, J., Chen, J., Shan, L., et al. (2008). Evaluation of antioxidant activities and total phenolic contents of typical malting barleys varieties. *Food Chem.*, 107:296–304.

Zheng, G.-Q., Kenney, P. M. and Lam, L. K. T. (1992). Sesquiterpenes from clove (*Eugenia caryophyllata*) as potential anticarcinogenic agents. *Journal of Natural Products (Lloydia)*, 55: 999–1003.

Zheng, W. and Wang, S. (2001). Antioxidant activity and phenolic composition in selected herbs. *Journal of Agricultural and Food Chemistry*, 49: 5165–5170.

Zhishen, J., Mengcheng, T., and Jianming, W. (1999). The determination of favonoid contents in mulberry and their scavenging effects on superoxide radicals. Food Chemistry, 64, 555-559.

Zhong, Y., and Shahidi, F. (2011). Lipophilized epigallocatechin gallate (EGCG) derivatives as novel antioxidants. *Journal of Agricultural and Food Chemistry*, *59*, 6526–6533.

Zhong, Y., and Shahidi, F. (2012a). Antioxidant behavior in bulk oil: Limitations of polar paradox theory. *Journal of Agricultural and Food Chemistry*, 60, 4–6.

Zhong, Y., and Shahidi, F. (2012b). Lipophilized epigallocatechin gallate (EGCG) derivatives and their antioxidant potential in food and biological systems. *Food Chemistry*, 131, 22–30.

Zhong, Y., Chiou, Y. S., Pan, M. H., Ho, C.T. and Shahidi, F. (2012). Protective effects of epigallocatechin gallate (EGCG) derivatives on azoxymethane-induced colonic carcinogenesis in mice. *Journal of Functional Foods*, 4: 323–330.

Zhong, Y., Lall, S. P., and Shahidi, F. (2007). Effects of oxidized dietary oil and vitamin E supplementation on lipid profile and oxidation of muscle and liver of juvenile Atlantic cod (*Gadus morhua*). *Journal of Agricultural and Food Chemistry*, *55*, 6379–6386.

Zhong, Y., Ma, C. M., and Shahidi, F. (2012). Antioxidant and antiviral activities of lipophilic epigallocatechin gallate (EGCG) derivatives. *Journal of Functional Foods*, *4*, 87–93.

Zhong, Y. and Shahidi, F. (2011). Lipophilized epigallocatechin gallate (EGCG) derivatives as novel antioxidants. *Journal of Agricultural and Food Chemistry*, 56: 6526–6533.

Zhou, B., Miao, Q., Yang, L. and Liu, Z. L. (2005). Antioxidative effects of flavonols and their glycosides against the freeradical-induced peroxidation of linoleic acid in solution and in micelles. *Chemistry – A European Journal*, 11: 680–691.

Zhou, D., Yu, H., He, F., Reilly, K. H., Zhang, J., Li, S., Zhang, T., Wang, B., Ding, Y., and Xi, B. (2014). Nut consumption in relation to cardiovascular disease risk and type 2 diabetes: A systematic review and meta-analysis of prospective studies. *The American Journal of Clinical Nutrition*, *100*, 270–277.

Zhou, J., Li, P., Cheng, N., Gao, H., Wang, B., Wei, Y. and Cao, W. (2012). Protective effects of buckwheat honey on DNA damage induced by hydroxyl radicals. *Food and Chemical Toxicology*, 50: 2766–2773.

Zhou, K., Su, L., and Yu, L. (2004). Phytochemicals and antioxidant properties in wheat bran. *J. Agric. Food Chem.,* 52:6108–6114.

Zhou, Z., Robards, K., Helliwell, S., and Blanchard, C. (2004). The distribution of phenolic acids in rice. *Food Chem.,* 87:401-406.

Zhu, B. T., Ezell, E. L. and Liehr, J. G. (1994). Catechol-*O*methyltransferase-catalyzed rapid *O*-methylation of mutagenic flavonoids. *Journal of Biological Chemistry*, 269: 292–299.

Zhu, M., Hu, J., Perez, E., Phillips, D., Kim, W., Ghaedian, R., Napora, J. K., and Zou, S. (2011). Effects of long-term cranberry supplementation on endocrine pancreas in aging rats. *Journals of Gerontology Series A: Biological Sciences and Medical Sciences*, *66*, 1139–1151.

Zhu, Q. Y., Huang, Y., Tsang, D., and Chen, Z-Y. (1999). Regeneration of a-tocopherol in human low-density lipoprotein by green tea catechin. Journal of Agricultural and Food Chemistry, 47, 2020-2025.

Zhu, Q., Liang, C. P., Cheng, K. W., Peng, X., Lo, C. Y., Shahidi, F., Chen, F., Ho, C.-T. and Wang, M. (2009). Trapping effects of green and black tea extracts on peroxidation-derived carbonyl substances of seal blubber oil. *Journal of Agricultural and Food Chemistry*, 57: 1065–1069.

Zia-Ul-Haq, M., Iqbal, S., Ahmad, S., Bhanger, M. I., Wiczkowski, W. and Amarowicz, R. (2008). Antioxidant potential of Desi chickpea varieties commonly consumed in Pakistan. *Journal of Food Lipids*, 15: 326–342.

Zielinska, D., Wiczkowski, W. and Piskula, M. K. (2008). Determination of the relative contribution of quercetin and its glucosides to the antioxidant capacity of onion by cyclic voltammetry and spectrophotometric methods. *Journal of Agricultural and Food Chemistry*, 56: 3524–3531.

Zielinski, H., Ceglinska, A., and Michalska, A. (2007). Antioxidant contents and properties as quality indices of rye cultivars. *Food Chem.,* 104:980-988.

Zielinski, H., Zielinska, D., and Kostyra, H. (2012). Antioxidant capacity of a new crispy type food products determined by updated analytical strategies. *Food Chemistry, 130,* 1098–1104.

Žilic, S., Serpen, A., Akillioglu, G., Jankovic, M. and Gökmen, V. (2012). Distribution of phenolic compounds, yellow pigments and oxidative enzymes in wheat grains and their relation to antioxidant capacity of bran and debranned flour. *Journal of Cereal Science*, 56(3), 652-658.

Zitko, V. and Rosik, J. (1962). Ein beitrag zur theorie-der tanningelatine-schonung von obstsaften, *Nahrung, 6,* 561.

Ziyatdinova, G., Nizamova, A., and Budnikov, H. (2011). Novel coulometric approach to evaluation of total free polyphenols in tea and coffee beverages in presence of milk proteins. *Food Analytical Methods*, *4*, 334–340.

Ziyatdinova, G., Salikhova, I., and Budnikov, H. (2014). Coulometric titration with electrogenerated oxidants as a tool for evaluation of cognac and brandy antioxidant properties. *Food Chemistry*, *150*(2014), 80–86.

Zucker, W.V. (1983). Tannins: does structure determine function? An ecological perspective. Am. Naturalist. 121: 335-365.

Zumbe, A. (1998). Polyphenols in cocoa: Are there health benefits? *British National Formulary. Nutrition Bulletin*, 23: 94–102.

Zuo, Y., Wang, C. and Zhan, J. (2002). Separation, characterization, and quantitation of benzoic and phenolic antioxidants in American cranberry fruit by GC–MS. *Journal of Agricultural and Food Chemistry*, 50: 3789–3794.

Zupfer, J. M., Churchill, K. E., Rasmusson, D. C., and Fulcher, R. G. (1998). Variation in ferulic acid concentration among diverse barley cultivars measured by HPLC and microspectrophotometry. *J. Agric. Food Chem.*, 46:1350-1354.

Zurita, J. L., Jos, A., Peso, A. D., Salguero, M., López-Artíguez, M. and Repetto, G. (2007). Ecotoxicological effects of the antioxidant additive propyl gallate in five aquatic systems.*Water Research*, 41: 2599–2611.

Zygadlo, J. A., Lamarque, A. L., Maestri, D. M., and Grosso, N. R. (1995). Empleo de aceites esenciales como antioxidantes naturales. Grasas y Aceites, 46, 285-288.

■■■■

Appendix

APPENDEX – I

Table 12.1: Conversion factors (weights and measures)

Length	1 m	10^{10} A	39.37 in	3.2808 ft
Area	1 m^2	10^4 cm^2	1550.0031 in^2	10.7639 ft^2
Volume	1 m^3	1000 L	10^6 cm^3	264.17 gal (US)
Mass	1 kg	2.2046 lbm	35.274 oz	1.1023 x 10^{-3} ton (short)
Density	1 g/cm^3	1000 kg/m^3	62.428 lbm/ft^3	8.3454 lbm/ gal (US)
Velocity	1 m/s	3.6 km/h	2.2369 min/h	196.85 ft/min
Force	1 kg-m/s^2 (N)	0.2248 lbf	10^5 g.cm/s^2 (dyn)	7.2330 lbm.ft/ s^2 (pdf)
Pressure	1 atm	101.325 kg/ms^2 (pa)	14.696 lbf/in^2 (psi)	760 mm Hg
Energy	1 kg.m^2/s^2 (J)	9.4782 x 10^{-4} Btu	0.2388 cal	0.7376 ft.lbf
Power	1 J/S (W)	1 kg/m^2/s^3	3.4121 Btu/h	1.341 x 10^{-3} hp
Thermal conductivity	1 W/m.k	0.5778 Btu/h.A.^{0}F	2.388 x 10^{-3} cal/s.cm.°C	6.9335 Btu.in/h.ft^2.^{0}F
Specific heat	1 KJ/kg.K	0.2388 cal/g.°C	0.2388 Btf/lbm.°F	185.86 ft.lbf/lbm.^{0}F
Viscosity (dynamic)	1 pa.S	1 kg/m.s	1000 cp	0.6720 lbm/ft.S
Temperature	°C=5/9 (F-32)	^{0}F = 9/5 (°C) + 32	K = ^{0}C + 273.15	^{0}R = ^{0}F + 45967

APPENDIX – II

INTERNATIONAL SYSTEM OF UNITS (SI)

With a few exceptions, most of the units used in this book are SI (System International). It is a system of units adopted in 1960 by an international organization to bring worldwide uniformity to scientific measurements. Some of the fundamental units are shown in Table 1 and the prefixes used to indicate their fractions and multiples are shown in Table 2. For example, the prefixes for 1 x 10^{-3} and 1 x 10^{-6} of a unit are milli- (m) and micro- (μ), respectively. For the mass unit gram (g), 1 milligram (mg) = 1 x 10^{-3} g and 1μg = 1 x 10^{-6} g. Similarly, for the amount unit mole (mol), 1 mmol = 1 x 10^{-3} mol, and 1 μmol = 1 x 10^{-6} mol.

Table 12.2: Selected SI Units

Quantity	Unit	Symbol
Mass	Kilogram	Kg
Volume	Liter	L
Amount of substance	Mole	mol
Length	Meter	m
Time	Second	s
Electric current	Ampere	A
Temperature	Kelvin	K

Table 12.3: Prefixes for SI Units

Prefix	Symbol	Multiples and submultiples
exa	E	10^{18}
peta	P	10^{15}
tera	T	10^{12}
giga	G	10^{9}
mega	M	10^{6}
kilo	K	10^{3}
deci	d	10^{-1}
centi	c	10^{-2}
milli	m	10^{-3}
micro	μ	10^{-6}
nano	n	10^{-9}
pico	p	10^{-12}
femto	f	10^{-15}
atto	a	10^{-18}

Table 12.4: Conversion factors for gram units

Source	**Kg**	**g**	**mg**	**µg**	**ng**	**pg**	***fg***
Kg	10^0	10^3	10^6	10^9	10^{12}	10^{15}	10^{18}
g	10^{-3}	10^0	10^3	10^6	10^9	10^{12}	10^{15}
mg	10^{-6}	10^{-3}	10^0	10^3	10^6	10^9	10^{12}
µg	10^{-9}	10^{-6}	10^{-3}	10^0	10^3	10^6	10^9
ng	10^{-12}	10^{-9}	10^{-6}	10^{-3}	10^0	10^3	10^6
pg	10^{-15}	10^{-12}	10^{-9}	10^{-6}	10^{-3}	10^0	10^3
*f*g	10^{-18}	10^{-15}	10^{-12}	10^{-9}	10^{-6}	10^{-3}	10^0

Table 12.5: Conversion factors for liter units

Source	**KL**	**L**	**dL**	**ml**	**µL**	**nL**
KL	10^0	10^3	10^4	10^6	10^9	10^{12}
L	10^{-3}	10^0	10^1	10^3	10^6	10^9
dL	10^{-4}	10^{-1}	10^0	10^2	10^5	10^8
mL	10^{-6}	10^{-3}	10^{-2}	10^0	10^3	10^6
µL	10^{-9}	10^{-6}	10^{-5}	10^{-3}	10^0	10^3
nL	10^{-12}	10^{-9}	10^{-8}	10^{-6}	10^{-3}	10^0

Table 12.6: Conversion factors for meter units

Source	**Km**	**m**	**cm**	**mm**	**µm**	**nm**	**Å**
Km	10^0	10^3	10^5	10^6	10^9	10^{12}	10^{13}
m	10^{-3}	10^0	10^2	10^3	10^6	10^9	10^{10}
cm	10^{-5}	10^{-2}	10^0	10^1	10^4	10^7	10^8
mm	10^{-6}	10^{-3}	10^{-1}	10^0	10^3	10^6	10^7
µm	10^{-9}	10^{-6}	10^{-4}	10^{-3}	10^0	10^3	10^4
Nm	10^{-12}	10^{-9}	10^{-7}	10^{-6}	10^{-3}	10^0	10^1
Å	10^{-13}	10^{-10}	10^{-8}	10^{-7}	10^{-4}	10^{-1}	10^0

Table 12.7: Conversion factors for mole-related units*

Source	**mol**	**mmol**	**µmol**	**nmol**	**pmol**	***f*mol**
mol	10^0	10^3	10^6	10^9	10^{12}	10^{15}
mmol	10^{-3}	10^0	10^3	10^6	10^9	10^{12}
µmol	10^{-6}	10^{-3}	10^0	10^3	10^6	10^9
nmol	10^{-9}	10^{-6}	10^{-3}	10^0	10^3	10^6
pmol	10^{-12}	10^{-9}	10^{-6}	10^{-3}	10^0	10^3
*f*mol	10^{-15}	10^{-12}	10^{-9}	10^{-6}	10^{-3}	10^0

*To use with molarity (M), Equivalence (equiv) or normality (N), substitute mol with M, equiv, or N.

Instructions for Tables 12.4 to 12.7.

Definitions are: source unit, the unit to be converted; target unit, the unit into which the source unit is to be converted. To convert from one unit to another, select the source unit from the boldfaced set (leftmost column). Move in a straight line across the corresponding row to a factor directly under the *italicized target unit* (topmost row). Multiply the data by this factor. For example, to convert 0.5 mg to µg the mg is selected from the boldfaced set in the leftmost column of units in Table 3. Moving straight across the corresponding row, the conversion factor 10^3 is directly under the *italicized µg* in the top row of units. This means that:

$$1 \text{ mg} = 10^3 \text{ µg}; 0.5 \text{ mg} = (0.5 \times 10^3) = 500 \text{ µg}$$

If the source unit is selected from the *italicized* set instead, follow the above instructions except that the target unit is selected from the boldfaced set, and the data are divided by the factor. Using the above example, the factor obtained is 10^{-3}.

$$0.5 \text{ mg} = 0.5 \div 10^3 = 500 \text{ µg}$$

Subject Index

3-glucoside 86
Absorbance 81
Acetone 74
a-glucosidase enzyme 181
Agricultural 217
Agronomical benefits 347
Albumin 108
Aldehydes 148
Aldose reductase 178
Alkaloids 6
Almond 173, 305
Amperometric 136
Animal foods 4
Anthocyanin 88, 92
Anthocyanin pigments 59
Anthocyanins 28, 165, 221, 226, 236, 242
Anthracnose 357
Antibiotic 275
Anticancer 274
Anti-carcinogenic 7
Anti-diabetic 276
Anti-glaucoma 274
Anti-inflammation 275, 361
Anti-microbial 361
Antimutagenic activity 12
Antimutagenic effects 7
Antinutritional 349
Antioxidant
Antioxidant acitivty 46, 80, 87, 141, 149, 216, 231, 265, 324, 333, 338
Antioxidant agents 15
Antioxidant capacity 79
Antioxidant potential 91
Antioxidant(s) 1, 2, 8, 10, 11, 12, 13, 54, 77, 139, 140, 143, 149, 187, 243, 267, 326, 341, 361
Antioxidative stability 20
Anti-pathogenic 275
Antiradical 203
Apples 245, 251
Apricots 254, 266
Aromatic amino acids 196
Aromatic nucleus 8
Ascorbate 332
Asparagus 167, 289
Assimilation 20
Atherosclerosis 161
Avenanthramides posses 91
Bananas 245
Barley 55, 217
Beach pea 68
Beer 55, 303
Beet 282
Beetroot 289
Beneficial effects 350
Benzoic acid 38
Berries 204
Beta carotene 171
Beverages 152
Bioaceessibility 268
Bioactive 268, 319
Bioavailability 225, 360
Biochemical 348
Biogenesis 61
Biological activity 117
Biological controls 351

Biological significance 65
Biomedical functions 164
Biosynthesis 23, 25, 39, 43
Biosynthetic pathways 23
Bird resistance 353
Bird-depredation 345
Bitter melon 285
Black currants 245
Black pepper 315
Blackberries 246
Blackcurrant 247
Bleaching 153
Blueberries 248
Body mass index 274
Boiling 337
Bone protection 276
Borage 297
Bound antioxidants 139
Bound phenolic compounds 15
Bound phenolics 218
Bovine serum 108
Brassica vegetables 284
Breast cancer 205
Broad bean 67, 73
Broccoli 282
Buckwheat 220, 321
Butylated hydroxytoluene 185
Calories 344
Cancer 7, 170
Canola 294
Carcinogenesis 163
Cardio protection 276
Caronary heart disease 170
Carotenoids 6, 20, 169, 264
Carrots 285
Casein 108
Cashew nuts 306
Catechin 52, 69, 226
Catechin equivalent 114
Cauliflower 282
Celery 290
Cereal grains 83, 91
Cereals 4, 86, 111, 342, 359
Chemiluminescence 133
Chemopreventive 361
Cherries 245
Chestnut 311
Chromatography 67, 215
Chromogen radical 133
Chronic diseases 360
Cinnamic acid 25, 263
Citric acid 166, 167
Citrus fruits 245, 253
Clove 314
Cocoa 302
Coffee 301
Collaborative 133
Colon cancer 163
Color 4
Condensed tannins 6, 43, 61, 64, 74, 75, 112, 135
Configuration 201
Conjugated dienes 77
Consumer 186
Consumption 174
Cooking 6, 20, 117, 320, 349
Co-products 7
Corn 221
Cottonseed 296
Coumarins 30, 39
Coumestan 6
Cranberries 244, 245, 248
Cyanidin 246
Cyanidin 3-glucoside 247
Cyanidin 3-monoglucoside 251
Cyanidin 86
Cyanidin-3-O-glucoside 52
Cyclization 40, 41
Cyclooxygenase 2
Cypridina luciferin 132
Decortication 224, 320

Degradation 208
Delocalization 201
Delphinidin 3-monoglucoside 251
Diabetes 7
Dietary antioxidants 84
Dietary carbohydrates 89
Dietary fibers 236
Dietary supplements 158
Dietary tannins 347
Differential scanning calorimetry 144
Digestion 51
Dihydroflavonols 199
Dried fruits 266, 279, 280
Drying 349
Electrochemical 140
Electron transfer 84
Electrophilic radical 98
Ellagic acid 33, 50, 163
Ellagitannns 31, 33
Embryo 239
Endosperm 239
Environment 157
Enzymes 108
Epicatechin 52, 69
Epidemiological 177
Epigallocatechin gallate 51
Eriodictyol 52
Essential amino acids 175
Esterification 182
Evening primrose 297
Extraction 56, 127
Faba bean 67, 73, 241
Fast foods 153
Fat-soluble bioactives 6
Fatty acid 151
Fatty acid metabolism 53
Fenugreek seeds 316
Fermentation 182, 183, 224, 225
Ferric chloride 79
Ferric ion 137
Ferrocyanide 137
Ferrous sulphate 141
Ferulic acid 25, 93, 218, 235, 238
Figs 245, 266
Finger millets 177, 181
Flaking 224
Flavan-3-ols 45
Flavones 12, 28, 96
Flavonoids 3, 4, 23, 29, 30, 87, 198, 211, 227, 257, 282
Flavonol 28, 87, 258
Flaxseed 295
Folin-Ciocalteu assay 142
Folin-Denis reagent 121
Food matrix 52
Forage plants 4
Formic acid 57
Free fatty acids 144
Free Radicals 167, 191
Freezing 327
Fruits 4, 158, 243, 281, 359
Functional foods 361
Galangin 86
Gallic acid 203
Gallotannins 31
Gastrointenstinal tract 85
Gelatin 108
Germination 3, 6, 349
Gingers 245, 315
Glucose 28
Glucosylation 39
Glutathione 9
Glutathione peroxidase 9
Glycosides 241
Goji berries 245
Grain production 351
Grains 319
Grape 250
Grapefrutis 166
Green pepper 282
Grinding 118, 224

Guavas 245
Harmful 358
Hazelnut 47, 307
Health benefits 167, 179
Health contribution 360
Heartnut 307
Heat treatment 224
Hemoglobin 108
High blood pressure 7
HPLC 154
HPLIC 215
Human health 341
Hydrocarbon 171
Hydrolysable tannins 207
Hydrolysis 218
Hydrolyzable tannins 4, 31
Hydroxy-benzoic 197
Hydroxycinnamates 197
Hydroxycinnamic acids 252
Hydroxyl radicals 188, 211
Hydrxylation 111
Hyperglycemia 178
Hypocholesterolaemic 361
Hypoglycaemic 276
Immune 178
Inhibition 158, 170
Inhibition assay 157
Inhibitors 64, 195
Initiator 159
Irradiation 208
Isoflavonoid 40, 198
Kaempferol 86
Ketones 148
Kinetic study 158
Kiwis 245
LDL oxidation 155
LDL-cholesterol 155
Legumes 29, 59, 106, 111, 240, 342
Lemons 166
Lettuce 290
Leucoanthocyanidins 96, 101
Lignocellulosic 48
Linoleic acid 152
Lipid hydroperoxides 147
Lipid oxidation 1
Lipid-lowering 275
Lipids 170
Low density lipoprotein 168
Lucigenin 132
Luminol 132
Lutein 264
Lycopene 169, 170, 171
Macadamia 47
Macadamia nut 311
Maillard reaction 15, 48
Malondialdehyde 148, 340
Malondialdehyde equivalents 48
Malt extract 89
Malting 56, 224, 320
Management 360
Mangoes 245
Mayonnaise 153
Melanoma 207
Metabolic 157
Metabolism 202, 360
Metabolizable energy 5
Metal chelating 179
Metals 95
Microwaving 336
Millets 175, 177, 223, 321
Milling 116
Mobility 278
Monomeric 125
Monomeric flavans 99
Monophenol 189
Mulberries 245
Mutagenesis 163
Natural antioxidants 19
Nectarines 245
Neurodegenerative diseases 7

Non hydrolyzable 31
Non-radical redox 137
Novel product 277
ntilipoperoxidant 203
Nutritive value 25
Nuts 7, 45
Oat 234
Oil stability index 150
Oils 293
Oilseeds 293
Oligomers 63
Olive oil 298
Onions 286
Optimisation 20
Oranges 166
Oregano 326
Organoleptic 25
Oxalates 132
Oxidation 151, 350
Oxidation reactions 10
p-anisidine value 145
Papayas 245
Passion fruits 245
Pasteurization 328
p-coumaric 27
p-coumarin 41
Pea 73
Pea hull 65
Peaches 245
Peanut 53, 309
Pearl millet 323
Pears 245
Peas 167
Pecan 53, 310
Peroxyl radicals 156
Phenolases 95
Phenolic 282
Phenolic acids 23, 25, 30, 235, 262
Phenolic compounds 47, 70, 120, 344, 346
Phenolic profiles 237
Phenolics 2, 92, 160, 345
Phenols 103
Phenomenon 30
Phenylpropanoid pathway 8
Phloroglucinol 99
Phosphenolpyruvate 38
Photocarcinogenesis 180
p-hydroxy benzaldehyde 236
Phytates 6
Phytochemicals 91, 256, 269
Phytoestrogens 264
Pigments 85
Pine nut 53, 311
Pineapples 245
Pistachio 204, 309, 340
Plant phenolics 23, 26
Polymeric flavans 125
Polymerization 29, 95, 101, 111, 119
Polymers 102
Polymethoxyflavones 253
Polyphenol 51, 93
Polyphenol content 355
Polyphenolic compounds 12
Polyphenols 8, 51, 55, 172, 318, 345
Pomegranates 254
Potato 287
Potatoes 167
Potent antioxidants 159
Prebiotics 173, 361
Primary oxidation 147
Proanthocyanidin 37, 46, 205
Processing 318, 326
Procyanidin 37, 63, 346
Prooxidants 152
Propagation 1
Prophylactic 253
Propyl gallate 185, 191
Proso millets 177
Prostate cancer 161
Protein digestibility 5

Protein matrix 112
Protein-precipitable phenol 127
Proteins 108, 170, 344
Prunes 245, 266
Prussian blue 137
Purple onion 282
Pyrogallol 132
Quercetin 86, 242
Radical 1, 54
Radical scavenging mechanism 133
Raisins 245, 266
Rapeseed 294
Raspberries 245
Reactive oxygen species 83
Red cabbage 204
Red raspberries 249
Refrigerated 328
Reliable 138
Resistance 357
Resistant 358
Rice 86, 221, 235
Rice bran oil 90
Risk toxicity 340
Roasted 343
Roasted nuts 339
Roasted nuts 7
Roasting 52
Rosemary 313
Rutin 52
Rye 238
Sage 313
Salicylic acid 219
Sauces 152
Secondary metabolites 3
Seed germination 356
Sensitive 138
Sensitivity 150
Serum-cholesterol 237
Sesame seeds 295
Sexual maturity 5
Shikimic aid pathway 38
Single antioxidant 160
Skin carcinoma 207
Soaking 224, 320, 337
Solvent 23
Sorghum 35, 232, 322, 352
Sorghum grain 227
Sorghum tannins 5, 126
Soy isoflavones 162
Soybean 293
Soybean molasses 14
Specificity 114
Spectrophotometric 110, 134
Spectrophotometry 145
Spinach 204, 282, 291
Star apples 25
Steaming 336, 337
Stoichiometric 105
Strawberries 245
Sugars 73, 74
Sulphydryl 2
Supercritical 23
Superoxide dismutase 9
Sweet cherries 251
Sweet potato 288
Synthetic antioxidants 15
Syringic acid 38, 60
Tamarind seeds 14
Tannic acid 4, 104
Tannin(s) 5, 29, 31, 60, 63, 97, 101, 102, 116, 205, 345, 350
Tannin-protein interactions 45
TBARS 154
TBHQ 190
Tea 299
Termination 1
Tertiary-butylhydroquinone 185
Theaflavin gallate 99
Thermal treatment 327
Threshability 354
Tocopherols 192

Tocotrienols 192
Tocotrienoxyl radicals 194
Tomato 292
Tomatoes 167
Total phenols 107
Trans-*p*-coumaric 60
Tree nuts 304
Triacylglycerols 19
Triethanolamine 109
Trolox equivalent 135
Turmeric 315
Unsaturated fatty acids 151
UV spectra 73
Vanillaldehyde 97
Vanillic acid 38
Vanillin 97, 100, 101, 114, 121
Vanillin assay 116
Vegetable protein 6
Vegetable tannins 4
Vegetables 4, 158, 281, 337, 359
Vitamin C 168, 335, 341
Vitamin E 196
Walnuts 47, 307, 340
Wheat 88, 212, 221
Wheat grains 214
Wilting 115
Wine 304
Zeaxanthin 264
α-carotene 264, 332
α-tocopherol 325, 332
β-carotene 143, 152, 153, 331, 332, 335
β-cryptoxanthin 264
β-glucosidases 161
γ-tocopherol 168, 332